INTUITIVE PROBABILITY AND RANDOM PROCESSES USING MATLAB®

INTUITIVE PROBABILITY AND RANDOM PROCESSES USING MATLAB®

STEVEN M. KAY
University of Rhode Island

 Springer

Steven M. Kay
University of Rhode Island
Dept. of Electrical & Computer
Engineering
Kingston, RI 02881

Kay, Steven M., 1951-
    Intuitive probability and random processes using MATLAB / Steven M. Kay.
        p. cm.
    ISBN-13: 978-0-387-24157-9 (acid-free paper)
    ISBN-10: 0-387-24157-4 (acid-free paper)
    ISBN-10: 0-387-24158-2 (e-book)
    1. Probabilities--Computer simulation--Textbooks. 2. Stochastic processes--Computer
simulation--Textbooks. 3. MATLAB--Textbooks. I. Title.

    QA273.K329 2005
    519.2'01'13--dc22

                                                          2005051721

Printed in the United States of America.

9 8 7 6 5 4 3 2

springer.com

To my wife
Cindy,
whose love and support
are without measure

and to my daughters
Lisa and Ashley,
who are a source of joy

NOTE TO INSTRUCTORS

As an aid to instructors interested in using this book for a course, the solutions to the exercises are available in electronic form. They may be obtained by contacting the author at kay@ele.uri.edu.

Preface

The subject of probability and random processes is an important one for a variety of disciplines. Yet, in the author's experience, a first exposure to this subject can cause difficulty in assimilating the material and even more so in applying it to practical problems of interest. The goal of this textbook is to lessen this difficulty. To do so we have chosen to present the material with an emphasis on conceptualization. As defined by Webster, a *concept* is "an abstract or generic idea generalized from particular instances." This embodies the notion that the "idea" is something we have formulated based on our past experience. This is in contrast to a *theorem*, which according to Webster is "an idea accepted or proposed as a demonstrable truth". A theorem then is the result of many *other* persons' past experiences, which may or may not coincide with our own. In presenting the material we prefer to first present "particular instances" or examples and then generalize using a definition/theorem. Many textbooks use the opposite sequence, which undeniably is cleaner and more compact, but omits the motivating examples that initially led to the definition/theorem. Furthermore, in using the definition/theorem-first approach, for the sake of mathematical correctness multiple concepts must be presented at once. This is in opposition to human learning for which "under most conditions, the greater the number of attributes to be bounded into a single concept, the more difficult the learning becomes"[1]. The philosophical approach of specific examples followed by generalizations is embodied in this textbook. It is hoped that it will provide an alternative to the more traditional approach for exploring the subject of probability and random processes.

To provide motivating examples we have chosen to use MATLAB[2], which is a very versatile scientific programming language. Our own engineering students at the University of Rhode Island are exposed to MATLAB as freshmen and continue to use it throughout their curriculum. Graduate students who have not been previously introduced to MATLAB easily master its use. The pedagogical utility of using MATLAB is that:

1. Specific computer generated examples can be constructed to provide motivation for the more general concepts to follow.

---

[1] Eli Saltz, *The Cognitive Basis of Human Learning*, Dorsey Press, Homewood, IL, 1971.

[2] Registered trademark of TheMathWorks, Inc.

2. Inclusion of computer code within the text allows the reader to interpret the mathematical equations more easily by seeing them in an alternative form.

3. Homework problems based on computer simulations can be assigned to illustrate and reinforce important concepts.

4. Computer experimentation by the reader is easily accomplished.

5. Typical results of probabilistic-based algorithms can be illustrated.

6. Real-world problems can be described and "solved" by implementing the solution in code.

Many MATLAB programs and code segments have been included in the book. In fact, most of the figures were generated using MATLAB. The programs and code segments listed within the book are available in the file `probbook_matlab_code.tex`, which can be found at http://www.ele.uri.edu/faculty/kay/New%20web/Books.htm. The use of MATLAB, along with a brief description of its syntax, is introduced early in the book in Chapter 2. It is then immediately applied to simulate outcomes of random variables and to estimate various quantities such as means, variances, probability mass functions, etc. *even though these concepts have not as yet been formally introduced.* This chapter sequencing is purposeful and is meant to expose the reader to some of the main concepts that will follow in more detail later. In addition, the reader will then immediately be able to simulate random phenomena to learn through doing, in accordance with our philosophy. In summary, we believe that the incorporation of MATLAB into the study of probability and random processes provides a "hands-on" approach to the subject and promotes better understanding.

Other pedagogical features of this textbook are the discussion of discrete random variables first to allow easier assimilation of the concepts followed by a parallel discussion for continuous random variables. Although this entails some redundancy, we have found less confusion on the part of the student using this approach. In a similar vein, we first discuss scalar random variables, then bivariate (or two-dimensional) random variables, and finally N-dimensional random variables, reserving separate chapters for each. All chapters, except for the introductory chapter, begin with a summary of the important concepts and point to the main formulas of the chapter, and end with a real-world example. The latter illustrates the utility of the material just studied and provides a powerful motivation for further study. It also will, hopefully, answer the ubiquitous question "Why do we have to study this?". We have tried to include real-world examples from many disciplines to indicate the wide applicability of the material studied. There are numerous problems in each chapter to enhance understanding with some answers listed in Appendix E. The problems consist of four types. There are "formula" problems, which are simple applications of the important formulas of the chapter; "word" problems, which require a problem-solving capability; and "theoretical" problems, which are more abstract and mathematically demanding; and finally, there are "computer" problems, which are either computer simulations or involve the application of computers to facilitate analytical solutions. A complete solutions manual for all the problems is available to instructors from the author upon request. Finally, we have provided warnings on how to avoid common errors as well as in-line explanations of equations within the derivations for clarification.

The book was written mainly to be used as a first-year graduate level course in probability and random processes. As such, we assume that the student has had some exposure to basic probability and therefore Chapters 3–11 can serve as a review and a summary of the notation. We then will cover Chapters 12–15 on probability and selected chapters from Chapters 16–22 on random processes. This book can also be used as a self-contained introduction to probability at the senior undergraduate or graduate level. It is then suggested that Chapters 1–7, 10, 11 be covered. Finally, this book is suitable for self-study and so should be useful to the practitioner as well as the student. The necessary background that has been assumed is a knowledge of calculus (a review is included in Appendix B); some linear/matrix algebra (a review is provided in Appendix C); and linear systems, which is necessary only for Chapters 18–20 (although Appendix D has been provided to summarize and illustrate the important concepts).

The author would like to acknowledge the contributions of the many people who over the years have provided stimulating discussions of teaching and research problems and opportunities to apply the results of that research. Thanks are due to my colleagues L. Jackson, R. Kumaresan, L. Pakula, and P. Swaszek of the University of Rhode Island. A debt of gratitude is owed to all my current and former graduate students. They have contributed to the final manuscript through many hours of pedagogical and research discussions as well as by their specific comments and questions. In particular, Lin Huang and Cuichun Xu proofread the entire manuscript and helped with the problem solutions, while Russ Costa provided feedback. Lin Huang also aided with the intricacies of LaTex while Lisa Kay and Jason Berry helped with the artwork and to demystify the workings of Adobe Illustrator 10.[3] The author is indebted to the many agencies and program managers who have sponsored his research, including the Naval Undersea Warfare Center, the Naval Air Warfare Center, the Air Force Office of Scientific Research, and the Office of Naval Research. As always, the author welcomes comments and corrections, which can be sent to kay@ele.uri.edu.

Steven M. Kay
University of Rhode Island
Kingston, RI 02881

---

[3]Registered trademark of Adobe Systems Inc.

Contents

Chapter 1

Introduction

1.1 What Is Probability?

Probability as defined by Webster's dictionary is "the chance that a given event will occur". Examples that we are familiar with are the probability that it will rain the next day or the probability that you will win the lottery. In the first example, there are many factors that affect the weather—so many, in fact, that we cannot be certain that it will or will not rain the following day. Hence, as a predictive tool we usually assign a number between 0 and 1 (or between 0% and 100%) indicating our degree of certainty that the event, rain, will occur. If we say that there is a 30% chance of rain, we believe that if identical conditions prevail, then 3 times out of 10, rain will occur the next day. Alternatively, we believe that the *relative frequency* of rain is 3/10. Note that if the science of meteorology had accurate enough models, then it is conceivable that we could determine exactly whether rain would or would not occur. Or we could say that the *probability* is either 0 or 1. Unfortunately, we have not progressed that far. In the second example, winning the lottery, our chance of success, assuming a fair drawing, is just one out of the number of possible lottery number sequences. In this case, we are uncertain of the outcome, not because of the inaccuracy of our model, but because the experiment has been designed to produce uncertain results.

The common thread of these two examples is the presence of a *random experiment*, a *set of outcomes*, and the *probabilities* assigned to these outcomes. We will see later that these attributes are common to all probabilistic descriptions. In the lottery example, the experiment is the drawing, the outcomes are the lottery number sequences, and the probabilities assigned are $1/N$, where N = total number of lottery number sequences. Another common thread, which justifies the use of probabilistic methods, is the concept of *statistical regularity*. Although we may never be able to predict with certainty the outcome of an experiment, we are, nonetheless, able to predict "averages". For example, the average rainfall in the summer in Rhode Island is 9.76 inches, as shown in Figure 1.1, while in Arizona it is only 4.40

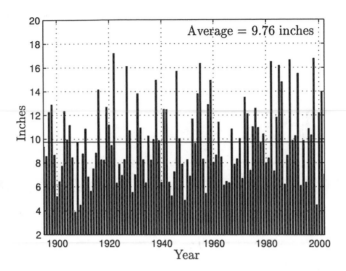

Figure 1.1: Annual summer rainfall in Rhode Island from 1895 to 2002 [NOAA/NCDC 2003].

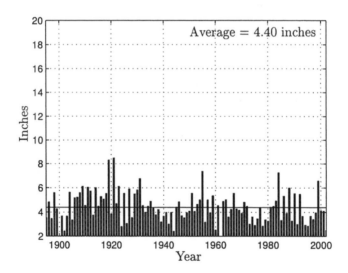

Figure 1.2: Annual summer rainfall in Arizona from 1895 to 2002 [NOAA/NCDC 2003].

inches, as shown in Figure 1.2. It is clear that the decision to plant certain types of crops could be made based on these averages. This is not to say, however, that we can predict the rainfall amounts for any given summer. For instance, in 1999 the summer rainfall in Rhode Island was only 4.5 inches while in 1984 the summer rainfall in Arizona was 7.3 inches. A somewhat more controlled experiment is the repeated tossing of a fair coin (one that is equally likely to come up heads or tails). We would expect about 50 heads out of 100 tosses, but of course, we could not predict the outcome of any one particular toss. An illustration of this is shown in Figure 1.3. Note that 53 heads were obtained in this particular experiment. This

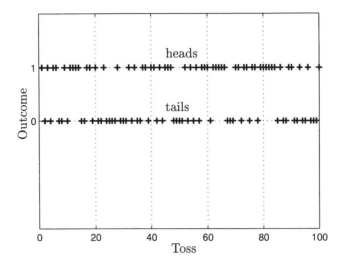

Figure 1.3: Outcomes for repeated fair coin tossings.

example, which is of seemingly little relevance to physical reality, actually serves as a good *model* for a variety of random phenomena. We will explore one example in the next section.

In summary, probability theory provides us with the ability to predict the behavior of random phenomena in the "long run." To the extent that this information is useful, probability can serve as a valuable tool for assessment and decision making. Its application is widespread, encountering use in all fields of scientific endeavor such as engineering, medicine, economics, physics, and others (see references at end of chapter).

1.2 Types of Probability Problems

Because of the mathematics required to determine probabilities, probabilistic methods are divided into two distinct types, *discrete* and *continuous*. A discrete approach is used when the number of experimental outcomes is finite (or infinite but countable as illustrated in Problem 1.7). For example, consider the number of persons at a business location that are talking on their respective phones anytime between 9:00 AM and 9:10 AM. Clearly, the possible outcomes are $0, 1, \ldots, N$, where N is the number of persons in the office. On the other hand, if we are interested in the length of time a particular caller is on the phone during that time period, then the outcomes may be anywhere from 0 to T minutes, where $T = 10$. Now the outcomes are infinite in number since they lie within the interval $[0, T]$. In the first case, since the outcomes are discrete (and finite), we can assign probabilities to the outcomes $\{0, 1, \ldots, N\}$. An equiprobable assignment would be to assign each outcome a probability of $1/(N + 1)$. In the second case, the outcomes are continuous (and therefore infinite) and so it is not possible to assign a nonzero probability *to each outcome* (see Problem 1.6).

We will henceforth delineate between probabilities assigned to discrete outcomes and those assigned to continuous outcomes, with the discrete case always discussed first. The discrete case is easier to conceptualize and to describe mathematically. It will be important to keep in mind which case is under consideration since otherwise, certain paradoxes may result (as well as much confusion on the part of the student!).

1.3   Probabilistic Modeling

Probability models are simplified approximations to reality. They should be detailed enough to capture important characteristics of the random phenomenon so as to be useful as a prediction device, but not so detailed so as to produce an unwieldy model that is difficult to use in practice. The example of the number of telephone callers can be modeled by assigning a probability p to each person being on the phone anytime in the given 10-minute interval and *assuming* that whether one or more persons are on the phone does not affect the probability of others being on the phone. One can thus liken the event of being on the phone to a coin toss— if heads, a person is on the phone and if tails, a person is not on the phone. If there are $N = 4$ persons in the office, then the experimental outcome is likened to 4 coin tosses (either in succession or simultaneously—it makes no difference in the modeling). We can then ask for the probability that 3 persons are on the phone by determining the probability of 3 heads out of 4 coin tosses. The solution to this problem will be discussed in Chapter 3, where it is shown that the probability of k heads out of N coin tosses is given by

$$P[k] = \binom{N}{k} p^k (1 - p)^{N-k} \tag{1.1}$$

where

$$\binom{N}{k} = \frac{N!}{(N - k)! k!}$$

for $k = 0, 1, \ldots, N$, and where $M! = 1 \cdot 2 \cdot 3 \cdots M$ for M a positive integer and by definition $0! = 1$. For our example, if $p = 0.75$ (we have a group of telemarketers) and $N = 4$ a compilation of the probabilities is shown in Figure 1.4. It is seen that the probability that three persons are on the phone is 0.42. Generally, the coin toss

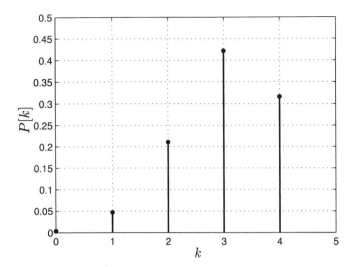

Figure 1.4: Probabilities for $N = 4$ coin tossings with $p = 0.75$.

model is a reasonable one for this type of situation. It will be poor, however, if the *assumptions are invalid*. Some practical objections to the model might be:

1. Different persons have different probabilities p (an eager telemarketer versus a not so eager one).

2. The probability of one person being on the phone is affected by whether his neighbor is on the phone (the two neighbors tend to talk about their planned weekends), i.e., the events are not "independent".

3. The probability p changes over time (later in the day there is less phone activity due to fatigue).

To accommodate these objections the model can be made more complex. In the end, however, the "more accurate" model may become a poorer predictor if the additional information used is not correct. It is generally accepted that a model should exhibit the property of "parsimony"—in other words, it should be as simple as possible.

The previous example had discrete outcomes. For continuous outcomes a frequently used probabilistic model is the *Gaussian* or "bell"-shaped curve. For the modeling of the length of time T a caller is on the phone it is not appropriate to ask for the probability that T will be *exactly*, for example, 5 minutes. This is because this probability will be zero (see Problem 1.6). Instead, we inquire as to the probability that T will be between 5 and 6 minutes. This question is answered by determining the area under the Gaussian curve shown in Figure 1.5. The form of the curve is given by

$$p_T(t) = \frac{1}{\sqrt{2\pi}} \exp\left[-\frac{1}{2}(t-7)^2\right] \qquad -\infty < t < \infty \qquad (1.2)$$

and although defined for all t, it is physically meaningful only for $0 \le t \le T_{\max}$,

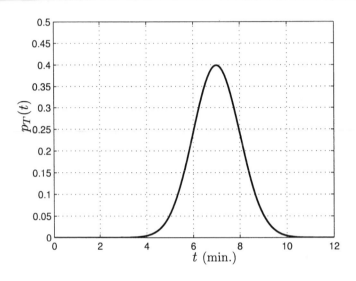

Figure 1.5: Gaussian or "bell"-shaped curve.

where $T_{\max} = 10$ for the current example. Since the area under the curve for times less than zero or greater than $T_{\max} = 10$ is nearly zero, this model is a reasonable approximation to physical reality. The curve has been chosen to be centered about $t = 7$ to relect an "average" time on the phone of 7 minutes for a given caller. Also, note that we let t denote the actual *value* of the *random* time T. Now, to determine the probability that the caller will be on the phone for between 5 and 6 minutes we integrate $p_T(t)$ over this interval to yield

$$P[5 \le T \le 6] = \int_5^6 p_T(t)dt = 0.1359. \qquad (1.3)$$

The value of the integral must be numerically determined. Knowing the function $p_T(t)$ allows us to determine the probability for any interval. (It is called the probability density function (PDF) and is the probability per unit length. The PDF will be discussed in Chapter 10.) Also, it is apparent from Figure 1.5 that phone usage of duration less than 4 minutes or greater than 10 minutes is highly unlikely. Phone usage in the range of 7 minutes, on the other hand, is most probable. As before, some objections might be raised as to the accuracy of this model. A particularly lazy worker could be on the phone for only 3 minutes, as an example.

In this book we will henceforth assume that the models, which are mathematical in nature, are perfect and thus can be used to determine probabilities. In practice, the user must ultimately choose a model that is a reasonable one for the application of interest.

1.4 Analysis versus Computer Simulation

In the previous section we saw how to compute probabilities once we were given certain probability functions such as (1.1) for the discrete case and (1.2) for the continuous case. For many practical problems it is not possible to determine these functions. However, if we have a model for the random phenomenon, then we may carry out the experiment a large number of times to obtain an approximate probability. For example, to determine the probability of 3 heads in 4 tosses of a coin with probability of heads being $p = 0.75$, we toss the coin four times and count the number of heads, say $x_1 = 2$. Then, we repeat the experiment by tossing the coin four more times, yielding $x_2 = 1$ head. Continuing in this manner we execute a succession of 1000 experiments to produce the sequence of number of heads as $\{x_1, x_2, \ldots, x_{1000}\}$. Then, to determine the probability of 3 heads we use a *relative frequency* interpretation of probability to yield

$$P[3 \text{ heads}] = \frac{\text{Number of times 3 heads observed}}{1000}. \tag{1.4}$$

Indeed, early on probabilists did exactly this, although it was extremely tedious. *It is therefore of utmost importance to be able to simulate this procedure.* With the advent of the modern digital computer this is now possible. A digital computer has no problem performing a calculation once, 100 times, or 1,000,000 times. What is needed to implement this approach is a means to simulate the toss of a coin. Fortunately, this is quite easy as most scientific software packages have built-in *random number generators*. In MATLAB, for example, a number in the interval $(0, 1)$ can be produced with the simple statement x=rand(1,1). The number is chosen "at random" so that it is equally likely to be anywhere in the $(0, 1)$ interval. As a result, a number in the interval $(0, 1/2]$ will be observed with probability $1/2$ and a number in the remaining part of the interval $(1/2, 1)$ also with probability $1/2$. Likewise, a number in the interval $(0, 0.75]$ will be observed with probability $p = 0.75$. A computer simulation of the number of persons in the office on the telephone can thus be implemented with the MATLAB code (see Appendix 2A for a brief introduction to MATLAB):

```
number=0;
for i=1:4 % set up simulation for 4 coin tosses
  if rand(1,1)<0.75 % toss coin with p=0.75
    x(i,1)=1; % head
  else
```

```
      x(i,1)=0; % tail
   end
number=number+x(i,1); % count number of heads
end
```

Repeating this code segment 1000 times will result in a simulation of the previous experiment.

Similarly, for a continuous outcome experiment we require a means to generate a continuum of outcomes on a digital computer. Of course, strictly speaking this is not possible since digital computers can only provide a finite set of numbers, which is determined by the number of bits in each word. But if the number of bits is large enough, then the approximation is adequate. For example, with 64 bits we could represent 2^{64} numbers between 0 and 1, so that neighboring numbers would be $2^{-64} = 5 \times 10^{-20}$ apart. With this ability MATLAB can produce numbers that follow a Gaussian curve by invoking the statement x=randn(1,1).

Throughout the text we will use MATLAB for examples and also exercises. However, any modern scientific software package can be used.

1.5 Some Notes to the Reader

The notation used in this text is summarized in Appendix A. Note that boldface type is reserved for vectors and matrices while regular face type will denote scalar quantities. All other symbolism is defined within the context of the discussion. Also, the reader will frequently be warned of potential "pitfalls". Common misconceptions leading to student errors will be described and noted. The pitfall or caution symbol shown below should be heeded.

The problems are of four types: computational or formula applications, word problems, computer exercises, and theoretical exercises. Computational or formula (denoted by **f**) problems are straightforward applications of the various formulas of the chapter, while word problems (denoted by **w**) require a more complete assimilation of the material to solve the problem. Computer exercises (denoted by **c**) will require the student to either use a computer to solve a problem or to simulate the analytical results. This will enhance understanding and can be based on MATLAB, although equivalent software may be used. Finally, theoretical exercises (denoted by **t**) will serve to test the student's analytical skills as well as to provide extensions to the material of the chapter. They are more challenging. Answers to selected problems are given in Appendix E. Those problems for which the answers are provided are noted in the problem section with the symbol ($\cdot\cdot$).

The version of MATLAB used in this book is 5.2, although newer versions should provide identical results. Many MATLAB outputs that are used for the text figures and for the problem solutions rely on random number generation. To match your results against those shown in the figures and the problem solutions, the same set of random numbers can be generated by using the MATLAB statements `rand('state',0)` and `randn('state',0)` at the beginning of each program. These statements will initialize the random number generators to produce the same set of random numbers. Finally, the MATLAB programs and code segments given in the book are indicated by the "typewriter" font, for example, `x=randn(1,1)`.

There are a number of other textbooks that the reader may wish to consult. They are listed in the following reference list, along with some comments on their contents.

Davenport, W.B., *Probability and Random Processes*, McGraw-Hill, New York, 1970. (Excellent introductory text.)

Feller, W., *An Introduction to Probability Theory and its Applications*, Vols. 1, 2, John Wiley, New York, 1950. (Definitive work on probability—requires mature mathematical knowledge.)

Hoel, P.G., S.C. Port, C.J. Stone, *Introduction to Probability Theory*, Houghton Mifflin Co., Boston, 1971. (Excellent introductory text but limited to probability.)

Leon-Garcia, A., *Probability and Random Processes for Electrical Engineering*, Addison-Wesley, Reading, MA, 1994. (Excellent introductory text.)

Parzen, E., *Modern Probability Theory and Its Applications*, John Wiley, New York, 1960. (Classic text in probability—useful for all disciplines).

Parzen, E., *Stochastic Processes*, Holden-Day, San Francisco, 1962. (Most useful for Markov process descriptions.)

Papoulis, A., *Probability, Random Variables, and Stochastic Processes*, McGraw-Hill, New York, 1965. (Classic but somewhat difficult text. Best used as a reference.)

Ross, S., *A First Course in Probability*, Prentice-Hall, Upper Saddle River, NJ, 2002. (Excellent introductory text covering only probability.)

Stark, H., J.W. Woods, *Probability and Random Processes with Applications to Signal Processing*, Third Ed., Prentice Hall, Upper Saddle River, NJ, 2002. (Excellent introductory text but at a somewhat more advanced level.)

References

Burdic, W.S., *Underwater Acoustic Systems Analysis*, Prentice-Hall, Englewood Cliffs, NJ, 1984.

Ellenberg, S.S., D.M. Finklestein, D.A. Schoenfeld, "Statistical Issues Arising in AIDS Clinical Trials," *J. Am. Statist. Asssoc.*, Vol. 87, pp. 562–569, 1992.

Gustafson, D.E., A.S. Willsky, J.Y. Wang, M.C. Lancaster, J.H. Triebwasser, "ECG/VCG Rhythm Diagnosis Using Statistical Signal Analysis, Part II: Identification of Transient Rhythms," *IEEE Trans. Biomed. Eng.*, Vol. BME-25, pp. 353–361, 1978.

Ives, R.B., "The Applications of Advanced Pattern Recognition Techniques for the Discrimination Between Earthquakes and Nuclear Detonations," *Pattern Recognition*, Vol. 14, pp. 155–161, 1981.

Justice, J.H., "Array Processing in Exploration Seismology," in *Array Signal Processing*, S. Haykin, Ed., Prentice-Hall, Englewood Cliffs, NJ, 1985.

Knight, W.S., R.G. Pridham, S.M. Kay, "Digital Signal Processing for Sonar," *Proc. IEEE*, Vol. 69, pp. 1451–1506, Nov. 1981.

NOAA/NCDC, Lawrimore, J., "Climate at a Glance," National Oceanic and Atmospheric Administration, http://www.ncdc.noaa.gov/oa/climate/resarch/cag3/NA.html, 2003.

Proakis, J., *Digital Communications*, Second Ed., McGraw-Hill, New York, 1989.

Skolnik, M.I., *Introduction to Radar Systems*, McGraw-Hill, New York, 1980.

Taylor, S., *Modelling Financial Time Series*, John Wiley, New York, 1986.

Problems

1.1 (☺) (w) A fair coin is tossed. Identify the random experiment, the set of outcomes, and the probabilities of each possible outcome.

1.2 (w) A card is chosen at random from a deck of 52 cards. Identify the random experiment, the set of outcomes, and the probabilities of each possible outcome.

1.3 (w) A fair die is tossed and the number of dots on the face noted. Identify the random experiment, the set of outcomes, and the probabilities of each possible outcome.

1.4 (w) It is desired to predict the annual summer rainfall in Rhode Island for 2010. If we use 9.76 inches as our prediction, how much in error might we be, based on the past data shown in Figure 1.1? Repeat the problem for Arizona by using 4.40 inches as the prediction.

1.5 (☺) (w) Determine whether the following experiments have discrete or continuous outcomes:

    a. Throw a dart with a point tip at a dartboard.

    b. Toss a die.

    c. Choose a lottery number.

    d. Observe the outdoor temperature using an analog thermometer.

    e. Determine the current time in hours, minutes, seconds, and AM or PM.

1.6 (w) An experiment has $N = 10$ outcomes that are equally probable. What is the probability of each outcome? Now let $N = 1000$ and also $N = 1,000,000$ and repeat. What happens as $N \to \infty$?

1.7 (☺) (f) Consider an experiment with possible outcomes $\{1, 2, 3, \ldots\}$. If we assign probabilities

$$P[k] = \frac{1}{2^k} \qquad k = 1, 2, 3, \ldots$$

to the outcomes, will these probabilties sum to one? Can you have an infinite number of outcomes but still assign nonzero probabilities to each outcome? Reconcile these results with that of Problem 1.6.

1.8 (w) An experiment consists of tossing a fair coin four times in succession. What are the possible outcomes? Now count up the number of outcomes with three heads. If the outcomes are equally probable, what is the probability of three heads? Compare your results to that obtained using (1.1).

1.9 (w) Perform the following experiment by *actually tossing* a coin of your choice. Flip the coin four times and observe the number of heads. Then, repeat this experiment 10 times. Using (1.1) determine the probability for $k = 0, 1, 2, 3, 4$ heads. Next use (1.1) to determine the number of heads that is most probable for a single experiment? In your 10 experiments which number of heads appeared most often?

1.10 (☺) (w) A coin is tossed 12 times. The sequence observed is the 12-tuple $(H, H, T, H, H, T, H, H, H, H, T, H)$. Is this a fair coin? Hint: Determine $P[k = 9]$ using (1.1) assuming a probability of heads of $p = 1/2$.

1.11 (t) Prove that $\sum_{k=0}^{N} P[k] = 1$, where $P[k]$ is given by (1.1). Hint: First prove the binomial theorem

$$(a + b)^N = \sum_{k=0}^{N} \binom{N}{k} a^k b^{N-k}$$

by induction (see Appendix B). Use Pascal's "triangle" rule

$$\binom{M}{k} = \binom{M-1}{k} + \binom{M-1}{k-1}$$

where

$$\binom{M}{k} = 0 \qquad k < 0 \text{ and } k > M.$$

1.12 (t) If $\int_a^b p_T(t)dt$ is the probability of observing T in the interval $[a, b]$, what is $\int_{-\infty}^{\infty} p_T(t)dt$?

1.13 (⌣) (f) Using (1.2) what is the probability of $T > 7$? Hint: Observe that $p_T(t)$ is symmetric about $t = 7$.

1.14 (⌣) (c) Evaluate the integral

$$\int_{-3}^{3} \frac{1}{\sqrt{2\pi}} \exp\left[-\frac{1}{2}t^2\right] dt$$

by using the approximation

$$\sum_{n=-L}^{L} \frac{1}{\sqrt{2\pi}} \exp\left[-\frac{1}{2}(n\Delta)^2\right] \Delta$$

where L is the integer closest to $3/\Delta$ (the rounded value), for $\Delta = 0.1$, $\Delta = 0.01$, $\Delta = 0.001$.

1.15 (c) Simulate a fair coin tossing experiment by modifying the code given in Section 1.4. Using 1000 repetitions of the experiment, count the number of times three heads occur. What is the simulated probability of obtaining three heads in four coin tosses? Compare your result to that obtained using (1.1).

1.16 (c) Repeat Problem 1.15 but instead consider a biased coin with $p = 0.75$. Compare your result to Figure 1.4.

Chapter 2

Computer Simulation

2.1 Introduction

Computer simulation of random phenomena has become an indispensable tool in
modern scientific investigations. So-called *Monte Carlo* computer approaches are
now commonly used to promote understanding of probabilistic problems. In this
chapter we continue our discussion of computer simulation, first introduced in Chapter 1, and set the stage for its use in later chapters. Along the way we will examine
some well known properties of random events in the process of simulating their
behavior. A more formal mathematical description will be introduced later but
careful attention to the details now, will lead to a better intuitive understanding of
the mathematical definitions and theorems to follow.

2.2 Summary

This chapter is an introduction to computer simulation of random experiments. In
Section 2.3 there are examples to show how we can use computer simulation to provide counterexamples, build intuition, and lend evidence to a conjecture. However,
it cannot be used to prove theorems. In Section 2.4 a simple MATLAB program is
given to simulate the outcomes of a discrete random variable. Section 2.5 gives many
examples of typical computer simulations used in probability, including probability
density function estimation, probability of an interval, average value of a random
variable, probability density function for a transformed random variable, and scatter diagrams for multiple random variables. Section 2.6 contains an application of
probability to the "real-world" example of a digital communication system. A brief
description of the MATLAB programming language is given in Appendix 2A.

2.3   Why Use Computer Simulation?

A computer simulation is valuable in many respects. It can be used

a. to provide counterexamples to proposed theorems

b. to build intuition by experimenting with random numbers

c. to lend evidence to a conjecture.

We now explore these uses by posing the following question: What is the effect of adding together the numerical outcomes of two or more experiments, i.e., what are the probabilities of the summed outcomes? Specifically, if U_1 represents the outcome of an experiment in which a number from 0 to 1 is chosen at random and U_2 is the outcome of an experiment in which another number is also chosen at random from 0 to 1, what are the probabilities of $X = U_1 + U_2$? The mathematical answer to this question is given in Chapter 12 (see Example 12.8), although at this point it is unknown to us. Let's say that someone asserts that there is a theorem that X is *equally likely* to be anywhere in the interval $[0, 2]$. To see if this is reasonable, we carry out a computer simulation by generating values of U_1 and U_2 and adding them together. Then we repeat this procedure M times. Next we plot a *histogram*, which gives the number of outcomes that fall in each subinterval within $[0, 2]$. As an example of a histogram consider the $M = 8$ possible outcomes for X of $\{1.7, 0.7, 1.2, 1.3, 1.8, 1.4, 0.6, 0.4\}$. Choosing the four subintervals (also called *bins*) $[0, 0.5], (0.5, 1], (1, 1.5], (1.5, 2]$, the histogram appears in Figure 2.1. In this

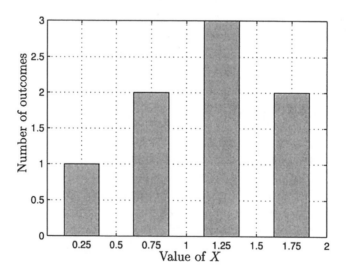

Figure 2.1: Example of a histogram for a set of 8 numbers in [0,2] interval.

example, 2 outcomes were between 0.5 and 1 and are therefore shown by the bar centered at 0.75. The other bars are similarly obtained. If we now increase the number of experiments to $M = 1000$, we obtain the histogram shown in Figure 2.2. Now it is clear that the values of X are *not equally likely*. Values near one appear

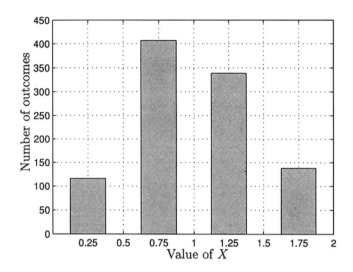

Figure 2.2: Histogram for sum of two equally likely numbers, both chosen in interval $[0, 1]$.

to be much more probable. Hence, we have generated a "counterexample" to the proposed theorem, or at least some evidence to the contrary.

We can build up our intuition by continuing with our experimentation. Attempting to justify the observed occurrences of X, we might suppose that the probabilities are higher near one because there are more ways to obtain these values. If we contrast the values of $X = 1$ versus $X = 2$, we note that $X = 2$ can only be obtained by choosing $U_1 = 1$ and $U_2 = 1$ but $X = 1$ can be obtained from $U_1 = U_2 = 1/2$ or $U_1 = 1/4, U_2 = 3/4$ or $U_1 = 3/4, U_2 = 1/4$, etc. We can lend credibility to this line of reasoning by supposing that U_1 and U_2 can only take on values in the set $\{0, 0.25, 0.5, 0.75, 1\}$ and finding all values of $U_1 + U_2$. In essence, we now look at a *simpler* problem in order to build up our intuition. An enumeration of the possible values is shown in Table 2.1 along with a "histogram" in Figure 2.3. It is clear now that the probability is highest at $X = 1$ because the number of combinations of U_1 and U_2 that will yield $X = 1$ is highest. Hence, we have learned about what happens when outcomes of experiments are added together by employing computer simulation.

We can now try to extend this result to the addition of three or more experimental outcomes via computer simulation. To do so define $X_3 = U_1 + U_2 + U_3$ and $X_4 = U_1 + U_2 + U_3 + U_4$ and repeat the simulation. A computer simulation with $M = 1000$ trials produces the histograms shown in Figure 2.4. It appears to

		U_2			
	0.00	0.25	0.50	0.75	1.00
0.00	0.00	0.25	0.50	0.75	1.00
0.25	0.25	0.50	0.75	1.00	1.25
U_1 0.50	0.50	0.75	1.00	1.25	1.50
0.75	0.75	1.00	1.25	1.50	1.75
1.00	1.00	1.25	1.50	1.75	2.00

Table 2.1: Possible values for $X = U_1 + U_2$ for intuition-building experiment.

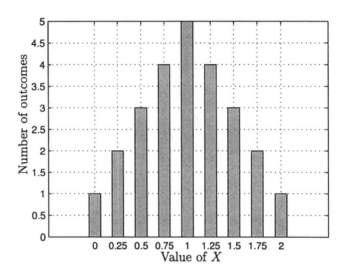

Figure 2.3: Histogram for X for intuition-building experiment.

bear out the conjecture that the most probable values are near the center of the $[0, 3]$ and $[0, 4]$ intervals, respectively. Additionally, the histograms appear more like a bell-shaped or Gaussian curve. Hence, we might now *conjecture*, based on these computer simulations, that as we add more and more experimental outcomes together, we will obtain a Gaussian-shaped histogram. This is in fact true, as will be proven later (see central limit theorem in Chapter 15). Note that we cannot *prove* this result using a computer simulation but only lend evidence to our theory. However, the use of computer simulations indicates *what* we need to prove, information that is invaluable in practice. In summary, computer simulation is a valuable tool for lending credibility to conjectures, building intuition, and uncovering new results.

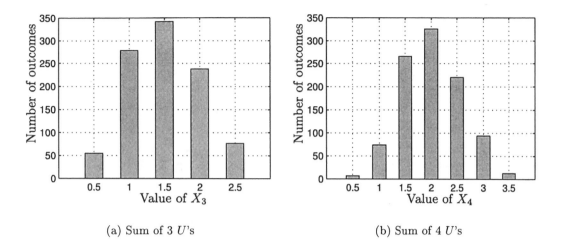

(a) Sum of 3 U's   (b) Sum of 4 U's

Figure 2.4: Histograms for addition of outcomes.

 Computer simulations cannot be used to prove theorems.

In Figure 2.2, which displayed the outcomes for 1000 trials, is it possible that the computer simulation could have produced 500 outcomes in [0,0.5], 500 outcomes in [1.5,2] and no outcomes in (0.5,1.5)? The answer is yes, although it is improbable. It can be shown that the probability of this occuring is

$$\binom{1000}{500} \left(\frac{1}{8}\right)^{1000} \approx 2.2 \times 10^{-604}$$

(see Problem 12.27).

2.4   Computer Simulation of Random Phenomena

In the previous chapter we briefly explained how to use a digital computer to simulate a random phenomenon. We now continue that discussion in more detail. Then, the following section applies the techniques to specific problems ecountered in probability. As before, we will distinguish between experiments that produce discrete outcomes from those that produce continuous outcomes.

We first define a *random variable X* as the *numerical outcome* of the random experiment. Typical examples are the number of dots on a die (discrete) or the distance of a dart from the center of a dartboard of radius one (continuous). The random variable X can take on the values in the set $\{1,2,3,4,5,6\}$ for the first example and in the set $\{r : 0 \leq r \leq 1\}$ for the second example. We denote the random variable by a *capital letter*, say X, and its possible *values* by a small letter, say x_i for the discrete case and x for the continuous case. The distinction is analogous to that between a function *defined* as $g(x) = x^2$ and the *values* $y = g(x)$ that $g(x)$ can take on.

Now it is of interest to determine various properties of X. To do so we use a computer simulation, performing many experiments and observing the outcome for each experiment. The number of experiments, which is sometimes referred to as the number of *trials*, will be denoted by M. To simulate a discrete random variable we use **rand**, which generates a number at random within the $(0,1)$ interval (see Appendix 2A for some MATLAB basics). Assume that in general the possible values of X are $\{x_1, x_2, \ldots, x_N\}$ with probabilities $\{p_1, p_2, \ldots, p_N\}$. As an example, if $N = 3$ we can generate M values of X by using the following code segment (which assumes M,x1,x2,x3,p1,p2,p3 have been previously assigned):

```
for i=1:M
  u=rand(1,1);
  if u<=p1
    x(i,1)=x1;
  elseif u>p1 & u<=p1+p2
    x(i,1)=x2;
  elseif u>p1+p2
    x(i,1)=x3;
  end
end
```

After this code is executed, we will have generated M values of the random variable X. Note that the values of X so obtained are termed the *outcomes* or *realizations* of X. The extension to any number N of possible values is immediate. For a continuous random variable X that is Gaussian we can use the code segment:

```
for i=1:M
  x(i,1)=randn(1,1);
end
```

or equivalently x=randn(M,1). Again at the conclusion of this code segment we will have generated M realizations of X. Later we will see how to generate realizations of random variables whose PDFs are not Gaussian (see Section 10.9).

2.5 Determining Characteristics of Random Variables

There are many ways to characterize a random variable. We have already alluded to the probability of the outcomes in the discrete case and the PDF in the continuous case. To be more precise consider a discrete random variable, such as that describing the outcome of a coin toss. If we toss a coin and let X be 1 if a head is observed and let X be 0 if a tail is observed, then the probabilities are defined to be p for $X = x_1 = 1$ and $1 - p$ for $X = x_2 = 0$. The probability p of $X = 1$ can be thought of as the relative frequency of the outcome of heads in a long succession of tosses. Hence, to determine the probability of heads we could toss a coin a large number of times and estimate p by the number of observed heads divided by the number of tosses. Using a computer to simulate this experiment, we might inquire as to the number of tosses that would be necessary to obtain an accurate estimate of the probability of heads. Unfortunately, this is not easily answered. A practical means, though, is to increase the number of tosses until the estimate so computed converges to a fixed number. A computer simulation is shown in Figure 2.5 where the estimate

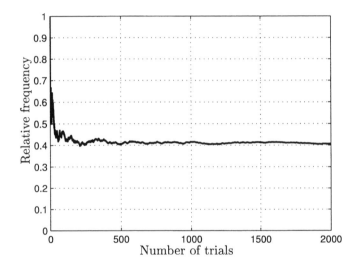

Figure 2.5: Estimate of probability of heads for various number of coin tosses.

appears to converge to about 0.4. Indeed, the true value (that value used in the simulation) was $p = 0.4$. It is also seen that the estimate of p is slightly higher than 0.4. This is due to the slight imperfections in the random number generator as well as computational errors. Increasing the number of trials will not improve the results. We next describe some typical simulations that will be useful to us. To illustrate the various simulations we will use a Gaussian random variable with realizations generated using `randn(1,1)`. Its PDF is shown in Figure 2.6.

Example 2.1 – Probability density function

A PDF may be estimated by first finding the histogram and then dividing the number of outcomes in each bin by M, the total number of realizations, to obtain the probability. Then to obtain the PDF $p_X(x)$ recall that the probability of X

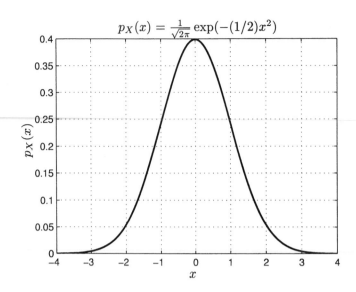

Figure 2.6: Gaussian probability density function.

taking on a value in an interval is found as the area under the PDF of that interval (see Section 1.3). Thus,

$$P[a \leq X \leq b] = \int_a^b p_X(x)dx \qquad (2.1)$$

and if $a = x_0 - \Delta x/2$ and $b = x_0 + \Delta x/2$, where Δx is small, then (2.1) becomes

$$P[x_0 - \Delta x/2 \leq X \leq x_0 + \Delta x/2] \approx p_X(x_0)\Delta x$$

and therefore the PDF at $x = x_0$ is approximately

$$p_X(x_0) \approx \frac{P[x_0 - \Delta x/2 \leq X \leq x_0 + \Delta x/2]}{\Delta x}.$$

Hence, we need only divide the estimated probability by the bin width Δx. Also, note that as claimed in Chapter 1, $p_X(x)$ is seen to be the *probability per unit length*. In Figure 2.7 is shown the estimated PDF for a Gaussian random variable as well as the true PDF as given in Figure 2.6. The MATLAB code used to generate the figure is also shown.

Example 2.2 – Probability of an interval
To determine $P[a \leq X \leq b]$ we need only generate M realizations of X, then count the number of outcomes that fall into the $[a, b]$ interval and divide by M. Of course

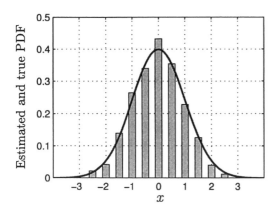

```
randn('state',0)
x=randn(1000,1);
bincenters=[-3.5:0.5:3.5]';
bins=length(bincenters);
h=zeros(bins,1);
for i=1:length(x)
  for k=1:bins
    if x(i)>bincenters(k)-0.5/2 ...
      & x(i)<=bincenters(k)+0.5/2
      h(k,1)=h(k,1)+1;
    end
  end
end
pxest=h/(1000*0.5);
xaxis=[-4:0.01:4]';
px=(1/sqrt(2*pi))*exp(-0.5*xaxis.^2);
```

Figure 2.7: Estimated and true probability density functions.

M should be large. In particular, if we let $a = 2$ and $b = \infty$, then we should obtain the value (which must be evaluated using numerical integration)

$$P[X > 2] = \int_2^\infty \frac{1}{\sqrt{2\pi}} \exp\left(-(1/2)x^2\right) dx = 0.0228$$

and therefore very few realizations can be expected to fall in this interval. The results for an increasing number of realizations are shown in Figure 2.8. This illustrates the problem with the simulation of small probability events. It requires a large number of realizations to obtain accurate results. (See Problem 11.47 on how to reduce the number of realizations required.)

◇

Example 2.3 – Average value

It is frequently important to measure characteristics of X in addition to the PDF. For example, we might only be interested in the average or *mean* or *expected value* of X. If the random variable is Gaussian, then from Figure 2.6 we would expect X to be zero on the average. This conjecture is easily "verified" by using the *sample mean* estimate

$$\frac{1}{M} \sum_{i=1}^{M} x_i$$

M	Estimated $P[X > 2]$	True $P[X > 2]$
100	0.0100	0.0228
1000	0.0150	0.0228
10,000	0.0244	0.0288
100,000	0.0231	0.0288

```
randn('state',0)
M=100;count=0;
x=randn(M,1);
for i=1:M
   if x(i)>2
      count=count+1;
   end
end
probest=count/M
```

Figure 2.8: Estimated and true probabilities.

of the mean. The results are shown in Figure 2.9.

M	Estimated mean	True mean
100	0.0479	0
1000	−0.0431	0
10,000	0.0011	0
100,000	0.0032	0

```
randn('state',0)
M=100;
meanest=0;
x=randn(M,1);
for i=1:M
    meanest=meanest+(1/M)*x(i);
end
meanest
```

Figure 2.9: Estimated and true mean.

Example 2.4 – A transformed random variable

One of the most important problems in probability is to determine the PDF for a transformed random variable, i.e., one that is a function of X, say X^2 as an example. This is easily accomplished by modifying the code in Figure 2.7 from `x=randn(1000,1)` to `x=randn(1000,1);x=x.^2;`. The results are shown in Figure 2.10.   Note that the shape of the PDF is completely different than the original Gaussian shape (see Example 10.7 for the true PDF). Additionally, we can obtain the mean of X^2 by using

$$\frac{1}{M}\sum_{i=1}^{M} x_i^2$$

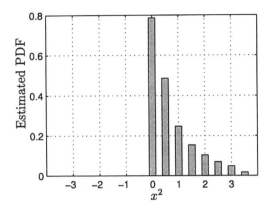

Figure 2.10: Estimated PDF of X^2 for X Gaussian.

as we did in Example 2.3. The results are shown in Figure 2.11.

M	Estimated mean	True mean
100	0.7491	1
1000	0.8911	1
10,000	1.0022	1
100,000	1.0073	1

```
randn('state',0)
M=100;
meanest=0;
x=randn(M,1);
for i=1:M
    meanest=meanest+(1/M)*x(i)^2;
end
meanest
```

Figure 2.11: Estimated and true mean.

Example 2.5 – Multiple random variables

Consider an experiment that yields two random variables or the *vector* random variable $[X_1\, X_2]^T$, where T denotes the transpose. An example might be the choice of a point in the square $\{(x, y) : 0 \leq x \leq 1, 0 \leq y \leq 1\}$ according to some procedure. This procedure may or may not cause the value of x_2 to depend on the value of x_1. For example, if the result of many repetitions of this experiment produced an even distribution of points indicated by the shaded region in Figure 2.12a, then we would say that there is no dependency between X_1 and X_2. On the other hand, if the points were evenly distributed within the shaded region shown in Figure 2.12b, then there is a strong dependency. This is because if, for example, $x_1 = 0.5$, then x_2 would have to lie in the interval $[0.25, 0.75]$. Consider next the random vector

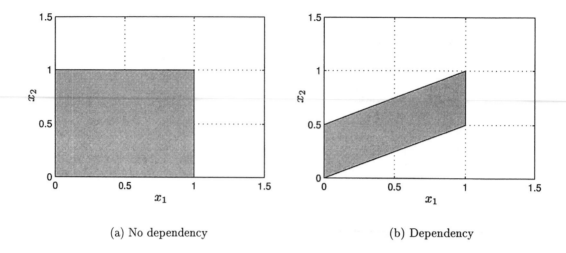

(a) No dependency   (b) Dependency

Figure 2.12: Relationships between random variables.

$$
\begin{bmatrix} X_1 \\ X_2 \end{bmatrix} = \begin{bmatrix} U_1 \\ U_2 \end{bmatrix}
$$

where each U_i is generated using **rand**. The result of $M = 1000$ realizations is shown in Figure 2.13a. We say that the random variables X_1 and X_2 are *independent*. Of course, this is what we expect from a good random number generator. If instead, we defined the new random variables,

$$
\begin{bmatrix} X_1 \\ X_2 \end{bmatrix} = \begin{bmatrix} U_1 \\ \frac{1}{2}U_1 + \frac{1}{2}U_2 \end{bmatrix}
$$

then from the plot shown in Figure 2.13b, we would say that the random variables are dependent. Note that this type of plot is called a *scatter diagram*.

◇

2.6   Real-World Example – Digital Communications

In a phase-shift keyed (PSK) digital communication system a binary digit (also termed a *bit*), which is either a "0" or a "1", is communicated to a receiver by sending either $s_0(t) = A\cos(2\pi F_0 t + \pi)$ to represent a "0" or $s_1(t) = A\cos(2\pi F_0 t)$ to represent a "1", where $A > 0$ [Proakis 1989]. The receiver that is used to decode the transmission is shown in Figure 2.14.   The input to the receiver is the noise

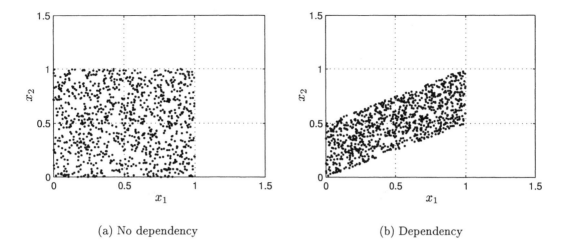

Figure 2.13: Relationships between random variables.

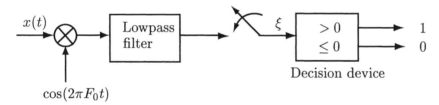

Figure 2.14: Receiver for a PSK digital communication system.

corrupted signal or $x(t) = s_i(t) + w(t)$, where $w(t)$ represents the channel noise. Ignoring the effect of noise for the moment, the output of the multiplier will be

$$s_0(t)\cos(2\pi F_0 t) = A\cos(2\pi F_0 t + \pi)\cos(2\pi F_0 t) = -A\left(\frac{1}{2} + \frac{1}{2}\cos(4\pi F_0 t)\right)$$

$$s_1(t)\cos(2\pi F_0 t) = A\cos(2\pi F_0 t)\cos(2\pi F_0 t) = A\left(\frac{1}{2} + \frac{1}{2}\cos(4\pi F_0 t)\right)$$

for a 0 and 1 sent, respectively. After the lowpass filter, which filters out the $\cos(4\pi F_0 t)$ part of the signal, and sampler, we have

$$\xi = \begin{cases} -\frac{A}{2} & \text{for a } 0 \\ \frac{A}{2} & \text{for a } 1. \end{cases}$$

The receiver decides a 1 was transmitted if $\xi > 0$ and a 0 if $\xi \le 0$. To model the channel noise we assume that the actual value of ξ observed is

$$\xi = \begin{cases} -\frac{A}{2} + W & \text{for a } 0 \\ \frac{A}{2} + W & \text{for a } 1 \end{cases}$$

where W is a Gaussian random variable. It is now of interest to determine how the error depends on the signal amplitude A. Consider the case of a 1 having been transmitted. Intuitively, if A is a large positive amplitude, then the chance that the noise will cause an error or equivalently, $\xi \leq 0$, should be small. This probability, termed the *probability of error* and denoted by P_e, is given by $P[A/2 + W \leq 0]$. Using a computer simulation we can plot P_e versus A with the result shown in Figure 2.15. Also, the true P_e is shown. (In Example 10.3 we will see how to analytically determine this probability.) As expected, the probability of error decreases as the

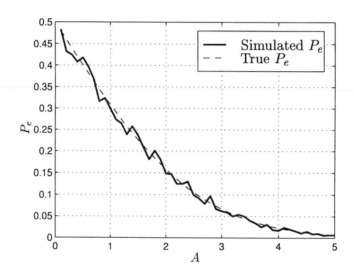

Figure 2.15: Probability of error for a PSK communication system.

signal amplitude increases. With this information we can design our system by choosing A to satisfy a given probability of error requirement. In actual systems this requirement is usually about $P_e = 10^{-7}$. Simulating this small probability would be exceedingly difficult due to the large number of trials required (but see also Problem 11.47). The MATLAB code used for the simulation is given in Figure 2.16.

References

Proakis, J., *Digitial Communications*, Second Ed., McGraw-Hill, New York, 1989.

Problems

Note: All the following problems require the use of a computer simulation. A realization of a *uniform* random variable is obtained by using `rand(1,1)` while a

```
A=[0.1:0.1:5]';
for k=1:length(A)
  error=0;
  for i=1:1000
    w=randn(1,1);
    if A(k)/2+w<=0
      error=error+1;
    end
  end
  Pe(k,1)=error/1000;
end
```

Figure 2.16: MATLAB code used to estimate the probability of error P_e in Figure 2.15.

realization of a *Gaussian* random variable is obtained by using `randn(1,1)`.

2.1 (☺) **(c)** An experiment consists of tossing a fair coin twice. If a head occurs on the first toss, we let $x_1 = 1$ and if a tail occurs we let $x_1 = 0$. The same assignment is used for the outcome x_2 of the second toss. Defining the random variable as $Y = X_1 X_2$, estimate the probabilities for the different possible values of Y. Explain your results.

2.2 **(c)** A pair of fair dice is tossed. Estimate the probability of "snake eyes" or a one for each die?

2.3 (☺) **(c)** Estimate $P[-1 \le X \le 1]$ if X is a Gaussian random variable. Verify the results of your computer simulation by numerically evaluating the integral

$$\int_{-1}^{1} \frac{1}{\sqrt{2\pi}} \exp\left(-\frac{1}{2}x^2\right) dx.$$

Hint: See Problem 1.14.

2.4 **(c)** Estimate the PDF of the random variable

$$X = \sum_{i=1}^{12} \left(U_i - \frac{1}{2}\right)$$

where U_i is a uniform random variable. Then, compare this PDF to the Gaussian PDF or

$$p_X(x) = \frac{1}{\sqrt{2\pi}} \exp\left(-\frac{1}{2}x^2\right).$$

2.5 (c) Estimate the PDF of $X = U_1 - U_2$, where U_1 and U_2 are uniform random variables. What is the most probable range of values?

2.6 ($\ddot{\smile}$) (c) Estimate the PDF of $X = U_1 U_2$, where U_1 and U_2 are uniform random variables. What is the most probable range of values?

2.7 (c) Generate realizations of a discrete random variable X, which takes on values 1, 2, and 3 with probabilities $p_1 = 0.1$, $p_2 = 0.2$ and $p_3 = 0.7$, respectively. Next based on the generated realizations estimate the probabilities of obtaining the various values of X.

2.8 ($\ddot{\smile}$) (c) Estimate the mean of U, where U is a uniform random variable. What is the true value?

2.9 (c) Estimate the mean of $X + 1$, where X is a Gaussian random variable. What is the true value?

2.10 (c) Estimate the mean of X^2, where X is a Gaussian random variable.

2.11 ($\ddot{\smile}$) (c) Estimate the mean of $2U$, where U is a uniform random variable. What is the true value?

2.12 (c) It is conjectured that if X_1 and X_2 are Gaussian random variables, then by subtracting them (let $Y = X_1 - X_2$), the probable range of values should be smaller. Is this true?

2.13 ($\ddot{\smile}$) (c) A large circular dartboard is set up with a "bullseye" at the center of the circle, which is at the coordinate $(0,0)$. A dart is thrown at the center but lands at (X, Y), where X and Y are two different Gaussian random variables. What is the average distance of the dart from the bullseye?

2.14 ($\ddot{\smile}$) (c) It is conjectured that the mean of $\sqrt{U}$, where U is a uniform random variable, is $\sqrt{\text{mean of } U}$. Is this true?

2.15 (c) The Gaussian random variables X_1 and X_2 are linearly transformed to the new random variables

$$\begin{aligned} Y_1 &= X_1 + 0.1X_2 \\ Y_2 &= X_1 + 0.2X_2. \end{aligned}$$

Plot a scatter diagram for Y_1 and Y_2. Could you approximately determine the value of Y_2 if you knew that $Y_1 = 1$?

2.16 (c,w) Generate a scatter diagram for the linearly transformed random variables

$$\begin{aligned} X_1 &= U_1 \\ X_2 &= U_1 + U_2 \end{aligned}$$

where U_1 and U_2 are uniform random variables. Can you explain why the scatter diagram looks like a parallelogram? Hint: Define the vectors

$$\mathbf{X} = \begin{bmatrix} X_1 \\ X_2 \end{bmatrix}$$

$$\mathbf{e}_1 = \begin{bmatrix} 1 \\ 1 \end{bmatrix}$$

$$\mathbf{e}_2 = \begin{bmatrix} 0 \\ 1 \end{bmatrix}$$

and express $\mathbf{X}$ as a linear combination of $\mathbf{e}_1$ and $\mathbf{e}_2$.

Appendix 2A

Brief Introduction to MATLAB

A brief introduction to the scientific software package MATLAB is contained in this appendix. Further information is available at the Web site www.mathworks.com. MATLAB is a scientific computation and data presentation language.

Overview of MATLAB

The chief advantage of MATLAB is its use of high-level instructions for matrix algebra and built-in routines for data processing. In this appendix as well as throughout the text a MATLAB command is indicated with the typewriter font such as `end`. MATLAB treats matrices of any size (which includes vectors and scalars as special cases) as *elements* and hence matrix multiplication is as simple as `C=A*B`, where `A` and `B` are conformable matrices. In addition to the usual matrix operations of addition `C=A+B`, multiplication `C=A*B`, and scaling by a constant `c` as `B=c*A`, certain matrix operators are defined that allow convenient manipulation. For example, assume we first define the column vector $\mathbf{x} = [1\,2\,3\,4]^T$, where T denotes transpose, by using `x=[1:4]'`. The vector starts with the element 1 and ends with the element 4 and the colon indicates that the intervening elements are found by incrementing the start value by one, which is the default. For other increments, say 0.5, we use `x=[1:0.5:4]'`. To define the vector $\mathbf{y} = [1^2\,2^2\,3^2\,4^2]^T$, we can use the matrix *element by element* exponentiation operator `.^` to form `y=x.^2` if `x=[1:4]'`. Similarly, the operators `.*` and `./` perform element by element multiplication and division of the matrices, respectively. For example, if

$$\mathbf{A} \;=\; \begin{bmatrix} 1 & 2 \\ 3 & 4 \end{bmatrix}$$

$$\mathbf{B} \;=\; \begin{bmatrix} 1 & 2 \\ 3 & 4 \end{bmatrix}$$

Character	Meaning
+	addition (scalars, vectors, matrices)
−	subtraction (scalars, vectors, matrices)
*	multiplication (scalars, vectors, matrices)
/	division (scalars)
^	exponentiation (scalars, square matrices)
.*	element by element multiplication
./	element by element division
.^	element by element exponentiation
;	suppress printed output of operation
:	specify intervening values
'	conjugate transpose (transpose for real vectors, matrices)
...	line continuation (when command must be split)
%	remainder of line interpreted as comment
==	logical equals
		logical or
&	logical and
~ =	logical not

Table 2A.1: Definition of common MATLAB characters.

then the statements C=A.*B and D=A./B produce the results

$$C \;=\; \begin{bmatrix} 1 & 4 \\ 9 & 16 \end{bmatrix}$$

$$D \;=\; \begin{bmatrix} 1 & 1 \\ 1 & 1 \end{bmatrix}$$

respectively. A listing of some common characters is given in Table 2A.1. MATLAB has the usual built-in functions of cos, sin, etc. for the trigonometric functions, sqrt for a square root, exp for the exponential function, and abs for absolute value, as well as many others. When a function is applied to a matrix, the function is applied to each element of the matrix. Other built-in symbols and functions and their meanings are given in Table 2A.2.

Matrices and vectors are easily specified. For example, to define the column vector $c_1 = [1\ 2]^T$, just use c1=[1 2].' or equivalently c1=[1;2]. To define the C matrix given previously, the construction C=[1 4;9 16] is used. Or we could first define $c_2 = [4\ 16]^T$ by c2=[4 16].' and then use C=[c1 c2]. It is also possible to extract portions of matrices to yield smaller matrices or vectors. For example, to extract the first column from the matrix C use c1=C(:,1). The colon indicates that all elements in the first column should be extracted. Many other convenient manipulations of matrices and vectors are possible.

Function	Meaning
`pi`	π
`i`	$\sqrt{-1}$
`j`	$\sqrt{-1}$
`round(x)`	rounds every element in $\mathbf{x}$ to the nearest integer
`floor(x)`	replaces every element in $\mathbf{x}$ by the nearest integer less than or equal to x
`inv(A)`	takes the inverse of the square matrix $\mathbf{A}$
`x=zeros(N,1)`	assigns an $N \times 1$ vector of all zeros to $\mathbf{x}$
`x=ones(N,1)`	assigns an $N \times 1$ vector of all ones to $\mathbf{x}$
`x=rand(N,1)`	generates an $N \times 1$ vector of all uniform random variables
`x=randn(N,1)`	generates an $N \times 1$ vector of all Gaussian random variables
`rand('state',0)`	initializes uniform random number generator
`randn('state',0)`	initializes Gaussian random number generator
`M=length(x)`	sets M equal to N if $\mathbf{x}$ is $N \times 1$
`sum(x)`	sums all elements in vector $\mathbf{x}$
`mean(x)`	computes the sample mean of the elements in $\mathbf{x}$
`flipud(x)`	flips the vector $\mathbf{x}$ upside down
`abs`	takes the absolute value (or complex magnitude) of every element of $\mathbf{x}$
`fft(x,N)`	computes the FFT of length N of $\mathbf{x}$ (zero pads if `N>length(x)`)
`ifft(x,N)`	computes the inverse FFT of length N of $\mathbf{x}$
`fftshift(x)`	interchanges the two halves of an FFT output
`pause`	pauses the execution of a program
`break`	terminates a loop when encountered
`whos`	lists all variables and their attributes in current workspace
`help`	provides help on commands, e.g., `help sqrt`

Table 2A.2: Definition of useful MATLAB symbols and functions.

Any vector that is generated whose dimensions are not explicitly specified is assumed to be a *row* vector. For example, if we say `x=ones(10)`, then it will be designated as the 1×10 row vector consisting of all ones. To yield a column vector use `x=ones(10,1)`.

Loops are implemented with the construction

```
for k=1:10
    x(k,1)=1;
end
```

which is equivalent to `x=ones(10,1)`. Logical flow can be accomplished with the construction

```
if x>0
    y=sqrt(x);
else
    y=0;
end
```

Finally, a good practice is to begin each program or script, which is called an "m" file (due to its syntax, for example, `pdf.m`), with a `clear all` command. This will clear all variables in the workspace, since otherwise the current program may inadvertently (on the part of the programmer) use previously stored variable data.

Plotting in MATLAB

Plotting in MATLAB is illustrated in the next section by example. Some useful functions are summarized in Table 2A.3.

Function	Meaning
`figure`	opens up a new figure window
`plot(x,y)`	plots the elements of **x** versus the elements of **y**
`plot(x1,y1,x2,y2)`	same as above except multiple plots are made
`plot(x,y,'.')`	same as `plot` except the points are not connected
`title('my plot')`	puts a title on the plot
`xlabel('x')`	labels the x axis
`ylabel('y')`	labels the y axis
`grid`	draws grid on the plot
`axis([0 1 2 4])`	plots only the points in range $0 \le x \le 1$ and $2 \le y \le 4$
`text(1,1,'curve 1')`	places the text "curve 1" at the point (1,1)
`hold on`	holds current plot
`hold off`	releases current plot

Table 2A.3: Definition of useful MATLAB plotting functions.

An Example Program

A complete MATLAB program is given below to illustrate how one might compute the samples of several sinusoids of different amplitudes. It also allows the sinusoids to be clipped. The sinusoid is $s(t) = A\cos(2\pi F_0 t + \pi/3)$, with $A = 1$, $A = 2$, and $A = 4$, $F_0 = 1$, and $t = 0, 0.01, 0.02, \ldots, 10$. The clipping level is set at ± 3, i.e., any sample above $+3$ is clipped to $+3$ and any sample less than -3 is clipped to -3.

```
%  matlabexample.m
%
%  This program computes and plots samples of a sinusoid
%  with amplitudes 1, 2, and 4.  If desired, the sinusoid can be
%  clipped to simulate the effect of a limiting device.
%  The frequency is 1 Hz and the time duration is 10 seconds.
%  The sample interval is 0.1 seconds. The code is not efficient but
%  is meant to illustrate MATLAB statements.
%
clear all % clear all variables from workspace
delt=0.01; % set sampling time interval
F0=1; % set frequency
t=[0:delt:10]'; % compute time samples 0,0.01,0.02,...,10
A=[1 2 4]'; % set amplitudes
clip='yes'; % set option to clip
for i=1:length(A) % begin computation of sinusoid samples
    s(:,i)=A(i)*cos(2*pi*F0*t+pi/3); % note that samples for sinusoid
                                     % are computed all at once and
                                     % stored as columns in a matrix
    if clip=='yes' % determine if clipping desired
        for k=1:length(s(:,i)) % note that number of samples given as
                               % dimension of column using length command
            if s(k,i)>3 % check to see if sinusoid sample exceeds 3
                s(k,i)=3; % if yes, then clip
            elseif s(k,i)<-3 % check to see if sinusoid sample is less
                s(k,i)=-3;   % than -3 if yes, then clip
            end
        end
    end
end
figure % open up a new figure window
plot(t,s(:,1),t,s(:,2),t,s(:,3)) % plot sinusoid samples versus time
                                 % samples for all three sinusoids
grid % add grid to plot
xlabel('time, t') % label x-axis
```

```
ylabel('s(t)') % label y-axis
axis([0 10 -4 4]) % set up axes using axis([xmin xmax ymin ymax])
legend('A=1','A=2','A=4') % display a legend to distinguish
                         % different sinusoids
```

The output of the program is shown in Figure 2A.1. Note that the different graphs will appear as different colors.

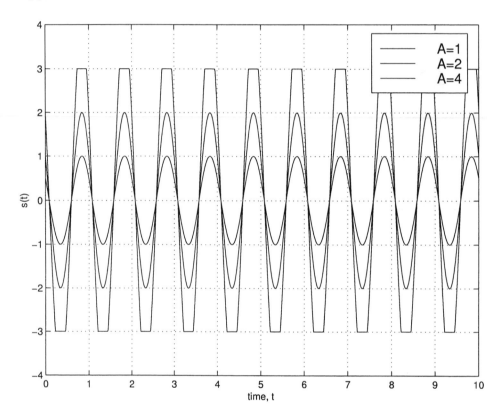

Figure 2A.1: Output of MATLAB program `matlabexample.m`.

Chapter 3

Basic Probability

3.1 Introduction

We now begin the formal study of probability. We do so by utilizing the properties of sets in conjunction with the *axiomatic* approach to probability. In particular, we will see how to solve a class of probability problems via *counting methods*. These are problems such as determining the probability of obtaining a royal flush in poker or of obtaining a defective item from a batch of mostly good items, as examples. Furthermore, the axiomatic approach will provide the basis for all our further studies of probability. Only the methods of determining the probabilities will have to be modified in accordance with the problem at hand.

3.2 Summary

Section 3.3 reviews set theory, with Figure 3.1 illustrating the standard definitions. Manipulation of sets can be facilitated using De Morgan's laws of (3.6) and (3.7). The application of set theory to probability is summarized in Table 3.1. Using the three axioms described in Section 3.4 a theory of probability can be formulated and a means for computing probabilities constructed. Properties of the probability function are given in Section 3.5. In addition, the probability for a union of three events is given by (3.20). An equally likely probability assignment for a continuous sample space is given by (3.22) and is shown to satisfy the basic axioms. Section 3.7 introduces the determination of probabilities for discrete sample spaces with equally likely outcomes. The basic formula is given by (3.24). To implement this approach for more complicated problems in which brute-force counting of outcomes is not possible, the subject of combinatorics is described in Section 3.8. Permutations and combinations are defined and applied to several examples for computing probabilities. Based on these counting methods the hypergeometric probability law of (3.27) and the binomial probability law of (3.28) are derived in Section 3.9. Finally, an example of the application of the binomial law to a quality control problem is given in Section 3.10.

3.3   Review of Set Theory

The reader has undoubtedly been introduced to set theory at some point in his/her education. We now summarize only the salient definitions and properties that are germane to probability. A set is defined as a collection of objects, for example, the set of students in a probability class. The set A can be defined either by the *enumeration* method, i.e., a listing of the students as

$$A = \{\text{Jane, Bill, Jessica, Fred}\} \tag{3.1}$$

or by the *description* method

$$A = \{\text{students: each student is enrolled in the probability class}\}$$

where the ":" is read as "such that". Another example would be the set of natural numbers or

$$
\begin{aligned}
B &= \{1, 2, 3, \ldots\} & \text{(enumeration)} & \tag{3.2} \\
B &= \{I : I \text{ is an integer and } I \geq 1\} & \text{(description)}.
\end{aligned}
$$

Each object in the set is called an *element* and each element is *distinct*. For example, the sets $\{1, 2, 3\}$ and $\{1, 2, 1, 3\}$ are equivalent. There is no reason to list an element in a set more than once. Likewise, the *ordering* of the elements within the set is not important. The sets $\{1, 2, 3\}$ and $\{2, 1, 3\}$ are equivalent. Sets are said to be *equal* if they contain the same elements. For example, if $C_1 = \{\text{Bill, Fred}\}$ and $C_2 = \{\text{male members in the probability class}\}$, then $C_1 = C_2$. Although the description may change, it is ultimately the contents of the set that is of importance. An element x of a set A is denoted using the symbolism $x \in A$, and is read as "x is contained in A", as for example, $1 \in B$ for the set B defined in (3.2). Some sets have no elements. If the instructor in the probability class does not give out any grades of "A", then the set of students receiving an "A" is $D = \{\,\}$. This is called the *empty set* or the *null set*. It is denoted by $\emptyset$ so that $D = \emptyset$. On the other hand, the instructor may be an easy grader and give out all "A"s. Then, we say that $D = \mathcal{S}$, where $\mathcal{S}$ is called the *universal set* or the set of *all* students enrolled in the probability class. These concepts, in addition to some others, are further illustrated in the next example.

Example 3.1 – Set concepts

Consider the set of *all outcomes* of a tossed die. This is

$$A = \{1, 2, 3, 4, 5, 6\}. \tag{3.3}$$

The numbers $1, 2, 3, 4, 5, 6$ are its elements, which are distinct. The set of integer numbers from 1 to 6 or $B = \{I : 1 \leq I \leq 6\}$ is equal to A. The set A is also the universal set $\mathcal{S}$ since it contains all the outcomes. This is in contrast to the set

$C = \{2, 4, 6\}$, which contains only the even outcomes. The set C is called a *subset* of A. A *simple set* is a set containing a *single* element, as for example, $C = \{1\}$.

$\diamond$

  Element vs. simple set

In the example of the probability class consider the set of instructors. Usually, there is only one instructor and so the set of instructors can be defined as the simple set $A = \{$Professor Laplace$\}$. However, this is not the same as the "element" given by Professor Laplace. A distinction is therefore made between the instructors teaching probability and an individual instructor. As another example, it is clear that sometimes elements in a set can be added, as, for example, $2 + 3 = 5$, but it makes no sense to add sets as in $\{2\} + \{3\} = \{5\}$.

More formally, a set B is defined as a subset of a set A if every element in B is also an element of A. We write this as $B \subset A$. This also includes the case of $B = A$. In fact, we can say that $A = B$ if $A \subset B$ *and* $B \subset A$.

Besides subsets, new sets may be derived from other sets in a number of ways. If $S = \{x : -\infty < x < \infty\}$ (called the set of *real numbers*), then $A = \{x : 0 < x \le 2\}$ is clearly a subset of S. The *complement* of A, denoted by A^c, is the set of elements in S but not in A. This is $A^c = \{x : x \le 0 \text{ or } x > 2\}$. Two sets can be combined together to form a new set. For example, if

$$
\begin{aligned}
A &= \{x : 0 \le x \le 2\} \\
B &= \{x : 1 \le x \le 3\}
\end{aligned}
\tag{3.4}
$$

then the *union* of A and B, denoted by $A \cup B$, is the set of elements that belong to A *or* B or both A *and* B (so-called *inclusive or*). Hence, $A \cup B = \{x : 0 \le x \le 3\}$. This definition may be extended to multiple sets $A_1, A_2, \ldots, A_N$ so that the union is the set of elements for which each element belongs to at least one of these sets. It is denoted by

$$
A_1 \cup A_2 \cup A_2 \cup \cdots \cup A_N = \bigcup_{i=1}^{N} A_i.
$$

The *intersection* of sets A and B, denoted by $A \cap B$, is defined as the set of elements that belong to *both A and B*. Hence, $A \cap B = \{x : 1 \le x \le 2\}$ for the sets of (3.4). We will sometimes use the shortened symbolism AB to denote $A \cap B$. This definition may be extended to multiple sets $A_1, A_2, \ldots, A_N$ so that the intersection is the set of elements for which each element belongs to *all* of these sets. It is denoted by

$$A_1 \cap A_2 \cap A_2 \cap \cdots \cap A_N = \bigcap_{i=1}^{N} A_i.$$

The *difference* between sets, denoted by $A - B$, is the set of elements in A but *not* in B. Hence, for the sets of (3.4) $A - B = \{x : 0 \leq x < 1\}$. These concepts can be illustrated pictorially using a *Venn diagram* as shown in Figure 3.1. The darkly

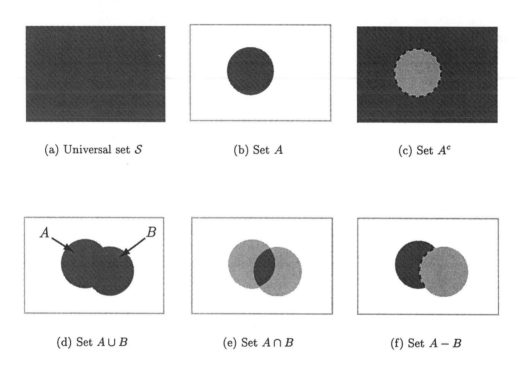

(a) Universal set $\mathcal{S}$ (b) Set A (c) Set A^c

(d) Set $A \cup B$ (e) Set $A \cap B$ (f) Set $A - B$

Figure 3.1: Illustration of set definitions – darkly shaded region indicates the set.

shaded regions are the sets described. The dashed portions are *not* included in the sets. A Venn diagram is useful for visualizing set operations. As an example, one might inquire whether the sets $A - B$ and $A \cap B^c$ are equivalent or if

$$A - B = A \cap B^c. \tag{3.5}$$

From Figures 3.2 and 3.1f we see that they appear to be. However, to formally prove that this relationship is true requires one to let $C = A - B$, $D = A \cap B^c$ and prove that (a) $C \subset D$ and (b) $D \subset C$. To prove (a) assume that $x \in A - B$. Then, by definition of the difference set (see Figure 3.1f) $x \in A$ but x is *not* an element of B. Hence, $x \in A$ *and* x must also be an element of B^c. Since $D = A \cap B^c$, x must be an element of D. Hence, $x \in A \cap B^c$ and since this is true for every $x \in A - B$,

Figure 3.2: Using Venn diagrams to "validate" set relationships.

we have that $A - B \subset A \cap B^c$. The reader is asked to complete the proof of (b) in Problem 3.6.

With the foregoing set definitions a number of results follow. They will be useful in manipulating sets to allow easier calculation of probabilities. We now list these.

1. $(A^c)^c = A$

2. $A \cup A^c = \mathcal{S}$, $A \cap A^c = \emptyset$

3. $A \cup \emptyset = A$, $A \cap \emptyset = \emptyset$

4. $A \cup \mathcal{S} = \mathcal{S}$, $A \cap \mathcal{S} = A$

5. $\mathcal{S}^c = \emptyset$, $\emptyset^c = \mathcal{S}$.

If two sets A and B have no elements in common, they are said to be *disjoint*. The condition for being disjoint is therefore $A \cap B = \emptyset$. If, furthermore, the sets contain between them all the elements of $\mathcal{S}$, then the sets are said to *partition* the universe. This latter additional condition is that $A \cup B = \mathcal{S}$. An example of sets that partition the universe is given in Figure 3.3. Note also that the sets A and A^c

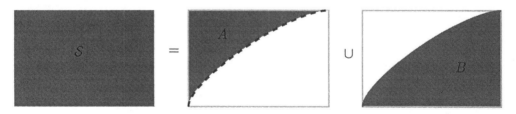

Figure 3.3: Sets that partition the universal set.

are always a partitioning of $\mathcal{S}$ (why?). More generally, *mutually* disjoint sets or sets $A_1, A_2, \ldots, A_N$ for which $A_i \cap A_j = \emptyset$ for all $i \neq j$ are said to partition the universe if $\mathcal{S} = \cup_{i=1}^{N} A_i$ (see also Problem 3.9 on how to construct these sets in general). For example, the set of students enrolled in the probability class, which is defined as the universe (although of course other universes may be defined such as the set of all students attending the given university), is partitioned by

$$A_1 = \{\text{males}\} = \{\text{Bill, Fred}\}$$
$$A_2 = \{\text{females}\} = \{\text{Jane, Jessica}\}.$$

Algebraic rules for manipulating multiple sets, which will be useful, are

1. $A \cup B = B \cup A$
   $A \cap B = B \cap A$ commutative properties

2. $A \cup (B \cup C) = (A \cup B) \cup C$
   $A \cap (B \cap C) = (A \cap B) \cap C$ associative properties

3. $A \cap (B \cup C) = (A \cap B) \cup (A \cap C)$
   $A \cup (B \cap C) = (A \cup B) \cap (A \cup C)$ distributive properties.

Another important relationship for manipulating sets is De Morgan's law. Referring

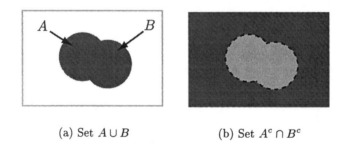

(a) Set $A \cup B$ (b) Set $A^c \cap B^c$

Figure 3.4: Illustration of De Morgan's law.

to Figure 3.4 it is obvious that

$$A \cup B = (A^c \cap B^c)^c \tag{3.6}$$

which allows one to convert from unions to intersections. To convert from intersections to unions we let $A = C^c$ and $B = D^c$ in (3.6) to obtain

$$C^c \cup D^c = (C \cap D)^c$$

and therefore

$$C \cap D = (C^c \cup D^c)^c. \tag{3.7}$$

In either case we can perform the conversion by the following set of rules:

1. Change the unions to intersections and the intersections to unions ($A \cup B \Rightarrow A \cap B$)

2. Complement each set ($A \cap B \Rightarrow A^c \cap B^c$)

3. Complement the overall expression $(A^c \cap B^c \Rightarrow (A^c \cap B^c)^c)$.

Finally, we discuss the *size* of a set. This will be of extreme importance in assigning probabilities. The set $\{2, 4, 6\}$ is a finite set, having a finite number of elements. The set $\{2, 4, 6, \ldots\}$ is an infinite set, having an infinite number of elements. In the latter case, although the set is infinite, it is said to be *countably infinite*. This means that "in theory" we can count the number of elements in the set. (We do so by pairing up each element in the set with an element in the set of natural numbers or $\{1, 2, 3, \ldots\}$). In either case, the set is said to be *discrete*. The set may be pictured as points on the real line. In contrast to these sets the set $\{x : 0 \leq x \leq 1\}$ is infinite and cannot be counted. This set is termed *continuous* and is pictured as a line segment on the real line. Another example follows.

Example 3.2 – Size of sets

The sets

$$A = \left\{\frac{1}{8}, \frac{1}{4}, \frac{1}{2}, 1\right\} \qquad \text{finite set - discrete}$$

$$B = \left\{1, \frac{1}{2}, \frac{1}{3}, \frac{1}{4}, \ldots\right\} \qquad \text{countably infinite set - discrete}$$

$$C = \{x : 0 \leq x \leq 1\} \qquad \text{infinite set - continuous}$$

are pictured in Figure 3.5.

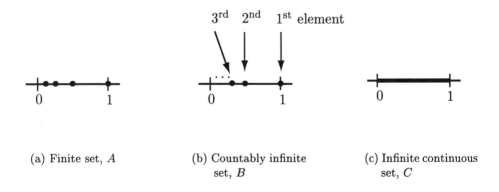

(a) Finite set, A

(b) Countably infinite set, B

(c) Infinite continuous set, C

Figure 3.5: Examples of sets of different sizes.

$\Diamond$

3.4   Assigning and Determining Probabilities

In the previous section we reviewed various aspects of set theory. This is because the concept of sets and operations on sets provide an ideal description for a probabilistic model and the means for determining the probabilites associated with the model. Consider the tossing of a fair die. The possible outcomes comprise the elements of the set $S = \{1, 2, 3, 4, 5, 6\}$. Note that this set is composed of *all* the possible outcomes, and as such is the universal set. In probability theory S is termed the *sample space* and its elements s are the *outcomes* or *sample points*. At times we may be interested in a particular outcome of the die tossing experiment. Other times we might not be interested in a particular outcome, but whether or not the outcome was an even number, as an example. Hence, we would inquire as to whether the outcome was *included* in the set $E_{\text{even}} = \{2, 4, 6\}$. Clearly, E_{even} is a subset of S and is termed an *event*. The simplest type of events are the ones that contain only a single outcome such as $E_1 = \{1\}$, $E_2 = \{2\}$, or $E_6 = \{6\}$, as examples. These are called *simple events*. Other events are S, the sample space itself, and $\emptyset = \{\}$, the set with no outcomes. These events are termed the *certain event* and the *impossible event*, respectively. This is because the outcome of the experiment must be an element of S so that S is certain to occur. Also, the event that does not contain any outcomes cannot occur so that this event is impossible. Note that we are saying that *an event occurs if the outcome is an element of the defining set of that event.* For example, the event that a tossed die produces an even number *occurs* if it comes up a 2 or a 4 or a 6. These numbers are just the elements of E_{even}. Disjoint sets such as $\{1, 2\}$ and $\{3, 4\}$ are said to be *mutually exclusive*, in that an outcome cannot be in both sets simultaneously and hence both events cannot occur. The events then are said to be mutually exclusive. It is seen that probabilistic questions can be formulated using set theory, albeit with its own terminology. A summary of the equivalent terms used is given in Table 3.1.

Set theory	Probability theory	Probability symbol
universe	sample space (certain event)	S
element	outcome (sample point)	s
subset	event	E
disjoint sets	mutually exclusive events	$E_1 \cap E_2 = \emptyset$
null set	impossible event	$\emptyset$
simple set	simple event	$E = \{s\}$

Table 3.1: Terminology for set and probability theory.

In order to develop a theory of probability we must next assign probabilities to events. For example, what is the probability that the tossed die will produce an even outcome? Denoting this probability by $P[E_{\text{even}}]$, we would intuitively say that it is $1/2$ since there are 3 chances out of 6 to produce an even outcome. Note that P is a *probability function* or a function that assigns a number between 0 and 1 to sets. It is sometimes called a *set function*. The reader is familiar with ordinary functions such as $g(x) = \exp(x)$, in which a number y, where $y = g(x)$, is assigned to each x

for $-\infty < x < \infty$, and where each x is a *distinct number*. The probability function must assign a number to every event, or to *every set*. For a coin toss whose outcome is either a head H or a tail T, all the events are $E_1 = \{H\}$, $E_2 = \{T\}$, $E_3 = \mathcal{S}$, and $E_4 = \emptyset$. For a die toss all the events are $E_0 = \emptyset$, $E_1 = \{1\}, \ldots, E_6 = \{6\}$, $E_{12} = \{1,2\}, \ldots, E_{56} = \{5,6\}, \ldots, E_{12345} = \{1,2,3,4,5\}, \ldots, E_{23456} = \{2,3,4,5,6\}$, $E_{123456} = \{1,2,3,4,5,6\} = \mathcal{S}$. There are a total of 64 events. In general, if the sample space has N simple events, the total number of events is 2^N (see Problem 3.15). We must be able to assign probabilities to all of these. In accordance with our intuitive notion of probability we assign a number, either zero or positive, to each event. Hence, we require that

Axiom 1 $P[E] \geq 0$ for every event E.

Also, since the die toss will always produce an outcome that is included in $\mathcal{S} = \{1,2,3,4,5,6\}$ we should require that

Axiom 2 $P[\mathcal{S}] = 1$.

Next we might inquire as to the assignment of a probability to the event that the die comes up *either* less than or equal to 2 *or* equal to 3. Intuitively, we would say that it is 3/6 since

$$
\begin{aligned}
P[\{1,2\} \cup \{3\}] &= P[\{1,2\}] + P[\{3\}] \\
&= \frac{2}{6} + \frac{1}{6} = \frac{1}{2}.
\end{aligned}
$$

However, we would *not* assert that the probability of the die coming up *either* less than or equal to 3 *or* equal to 3 is

$$
\begin{aligned}
P[\{1,2,3\} \cup \{3\}] &= P[\{1,2,3\}] + P[\{3\}] \\
&= \frac{3}{6} + \frac{1}{6} = \frac{4}{6}.
\end{aligned}
$$

This is because the event $\{1,2,3\} \cup \{3\}$ is just $\{1,2,3\}$ (we should not count the 3 twice) and so the probability should be 1/2. In the first example, the events are *mutually exclusive* (the sets are disjoint) while in the second example they are not. Hence, the probability of an event that is the union of two mutually exclusive events should be the sum of the probabilities. Combining this axiom with the previous ones produces the full set of axioms, which we summarize next for convenience.

Axiom 1 $P[E] \geq 0$ for every event E

Axiom 2 $P[\mathcal{S}] = 1$

Axiom 3 $P[E \cup F] = P[E] + P[F]$ for E and F mutually exclusive.

Using induction (see Problem 3.17) the third axiom may be extended to

Axiom 3′ $P[\bigcup_{i=1}^{N} E_i] = \sum_{i=1}^{N} P[E_i]$ for all E_i's mutually exclusive.

The acceptance of these axioms as the basis for probability is called the *axiomatic approach to probability*. It is remarkable that these three axioms, along with a fourth axiom to be introduced later, are adequate to formulate the entire theory. We now illustrate the application of these axioms to probability calculations.

Example 3.3 – Die toss

Determine the probability that the outcome of a fair die toss is even. The event is $E_{\text{even}} = \{2, 4, 6\}$. The assumption that the die is fair means that each outcome must be *equally likely*. Defining E_i as the simple event $\{i\}$ we note that

$$S = \bigcup_{i=1}^{6} E_i$$

and from Axiom 2 we must have

$$P\left[\bigcup_{i=1}^{6} E_i\right] = P[S] = 1. \tag{3.8}$$

But since each E_i is a simple event and by definition the simple events are mutually exclusive (only one outcome or simple event can occur), we have from Axiom 3′ that

$$P\left[\bigcup_{i=1}^{6} E_i\right] = \sum_{i=1}^{6} P[E_i]. \tag{3.9}$$

Next we note that the outcomes are assumed to be equally likely which means that $P[E_1] = P[E_2] = \cdots = P[E_6] = p$. Hence, we must have from (3.8) and (3.9) that

$$\sum_{i=1}^{6} P[E_i] = 6p = 1$$

or $P[E_i] = 1/6$ for all i. We can now finally determine $P[E_{\text{even}}]$ since $E_{\text{even}} = E_2 \cup E_4 \cup E_6$. By applying Axiom 3′ once again we have

$$P[E_{\text{even}}] = P[E_2 \cup E_4 \cup E_6] = P[E_2] + P[E_4] + P[E_6] = \frac{1}{6} + \frac{1}{6} + \frac{1}{6} = \frac{1}{2}.$$

$\Diamond$

In general, the probabilities assigned to each simple event need not be the same, i.e., the outcomes of a die toss may not have equal probabilities. One might have weighted the die so that the number 6 comes up twice as often as all the others. The numbers $1, 2, 3, 4, 5$ could still be equally likely. In such a case, since the probabilities of the all the simple events must sum to one, we would have the assignment $P[\{i\}] =$

1/7 for $i = 1, 2, 3, 4, 5$ and $P[\{6\}] = 2/7$. In either case, to compute the probability of *any event* it is only necessary to sum the probabilities of the simple events that make up that event. Letting $P[\{s_i\}]$ be the probability of the ith simple event we have that

$$P[E] = \sum_{\{i:S_i \in E\}} P[\{s_i\}]. \tag{3.10}$$

We now simplify the notation by omitting the $\{\}$ when referring to events. Instead of $P[\{1\}]$ we will use $P[1]$. Another example follows.

Example 3.4 – Defective die toss

A defective die is tossed whose sides have been mistakenly manufactured with the number of dots being $1, 1, 2, 2, 3, 4$. The simple events are $s_1 = 1$, $s_2 = 1$, $s_3 = 2$, $s_4 = 2$, $s_5 = 3$, $s_6 = 4$. Even though some of the outcomes have the same number of dots, they are actually different in that a *different side* is being observed. Each side is equally likely to appear. What is the probability that the outcome is less than 3? Noting that the event of interest is $\{s_1, s_2, s_3, s_4\}$, we use (3.10) to obtain

$$P[E] = P[\text{outcome} < 3] = \sum_{i=1}^{4} P[s_i] = \frac{4}{6}.$$

$\Diamond$

The formula given by (3.10) also applies to probability problems for which the sample space is countably infinite. Therefore, it applies to all *discrete* sample spaces (see also Example 3.2).

Example 3.5 – Countably infinite sample space

A habitually tardy person arrives at the theater late by s_i minutes, where

$$s_i = i \quad i = 1, 2, 3 \dots .$$

If $P[s_i] = (1/2)^i$, what is the probability that he will be more than 1 minute late? The event is $E = \{2, 3, 4, \dots\}$. Using (3.10) we have

$$P[E] = \sum_{i=2}^{\infty} \left(\frac{1}{2}\right)^i .$$

Using the formula for the sum of a geometric progression (see Appendix B)

$$\sum_{i=k}^{\infty} a^i = \frac{a^k}{1 - a} \quad \text{for } |a| < 1$$

we have that

$$P[E] = \frac{\left(\frac{1}{2}\right)^2}{1 - \frac{1}{2}} = \frac{1}{2}.$$

In the above example we have implicitly used the relationship

$$P\left[\bigcup_{i=1}^{\infty} E_i\right] = \sum_{i=1}^{\infty} P[E_i] \qquad (3.11)$$

where $E_i = \{s_i\}$ and hence the E_i's are mutually exclusive. This does not automatically follow from Axiom 3' since N is now infinite. However, we will assume for our problems of interest that it does. Adding (3.11) to our list of axioms we have

Axiom 4 $P[\bigcup_{i=1}^{\infty} E_i] = \displaystyle\sum_{i=1}^{\infty} P[E_i]$ for all E_i's mutually exclusive.

See [Billingsley 1986] for further details.

3.5   Properties of the Probability Function

From the four axioms we may derive many useful properties for evaluating probabilities. We now summarize these properties.

Property 3.1 – Probability of complement event

$$P[E^c] = 1 - P[E]. \qquad (3.12)$$

<u>Proof:</u> By definition $E \cup E^c = \mathcal{S}$. Also, by definition E and E^c are mutually exclusive. Hence,

$$
\begin{aligned}
1 &= P[\mathcal{S}] & \text{(Axiom 2)} \\
  &= P[E \cup E^c] & \text{(definition of complement set)} \\
  &= P[E] + P[E^c] & \text{(Axiom 3)}
\end{aligned}
$$

from which (3.12) follows.

$\square$

We could have determined the probability in Example 3.5 without the use of the geometric progression formula by using $P[E] = 1 - P[E^c] = 1 - P[1] = 1/2$.

Property 3.2 – Probability of impossible event

$$P[\emptyset] = 0. \qquad (3.13)$$

<u>Proof:</u> Since $\emptyset = \mathcal{S}^c$ we have

$$
\begin{aligned}
P[\emptyset] &= P[\mathcal{S}^c] \\
             &= 1 - P[\mathcal{S}] & \text{(from Property 3.1)} \\
             &= 1 - 1 & \text{(from Axiom 2)} \\
             &= 0.
\end{aligned}
$$

$\square$

We will see later that there are other events for which the probability can be zero. Thus, the converse is *not true*.

Property 3.3 – All probabilities are between 0 and 1.

Proof:

$$S = E \cup E^c \qquad \text{(definition of complement set)}$$
$$P[S] = P[E] + P[E^c] \quad \text{(Axiom 3)}$$
$$1 = P[E] + P[E^c] \quad \text{(Axiom 2)}$$

But from Axiom 1 $P[E^c] \geq 0$ and therefore

$$P[E] = 1 - P[E^c] \leq 1. \tag{3.14}$$

Combining this result with Axiom 1 proves Property 3.3.

□

Property 3.4 – Formula for $P[E \cup F]$ where E and F are not mutually exclusive

$$P[E \cup F] = P[E] + P[F] - P[EF]. \tag{3.15}$$

(We have shortened $E \cap F$ to EF.)

Proof: By the definition of $E - F$ we have that $E \cup F = (E - F) \cup F$ (see Figure 3.1d,f). Also, the events $E - F$ and F are by definition mutually exclusive. It follows that

$$P[E \cup F] = P[E - F] + P[F] \quad \text{(Axiom 3)}. \tag{3.16}$$

But by definition $E = (E - F) \cup EF$ (draw a Venn diagram) and $E - F$ and EF are mutually exclusive. Thus,

$$P[E] = P[E - F] + P[EF] \qquad \text{(Axiom 3)}. \tag{3.17}$$

Combining (3.16) and (3.17) produces Property 3.4.

□

The effect of this formula is to make sure that the intersection EF is not counted twice in the probability calculation. This would be the case if Axiom 3 were mistakenly applied to sets that were not mutually exclusive. In the die example, if we wanted the probability of the die coming up either *less than or equal to 3* or *equal to 3*, then we would first define

$$E = \{1, 2, 3\}$$
$$F = \{3\}$$

so that $EF = \{3\}$. Using Property 3.4, we have that

$$P[E \cup F] = P[E] + P[F] - P[EF] = \frac{3}{6} + \frac{1}{6} - \frac{1}{6} = \frac{3}{6}.$$

Of course, we could just as easily have noted that $E \cup F = \{1, 2, 3\} = E$ and then applied (3.10). Another example follows.

Example 3.6 – Switches in parallel

A switching circuit shown in Figure 3.6 consists of two potentially faulty switches in parallel. In order for the circuit to operate properly at least one of the switches must

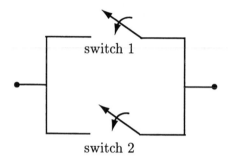

Figure 3.6: Parallel switching circuit.

close to allow the overall circuit to be closed. Each switch has a probability of 1/2 of closing. The probability that both switches close simultaneously is 1/4. What is the probability that the switching circuit will operate correctly? To solve this problem we first define the events $E_1 = \{\text{switch 1 closes}\}$ and $E_2 = \{\text{switch 2 closes}\}$. The event that *at least one* switch closes is $E_1 \cup E_2$. This includes the possibility that both switches close. Then using Property 3.4 we have

$$\begin{aligned} P[E_1 \cup E_2] &= P[E_1] + P[E_2] - P[E_1 E_2] \\ &= \frac{1}{2} + \frac{1}{2} - \frac{1}{4} = \frac{3}{4}. \end{aligned}$$

Note that by using two switches in parallel as opposed to only one switch, the probability that the circuit will operate correctly has been increased. What do you think would happen if we had used three switches in parallel? Or if we had used N switches? Could you ever be assured that the circuit would operate flawlessly? (See Problem 3.26.)

Property 3.5 – Monotonicity of probability function

Monotonicity asserts that the larger the set, the larger the probability of that set. Mathematically, this translates into the statement that if $E \subset F$, then $P[E] \leq P[F]$.

<u>Proof:</u> If $E \subset F$, then by definition $F = E \cup (F - E)$, where E and $F - E$ are mutually exclusive by definition. Hence,

$$
\begin{aligned}
P[F] &= P[E] + P[F - E] \quad \text{(Axiom 3)} \\
&\geq P[E] \quad\quad\quad\quad\;\; \text{(Axiom 1)}.
\end{aligned}
$$

$\square$

Note that since $EF \subset F$ and $EF \subset E$, we have that $P[EF] \leq P[E]$ and also that $P[EF] \leq P[F]$. The probability of an intersection is always less than or equal to the probability of the set with the smallest probability.

Example 3.7 − Switches in series

A switching circuit shown in Figure 3.7 consists of two potentially faulty switches in series. In order for the circuit to operate properly *both* switches must close. For the

switch 1   switch 2

Figure 3.7: Series switching circuit.

same switches as described in Example 3.6 what is the probability that the circuit will operate properly? Now we need to find $P[E_1 E_2]$. This was given as $1/4$ so that

$$
\frac{1}{4} = P[E_1 E_2] \leq P[E_1] = \frac{1}{2}
$$

Could the series circuit ever outperform the parallel circuit? (See Problem 3.27.)

$\Diamond$

One last property that is often useful is the probability of a union of more than two events. This extends Property 3.4. Consider first three events so that we wish to derive a formula for $P[E_1 \cup E_2 \cup E_3]$, which is equivalent to $P[(E_1 \cup E_2) \cup E_3]$ or $P[E_1 \cup (E_2 \cup E_3)]$ by the associative property. Writing this as $P[E_1 \cup (E_2 \cup E_3)]$, we have

$$
\begin{aligned}
P[E_1 \cup E_2 \cup E_3] &= P[E_1 \cup (E_2 \cup E_3)] \\
&= P[E_1] + P[E_2 \cup E_3] - P[E_1(E_2 \cup E_3)] \quad \text{(Property 3.4)} \\
&= P[E_1] + (P[E_2] + P[E_3] - P[E_2 E_3]) \\
&\quad\; -P[E_1(E_2 \cup E_3)] \quad\quad\quad\quad\quad\quad \text{(Property 3.4)}
\end{aligned}
$$
(3.18)

But $E_1(E_2 \cup E_3) = E_1 E_2 \cup E_1 E_3$ by the distributive property (draw a Venn diagram) so that

$$
\begin{aligned}
P[E_1(E_2 \cup E_3)] &= P[E_1 E_2 \cup E_1 E_3] \\
&= P[E_1 E_2] + P[E_1 E_3] - P[E_1 E_2 E_3] \quad \text{(Property 3.4). (3.19)}
\end{aligned}
$$

Substituting (3.19) into (3.18) produces

$$P[E_1 \cup E_2 \cup E_3] = P[E_1] + P[E_2] + P[E_3] - P[E_2 E_3] - P[E_1 E_2] - P[E_1 E_3] + P[E_1 E_2 E_3]$$
$$(3.20)$$

which is the desired result. It can further be shown that (see Problem 3.29)

$$P[E_1 E_2] + P[E_1 E_3] + P[E_2 E_3] \geq P[E_1 E_2 E_3]$$

so that

$$P[E_1 \cup E_2 \cup E_3] \leq P[E_1] + P[E_2] + P[E_3] \qquad (3.21)$$

which is known as *Boole's inequality* or the *union bound*. Clearly, equality holds if and only if the E_i's are mutually exclusive. Both (3.20) and (3.21) can be extended to any finite number of unions [Ross 2002].

3.6   Probabilities for Continuous Sample Spaces

We have introduced the axiomatic approach to probability and illustrated the approach with examples from a discrete sample space. The axiomatic approach is completely general and applies to continuous sample spaces as well. However, (3.10) cannot be used to determine probabilities of events. This is because the simple events of the continuous sample space are not countable. For example, suppose one throws a dart at a "linear" dartboard as shown in Figure 3.8 and measures the horizontal distance from the "bullseye" or center at $x = 0$. We will then have a sample space

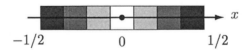

Figure 3.8: "Linear" dartboard.

$\mathcal{S} = \{x : -1/2 \leq x \leq 1/2\}$, which is not countable. A possible approach is to assign probabilities to *intervals* as opposed to sample points. If the dart is equally likely to land anywhere, then we could assign the interval $[a, b]$ a probability equal to the length of the interval or

$$P[a \leq x \leq b] = b - a \quad -1/2 \leq a \leq b \leq 1/2. \qquad (3.22)$$

Also, we will *assume* that the probability of disjoint intervals is the sum of the probabilities for each interval. This assignment is entirely consistent with our axioms

since

$$P[E] = P[a \leq x \leq b] = b - a \geq 0. \quad \text{(Axiom 1)}$$

$$P[\mathcal{S}] = P[-1/2 \leq x \leq 1/2] = 1/2 - (-1/2) = 1. \quad \text{(Axiom 2)}$$

$$\begin{aligned} P[E \cup F] &= P[a \leq x \leq b \cup c \leq x \leq d] \\ &= (b-a) + (d-c) \quad \text{(assumption)} \\ &= P[a \leq x \leq b] + P[c \leq x \leq d] \\ &= P[E] + P[F] \quad \text{(Axiom 3)} \end{aligned}$$

for $a \leq b < c \leq d$ so that E and F are mutually exclusive. Hence, an equally likely type probability assignment for a continuous sample space is a valid one and produces a probability equal to the length of the interval. If the sample space does not have unity length, as for example, a dartboard with a length L, then we should use

$$P[E] = \frac{\text{Length of interval}}{\text{Length of dartboard}} = \frac{\text{Length of interval}}{L}. \quad (3.23)$$

 Probability of a bullseye

It is an inescapable fact that the probability of the dart landing at say $x = 0$ is zero since the length of this interval is zero. For that matter the probability of the dart landing at any one particular point x_0 is zero as follows from (3.22) with $a = b = x_0$. The first-time reader of probability will find this particularly disturbing and argue that "How can the probability of landing at every point be zero if indeed the dart had to land at *some* point?" From a pragmatic viewpoint we will seldom be interested in probabilities of points in a continuous sample space but only in those of intervals. How many darts are there whose tips have width zero and so can be said to land at a point? It is more realistic in practice then to ask for the probability that the dart lands in the bullseye, which is a small interval with some nonzero length. That probability is found by using (3.22). From a mathematical viewpoint it is not possible to "sum" up an infinite number of positive numbers of *equal value* and not obtain infinity, as opposed to one, as assumed in Axiom 2. The latter is true for continuous sample spaces, in which we have an uncountably infinite set, and also for discrete sample spaces, which is composed of a infinite but countable set. (Note that in Example 3.5 we had a countably infinite sample space *but* the probabilities were not equal.)

Since the probability of a point event occurring is zero, the probability of any interval is the same whether or not the endpoints are included. Thus, for our example

$$P[a \leq x \leq b] = P[a < x \leq b] = P[a \leq x < b] = P[a < x < b].$$

3.7   Probabilities for Finite Sample Spaces – Equally Likely Outcomes

We now consider in more detail a discrete sample space with a finite number of outcomes. Some examples that we are already familiar with are a coin toss, a die toss, or the students in a class. Furthermore, we assume that the simple events or outcomes are *equally likely*. Many problems have this structure and can be approached using *counting methods* or *combinatorics*. For example, if two dice are tossed, then the sample space is

$$\mathcal{S} = \{(i, j) : i = 1, \ldots, 6; j = 1, \ldots, 6\}$$

which consists of 36 outcomes with each outcome or simple event denoted by an ordered pair of numbers. If we wish to assign probabilities to events, then we need only assign probabilities to the simple events and then use (3.10). But if all the simple events, denoted by $\mathcal{S}_{ij}$, are equally likely, then

$$P[\mathcal{S}_{ij}] = \frac{1}{N_{\mathcal{S}}} = \frac{1}{36}$$

where $N_{\mathcal{S}}$ is the number of outcomes in $\mathcal{S}$. Now using (3.10) we have for any event that

$$
\begin{aligned}
P[E] \;&=\; \sum_{\{(i,j):\, \mathcal{S}_{ij} \in E\}} \sum P[\mathcal{S}_{ij}] \\
&=\; \sum_{\{(i,j):\, \mathcal{S}_{ij} \in E\}} \sum \frac{1}{N_{\mathcal{S}}} \\
&=\; \frac{N_E}{N_{\mathcal{S}}} \\
&=\; \frac{\text{Number of outcomes in } E}{\text{Number of outcomes in } \mathcal{S}}.
\end{aligned}
\qquad (3.24)
$$

We will use combinatorics to determine N_E and $N_{\mathcal{S}}$ and hence $P[E]$.

Example 3.8 –   Probability of equal values for two-dice toss
Each outcome with equal values is of the form (i, i) so that

$$P[E] = \frac{\text{Number of outcomes with } (i, i)}{\text{Total number of outcomes}}.$$

There are 6 outcomes with equal values or (i, i) for $i = 1, 2, \ldots, 6$. Thus,

$$P[E] = \frac{6}{36} = \frac{1}{6}.$$

Example 3.9 – A more challenging problem - urns
An urn contains 3 red balls and 2 black balls. Two balls are chosen in succession. The first ball is returned to the urn before the second ball is chosen. Each ball is chosen *at random*, which means that we are equally likely to choose any ball. What is the probability of choosing first a red ball and then a black ball? To solve this problem we first need to define the sample space. To do so we assign numbers to the balls as follows. The red balls are numbered $1, 2, 3$ and the black balls are numbered $4, 5$. The sample space is then $S = \{(i, j) : i = 1, 2, 3, 4, 5; j = 1, 2, 3, 4, 5\}$. The event of interest is $E = \{(i, j) : i = 1, 2, 3; j = 4, 5\}$. We assume that all the simple events are equally likely. An enumeration of the outcomes is shown in Table 3.2. The outcomes with the asterisks comprise E. Hence, the probability is $P[E] = 6/25$. This problem could also have been solved using combinatorics as follows. Since there

	$j = 1$	$j = 2$	$j = 3$	$j = 4$	$j = 5$
$i = 1$	$(1, 1)$	$(1, 2)$	$(1, 3)$	$(1, 4)^*$	$(1, 5)^*$
$i = 2$	$(2, 1)$	$(2, 2)$	$(2, 3)$	$(2, 4)^*$	$(2, 5)^*$
$i = 3$	$(3, 1)$	$(3, 2)$	$(3, 3)$	$(3, 4)^*$	$(3, 5)^*$
$i = 4$	$(4, 1)$	$(4, 2)$	$(4, 3)$	$(4, 4)$	$(4, 5)$
$i = 5$	$(5, 1)$	$(5, 2)$	$(5, 3)$	$(5, 4)$	$(5, 5)$

Table 3.2: Enumeration of outcomes for urn problem of Example 3.9.

are 5 possible choices for each ball, there are a total of $5^2 = 25$ outcomes in the sample space. There are 3 possible ways to choose a red ball on the first draw and 2 possible ways to choose a black ball on the second draw, yielding a total of $3 \cdot 2 = 6$ possible ways of choosing a red ball *followed* by a black ball. We thus arrive at the same probability.

3.8 Combinatorics

Combinatorics is the study of counting. As illustrated in Example 3.9, we often have an outcome that can be represented as a 2-tuple or (z_1, z_2), where z_1 can take on one of N_1 values and z_2 can take on one of N_2 values. For that example, the total number of 2-tuples in S is $N_1 N_2 = 5 \cdot 5 = 25$, while that in E is $N_1 N_2 = 3 \cdot 2 = 6$, as can be verified by referring to Table 3.2. It is important to note that *order matters* in the description of a 2-tuple. For example, the 2-tuple $(1, 2)$ is not the same as the 2-tuple $(2, 1)$ since each one describes a different outcome of the experiment. We will frequently be using 2-tuples and more generally r-tuples denoted by $(z_1, z_2, \ldots, z_r)$ to describe the outcomes of urn experiments.

In drawing balls from an urn there are two possible strategies. One method is to draw a ball, note which one it is, return it to the urn, and then draw a second ball. This is called *sampling with replacement* and was used in Example 3.9. However, it is also possible that the first ball is not returned to the urn before the second one is chosen. This method is called *sampling without replacement*. The contrast between the two strategies is illustrated next.

Example 3.10 – Computing probabilities of drawing balls from urns - with and without replacement

An urn has k red balls and $N - k$ black balls. If two balls are chosen in succession and at random *with replacement*, what is the probability of a red ball *followed* by a black ball? We solve this problem by first labeling the k red balls with $1, 2, \ldots, k$ and the black balls with $k + 1, k + 2, \ldots, N$. In doing so the possible outcomes of the experiment can be represented by a 2-tuple (z_1, z_2), where $z_1 \in \{1, 2, \ldots, N\}$ and $z_2 \in \{1, 2, \ldots, N\}$. A successful outcome is a red ball followed by a black one so that the successful event is $E = \{(z_1, z_2) : z_1 = 1, \ldots, k; z_2 = k + 1, \ldots, N\}$. The total number of 2-tuples in the sample space is $N_S = N^2$, while the total number of 2-tuples in E is $N_E = k(N - k)$ so that

$$
\begin{aligned}
P[E] &= \frac{N_E}{N_S} \\
&= \frac{k(N - k)}{N^2} \\
&= \frac{k}{N}\left(1 - \frac{k}{N}\right).
\end{aligned}
$$

Note that if we let $p = k/N$ be the proportion of red balls, then $P[E] = p(1 - p)$. Next consider the case of sampling *without replacement*. Now since the same ball cannot be chosen twice in succession, and therefore, $z_1 \neq z_2$, we have one fewer choice for the second ball. Therefore, $N_S = N(N - 1)$. As before, the number of successful 2-tuples is $N_E = k(N - k)$, resulting in

$$
\begin{aligned}
P[E] &= \frac{k(N - k)}{N(N - 1)} = \frac{k}{N}\frac{N - k}{N}\frac{N}{N - 1} \\
&= p(1 - p)\frac{N}{N - 1}.
\end{aligned}
$$

The probability is seen to be higher. Can you explain this? (It may be helpful to think about the effect of a successful first draw on the probability of a success on the second draw.) Of course, for large N the probabilities for sampling with and without replacement are seen to be approximately the same, as expected.

◇

If we now choose r balls *without replacement* from an urn containing N balls, then all the possible outcomes are of the form $(z_1, z_2, \ldots, z_r)$, where the z_i's must be different. On the first draw we have N possible balls, on the second draw we have $N - 1$ possible balls, etc. Hence, the total number of possible outcomes or number of r-tuples is $N(N - 1) \cdots (N - r + 1)$. We denote this by $(N)_r$. If all the balls are selected, forming an N-tuple, then the number of outcomes is

$$(N)_N = N(N - 1) \cdots 1$$

which is defined as $N!$ and is termed N *factorial*. As an example, if there are 3 balls labeled A,B,C, then the number of 3-tuples is $3! = 3 \cdot 2 \cdot 1 = 6$. To verify this we have by enumeration that the possible 3-tuples are (A,B,C), (A,C,B), (B,A,C), (B,C,A), (C,A,B), (C,B,A). Note that $3!$ is the number of ways that 3 objects can be arranged. These arrangements are termed the *permutations* of the letters A, B, and C. Note that with the definition of a factorial we have that $(N)_r = N!/(N-r)!$. Another example follows.

Example 3.11 – More urns - using permutations

Five balls numbered $1, 2, 3, 4, 5$ are drawn from an urn *without replacement*. What is the probability that they will be drawn in the same order as their number? Each outcome is represented by the 5-tuple $(z_1, z_2, z_3, z_4, z_5)$. The only outcome in E is $(1, 2, 3, 4, 5)$ so that $N_E = 1$. To find N_S we require the number of ways that the numbers $1, 2, 3, 4, 5$ can be arranged or the number of permutations. This is $5! = 120$. Hence, the desired probability is $P[E] = 1/120$.

◇

Before continuing, we give one more example to explain our fixation with drawing balls out of urns.

Example 3.12 – The birthday problem

A probability class has N students enrolled. What is the probability that at least two of the students will have the same birthday? We first assume that each student in the class is equally likely to be born on any day of the year. To solve this problem consider a "birthday urn" that contains 365 balls. Each ball is labeled with a different day of the year. Now allow each student to select a ball at random, note its date, and return it to the urn. The day of the year on the ball becomes his/her birthday. The probability desired is of the event that two or more students choose the same ball. It is more convenient to determine the probability of the complement event or that no two students have the same birthday. Then, using Property 3.1

$P[$at least 2 students have same birthday$] = 1 - P[$no students have same birthday$]$.

The sample space is composed of $N_S = 365^N$ N-tuples (sampling with replacement). The number of N-tuples for which all the outcomes are different is $N_E = (365)_N$. This is because the event that no two students have the same birthday occurs if the first student chooses any of the 365 balls, the second student chooses any of the remaining 364 balls, etc., which is the same as if sampling without replacement were used. The probability is then

$$P[\text{at least 2 students have same birthday}] = 1 - \frac{(365)_N}{365^N}.$$

This probability is shown in Figure 3.9 as a function of the number of students. It is seen that if the class has 23 or more students, there is a probability of 0.5 or greater that two students will have the same birthday.

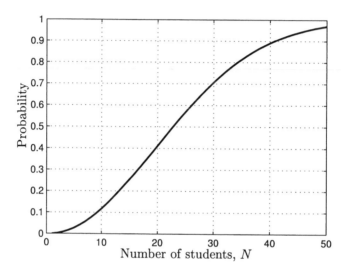

Figure 3.9: Probability of at least two students having the same birthday.

   Why this doesn't appear to make sense.

This result may seem counterintuitive at first, but this is only because the reader is misinterpreting the question. Most persons would say that you need about 180 people for a 50% chance of two identical birthdays. In contrast, if the question was posed as to the probability that at least two persons were born on January 1, then the event would be at least two persons choose the ball labeled "January 1" from the birthday urn. For 23 people this probability is considerably smaller (see Problem 3.38). It is the possibility that the two identical birthdays *can occur on any day of the year* (365 possibilities) that leads to the unexpected large probability. To verify this result the MATLAB program given below can be used. When run, the estimated probability for 10,000 repeated experiments was 0.5072. The reader may wish to reread Section 2.4 at this point.

```
% birthday.m
%
clear all
rand('state',0)
BD=[0:365]';
event=zeros(10000,1); % initialize to no successful events
for ntrial=1:10000
for i=1:23
   x(i,1)=ceil(365*rand(1,1)); % chooses birthdays at random
                              % (ceil rounds up to nearest integer)
end
y=sort(x); % arranges birthdays in ascending order
z=y(2:23)-y(1:22); % compares successive birthdays to each other
w=find(z==0); % flags same birthdays
if length(w)>0
   event(ntrial)=1; % event occurs if one or more birthdays the same
end
end
prob=sum(event)/10000
```

We summarize our counting formulas so far. Each outcome of an experiment produces an r-tuple, which can be written as $(z_1, z_2, \ldots, z_r)$. If we are choosing balls in succession from an urn containing N balls, then with replacement each z_i can take on one of N possible values. The number of possible r-tuples is then N^r. If we sample without replacement, then the number of r-tuples is only $(N)_r = N(N-1)\cdots(N-r+1)$. If we sample without replacement and $r = N$ or all the balls are chosen, then the number of r-tuples is $N!$. In arriving at these formulas we have used the r-tuple representation in which the *ordering* is used in the counting. For example, the 3-tuple (A,B,C) is different than (C,A,B), which is different than (C,B,A), etc. In fact, there are 3! possible orderings or permutations of the letters A, B, and C. We are frequently not interested in the ordering but only in the number of *distinct* elements. An example might be to determine the number of possible sum-values that can be made from one penny (p), one nickel (n), and one dime (d) if two coins are chosen. To determine this we use a *tree diagram* as shown in Figure 3.10. Note that since this is essentially sampling without replacement, we cannot have the outcomes pp, nn, or dd (shown in Figure 3.10 as dashed). The number of possible outcomes are 3 for the first coin and 2 for the second so that as usual there are $(3)_2 = 3 \cdot 2 = 6$ outcomes. However, *only 3 of these are distinct* or produce different sum-values for the two coins. The outcome (p,n) is counted the same as (n,p) for example. Hence, the ordering of the outcome does not matter. Both orderings are treated as the *same outcome*. To remind us that

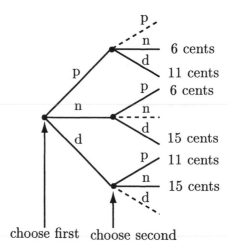

Figure 3.10: Tree diagram enumerating possible outcomes.

ordering is immaterial we will replace the 2-tuple description by the set description (recall that the elements of a set may be arranged in any order to yield the same set). The outcomes of this experiment are therefore {p,n}, {p,d}, {n,d}. In effect, all permutations are considered as a *single combination*. Thus, to find the number of combinations:

Number of combinations × Number of permutations   =   Total number of
                                                      r-tuple outcomes

or for this example,

$$\text{Number of combinations} \times 2! = (3)_2$$

which yields

$$\text{Number of combinations} = \frac{(3)_2}{2!} = \frac{3!}{1!2!} = 3.$$

The number of combinations is given by the symbol $\binom{3}{2}$ and is said to be "3 things taken 2 at a time". Also, $\binom{3}{2}$ is termed the binomial coefficient due to its appearance in the binomial expansion (see Problem 3.43). In general the number of combinations of N things taken k at a time, i.e., order does not matter, is

$$\binom{N}{k} = \frac{(N)_k}{k!} = \frac{N!}{(N-k)!k!}.$$

Example 3.13 – Correct change

If a person has a penny, nickel, and dime in his pocket and selects two coins at random, what is the probability that the sum-value will be 6 cents? The sample space is now $\mathcal{S} = \{\{p, n\}, \{p, d\}, \{n, d\}\}$ and $E = \{\{p, n\}\}$. Thus,

$$
\begin{aligned}
P[\text{6 cents}] &= P[\{p, n\}] = \frac{N_E}{N_{\mathcal{S}}} \\
&= \frac{1}{3}.
\end{aligned}
$$

Note that each simple event is of the form $\{\cdot, \cdot\}$. Also, $N_{\mathcal{S}}$ can be found from the original problem statement as $\binom{3}{2} = 3$.

◇

Example 3.14 – How probable is a royal flush?

A person draws 5 cards from a deck of 52 freshly shuffled cards. What is the probability that he obtains a royal flush? To obtain a royal flush he must draw an ace, king, queen, jack, and ten of the same suit *in any order*. There are 4 possible suits that will be produce the flush. The total number of combinations of cards or "hands" that can be drawn is $\binom{52}{5}$ and a royal flush will result from 4 of these combinations. Hence,

$$
P[\text{royal flush}] = \frac{4}{\binom{52}{5}} \approx 0.00000154.
$$

◇

   Ordered vs. unordered

It is sometimes confusing that $\binom{52}{5}$ is used for $N_{\mathcal{S}}$. It might be argued that the first card can be chosen in 52 ways, the second card in 51 ways, etc. for a total of $(52)_5$ possible outcomes. Likewise, for a royal flush in hearts we can choose any of 5 cards, followed by any of 4 cards, etc. for a total of 5! possible outcomes. Hence, the probability of a royal flush in hearts should be

$$
P[\text{royal flush in hearts}] = \frac{5!}{(52)_5}.
$$

But this is just the same as $1/\binom{52}{5}$ which is the same as obtained by counting combinations. In essence, we have reduced the sample space by a factor of 5! but additionally each event is commensurately reduced by 5!, yielding the same probability. Equivalently, we have grouped together each set of 5! permutations to yield a single combination.

3.9   Binomial Probability Law

In Chapter 1 we cited the binomial probability law for the number of heads obtained for N tosses of a coin. The same law also applies to the problem of drawing balls from an urn. First, however, we look at a related problem that is of considerable practical interest. Specifically, consider an urn consisting of a *proportion p* of red balls and the remaining proportion $1 - p$ of black balls. What is the probability of drawing k red balls in M drawings *without replacement*? Note that we can associate the drawing of a red ball as a "success" and the drawing of a black ball as a "failure". Hence, we are equivalently asking for the probability of k successes out of a maximum of M successes. To determine this probability we first assume that the urn contains N balls, of which N_R are red and N_B are black. We sample the urn by drawing M balls without replacement. To make the balls distinguishable we label the red balls as $1, 2, \ldots, N_R$ and the black ones as $N_R + 1, N_R + 2, \ldots, N$. The sample space is

$$S = \{(z_1, z_2, \ldots, z_M) : z_i = 1, \ldots, N \text{ and no two } z_i\text{'s are the same}\}.$$

We assume that the balls are selected at random so that the outcomes are equally likely. The total number of outcomes is $N_S = (N)_M$. Hence, the probability of obtaining k red balls is

$$P[k] = \frac{N_E}{(N)_M}. \tag{3.25}$$

N_E is the number of M-tuples that contain k distinct integers in the range from 1 to N_R and $M - k$ distinct integers in the range $N_R + 1$ to N. For example, if $N_R = 3$, $N_B = 4$ (and hence $N = 7$), $M = 4$, and $k = 2$, the red balls are contained in $\{1, 2, 3\}$, the black balls are contained in $\{4, 5, 6, 7\}$ and we choose 4 balls without replacement. A successful outcome has two red balls and two black balls. Some successful outcomes are $(1, 4, 2, 5)$, $(1, 4, 5, 2)$, $(1, 2, 4, 5)$, etc. or $(2, 3, 4, 6)$, $(2, 4, 3, 6)$, $(2, 6, 3, 4)$, etc. Hence, N_E is the total number of outcomes for which two of the z_i's are elements of $\{1, 2, 3\}$ and two of the z_i's are elements of $\{4, 5, 6, 7\}$. To determine this number of successful M-tuples we

1. Choose the k positions of the M-tuple to place the red balls. (The remaining positions will be occupied by the black balls.)

2. Place the N_R red balls in the k positions obtained from step 1.

3. Place the N_B black balls in the remaining $M - k$ positions.

Step 1 is accomplished in $\binom{M}{k}$ ways since any permutation of the chosen positions produces the same set of positions. Step 2 is accomplished in $(N_R)_k$ ways and step

3 is accomplished in $(N_B)_{M-k}$ ways. Thus, we have that

$$
\begin{aligned}
N_E &= \binom{M}{k} (N_R)_k (N_B)_{M-k} \\
&= \frac{M!}{(M-k)!k!} (N_R)_k (N_B)_{M-k} \\
&= M! \binom{N_R}{k} \binom{N_B}{M-k}
\end{aligned}
\tag{3.26}
$$

so that finally we have from (3.25)

$$
\begin{aligned}
P[k] &= \frac{M! \binom{N_R}{k} \binom{N_B}{M-k}}{(N)_M} \\
&= \frac{\binom{N_R}{k} \binom{N_B}{M-k}}{\binom{N}{M}}.
\end{aligned}
\tag{3.27}
$$

This law is called the *hypergeometric law* and describes the probability of k successes when sampling *without replacement* is used. If sampling *with replacement* is used, then the binomial law results. However, instead of repeating the entire derivation for sampling with replacement, we need only assume that N is large. Then, whether the balls are replaced or not will not affect the probability. To show that this is indeed the case, we start with the expression given by (3.26) and note that for N large and $M \ll N$, then $(N)_M \approx N^M$. Similarly, we assume that $M \ll N_R$ and $M \ll N_B$ and make similar approximations. As a result we have from (3.25) and (3.26)

$$
\begin{aligned}
P[k] &\approx \binom{M}{k} \frac{N_R^k N_B^{M-k}}{N^M} \\
&= \binom{M}{k} \left(\frac{N_R}{N} \right)^k \left(\frac{N_B}{N} \right)^{M-k}.
\end{aligned}
$$

Letting $N_R/N = p$ and $N_B/N = (N - N_R)/N = 1 - p$, we have at last the *binomial law*

$$
P[k] = \binom{M}{k} p^k (1-p)^{M-k}.
\tag{3.28}
$$

To summarize, the binomial law not only applies to the drawing of balls from urns *with replacement* but also applies to the drawing of balls *without replacement if the number of balls in the urn is large*. We next use our results in a quality control application.

3.10   Real-World Example – Quality Control

A manufacturer of electronic memory chips produces batches of 1000 chips for ship-ment to computer companies. To determine if the chips meet specifications the manufacturer initially tests all 1000 chips in each batch. As demand for the chips grows, however, he realizes that it is impossible to test all the chips and so proposes that only a subset or sample of the batch be tested. The criterion for acceptance of the batch is that at least 95% of the sample chips tested meet specifications. If the criterion is met, then the batch is accepted and shipped. This criterion is based on past experience of what the computer companies will find acceptable, i.e., if the batch "yield" is less than 95% the computer companies will not be happy. The production manager proposes that a sample of 100 chips from the batch be tested and if 95 or more are deemed to meet specifications, then the batch is judged to be acceptable. However, a quality control supervisor argues that even if only 5 of the sample chips are defective, then it is still quite probable that the batch will not have a 95% yield and thus be defective.

The quality control supervisor wishes to convince the production manager that a defective batch can frequently produce 5 or fewer defective chips in a chip sample of size 100. He does so by determining the probability that a *defective batch* will have a chip sample with 5 or fewer defective chips as follows. He first needs to assume the proportion of chips in the defective batch that will be good. Since a good batch has a proportion of good chips of 95%, a defective batch will have a proportion of good chips of less than 95%. Since he is quite conservative, he chooses this proportion as exactly $p = 0.94$, although it may actually be less. Then, according to the production manager a batch is judged to be acceptable if the sample produces $95, 96, 97, 98, 99$, or 100 good chips. The quality control supervisor likens this problem to the drawing of 100 balls from an "chip urn" containing 1000 balls. In the urn there are $1000p$ good balls and $1000(1-p)$ bad ones. The probability of drawing 95 or more good balls from the urn is given *approximately* by the binomial probability law. We have assumed that the true law, which is hypergeometric due to the use of sampling without replacement, can be approximated by the binomial law, which assumes sampling with replacement. See Problem 3.48 for the accuracy of this approximation.

Now the defective batch will be judged as acceptable if there are 95 or more successes out of a possible 100 draws. The probability of this occurring is

$$P[k \geq 95] = \sum_{k=95}^{100} \binom{100}{k} p^k (1-p)^{100-k}$$

where $p = 0.94$. The probability $P[k \geq 95]$ versus p is plotted in Figure 3.11. For $p = 0.94$ we see that the defective batch will be accepted with a probability of about 0.45 or almost half of the defective batches will be shipped. The quality control supervisor is indeed correct. The production manager does not believe the

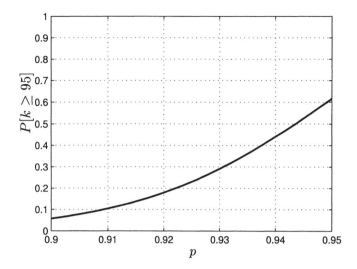

Figure 3.11: Probability of accepting a defective batch versus proportion of good chips in the defective batch – accept if 5 or fewer bad chips in a sample of 100.

result since it appears to be too high. Using sampling with replacement, which will produce results in accordance with the binomial law, he performs a computer simulation (see Problem 3.49). Based on the simulated results he reluctantly accepts the supervisor's conclusions. In order to reduce this probability the quality control supervisor suggests changing the acceptance strategy to one in which the batch is accepted only if 98 or more of the samples meet the specifications. Now the probability that the defective batch will be judged as acceptable is

$$P[k \geq 98] = \sum_{k=98}^{100} \binom{100}{k} p^k (1-p)^{100-k}$$

where $p = 0.94$, the assumed proportion of good chips in the defective batch. This produces the results shown in Figure 3.12. The acceptance probability for a defective batch is now reduced to only about 0.05.

There is a price to be paid, however, for only accepting a batch if 98 or more of the samples are good. Many more good batches will be rejected than if the previous strategy were used (see Problem 3.50). This is deemed to be a reasonable tradeoff. Note that the supervisor may well be advised to examine his initial assumption about p for the defective batch. If, for instance, he assumed that a defective batch could be characterized by $p = 0.9$, then according to Figure 3.11, the production manager's original strategy would produce a probability of less than 0.1 of accepting a defective batch.

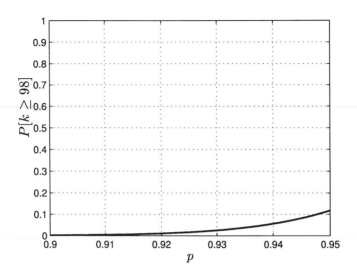

Figure 3.12: Probability of accepting a defective batch versus proportion of good chips in the defective batch – accept if 2 or fewer bad chips in a sample of 100.

References

Billingsley, P., *Probability and Measure*, John Wiley & Sons, New York, 1986.

Ross, S., *A First Course in Probability*, Prentice-Hall, Upper Saddle River, NJ, 2002.

Problems

3.1 ($\ddot\smile$) (w) The universal set is given by $\mathcal{S} = \{x : -\infty < x < \infty\}$ (the real line). If $A = \{x : x > 1\}$ and $B = \{x : x \leq 2\}$, find the following:

a. A^c and B^c

b. $A \cup B$ and $A \cap B$

c. $A - B$ and $B - A$

3.2 (w) Repeat Problem 3.1 if $\mathcal{S} = \{x : x \geq 0\}$.

3.3 (w) A group of voters go to the polling place. Their names and ages are Lisa, 21, John, 42, Ashley, 18, Susan, 64, Phillip, 58, Fred, 48, and Brad, 26. Find the following sets:

      a. Voters older than 30

      b. Voters younger than 30

      c. Male voters older than 30

      d. Female voters younger than 30

      e. Voters that are male or younger than 30

      f. Voters that are female and older than 30

Next find any two sets that partition the universe.

3.4 (w) Given the sets $A_i = \{x : 0 \leq x \leq i\}$ for $i = 1, 2, \ldots, N$, find $\cup_{i=1}^{N} A_i$ and $\cap_{i=1}^{N} A_i$. Are the A_i's disjoint?

3.5 (w) Prove that the sets $A = \{x : x \geq -1\}$ and $B = \{x : 2x + 2 \geq 0\}$ are equal.

3.6 (t) Prove that if $x \in A \cap B^c$, then $x \in A - B$.

3.7 ($\smile$) (w) If $\mathcal{S} = \{1, 2, 3, 4, 5, 6\}$, find sets A and B that are disjoint. Next find sets C and D that partition the universe.

3.8 (w) If $\mathcal{S} = \{(x, y) : 0 \leq x \leq 1 \text{ and } 0 \leq y \leq 1\}$, find sets A and B that are disjoint. Next find sets C and D that partition the universe.

3.9 (t) In this problem we see how to construct disjoint sets from ones that are not disjoint so that their unions will be the same. We consider only three sets and ask the reader to generalize the result. Calling the nondisjoint sets A, B, C and the union $D = A \cup B \cup C$, we wish to find three disjoint sets E_1, E_2, and E_3 so that $D = E_1 \cup E_2 \cup E_3$. To do so let

$$
\begin{aligned}
E_1 &= A \\
E_2 &= B - E_1 \\
E_3 &= C - (E_1 \cup E_2).
\end{aligned}
$$

Using a Venn diagram explain this procedure. If we now have sets $A_1, A_2, \ldots, A_N$, explain how to construct N disjoint sets with the same union.

3.10 ($\smile$) (f) Replace the set expression $A \cup B \cup C$ with one using intersections and complements. Replace the set expression $A \cap B \cap C$ with one using unions and complements.

3.11 (w) The sets A, B, C are subsets of $\mathcal{S} = \{(x, y) : 0 \leq x \leq 1 \text{ and } 0 \leq y \leq 1\}$. They are defined as

$$
\begin{aligned}
A &= \{(x, y) : x \leq 1/2, 0 \leq y \leq 1\} \\
B &= \{(x, y) : x \geq 1/2, 0 \leq y \leq 1\} \\
C &= \{(x, y) : 0 \leq x \leq 1, y \leq 1/2\}.
\end{aligned}
$$

Explicitly determine the set $A \cup (B \cap C)^c$ by drawing a picture of it as well as pictures of all the individual sets. For simplicity you can ignore the edges of the sets in drawing any diagrams. Can you represent the resultant set using only unions and complements?

3.12 ($\smile$) (w) Give the size of each set and also whether it is discrete or continuous. If the set is infinite, determine if it is countably infinite or not.

    a. $A = \{\text{seven-digit numbers}\}$

    b. $B = \{x : 2x = 1\}$

    c. $C = \{x : 0 \le x \le 1 \text{ and } 1/2 \le x \le 2\}$

    d. $D = \{(x,y) : x^2 + y^2 = 1\}$

    e. $E = \{x : x^2 + 3x + 2 = 0\}$

    f. $F = \{\text{positive even integers}\}$

3.13 (w) Two dice are tossed and the number of dots on each side that come up are added together. Determine the sample space, outcomes, impossible event, three different events including a simple event, and two mutually exclusive events. Use appropriate set notation.

3.14 ($\smile$) (w) The temperature in Rhode Island on a given day in August is found to always be in the range from 30° F to 100° F. Determine the sample space, outcomes, impossible event, three different events including a simple event, and two mutually exclusive events. Use appropriate set notation.

3.15 (t) Prove that if the sample space has size N, then the total number of events (including the impossible event and the certain event) is 2^N. Hint: There are $\binom{N}{k}$ ways to choose an event with k outcomes from a total of N outcomes. Also, use the binomial formula

$$(a + b)^N = \sum_{k=0}^{N} \binom{N}{k} a^k b^{N-k}$$

which was proven in Problem 1.11.

3.16 (w) An urn contains 2 red balls and 3 black balls. The red balls are labeled with the numbers 1 and 2 and the black balls are labeled as 3, 4, and 5. Three balls are drawn without replacement. Consider the events that

$$\begin{aligned} A &= \{\text{a majority of the balls drawn are black}\} \\ B &= \{\text{the sum of the numbers of the balls drawn} \ge 10\}. \end{aligned}$$

Are these events mutually exclusive? Explain your answer.

3.17 (t) Prove Axiom 3′ by using mathematical induction (see Appendix B) and Axiom 3.

3.18 (☺) (w) A roulette wheel has numbers 1 to 36 equally spaced around its perimeter. The odd numbers are colored red while the even numbers are colored black. If a spun ball is equally likely to yield any of the 36 numbers, what is the probability of a black number, of a red number? What is the probability of a black number that is greater than 24? What is the probability of a black number or a number greater than 24?

3.19 (☺) (c) Use a computer simulation to simulate the tossing of a fair die. Based on the simulation what is the probability of obtaining an even number? Does it agree with the theoretical result? Hint: See Section 2.4.

3.20 (w) A fair die is tossed. What is the probability of obtaining an even number, an odd number, a number that is even or odd, a number that is even and odd?

3.21 (☺) (w) A die is tossed that yields an even number with twice the probability of yielding an odd number. What is the probability of obtaining an even number, an odd number, a number that is even or odd, a number that is even and odd?

3.22 (w) If a single letter is selected at random from $\{A, B, C\}$, find the probability of all events. Recall that the total number of events is 2^N, where N is the number of simple events. Do these probabilities sum to one? If not, why not? Hint: See Problem 3.15.

3.23 (☺) (w) A number is chosen from $\{1, 2, 3, \ldots\}$ with probability

$$
P[i] = \begin{cases}
\frac{4}{7} & i = 1 \\
\frac{2}{7} & i = 2 \\
\left(\frac{1}{8}\right)^{i-2} & i \geq 3
\end{cases}
$$

Find $P[i \geq 4]$.

3.24 (f) For a sample space $\mathcal{S} = \{0, 1, 2, \ldots\}$ the probability assignment

$$
P[i] = \exp(-2)\frac{2^i}{i!}
$$

is proposed. Is this a valid assignment?

3.25 (☺) (w) Two fair dice are tossed. Find the probability that only one die comes up a 6.

3.26 (w) A circuit consists of N switches in parallel (see Example 3.6 for $N = 2$). The sample space can be summarized as $S = \{(z_1, z_2, \ldots, z_N) : z_i = \text{s or f}\}$, where s indicates a success or the switch closes and f indicates a failure or the switch fails to close. Assuming that all the simple events are equally likely, what is the probability that a circuit is closed when all the switches are activated to close? Hint: Consider the complement event.

3.27 (☺) (w) Can the series circuit of Figure 3.7 ever outperform the parallel circuit of Figure 3.6 in terms of having a higher probability of closing when both switches are activated to close? Assume that switch 1 closes with probability p, switch 2 closes with probability p, and both switches close with probability p^2.

3.28 (w) Verify the formula (3.20) for $P[E_1 \cup E_2 \cup E_3]$ if E_1, E_2, E_3 are events that are not necessarily mutually exclusive. To do so use a Venn diagram.

3.29 (t) Prove that

$$P[E_1 E_2] + P[E_1 E_3] + P[E_2 E_3] \geq P[E_1 E_2 E_3].$$

3.30 (w) A person always arrives at his job between 8:00 AM and 8:20 AM. He is equally likely to arrive anytime within that period. What is the probability that he will arrive at 8:10 AM? What is the probability that he will arrive between 8:05 and 8:10 AM?

3.31 (w) A random number generator produces a number that is equally likely to be anywhere in the interval $(0, 1)$. What are the simple events? Can you use (3.10) to find the probability that a generated number will be less than $1/2$? Explain.

3.32 (w) If two fair dice are tossed, find the probability that the same number will be observed on each one. Next, find the probability that different numbers will be observed.

3.33 (☺) (w) Three fair dice are tossed. Find the probability that 2 of the numbers will be the same and the third will be different.

3.34 (w,c) An urn contains 4 red balls and 2 black balls. Two balls are chosen at random and without replacement. What is the probability of obtaining one red ball and one black ball in any order? Verify your results by enumerating all possibilities using a computer evaluation.

3.35 (☺) (f) Rhode Island license plate numbers are of the form GR315 (2 letters followed by 3 digits). How many different license plates can be issued?

3.36 (f) A baby is to be named using four letters of the alphabet. The letters can be used as often as desired. How many different names are there? (Of course, some of the names may not be pronounceable).

3.37 (c) It is difficult to compute $N!$ when N is large. As an approximation, we can use Stirling's formula, which says that for large N

$$N! \approx \sqrt{2\pi} N^{N+1/2} \exp(-N).$$

Compare Stirling's approximation to the true value of $N!$ for $N = 1, 2, \ldots, 100$ using a digital computer. Next try calculating the exact value of $N!$ for $N = 200$ using a computer. Hint: Try printing out the logarithm of $N!$ and compare it to the logarithm of its approximation.

3.38 (⌣) (t) Determine the probability that in a class of 23 students two or more students have birthdays on January 1.

3.39 (c) Use a computer simulation to verify your result in Problem 3.38.

3.40 (⌣) (w) A pizza can be ordered with up to four different toppings. Find the total number of different pizzas (including no toppings) that can be ordered. Next, if a person wishes to pay for only two toppings, how many two-topping pizzas can he order?

3.41 (f) How many subsets of size three can be made from $\{A, B, C, D, E\}$?

3.42 (w) List all the combinations of two coins that can be chosen from the following coins: one penny (p), one nickel (n), one dime (d), one quarter (q). What are the possible sum-values?

3.43 (f) The binomial theorem states that

$$(a + b)^N = \sum_{k=0}^{N} \binom{N}{k} a^k b^{N-k}.$$

Expand $(a + b)^3$ and $(a + b)^4$ into powers of a and b and compare your results to the formula.

3.44 (⌣) (w) A deck of poker cards contains an ace, king, queen, jack, 10, 9, 8, 7, 6, 5, 4, 3, 2 in each of the four suits, hearts (h), clubs (c), diamonds (d), and spades (s), for a total of 52 cards. If 5 cards are chosen at random from a deck, find the probability of obtaining 4 of a kind, as for example, 8-h, 8-c, 8-d, 8-s, 9-c. Next find the probability of a flush, which occurs when all five cards have the same suit, as for example, 8-s, queen-s, 2-s, ace-s, 5-s.

3.45 (w) A class consists of 30 students, of which 20 are freshmen and 10 are sophomores. If 5 students are selected at random, what is the probability that they will all be sophomores?

3.46 (w) An urn containing an infinite number of balls has a proportion p of red balls, and the remaining portion $1 - p$ of black balls. Two balls are chosen at random. What value of p will yield the highest probability of obtaining one red ball and one black ball in any order?

3.47 (w) An urn contains an infinite number of coins that are either two-headed or two-tailed. The proportion of each kind is the same. If we choose M coins at random, explain why the probability of obtaining k heads is given by (3.28) with $p = 1/2$. Also, how does this experiment compare to the tossing of a fair coin M times?

3.48 (c) Compare the hypergeometric law to the binomial law if $N = 1000$, $M = 100$, $p = 0.94$ by calculating the probability $P[k]$ for $k = 95, 96, \ldots, 100$. Hint: To avoid computational difficulties of calculating $N!$ for large N, use the following strategy to find $x = 1000!/900!$ as an example.

$$y = \ln(x) = \ln(1000!) - \ln(900!) = \sum_{i=1}^{1000} \ln(i) - \sum_{i=1}^{900} \ln(i)$$

and then $x = \exp(y)$. Alternatively, for this example you can cancel out the common factors in the quotient of x and write it as $x = (1000)_{100}$, which is easier to compute. But in general, this may be more difficult to set up and program.

3.49 (☺) (c) A defective batch of 1000 chips contains 940 good chips and 60 bad chips. If we choose a sample of 100 chips, find the probability that there will be 95 or more good chips by using a computer simulation. To simpify the problem assume sampling with replacement for the computer simulation and the theoretical probability. Compare your result to the theoretical prediction in Section 3.10.

3.50 (c) For the real-world problem discussed in Section 3.10 use a computer simulation to determine the probability of rejecting a good batch. To simpify your code assume sampling *with replacement*. A good batch is defined as one with a probability of obtaining a good chip of $p = 0.95$. The two strategies are to accept the batch if 95 or more of the 100 samples are good and if 98 or more of the 100 samples are good. Explain your results. Can you use Figures 3.11 and 3.12 to determine the theoretical probabilities?

Chapter 4

Conditional Probability

4.1 Introduction

In the previous chapter we determined the probabilities for some simple experiments. An example was the die toss that produced a number from 1 to 6 "at random". Hence, a probability of 1/6 was assigned to each possible outcome. In many real-world "experiments", the outcomes are not completely random since we have some prior knowledge. For instance, knowing that it has rained the previous 2 days might influence our assignment of the probability of sunshine for the following day. Another example is to determine the probability that an individual chosen from some general population weighs more than 200 lbs., knowing that his height exceeds 6 ft. This motivates our interest in how to determine the probability of an event, given that we have some prior knowledge. For the die tossing experiment we might inquire as to the probability of obtaining a 4, if it is known that the outcome is an even number. The additional knowledge should undoubtedly change our probability assignments. For example, if it is known that the outcome is an even number, then the probability of any odd-numbered outcome must be zero. It is this interaction between the original probabilities and the probabilities in light of prior knowledge that we wish to describe and quantify, leading to the concept of a *conditional probability*.

4.2 Summary

Section 4.3 motivates and then defines the conditional probability as (4.1). In doing so the concept of a joint event and its probability are introduced as well as the marginal probability of (4.3). Conditional probabilities can be greater than, less than, or equal to the ordinary probability as illustrated in Figure 4.2. Also, conditional probabilities are true probabilities in that they satisfy the basic axioms and so can be manipulated in the usual ways. Using the law of total probability (4.4), the probabilities for compound experiments are easily determined. When the conditional probability is equal to the ordinary probability, the events are said to be statistically independent. Then, knowledge of the occurrence of one event does not change the probability of the other event. The condition for two events to be independent is given by (4.5). Three events are statistically independent if the conditions (4.6)–(4.9) hold. Bayes' theorem is defined by either (4.13) or (4.14). Embodied in the theorem are the concepts of a prior probability (before the experiment is conducted) and a posterior probability (after the experiment is conducted). Conclusions may be drawn based on the outcome of an experiment as to whether certain hypotheses are true. When an experiment is repeated multiple times and the experiments are independent, the probability of a joint event is easily found via (4.15). Some probability laws that result from the independent multiple experiment assumption are the binomial (4.16), the geometric (4.17), and the multinomial (4.19). For dependent multiple experiments (4.20) must be used to determine probabilities of joint events. If, however, the experimental outcomes probabilities only depend on the previous experimental outcome, then the Markov condition is satisfied. This results in the simpler formula for determining joint probabilities given by (4.21). Also, this assumption leads to the concept of a Markov chain, an example of which is shown in Figure 4.8. Finally, in Section 4.7 an example of the use of Bayes' theorem to detect the presence of a cluster is investigated.

4.3   Joint Events and the Conditional Probability

In formulating a useful theory of conditional probability we are led to consider two events. Event A is our event of interest while event B represents the event that embodies our prior knowledge. For the fair die toss example described in the introduction, the event of interest is $A = \{4\}$ and the event describing our prior knowledge is an even outcome or $B = \{2, 4, 6\}$. Note that when we say that the outcome must be even, we do not elaborate on why this is the case. It may be because someone has observed the outcome of the experiment and conveyed this partial information to us. Alternatively, it may be that the experimenter loathes odd outcomes, and therefore keeps tossing the die until an even outcome is obtained. Conditional probability does not address the reasons for the prior information, only how to accommodate it into a probabilistic framework. Continuing with the fair die example, a typical sequence of outcomes for a repeated experiment is shown in Figure 4.1. The odd outcomes are shown as dashed lines and are to be ignored. From the figure we see that the probability of a 4 is about $9/25 = 0.36$, or about $1/3$, using a relative frequency interpretation of probability. This has been found by taking the total number of 4's and dividing by the total number of 2's, 4's, and 6's. Specifically, we have that

$$\frac{N_A}{N_B} = \frac{9}{25}.$$

Another problem might be to determine the probability of $A = \{1, 4\}$, knowing that the outcome is even. In this case, we should use $N_{A \cap B}/N_B$ to make sure we

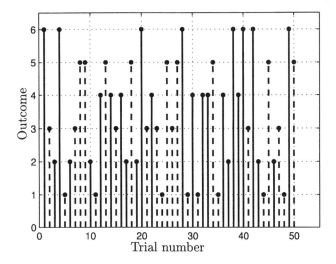

Figure 4.1: Outcomes for repeated tossing of a fair die.

only count the outcomes that can occur in light of our knowledge of B. For this example, only the 4 in $\{1, 4\}$ could have occurred. If an outcome is not in B, then that outcome will not be included in $A \cap B$ and will not be counted in $N_{A \cap B}$. Now letting $\mathcal{S} = \{1, 2, 3, 4, 5, 6\}$ be the sample space and $N_{\mathcal{S}}$ its size, the probability of A given B is

$$\frac{N_{A \cap B}}{N_B} = \frac{\frac{N_{A \cap B}}{N_{\mathcal{S}}}}{\frac{N_B}{N_{\mathcal{S}}}} \approx \frac{P[A \cap B]}{P[B]}.$$

This is termed the *conditional probability* and is denoted by $P[A|B]$ so that we have as our definition

$$P[A|B] = \frac{P[A \cap B]}{P[B]}. \tag{4.1}$$

Note that to determine it, we require $P[A \cap B]$ which is the probability of both A and B occurring or the probability of the intersection. Intuitively, the conditional probability is the proportion of time A and B occurs divided by the proportion of time that B occurs. The event $B = \{2, 4, 6\}$ comprises a new sample space and is sometimes called the *reduced sample space*. The denominator term in (4.1) serves to normalize the conditional probabilities so that the probability of the reduced sample space is one (set $A = B$ in (4.1)). Returning to the die toss, the probability of a 4, given that the outcome is even, is found as

$$\begin{aligned} A \cap B &= \{4\} \cap \{2, 4, 6\} = \{4\} = A \\ B &= \{2, 4, 6\} \end{aligned}$$

	W_1 100–130	W_2 130–160	W_3 160–190	W_4 190–220	W_5 220–250	$P[H_i]$
H_1 $5' - 5'4''$	0.08	0.04	0.02	0	0	0.14
H_2 $5'4'' - 5'8''$	0.06	0.12	0.06	0.02	0	0.26
H_3 $5'8'' - 6'$	0	0.06	0.14	0.06	0	0.26
H_4 $6' - 6'4''$	0	0.02	0.06	0.10	0.04	0.22
H_5 $6'4'' - 6'8''$	0	0	0	0.08	0.04	0.12

Table 4.1: Joint probabilities for heights and weights of college students.

and therefore

$$P[A|B] = \frac{P[A \cap B]}{P[B]} = \frac{P[A]}{P[B]}$$
$$= \frac{1/6}{3/6} = \frac{1}{3}$$

as expected. Note that $P[A \cap B]$ and $P[B]$ are computed based on the *original sample space*, S.

The event $A \cap B$ is usually called the *joint event* since both events must occur for a nonempty intersection. Likewise, $P[A \cap B]$ is termed the *joint probability*, but of course, it is nothing more than the probability of an intersection. Also, $P[A]$ is called the *marginal probability* to distinguish it from the joint and conditional probabilities. The reason for this terminology will be discussed shortly.

In defining the conditional probability of (4.1) it is assumed that $P[B] \neq 0$. Otherwise, theoretically and practically, the definition would not make sense. Another example follows.

Example 4.1 – Heights and weights of college students

A population of college students have heights H and weights W which are grouped into ranges as shown in Table 4.1. The table gives the joint probability of a student having a given height and weight, which can be denoted as $P[H_i \cap W_j]$. For example, if a student is selected, the probability of his/her height being between $5'4''$ and $5'8''$ and also his/her weight being between 130 lbs. and 160 lbs. is 0.12. Now consider the event that the student has a weight in the range 130–160 lbs. Calling this event A we next determine its probability. Since $A = \{(H, W) : H = H_1, \ldots, H_5; W = W_2\}$, it is explicitly

$$A = \{(H_1, W_2), (H_2, W_2), (H_3, W_2), (H_4, W_2), (H_5, W_2)\}$$

and since the simple events are by definition mutually exclusive, we have by Axiom

3′ (see Section 3.4)

$$P[A] = \sum_{i=1}^{5} P[(H_i, W_2)] = 0.04 + 0.12 + 0.06 + 0.02 + 0$$
$$= 0.24.$$

Next we determine the probability that a student's weight is in the range of 130–160 lbs., *given* that the student has height less than 6′. The event of interest A is the same as before. The conditioning event is $B = \{(H, W) : H = H_1, H_2, H_3; W = W_1, \ldots, W_5\}$ so that $A \cap B = \{(H_1, W_2), (H_2, W_2), (H_3, W_2)\}$ and

$$P[A|B] = \frac{P[A \cap B]}{P[B]} = \frac{0.04 + 0.12 + 0.06}{0.14 + 0.26 + 0.26}$$
$$= 0.33.$$

We see that it is more probable that the student has weight between 130 and 160 lbs. if it is known beforehand that his/her height is less than 6′. Note that in finding $P[B]$ we have used

$$P[B] = \sum_{i=1}^{3} \sum_{j=1}^{5} P[(H_i, W_j)] \tag{4.2}$$

which is determined by first summing along each row to produce the entries shown in Table 4.1 as $P[H_i]$. These are given by

$$P[H_i] = \sum_{j=1}^{5} P[(H_i, W_j)] \tag{4.3}$$

and then summing the $P[H_i]$'s for $i = 1, 2, 3$. Hence, we could have written (4.2) equivalently as

$$P[B] = \sum_{i=1}^{3} P[H_i].$$

The probabilities $P[H_i]$ are called the *marginal probabilities* since they are written in the *margin* of the table. If we were to sum along the columns, then we would obtain the marginal probabilities for the weights or $P[W_j]$. These are given by

$$P[W_j] = \sum_{i=1}^{5} P[(H_i, W_j)].$$

It is important to observe that by utilizing the information that the student's height is less than 6′, the probability of the event has changed; in this case, it has increased from 0.24 to 0.33. It is also possible that the opposite may occur. If we were to determine the probability that the student's weight is in the range

130–160 lbs., given that he/she has a height *greater* than $6'$, then defining the conditioning event as $B = \{(H, W) : H = H_4, H_5; W = W_1, \ldots, W_5\}$ and noting that $A \cap B = \{(H_4, W_2), (H_5, W_2\}$ we have

$$P[A|B] = \frac{0.02 + 0}{0.22 + 0.12}$$
$$= 0.058.$$

Hence, the conditional probability has now decreased with respect to the unconditional probability or $P[A]$.

$\diamond$

In general we may have

$$P[A|B] > P[A]$$
$$P[A|B] < P[A]$$
$$P[A|B] = P[A].$$

See Figure 4.2 for another example. The last possibility is of particular interest since

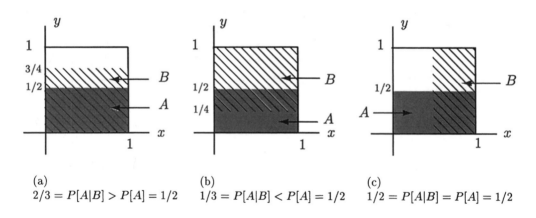

(a)   (b)   (c)
$2/3 = P[A|B] > P[A] = 1/2$ $1/3 = P[A|B] < P[A] = 1/2$ $1/2 = P[A|B] = P[A] = 1/2$

Figure 4.2: Illustration of possible relationships of conditional probability to ordinary probability.

it states that the probability of an event A is the same whether or not we know that B has occurred. In this case, the event A is said to be *statistically independent* of the event B. In the next section, we will explore this further.

Before proceeding, we wish to emphasize that a conditional probability is a true probability in that it satisfies the axioms described in Chapter 3. As a result, all the rules that allow one to manipulate probabilities also apply to conditional probabilities. For example, since Property 3.1 must hold, it follows that $P[A^c|B] = 1 - P[A|B]$ (see also Problem 4.10). To prove that the axioms are satisfied for conditional probabilities we first assume that the axioms hold for ordinary probabilities. Then,

Axiom 1

$$P[A|B] = \frac{P[A \cap B]}{P[B]} \geq 0$$

since $P[A \cap B] \geq 0$ and $P[B] \geq 0$.

Axiom 2

$$P[\mathcal{S}|B] = \frac{P[\mathcal{S} \cap B]}{P[B]} = \frac{P[B]}{P[B]} = 1.$$

Axiom 3 If A and C are mutually exclusive events, then

$$
\begin{aligned}
P[A \cup C|B] &= \frac{P[(A \cup C) \cap B]}{P[B]} && \text{(definition)} \\
&= \frac{P[(A \cap B) \cup (C \cap B)]}{P[B]} && \text{(distributive property)} \\
&= \frac{P[A \cap B] + P[C \cap B]}{P[B]} && \text{(Axiom 3 for ordinary probability,} \\
& && A \cap C = \emptyset \Rightarrow (A \cap B) \cap (C \cap B) = \emptyset) \\
&= P[A|B] + P[C|B] && \text{(definition of conditional probability).}
\end{aligned}
$$

Conditional probabilities are useful in that they allow us to simplify probability calculations. One particularly important relationship based on conditional probability is described next. Consider a partitioning of the sample space $\mathcal{S}$. Recall that a partition is defined as a group of sets $B_1, B_2, \ldots, B_N$ such that $\mathcal{S} = \cup_{i=1}^{N} B_i$ and $B_i \cap B_j = \emptyset$ for $i \neq j$. Then we can rewrite the probability $P[A]$ as

$$P[A] = P[A \cap \mathcal{S}] = P\left[A \cap \left(\cup_{i=1}^{N} B_i\right)\right].$$

But by a slight extension of the distributive property of sets, we can express this as

$$P[A] = P[(A \cap B_1) \cup (A \cap B_2) \cup \cdots \cup (A \cap B_N)].$$

Since the B_i's are mutually exclusive, then so are the $A \cap B_i$'s, and therefore

$$P[A] = \sum_{i=1}^{N} P[A \cap B_i]$$

or finally

$$P[A] = \sum_{i=1}^{N} P[A|B_i]P[B_i]. \tag{4.4}$$

This relationship is called the *law of total probability*. Its utility is illustrated next.

Example 4.2 – A compound experiment

Two urns contain different proportions of red and black balls. Urn 1 has a proportion p_1 of red balls and a proportion $1 - p_1$ of black balls whereas urn 2 has proportions of p_2 and $1 - p_2$ of red balls and black balls, respectively. A *compound* experiment is performed in which an urn is chosen at random, followed by the selection of a ball. We would like to find the probability that a red ball is selected. To do so we use (4.4) with $A = \{$red ball selected$\}$, $B_1 = \{$urn 1 chosen$\}$, and $B_2 = \{$urn 2 chosen$\}$. Then

$$
\begin{aligned}
P[\text{red ball selected}] &= P[\text{red ball selected}|\text{urn 1 chosen}]P[\text{urn 1 chosen}] \\
&\quad + P[\text{red ball selected}|\text{urn 2 chosen}]P[\text{urn 2 chosen}] \\
&= p_1\tfrac{1}{2} + p_2\tfrac{1}{2} = \tfrac{1}{2}(p_1 + p_2).
\end{aligned}
$$

 Do B_1 and B_2 really partition the sample space?

To verify that the application of the law of total probability is indeed valid for this problem, we need to show that $B_1 \cup B_2 = \mathcal{S}$ and $B_1 \cap B_2 = \emptyset$. In our description of B_1 and B_2 we refer to the choice of an urn. In actuality, this is shorthand for all the balls in the urn. If urn 1 contains balls numbered 1 to N_1, then by choosing urn 1 we are really saying that the event is that one of the balls numbered 1 to N_1 is chosen and similarly for urn 2 being chosen. Hence, since the sample space consists of all the numbered balls in urns 1 and 2, it is observed that the union of B_1 and B_2 is the set of all possible outcomes or the sample space. Also, B_1 and B_2 are mutually exclusive since we choose urn 1 *or* urn 2 but not both.

Some more examples follow.

Example 4.3 – Probability of error in a digital communication system

In a digital communication system a "0" or "1" is transmitted to a receiver. Typically, either *bit* is equally likely to occur so that a *prior probability* of 1/2 is assumed. At the receiver a decoding error can be made due to channel noise, so that a 0 may be mistaken for a 1 and vice versa. Defining the probability of decoding a 1 when a 0 is transmitted as ϵ and a 0 when a 1 is transmitted also as ϵ, we are interested in the overall probability of an error. A probabilistic model summarizing the relevant features is shown in Figure 4.3. Note that the problem at hand is essentially the same as the previous one. If urn 1 is chosen, then we transmit a 0 and if urn 2 is chosen, we transmit a 1. The effect of the channel is to introduce an error so that even if we know which bit was transmitted, we do not know the received bit. This is analogous to not knowing which ball was chosen from the given urn. The

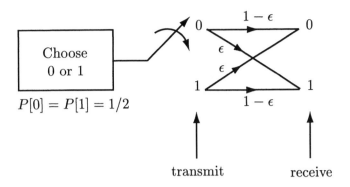

Figure 4.3: Probabilistic model of a digital communication system.

probability of error is from (4.4)

$$
\begin{aligned}
P[\text{error}] &= P[\text{error}|0 \text{ transmitted}]P[0 \text{ transmitted}] \\
&\quad + P[\text{error}|1 \text{ transmitted}]P[1 \text{ transmitted}] \\
&= \epsilon\tfrac{1}{2} + \epsilon\tfrac{1}{2} = \epsilon.
\end{aligned}
$$

$\diamondsuit$

Conditional probabilities can be quite tricky, in that they sometimes produce counterintuitive results. A famous instance of this is the Monty Hall or Let's Make a Deal problem.

Example 4.4 – Monty Hall problem

About 40 years ago there was a television game show called "Let's Make a Deal". The game show host, Monty Hall, would present the contestant with three closed doors. Behind one door was a new car, while the others concealed less desireable prizes, for instance, farm animals. The contestant would first have the opportunity to choose a door, but it would not be opened. Monty would then choose one of the remaining doors and open it. Since he would have knowledge of which door led to the car, he would always choose a door to reveal one of the farm animals. Hence, if the contestant had chosen one of the farm animals, Monty would then choose the door that concealed the other farm animal. If the contestant had chosen the door behind which was the car, then Monty would choose one of the other doors, both concealing farm animals, at random. At this point in the game, the contestant was faced with two closed doors, one of which led to the car and the other to a farm animal. The contestant was given the option of either opening the door she had originally chosen or deciding to open the other door. What should she do? The answer, surprisingly, is that by choosing to switch doors she has a probability of 2/3 of winning the car! If she stays with her original choice, then the probability is only 1/3. Most people would say that irregardless of which strategy she decided upon, her probability of winning the car is 1/2.

		M_j		
		1	2	3
	1	0	$\frac{1}{6}$	$\frac{1}{6}$
C_i 2		0	0	$\frac{1}{3}$*
	3	0	$\frac{1}{3}$*	0

Table 4.2: Joint probabilities $(P[C_i, M_j] = P[M_j|C_i]P[C_i])$ for contestant's initial and Monty's choice of doors. Winning door is 1.

To see how these probabilities are determined first assume she stays with her original choice. Then, since the car is equally likely to be placed behind any of the three doors, the probability of the contestant's winning the car is 1/3. Monty's choice of a door is irrelevant since her final choice is always the same as her initial choice. However, if as a result of Monty's action a different door is selected by the contestant, then the probability of winning becomes a *conditional probability*. We now compute this by assuming that the car is behind door one. Define the events $C_i = \{$contestant initially chooses door $i\}$ for $i = 1, 2, 3$ and $M_j = \{$Monty opens door $j\}$ for $j = 1, 2, 3$. Next we determine the joint probabilities $P[C_i, M_j]$ by using

$$P[C_i, M_j] = P[M_j|C_i]P[C_i].$$

Since the winning door is never chosen by Monty, we have $P[M_1|C_i] = 0$. Also, Monty never opens the door initially chosen by the contestant so that $P[M_i|C_i] = 0$. Then, it is easily verified that

$$P[M_2|C_3] = P[M_3|C_2] = 1 \qquad \text{(contestant chooses losing door)}$$
$$P[M_3|C_1] = P[M_2|C_1] = \frac{1}{2} \qquad \text{(contestant chooses winning door)}$$

and $P[C_i] = 1/3$. The joint probabilities are summarized in Table 4.2. Since the contestant always switches doors, the winning events are $(2, 3)$ (the contestant initially chooses door 2 and Monty chooses door 3) and $(3, 2)$ (the contestant initially chooses door 3 and Monty chooses door 2). As shown in Table 4.2 (the entries with asterisks), the total probability is 2/3. This may be verified directly using

$$P[\text{final choice is door 1}] = P[M_3|C_2]P[C_2] + P[M_2|C_3]P[C_3]$$
$$= P[C_2, M_3] + P[C_3, M_2].$$

Alternatively, the only way she can *lose* is if she initially chooses door one since she always switches doors. This has a probability of 1/3 and hence her probability of winning is 2/3. In effect, Monty, by eliminating a door, has improved her odds!

4.4 Statistically Independent Events

Two events A and B are said to be *statistically independent* (or sometimes just *independent*) if $P[A|B] = P[A]$. If this is true, then

$$P[A|B] = \frac{P[A \cap B]}{P[B]} = P[A]$$

which results in the condition for statistical independence of

$$P[A \cap B] = P[A]P[B]. \tag{4.5}$$

An example is shown in Figure 4.2c. There, the probability of A is unchanged if we know that the outcome is contained in the event B. Note, however, that once we know that B has occurred, the outcome could not have been in the uncross-hatched region of A but must be in the cross-hatched region. Knowing that B has occurred does in fact affect the possible outcomes. However, it is the *ratio* of $P[A \cap B]$ to $P[B]$ that remains the same.

Example 4.5 – Statistical independence does not mean one event does not affect another event.

If a fair die is tossed, the probability of a 2 or a 3 is $P[A = \{2,3\}] = 1/3$. Now assume we know that the outcome is an even number or $B = \{2,4,6\}$. Recomputing the probability

$$
\begin{aligned}
P[A|B] &= \frac{P[A \cap B]}{P[B]} = \frac{P[\{2\}]}{P[\{2,4,6\}]} \\
&= \frac{1}{3} = P[A].
\end{aligned}
$$

Hence, A and B are independent. Yet, knowledge of B occurring has affected the possible outcomes. In particular, the event $A \cap B = \{2\}$ has half as many elements as A, but the reduced sample space $\mathcal{S}' = B$ also has half as many elements.

$\Diamond$

The condition for the event A to be independent of the event B is $P[A \cap B] = P[A]P[B]$. Hence, we need only know the *marginal probabilities* or $P[A], P[B]$ to determine the *joint probability* $P[A \cap B]$. In practice, this property turns out to be very useful. Finally, it is important to observe that statistical independence has a symmetry property, as we might expect. If A is independent of B, then B must be independent of A since

$$
\begin{aligned}
P[B|A] &= \frac{P[B \cap A]}{P[A]} && \text{(definition)} \\
&= \frac{P[A \cap B]}{P[A]} && \text{(commutative property)} \\
&= \frac{P[A]P[B]}{P[A]} && (A \text{ is independent of } B) \\
&= P[B]
\end{aligned}
$$

and therefore B is independent of A. Henceforth, we can say that the events A and B are statistically independent of each other, without further elaboration.

⚠️ **Statistically independent events are different than mutually exclusive events.**

If A and B are mutually exclusive and B occurs, then A cannot occur. Thus, $P[A|B] = 0$. If A and B are statistically independent and B occurs, then $P[A|B] = P[A]$. Clearly, the probabilities $P[A|B]$ are only the same if $P[A] = 0$. In general then, the conditions of mutually exclusivity and independence must be different since they lead to different values of $P[A|B]$. A specific example of events that

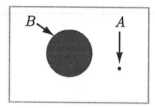

Figure 4.4: Events that are mutually exclusive (since $A \cap B = \emptyset$) and independent (since $P[A \cap B] = P[\emptyset] = 0$ and $P[A]P[B] = 0 \cdot P[B] = 0$).

are both mutually exclusive and statistically independent is shown in Figure 4.4. Finally, the two conditions produce different relationships, namely

$$P[A \cup B] = P[A] + P[B] \qquad \text{mutually exclusive events}$$
$$P[A \cap B] = P[A]P[B] \qquad \text{statistically independent events.}$$

See also Figure 4.2c for statistically independent but not mutually exclusive events. Can you think of a case of mutually exclusive but not independent events?

Consider now the extension of the idea of statistical independence to three events. Three events are defined to be independent if the knowledge that any one or two of the events has occurred does not affect the probability of the third event. For example, one condition is that $P[A|B \cap C] = P[A]$. We will use the shorthand notation $P[A|B,C]$ to indicate that this is the probability of A given that B *and* C has occurred. Note that if B and C has occurred, then by definition $B \cap C$ has occurred. The full set of conditions is

$$P[A|B] = P[A|C] = P[A|B,C] = P[A]$$
$$P[B|A] = P[B|C] = P[B|A,C] = P[B]$$
$$P[C|A] = P[C|B] = P[C|A,B] = P[C].$$

These conditions are satisfied if and only if

$$P[AB] = P[A]P[B] \tag{4.6}$$
$$P[AC] = P[A]P[C] \tag{4.7}$$
$$P[BC] = P[B]P[C] \tag{4.8}$$
$$P[ABC] = P[A]P[B]P[C]. \tag{4.9}$$

If the first three conditions (4.6)–(4.8) are satisfied, then the events are said to be *pairwise* independent. They are not enough, however, to ensure independence. The last condition (4.9) is also required since without it we could not assert that

$$
\begin{aligned}
P[A|B,C] &= P[A|BC] && \text{(definition of } B \text{ and } C \text{ occurring)} \\
&= \frac{P[ABC]}{P[BC]} && \text{(definition of conditional probability)} \\
&= \frac{P[ABC]}{P[B]P[C]} && \text{(from (4.8))} \\
&= \frac{P[A]P[B]P[C]}{P[B]P[C]} && \text{(from (4.9))} \\
&= P[A]
\end{aligned}
$$

and similarly for the other conditions (see also Problem 4.20 for an example). In general, events $E_1, E_2, \ldots, E_N$ are defined to be statistically independent if

$$
\begin{aligned}
P[E_i E_j] &= P[E_i]P[E_j] && i \neq j \\
P[E_i E_j E_k] &= P[E_i]P[E_j]P[E_k] && i \neq j \neq k \\
&\;\;\vdots && \vdots \\
P[E_1 E_2 \cdots E_N] &= P[E_1]P[E_2] \cdots P[E_N].
\end{aligned}
$$

Although statistically independent events allow us to compute joint probabilities based on only the marginal probabilities, we can still determine joint probabilities without this property. Of course, it becomes much more difficult. Consider three events as an example. Then, the joint probability is

$$
\begin{aligned}
P[ABC] &= P[A|B,C]P[BC] \\
&= P[A|B,C]P[B|C]P[C]. \tag{4.10}
\end{aligned}
$$

This relationship is called the *probability chain rule*. One is required to determine conditional probabilities, not always an easy matter. A simple example follows.

Example 4.6 –   Tossing a fair die - once again

If we toss a fair die, then it is clear that the probability of the outcome being 4 is 1/6. We can, however, rederive this result by using (4.10). Letting

$$
\begin{aligned}
A &= \{\text{even number}\} = \{2,4,6\} \\
B &= \{\text{numbers} > 2\} = \{3,4,5,6\} \\
C &= \{\text{numbers} < 5\} = \{1,2,3,4\}
\end{aligned}
$$

we have that $ABC = \{4\}$. These events can be shown to be dependent (see Problem 4.21). Now making use of (4.10) and noting that $BC = \{3,4\}$ it follows that

$$
\begin{aligned}
P[ABC] &= P[A|B,C]P[B|C]P[C] \\
&= \left(\frac{1/6}{2/6}\right)\left(\frac{2/6}{4/6}\right)\left(\frac{4}{6}\right) = \frac{1}{6}.
\end{aligned}
$$

4.5   Bayes' Theorem

The definition of conditional probability leads to a famous and sometimes controversial formula for computing conditional probabilities. Recalling the definition, we have that

$$
P[A|B] = \frac{P[AB]}{P[B]} \tag{4.11}
$$

and

$$
P[B|A] = \frac{P[AB]}{P[A]}. \tag{4.12}
$$

Upon substitution of $P[AB]$ from (4.11) into (4.12)

$$
P[B|A] = \frac{P[A|B]P[B]}{P[A]}. \tag{4.13}
$$

This is called *Bayes' theorem*. By knowing the marginal probabilities $P[A], P[B]$ and the conditional probability $P[A|B]$, we can determine the other conditional probability $P[B|A]$. The theorem allows us to perform "inference" or to assess (with some probability) the validity of an event when some other event has been observed. For example, if an urn containing an unknown composition of balls is sampled with replacement and produces an outcome of 10 red balls, what are we to make of this? One might conclude that the urn contains only red balls. Yet, another individual might claim that the urn is a "fair" one, containing half red balls and half black balls, and attribute the outcome to luck. To test the latter conjecture we now determine the probability of a fair urn given that 10 red balls have just been drawn. The reader should note that we are essentially going "backwards" – usually we compute the probability of choosing 10 red balls *given* a fair urn. Now we are *given* the outcomes and wish to determine the probability of a fair urn. In doing so we believe that the urn is fair with probability 0.9. This is due to our past experience with our purchases from urn.com. In effect, we assume that the prior probability of $B = \{\text{fair urn}\}$ is $P[B] = 0.9$. If $A = \{10 \text{ red balls drawn}\}$, we wish to determine $P[B|A]$, which is the probability of the urn being fair *after* the experiment has been performed or the *posterior* probability. This probability is *our reassessment of the fair urn in light of the new evidence (10 red balls drawn)*. Let's compute $P[B|A]$ which according to (4.13) requires knowledge of the *prior probability* $P[B]$ and the conditional probability $P[A|B]$. The former was assumed to be 0.9 and the latter is the probability of drawing 10 successive red balls from an urn with $p = 1/2$. From our previous work this is given by the binomial law as

$$P[A|B] \;=\; P[k = 10] = \binom{M}{k} p^k (1-p)^{M-k}$$

$$=\; \binom{10}{10} \left(\frac{1}{2}\right)^{10} \left(\frac{1}{2}\right)^{0} = \left(\frac{1}{2}\right)^{10}.$$

We still need to find $P[A]$. But this is easily found using the law of total probability as

$$P[A] \;=\; P[A|B]P[B] + P[A|B^c]P[B^c]$$

$$=\; P[A|B]P[B] + P[A|B^c](1 - P[B])$$

and thus only $P[A|B^c]$ needs to be determined (and which is *not equal to* $1 - P[A|B]$ as is shown in Problem 4.9). This is the conditional probability of drawing 10 red balls from a *unfair urn*. For simplicity we will assume that an unfair urn has all red balls and thus $P[A|B^c] = 1$. Now we have that

$$P[A] = \left(\frac{1}{2}\right)^{10} (0.9) + (1)(0.1)$$

and using this in (4.13) yields

$$P[B|A] = \frac{\left(\frac{1}{2}\right)^{10}(0.9)}{\left(\frac{1}{2}\right)^{10}(0.9) + (1)(0.1)} = 0.0087.$$

The posterior probability (after 10 red balls have been drawn) that the urn is fair is only 0.0087. Our conclusion would be to reject the assumption of a fair urn.

Another way to quantify the result is to compare the posterior probability of the unfair urn to the probability of the fair urn by the ratio of the former to the latter. This is called the *odds ratio* and it is interpreted as the odds *against* the hypothesis of a fair urn. In this case it is

$$\text{odds} = \frac{P[B^c|A]}{P[B|A]} = \frac{1 - 0.0087}{0.0087} = 113.$$

It is seen from this example that based on observed "data", prior beliefs embodied in $P[B] = 0.9$ can be modified to yield posterior beliefs or $P[B|A] = 0.0087$. This is an important concept in statistical inference [Press 2003].

In the previous example, we used the law of total probability to determine the posterior probability. More generally, if a set of B_i's partition the sample space, then Bayes' theorem can be expressed as

$$P[B_k|A] = \frac{P[A|B_k]P[B_k]}{\sum_{i=1}^{N} P[A|B_i]P[B_i]} \qquad k = 1, 2, \ldots, N. \qquad (4.14)$$

The denominator in (4.14) serves to normalize the posterior probability so that the conditional probabilities sum to one or

$$\sum_{k=1}^{N} P[B_k|A] = 1.$$

In many problems one is interested in determining whether an observed event or *effect* is the result of some *cause*. Again the backwards or *inferential* reasoning is implicit. Bayes' theorem can be used to quantify this connection as illustrated next.

Example 4.7 – Medical diagnosis

Suppose it is known that 0.001% of the general population has a certain type of cancer. A patient visits a doctor complaining of symptoms that might indicate the presence of this cancer. The doctor performs a blood test that will confirm the cancer with a probability of 0.99 if the patient does indeed have cancer. However, the test also produces *false positives* or says a person has cancer when he does not. This occurs with a probability of 0.2. If the test comes back positive, what is the probability that the person has cancer?

To solve this problem we let $B = \{$person has cancer$\}$, the causitive event, and $A = \{$test is positive$\}$, the effect of that event. Then, the desired probability is

$$
\begin{aligned}
P[B|A] &= \frac{P[A|B]P[B]}{P[A|B]P[B] + P[A|B^c]P[B^c]} \\
&= \frac{(0.99)(0.00001)}{(0.99)(0.00001) + (0.2)(0.99999)} = 4.95 \times 10^{-5}.
\end{aligned}
$$

The prior probability of the person having cancer is $P[B] = 10^{-5}$ while the posterior probability of the person having cancer (after the test is performed and found to be positive) is $P[B|A] = 4.95 \times 10^{-5}$. With these results the doctor might be hard pressed to order additional tests. This is quite surprising, and is due to the prior probability assumed, which is quite small and therefore tends to nullify the test results. If we had assumed that $P[B] = 0.5$, for indeed the doctor is seeing a patient who is complaining of symptoms consistent with cancer and not some person chosen at random from the general population, then

$$P[B|A] = \frac{(0.99)(0.5)}{(0.99)(0.5) + (0.2)(0.5)} = 0.83$$

which seems more reasonable (see also Problem 4.23). The controversy surrounding the use of Bayes' theorem in probability calculations can almost always be traced back to the prior probability assumption. Bayes' theorem is mathematically correct – only its application is sometimes in doubt!

4.6 Multiple Experiments

4.6.1 Independent Subexperiments

An experiment that was discussed in Chapter 1 was the repeated tossing of a coin. We can alternatively view this experiment as a succession of *subexperiments*, with each subexperiment being a single toss of the coin. It is of interest to investigate the relationship between the probabilities defined on the experiment and those defined on the subexperiments. To be more concrete, assume a coin is tossed twice in succession and we wish to determine the probability of the event $A = \{(H, T)\}$. Recall that the notation (H, T) denotes an *ordered* 2-tuple and represents a head on toss 1 and a tail on toss 2. For a fair coin it was determined to be 1/4 since we assumed that all 4 possible outcomes were equally likely. This seemed like a reasonable assumption. However, if the coin had a probability of heads of 0.99, we might not have been so quick to agree with the equally likely assumption. How then are we to determine the probabilities? Let's first consider the experiment to be composed of two separate subexperiments with each subexperiment having a sample space $\mathcal{S}^1 = \{H, T\}$. The sample space of the overall experiment is obtained by forming the *cartesian product*, which for this example is defined as

$$
\begin{aligned}
\mathcal{S} &= \mathcal{S}^1 \times \mathcal{S}^1 \\
&= \{(i, j) : i \in \mathcal{S}^1; j \in \mathcal{S}^1\} \\
&= \{(H, H), (H, T), (T, H), (T, T)\}.
\end{aligned}
$$

It is formed by taking an outcome from $\mathcal{S}^1$ for the first element of the 2-tuple and an outcome from $\mathcal{S}^1$ for the second element of the 2-tuple and doing this for all possible outcomes. It would be exceedingly useful if we could determine probabilities for events defined on $\mathcal{S}$ from those probabilities for events defined on $\mathcal{S}^1$. In this way the determination of probabilities of very complicated events could be simplified. Such is the case if we assume that the *subexperiments are independent*. Continuing on, we next calculate $P[A] = P[(H, T)]$ for a coin with an arbitrary probability of heads p. This event is defined on the sample space of 2-tuples, which is S. We can, however, express it as an intersection

$$\begin{aligned}
\{(H,T)\} &= \{(H,H),(H,T)\} \cap \{(H,T),(T,T)\} \\
&= \{\text{heads on toss 1}\} \cap \{\text{tails on toss 2}\} \\
&= H_1 \cap T_2.
\end{aligned}$$

We would expect the events H_1 and T_2 to be independent of each other. Whether a head or tail appears on the first toss should not affect the probability of the outcome of the second toss and vice versa. Hence, we will let $P[(H,T)] = P[H_1]P[T_2]$ in accordance with the definition of statistically independent events. We can determine $P[H_1]$ either as $P[(H,H),(H,T)]$, which is defined on S or *equivalently due to the independence assumption* as $P[H]$, which is defined on S^1. Note that $P[H]$ is the marginal probability and is equal to $P[(H,H)] + P[(H,T)]$. But the latter was specified to be p and therefore we have that

$$\begin{aligned}
P[H_1] &= p \\
P[T_2] &= 1 - p
\end{aligned}$$

and finally,

$$P[(H,T)] = p(1-p).$$

For a fair coin we recover the previous value of 1/4, but not otherwise.

Experiments that are composed of subexperiments whose probabilities of the outcomes do not depend on the outcomes of any of the other subexperiments are defined to be *independent subexperiments*. Their utility is to allow calculation of joint probabilities from marginal probabilities. More generally, if we have M independent subexperiments, with A_i an event described for experiment i, then the joint event $A = A_1 \cap A_2 \cap \cdots \cap A_M$ has probability

$$P[A] = P[A_1]P[A_2]\cdots P[A_M]. \qquad (4.15)$$

Apart from the differences in sample spaces upon which the probabilities are defined, independence of subexperiments is equivalent to statistical independence of events defined on the *same sample space*.

4.6.2   Bernoulli Sequence

The single tossing of a coin with probability p of heads is an example of a *Bernoulli trial*. Consecutive *independent* Bernoulli trials comprise a *Bernoulli sequence*. More generally, any sequence of M independent subexperiments with each subexperiment producing two possible outcomes is called a Bernoulli sequence. Typically, the subexperiment outcomes are labeled as 0 and 1 with the probability of a 1 being p. Hence, for a Bernoulli trial $P[0] = 1 - p$ and $P[1] = p$. Several important probability laws are based on this model.

Binomial Probability Law

Assume that M independent Bernoulli trials are carried out. We wish to determine the probability of k 1's (or successes). Each outcome is an M-tuple and a successful outcome would consist of k 1's and $M - k$ 0's in any order. Thus, each successful outcome has a probability of $p^k(1-p)^{M-k}$ due to independence. The total number of successful outcomes is the number of ways k 1's may be placed in the M-tuple. This is known from combinatorics to be $\binom{M}{k}$ (see Section 3.8). Hence, by summing up the probabilities of the successful simple events, which are mutually exclusive, we have

$$P[k] = \binom{M}{k} p^k (1-p)^{M-k} \qquad k = 0, 1, \ldots, M \qquad (4.16)$$

which we immediately recognize as the binomial probability law. We have previously encountered the same law when we chose M balls at random from an urn with replacement and desired the probability of obtaining k red balls. The proportion of red balls was p. In that case, each subexperiment was the choosing of a ball and all the subexperiments were *independent* of each other. The binomial probabilities are shown in Figure 4.5 for various values of p.

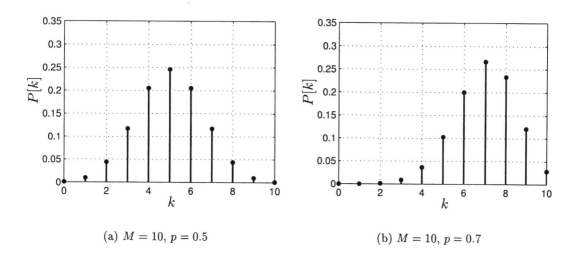

(a) $M = 10$, $p = 0.5$ (b) $M = 10$, $p = 0.7$

Figure 4.5: The binomial probability law for different values of p.

Geometric Probability Law

Another important aspect of a Bernoulli sequence is the appearance of the first success. If we let k be the Bernoulli trial for which the first success is observed, then the event of interest is the simple event $(f, f, \ldots, f, s)$, where s, f denote success and failure, respectively. This is a k-tuple with the first $k - 1$ elements all f's. The probability of the first success at trial k is therefore

$$P[k] = (1-p)^{k-1}p \qquad k = 1, 2, \ldots \tag{4.17}$$

where $0 < p < 1$. This is called the *geometric probability law*. The geometric probabilities are shown in Figure 4.6 for various values of p. It is interesting to note that the first success is always most likely to occur on the first trial or for $k = 1$. This is true even for small values of p, which is somewhat counterintuitive. However, upon further reflection, for the first success to occur on trial $k = 1$ we must have a success on trial 1 and the outcomes of the remaining trials are arbitrary. For a success on trial $k = 2$, for example, we must have a failure on trial 1 followed by a success on trial 2, with the remaining outcomes arbitrary. This additional constraint reduces the probability. It will be seen later, though, that the average number of trials required for a success is $1/p$, which is more in line with our intuition. An

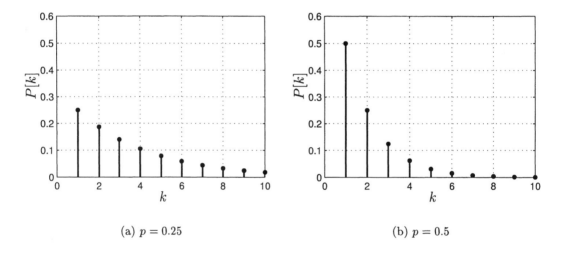

(a) $p = 0.25$ (b) $p = 0.5$

Figure 4.6: The geometric probability law for different values of p.

example of its use follows.

Example 4.8 – Telephone calling

A fax machine dials a phone number that is typically busy 80% of the time. The machine dials it every 5 minutes until the line is clear and the fax is able to be transmitted. What is the probability that the fax machine will have to dial the number 9 times? The number of times the line is busy can be considered the number of failures with each failure having a probability of $1 - p = 0.8$. If the number is dialed 9 times, then the first success occurs for $k = 9$ and

$$P[9] = (0.8)^8(0.2) = 0.0336.$$

$\Diamond$

A useful property of the geometric probability law is that it is memoryless. Assume it is known that no successes occurred in the first m trials. Then, the probability of the first success at trial $m+l$ is the same as if we had started the Bernoulli sequence experiment over again and determined the probability of the first success at trial l (see Problem 4.34).

4.6.3 Multinomial Probability Law

Consider an extension to the Bernoulli sequence in which the trials are still independent but the outcomes for each trial may take on more than two values. For example, let $\mathcal{S}^1 = \{1, 2, 3\}$ and denote the probabilities of the outcomes 1, 2, and 3 by p_1, p_2, and p_3, respectively. As usual, the assignment of these probabilities must satisfy $\sum_{i=1}^{3} p_i = 1$. Also, let the number of trials be $M = 6$ so that a possible outcome might be $(2, 1, 3, 1, 2, 2)$, whose probability is $p_2 p_1 p_3 p_1 p_2 p_2 = p_1^2 p_2^3 p_3^1$. The multinomial probability law specifies the probability of obtaining k_1 1's, k_2 2's, and k_3 3's, where $k_1 + k_2 + k_3 = M = 6$. In the current example, $k_1 = 2$, $k_2 = 3$, and $k_3 = 1$. Some outcomes with the same number of 1's, 2's', and 3's are $(2, 1, 3, 1, 2, 2)$, $(1, 2, 3, 1, 2, 2)$, $(1, 2, 1, 2, 2, 3)$, etc., with each outcome having a probability of $p_1^2 p_2^3 p_3^1$. The total number of these outcomes will be the total number of distinct 6-tuples that can be made with the numbers $1, 1, 2, 2, 2, 3$. If the numbers to be used were all different, then the total number of 6-tuples would be 6! , or all permutations. However, since they are not, some of these permutations will be the same. For example, we can arrange the 2's 3! ways and still have the same 6-tuple. Likewise, the 1's can be arranged 2! ways without changing the 6-tuple. As a result, the total number of *distinct* 6-tuples is

$$\frac{6!}{2!3!1!} \tag{4.18}$$

which is called the *multinomial coefficient*. (See also Problem 4.36 for another way to derive this.) It is sometimes denoted by

$$\binom{6}{2, 3, 1}.$$

Finally, for our example the probability of the sequence exhibiting two 1's, three 2's, and one 3 is

$$\frac{6!}{2!3!1!} p_1^2 p_2^3 p_3^1.$$

This can be generalized to the case of M trials with N possible outcomes for each trial. The probability of k_1 1's, k_2 2's,..., k_N N's is

$$P[k_1, k_2, \ldots, k_N] = \binom{M}{k_1, k_2, \ldots, k_N} p_1^{k_1} p_2^{k_2} \cdots p_N^{k_N} \qquad k_1 + k_2 + \cdots + k_N = M \tag{4.19}$$

and where $\sum_{i=1}^{N} p_i = 1$. This is termed the *multinomial probability law*. Note that if $N = 2$, then it reduces to the binomial law (see Problem 4.37). An example follows.

Example 4.9 – A version of scrabble

A person chooses 9 letters at random from the English alphabet with replacement. What is the probability that she will be able to make the word "committee"? Here we have that the outcome on each trial is one of 26 letters. To be able to make the word she needs $k_c = 1, k_e = 2, k_i = 1, k_m = 2, k_o = 1, k_t = 2$, and $k_{\text{other}} = 0$. We have denoted the outcomes as c, e, i, m, o, t, and "other". "Other" represents the remaining 20 letters so that $N = 7$. Thus, the probability is from (4.19)

$$P[k_c = 1, k_e = 2, k_i = 1, k_m = 2, k_o = 1, k_t = 2, k_{\text{other}} = 0] =$$

$$\binom{9}{1,2,1,2,1,2,0} \left(\frac{1}{26}\right)^9 \left(\frac{20}{26}\right)^0$$

since $p_c = p_e = p_i = p_m = p_o = p_t = 1/26$ and $p_{\text{other}} = 20/26$ due to the assumption of "at random" sampling and with replacement. This becomes

$$P[k_c = 1, k_e = 2, k_i = 1, k_m = 2, k_o = 1, k_t = 2, k_{\text{other}} = 0] =$$

$$\frac{9!}{1!2!1!2!1!2!0!} \left(\frac{1}{26}\right)^9 = 8.35 \times 10^{-9}.$$

4.6.4   Nonindependent Subexperiments

When the subexperiments are independent, the calculation of probabilities can be greatly simplified. An event that can be written as $A = A_1 \cap A_2 \cap \cdots \cap A_M$ can be found via

$$P[A] = P[A_1]P[A_2] \cdots P[A_M]$$

where each $P[A_i]$ can be found by considering only the individual subexperiment. However, the assumption of independence can sometimes be unreasonable. In the absence of independence, the probability would be found by using the chain rule (see (4.10) for $M = 3$)

$$P[A] = P[A_M|A_{M-1}, \ldots, A_1]P[A_{M-1}|A_{M-2}, \ldots, A_1] \cdots P[A_2|A_1]P[A_1]. \quad (4.20)$$

Such would be the case if a Bernoulli sequence were composed of nonindependent trials as illustrated next.

Example 4.10 – Dependent Bernoulli trials

Assume that we have two coins. One is fair and the other is weighted to have a probability of heads of $p \neq 1/2$. We begin the experiment by first choosing at random one of the two coins and then tossing it. If it comes up heads, we choose the fair coin to use on the next trial. If it comes up tails, we choose the weighted coin to use on the next trial. We repeat this procedure for all the succeeding trials. One possible sequence of outcomes is shown in Figure 4.7a for the weighted coin having $p = 1/4$. Also shown is the case when $p = 1/2$ or a fair coin is always used,

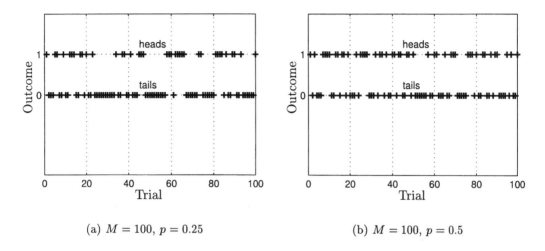

(a) $M = 100, p = 0.25$ (b) $M = 100, p = 0.5$

Figure 4.7: Dependent Bernoulli sequence for different values of p.

so that we are equally likely to observe a head or a tail on each trial. Note that in the case of $p = 1/4$ (see Figure 4.7a), if the outcome is a tail on any trial, then we use the weighted coin for the next trial. Since the weighted coin is biased towards producing a tail, we would expect to again see a tail, and so on. This accounts for the long run of tails observed. Clearly, the trials are not independent.

$\diamond$

If we think some more about the previous experiment, we realize that the dependency between trials is due only to the outcome of the $(i - 1)^{\text{st}}$ trial affecting the outcome of the ith trial. In fact, once the coin has been chosen, the probabilities for the next trial are either $P[0] = P[1] = 1/2$ if a head occurred on the previous trial or $P[0] = 3/4, P[1] = 1/4$ if the previous trial produced a tail. The previous outcome is called the *state* of the sequence. This behavior may be summarized by the *state probability diagram* shown in Figure 4.8. The probabilities shown are actually conditional probabilities. For example, 3/4 is the probability $P[\text{tail on } i\text{th toss}|\text{tail on } i - 1^{\text{st}} \text{ toss}] = P[0|0]$, and similarly for the others. This type of Bernoulli sequence, in which the probabilities for trial i depend only on the outcome of the previous trial, is called a *Markov sequence*. Mathematically, the probability of the event A_i on the ith trial given all the previous outcomes can be written as

$$P[A_i|A_{i-1}, A_{i-2}, \ldots, A_1] = P[A_i|A_{i-1}].$$

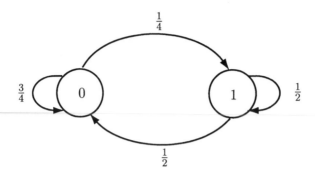

Figure 4.8: Markov state probability diagram.

Using this in (4.20) produces

$$P[A] = P[A_M|A_{M-1}]P[A_{M-1}|A_{M-2}] \cdots P[A_2|A_1]P[A_1]. \qquad (4.21)$$

The conditional probabilities $P[A_i|A_{i-1}]$ are called the *state transition probabilities*, and along with the initial probability $P[A_1]$, the probability of any joint event can be determined. For example, we might wish to determine the probability of $N = 10$ tails in succession or of the event $A = \{(0,0,0,0,0,0,0,0,0,0)\}$. If the weighted coin was actually fair, then $P[A] = (1/2)^{10} = 0.000976$, but if $p = 1/4$, we have by letting $A_i = \{0\}$ for $i = 1, 2, \ldots, 10$ in (4.21)

$$P[A] = \left(\prod_{i=2}^{10} P[A_i|A_{i-1}]\right) P[A_1].$$

But $P[A_i|A_{i-1}] = P[0|0] = P[\text{tails}|\text{weighted coin}] = 3/4$ for $i = 2, 3, \ldots, 10$. Since we initially choose one of the coins at random, we have

$$
\begin{aligned}
P[A_1] &= P[0] = P[\text{tail}|\text{weighted coin}]P[\text{weighted coin}] \\
&\quad + P[\text{tail}|\text{fair coin}]P[\text{fair coin}] \\
&= \left(\frac{3}{4}\right)\left(\frac{1}{2}\right) + \left(\frac{1}{2}\right)\left(\frac{1}{2}\right) = \frac{5}{8}.
\end{aligned}
$$

Thus, we have that

$$P[A] = \left(\prod_{i=2}^{10} \frac{3}{4}\right)\left(\frac{5}{8}\right) = 0.0469$$

or about 48 times more probable than if the weighted coin were actually fair. Note that we could also represent the process by using a *trellis diagram* as shown in Figure 4.9. The probability of any sequence is found by tracing the sequence values through the trellis and multiplying the probabilities for each branch together, along with the initial probability. Referring to Figure 4.9 the sequence $1, 0, 0$ has a probability of $(3/8)(1/2)(3/4)$. The foregoing example is a simple case of a *Markov chain*. We will study this modeling in much more detail in Chapter 22.

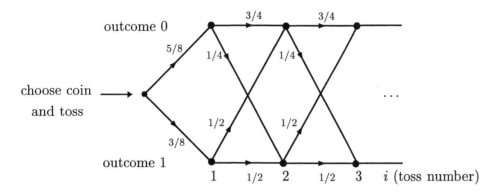

Figure 4.9: Trellis diagram.

4.7 Real-World Example – Cluster Recognition

In many areas an important problem is the detection of a "cluster." Epidemiology is concerned with the incidence of a greater than expected number of disease cases in a given geographic area. If such a situation is found to exist, then it may indicate a problem with the local water supply, as an example. Police departments may wish to focus their resources on areas of a city that exhibit an unusually high incidence of crime. Portions of a remotely sensed image may exhibit an increased number of noise bursts. This could be due to a group of sensors that are driven by a faulty power source. In all these examples, we wish to determine if a cluster of events has occurred. By cluster, we mean that more occurrences of an event are observed than would normally be expected. An example could be a geographic area which is divided into a grid of 50 × 50 cells as shown in Figure 4.10. It is seen that an event or "hit", which is denoted by a black square, occurs rather infrequently. In this example, it occurs $29/2500 = 1.16\%$ of the time. Now consider Figure 4.11. We see that the shaded area appears to exhibit more hits than the expected $145 \times 0.0116 = 1.68$ number. One might be inclined to call this shaded area a cluster. But how probable is this cluster? And how can we make a decision to either accept the hypothesis that this area is a cluster or to reject it? To arrive at a decision we use a Bayesian approach. It computes the *odds ratio against* the occurrence of a cluster (or in favor of no cluster), which is defined as

$$\text{odds} = \frac{P[\text{no cluster}|\text{observed data}]}{P[\text{cluster}|\text{observed data}]}.$$

If this number is large, typically much greater than one, we would be inclined to reject the hypothesis of a cluster, and otherwise, to accept it. We can use Bayes' theorem to evaluate the odds ratio by letting $B = \{\text{cluster}\}$ and $A = \{\text{observed data}\}$. Then,

$$\text{odds} = \frac{P[B^c|A]}{P[B|A]} = \frac{P[A|B^c]P[B^c]}{P[A|B]P[B]}.$$

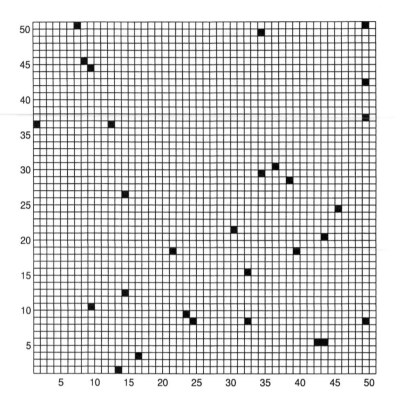

Figure 4.10: Geographic area with incidents shown as black squares – no cluster present.

Note that $P[A]$ is not needed since it cancel outs in the ratio. To evaluate this we need to determine $P[B], P[A|B^c], P[A|B]$. The first probability $P[B]$ is the prior probability of a cluster. Since we believe a cluster is quite unlikely, we assign a probability of 10^{-6} to this. Next we need $P[A|B^c]$ or the probability of the observed data if there is no cluster. Since each cell can take on only one of two values, either a hit or no hit, and if we assume that the outcomes of the various cells are independent of each other, we can model the data as a Bernoulli sequence. For this problem, we might be tempted to call it a Bernoulli *array* but the determination of the probabilities will of course proceed as usual. If M cells are contained in the supposed cluster area (shown as shaded in Figure 4.11 with $M = 145$), then the probability of k hits is given by the binomial law

$$P[k] = \binom{M}{k} p^k (1-p)^{M-k}.$$

Next must assign values to p under the hypothesis of a cluster present and no cluster present. From Figure 4.10 in which we did not suspect a cluster, the relative

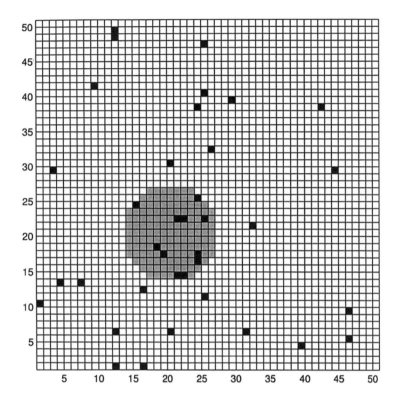

Figure 4.11: Geographic area with incidents shown as black squares – possible cluster present.

frequency of hits was about 0.0116 so that we assume $p_{nc} = 0.01$ when there is no cluster. When we believe a cluster is present, we assume that $p_c = 0.1$ in accordance with the relative frequency of hits in the shaded area of Figure 4.11, which is $11/145 = 0.07$. Thus,

$$
\begin{aligned}
P[A|B^c] &= P[\text{observed data}|\text{no cluster}] = \binom{M}{k} p_{nc}^k (1 - p_{nc})^{M-k} \\
&= P[k = 11|\text{no cluster}] = \binom{145}{11} (0.01)^{11}(0.99)^{134} \\
P[A|B] &= P[\text{observed data}|\text{cluster}] = \binom{M}{k} p_c^k (1 - p_c)^{M-k} \\
&= P[k = 11|\text{cluster}] = \binom{145}{11} (0.1)^{11}(0.9)^{134}
\end{aligned}
$$

which results in an odds ratio of

$$
\text{odds} = \frac{(0.01)^{11}(0.99)^{134}(1 - 10^{-6})}{(0.1)^{11}(0.9)^{134}(10^{-6})} = 3.52.
$$

Since the posterior probability of no cluster is 3.52 times larger than the posterior probability of a cluster, we would reject the hypothesis of a cluster present. However, the odds against a cluster being present are not overwhelming. In fact, the computer simulation used to generate Figures 4.11 employed $p = 0.01$ for the unshaded region and $p = 0.1$ for the shaded cluster region. The reader should be aware that it is mainly the influence of the small prior probability of a cluster, $P[B] = 10^{-6}$, that has resulted in the greater than unity odds ratio and a decision to reject the cluster present hypothesis.

References

S. Press, *Subjective and Objective Bayesian Statistics*, John Wiley & Sons, New York, 2003.

D. Salsburg, *The Lady Tasting Tea: How Statistics Revolutionized Science in the Twentieth Century*, W.H. Freeman, New York, 2001.

Problems

4.1 (f) If $B \subset A$, what is $P[A|B]$? Explain your answer.

4.2 ($\smile$) (f) A point x is chosen at random within the interval $(0, 1)$. If it is known that $x \geq 1/2$, what is the probability that $x \geq 7/8$?

4.3 (w) A coin is tossed three times with each 3-tuple outcome being equally likely. Find the probability of obtaining (H, T, H) if it is known that the outcome has 2 heads. Do this by 1) using the idea of a reduced sample space and 2) using the definition of conditional probability.

4.4 (w) Two dice are tossed. Each 2-tuple outcome is equally likely. Find the probability that the number that comes up on die 1 is the same as the number that comes up on die 2 if it is known that the sum of these numbers is even.

4.5 ($\smile$) (f) An urn contains 3 red balls and 2 black balls. If two balls are chosen without replacement, find the probability that the second ball is black if it is known that the first ball chosen is black.

4.6 (f) A coin is tossed 11 times in succession. Each 11-tuple outcome is equally likely to occur. If the first 10 tosses produced all heads, what is the probability that the 11th toss will also be a head?

4.7 ($\smile$) (w) Using Table 4.1, determine the probability that a college student will have a weight greater than 190 lbs. if he/she has a height exceeding $5'8''$. Next, find the probability that a student's weight will exceed 190 lbs.

4.8 (w) Using Table 4.1, find the probability that a student has weight less than 160 lbs. if he/she has height *greater* than $5'4''$. Also, find the probability that a student's weight is less than 160 lbs. if he/she has height *less* than $5'4''$. Are these two results related?

4.9 (t) Show that the statement $P[A|B] + P[A|B^c] = 1$ is false. Use Figure 4.2a to provide a counterexample.

4.10 (t) Prove that for the events A, B, C, which are not necessarily mutually exclusive,

$$P[A \cup B|C] = P[A|C] + P[B|C] - P[AB|C].$$

4.11 ($\ddot{\smile}$) (w) A group of 20 patients afflicted with a disease agree to be part of a clinical drug trial. The group is divided up into two groups of 10 subjects each, with one group given the drug and the other group given sugar water, i.e., this is the control group. The drug is 80% effective in curing the disease. If one is not given the drug, there is still a 20% chance of a cure due to remission. What is the probability that a randomly selected subject will be cured?

4.12 (w) A new bus runs on Sunday, Tuesday, Thursday, and Saturday while an older bus runs on the other days. The new bus has a probability of being on time of 2/3 while the older bus has a probability of only 1/3. If a passenger chooses an arbitrary day of the week to ride the bus, what is the probability that the bus will be on time?

4.13 (w) A digital communication system transmits one of the three values $-1, 0, 1$. A channel adds noise to cause the decoder to sometimes make an error. The error rates are 12.5% if a -1 is transmitted, 75% if a 0 is transmitted, and 12.5% if a 1 is transmitted. If the probabilities for the various symbols being transmitted are $P[-1] = P[1] = 1/4$ and $P[0] = 1/2$, find the probability of error. Repeat the problem if $P[-1] = P[0] = P[1]$ and explain your results.

4.14 ($\ddot{\smile}$) (w) A sample space is given by $\mathcal{S} = \{(x,y) : 0 \le x \le 1, 0 \le y \le 1\}$. Determine $P[A|B]$ for the events

$$
\begin{aligned}
A &= \{(x,y) : y \le 2x, 0 \le x \le 1/2 \text{ and } y \le 2 - 2x, 1/2 \le x \le 1\} \\
B &= \{(x,y) : 1/2 \le x \le 1, 0 \le y \le 1\}.
\end{aligned}
$$

Are A and B independent?

4.15 (w) A sample space is given by $\mathcal{S} = \{(x,y) : 0 \le x \le 1, 0 \le y \le 1\}$. Are the events

$$
\begin{aligned}
A &= \{(x,y) : y \le x\} \\
B &= \{(x,y) : y \le 1 - x\}
\end{aligned}
$$

independent? Repeat if $B = \{(x,y) : x \le 1/4\}$.

4.16 (t) Give an example of two events that are mutually exclusive but not independent. Hint: See Figure 4.4.

4.17 (t) Consider the sample space $\mathcal{S} = \{(x,y,z) : 0 \le x \le 1, 0 \le y \le 1, 0 \le z \le 1\}$, which is the unit cube. Can you find three events that are independent? Hint: See Figure 4.2c.

4.18 (t) Show that if (4.9) is satisfied for *all* possible events, then pairwise independence follows. In this case all events are independent.

4.19 (⌣) (f) It is known that if it rains, there is a 50% chance that a sewer will overflow. Also, if the sewer overflows, then there is a 30% chance that the road will flood. If there is a 20% chance that it will rain, what is the probability that the road will flood?

4.20 (w) Consider the sample space $\mathcal{S} = \{1,2,3,4\}$. Each simple event is equally likely. If $A = \{1,2\}, B = \{1,3\}, C = \{1,4\}$, are these events pairwise independent? Are they independent?

4.21 (⌣) (w) In Example 4.6 determine if the events are pairwise independent. Are they independent?

4.22 (⌣) (w) An urn contains 4 red balls and 2 black balls. Two balls are chosen in succession without replacement. If it is known that the first ball drawn is black, what are the odds in favor of a red ball being chosen on the second draw?

4.23 (w) In Example 4.7 plot the probability that the person has cancer given that the test results are positive, i.e., the posterior probability, as a function of the prior probability $P[B]$. How is the posterior probability that the person has cancer related to the prior probability?

4.24 (w) An experiment consists of two subexperiments. First a number is chosen at random from the interval $(0, 1)$. Then, a second number is chosen at random from the same interval. Determine the sample space $\mathcal{S}^2$ for the overall experiment. Next consider the event $A = \{(x,y) : 1/4 \le x \le 1/2, 1/2 \le y \le 3/4\}$ and find $P[A]$. Relate $P[A]$ to the probabilities defined on $\mathcal{S}^1 = \{u : 0 < u < 1\}$, where $\mathcal{S}^1$ is the sample space for each subexperiment.

4.25 (w,c) A fair coin is tossed 10 times. What is the probability of a run of exactly 5 heads in a row? Do not count runs of 6 or more heads in a row. Now verify your solution using a computer simulation.

4.26 (⌣) (w) A lady claims that she can tell whether a cup of tea containing milk had the tea poured first or the milk poured first. To test her claim an experiment is set up whereby at random the milk or tea is added first to an empty cup. This experiment is repeated 10 times. If she correctly identifies which liquid was poured first 8 times out of 10, how likely is it that she is guessing? See [Salsburg 2001] for a further discussion of this famous problem.

4.27 (f) The probability $P[k]$ is given by the binomial law. If $M = 10$, for what value of p is $P[3]$ maximum? Explain your answer.

4.28 ($\smile$) (f) A sequence of independent subexperiments is conducted. Each subexperiment has the outcomes "success", "failure", or "don't know". If $P[\text{success}] = 1/2$ and $P[\text{failure}] = 1/4$, what is the probability of 3 successes in 5 trials?

4.29 (c) Verify your results in Problem 4.28 by using a computer simulation.

4.30 (w) A drunk person wanders aimlessly along a path by going forward one step with probability $1/2$ and going backward one step with probability $1/2$. After 10 steps what is the probability that he has moved 2 steps forward?

4.31 (f) Prove that the geometric probability law (4.17) is a valid probability assignment.

4.32 (w) For a sequence of independent Bernoulli trials find the probability of the first failure at the kth trial for $k = 1, 2, \ldots$.

4.33 ($\smile$) (w) For a sequence of independent Bernoulli trials find the probability of the second success occurring at the kth trial.

4.34 (t) Consider a sequence of independent Bernoulli trials. If it is known that the first m trials resulted in failures, prove that the probability of the first success occurring at $m + l$ is given by the geometric law with k replaced by l. In other words, the probability is the same as if we had started the process over again after the mth failure. There is no memory of the first m failures.

4.35 (f) An urn contains red, black, and white balls. The proportion of red is 0.4, the proportion of black is 0.4, and the proportion of white is 0.2. If 5 balls are drawn with replacement, what is the probability of 2 red, 2 black, and 1 white in any order?

4.36 (t) We derive the multinomial coefficient for $N = 3$. This will yield the number of ways that an M-tuple can be formed using k_1 1's, k_2 2's and k_3 3's. To do so choose k_1 places in the M-tuple for the 1's. There will be $M - k_1$ positions remaining. Of these positions choose k_2 places for the 2's. Fill in the remaining $k_3 = M - k_1 - k_2$ positions using the 3's. Using this result, determine the number of different M digit sequences with k_1 1's, k_2 2's, and k_3 3's.

4.37 (t) Show that the multinomial probability law reduces to the binomial law for $N = 2$.

4.38 (☺) (w,c) An urn contains 3 red balls, 3 black balls, and 3 white balls. If 6 balls are chosen with replacement, how many of each color is most likely? Hint: You will need a computer to evaluate the probabilities.

4.39 (w,c) For the problem discussed in Example 4.10 change the probability of heads for the weighted coin from $p = 0.25$ to $p = 0.1$. Redraw the Markov state probability diagram. Next, using a computer simulation generate a sequence of length 100. Explain your results.

4.40 (☺) (f) For the Markov state diagram shown in Figure 4.8 with an initial state probability of $P[0] = 3/4$, find the probability of the sequence $0, 1, 1, 0$.

4.41 (f) A *two-state* Markov chain (see Figure 4.8) has the *state transition probabilities* $P[0|0] = 1/4, P[0|1] = 3/4$ and the initial state probability of $P[0] = 1/2$. What is the probability of the sequence $0, 1, 0, 1, 0$?

4.42 (w) A digital communication system model is shown in Figure 4.12. It consists of two sections with each one modeling a different portion of the communication channel. What is the probability of a bit error? Compare this to the probability of error for the single section model shown in Figure 4.3, assuming that $\epsilon < 1/2$, which is true in practice. Note that Figure 4.12 is a trellis.

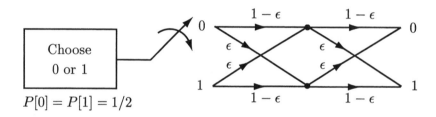

Figure 4.12: Probabilistic model of a digital communication system with two sections.

4.43 (☺) (f) For the trellis shown in Figure 4.9 find the probability of the event $A = \{(0, 1, 0, 0), (0, 0, 0, 0)\}$.

Chapter 5

Discrete Random Variables

5.1 Introduction

Having been introduced to the basic probabilistic concepts in Chapters 3 and 4, we now begin their application to solving problems of interest. To do so we define the random variable. It will be seen to be a function, also called a *mapping*, of the outcomes of a random experiment to the set of real numbers. With this association we are able to use the real number description to quantify items of interest. In this chapter we describe the *discrete* random variable, which is one that takes on a finite or countably infinite number of values. Later we will extend the definition to a random variable that takes on a continuum of values, the *continuous* random variable. The mathematics associated with a discrete random variable are inherently simpler and so conceptualization is facilitated by first concentrating on the discrete problem. The reader has already been introduced to the concept of a random variable in Chapter 2 in an informal way and hence may wish to review the computer simulation methodology described therein.

5.2 Summary

The random variable, which is a mapping from the sample space into the set of real numbers, is formally discussed and illustrated in Section 5.3. In Section 5.4 the probability of a random variable taking on its possible values is given by (5.2). Next the probability mass function is defined by (5.3). Some important probability mass functions are summarized in Section 5.5. They include the Bernoulli (5.5), the binomial (5.6), the geometric (5.7), and the Poisson (5.8). The binomial probability mass function can be approximated by the Poisson as shown in Figure 5.8 if $M \to \infty$ and $p \to 0$, with Mp remaining constant. This motivates the use of the Poisson probability mass function for traffic modeling. If a random variable is transformed to a new one via a mapping, then the new random variable has a probability mass function given by (5.9). Next the cumulative distribution function is introduced and is given by (5.10). It can be used as an equivalent description for the probability of a discrete random variable. Its properties are summarized in Section 5.8. The computer simulation of discrete random variables is revisited in Section 5.9 with the estimate of the probability mass function and the cumulative distribution function given by (5.14) and (5.15),(5.16), respectively. Finally, the application of the Poisson probability model to determining the resources required to service customers is described in Section 5.10.

5.3   Definition of Discrete Random Variable

We have previously used a coin toss and a die toss as examples of a random experiment. In the case of a die toss the outcomes comprised the sample space $S = \{1, 2, 3, 4, 5, 6\}$. This was because each face of a die has a dot pattern consisting of 1, 2, 3, 4, 5, or 6 dots. A natural description of the outcome of a die toss is therefore the number of dots observed on the face that appears upward. In effect, we have mapped the *dot pattern* into the *number of dots* in describing the outcome. This type of experiment is called a *numerically valued* random phenomenon since the basic output is a real number. In the case of a coin toss the outcomes comprise the *nonnumerical* sample space $S = \{\text{head, tail}\}$. We have, however, at times replaced the sample space by one consisting only of real numbers such as $S_X = \{0, 1\}$, where a head is mapped into a 1 and a tail is mapped into a 0. This mapping is shown in Figure 5.1. For many applications this is a convenient mapping. For example, in

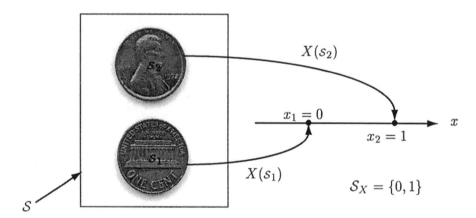

Figure 5.1: Mapping of the outcome of a coin toss into the set of real numbers.

a succession of M coin tosses, we might be interested in the total number of heads observed. With the defined mapping of

$$X(s_i) = \begin{cases} 0 & s_1 = \text{tail} \\ 1 & s_2 = \text{head} \end{cases}$$

we could represent the number of heads as $\sum_{i=1}^{M} X(s_i)$, where s_i is the outcome of the ith toss. The function that maps S into S_X and which is denoted by $X(\cdot)$ is called a *random variable*. It is a function that takes each outcome of S (which may not necessarily be a set of numbers) and maps it into the subset of the set of real numbers. Note that as previously mentioned in Chapter 2, a *capital* letter X will denote the random variable and a *lowercase* letter x its value. This convention for the coin toss example produces the assignment

$$X(s_i) = x_i \qquad i = 1, 2$$

where $s_1 = $ tail and thus $x_1 = 0$, and $s_2 = $ head and thus $x_2 = 1$. The name *random variable* is a poor one in that the function $X(\cdot)$ *is not random* but a known one and usually one of our own choosing. What is random is the input argument s_i and hence the output of the function is random. However, due to its long-standing usage in probability we will retain this terminology.

Sometimes it is more convenient to use a particular random variable for a given experiment. For example, in Chapter 2 we described a digital communication system called a PSK system. A bit is communicated using the transmitted signals

$$s(t) = \begin{cases} -A \cos 2\pi F_0 t & \text{for a } 0 \\ A \cos 2\pi F_0 t & \text{for a } 1. \end{cases}$$

Usually a 1 or a 0 occurs with equal probability so that the choice of a bit can be modeled as the outcome of a fair coin tossing experiment. If a head is observed, then a 1 is transmitted and a 0 otherwise. As a result, we could represent the transmitted signal with the model

$$s_i(t) = X(s_i) A \cos 2\pi F_0 t$$

where $s_1 = $ tail and $s_2 = $ head and hence we have the defined random variable

$$X(s_i) = \begin{cases} -1 & s_1 = \text{tail} \\ +1 & s_2 = \text{head}. \end{cases}$$

This random variable is a convenient one for this application.

In general, a *random variable* is a function that maps the sample space S into a subset of the real line. The real line will be denoted by R ($R = \{x : -\infty < x < \infty\}$). For a discrete random variable this subset is a finite or countably infinite set of points. The subset forms a new sample space which we will denote by S_X, and which is illustrated in Figure 5.2. A discrete random variable may also map *multiple elements* of the sample space into the *same number* as illustrated in Figure 5.3. An example would be a die toss experiment in which we were only interested in whether the outcome is even or odd. To quantify this outcome we could define

$$X(s_i) = \begin{cases} 0 & \text{if } s_i = 1, 3, 5 \text{ dots} \\ 1 & \text{if } s_i = 2, 4, 6 \text{ dots}. \end{cases}$$

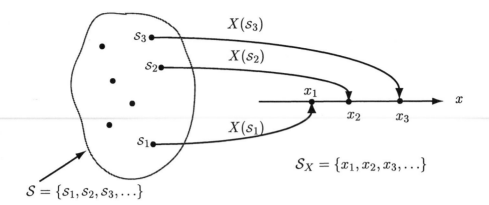

Figure 5.2: Discrete random variable as a *one-to-one* mapping of a countably infinite sample space into set of real numbers.

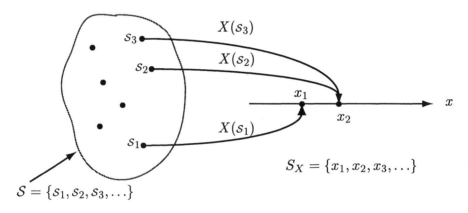

Figure 5.3: Discrete random variable as a *many-to-one* mapping of a countably infinite sample space into set of real numbers.

This type of mapping is usually called a *many-to-one* mapping while the previous one is called a *one-to-one* mapping. Note that for a many-to-one mapping we cannot recover the outcome of S if we know the value of $X(s)$. But as already explained, this is of little concern since we initially defined the random variable to output the item of interest. Lastly, for numerically valued random experiments in which s is contained in R, we can still use the random variable approach if we define $X(s) = s$ for all s. This allows the concept of a random variable to be used for all random experiments, with either numerical or nonnumerical outputs.

5.4   Probability of Discrete Random Variables

We would next like to determine the probabilities of the random variable taking on its possible values. In other words, what is the probability $P[X(s_i) = x_i]$ for each

$x_i \in \mathcal{S}_X$? Since the sample space $\mathcal{S}$ is discrete, the random variable can take on at most a countably infinite number of values or $X(s_i) = x_i$ for $i = 1, 2, \ldots$. It should be clear that if $X(\cdot)$ maps each s_i into a *different* x_i (or $X(\cdot)$ is one-to-one), then because s_i and x_i are just two different names for the same event

$$P[X(s) = x_i] = P[\{s_j : X(s_j) = x_i\}] = P[\{s_i\}] \tag{5.1}$$

or we assign a probability to the value of the random variable equal to that of the simple event in $\mathcal{S}$ that yields that value. If, however, there are multiple outcomes in $\mathcal{S}$ that map into the same value x_i (or $X(\cdot)$ is many-to-one) then

$$\begin{aligned} P[X(s) = x_i] &= P[\{s_j : X(s_j) = x_i\}] \\ &= \sum_{\{j : X(\mathcal{S}_j) = x_i\}} P[\{s_j\}] \end{aligned} \tag{5.2}$$

since the s_j's are simple events in $\mathcal{S}$ and are therefore mutually exclusive. It is said that the events $\{X = x_i\}$, defined on $\mathcal{S}_X$, and $\{s_j : X(s_j) = x_i\}$, defined on $\mathcal{S}$, are *equivalent events*. As such they are assigned the same probability. Note that the probability assignment (5.2) subsumes that of (5.1) and that in either case we can summarize the probabilities that the random variable values take on by defining the *probability mass function* (PMF) as

$$p_X[x_i] = P[X(s) = x_i] \tag{5.3}$$

and use (5.2) to evaluate it from a knowledge of the mapping. It is important to observe that in the notation $p_X[x_i]$ the subscript X refers to the random variable and also the $[\cdot]$ notation is meant to remind the reader that the argument is a discrete one. Later, we will use $(\cdot)$ for continuous arguments. In summary, the probability mass function is the probability that the random variable X takes on the value x_i for each possible x_i. An example follows.

Example 5.1 – Coin toss – one-to-one mapping

The experiment consists of a single coin toss with a probability of heads equal to p. The sample space is $\mathcal{S} = \{\text{head, tail}\}$ and we define the random variable as

$$X(s_i) = \begin{cases} 0 & s_i = \text{tail} \\ 1 & s_i = \text{head.} \end{cases}$$

The PMF is therefore from from (5.3) and (5.1)

$$\begin{aligned} p_X[0] &= P[X(s) = 0] = 1 - p \\ p_X[1] &= P[X(s) = 1] = p. \end{aligned}$$

Example 5.2 – Die toss – many-to-one mapping

The experiment consists of a single fair die toss. With a sample space of $\mathcal{S} = \{1, 2, 3, 4, 5, 6\}$ and an interest only in whether the outcome is even or odd we define the random variable

$$X(s_i) = \begin{cases} 0 & \text{if } i = 1, 3, 5 \\ 1 & \text{if } i = 2, 4, 6. \end{cases}$$

Thus, using (5.3) and (5.2) we have the PMF

$$p_X[0] = P[X(s) = 0] = \sum_{j=1,3,5} P[\{s_j\}] = \frac{3}{6}$$

$$p_X[1] = P[X(s) = 1] = \sum_{j=2,4,6} P[\{s_j\}] = \frac{3}{6}.$$

$\diamond$

The use of (5.2) may seem familiar and indeed it should. We have summed the probabilities of simple events in $\mathcal{S}$ to obtain the probability of an event in $\mathcal{S}$ using (3.10). Here, the event is just the subset of $\mathcal{S}$ for which $X(s) = x_i$ holds. The introduction of a random variable has quantified the events of interest!

Finally, because PMFs $p_X[x_i]$ are just new names for the probabilities $P[X(s) = x_i]$ they must satisfy the usual properties:

Property 5.1 – Range of values

$$0 \le p_X[x_i] \le 1$$

$\square$

Property 5.2 – Sum of values

$$\sum_{i=1}^{M} p_X[x_i] = 1 \text{ if } \mathcal{S}_X \text{ consists of } M \text{ outcomes}$$

$$\sum_{i=1}^{\infty} p_X[x_i] = 1 \text{ if } \mathcal{S}_X \text{ is countably infinite.}$$

$\square$

We will frequently omit the s argument of X to write $p_X[x_i] = P[X = x_i]$.

Once the PMF has been specified all subsequent probability calculations can be based on it, without referring back to the original sample space $\mathcal{S}$. Specifically, for an event A defined on $\mathcal{S}_X$ the probability is given by

$$P[X \in A] = \sum_{\{i : x_i \in A\}} p_X[x_i]. \tag{5.4}$$

An example follows.

Example 5.3 – Calculating probabilities based on the PMF

Consider a die whose sides have been labeled with two sides having 1 dot, two sides having 2 dots, and two sides having 3 dots. Hence, $S = \{s_1, s_2, \ldots, s_6\} = $ {side 1, side 2, side 3, side 4, side 5, side 6}. Then if we are interested in the probabilities of the outcomes displaying either 1, 2, or 3 dots, we would define a random variable as

$$X(s_i) = \begin{cases} 1 & i = 1, 2 \\ 2 & i = 3, 4 \\ 3 & i = 5, 6. \end{cases}$$

It easily follows then that the PMF is from (5.2)

$$p_X[1] = p_X[2] = p_X[3] = \frac{1}{3}.$$

Now assume we are interested in the probability that a 2 or 3 occurs or $A = \{2, 3\}$. Then from (5.4) we have

$$P[X \in \{2, 3\}] = p_X[2] + p_X[3] = \frac{2}{3}.$$

There is no need to reconsider the original sample space S and all probability calculations of interest are obtainable from the PMF.

$\diamond$

5.5 Important Probability Mass Functions

We have already encountered many of these in Chapter 4. We now summarize these in our new notation. Since the sample spaces S_X consist of integer values we will replace the notation x_i by k, which indicates an integer.

5.5.1 Bernoulli

$$p_X[k] = \begin{cases} 1 - p & k = 0 \\ p & k = 1. \end{cases} \tag{5.5}$$

The PMF is shown in Figure 5.4 and is recognized as a *sequence* of numbers that is nonzero only for the indices $k = 0, 1$. It is convenient to represent the Bernoulli PMF using the shorthand notation Ber(p). With this notation we replace the description that "X is distributed according to a Bernoulli random variable with PMF Ber(p)" by the shorthand notation $X \sim$ Ber(p), where $\sim$ means "is distributed according to".

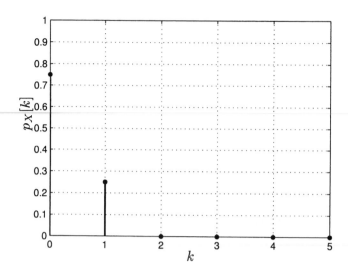

Figure 5.4: Bernoulli probability mass function for $p = 0.25$.

5.5.2 Binomial

$$p_X[k] = \binom{M}{k} p^k (1-p)^{M-k} \qquad k = 0, 1, \ldots, M. \tag{5.6}$$

The PMF is shown in Figure 5.5. The shorthand notation for the binomial PMF is

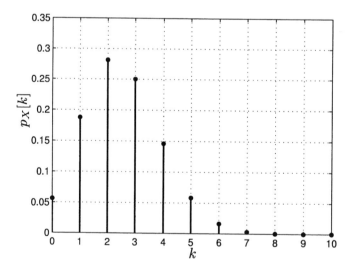

Figure 5.5: Binomial probability mass function for $M = 10, p = 0.25$.

$\text{bin}(M, p)$. The location of the maximum of the PMF can be shown to be given by $[(M+1)p]$, where $[x]$ denotes the largest integer less than or equal to x (see Problem 5.7).

5.5.3 Geometric

$$p_X[k] = (1-p)^{k-1}p \qquad k = 1, 2, \ldots. \tag{5.7}$$

The PMF is shown in Figure 5.6. The shorthand notation for the geometric PMF is geom(p).

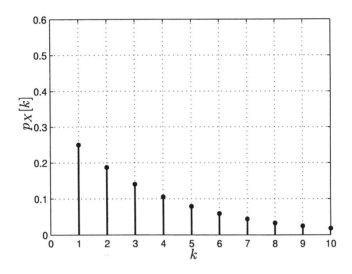

Figure 5.6: Geometric probability mass function for $M = 10, p = 0.25$.

5.5.4 Poisson

$$p_X[k] = \exp(-\lambda)\frac{\lambda^k}{k!} \qquad k = 0, 1, 2, \ldots \tag{5.8}$$

where $\lambda > 0$. The PMF is shown in Figure 5.7 for several values of λ. Note that the maximum occurs at $[\lambda]$ (see Problem 5.11). The shorthand notation is Pois(λ).

5.6 Approximation of Binomial PMF by Poisson PMF

The binomial and Poisson PMFs are related to each other under certain conditions. This relationship helps to explain why the Poisson PMF is used in various applications, primarily traffic modeling as described further in Section 5.10. The relationship is as follows. If in a binomial PMF, we let $M \to \infty$ as $p \to 0$ such that the product $\lambda = Mp$ remains constant, then bin(M, p) $\to$ Pois(λ). Note that $\lambda = Mp$ represents the expected or average number of successes in M Bernoulli trials (see Chapter 6 for definition of expectation). Hence, by keeping the average number of successes fixed but assuming more and more trials with smaller and smaller probabilities of success on each trial, we are led to a Poisson PMF. As an example, a comparison is shown in Figure 5.8 for $M = 10, p = 0.5$ and $M = 100, p = 0.05$. This

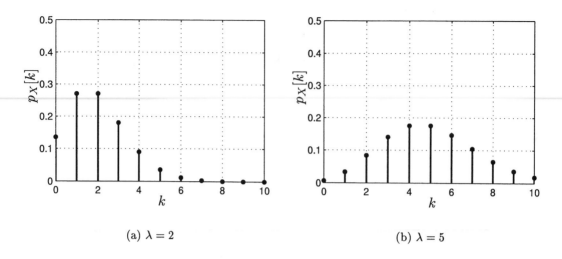

(a) $\lambda = 2$ (b) $\lambda = 5$

Figure 5.7: The Poisson probability mass function for different values of λ.

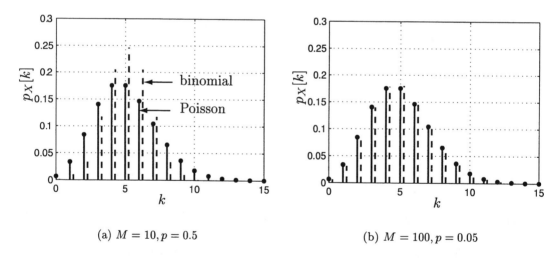

(a) $M = 10, p = 0.5$ (b) $M = 100, p = 0.05$

Figure 5.8: The Poisson approximation to the binomial probability mass function.

result is primarily useful since Poisson PMFs are easier to manipulate and also arise in the modeling of point processes as described in Chapter 21.

To make this connection we have for the binomial PMF with $p = \lambda/M \to 0$ as

$M \to \infty$ (and λ fixed)

$$
\begin{aligned}
p_X[k] &= \binom{M}{k} p^k (1-p)^{M-k} \\
&= \frac{M!}{(M-k)!k!} \left(\frac{\lambda}{M}\right)^k \left(1 - \frac{\lambda}{M}\right)^{M-k} \\
&= \frac{(M)_k}{k!} \frac{\lambda^k}{M^k} \frac{(1-\lambda/M)^M}{(1-\lambda/M)^k} \\
&= \frac{\lambda^k}{k!} \frac{(M)_k}{M^k} \frac{(1-\lambda/M)^M}{(1-\lambda/M)^k}.
\end{aligned}
$$

But for a fixed k, as $M \to \infty$, we have that $(M)_k/M^k \to 1$. Also, for a fixed k, $(1-\lambda/M)^k \to 1$ so that we need only find the limit of $g(M) = (1-\lambda/M)^M$ as $M \to \infty$. This is shown in Problem 5.15 to be $\exp(-\lambda)$ and therefore

$$
p_X[k] \to \frac{\lambda^k}{k!} \exp(-\lambda).
$$

Also, since the binomial PMF is defined for $k = 0, 1, \ldots, M$, as $M \to \infty$ the limiting PMF is defined for $k = 0, 1, \ldots$. This result can also be found using characteristic functions as shown in Chapter 6.

5.7 Transformation of Discrete Random Variables

It is frequently of interest to be able to determine the PMF of a *transformed* random variable. Mathematically, we desire the PMF of the new random variable $Y = g(X)$, where X is a discrete random variable. For example, consider a die whose faces are labeled with the numbers $0, 0, 1, 1, 2, 2$. We wish to find the PMF of the number observed when the die is tossed, assuming all sides are equally likely to occur. If the original sample space is composed of the possible cube sides that can occur, so that $\mathcal{S}_X = \{1, 2, 3, 4, 5, 6\}$, then the transformation appears as shown in Figure 5.9. Specifically, we have that

$$
Y = \begin{cases}
y_1 = 0 & \text{if } x = x_1 = 1 \text{ or } x = x_2 = 2 \\
y_2 = 1 & \text{if } x = x_3 = 3 \text{ or } x = x_4 = 4 \\
y_3 = 2 & \text{if } x = x_5 = 5 \text{ or } x = x_6 = 6.
\end{cases}
$$

Note that the transformation is many-to-one. Since events such as $\{y : y = y_1 = 0\}$ and $\{x : x = x_1 = 1, x = x_2 = 2\}$, for example, are equivalent, they should be assigned the same probability. Thus, using the property that the events $\{X = x_i\}$ are simple events defined on $\mathcal{S}_X$, we have that

$$
p_Y[y_i] = \begin{cases}
p_X[1] + p_X[2] = \frac{1}{3} & i = 1 \\
p_X[3] + p_X[4] = \frac{1}{3} & i = 2 \\
p_X[5] + p_X[6] = \frac{1}{3} & i = 3.
\end{cases}
$$

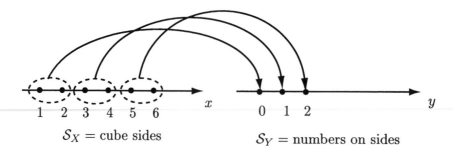

$$S_X = \text{cube sides} \qquad\qquad S_Y = \text{numbers on sides}$$

Figure 5.9: Transformation of discrete random variable.

In general, we have that

$$p_Y[y_i] = \sum_{\{j:g(x_j)=y_i\}} p_X[x_j]. \qquad (5.9)$$

We just sum up the probabilities for all the values of $X = x_j$ that are mapped into $Y = y_i$. This is reminiscent of (5.2) in which the transformation was from the objects s_j defined on S to the numbers x_i defined on S_X. In fact, it is nearly identical except that we have replaced the objects that are to be transformed by numbers, i.e., the x_j's. Some examples of this procedure follow.

Example 5.4 – One-to-one transformation of Bernoulli random variable
If $X \sim \text{Ber}(p)$ and $Y = 2X - 1$, determine the PMF of Y. The sample space for X is $S_X = \{0, 1\}$ and consequently that for Y is $S_Y = \{-1, 1\}$. It follows that $x_1 = 0$ maps into $y_1 = -1$ and $x_2 = 1$ maps into $y_2 = 1$. As a result, we have from (5.9)

$$\begin{aligned} p_Y[-1] &= p_X[0] = 1 - p \\ p_Y[1] &= p_X[1] = p. \end{aligned}$$

Note that this mapping is particularly simple since it is one-to-one. A slightly more complicated example is next.

◊

Example 5.5 – Many-to-one transformation
Let the transformation be $Y = g(X) = X^2$ which is defined on the sample spaces $S_X = \{-1, 0, 1\}$ so that $S_Y = \{0, 1\}$. Clearly, $g(x_j) = x_j^2 = 0$ only for $x_j = 0$. Hence,

$$p_Y[0] = p_X[0].$$

However, $g(x_j) = x_j^2 = 1$ for $x_j = -1$ and $x_j = 1$. Thus, using (5.9) we have

$$\begin{aligned} p_Y[1] &= \sum_{\{x_j : x_j^2 = 1\}} p_X[x_j] \\ &= p_X[-1] + p_X[1]. \end{aligned}$$

Note that we have determined $p_Y[y_i]$ by summing the probabilities of all the x_j's that map into y_i via the transformation $y = g(x)$. This is in essence the meaning of (5.9).

$\diamond$

Example 5.6 – Many-to-one transformation of Poisson random variable

Now consider $X \sim \text{Pois}(\lambda)$ and define the transformation $Y = g(X)$ as

$$Y = \begin{cases} 1 & \text{if } X = k \text{ is even} \\ -1 & \text{if } X = k \text{ is odd.} \end{cases}$$

To find the PMF for Y we use

$$p_Y[k] = P[Y = k] = \begin{cases} P[X \text{ is even}] & k = 1 \\ P[X \text{ is odd}] & k = -1. \end{cases}$$

We need only determine $p_Y[1]$ since $p_Y[-1] = 1 - p_Y[1]$. Thus, from (5.9)

$$\begin{aligned} p_Y[1] &= \sum_{\substack{j=0 \text{ and even}}}^{\infty} p_X[j] \\ &= \sum_{\substack{j=0 \text{ and even}}}^{\infty} \exp(-\lambda)\frac{\lambda^j}{j!}. \end{aligned}$$

To evaluate the infinite sum in closed form we use the following "trick"

$$\begin{aligned} \sum_{\substack{j=0 \text{ and even}}}^{\infty} \frac{\lambda^j}{j!} &= \frac{1}{2}\sum_{j=0}^{\infty} \frac{\lambda^j}{j!} + \frac{1}{2}\sum_{j=0}^{\infty} \frac{(-\lambda)^j}{j!} \\ &= \frac{1}{2}\exp(\lambda) + \frac{1}{2}\exp(-\lambda) \end{aligned}$$

since the Taylor expansion of $\exp(x)$ is known to be $\sum_{j=0}^{\infty} x^j/j!$ (see Problem 5.22). Finally, we have that

$$\begin{aligned} p_Y[1] &= \exp(-\lambda)\left[\frac{1}{2}\exp(\lambda) + \frac{1}{2}\exp(-\lambda)\right] = \frac{1}{2}(1 + \exp(-2\lambda)) \\ p_Y[-1] &= 1 - p_Y[1] = \frac{1}{2}(1 - \exp(-2\lambda)). \end{aligned}$$

$\diamond$

5.8 Cumulative Distribution Function

An alternative means of summarizing the probabilities of a discrete random variable is the *cumulative distribution function* (CDF). It is sometimes referred to more succinctly as the *distribution function*. The CDF for a random variable X and evaluated at x is given by $P[\{\text{real numbers } x' : x' \leq x\}]$, which is the probability that X lies in the semi-infinite interval $(-\infty, x]$. It is therefore defined as

$$F_X(x) = P[X \leq x] \quad -\infty < x < \infty. \tag{5.10}$$

It is important to observe that the value $X = x$ is included in the interval. As an example, if $X \sim \text{Ber}(p)$, then the PMF and the corresponding CDF are shown in Figure 5.10. Because the random variable takes on only the values 0 and 1, the CDF

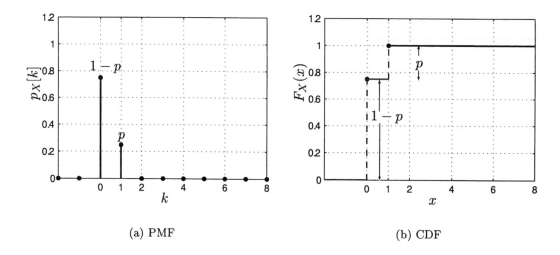

(a) PMF   (b) CDF

Figure 5.10: The Bernoulli probability mass function and cumulative distribution function for $p = 0.25$.

changes its value only at these points, where it jumps. The CDF can be thought of as a "running sum" which adds up the probabilities of the PMF starting at $-\infty$ and ending at $+\infty$. When the value x of $F_X(x)$ encounters a nonzero value of the PMF, the additional mass causes the CDF to jump, with the size of the jump equal to the value of the PMF at that point. For example, referring to Figure 5.10b, at $x = 0$ we have $F_X(0) = p_X[0] = 1 - p = 3/4$ and at $x = 1$ we have $F_X(1) = p_X[0] + p_X[1] = 1$, with the jump having size $p_X[1] = p = 1/4$. Another example follows.

Example 5.7 – CDF for geometric random variable

Since $p_X[k] = (1 - p)^{k-1}p$ for $k = 1, 2, \ldots$, we have the CDF

$$F_X(x) = \begin{cases} 0 & x < 1 \\ \displaystyle\sum_{i=1}^{[x]}(1 - p)^{i-1}p & x \geq 1 \end{cases}$$

where $[x]$ denotes the largest integer less than or equal to x. This evaluates to

$$F_X(x) = \begin{cases} 0 & x < 1 \\ p & 1 \le x < 2 \\ p + (1-p)p & 2 \le x < 3 \\ \text{etc.} \end{cases}$$

The PMF and CDF are plotted in Figure 5.11 for $p = 0.5$. Since the CDF jumps at

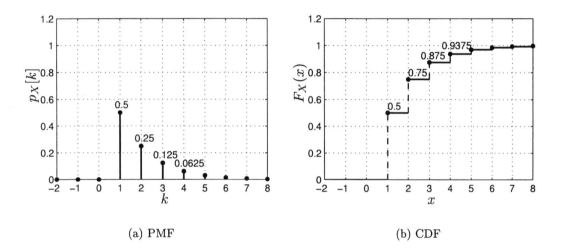

(a) PMF   (b) CDF

Figure 5.11: The geometric probability mass function and cumulative distribution function for $p = 0.5$.

each nonzero value of the PMF and the jump size is that value of the PMF, we can recover the PMF from the CDF. In particular, we have that

$$p_X[x] = F_X(x^+) - F_X(x^-)$$

where x^+ denotes a value just slightly larger than x and x^- denotes a value just slightly smaller than x. Thus, if $F_X(x)$ does not have a discontinuity at x the value of the PMF is zero. At a discontinuity the value of the PMF is just the jump size as previously asserted. Also, because of the definition of the CDF, i.e., that $F_X(x) = P[X \le x] = P[X < x \text{ or } X = x]$, the value of $F_X(x)$ is the value *after* the jump. The CDF is said to be *right-continuous* which is sometimes stated mathematically as $\lim_{x \to x_0^+} F_X(x) = F_X(x_0)$ at the point $x = x_0$.

$\diamond$

From the previous example we see that the PMF and CDF are equivalent descriptions of the probability assignment for X. Either one can be used to find the probability of X being in an interval (even an interval of length zero). For example, to determine $P[3/2 < X \leq 7/2]$ for the geometric random variable

$$P\left[\frac{3}{2} < X \leq \frac{7}{2}\right] = p_X[2] + p_X[3]$$

$$= F_X\left(\frac{7}{2}\right) - F_X\left(\frac{3}{2}\right)$$

as is evident by referring to Figure 5.11b. We need to be careful, however, to note whether the endpoints of the interval are included or not. This is due to the discontinuities of the CDF. Because of the definition of the CDF as the probability of X being within the interval $(-\infty, x]$, which *includes* the right-most point, we have for the interval $(a, b]$

$$P[a < X \leq b] = F_X(b^+) - F_X(a^+). \tag{5.11}$$

Also, the other intervals (a, b), $[a, b)$, and $[a, b]$ will in general have different probabilities than that given by (5.11). From Figure 5.11b and (5.11) we have as an example that

$$P[2 < X \leq 3] = F_X(3^+) - F_X(2^+) = p_X[3] = (1-p)^2 p = 0.125$$

but

$$P[2 \leq X \leq 3] = F_X(3^+) - F_X(2^-) = (1-p)p + (1-p)^2 p = 0.375.$$

From the definition of the CDF and as further illustrated in Figures 5.10 and 5.11 the CDF has several important properties. They are now listed and proven.

Property 5.3 –   CDF is between 0 and 1.

$$0 \leq F_X(x) \leq 1 \qquad -\infty < x < \infty$$

Proof: Since by definition $F_X(x) = P[X \leq x]$ is a probability for all x, it must lie between 0 and 1.

$\qquad\qquad\qquad\qquad\qquad\qquad\qquad\qquad\qquad\qquad\qquad\qquad\qquad\qquad$ □

Property 5.4 –   Limits of CDF as $x \to -\infty$ and as $x \to \infty$

$$\lim_{x \to -\infty} F_X(x) = 0$$
$$\lim_{x \to +\infty} F_X(x) = 1.$$

Proof:

$$\lim_{x \to -\infty} F_X(x) = P[\{s : X(s) < -\infty\}] = P[\emptyset] = 0$$

since the values that $X(s)$ can take on do not include $-\infty$. Also,

$$\lim_{x \to +\infty} F_X(x) = P[\{s : X(s) < +\infty\}] = P[\mathcal{S}] = 1$$

since the values that $X(s)$ can take on are all included on the real line.

$\square$

Property 5.5 – CDF is monotonically increasing.

A monotonically increasing function $g(\cdot)$ is one in which for every x_1 and x_2 with $x_1 \leq x_2$, it follows that $g(x_1) \leq g(x_2)$ or the function increases or stays the same as the argument increases (see also Problem 5.29).
Proof:

$$
\begin{aligned}
F_X(x_2) &= P[X \leq x_2] && \text{(definition)} \\
&= P[(X \leq x_1) \cup (x_1 < X \leq x_2)] \\
&= P[X \leq x_1] + P[x_1 < X \leq x_2] && \text{(Axiom 3)} \\
&= F_X(x_1) + P[x_1 < X \leq x_2] \geq F_X(x_1). && \text{(definition and Axiom 1)}
\end{aligned}
$$

Alternatively, if $A = \{-\infty < X \leq x_1\}$ and $B = \{-\infty < X \leq x_2\}$ with $x_1 \leq x_2$, then $A \subset B$. From Property 3.5 (montonicity) $F_X(x_2) = P[B] \geq P[A] = F_X(x_1)$.

$\square$

Property 5.6 – CDF is right-continuous.

By right-continuous it is meant that as we approach the point x_0 from the right, the limiting value of the CDF should be the value of the CDF at that point. Mathematically, it is expressed as

$$\lim_{x \to x_0^+} F_X(x) = F_X(x_0).$$

Proof:
The proof relies on the continuity property of the probability function. It can be found in [Ross 2002].

$\square$

Property 5.7 – Probability of interval found using the CDF

$$P[a < X \leq b] = F_X(b) - F_X(a) \tag{5.12}$$

or more explicitly to remind us of possible discontinuities

$$P[a < X \leq b] = F_X(b^+) - F_X(a^+). \tag{5.13}$$

<u>Proof:</u>

Since for $a < b$

$$\{-\infty < X \leq b\} = \{-\infty < X \leq a\} \cup \{a < X \leq b\}$$

and the intervals on the right-hand-side are disjoint (mutually exclusive events), by Axiom 3

$$P[-\infty < X \leq b] = P[-\infty < X \leq a] + P[a < X \leq b]$$

or rearranging terms we have that

$$P[a < X \leq b] = P[-\infty < X \leq b] - P[-\infty < X \leq a] = F_X(b) - F_X(a).$$

$\square$

5.9   Computer Simulation

In Chapter 2 we discussed how to simulate a discrete random variable on a digital computer. In particular, Section 2.4 presented some MATLAB code. We now continue that discussion to show how to simulate a discrete random variable and estimate its PMF and CDF. Assume that X can take on values in $\mathcal{S}_X = \{1, 2, 3\}$ with a PMF

$$p_X[x] = \begin{cases} p_1 = 0.2 & \text{if } x = x_1 = 1 \\ p_2 = 0.6 & \text{if } x = x_2 = 2 \\ p_3 = 0.2 & \text{if } x = x_3 = 3. \end{cases}$$

The PMF and CDF are shown in Figure 5.12. The code from Section 2.4 for generating M realizations of X is

```
for i=1:M
  u=rand(1,1);
  if u<=0.2
    x(i,1)=1;
  elseif u>0.2 & u<=0.8
    x(i,1)=2;
  elseif u>0.8
    x(i,1)=3;
  end
end
```

Recall that U is a random variable whose values are equally likely to fall within the interval $(0, 1)$. It is called the *uniform random variable* and is described further in Chapter 10. Now to estimate the PMF $p_X[k] = P[X = k]$ for $k = 1, 2, 3$ we use the relative frequency interpretation of probability to yield

$$\hat{p}_X[k] = \frac{\text{Number of outcomes equal to } k}{M} \qquad k = 1, 2, 3. \tag{5.14}$$

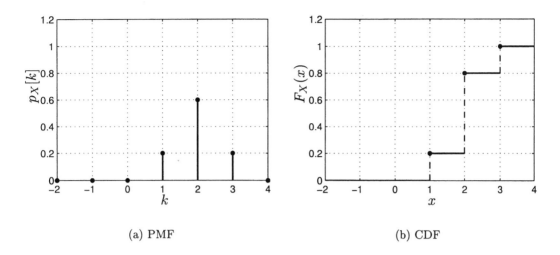

(a) PMF   (b) CDF

Figure 5.12: The probability mass function and cumulative distribution function for computer simulation example.

For $M = 100$ this is shown in Figure 5.13a. Also, the CDF is estimated for all x via

$$\hat{F}_X(x) = \frac{\text{Number of outcomes } \leq x}{M} \tag{5.15}$$

or equivalently by

$$\hat{F}_X(x) = \sum_{\{k:k\leq x\}} \hat{p}_X[k] \tag{5.16}$$

and is shown in Figure 5.13b. For finite sample spaces this approach to simulate a discrete random variable is adequate. But for infinite sample spaces such as for the geometric and Poisson random variables a different approach is needed. See Problem 5.30 for a further discussion.

Before concluding our discussion we wish to point out a useful property of CDFs that simplifies the computer generation of random variable outcomes. Note from Figure 5.12b with $u = F_X(x)$ that we can define an inverse CDF as $x = F_X^{-1}(u)$ where

$$x = F_X^{-1}(u) = \begin{cases} 1 & \text{if } 0 < u \leq 0.2 \\ 2 & \text{if } 0.2 < u \leq 0.8 \\ 3 & \text{if } 0.8 < u < 1 \end{cases}$$

or we choose the value of x as shown in Figure 5.14. But if u is the outcome of a uniform random variable U on $(0,1)$, then this procedure is identical to that implemented in the previous MATLAB program used to generate realizations of X. A more general program is given in Appendix 6B as PMFdata.m. This is not merely a coincidence but can be shown to follow from the definition of the CDF.

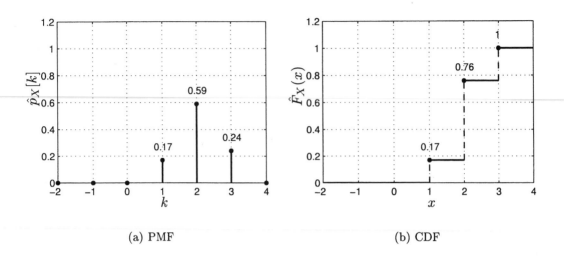

(a) PMF   (b) CDF

Figure 5.13: The estimated probability mass function and corresponding estimated cumulative distribution function.

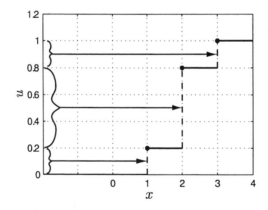

Figure 5.14: Relationship of inverse CDF to generation of discrete random variable. Value of u is mapped into value of x.

Although little more than a curiousity now, it will become important when we simulate continuous random variables in Chapter 10.

5.10   Real-World Example – Servicing Customers

A standard problem in many disciplines is the allocation of resources to service customers. It occurs in determining the number of cashiers needed at a store, the computer capacity needed to service download requests, and the amount of equipment necessary to service phone customers, as examples. In order to service these customers in a timely manner, it is necessary to know the distribution of arrival times of their requests. Since this will vary depending on many factors such as time of day, popularity of a file request, etc., the best we can hope for is a determination of the probabilities of these arrivals. As we will see shortly, the Poisson probability PMF is particularly suitable as a model. We now focus on the problem of determining the number of cashiers needed in a supermarket.

A supermarket has one express lane open from 5 to 6 PM on weekdays (Monday through Friday). This time of the day is usually the busiest since people tend to stop on their way home from work to buy groceries. The number of items allowed in the express lane is limited to 10 so that the average time to process an order is fairly constant at about 1 minute. The manager of the supermarket notices that there is frequently a long line of people waiting and hears customers grumbling about the wait. To improve the situation he decides to open additional express lanes during this time period. If he does, however, he will have to "pull" workers from other jobs around the store to serve as cashiers. Hence, he is reluctant to open more lanes than necessary. He hires Professor Poisson to study the problem and tell him how many lanes should be opened. The manager tells Professor Poisson that there should be no more than one person waiting in line 95% of the time. Since the processing time is 1 minute, there can be at most two arrivals in each time slot of 1 minute length. He reasons that one will be immediately serviced and the other will only have to wait a maximum of 1 minute. After a week of careful study, Professor Poisson tells the manager to open two lanes from 5 to 6 PM. Here is his reasoning.

First Professor Poisson observes the arrivals of customers in the express lane on a Monday from 5 to 6 PM. The observed arrivals are shown in Figure 5.15, where the arrival times are measured in seconds. On Monday there are a total of

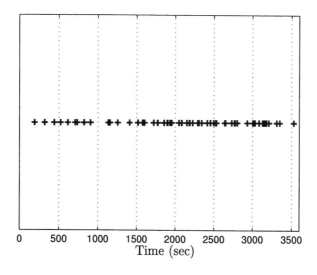

Figure 5.15: Arrival times at one express lane on Monday (a '+' indicates an arrival).

80 arrivals. He repeats his experiment on the following 4 days (Tuesday through Friday) and notes total arrivals of 68, 70, 59, and 66 customers, respectively. On the average there are 68.6 arrivals, which he rounds up to 70. Thus, the arrival rate is 1.167 customers per minute. He then likens the arrival process to one in which the 5 to 6 PM time interval is broken up into 3600 time slots of 1 second each. He reasons that there is at most 1 arrival in a given time slot and there may be no arrivals in that time slot. (This of course would not be valid if for instance, two friends did their shopping together and arrived at the same time.) Hence, Professor Poisson reasons that a good arrival model is a sequence of independent Bernoulli trials, where 0 indicates no arrival and 1 indicates an arrival in each 1-second time slot. The probability p of a 1 is estimated from his observed data as the number of arrivals from 5 to 6 PM divided by the total number of time slots in seconds. This yields $\hat{p} = 70/3600 = 0.0194$ for each 1-second time slot. Instead of using the binomial PMF to describe the number of arrivals in each 1-minute time slot (for which $p = 0.0194$ and $M = 60$), he decides to approximate it using his favorite PMF, the Poisson model. Therefore, the probability of k arrivals (or successes) in a time interval of 60 seconds would be

$$p_{X_1}[k] = \exp(-\lambda_1)\frac{\lambda_1^k}{k!} \qquad k = 0, 1, \ldots \tag{5.17}$$

where the subscripts on X and λ are meant to remind us that we will initially consider the arrivals at *one* express lane. The value of λ_1 to be used is $\lambda_1 = Mp$, which is estimated as $\hat{\lambda}_1 = M\hat{p} = 60(70/3600) = 7/6$. This represents the expected number of customers arriving in the 1-minute interval. According to the manager's requirements, within this time interval there should be at most 2 customers arriving 95% of the time. Hence, we require that

$$P[X_1 \leq 2] = \sum_{k=0}^{2} p_{X_1}[k] \geq 0.95.$$

But from (5.17) this becomes

$$P[X_1 \leq 2] = \exp(-\lambda_1)\left(1 + \lambda_1 + \frac{1}{2}\lambda_1^2\right) = 0.88$$

using $\lambda_1 = 7/6$. Hence, the probability of 2 or fewer customers arriving at the express lane is not greater than 0.95. If a second express lane is opened, then the average number of arrivals at each lane during the 1-minute time interval will be halved to 35. Therefore, the Poisson PMF for the number of arrivals at each lane will be characterized by $\lambda_2 = 7/12$. Now, however, there are two lanes and two sets of arrivals. Since the arrivals are modeled as *independent* Bernoulli trials, we can assert that

$$
\begin{aligned}
P[2 \text{ or fewer arrivals at both lanes}] &= P[2 \text{ or fewer arrivals at lane 1}] \\
&\quad \cdot P[2 \text{ or fewer arrivals at lane 2}] \\
&= P[2 \text{ or fewer arrivals at lane 1}]^2 \\
&= P[X_1 \leq 2]^2
\end{aligned}
$$

so that

$$
\begin{aligned}
P[2 \text{ or fewer arrivals at both lanes}] &= \left(\sum_{k=0}^{2} p_{X_1}[k] \right)^2 \\
&= \left[\exp(-\lambda_2) \left(1 + \lambda_2 + \frac{1}{2}\lambda_2^2 \right) \right]^2 = 0.957
\end{aligned}
$$

which meets the requirement. An example is shown for one of the two express lanes with an average number of customer arrivals per minute of 7/12 in Figures 5.16 and 5.17, with the latter an expanded version of the former. The dashed vertical lines

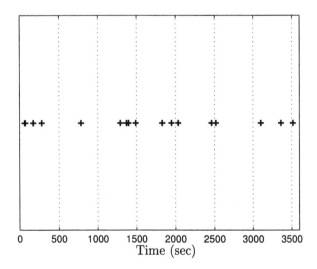

Figure 5.16: Arrival times at one of the two express lanes (a '+' indicates an arrival).

in Figure 5.17 indicate 1-minute intervals. There are no 1-minute intervals with more than 2 arrivals, as we expect.

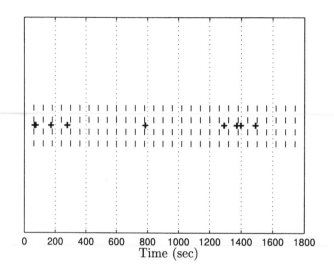

Figure 5.17: Expanded version of Figure 5.16 (a '+' indicates an arrival). Time slots of 60 seconds are shown by dashed lines.

References

Ross, S., *A First Course in Probability*, Prentice-Hall, Upper Saddle River, NJ, 2002.

Problems

5.1 (w) Draw a picture depicting a mapping of the outcome of a die toss, i.e., the pattern of dots that appear, to the numbers $1, 2, 3, 4, 5, 6$.

5.2 (w) Repeat Problem 5.1 for a mapping of the sides that display 1, 2, or 3 dots to the number 0 and the remaining sides to the number 1.

5.3 (w) Consider a random experiment for which $\mathcal{S} = \{s_i : s_i = i, i = 1, 2, \ldots, 10\}$ and the outcomes are equally likely. If a random variable is defined as $X(s_i) = s_i^2$, find $\mathcal{S}_X$ and the PMF.

5.4 (☺) (w) Consider a random experiment for which $\mathcal{S} = \{s_i : s_i = -3, -2, -1, 0, 1, 2, 3\}$ and the outcomes are equally likely. If a random variable is defined as $X(s_i) = s_i^2$, find $\mathcal{S}_X$ and the PMF.

5.5 (w) A man is late for his job by $s_i = i$ minutes, where $i = 1, 2, \ldots$. If $P[s_i] = (1/2)^i$ and he is fined \$0.50 per minute, find the PMF of his fine. Next find the probability that he will be fined more than \$10.

5.6 (☺) (w) If $p_X[k] = \alpha p^k$ for $k = 2, 3, \ldots$ is to be a valid PMF, what are the possible values for α and p?

5.7 (t) The maximum value of the binomial PMF occurs for the unique value $k = [(M + 1)p]$, where $[x]$ denotes the largest integer less than or equal to x, if $(M + 1)p$ is not an integer. If, however, $(M + 1)p$ is an integer, then the PMF will have the same maximum value at $k = (M + 1)p$ and $k = (M + 1)p - 1$. For the latter case when $(M + 1)p$ is an integer you are asked to prove this result. To do so first show that

$$p_X[k]/p_X[k - 1] = 1 + \frac{(M + 1)p - k}{k(1 - p)}.$$

5.8 (☺) (w) At a party a large barrel is filled with 99 gag gifts and 1 diamond ring, all enclosed in identical boxes. Each person at the party is given a chance to pick a box from the barrel, open the box to see if the diamond is inside, and if not, to close the box and return it to the barrel. What is the probability that at least 19 persons will choose gag gifts before the diamond ring is selected?

5.9 (f,c) If X is a geometric random variable with $p = 0.25$, what is the probability that $X \geq 4$? Verify your result by performing a computer simulation.

5.10 (c) Using a computer simulation to generate a geom(0.25) random variable, determine the average value for a large number of realizations. Relate this to the value of p and explain the results.

5.11 (t) Prove that the maximum value of a Poisson PMF occurs at $k = [\lambda]$. Hint: See Problem 5.7 for the approach.

5.12 (w,c) If $X \sim \text{Pois}(\lambda)$, plot $P[X \geq 2]$ versus λ and explain your results.

5.13 (☺) (c) Use a computer simulation to generate realizations of a $\text{Pois}(\lambda)$ random variable with $\lambda = 5$ by approximating it with a bin(100,0.05) random variable. What is the average value of X?

5.14 (☺) (w) If $X \sim \text{bin}(100, 0.01)$, determine $p_X[5]$. Next compare this to the value obtained using a Poisson approximation.

5.15 (t) Prove the following limit:

$$\lim_{M \to \infty} g(M) = \lim_{M \to \infty} \left(1 + \frac{x}{M}\right)^M = \exp(x).$$

To do so note that the same limit is obtained if M is replaced by a continuous variable, say u, and that one can consider $\ln g(u)$ since the logarithm is a continuous function. Hint: Use L'Hospital's rule.

5.16 (f,c) Compare the PMFs for Pois(1) and bin(100,0.01) random variables.

5.17 (c) Generate realizations of a Pois(1) random variable by using a binomial approximation.

5.18 (☺) (c) Compare the theoretical value of $P[X = 3]$ for the Poisson random variable to the estimated value obtained from the simulation of Problem 5.17.

5.19 (f) If $X \sim \text{Ber}(p)$, find the PMF for $Y = -X$.

5.20 (☺) (f) If $X \sim \text{Pois}(\lambda)$, find the PMF for $Y = 2X$.

5.21 (f) A discrete random variable X has the PMF

$$
p_X[x_i] = \begin{cases}
\frac{1}{2} & x_1 = -1 \\
\frac{1}{4} & x_2 = -\frac{1}{2} \\
\frac{1}{8} & x_3 = 0 \\
\frac{1}{16} & x_4 = \frac{1}{2} \\
\frac{1}{16} & x_5 = 1.
\end{cases}
$$

If $Y = \sin \pi X$, find the PMF for Y.

5.22 (t) In this problem we derive the Taylor expansion for the function $g(x) = \exp(x)$. To do so note that the expansion about the point $x = 0$ is given by

$$
g(x) = \sum_{n=0}^{\infty} \frac{g^{(n)}(0)}{n!} x^n
$$

where $g^{(0)}(0) = g(0)$ and $g^{(n)}(0)$ is the nth derivative of $g(x)$ evaluated at $x = 0$. Prove that it is given by

$$
\exp(x) = \sum_{n=0}^{\infty} \frac{x^n}{n!}.
$$

5.23 (f) Plot the CDF for

$$
p_X[k] = \begin{cases}
\frac{1}{4} & k = 1 \\
\frac{1}{2} & k = 2 \\
\frac{1}{4} & k = 3.
\end{cases}
$$

5.24 (w) A horizontal bar of negligible weight is loaded with three weights as shown in Figure 5.18. Assuming that the weights are concentrated at their center locations, plot the total mass of the bar starting at the left end (where $x = 0$ meters) to any point on the bar. How does this relate to a PMF and a CDF?

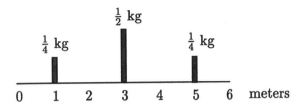

Figure 5.18: Weightless bar supporting three weights.

5.25 (f) Find and plot the CDF of $Y = -X$ if $X \sim \text{Ber}(\frac{1}{4})$.

5.26 (⌣) (w) Find the PMF if X is a discrete random variable with the CDF

$$F_X(x) = \begin{cases} 0 & x < 0 \\ \frac{[x]}{5} & 0 \le x \le 5 \\ 1 & x > 5. \end{cases}$$

5.27 (w) Is the following a valid CDF? If not, why not, and how could you modify it to become a valid one?

$$F_X(x) = \begin{cases} 0 & x < 2 \\ \frac{1}{2} & 2 \le x \le 3 \\ \frac{3}{4} & 3 < x \le 4 \\ 1 & x \ge 4. \end{cases}$$

5.28 (⌣) (f) If X has the CDF shown in Figure 5.11b, determine $P[2 \le X \le 4]$ from the CDF.

5.29 (t) Prove that the function $g(x) = \exp(x)$ is a monotonically increasing function by showing that $g(x_2) \ge g(x_1)$ if $x_2 \ge x_1$.

5.30 (c) Estimate the PMF for a geom(0.25) random variable for $k = 1, 2, \ldots, 20$ using a computer simulation and compare it to the true PMF. Also, estimate the CDF from your computer simulation.

5.31 (⌣) (f,c) The arrival rate of calls at a mobile switching station is 1 per second. The probability of k calls in a T second interval is given by a Poisson PMF with $\lambda = $ arrival rate $\times T$. What is the probability that there will be more than 100 calls placed in a 1-minute interval?

Chapter 6

Expected Values for Discrete Random Variables

6.1 Introduction

The probability mass function (PMF) discussed in Chapter 5 is a complete description of a discrete random variable. As we have seen, it allows us to determine probabilities of any event. Once the probability of an event of interest is determined, however, the question of its interpretation arises. Consider, for example, whether there is adequate rainfall in Rhode Island to sustain a farming endeavor. The past history of yearly summer rainfall was shown in Figure 1.1 and is repeated in Figure 6.1a for convenience. Along with it, the estimated PMF of this yearly data is shown in Figure 6.1b (see Section 5.9 for a discussion on how to estimate the PMF). For a particular crop we might need a rainfall of between 8 and 12 inches. This event has probability 0.5278, obtained by $\sum_{k=8}^{12} \hat{p}_X[k]$ for the estimated PMF shown in Figure 6.1b. Is this adequate or should the probability be higher? Answers to such questions are at best problematic. Rather we might be better served by ascertaining the *average* rainfall since this is closer to the requirement of an adequate amount of rainfall. In the case of Figure 6.1a the average is 9.76 inches, and is obtained by summing all the yearly rainfalls and dividing by the number of years. Based on the given data it is a simple matter to estimate the average value of a random variable (the rainfall in this case). Some computer simulation results pertaining to averages have already been presented in Example 2.3. In this chapter we address the topic of the *average* or *expected value* of a discrete random variable and study its properties.

6.2 Summary

The expected value of a random variable is the average value of the outcomes of a large number of experimental trials. It is formally defined by (6.1). For discrete random variables with integer values it is given by (6.2) and some examples of its

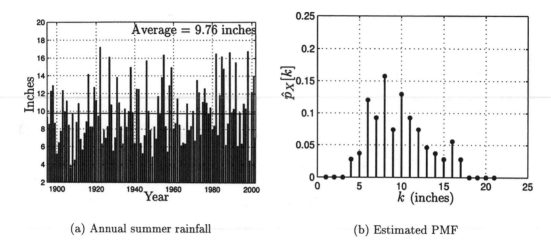

(a) Annual summer rainfall   (b) Estimated PMF

Figure 6.1: Annual summer rainfall in Rhode Island and its estimated probability mass function.

determination given in Section 6.4. The expected value does not exist for all PMFs as illustrated in Section 6.4. For functions of a random variable the expected value is easily computed via (6.5). It is shown to be a linear operation in Section 6.5. Another interpretation of the expected value is as the best predictor of the outcome of an experiment as shown in Example 6.3. The variability of the values exhibited by a random variable is quantified by the variance. It is defined in (6.6) with examples given in Section 6.6. Some properties of the variance are summarized in Section 6.6 as Properties 1 and 2. An alternative way to determine means and variances of a discrete random variable is by using the characteristic function. It is defined by (6.10) and for integer valued random variables it is evaluated using (6.12), which is a Fourier transform of the PMF. Having determined the characteristic function, one can easily determine the mean and variance by using (6.13). Some examples of this procedure are given in Section 6.7, as are some further important properties of the characteristic function. An important property is that the PMF may be obtained from the characteristic function as an inverse Fourier transform as expressed by (6.19). In Section 6.8 an example is given to illustrate how to estimate the mean and variance of a discrete random variable. Finally, Section 6.9 describes the use of the expected value to reduce the average code length needed to store symbols in a digital format. This is called data compression.

6.3   Determining Averages from the PMF

We now discuss how the average of a discrete random variable can be obtained from the PMF. To motivate the subsequent definition we consider the following game of chance. A barrel is filled with US dollar bills with denominations of \$1, \$5, \$10, and \$20. The proportion of each denomination bill is the same. A person playing the game gets to choose a bill from the barrel, but must do so while blindfolded. He pays \$10 to play the game, which consists of a single draw from the barrel. After he observes the denomination of the bill, the bill is returned to the barrel and he wins that amount of money. Will he make a profit by playing the game many times? A typical sequence of outcomes for the game is shown in Figure 6.2. His average

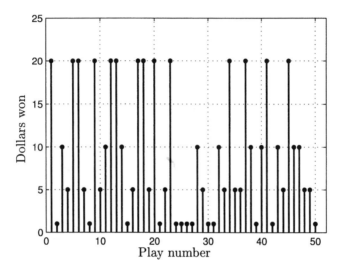

Figure 6.2: Dollar winnings for each play.

winnings per play is found by adding up all his winnings and dividing by the number of plays N. This is computed by

$$\bar{x} = \frac{1}{N} \sum_{i=1}^{N} x_i$$

where x_i is his winnings for play i. Alternatively, we can compute $\bar{x}$ using a slightly different approach. From Figure 6.2 the number of times he wins k dollars (where $k = 1, 5, 10, 20$) is given by N_k, where

$$
\begin{aligned}
N_1 &= 13 \\
N_5 &= 13 \\
N_{10} &= 10 \\
N_{20} &= 14.
\end{aligned}
$$

As a result, we can determine the average winnings per play by

$$
\begin{aligned}
\bar{x} &= \frac{1 \cdot N_1 + 5 \cdot N_5 + 10 \cdot N_{10} + 20 \cdot N_{20}}{N_1 + N_5 + N_{10} + N_{20}} \\
&= 1 \cdot \frac{N_1}{N} + 5 \cdot \frac{N_5}{N} + 10 \cdot \frac{N_{10}}{N} + 20 \cdot \frac{N_{20}}{N} \\
&= 1 \cdot \frac{13}{50} + 5 \cdot \frac{13}{50} + 10 \cdot \frac{10}{50} + 20 \cdot \frac{14}{50} \\
&= 9.16
\end{aligned}
$$

since $N = N_1 + N_5 + N_{10} + N_{20} = 50$. If he were to play the game a large number of times, then as $N \to \infty$ we would have $N_k/N \to p_X[k]$, where the latter is just the PMF for choosing a bill with denomination k, and results from the relative frequency interpretation of probability. Then, his average winnings per play would be found as

$$
\begin{aligned}
\bar{x} &\to 1 \cdot p_X[1] + 5 \cdot p_X[5] + 10 \cdot p_X[10] + 20 \cdot p_X[20] \\
&= 1 \cdot \frac{1}{4} + 5 \cdot \frac{1}{4} + 10 \cdot \frac{1}{4} + 20 \cdot \frac{1}{4} \\
&= 9
\end{aligned}
$$

where $p_X[k] = 1/4$ for $k = 1, 5, 10, 20$ since the proportion of bill denominations in the barrel is the same for each denomination. It is now clear that "on the average" he will lose \$1 per play. The value that the average converges to is called the *expected value* of X, where X is the random variable that describes his winnings for a single play and takes on the values $1, 5, 10, 20$. The expected value is denoted by $E[X]$. For this example, the PMF as well as the expected value is shown in Figure 6.3. The expected value is also called the *expectation* of X, the *average* of X, and the *mean* of X. With this example as motivation we now define the expected value of a discrete random variable X as

$$
E[X] = \sum_i x_i p_X[x_i] \tag{6.1}
$$

where the sum is over all values of x_i for which $p_X[x_i]$ is nonzero. It is determined from the PMF and as we have seen coincides with our notion of the outcome of an experiment in the "long run" or "on the average." The expected value may also be intepreted as the best prediction of the outcome of a random experiment for a single trial (to be described in Example 6.3). Finally, the expected value is analogous to the center of mass of a system of linearly arranged masses as illustrated in Problem 6.1.

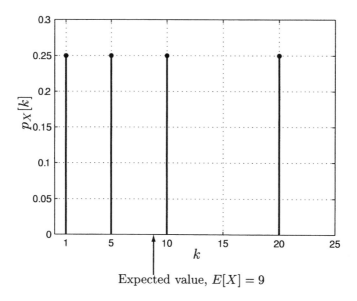

Figure 6.3: PMF and expected value of dollar bill denomination chosen.

6.4 Expected Values of Some Important Random Variables

The definition of the expected value was given by (6.1). When the random variable takes on only integer values, we can rewrite it as

$$E[X] = \sum_{k=-\infty}^{\infty} k p_X[k]. \tag{6.2}$$

We next determine the expected values for some important discrete random variables (see Chapter 5 for a definition of the PMFs).

6.4.1 Bernoulli

If $X \sim \text{Ber}(p)$, then the expected value is

$$
\begin{aligned}
E[X] &= \sum_{k=0}^{1} k p_X[k] \\
&= 0 \cdot (1-p) + 1 \cdot p \\
&= p.
\end{aligned}
$$

Note that $E[X]$ need not be a value that the random variable takes on. In this case, it is between $X = 0$ and $X = 1$.

6.4.2 Binomial

If $X \sim \text{bin}(M, p)$, then the expected value is

$$
\begin{aligned}
E[X] &= \sum_{k=0}^{M} k p_X[k] \\
&= \sum_{k=0}^{M} k \binom{M}{k} p^k (1-p)^{M-k}.
\end{aligned}
$$

To evaluate this in closed form we will need to find an expression for the sum. Continuing, we have that

$$
\begin{aligned}
E[X] &= \sum_{k=0}^{M} k \frac{M!}{(M-k)!k!} p^k (1-p)^{M-k} \\
&= Mp \sum_{k=1}^{M} \frac{(M-1)!}{(M-k)!(k-1)!} p^{k-1} (1-p)^{M-1-(k-1)}
\end{aligned}
$$

and letting $M' = M - 1$, $k' = k - 1$, this becomes

$$
\begin{aligned}
E[X] &= Mp \sum_{k'=0}^{M'} \frac{M'!}{(M'-k'!)k'!} p^{k'} (1-p)^{M'-k'} \\
&= Mp \sum_{k'=0}^{M'} \binom{M'}{k'} p^{k'} (1-p)^{M'-k'} \\
&= Mp
\end{aligned}
$$

since the summand is just the PMF of a $\text{bin}(M', p)$ random variable. Therefore, we have that $E[X] = Mp$ for a binomial random variable. This derivation is typical in that we attempt to manipulate the sum into one whose summands are the values of a PMF and so the sum must evaluate to one. Intuitively, we expect that if p is the probability of success for a Bernoulli trial, then the expected number of successes for M independent Bernoulli trials (which is binomially distributed) is Mp.

6.4.3 Geometric

If $X \sim \text{geom}(p)$, the the expected value is

$$
E[X] = \sum_{k=1}^{\infty} k(1-p)^{k-1} p.
$$

To evaluate this in closed form, we need to modify the summand to be a PMF, which in this case will produce a geometric series. To do so we use differentiation

by first letting $q = 1 - p$ to produce

$$E[X] \;=\; p\sum_{k=1}^{\infty} \frac{d}{dq}q^k$$

$$=\; p\frac{d}{dq}\sum_{k=1}^{\infty}q^k.$$

But since $0 < q < 1$ we have upon using the formula for the sum of a geometric series or $\sum_{k=1}^{\infty} q^k = q/(1-q)$ that

$$E[X] \;=\; p\frac{d}{dq}\left(\frac{q}{1-q}\right)$$

$$=\; p\frac{(1-q)-q(-1)}{(1-q)^2}$$

$$=\; p\frac{1}{(1-q)^2}$$

$$=\; \frac{1}{p}.$$

The expected number of Bernoulli trials until the first success (which is geometrically distributed) is $E[X] = 1/p$. For example, if $p = 1/10$, then on the average it takes 10 trials for a success, an intuitively pleasing result.

6.4.4   Poisson

If $X \sim \text{Pois}(\lambda)$, then it can be shown that $E[X] = \lambda$. The reader is asked to verify this in Problem 6.5. Note that this result is consistent with the Poisson approximation to the binomial PMF since the approximation constrains Mp (the expected value of the binomial random variable) to be λ (the expected value of the Poisson random variable).

   Not all PMFs have expected values.

Discrete random variables with a finite number of values always have expected values. In the case of a countably infinite number of values, a discrete random variable *may not have an expected value.* As an example of this, consider the PMF

$$p_X[k] = \frac{4/\pi^2}{k^2} \qquad k = 1, 2, \dots. \tag{6.3}$$

This is a valid PMF since it can be shown to sum to one. Attempting to find the

expected value produces

$$E[X] = \sum_{k=1}^{\infty} kp_X[k]$$

$$= \frac{4}{\pi^2} \sum_{k=1}^{\infty} \frac{1}{k} \to \infty$$

since $1/k$ is a *harmonic* series which is known not to summable (meaning that the partial sums do not converge). Hence, the random variable described by the PMF of (6.3) does not have a finite expected value. It is even possible for a sum $\sum_{k=-\infty}^{\infty} kp_X[k]$ that is composed of positive and negative terms to produce different results depending upon the order in which the terms are added together. In this case the value of the sum is said to be ambiguous. These difficulties can be avoided, however, if we require the sum to be absolutely summable or if the sum of the absolute values of the terms is finite [Gaughan 1975]. Hence we will say that the expected value *exists* if

$$E[|X|] = \sum_{k=\infty}^{\infty} |k|p_x[k] < \infty.$$

In Problem 6.6 a further discussion of this point is given.

Lastly, note the following properties of the expected value.

1. It is located at the "center" of the PMF if the PMF is symmetric about some point (see Problem 6.7).

2. It does not generally indicate the most probable value of the random variable (see Problem 6.8).

3. More than one PMF may have the same expected value (see Problem 6.9).

6.5   Expected Value for a Function of a Random Variable

The expected value may easily be found for a *function* of a random variable X if the PMF $p_X[x_i]$ is known. If the function of interest is $Y = g(X)$, then by the definition of expected value

$$E[Y] = \sum_i y_i p_Y[y_i]. \tag{6.4}$$

But as shown in Appendix 6A we can avoid having to find the PMF for Y by using the much more convenient form

$$E[g(X)] = \sum_i g(x_i)p_X[x_i]. \tag{6.5}$$

Otherwise, we would be forced to determine $p_Y[y_i]$ from $p_X[x_i]$ and $g(X)$ using (5.9). This result proves to be very useful, especially when the function is a complicated one such as $g(x) = \sin[(\pi/2)x]$ (see Problem 6.10). Some examples follow.

Example 6.1 – A linear function

If $g(X) = aX + b$, where a and b are constants, then

$$
\begin{aligned}
E[g(X)] &= E[aX + b] \\
&= \sum_i (ax_i + b)p_X[x_i] \qquad \text{(from (6.5))} \\
&= a\sum_i x_i p_X[x_i] + b\sum_i p_X[x_i] \\
&= aE[X] + b \qquad \text{(definition of } E[X] \text{ and PMF values sum to one.)}
\end{aligned}
$$

In particular, if we set $a = 1$, then $E[X + b] = E[X] + b$. This allows us to set the expected value of a random variable to any desired value by adding the appropriate constant to X. Finally, a simple extension of this example produces

$$
E[a_1 g_1(X) + a_2 g_2(X)] = a_1 E[g_1(X)] + a_2 E[g_2(X)]
$$

for any two constants a_1 and a_2 and any two functions g_1 and g_2 (see Problem 6.11). It is said that the *expectation operator E is linear.*

$\Diamond$

Example 6.2 – A nonlinear function

Assume that X has a PMF given by

$$
p_X[k] = \frac{1}{5} \qquad k = 0, 1, 2, 3, 4
$$

and determine $E[Y]$ for $Y = g(X) = X^2$. Then, using (6.5) produces

$$
\begin{aligned}
E[X^2] &= \sum_{k=0}^{4} k^2 p_X[k] \\
&= \sum_{k=0}^{4} k^2 \frac{1}{5} \\
&= 6.
\end{aligned}
$$

$\Diamond$

 It is not true that $E[g(X)] = g(E[X])$.

From the previous example with $g(X) = X^2$, we had that $E[g(X)] = E[X^2] = 6$ but $g(E[X]) = (E[X])^2 = 2^2 = 4 \neq E[g(X)]$. It is said that the expectation operator *does not commute* (or we cannot just take $E[g(X)]$ and interchange the E and g) for nonlinear functions. This manipulation is valid, however, for *linear* (actually affine) functions as Example 6.1 demonstrates. Henceforth, we will use the notation $E^2[X]$ to replace the more cumbersome $(E[X])^2$.

Example 6.3 – Predicting the outcome of an experiment

It is always of great interest to be able to predict the outcome of an experiment before it has occurred. For example, if the experiment were the summer rainfall in Rhode Island in the coming year, then a farmer would like to have this information before he decides upon which crops to plant. One way to do this is to check the Farmer's almanac, but its accuracy may be in dispute! Another approach would be to *guess* this number based on the PMF (statisticians, however, use the more formal term "predict" or "estimate" which sounds better). Denoting the prediction by the number b, we would like to choose a number so that *on the average* it is close to the true outcome of the random variable X. To measure the error we could use $x - b$, where x is the outcome, and to account for positive and negative errors equally we could use $(x - b)^2$. This *squared error* may at times be small and at other times large, depending on the outcome of X. What we want is the *average value* of the squared error. This is measured by $E[(X - b)^2]$, and is termed the *mean square error* (MSE). We denote it by mse(b) since it will depend on our choice of b. A reasonable method for choosing b is to choose the value that *minimizes* the MSE. We now proceed to find that value of b.

$$
\begin{aligned}
\text{mse}(b) &= E[(X - b)^2] \\
&= E[X^2 - 2bX + b^2] \\
&= E[X^2] - 2bE[X] + E[b^2] \qquad \text{(linearity of } E(\cdot)\text{)} \\
&= E[X^2] - 2bE[X] + b^2 \qquad \text{(expected value of constant is the constant)}.
\end{aligned}
$$

To find the value of b that minimizes the MSE we need only differentiate the MSE, set the derivative equal to zero, and solve for b. This is because the MSE is a quadratic function of b whose minimum is located at the stationary point. Thus, we have

$$
\frac{d\text{mse}(b)}{db} = -2E[X] + 2b = 0
$$

which produces the minimizing or optimal value of b given by $b_{\text{opt}} = E[X]$. Hence, the best predictor of the outcome of an experiment is the expected value or mean of the random variable. For example, the best predictor of the outcome of a die toss would be 3.5. This result provides another interpretation of the expected value. *The expected value of a random variable is the best predictor of the outcome of the experiment*, where "best" is to be interpreted as the value that minimizes the MSE.

6.6 Variance and Moments of a Random Variable

Another function of a random variable that yields important information about its behavior is that given by $g(X) = (X - E[X])^2$. Whereas $E[X]$ measures the mean of a random variable, $E[(X - E[X])^2]$ measures the *average squared deviation from the mean*. For example, a uniform discrete random variable whose PMF is

$$p_X[k] = \frac{1}{2M + 1} \qquad k = -M, -M + 1, \ldots, M$$

is easily shown to have a mean of zero for any M. However, as seen in Figure 6.4 the variability of the outcomes of the random variable becomes larger as M increases. This is because the PMF for $M = 10$ can have values exceeding those for $M = 2$. The variability is measured by the *variance* which is defined as

$$\text{var}(X) = E[(X - E[X])^2]. \tag{6.6}$$

Note that the variance is *always greater than or equal to zero*. It is determined from the PMF using (6.5) with $g(X) = (X - E[X])^2$ to yield

$$\text{var}(X) = \sum_i (x_i - E[X])^2 p_X[x_i]. \tag{6.7}$$

For the current example, $E[X] = 0$ due to the symmetry of the PMF about $k = 0$ so that

$$\begin{aligned}
\text{var}(X) &= \sum_i x_i^2 p_X[x_i] \\
&= \sum_{k=-M}^{M} k^2 \frac{1}{2M + 1} \\
&= \frac{2}{2M + 1} \sum_{k=1}^{M} k^2.
\end{aligned}$$

But it can be shown that

$$\sum_{k=1}^{M} k^2 = \frac{M(M + 1)(2M + 1)}{6}$$

which yields

$$\begin{aligned}
\text{var}(X) &= \frac{2}{2M + 1} \frac{M(M + 1)(2M + 1)}{6} \\
&= \frac{M(M + 1)}{3}.
\end{aligned}$$

Clearly, the variance increases with M, or equivalently with the *width* of the PMF, as is also evident from Figure 6.4. We next give another example of the determination of the variance and then summarize the results for several important PMFs.

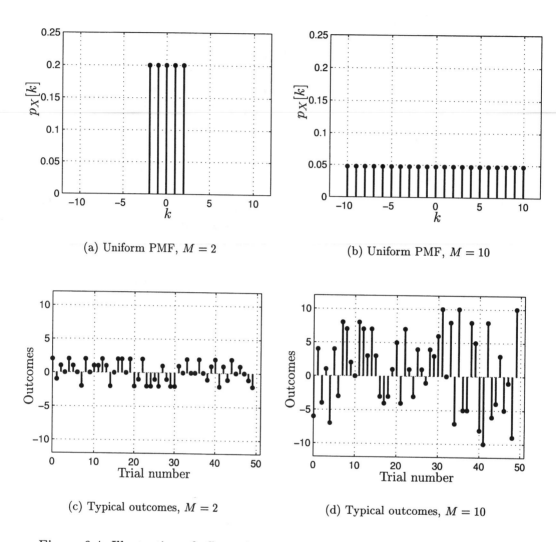

(a) Uniform PMF, $M = 2$ (b) Uniform PMF, $M = 10$

(c) Typical outcomes, $M = 2$ (d) Typical outcomes, $M = 10$

Figure 6.4: Illustration of effect of width of PMF on variability of outcomes.

Example 6.4 – Variance of Bernoulli random variable

If $X \sim \text{Ber}(p)$, then since $E[X] = p$, we have

$$\begin{aligned}
\text{var}(X) &= \sum_{i}(x_i - E[X])^2 p_X[x_i] \\
&= \sum_{k=0}^{1}(k - p)^2 p_X[k] \\
&= (0 - p)^2(1 - p) + (1 - p)^2 p \\
&= p(1 - p).
\end{aligned}$$

$\Diamond$

	Values	PMF	$E[X]$	$\text{var}(X)$	$\phi_X(\omega)$
Uniform	$k=-M,...,M$	$\frac{1}{2M+1}$	0	$\frac{M(M+1)}{3}$	$\frac{\sin[(2M+1)\omega/2]}{(2M+1)\sin[\omega/2]}$
Bernoulli	$k=0,1$	$p^k(1-p)^{1-k}$	p	$p(1-p)$	$p\exp(j\omega)+(1-p)$
Binomial	$k=0,1,...,M$	$\binom{M}{k}p^k(1-p)^{M-k}$	Mp	$Mp(1-p)$	$[p\exp(j\omega)+(1-p)]^M$
Geometric	$k=1,2,...$	$(1-p)^{k-1}p$	$\frac{1}{p}$	$\frac{1-p}{p^2}$	$\frac{p}{\exp(-j\omega)-(1-p)}$
Poisson	$k=0,1,...$	$\exp(-\lambda)\frac{\lambda^k}{k!}$	λ	λ	$\exp[\lambda(\exp(j\omega)-1)]$

Table 6.1: Properties of discrete random variables.

It is interesting to note that the variance is minimized and equals zero if $p = 0$ or $p = 1$. Also, it is maximized for $p = 1/2$. Can you explain this? Important PMFs with their means, variances, and characteristic functions (to be discussed in Section 6.7) are listed in . The reader is asked to derive some of these entries in the Problems.

An alternative useful expression for the variance can be developed based on the properties of the expectation operator. We have that

$$
\begin{aligned}
\text{var}(X) &= E[(X - E[X])^2] \\
&= E[X^2 - 2XE[X] + E^2[X]] \\
&= E[X^2] - 2E[X]E[X] + E^2[X]
\end{aligned}
$$

where the last step is due to linearity of the expectation operator and the fact that $E[X]$ is a constant. Hence

$$
\text{var}(X) = E[X^2] - E^2[X]
$$

and is seen to depend on $E[X]$ and $E[X^2]$. In the case where $E[X] = 0$, we have the simple result that $\text{var}(X) = E[X^2]$. This property of the variance along with some others is now summarized.

Property 6.1 – Alternative expression for variance

$$
\text{var}(X) = E[X^2] - E^2[X] \tag{6.8}
$$

$\square$

Property 6.2 – Variance for random variable modified by a constant
For c a constant

$$
\begin{aligned}
\text{var}(c) &= 0 \\
\text{var}(X + c) &= \text{var}(X) \\
\text{var}(cX) &= c^2 \text{var}(X)
\end{aligned}
$$

$\square$

The reader is asked to verify Property 6.2 in Problem 6.21.

The expectations $E[X]$ and $E[X^2]$ are called the *first* and *second moments* of X, respectively. The term moment has been borrowed from physics, where $E[X]$ is called the center of mass or moment of mass (see also Problem 6.1). In general, the nth moment is defined as $E[X^n]$ and exists (meaning that the value can be determined unambiguously and is finite) if $E[|X|^n]$ is finite. The latter is called the n absolute moment. It can be shown that if $E[X^s]$ exists, then $E[X^r]$ exists for $r < s$ (see Problem 6.23). As a result, if $E[X^2]$ is finite, then $E[X]$ exists and by (6.8) the variance will also exist. In summary, *the mean and variance of a discrete random variable will exist if the second moment is finite.*

A variant of the notion of moments is that of the *central moments*. They are defined as $E[(X - E[X])^n]$, in which the mean is first subtracted from X before the n moment is computed. They are useful in assessing the average deviations from the mean. In particular, for $n = 2$ we have the usual definition of the variance. See also Problem 6.26 for the relationship between the moments and central moments.

   Variance is a nonlinear operator.

The variance of a random variable *does not* have the linearity property of the expectation operator. Hence, in general

$$
\text{var}(g_1(X) + g_2(X)) = \text{var}(g_1(X)) + \text{var}(g_2(X)) \qquad \textit{is not true.}
$$

Just consider $\text{var}(X + X)$, where $E[X] = 0$ as a simple example.

As explained previously, an alternative interpretation of $E[X]$ is as the best predictor of X. Recall that this predictor is the constant $b_{\text{opt}} = E[X]$ when the mean square error is used as a measure of error. We wish to point out that the *minimum* mse is then

$$
\begin{aligned}
\text{mse}_{\min} &= E[(X - b_{\text{opt}})^2] \\
&= E[(X - E[X])^2] \\
&= \text{var}(X).
\end{aligned}
\tag{6.9}
$$

Thus, how well we can predict the outcome of an experiment depends on the variance of the random variable. As an example, consider a coin toss with a probability of heads ($X = 1$) of p and of tails ($X = 0$) of $1 - p$, i.e., a Bernoulli random variable. We would predict the outcome of X to be $b_{\text{opt}} = E[X] = p$ and the minimum mse is the variance which from Example 6.4 is $\text{mse}_{\text{min}} = p(1-p)$. This is plotted in Figure 6.5 versus p. It is seen that the minimum mse is smallest when $p = 0$ or $p = 1$ and largest when $p = 1/2$, or most predictable for $p = 0$ and $p = 1$ and least predictable for $p = 1/2$. Can you explain this?

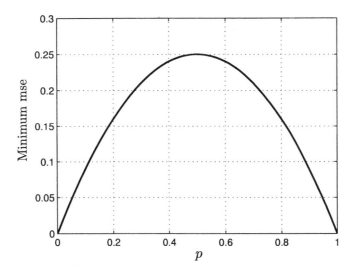

Figure 6.5: Measure of predictability of the outcome of a coin toss.

6.7 Characteristic Functions

Determining the moments $E[X^n]$ of a random variable can be a difficult task for some PMFs. An alternative method that can be considerably easier is based on the *characteristic function*. In addition, the characteristic function can be used to examine convergence of PMFs, as, for example, in the convergence of the binomial PMF to the Poisson PMF, and to determine the PMF for a sum of independent random variables, which will be examined in Chapter 7. In this section we discuss the use of the characteristic function for the calculation of moments and to investigate the convergence of a PMF.

The *characteristic function* of a random variable X is defined as

$$\phi_X(\omega) = E[\exp(j\omega X)] \tag{6.10}$$

where j is the square root of -1 and where ω takes on a suitable range of values. Note that the function $g(X) = \exp(j\omega X)$ is complex but by defining $E[g(X)] =$

$E[\cos(\omega X) + j\sin(\omega X)] = E[\cos(\omega X)] + jE[\sin(\omega X)]$, we can apply (6.5) to the real and imaginary parts of $\phi_X(\omega)$ to yield

$$
\begin{aligned}
\phi_X(\omega) &= E[\exp(j\omega X)] \\
&= E[\cos(\omega X) + j\sin(\omega X)] \\
&= E[\cos(\omega X)] + jE[\sin(\omega X)] \\
&= \sum_i \cos(\omega x_i)p_X[x_i] + j\sum_i \sin(\omega x_i)p_X[x_i] \\
&= \sum_i \exp(j\omega x_i)p_X[x_i]. \tag{6.11}
\end{aligned}
$$

To simplify the discussion, yet still be able to apply our results to the important PMFs, we assume that the sample space $\mathcal{S}_X$ is a subset of the integers. Then (6.11) becomes

$$
\phi_X(\omega) = \sum_{k=-\infty}^{\infty} \exp(j\omega k)p_X[k]
$$

or rearranging

$$
\phi_X(\omega) = \sum_{k=-\infty}^{\infty} p_X[k]\exp(j\omega k) \tag{6.12}
$$

where $p_X[k] = 0$ for those integers not included in $\mathcal{S}_X$. For example, in the Poisson PMF the range of summation in (6.12) would be $k \geq 0$. In this form, the characteristic function is immediately recognized as being the Fourier transform of the *sequence $p_X[k]$* for $-\infty < k < \infty$. Its definition is slightly different than the usual Fourier transform, called the discrete-time Fourier transform, which uses the function $\exp(-j\omega k)$ in its definition [Jackson 1991]. As a Fourier transform, it exhibits all the usual properties. In particular, the *Fourier transform* of a sequence is periodic with period of 2π (see Property 6.4 for a proof). As a result, we need only examine the characteristic function over the interval $-\pi \leq \omega \leq \pi$, which is defined to be the fundamental period. For our purposes the most useful property is that we can differentiate the sum in (6.12) "term by term" or

$$
\begin{aligned}
\frac{d\phi_X(\omega)}{d\omega} &= \frac{d}{d\omega}\sum_{k=-\infty}^{\infty} p_X[k]\exp(j\omega k) \\
&= \sum_{k=-\infty}^{\infty} p_X[k]\frac{d}{d\omega}\exp(j\omega k).
\end{aligned}
$$

The utility in doing so is to produce a formula for $E[X]$. Carrying out the differentiation

$$
\frac{d\phi_X(\omega)}{d\omega} = \sum_{k=-\infty}^{\infty} p_X[k]jk\exp(j\omega k)
$$

so that

$$\frac{1}{j}\frac{d\phi_X(\omega)}{d\omega}\bigg|_{\omega=0} = \sum_{k=-\infty}^{\infty} k p_X[k]$$

$$= E[X].$$

In fact, repeated differentiation produces the formula for the nth moment as

$$E[X^n] = \frac{1}{j^n}\frac{d^n \phi_X(\omega)}{d\omega^n}\bigg|_{\omega=0}. \tag{6.13}$$

All the moments that exist may be found by repeated differentiation of the characteristic function. An example follows.

Example 6.5 – First two moments of geometric random variable
Since the PMF for a geometric random variable is given by $p_X[k] = (1-p)^{k-1}p$ for $k = 1, 2, \ldots$, we have that

$$\phi_X(\omega) = \sum_{k=1}^{\infty} p_X[k]\exp(j\omega k)$$

$$= \sum_{k=1}^{\infty}(1-p)^{k-1}p\exp(j\omega k)$$

$$= p\exp(j\omega)\sum_{k=1}^{\infty}[(1-p)\exp(j\omega)]^{k-1}.$$

But since $|(1-p)\exp(j\omega)| < 1$, we can use the result

$$\sum_{k=1}^{\infty} z^{k-1} = \sum_{k=0}^{\infty} z^k = \frac{1}{1-z}$$

for z a complex number with $|z| < 1$ to yield the characteristic function

$$\phi_X(\omega) = \frac{p\exp(j\omega)}{1-[(1-p)\exp(j\omega)]}$$

$$= \frac{p}{\exp(-j\omega)-(1-p)}. \tag{6.14}$$

Note that as claimed the characteristic function is periodic with period 2π. To find the mean we use (6.13) with $n = 1$ to produce

$$E[X] = \frac{1}{j}\frac{d\phi_X(\omega)}{d\omega}\bigg|_{\omega=0}$$

$$= \frac{1}{j}p(-1)\frac{-j\exp(-j\omega)}{[\exp(-j\omega)-(1-p)]^2}\bigg|_{\omega=0} \tag{6.15}$$

$$= \frac{1}{j}p\frac{j}{p^2} = \frac{1}{p} \tag{6.16}$$

which agrees with our earlier results based on using the definition of expected value. To find the second moment and hence the variance using (6.8)

$$
\begin{aligned}
E[X^2] &= \frac{1}{j^2} \frac{d^2\phi_X(\omega)}{d\omega^2}\bigg|_{\omega=0} \\
&= \frac{p}{j} \frac{d}{d\omega} \frac{\exp(-j\omega)}{[\exp(-j\omega) - (1-p)]^2}\bigg|_{\omega=0} \qquad \text{(from (6.15))} \\
&= \frac{p}{j} \frac{D^2(-j)\exp(-j\omega) - \exp(-j\omega)2D(-j)\exp(-j\omega)}{D^4}\bigg|_{\omega=0}
\end{aligned}
$$

where $D = \exp(-j\omega) - (1-p)$. Since $D|_{\omega=0} = p$, we have that

$$
\begin{aligned}
E[X^2] &= \left(\frac{p}{j}\right)\left(\frac{-jp^2 + 2jp}{p^4}\right) \\
&= \frac{2p - p^2}{p^3} \\
&= \frac{2}{p^2} - \frac{1}{p}
\end{aligned}
$$

so that finally we have

$$
\begin{aligned}
\text{var}(X) &= E[X^2] - E^2[X] \\
&= \frac{2}{p^2} - \frac{1}{p} - \frac{1}{p^2} \\
&= \frac{1-p}{p^2}.
\end{aligned}
$$

$\Diamond$

As a second example, we consider the binomial PMF.

Example 6.6 – Expected value of binomial PMF

We first determine the characteristic function as

$$
\begin{aligned}
\phi_X(\omega) &= \sum_{k=-\infty}^{\infty} p_X[k]\exp(j\omega k) \\
&= \sum_{k=0}^{M}\binom{M}{k} p^k (1-p)^{M-k}\exp(j\omega k) \\
&= \sum_{k=0}^{M}\binom{M}{k}\left[\underbrace{p\exp(j\omega)}_{a}\right]^k \left[\underbrace{1-p}_{b}\right]^{M-k} \qquad (6.17) \\
&= (a+b)^M \qquad \text{(binomial theorem)} \\
&= [p\exp(j\omega) + (1-p)]^M. \qquad (6.18)
\end{aligned}
$$

The expected value then follows as

$$
\begin{aligned}
E[X] &= \frac{1}{j} \left. \frac{d\phi_X(\omega)}{d\omega} \right|_{\omega=0} \\
&= \frac{1}{j} M \left[p \exp(j\omega) + (1-p) \right]^{M-1} pj \exp(j\omega) \Big|_{\omega=0} \\
&= Mp
\end{aligned}
$$

which is in agreement with our earlier results. The variance can be found by using (6.8) and (6.13) for $n = 2$. It is left as an exercise to the reader to show that (see Problem 6.29)

$$
\text{var}(X) = Mp(1-p).
$$

$\diamond$

The characteristic function for the other important PMFs are given in Table 6.1. Some important properties of the characteristic function are listed next.

Property 6.3 – Characteristic function always exists since $|\phi_X(\omega)| < \infty$
Proof:

$$
\begin{aligned}
|\phi_X(\omega)| &= \left| \sum_{k=-\infty}^{\infty} p_X[k] \exp(j\omega k) \right| \\
&\leq \sum_{k=-\infty}^{\infty} |p_X[k] \exp(j\omega k)| \qquad \text{(magnitude of sum of complex numbers} \\
&\qquad\qquad\qquad\qquad\qquad\qquad \text{cannot exceed sum of magnitudes)} \\
&= \sum_{k=-\infty}^{\infty} |p_X[k]| \qquad\qquad (|\exp(j\omega k)| = 1) \\
&= \sum_{k=-\infty}^{\infty} p_X[k] \qquad\qquad (p_X[k] \geq 0) \\
&= 1.
\end{aligned}
$$

$\square$

Property 6.4 – Characteristic function is periodic with period 2π.
Proof: For m an integer

$$
\begin{aligned}
\phi_X(\omega + 2\pi m) &= \sum_{k=-\infty}^{\infty} p_X[k] \exp[j(\omega + 2\pi m)k] \\
&= \sum_{k=-\infty}^{\infty} p_X[k] \exp[j\omega k] \exp[j2\pi m k] \\
&= \sum_{k=-\infty}^{\infty} p_X[k] \exp[j\omega k] \qquad \text{(since } \exp(j2\pi m k) = 1 \\
&\qquad\qquad\qquad\qquad\qquad\qquad \text{for } mk \text{ an integer)} \\
&= \phi_X(\omega).
\end{aligned}
$$

$\square$

Property 6.5 –   The PMF may be recovered from the characteristic function.

Given the characteristic function, we may determine the PMF using

$$p_X[k] = \int_{-\pi}^{\pi} \phi_X(\omega) \exp(-j\omega k) \frac{d\omega}{2\pi} \qquad -\infty < k < \infty. \qquad (6.19)$$

Proof: Since the characteristic function is the Fourier transform of a sequence (although its definition uses a $+j$ instead of the usual $-j$), it has an inverse Fourier transform. Although any interval of length 2π may be used to perform the integration in the inverse Fourier transform, it is customary to use $[-\pi, \pi]$ which results in (6.19).

□

Property 6.6 –   Convergence of characteristic functions guarantees convergence of PMFs.

This property says that if we have a sequence of characteristic functions, say $\phi_X^{(n)}(\omega)$, which converges to a given characteristic function, say $\phi_X(\omega)$, then the corresponding sequence of PMFs, say $p_X^{(n)}[k]$, must converge to a given PMF say $p_X[k]$, where $p_X[k]$ is given by (6.19). The importance of this theorem is that it allows us to approximate PMFs by simpler ones if we can show that the characteristic functions are approximately equal. An illustration is given next. This theorem is known at the *continuity theorem of probability*. Its proof is beyond the scope of this text but can be found in [Pollard 2002].

□

We recall the approximation of the binomial PMF by the Poisson PMF under the conditions that $p \to 0$ and $M \to \infty$ with $Mp = \lambda$ fixed (see Section 5.6). To show this using the characteristic function approach (based on Property 6.6) we let X_b denote a binomial random variable. Its characteristic function is from (6.18)

$$\phi_{X_b}(\omega) = [p \exp(j\omega) + (1 - p)]^M$$

and replacing p by λ/M we have

$$\begin{aligned} \phi_{X_b}(\omega) &= \left[\frac{\lambda}{M} \exp(j\omega) + \left(1 - \frac{\lambda}{M}\right)\right]^M \\ &= \left[1 + \frac{\lambda(\exp(j\omega) - 1)}{M}\right]^M \\ &\to \exp[\lambda(\exp(j\omega) - 1)] \qquad \text{(see Problem 5.15, results are also valid for a complex variable)} \end{aligned}$$

as $M \to \infty$. For a Poisson random variable X_P we have that

$$
\begin{aligned}
\phi_{X_P}(\omega) &= \sum_{k=0}^{\infty} \exp(-\lambda) \frac{\lambda^k}{k!} \exp(j\omega k) \\
&= \exp(-\lambda) \sum_{k=0}^{\infty} \frac{[\lambda \exp(j\omega)]^k}{k!} \\
&= \exp(-\lambda) \exp[\lambda \exp(j\omega)] \qquad \text{(using results from Problem 5.22 which also hold for a complex variable)} \\
&= \exp[\lambda(\exp(j\omega) - 1)].
\end{aligned}
$$

Since $\phi_{X_b}(\omega) \to \phi_{X_P}(\omega)$ as $M \to \infty$, by Property 6.6, we must have that $p_{X_b}[k] \to p_{X_P}[k]$ for all k. Hence, under the stated conditions the binomial PMF becomes the Poisson PMF as $M \to \infty$. This was previously proven by other means in Section 5.6. Our derivation here though is considerably simpler.

6.8 Estimating Means and Variances

As alluded to earlier, an important aspect of the mean and variance of a PMF is that they are easily estimated in practice. We have already briefly discussed this in Chapter 2 where it was demonstrated how to do this with computer simulated data (see Example 2.3). We now continue that discussion in more detail. To illustrate the approach we will consider the PMF shown in Figure 6.6a. Since the theoretical

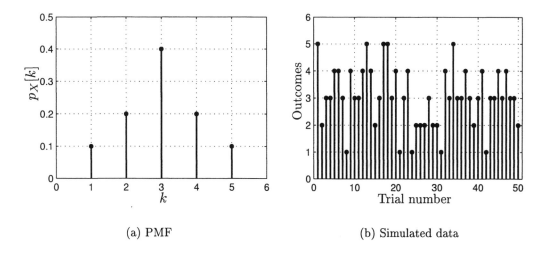

(a) PMF   (b) Simulated data

Figure 6.6: PMF and computer generated data used to illustrate estimation of mean and variance.

expected value or mean is given by

$$E[X] = \sum_{k=1}^{5} k p_X[k]$$

then by the relative frequency interpretation of probability we can use the approximation

$$p_X[k] \approx \frac{N_k}{N}$$

where N_k is the number of trials in which a k was the outcome and N is the total number of trials. As a result, we can estimate the mean by

$$\widehat{E[X]} = \sum_{k=1}^{5} \frac{k N_k}{N}.$$

The "hat" will always denote an estimated quantity. But $k N_k$ is just the sum of all the k outcomes that appear in the N trials and therefore $\sum_{k=1}^{5} k N_k$ is the sum of *all* the outcomes in the N trials. Denoting the latter by $\sum_{i=1}^{N} x_i$, we have as our estimate of the mean

$$\widehat{E[X]} = \frac{1}{N} \sum_{i=1}^{N} x_i \tag{6.20}$$

where x_i is the outcome of the ith trial. Note that we have just reversed our line of reasoning used in the introduction to motivate the use of $E[X]$ as the definition of the expected value of a random variable. Also, we have previously seen this type of estimate in Example 2.3 where it was referred to as the *sample mean*. It is usually denoted by $\bar{x}$. For the data shown in Figure 6.6b we plot the sample mean in Figure 6.7a versus N. Note that as N becomes larger, we have that $\widehat{E[X]} \to 3 = E[X]$.

The true variance of the PMF shown in Figure 6.6a is computed as

$$
\begin{aligned}
\text{var}(X) &= E[X^2] - E^2[X] \\
&= \sum_{k=1}^{5} k^2 p_X[k] - E^2[X]
\end{aligned}
$$

which is easily shown to be $\text{var}(X) = 1.2$. It is estimated as

$$\widehat{\text{var}(X)} = \widehat{E[X^2]} - (\widehat{E[X]})^2$$

and by the same rationale as before we use

$$\widehat{E[X^2]} = \frac{1}{N} \sum_{i=1}^{N} x_i^2$$

so that our estimate of the variance becomes

$$\widehat{\text{var}(X)} = \frac{1}{N}\sum_{i=1}^{N} x_i^2 - \left(\frac{1}{N}\sum_{i=1}^{N} x_i\right)^2. \tag{6.21}$$

This estimate is shown in Figure 6.7b as a function of N. Note that as the number of

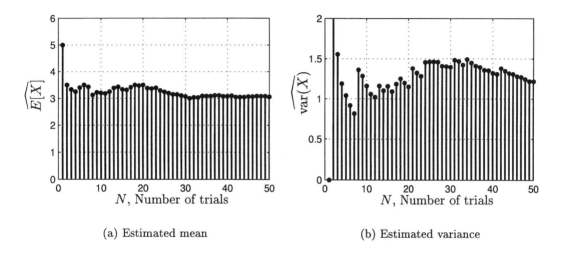

(a) Estimated mean   (b) Estimated variance

Figure 6.7: Estimated mean and variance for computer data shown in Figure 6.6.

trials increases the estimate of variance converges to the true value of $\text{var}(X) = 1.2$. The MATLAB code used to generate the data and estimate the mean and variance is given in Appendix 6B. Also, in that appendix is listed the MATLAB subprogram PMFdata.m which allows easier generation of the outcomes of a discrete random variable. In practice, it is customary to use (6.20) and (6.21) to analyze real-world data as a first step in assessing the characteristics of an unknown PMF.

6.9   Real-World Example – Data Compression

The digital revolution of the past 20 years has made it commonplace to record and store information in a digital format. Such information consists of speech data in telephone transmission, music data stored on compact discs, video data stored on digital video discs, and facsimile data, to name but a few. The amount of data can become quite large so that it is important to be able to reduce the amount of storage required. The process of storage reduction is called *data compression*. We now illustrate how this is done. To do so we simplify the discussion by assuming that the data consists of a sequence of the letters A, B, C, D. One could envision these letters as representing the chords of a rudimentary musical instrument, for example. The extension to the entire English alphabet consisting of 26 letters will be apparent. Consider a typical sequence of 50 letters

$$\text{AAAAAAAAAAABAAAAAAAAAAAA} \\ \text{AAAAAACABADAABAAABAAAAAAD}. \qquad (6.22)$$

To encode these letters for storage we could use the two-bit code

$$
\begin{array}{ccc}
A & \rightarrow & 00 \\
B & \rightarrow & 01 \\
C & \rightarrow & 10 \\
D & \rightarrow & 11
\end{array}
\qquad (6.23)
$$

which would then require a storage of 2 bits per letter for a total storage of 100 bits. However, as seen above the typical sequence is characterized by a much larger probability of observing an "A" as opposed to the other letters. In fact, there are 43 A's, 4 B's, 1 C, and 2 D's. It makes sense then to attempt a reduction in storage by assigning *shorter code words* to the letters that *occur more often*, in this case, to the "A". As a possible strategy, consider the code assignment

$$
\begin{array}{ccc}
A & \rightarrow & 0 \\
B & \rightarrow & 10 \\
C & \rightarrow & 110 \\
D & \rightarrow & 111.
\end{array}
\qquad (6.24)
$$

Using this code assignment for our typical sequence would require only $1 \cdot 43 + 2 \cdot 4 + 3 \cdot 1 + 3 \cdot 2 = 60$ bits or 1.2 bits per letter. The code given by (6.24) is called a *Huffman code*. It can be shown to produce less bits per letter "on the average" [Cover, Thomas 1991].

   To determine actual storage savings we need to determine the *average length* of the code word per letter. First we define a discrete random variable that measures the length of the code word. For the sample space $\mathcal{S} = \{A, B, C, D\}$ we define the random variable

$$
X(s_i) = \begin{cases}
1 & s_1 = A \\
2 & s_2 = B \\
3 & s_3 = C \\
3 & s_4 = D
\end{cases}
$$

which yields the code length for each letter. The probabilities used to generate the sequence of letters shown in (6.22) are $P[A] = 7/8$, $P[B] = 1/16$, $P[C] = 1/32$, $P[D] = 1/32$. As a result the PMF for X is

$$
p_X[k] = \begin{cases}
\frac{7}{8} & k = 1 \\
\frac{1}{16} & k = 2 \\
\frac{1}{16} & k = 3.
\end{cases}
$$

The average code length is given by

$$
\begin{aligned}
E[X] &= \sum_{k=1}^{3} k p_X[k] \\
&= 1 \cdot \frac{7}{8} + 2 \cdot \frac{1}{16} + 3 \cdot \frac{1}{16} \\
&= 1.1875 \text{ bits per letter.}
\end{aligned}
$$

This results in a compression ratio of $2 : 1.1875 = 1.68$ or we require about 40% less storage.

It is also of interest to note that the average code word length per letter can be reduced even further. However, it requires more complexity in coding (and of course in decoding). A fundamental theorem due to Shannon, who in many ways laid the groundwork for the digital revolution, says that the average code word length per letter can be no less than [Shannon 1948]

$$
H = \sum_{i=1}^{4} P[s_i] \log_2 \frac{1}{P[s_i]} \qquad \text{bits per letter.} \tag{6.25}
$$

This quantity is termed the *entropy* of the source. In addition, he showed that a code exists that can attain, to within any small deviation, this minimum average code length. For our example, the entropy is

$$
\begin{aligned}
H &= \frac{7}{8} \log_2 \frac{1}{7/8} + \frac{1}{16} \log_2 \frac{1}{1/16} + \frac{1}{32} \log_2 \frac{1}{1/32} + \frac{1}{32} \log_2 \frac{1}{1/32} \\
&= 0.7311 \qquad \text{bits per letter.}
\end{aligned}
$$

Hence, the potential compression ratio is $2 : 0.7311 = 2.73$ for about a 63% reduction.

Clearly, it is seen from this example that the amount of reduction will depend critically upon the probabilities of the letters occuring. If they are all equally likely to occur, then the minimum average code length is from (6.25) with $P[s_i] = 1/4$

$$
H = 4 \left(\frac{1}{4} \log_2 \frac{1}{1/4} \right) = 2 \qquad \text{bits per letter.}
$$

In this case no compression is possible and the original code given by (6.23) will be optimal. The interested reader should consult [Cover and Thomas 1991] for further details.

References

Cover, T.M., J.A. Thomas, *Elements of Information Theory*, John Wiley & Sons, New York, 1991.

Gaughan, E.D., *Introduction to Analysis*, Brooks/Cole, Monterey, CA, 1975.

Jackson, L.B., *Signals, Systems, and Transforms*, Addison-Wesley, Reading, MA, 1991.

Pollard., D. *A User's Guide to Measure Theoretic Probability*, Cambridge University Press, New York, 2002.

Shannon, C.E., "A Mathematical Theory of Communication," *Bell System Tech. Journal*, Vol. 27, pp. 379–423, 623–656, 1948.

Problems

6.1 (w) The center of mass of a system of masses situated on a line is the point at which the system is balanced. That is to say that at this point the sum of the moments, where the moment is the distance from center of mass times the mass, is zero. If the center of mass is denoted by CM, then

$$\sum_{i=1}^{M}(x_i - \text{CM})m_i = 0$$

where x_i is the position of the ith mass along the x direction and m_i is its corresponding mass. First solve for CM. Then, for the system of weights shown in Figure 6.8 determine the center of mass. How is this analogous to the expected value of a discrete random variable?

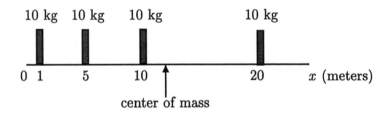

Figure 6.8: Weightless bar supporting four weights.

6.2 (☺) (f) For the discrete random variable with PMF

$$p_X[k] = \frac{1}{10} \qquad k = 0, 1, \ldots, 9$$

find the expected value of X.

6.3 (w) A die is tossed. The probability of obtaining a 1, 2, or 3 is the same. Also, the probability of obtaining a 4, 5, or 6 is the same. However, a 5 is twice as likely to be observed as a 1. For a large number of tosses what is the average value observed?

6.4 (☺) (f) A coin is tossed with the probability of heads being 2/3. A head is mapped into $X = 1$ and a tail into $X = 0$. What is the expected outcome of this experiment?

6.5 (f) Determine the expected value of a Poisson random variable. Hint: Differentiate $\sum_{k=0}^{\infty} \lambda^k / k!$ with respect to λ.

6.6 (t) Consider the PMF $p_X[k] = (2/\pi)/k^2$ for $k = \ldots, -1, 0, 1, \ldots$. The expected value is defined as

$$E[X] = \sum_{k=-\infty}^{\infty} k p_X[k]$$

which is actually shorthand for

$$E[X] = \lim_{\substack{N_L \to -\infty \\ N_U \to \infty}} \sum_{k=N_L}^{N_U} k p_X[k]$$

where the L and U represent "lower" and "upper", respectively. This may be written as

$$E[X] = \lim_{N_L \to -\infty} \sum_{k=N_L}^{-1} k p_X[k] + \lim_{N_U \to \infty} \sum_{k=1}^{N_U} k p_X[k]$$

where the limits are taken *independently* of each other. For $E[X]$ to be unambiguous and finite both limits must be finite. As a result, show that the expected value for the given PMF does not exist. If, however, we were to constrain $N_L = N_U$, show that the expected value is zero. Note that if $N_L = N_U$, we are reordering the terms before performing the sum since the partial sums become $\sum_{k=-1}^{1} k p_X[k]$, $\sum_{k=-2}^{2} k p_X[k]$, etc. But for the expected value to be unambiguous, the value should not depend on the ordering. If a sum is *absolutely* summable, any ordering will produce the same result [Gaughan 1975], hence our requirement for the existence of the expected value.

6.7 (t) Assume that a discrete random variable takes on the values $k = \ldots, -1, 0, 1, \ldots$ and that its PMF satisfies $p_X[m + i] = p_X[m - i]$, where m is a fixed integer and $i = 1, 2, \ldots$. This says that the PMF is symmetric about the point $x = m$. Prove that the expected value of the random variable is $E[X] = m$.

6.8 (☺) (t) Give an example where the expected value of a random variable is *not* its most probable value.

6.9 (t) Give an example of two PMFs that have the same expected value.

6.10 (f) A discrete random variable X has the PMF $p_X[k] = 1/5$ for $k = 0, 1, 2, 3, 4$. If $Y = \sin[(\pi/2)X]$, find $E[Y]$ using (6.4) and (6.5). Which way is easier?

6.11 (t) Prove the linearity property of the expectation operator

$$E[a_1 g_1(X) + a_2 g_2(X)] = a_1 E[g_1(X)] + a_2 E[g_2(X)]$$

where a_1 and a_2 are constants.

6.12 (☺) (f) Determine $E[X^2]$ for a geom(p) random variable using (6.5). Hint: You will need to differentiate twice.

6.13 (☺) (t) Can $E[X^2]$ ever be equal to $E^2[X]$? If so, when?

6.14 (☺) (w) A discrete random variable X has the PMF

$$p_X[k] = \begin{cases} \frac{1}{8} & k = 1 \\ \frac{2}{8} & k = 2 \\ \frac{4}{8} & k = 3 \\ \frac{1}{8} & k = 4. \end{cases}$$

If the experiment that produces a value of X is conducted, find the minimum mean square error predictor of the outcome. What is the minimum mean square error of the predictor?

6.15 (☺) (c) For Problem 6.14 use a computer to simulate the experiment for many trials. Compare the estimate to the actual outcomes of the computer experiment. Also, compute the minimum mean square error and compare it to the theoretical value obtained in Problem 6.14.

6.16 (w) Of the three PMFs shown in Figure 6.9, which one has the smallest variance? Hint: You do not need to actually calculate the variances.

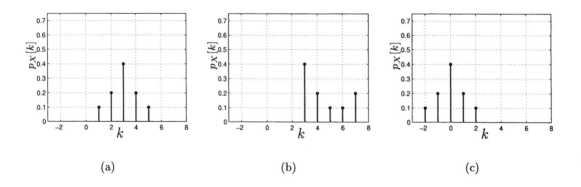

Figure 6.9: PMFs for Problem 6.16.

6.17 (w) If $Y = aX + b$, what is the variance of Y in terms of the variance of X?

6.18 (f) Find the variance of a Poisson random variable. See the hint for Problem 6.12.

6.19 (f) For the PMF given in Problem 6.2 find the variance.

6.20 (☺) (f) Find the second moment for a Poisson random variable by using the characteristic function, which is given in Table 6.1.

6.21 (t) If X is a discrete random variable and c is a constant, prove the following properties of the variance:

$$\begin{aligned} \text{var}(c) &= 0 \\ \text{var}(X + c) &= \text{var}(X) \\ \text{var}(cX) &= c^2\text{var}(X). \end{aligned}$$

6.22 (t) If a discrete random variable X has $\text{var}(X) = 0$, prove that X must be a constant c. This provides a converse to the property that if $X = c$, then $\text{var}(X) = 0$.

6.23 (t) In this problem we prove that if $E[X^s]$ exists, meaning that $E[|X|^s] < \infty$, then $E[X^r]$ also exists for $0 < r < s$. Provide the explanations for the following steps:

a. For $|x| \leq 1$, $|x|^r \leq 1$

b. For $|x| > 1$, $|x|^r \leq |x|^s$

c. For all $|x|$, $|x|^r \leq |x|^s + 1$

d. $E[|X|^r] = \sum_i |x_i|^r p_X[x_i] \leq \sum_i (|x_i|^s + 1) p_X[x_i] = E[|X|^s] + 1 < \infty.$

6.24 (f) If a discrete random variable has the PMF $p_X[k] = 1/4$ for $k = -1$ and $p_X[k] = 3/4$ for $k = 1$, find the mean and variance.

6.25 (t) A symmetric PMF satisfies the relationship $p_X[-k] = p_X[k]$ for $k = \ldots, -1, 0, 1, \ldots$. Prove that all the odd order moments, $E[X^n]$ for n odd, are zero.

6.26 (☺) (t) A central moment of a discrete random variable is defined as $E[(X - E[X])^n]$, for n a positive integer. Derive a formula that relates the central moment to the usual moments. Hint: You will need the binomial formula.

6.27 (☺) (t) If $Y = aX + b$, find the characteristic function of Y in terms of that for X. Next use your result to prove that $E[Y] = aE[X] + b$.

6.28 (☺) (f) Find the characteristic function for the PMF $p_X[k] = 1/5$ for $k = -2, -1, 0, 1, 2$.

6.29 (f) Determine the variance of a binomial random variable by using the properties of the characteristic function. You can assume knowledge of the characteristic function for a binomial random variable.

6.30 (f) Determine the mean and variance of a Poisson random variable by using the properties of the characteristic function. You can assume knowledge of the characteristic function for a Poisson random variable.

6.31 (f) Which PMF $p_X[k]$ for $k = \ldots, -1, 0, 1, \ldots$ has the characteristic function $\phi_X(\omega) = \cos\omega$?

6.32 ($\smile$) (c) For the random variable described in Problem 6.24 perform a computer simulation to estimate its mean and variance. How does it compare to the true mean and variance?

Appendix 6A

Derivation of $E[g(X)]$ Formula

Assume that X is a discrete random variable taking on values in $\mathcal{S}_X = \{x_1, x_2, \ldots\}$ with PMF $p_X[x_i]$. Then, if $Y = g(X)$ we have from the definition of expected value

$$E[Y] = \sum_i y_i p_Y[y_i] \tag{6A.1}$$

where the sum is over all $y_i \in \mathcal{S}_Y$. Note that it is assumed that the y_i are *distinct* (all different). But from (5.9)

$$p_Y[y_i] = \sum_{\{x_j : g(x_j) = y_i\}} p_X[x_j]. \tag{6A.2}$$

To simplify the notation we will define the *indicator function*, which indicates whether a number x is within a given set A, as

$$I_A(x) = \begin{cases} 1 & x \in A \\ 0 & \text{otherwise.} \end{cases}$$

Then (6A.2) can be rewritten as

$$p_Y[y_i] = \sum_{j=1}^{\infty} p_X[x_j] I_{\{0\}}(y_i - g(x_j))$$

since the sum will include the term $p_X[x_j]$ only if $y_i - g(x_j) = 0$. Using this, we have from (6A.1)

$$
\begin{aligned}
E[Y] &= \sum_i y_i \sum_{j=1}^{\infty} p_X[x_j] I_{\{0\}}(y_i - g(x_j)) \\
&= \sum_{j=1}^{\infty} \left[\sum_i y_i I_{\{0\}}(y_i - g(x_j)) \right] p_X[x_j].
\end{aligned}
$$

Now for a given j, $g(x_j)$ is a fixed number and since the y_i's are distinct, there is only one y_i for which $y_i = g(x_j)$. Thus, we have that

$$\sum_i y_i I_{\{0\}}(y_i - g(x_j)) = g(x_j)$$

and finally

$$E[Y] = E[g(X)] = \sum_{j=1}^{\infty} g(x_j) p_X[x_j].$$

Appendix 6B

MATLAB Code Used to Estimate Mean and Variance

Figures 6.6 and 6.7 are based on the following MATLAB code.

```
%  PMFdata.m
%
%  This program generates the outcomes for N trials
%  of an experiment for a discrete random variable.
%  Uses the method of Section 5.9.
%  It is a function subprogram.
%
%  Input parameters:
%
%     N   - number of trials desired
%     xi  - values of x_i's of discrete random variable (M x 1 vector)
%     pX  - PMF of discrete random variable (M x 1 vector)
%
%  Output parameters:
%
%     x   - outcomes of N trials (N x 1 vector)
%
function x=PMFdata(N,xi,pX)
M=length(xi);M2=length(pX);
if M~=M2
   message='xi and pX must have the same dimension'
end
for k=1:M ; % see Section 5.9 and Figure 5.14 for approach used here
   if k==1
```

```
      bin(k,1)=pX(k); % set up first interval of CDF as [0,pX(1)]

   else
      bin(k,1)=bin(k-1,1)+pX(k); % set up succeeding intervals
                                 % of CDF
   end
end
u=rand(N,1); % generate N outcomes of uniform random variable
for i=1:N % determine which interval of CDF the outcome lies in
          % and map into value of xi
   if u(i)>0&u(i)<=bin(1)
      x(i,1)=xi(1);
   end
   for k=2:M
      if u(i)>bin(k-1)&u(i)<=bin(k)
         x(i,1)=xi(k);
      end
   end
end
```

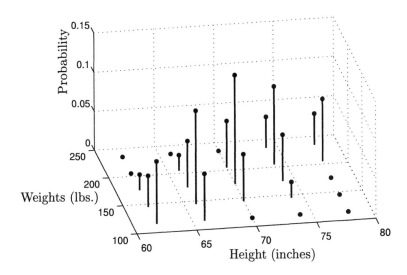

Figure 7.1: Joint probabilities for heights and weights of college students.

tended to any finite number of random variables (see Chapter 9 for this extension). As we will see throughout our discussions, the new and very important concept will be the *dependencies* between the multiple random variables. Questions such as "Can we predict a person's height from his weight?" naturally arise and can be addressed once we extend our description of a single random variable to multiple random variables.

7.2   Summary

The concept of jointly distributed discrete random variables is illustrated in Figure 7.2. Two random variables can be thought of as a random vector and assigned a joint PMF $p_{X,Y}[x_i, y_j]$ as described in Section 7.3, and which has Properties 7.1 and 7.2. The joint PMF may be obtained if the probabilities on the original experimental sample space is known by using (7.2), and is illustrated in Example 7.1. Once the joint PMF is specified, the probability of any event concerning the random variables is determined via (7.3). The marginal PMFs of the two random variables, which are the probabilities of each random variable taking on its possible values, is obtained from the joint PMF using (7.5) and (7.6). However, the joint PMF is not uniquely determined from the marginal PMFs. The joint CDF is defined by (7.7) and evaluated using (7.8). It has the usual properties as summarized via Properties 7.3–7.6. Random variables are defined to be independent if the probabilities of all the joint events can be found as the product of the probabilities of the single events. If the random variables are independent, then the joint PMF factors as in (7.11). Given a joint PMF, independence can be established by determining if the PMF factors. Conversely, if we know the random variables are independent, and

Chapter 7

Multiple Discrete Random Variables

7.1 Introduction

In Chapter 5 we introduced the concept of a discrete random variable as a mapping from the sample space $\mathcal{S} = \{s_i\}$ to a countable set of real numbers (either finite or countably infinite) via a mapping $X(s_i)$. In effect, the mapping yields useful *numerical* information about the outcome of the random phenomenon. In some instances, however, we would like to measure more than just one attribute of the outcome. For example, consider the choice of a student at random from a population of college students. Then, for the purpose of assessing the student's health we might wish to know his/her height, weight, blood pressure, pulse rate, etc. All these measurements and others are used by a physician for a disease risk assessment. Hence, the mapping from the sample space of college students to the important measurements of height and weight, for example, would be $H(s_i) = h_i$ and $W(s_i) = w_i$, where H and W represent the height and weight of the student selected. In Table 4.1 we summarized a hypothetical set of probabilities for heights and weights. The table is a two-dimensional array that lists the probabilities $P[H = h_i \text{ and } W = w_j]$. This information can also be displayed in a three-dimensional format as shown in Figure 7.1, where we have associated the center point of each interval of height and weight given in Table 4.1 with the probability displayed. These probabilities were termed *joint probabilities*. In this chapter we discuss the case of *multiple random variables*. For example, the height and weight could be represented as a 2×1 *random vector*

$$\begin{bmatrix} H \\ W \end{bmatrix}$$

and as such, its value is located in the plane (also called R^2). We will initially describe the simplest case of two random variables but all concepts are easily exwe are given the marginal PMFs, then the joint PMF is found as the product of the marginals. The joint PMF of a transformed vector random variable is given by (7.12) and illustrated in Example 7.6. The PMF for the sum of two independent discrete random variables can be found using (7.22) or via characteristic functions using (7.24). The expected value of a function of two random variables is found from (7.28). Also, the variance of the sum of two random variables is given by (7.33) and involves the covariance, which is defined by (7.34). The interpretation of the covariance is given in Section 7.8 and is seen to provide a quantification of the knowledge of the outcome of one random variable on the probability of the other. Independent random variables have a covariance of zero, but the converse is not true. In Section 7.9 linear prediction of one random variable based on observation of another random variable is explored. The optimal linear predictor is given by (7.41). A variation of this prediction equation results in the important parameter called the correlation coefficient (7.43). It quantifies the relationship of one random variable with another. However, a nonzero correlation does not indicate a causal relationship. The joint characteristic function is introduced in Section 7.10 and is defined by (7.45) and evaluated by (7.46). It is shown to provide a convenient means of determining the PMF for a sum of independent random variables. In Section 7.11 a method to simulate a random vector is described. Also, methods to estimate joint PMFs, marginal PMFs, and other quantities of interest are given. Finally, in Section 7.12 an application of the methods of the chapter to disease risk assessment is described.

7.3 Jointly Distributed Random Variables

We consider two discrete random variables that will be denoted by X and Y. As alluded to in the introduction, they represent the functions that map an outcome of an experiment s_i to a value in the plane. Hence, we have the mapping

$$
\left[\begin{array}{c} X(s_i) \\ Y(s_i) \end{array} \right] = \left[\begin{array}{c} x_i \\ y_i \end{array} \right]
$$

for all $s_i \in S$. An example is shown in Figure 7.2 in which the experiment consists of the simultaneous tossing of a penny and a nickel. The outcome in the sample space S is represented by a TH, for example, if the penny comes up tails and the nickel comes up heads. Explicitly, the mapping is

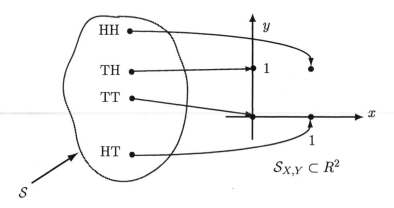

Figure 7.2: Example of mapping for jointly distributed discrete random variables.

$$
\begin{bmatrix} X(s_i) \\ Y(s_i) \end{bmatrix} = \begin{cases} \begin{bmatrix} 0 \\ 0 \end{bmatrix} & \text{if } s_i = \text{TT} \\[2ex] \begin{bmatrix} 0 \\ 1 \end{bmatrix} & \text{if } s_i = \text{TH} \\[2ex] \begin{bmatrix} 1 \\ 0 \end{bmatrix} & \text{if } s_i = \text{HT} \\[2ex] \begin{bmatrix} 1 \\ 1 \end{bmatrix} & \text{if } s_i = \text{HH.} \end{cases}
$$

Two random variables that are defined on the *same sample space S* are said to be *jointly distributed*. In this example, the random variables are also *discrete* random variables in that the possible values (which are actually 2×1 vectors) are countable. In this case there are just four vector values. These values comprise the sample space which is the subset of the plane given by

$$
\mathcal{S}_{X,Y} = \left\{ \begin{bmatrix} 0 \\ 0 \end{bmatrix}, \begin{bmatrix} 0 \\ 1 \end{bmatrix}, \begin{bmatrix} 1 \\ 0 \end{bmatrix}, \begin{bmatrix} 1 \\ 1 \end{bmatrix} \right\}.
$$

We can also refer to the two random variables as the single *random vector* $[X\,Y]^T$, where T denotes the vector transpose. Hence, we will use the terms *multiple random variables* and *random vector* interchangeably. The values of the random vector will be denoted either by (x, y), which is an ordered pair or a point in the plane, or by $[x\,y]^T$, which denotes a two-dimensional vector. These notations will be synonomous.

The size of the sample space for discrete random variables can be finite or countably infinite. In the example of Figure 7.2, since X can take on 2 values, denoted by $N_X = 2$, and Y can take on 2 values, denoted by $N_Y = 2$, the total number of elements in $\mathcal{S}_{X,Y}$ is $N_X N_Y = 4$. More generally, if X can take on values in $\mathcal{S}_X = \{x_1, x_2, \ldots, x_{N_X}\}$ and Y can take on values in $\mathcal{S}_Y = \{y_1, y_2, \ldots, y_{N_Y}\}$, then the random vector can take on values in

$$\mathcal{S}_{X,Y} = \mathcal{S}_X \times \mathcal{S}_Y = \{(x_i, y_j) : i = 1, 2, \ldots, N_X; j = 1, 2, \ldots, N_Y\}$$

for a total of $N_{X,Y} = N_X N_Y$ values. This is shown in Figure 7.3 for the case of $N_X = 4$ and $N_Y = 3$. The notation $A \times B$, where A and B are sets, denotes a *cartesian product set*. It consists of all ordered pairs (a_i, b_j), where $a_i \in A$ and $b_j \in B$. If either $\mathcal{S}_X$ or $\mathcal{S}_Y$ is countably infinite, then the random vector will also have a countably infinite set of values.

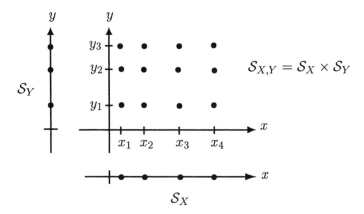

Figure 7.3: Example of sample space for jointly distributed discrete random variables.

Just as we defined the PMF for a single discrete random variable in Chapter 5 as $p_X[x_i] = P[X(s) = x_i]$, we can define the *joint PMF* (or sometimes called the *bivariate* PMF) as

$$p_{X,Y}[x_i, y_j] = P[X(s) = x_i, Y(s) = y_j] \quad i = 1, 2, \ldots, N_X; j = 1, 2, \ldots, N_Y.$$

Note that the set of all outcomes s for which $X(s) = x_i, Y(s) = y_j$ is the same as the set of outcomes for which

$$\begin{bmatrix} X(s) \\ Y(s) \end{bmatrix} = \begin{bmatrix} x_i \\ y_j \end{bmatrix}$$

so that for the random vector to have the value $[x_i \, y_j]^T$, both $X(s) = x_i$ and $Y(s) = y_j$ must be satisfied. Thus, the *comma* used in the statement $X(s) = x_i, Y(s) = y_j$ is to be read as "and". An example of the joint PMF for students' heights and weights is given in Figure 7.1 in which we set $X = $ height and $Y = $ weight and the vertical axis represents $p_{X,Y}[x_i, y_j]$. To verify that a set of probabilities as in Figure 7.1 can be viewed as a joint PMF we need only verify the usual properties of probability. Assuming N_X and N_Y are finite, these are:

Property 7.1 – Range of values of joint PMF

$$0 \le p_{X,Y}[x_i, y_j] \le 1 \qquad i = 1, 2, \ldots, N_X; j = 1, 2, \ldots, N_Y.$$

$\square$

Property 7.2 – Sum of values of joint PMF

$$\sum_{i=1}^{N_X} \sum_{j=1}^{N_Y} p_{X,Y}[x_i, y_j] = 1$$

$\square$

and similarly for a countably infinite sample space. For the coin toss example of Figure 7.2 we require that

$$0 \le p_{X,Y}[0,0] \le 1$$
$$0 \le p_{X,Y}[0,1] \le 1$$
$$0 \le p_{X,Y}[1,0] \le 1$$
$$0 \le p_{X,Y}[1,1] \le 1$$

and

$$\sum_{i=0}^{1} \sum_{j=0}^{1} p_{X,Y}[i,j] = 1.$$

Many possibilities exist. For two fair coins that do not interact as they are tossed (i.e., they are independent) we might assign $p_{X,Y}[i,j] = 1/4$ for all i and j. For two coins that are weighted but again do not interact with each other as they are tossed, we might assign

$$p_{X,Y}[i,j] = \begin{cases} (1-p)^2 & i = 0, j = 0 \\ (1-p)p & i = 0, j = 1 \\ p(1-p) & i = 1, j = 0 \\ p^2 & i = 1, j = 1 \end{cases}$$

if each coin has a probability of heads of p. It is easily shown that the joint PMF satisfies Properties 7.1 and 7.2 for any $0 \le p \le 1$. In obtaining these values for the joint PMF we have used the concept of equivalent events, which allows us to determine probabilities for events defined on $\mathcal{S}_{X,Y}$ from those defined on the original sample space $\mathcal{S}$. For example, since the events TH and $(0, 1)$ are equivalent as seen in Figure 7.2, we have that

$$
\begin{aligned}
p_{X,Y}[0,1] &= P[X(s) = 0, Y(s) = 1] \\
&= P[\{s_i : X(s_i) = 0, Y(s_i) = 1\}] \quad &\text{(equivalent event in } \mathcal{S}) \\
&= P[s_i = \text{TH}] \quad &\text{(mapping is one-to-one)} \\
&= (1-p)p \quad &\text{(independence)}
\end{aligned}
$$

where we have assumed independence of the penny and nickel toss subexperiments as described in Section 4.6.1.

In general, the procedure to determine the joint PMF from the probabilities defined on $\mathcal{S}$ depends on whether the random variable mapping is one-to-one or many-to-one. For a one-to-one mapping from $\mathcal{S}$ to $\mathcal{S}_{X,Y}$ we have

$$
\begin{aligned}
p_{X,Y}[x_i, y_j] &= P[X(s) = x_i, Y(s) = y_j] \\
&= P[\{s : X(s) = x_i, Y(s) = y_j\}] \\
&= P[\{s_k\}] \quad\quad (7.1)
\end{aligned}
$$

where it is assumed that s_k is the only solution to $X(s) = x_i$ and $X(s) = y_j$. For a many-to-one transformation the joint PMF is found as

$$
p_{X,Y}[x_i, y_j] = \sum_{\{k : X(\mathcal{S}_k) = x_i, Y(\mathcal{S}_k) = y_j\}} P[\{s_k\}]. \quad\quad (7.2)
$$

This is the extension of (5.1) and (5.2) to a two-dimensional random vector. An example follows.

Example 7.1 – Two dice toss with different colored dice

A red die and a blue die are tossed. The die that yields the larger number of dots is chosen. If they both display the same number of dots, the red die is chosen. The numerical outcome of the experiment is defined to be 0 if the blue die is chosen and 1 if the red die is chosen, along with its corresponding number of dots. The random vector is therefore defined as

$$
\begin{aligned}
X &= \begin{cases} 0 & \text{blue die chosen} \\ 1 & \text{red die chosen} \end{cases} \\
Y &= \text{number of dots on chosen die.}
\end{aligned}
$$

The outcomes of the experiment can be represented by (i, j) where $i = 0$ for blue, $i = 1$ for red, and j is the number of dots observed. What then is $p_{X,Y}[1,3]$, for example? To determine this we first list all outcomes in Table 7.1 for each number of dots observed on the red and blue dice. It is seen that the mapping is many-to-one. For example, if the red die displays 6 dots, then the outcome is the same, which is $(1, 6)$, for all possible blue outcomes. To determine the desired value of the PMF,

	blue=1	blue=2	blue=3	blue=4	blue=5	blue=6
red=1	$(1,1)$	$(0,2)$	$(0,3)$	$(0,4)$	$(0,5)$	$(0,6)$
red=2	$(1,2)$	$(1,2)$	$(0,3)$	$(0,4)$	$(0,5)$	$(0,6)$
red=3	$(1,3)$	$(1,3)$	$(1,3)$	$(0,4)$	$(0,5)$	$(0,6)$
red=4	$(1,4)$	$(1,4)$	$(1,4)$	$(1,4)$	$(0,5)$	$(0,6)$
red=5	$(1,5)$	$(1,5)$	$(1,5)$	$(1,5)$	$(1,5)$	$(0,6)$
red=6	$(1,6)$	$(1,6)$	$(1,6)$	$(1,6)$	$(1,6)$	$(1,6)$

Table 7.1: Mapping of outcomes in S to outcomes in $S_{X,Y}$. The outcomes of (X,Y) are (i,j), where i indicates the color of the die with more dots (red=1, blue=0), j indicates the number of dots on that die.

we assume that each outcome in S is equally likely and therefore is equal to $1/36$. Then, from (7.2)

$$
p_{X,Y}[1,3] = \sum_{\{k:X(S_k)=1,Y(S_k)=3\}} P[\{S_k\}]
$$

$$
= \sum_{\{k:X(S_k)=1,Y(S_k)=3\}} \frac{1}{36}
$$

$$
= \frac{3}{36} = \frac{1}{12}
$$

since there are three outcomes of the experiment in S that map into $(1,3)$. They are (red=3,blue=1), (red=3,blue=2), and (red=3,blue=3).

◇

In general, as in the case of a single random variable we can use the joint PMF to compute probabilities of all events defined on $S_{X,Y} = S_X \times S_Y$. For the event $A \subset S_{X,Y}$, the probability is

$$
P[(X,Y) \in A] = \sum_{\{(i,j):(x_i,y_j) \in A\}} p_{X,Y}[x_i, y_j]. \tag{7.3}
$$

Once we have knowledge of the joint PMF, we no longer need to retain the underlying sample space S of the experiment. All our probability calculations can be made concerning values of (X,Y) by using (7.3).

7.4 Marginal PMFs and CDFs

If the joint PMF is known, then the PMF for X, i.e., $p_X[x_i]$, and the PMF for Y, i.e., $p_Y[y_j]$, can be determined. These are termed the *marginal PMFs*. Consider first the determination of $p_X[x_i]$. Since $\{X = x_i\}$ does not specify any particular value for Y, the event $\{X = x_i\}$ is equivalent to the joint event $\{X = x_i, Y \in S_Y\}$.

To determine the probability of the latter event we assume the general case of a countably infinite sample space. Then, (7.3) becomes

$$P[(X, Y) \in A] = \sum_{\substack{i=1 \\ \{(i,j):(x_i,y_j)\in A\}}}^{\infty} \sum_{j=1}^{\infty} p_{X,Y}[x_i, y_j]. \tag{7.4}$$

Next let $A = \{x_k\} \times S_Y$, which is illustrated in Figure 7.4 for $k = 3$. Then, we have

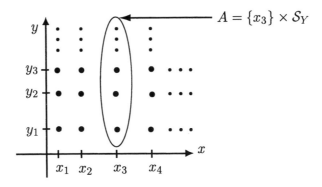

Figure 7.4:   Determination of marginal PMF value $p_X[x_3]$ from joint PMF $p_{X,Y}[x_i, y_j]$ by summing along y direction.

$$
\begin{aligned}
P[(X, Y) \in \{x_k\} \times S_Y] &= P[X = x_k, Y \in S_Y] \\
&= P[X = x_k] \\
&= p_X[x_k]
\end{aligned}
$$

so that from (7.4) with $i = k$ only

$$p_X[x_k] = \sum_{j=1}^{\infty} p_{X,Y}[x_k, y_j] \tag{7.5}$$

and is obtained for $k = 3$ by summing the probabilities along the column shown in Figure 7.4. The terminology "marginal" PMF originates from the process of summing the probabilities along each column and writing the results in the margin (below the x axis), much the same as the process for computing the marginal probability discussed in Section 4.3. Likewise, by summing along each row or in the x direction we obtain the marginal PMF for Y as

$$p_Y[y_k] = \sum_{i=1}^{\infty} p_{X,Y}[x_i, y_k]. \tag{7.6}$$

In summary, we see that *from the joint PMF we can obtain the marginal PMFs*. Another example follows.

Example 7.2 – Two coin toss

As before we toss a penny and a nickel and map the outcomes into a 1 for a head and a 0 for a tail. The random vector is (X, Y), where X is the random variable representing the penny outcome and Y is the random variable representing the nickel outcome. The mapping is shown in Figure 7.2. Consider the joint PMF

$$p_{X,Y}[i,j] = \begin{cases} \frac{1}{8} & i = 0, j = 0 \\ \frac{1}{8} & i = 0, j = 1 \\ \frac{1}{4} & i = 1, j = 0 \\ \frac{1}{2} & i = 1, j = 1. \end{cases}$$

Then, the marginal PMFs are given as

$$p_X[i] = \sum_{j=0}^{1} p_{X,Y}[i,j] = \begin{cases} \frac{1}{8} + \frac{1}{8} = \frac{1}{4} & i = 0 \\ \frac{1}{4} + \frac{1}{2} = \frac{3}{4} & i = 1 \end{cases}$$

$$p_Y[j] = \sum_{i=0}^{1} p_{X,Y}[i,j] = \begin{cases} \frac{1}{8} + \frac{1}{4} = \frac{3}{8} & j = 0 \\ \frac{1}{8} + \frac{1}{2} = \frac{5}{8} & j = 1. \end{cases}$$

As expected, $\sum_{i=0}^{1} p_X[i] = 1$ and $\sum_{j=0}^{1} p_Y[j] = 1$. We could also have arranged the joint PMF and marginal PMF values in a table as shown in Table 7.2. Note that

	$j = 0$	$j = 1$	$p_X[i]$
$i = 0$	$\frac{1}{8}$	$\frac{1}{8}$	$\frac{1}{4}$
$i = 1$	$\frac{1}{4}$	$\frac{1}{2}$	$\frac{3}{4}$
$p_Y[j]$	$\frac{3}{8}$	$\frac{5}{8}$	

Table 7.2: Joint PMF and marginal PMF values for Examples 7.2 and 7.4.

the marginal PMFs are found by summing across a row (for p_X) or a column (for p_Y) and are written in the "margins".

   Joint PMF cannot be determined from marginal PMFs.

Having obtained the marginal PMFs from the joint PMF, we might suppose we could reverse the process to find the joint PMF from the marginal PMFs. However, this is *not possible* in general. To see why, consider the joint PMF summarized in Table 7.3. The marginal PMFs are the same as the ones shown in Table 7.2. In

	$j=0$	$j=1$	$p_X[i]$
$i=0$	$\frac{1}{16}$	$\frac{3}{16}$	$\frac{1}{4}$
$i=1$	$\frac{5}{16}$	$\frac{7}{16}$	$\frac{3}{4}$
$p_Y[j]$	$\frac{3}{8}$	$\frac{5}{8}$	

Table 7.3: Joint PMF values for "caution" example.

fact, *there are an infinite number of joint PMFs that have the same marginal PMFs.* Hence,

$$\text{joint PMF} \Rightarrow \text{marginal PMFs}$$

but

$$\text{marginal PMFs} \nRightarrow \text{joint PMF}.$$

A joint cumulative distribution function (CDF) can also be defined for a random vector. It is given by

$$F_{X,Y}(x,y) = P[X \le x, Y \le y] \tag{7.7}$$

and can be found explicitly by summing the joint PMF as

$$F_{X,Y}(x,y) = \sum\sum_{\{(i,j):x_i \le x, y_j \le y\}} p_{X,Y}[x_i, y_j]. \tag{7.8}$$

An example is shown in Figure 7.5, along with the joint PMF. The *marginal* CDFs can be easily found from the joint CDF as

$$
\begin{aligned}
F_X(x) &= P[X \le x] = P[X \le x, Y < \infty] = F_{X,Y}(x,\infty) \\
F_Y(y) &= P[Y \le y] = P[X < \infty, Y \le y] = F_{X,Y}(\infty,y).
\end{aligned}
$$

The joint CDF has the usual properties which are:

Property 7.3 – Range of values

$$0 \le F_{X,Y}(x,y) \le 1$$

□

Property 7.4 – Values at "endpoints"

$$
\begin{aligned}
F_{X,Y}(-\infty,-\infty) &= 0 \\
F_{X,Y}(\infty,\infty) &= 1
\end{aligned}
$$

□

Property 7.5 – Monotonically increasing

$F_{X,Y}(x,y)$ monotonically increases as x and/or y increases.

$\square$

Property 7.6 – "Right" continuous

As expected, the joint CDF takes the value *after* the jump. However, in this case the jump is a line discontinuity as seen, for example, in Figure 7.5b. *After* the jump means as we move in the northeast direction in the x-y plane.

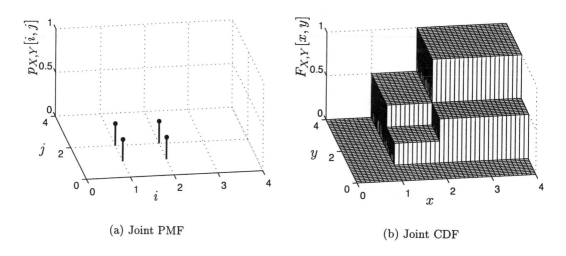

(a) Joint PMF   (b) Joint CDF

Figure 7.5: Joint PMF and corresponding joint CDF.

$\square$

The reader is asked to verify some of these properties in Problem 7.17. Finally, to recover the PMF we can use

$$p_{X,Y}[x_i, y_j] = F_{X,Y}(x_i^+, y_j^+) - F_{X,Y}(x_i^+, y_j^-) - F_{X,Y}(x_i^-, y_j^+) + F_{X,Y}(x_i^-, y_j^-). \quad (7.9)$$

The reader should verify this formula for the joint CDF shown in Figure 7.5b. In particular, consider the joint PMF at the point $(x_i, y_j) = (2, 2)$ to see why we need four terms.

7.5   Independence of Multiple Random Variables

Consider the experiment of tossing a coin and then a die. The outcome of the coin toss is denoted by X and equals 0 for a tail and 1 for a head. The outcome for the die is denoted by Y, which takes on the usual values $1, 2, 3, 4, 5, 6$. In determining the probability of the random vector (X, Y) taking on a value, there is no reason to believe that the probability of $Y = y_j$ should depend on the outcome of the coin toss. Likewise, the probability of $X = x_i$ should not depend on the outcome of the die toss (especially since the die toss occurs at a later time). We expect that these two events are *independent*. The formal definition of *independent random variables* X and Y is that they are independent if *all* the joint events on $\mathcal{S}_{X,Y}$ are independent. Mathematically X and Y are independent random variables if for all events $A \subset \mathcal{S}_X$ and $B \subset \mathcal{S}_Y$

$$P[X \in A, Y \in B] = P[X \in A]P[Y \in B]. \tag{7.10}$$

The probabilities on the right-hand-side of (7.10) are defined on $\mathcal{S}_X$ and $\mathcal{S}_Y$, respectively (see Figure 7.3 for an example of the relationship of $\mathcal{S}_X, \mathcal{S}_Y$ to $\mathcal{S}_{X,Y}$). The utility of the independence property is that the probabilities of joint events may be reduced to probabilities of "marginal events" (defined on $\mathcal{S}_X$ and $\mathcal{S}_Y$), which are always easier to determine. Specifically, if X and Y are independent random variables, then it follows from (7.10) that

$$p_{X,Y}[x_i, y_j] = p_X[x_i]p_Y[y_j] \tag{7.11}$$

as we now show. If $A = \{x_i\}$ and $B = \{y_j\}$, then the left-hand-side of (7.10) becomes

$$
\begin{aligned}
P[X \in A, Y \in B] &= P[X = x_i, Y = y_j] \\
&= p_{X,Y}[x_i, y_j]
\end{aligned}
$$

and the right-hand-side of (7.10) becomes

$$P[X \in A]P[Y \in B] = p_X[x_i]p_Y[y_j].$$

Hence, *if X and Y are independent random variables, the joint PMF factors into the product of the marginal PMFs.* Furthermore, the converse is true—*if the joint PMF factors, then X and Y are independent.* To prove the converse assume that the joint PMF factors according to (7.11). Then for all A and B we have

$$
\begin{aligned}
P[X \in A, Y \in B] &= \sum_{\{i : x_i \in A\}} \sum_{\{j : y_j \in B\}} p_{X,Y}[x_i, y_j] \quad &\text{(from (7.3))} \\
&= \sum_{\{i : x_i \in A\}} \sum_{\{j : y_j \in B\}} p_X[x_i]p_Y[y_j] \quad &\text{(assumption)} \\
&= \sum_{\{i : x_i \in A\}} p_X[x_i] \sum_{\{j : y_j \in B\}} p_Y[y_j] \\
&= P[X \in A]P[Y \in B].
\end{aligned}
$$

We now illustrate the concept of independent random variables with some examples.

Example 7.3 – Two coin toss – independence

Assume that we toss a penny and a nickel and that as usual a tail is mapped into a 0 and a head into a 1. If all outcomes are equally likely or equivalently the joint PMF is given in Table 7.4, then the random variables must be independent. This is

	$j=0$	$j=1$	$p_X[i]$
$i=0$	$\frac{1}{4}$	$\frac{1}{4}$	$\frac{1}{2}$
$i=1$	$\frac{1}{4}$	$\frac{1}{4}$	$\frac{1}{2}$
$p_Y[j]$	$\frac{1}{2}$	$\frac{1}{2}$	

Table 7.4: Joint PMF and marginal PMF values for Example 7.3.

because we can factor the joint PMF as

$$p_{X,Y}[i,j] = \left(\frac{1}{2}\right)\left(\frac{1}{2}\right) = p_X[i]p_Y[j]$$

for all i and j for which $p_{X,Y}[i,j]$ is nonzero. Furthermore, the marginal PMFs indicate that each coin is fair since $p_X[0] = p_X[1] = 1/2$ and $p_Y[0] = p_Y[1] = 1/2$.

$\Diamond$

Example 7.4 – Two coin toss – dependence

Now consider the same experiment but with a joint PMF given in Table 7.2. We see that $p_{X,Y}[0,0] = 1/8 \neq (1/4)(3/8) = p_X[0]p_Y[0]$ and hence X and Y cannot be independent. If two random variables are not independent, they are said to be *dependent*.

$\Diamond$

Example 7.5 – Two coin toss – dependent but fair coins

Consider the same experiment again but with the joint PMF given in Table 7.5. Since $p_{X,Y}[0,0] = 3/8 \neq (1/2)(1/2) = p_X[0]p_Y[0]$, X and Y are *dependent*. However,

	$j=0$	$j=1$	$p_X[i]$
$i=0$	$\frac{3}{8}$	$\frac{1}{8}$	$\frac{1}{2}$
$i=1$	$\frac{1}{8}$	$\frac{3}{8}$	$\frac{1}{2}$
$p_Y[j]$	$\frac{1}{2}$	$\frac{1}{2}$	

Table 7.5: Joint PMF and marginal PMF values for Example 7.5.

by examining the marginal PMFs we see that the coins are in some sense fair since $P[\text{heads}] = 1/2$, and therefore we might conclude that the random variables were independent. This is incorrect and underscores the fact that the marginal PMFs do not tell us much about the joint PMF. The joint PMF of Table 7.4 also has the same marginal PMFs but there X and Y were independent.

$\Diamond$

Finally, note that if the random variables are independent, the joint CDF factors as well. This is left as an exercise for the student (see Problem 7.20). Intuitively, if X and Y are independent random variables, then knowledge of the outcome of X does not change the probabilities of the outcomes of Y. This means that we cannot predict Y based on knowing that $X = x_i$. Our best predictor of Y is just $E[Y]$, as described in Example 6.3. When X and Y are dependent, however, we can improve upon the predictor $E[Y]$ by using the knowledge that $X = x_i$. How we actually do this is described in Section 7.9.

7.6 Transformations of Multiple Random Variables

In Section 5.7 we have seen how to find the PMF of $Y = g(X)$ if the PMF of X is given. It is determined using

$$p_Y[y_i] = \sum_{\{j:g(x_j)=y_i\}} p_X[x_j].$$

We need only sum the probabilities of the x_j's that map into y_i. In the case of two discrete random variables X and Y that are transformed into $W = g(X,Y)$ and $Z = h(X,Y)$, we have the similar result

$$p_{W,Z}[w_i, z_j] = \sum_{\left\{(k,l): \begin{smallmatrix} g(x_k,y_l)=w_i \\ h(x_k,y_l)=z_j \end{smallmatrix} \right\}} \sum p_{X,Y}[x_k, y_l] \qquad i = 1, 2, \ldots, N_W; j = 1, 2, \ldots, N_Z$$

(7.12)

where N_W and/or N_Z may be infinite. An example follows.

Example 7.6 – Independent Poisson random variables

Assume that the joint PMF is given as the product of the marginal PMFs, where each marginal PMF is a Poisson PMF. Then,

$$p_{X,Y}[k, l] = \exp[-(\lambda_X + \lambda_Y)]\frac{\lambda_X^k \lambda_Y^l}{k! \, l!} \qquad k = 0, 1, \ldots; l = 0, 1, \ldots \qquad (7.13)$$

Note that $X \sim \text{Pois}(\lambda_X)$, $Y \sim \text{Pois}(\lambda_Y)$, and X and Y are independent random variables. Consider the transformation

$$\begin{aligned} W &= g(X,Y) = X \\ Z &= h(X,Y) = X + Y. \end{aligned} \qquad (7.14)$$

The possible values of W are those of X, which are $0, 1, \ldots$, and the possible values of Z are also $0, 1, \ldots$. According to (7.12), we need to determine all (k, l) so that

$$
\begin{aligned}
g(x_k, y_l) &= w_i \\
h(x_k, y_l) &= z_j.
\end{aligned}
\tag{7.15}
$$

But x_k and y_l can be replaced by k and l, respectively, for $k = 0, 1, \ldots$ and $l = 0, 1, \ldots$. Also, w_i and z_j can be replaced by i and j, respectively, for $i = 0, 1, \ldots$ and $j = 0, 1, \ldots$. The transformation equations become

$$
\begin{aligned}
g(k, l) &= i \\
h(k, l) &= j
\end{aligned}
$$

which from (7.14) become

$$
\begin{aligned}
i &= k \\
j &= k + l.
\end{aligned}
$$

Solving for (k, l) for the *given* (i, j) desired, we have that $k = i$ and $l = j - i \geq 0$, which is the only solution. Note that from (7.13) the joint PMF for X and Y is nonzero only if $l = 0, 1, \ldots$. Therefore, we must have $l \geq 0$ so that $l = j - i \geq 0$. From (7.12) we now have

$$
\begin{aligned}
p_{W,Z}[i, j] &= \sum_{\substack{k=0 \\ \{(k,l):k=i,l=j-i\geq 0\}}}^{\infty} \sum_{l=0}^{\infty} p_{X,Y}[k, l] \\
&= p_{X,Y}[i, j - i]u[i]u[j - i]
\end{aligned}
\tag{7.16}
$$

where $u[n]$ is the discrete unit step sequence defined as

$$
u[n] = \begin{cases} 0 & n = \ldots, -2, -1 \\ 1 & n = 0, 1, \ldots \end{cases}.
$$

Finally, we have upon using (7.13)

$$
p_{W,Z}[i, j] = \exp[-(\lambda_X + \lambda_Y)] \frac{\lambda_X^i \lambda_Y^{j-i}}{i!(j-i)!} u[i]u[j - i]
\tag{7.17}
$$

$$
= \exp[-(\lambda_X + \lambda_Y)] \frac{\lambda_X^i \lambda_Y^{j-i}}{i!(j-i)!} \quad \begin{matrix} i = 0, 1, \ldots \\ j = i, i+1, \ldots \end{matrix}.
\tag{7.18}
$$

Use the discrete unit step sequence to avoid mistakes.

As we have seen in the preceding example, the discrete unit step sequence was introduced to designate the region of the w-z plane over which $p_{W,Z}[i,j]$ is nonzero. A common mistake in problems of this type is to disregard this region and assert that the joint PMF given by (7.18) is nonzero over $i = 0, 1, \ldots ; j = 0, 1, \ldots$. Note, however, that the transformation will generally change the region over which the new joint PMF is nonzero. It is as important to determine this region as it is to find the analytical form of $p_{W,Z}$. To avoid possible errors it is advisable to replace (7.13) at the outset by

$$p_{X,Y}[k,l] = \exp[-(\lambda_X + \lambda_Y)]\frac{\lambda_X^k \lambda_Y^l}{k!l!}u[k]u[l].$$

Then, the use of the unit step functions will serve to keep track of the nonzero PMF regions before and after the transformation. See also Problem 7.25 for another example.

We sometimes wish to determine the PMF of $Z = h(X,Y)$ only, which is a transformation from (X,Y) to Z. In this case, we can use an *auxiliary* random variable. That is to say, we add another random variable W so that the transformation becomes a transformation from (X,Y) to (W,Z) as before. We can then determine $p_{W,Z}[w_i, z_j]$ by once again using (7.12), and then p_Z, which is the marginal PMF, can be found as

$$p_Z[z_j] = \sum_{\{i:w_i \in \mathcal{S}_W\}} p_{W,Z}[w_i, z_j]. \tag{7.19}$$

As we have seen in the previous example, we will first need to solve (7.15) for x_k and y_l. To facilitate the solution we usually define a simple auxiliary random variable such as $W = X$.

Example 7.7 – PMF for sum of independent Poisson random variables (continuation of previous example)

To find the PMF of $Z = X + Y$ from the joint PMF given by (7.13), we use (7.19) with $W = X$. We then have $\mathcal{S}_W = \mathcal{S}_X = \{0, 1, \ldots\}$ and

$$
\begin{aligned}
p_Z[j] &= \sum_{i=0}^{\infty} p_{W,Z}[i,j] && \text{(from (7.19))} \quad (7.20)\\
&= \sum_{i=0}^{\infty} \exp[-(\lambda_X + \lambda_Y)]\frac{\lambda_X^i \lambda_Y^{j-i}}{i!(j-i)!}u[i]u[j-i] && \text{(from (7.17))}
\end{aligned}
$$

and since $u[i] = 1$ for $i = 0, 1, \ldots$ and $u[j - i] = 1$ for $i = 0, 1, \ldots, j$ and $u[j - i] = 0$ for $i > j$, this reduces to

$$p_Z[j] = \sum_{i=0}^{j} \exp[-(\lambda_X + \lambda_Y)] \frac{\lambda_X^i \lambda_Y^{j-i}}{i!(j-i)!} \qquad j = 0, 1, \ldots.$$

Note that Z can take on values $j = 0, 1, \ldots$ since $Z = X + Y$ and both X and Y take on values in $\{0, 1, \ldots\}$. To evaluate this sum we can use the binomial theorem as follows:

$$
\begin{aligned}
p_Z[j] &= \exp[-(\lambda_X + \lambda_Y)] \frac{1}{j!} \sum_{i=0}^{j} \frac{j!}{(j-i)!i!} \lambda_X^i \lambda_Y^{j-i} \\
&= \exp[-(\lambda_X + \lambda_Y)] \frac{1}{j!} \sum_{i=0}^{j} \binom{j}{i} \lambda_X^i \lambda_Y^{j-i} \\
&= \exp[-(\lambda_X + \lambda_Y)] \frac{1}{j!} (\lambda_X + \lambda_Y)^j \qquad \text{(use binomial theorem)} \\
&= \exp(-\lambda) \frac{\lambda^j}{j!} \qquad \text{(let } \lambda = \lambda_X + \lambda_Y)
\end{aligned}
$$

for $j = 0, 1, \ldots$. This is recognized as a Poisson PMF with $\lambda = \lambda_X + \lambda_Y$. By this example then, we have shown that if $X \sim \text{Pois}(\lambda_X)$, $Y \sim \text{Pois}(\lambda_Y)$, and X and Y are independent, then $X + Y \sim \text{Pois}(\lambda_X + \lambda_Y)$. This is called the *reproducing* PMF property. It is also extendible to any number of independent Poisson random variables that are added together.

$$\diamondsuit$$

The formula given by (7.20) when we let $p_{W,Z}[i, j] = p_{X,Y}[i, j - i]$ from (7.16) is valid for the PMF of the sum of any two discrete random variables, whether they are independent or not. Summarizing, if X and Y are random variables that take on integer values from $-\infty$ to $+\infty$, then $Z = X + Y$ has the PMF

$$p_Z[j] = \sum_{i=-\infty}^{\infty} p_{X,Y}[i, j - i]. \qquad (7.21)$$

This result says that we should sum all the values of the joint PMF such that the x value, which is i, and the y value, which is $j - i$, sums to the z value of j. In particular, if the random variables are *independent*, then since the joint PMF must factor, we have the result

$$p_Z[j] = \sum_{i=-\infty}^{\infty} p_X[i] p_Y[j - i]. \qquad (7.22)$$

But this summation operation is a *discrete convolution* [Jackson 1991]. It is usually written succinctly as $p_Z = p_X \star p_Y$, where $\star$ denotes the convolution operator. This result suggests that the use of Fourier transforms would be a useful tool since a convolution can be converted into a simple multiplication in the Fourier domain. We have already seen in Chapter 6 that the Fourier transform (defined with a $+j$) of a PMF $p_X[k]$ is the characteristic function $\phi_X(\omega) = E[\exp(j\omega X)]$. Therefore, taking the Fourier transform of both sides of (7.22) produces

$$\phi_Z(\omega) = \phi_X(\omega)\phi_Y(\omega) \qquad (7.23)$$

and by converting back to the original sequence domain, the PMF becomes

$$p_Z[j] = \mathcal{F}^{-1}\{\phi_X(\omega)\phi_Y(\omega)\} \qquad (7.24)$$

where $\mathcal{F}^{-1}$ denotes the inverse Fourier transform. An example follows.

Example 7.8 – PMF for sum of independent Poisson random variables using characteristic function approach

From Section 6.7 we showed that if $X \sim \text{Pois}(\lambda)$, then

$$\phi_X(\omega) = \exp\left[\lambda(\exp(j\omega) - 1)\right]$$

and thus using (7.23) and (7.24)

$$
\begin{aligned}
p_Z[j] &= \mathcal{F}^{-1}\left\{\exp\left[\lambda_X(\exp(j\omega) - 1)\right]\exp\left[\lambda_Y(\exp(j\omega) - 1)\right]\right\} \\
&= \mathcal{F}^{-1}\left\{\exp\left[(\lambda_X + \lambda_Y)(\exp(j\omega) - 1)\right]\right\}.
\end{aligned}
$$

But the characteristic function in the braces is that of a Poisson random variable. Using Property 6.5 we see that $Z \sim \text{Pois}(\lambda_X + \lambda_Y)$. The use of characteristic functions for the determination of the PMF for a sum of independent random variables has *considerably* simplified the derivation.

$\diamondsuit$

In summary, if X and Y are independent random variables with integer values, then the PMF of $Z = X + Y$ is given by

$$
\begin{aligned}
p_Z[k] &= \mathcal{F}^{-1}\{\phi_X(\omega)\phi_Y(\omega)\} \\
&= \int_{-\pi}^{\pi} \phi_X(\omega)\phi_Y(\omega)\exp(-j\omega k)\frac{d\omega}{2\pi}. \qquad (7.25)
\end{aligned}
$$

When the sample space $\mathcal{S}_{X,Y}$ is finite, it is sometimes possible to obtain the PMF of $Z = g(X,Y)$ by a direct calculation, thus avoiding the need to use (7.19). The latter requires one to first find the transformed joint PMF $p_{W,Z}$. To do so we

1. Determine the finite sample space $\mathcal{S}_Z$.

2. Determine which sample points (x_i, y_j) in $\mathcal{S}_{X,Y}$ map into each $z_k \in \mathcal{S}_Z$.

3. Sum the probabilities of those (x_i, y_j) sample points to yield $p_Z[z_k]$.

Mathematically, this is equivalent to

$$p_Z[z_k] = \sum_{\{(i,j):z_k=g(x_i,y_j)\}} \sum p_{X,Y}[x_i,y_j]. \tag{7.26}$$

An example follows.

Example 7.9 – Direct computation of PMF for transformed random variable, $Z = g(X,Y)$

Consider the transformation of the random vector (X,Y) into the scalar random variable $Z = X^2 + Y^2$. The joint PMF is given by

$$p_{X,Y}[i,j] = \begin{cases} \frac{3}{8} & i=0, j=0 \\ \frac{1}{8} & i=1, j=0 \\ \frac{1}{8} & i=0, j=1 \\ \frac{3}{8} & i=1, j=1. \end{cases}$$

To find the PMF for Z first note that (X,Y) takes on the values $(i,j) = (0,0), (1,0)$, $(0,1), (1,1)$. Therefore, Z must take on the values $z_k = i^2 + j^2 = 0, 1, 2$. Then from (7.26)

$$\begin{aligned} p_Z[0] &= \sum_{\{(i,j):0=i^2+j^2\}} \sum p_{X,Y}[i,j] \\ &= \sum_{i=0}^{0} \sum_{j=0}^{0} p_{X,Y}[i,j] \\ &= p_{X,Y}[0,0] = \frac{3}{8} \end{aligned}$$

and similarly

$$\begin{aligned} p_Z[1] &= p_{X,Y}[0,1] + p_{X,Y}[1,0] = \frac{2}{8} \\ p_Z[2] &= p_{X,Y}[1,1] = \frac{3}{8}. \end{aligned}$$

7.7   Expected Values

In addition to determining the PMF of a function of two random variables, we are frequently interested in the average value of that function. Specifically, if $Z = g(X,Y)$, then by definition its expected value is

$$E[Z] = \sum_i z_i p_Z[z_i]. \tag{7.27}$$

To determine $E[Z]$ according to (7.27) we need to first find the PMF of Z and then perform the summation. Alternatively, by a similar derivation to that given in Appendix 6A, we can show that a more direct approach is

$$E[Z] = \sum_i \sum_j g(x_i, y_j) p_{X,Y}[x_i, y_j]. \tag{7.28}$$

To remind us that we are using $p_{X,Y}$ as the averaging PMF, we will modify our previous notation from $E[Z]$ to $E_{X,Y}[Z]$, where of course, Z depends on X and Y. We therefore have the useful result that the expected value of a function of two random variables is

$$E_{X,Y}[g(X,Y)] = \sum_i \sum_j g(x_i, y_j) p_{X,Y}[x_i, y_j]. \tag{7.29}$$

Some examples follow.

Example 7.10 – Expected value of a sum of random variables
If $Z = g(X,Y) = X + Y$, then

$$
\begin{aligned}
E_{X,Y}[X + Y] &= \sum_i \sum_j (x_i + y_j) p_{X,Y}[x_i, y_j] \\
&= \sum_i \sum_j x_i p_{X,Y}[x_i, y_j] + \sum_i \sum_j y_j p_{X,Y}[x_i, y_j] \\
&= \sum_i x_i \underbrace{\sum_j p_{X,Y}[x_i, y_j]}_{p_X[x_i]} + \sum_j y_j \underbrace{\sum_i p_{X,Y}[x_i, y_j]}_{p_Y[y_j]} \quad \text{(from (7.6))} \\
&= E_X[X] + E_Y[Y] \qquad \text{(definition of expected value).}
\end{aligned}
$$

Hence, the expected value of a sum of random variables is the sum of the expected values. Note that we now use the more descriptive notation $E_X[X]$ to replace $E[X]$ used previously.

$\Diamond$

Similarly

$$E_{X,Y}[aX + bY] = aE_X[X] + bE_Y[Y]$$

and thus as we have seen previously for a single random variable, *the expectation $E_{X,Y}$ is a linear operation.*

Example 7.11 – Expected value of a product of random variables
If $g(X,Y) = XY$, then

$$E_{X,Y}[XY] = \sum_i \sum_j x_i y_j p_{X,Y}[x_i, y_j].$$

We cannot evaluate this further without specifying $p_{X,Y}$. If, however, X and Y are independent, then since the joint PMF factors, we have

$$
\begin{aligned}
E_{X,Y}[XY] &= \sum_i \sum_j x_i y_j p_X[x_i] p_Y[y_j] \\
&= \sum_i x_i p_X[x_i] \sum_j y_j p_Y[y_j] \\
&= E_X[X] E_Y[Y].
\end{aligned}
\tag{7.30}
$$

More generally, we can show by using (7.29) that if X and Y are independent, then (see Problem 7.30)

$$
E_{X,Y}[g(X)h(Y)] = E_X[g(X)]E_Y[h(Y)].
\tag{7.31}
$$

$\Diamond$

Example 7.12 – Variance of a sum of random variables

Consider the calculation of $\text{var}(X + Y)$. Then, letting $Z = g(X,Y) = (X + Y - E_{X,Y}[X+Y])^2$, we have

$$
\begin{aligned}
&\text{var}(X + Y) \\
&= E_Z[Z] && \text{(definition of variance)} \\
&= E_{X,Y}[g(X,Y)] && \text{(from (7.28))} \\
&= E_{X,Y}[(X + Y - E_{X,Y}[X+Y])^2] \\
&= E_{X,Y}[[(X - E_X[X]) + (Y - E_Y[Y])]^2] \\
&= E_{X,Y}[(X - E_X[X])^2 + 2(X - E_X[X])(Y - E_Y[Y]) \\
&\qquad + (Y - E_Y[Y])^2] \\
&= E_X[(X - E_X[X])^2] + 2E_{X,Y}[(X - E_X[X])(Y - E_Y[Y])] \\
&\qquad + E_Y[(Y - E_Y[Y])^2] && \text{(linearity of expectation)} \\
&= \text{var}(X) + 2E_{X,Y}[(X - E_X[X])(Y - E_Y[Y])] + \text{var}(Y) && \text{(definition of variance)}
\end{aligned}
$$

where we have also used $E_{X,Y}[g(X)] = E_X[g(X)]$ and $E_{X,Y}[h(Y)] = E_Y[h(Y)]$ (see Problem 7.28). The cross-product term is called the *covariance* and is denoted by $\text{cov}(X,Y)$ so that

$$
\text{cov}(X,Y) = E_{X,Y}[(X - E_X[X])(Y - E_Y[Y])].
\tag{7.32}
$$

Its interpretation is discussed in the next section. Hence, we finally have that the variance of a sum of random variables is

$$
\text{var}(X + Y) = \text{var}(X) + \text{var}(Y) + 2\text{cov}(X,Y).
\tag{7.33}
$$

Unlike the expected value or mean, *the variance of a sum is not in general the sum of the variances.* It will only be so when $\text{cov}(X, Y) = 0$. An alternative expression for the covariance is (see Problem 7.34)

$$\text{cov}(X, Y) = E_{X,Y}[XY] - E_X[X]E_Y[Y] \tag{7.34}$$

which is analogous to Property 6.1 for the variance.

7.8 Joint Moments

Joint PMFs describe the probabilistic behavior of two random variables completely. At times it is important to answer questions such as "If the outcome of one random variable is a given value, what can we say about the outcome of the other random variable? Will it be about the same or have the same magnitude or have no relationship to the other random variable?" For example, in Table 4.1, which lists the joint probabilities of college students having various heights and weights, there is clearly some type of relationship between height and weight. It is our intention to quantify this type of relationship in a succinct and meaningful way as opposed to a listing of probabilities of the various height-weight pairs. The concept of the covariance allows us to accomplish this goal. Note from (7.32) that the covariance is a joint *central* moment. To appreciate the information that it can provide we refer to the three possible joint PMFs depicted in Figure 7.6. The possible values of each joint PMF are shown as solid circles and each possible outcome has a probability of 1/2. In Figure 7.6a if $X = 1$, then $Y = 1$, and if $X = -1$, then $Y = -1$. The relationship

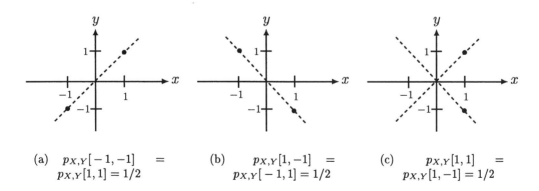

(a) $p_{X,Y}[-1, -1] = $
    $p_{X,Y}[1, 1] = 1/2$

(b) $p_{X,Y}[1, -1] = $
    $p_{X,Y}[-1, 1] = 1/2$

(c) $p_{X,Y}[1, 1] = $
    $p_{X,Y}[1, -1] = 1/2$

Figure 7.6: Joint PMFs depicting different relationships between the random variables X and Y.

is $Y = X$. Note, however, that we cannot determine the value of Y until after the experiment is performed and we are told the value of X. If $X = x_1$, then we know that $Y = X = x_1$. Likewise, in Figure 7.6b we have that $Y = -X$ and so if $X = x_1$, then $Y = -x_1$. However, in Figure 7.6c if $X = 1$, then Y can equal either $+1$ or -1. On the average if $X = 1$, we will have that $Y = 0$ since $Y = \pm 1$ with equal probability. To quantify these relationships we form the product XY, which can take on the values $+1$, -1, and ± 1 for the joint PMFs of Figures 7.6a, 7.6b, and 7.6c, respectively. To determine the value of XY *on the average* we define the *joint moment* as $E_{X,Y}[XY]$. From (7.29) this is evaluated as

$$E_{X,Y}[XY] = \sum_i \sum_j x_i y_j p_{X,Y}[x_i, y_j]. \tag{7.35}$$

The reader should compare the joint moment with the usual moment for a single random variable $E_X[X] = \sum_i x_i p_X[x_i]$. For the joint PMFs of Figure 7.6 the joint moment is

$$
\begin{aligned}
E_{X,Y}[XY] &= \sum_{i=1}^{2} \sum_{j=1}^{2} x_i y_j p_{X,Y}[x_i, y_j] \\
&= (1)(1)\frac{1}{2} + (-1)(-1)\frac{1}{2} = 1 \qquad \text{(for PMF of Figure 7.6a)} \\
&= (1)(-1)\frac{1}{2} + (-1)(1)\frac{1}{2} = -1 \qquad \text{(for PMF of Figure 7.6b)} \\
&= (1)(-1)\frac{1}{2} + (1)(1)\frac{1}{2} = 0 \qquad \text{(for PMF of Figure 7.6c)}
\end{aligned}
$$

as we might have expected.

In Figure 7.6a note that $E_X[X] = E_Y[Y] = 0$. If they are not zero, as for the joint PMF shown in Figure 7.7 in which $E_{X,Y}[XY] = 2$, then the joint moment will

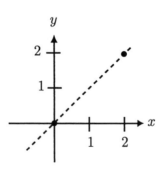

Figure 7.7: Joint PMF for nonzero means with equally probable outcomes.

depend on the values of the means. It is seen that even though the relationship $Y = X$ is preserved, the joint moment has changed. To nullify this effect of having nonzero means influence the joint moment it is more convenient to use the *joint central moment*

$$E_{X,Y}[(X - E_X[X])(Y - E_Y[Y])] \tag{7.36}$$

which will produce the desired $+1$ for the joint PMF of Figure 7.7. This quantity is recognized as the *covariance* of X and Y so that we denote it by $\text{cov}(X, Y)$. As we have just seen, the covariance may be positive, negative, or zero. Note that the covariance is a measure of how the random variables *covary* with respect to each other. If they vary in the same direction, i.e., both positive or negative at the same time, then the covariance will be positive. If they vary in opposite directions, the covariance will be negative. This explains why $\text{var}(X + Y)$ may be greater than $\text{var}(X) + \text{var}(Y)$, for the case of a positive covariance. Similarly, the variance of the sum of the random variables will be less than the sum of the variances if the covariance is negative.

If X and Y are independent random variables, then from (7.31) we have

$$
\begin{aligned}
\text{cov}(X, Y) &= E_{X,Y}[(X - E_X[X])(Y - E_Y[Y])] \\
&= E_X[X - E_X[X]]E_Y[Y - E_Y[Y]] = 0. \qquad (7.37)
\end{aligned}
$$

Hence, independent random variables have a covariance of zero. This also says that for independent random variables the *variance of the sum of random variables is the sum of the variances*, i.e., $\text{var}(X + Y) = \text{var}(X) + \text{var}(Y)$ (see (7.33)). However, the covariance may still be zero even if the random variables are not independent – the converse is not true. Some other properties of the covariance are given in Problem 7.34.

⚠️ **Independence implies zero covariance but zero covariance does not imply independence.**

Consider the joint PMF which assigns equal probability of $1/4$ to each of the four points shown in Figure 7.8. The joint and marginal PMFs are listed in Table 7.6.

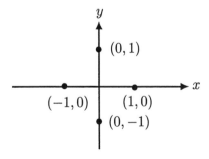

Figure 7.8: Joint PMF of random variables having zero covariance but that are dependent.

For this joint PMF the covariance is zero since

$$
E_X[X] = -1\left(\frac{1}{4}\right) + 0\left(\frac{1}{2}\right) + 1\left(\frac{1}{4}\right) = 0
$$

	$j = -1$	$j = 0$	$j = 1$	$p_X[i]$
$i = -1$	0	$\frac{1}{4}$	0	$\frac{1}{4}$
$i = 0$	$\frac{1}{4}$	0	$\frac{1}{4}$	$\frac{1}{2}$
$i = 1$	0	$\frac{1}{4}$	0	$\frac{1}{4}$
$p_Y[j]$	$\frac{1}{4}$	$\frac{1}{2}$	$\frac{1}{4}$	

Table 7.6: Joint PMF values.

and thus from (7.34)

$$\begin{aligned}
\text{cov}(X,Y) &= E_{X,Y}[XY] \\
&= \sum_{i=-1}^{1}\sum_{j=-1}^{1} ij p_{X,Y}[i,j] = 0
\end{aligned}$$

since either x or y is always zero. However, X are Y are dependent because $p_{X,Y}[1,0] = 1/4$ but $p_X[1]p_Y[0] = (1/4)(1/2) = 1/8$. Alternatively, we may argue that the random variables must be dependent since Y can be predicted from X. For example, if $X = 1$, then surely we must have $Y = 0$.

More generally the joint k-lth moment is defined as

$$E_{X,Y}[X^k Y^l] = \sum_{i}\sum_{j} x_i^k y_j^l p_{X,Y}[x_i, y_j] \tag{7.38}$$

for $k = 1, 2, \ldots; l = 1, 2, \ldots$, when it exists. The joint k-lth central moment is defined as

$$E_{X,Y}[(X - E_X[X])^k (Y - E_Y[Y])^l] = \sum_{i}\sum_{j}(x_i - E_X[X])^k (y_j - E_Y[Y])^l p_{X,Y}[x_i, y_j]$$
$$\tag{7.39}$$

for $k = 1, 2, \ldots; l = 1, 2, \ldots$, when it exists.

7.9   Prediction of a Random Variable Outcome

The covariance between two random variables has an important bearing on the predictability of Y based on knowledge of the outcome of X. We have already seen in Figures 7.6a,b that Y can be perfectly predicted from X as $Y = X$ (see Figure 7.6a) or as $Y = -X$ (see Figure 7.6b). These are extreme cases. More generally, we seek a predictor of Y that is *linear* (actually affine) in X or

$$\hat{Y} = aX + b$$

where the "hat" indicates an estimator. The constants a and b are to be chosen so that "on the average" the observed value of $\hat{Y}$, which is $ax + b$ if the experimental outcome is (x, y), is close to the observed value of Y, which is y. To determine these constants we shall adopt as our measure of closeness the mean square error (MSE) criterion described previously in Example 6.3. It is given by

$$\text{mse}(a, b) = E_{X,Y}[(Y - \hat{Y})^2]. \tag{7.40}$$

Note that since the predictor $\hat{Y}$ depends on X, we need to average with respect to X and Y. Previously, we let $\hat{Y} = b$, not having the additional information of the outcome of another random variable. It was found in Example 6.3 that the *optimal* value of b, i.e., the value that minimized the MSE, was $b_{\text{opt}} = E_Y[Y]$ and therefore $\hat{Y} = E_Y[Y]$. Now, however, we presume to know the outcome of X. With the additional knowledge of the outcome of X we should be able to find a better predictor. To find the optimal values of a and b we minimize (7.40) over a and b. Before doing so we simplify the expression for the MSE. Starting with (7.40)

$$
\begin{aligned}
\text{mse}(a, b) &= E_{X,Y}[(Y - aX - b)^2] \\
&= E_{X,Y}[(Y - aX)^2 - 2b(Y - aX) + b^2] \\
&= E_{X,Y}[Y^2 - 2aXY + a^2X^2 - 2bY + 2abX + b^2] \\
&= E_Y[Y^2] - 2aE_{X,Y}[XY] + a^2 E_X[X^2] - 2bE_Y[Y] + 2abE_X[X] + b^2.
\end{aligned}
$$

To find the values of a and b that minimize the function $\text{mse}(a, b)$, we determine a stationary point by partial differentiation. Since the function is quadratic in a and b, this will yield the minimizing values of a and b. Using partial differentiation and setting each partial derivative equal to zero produces

$$
\begin{aligned}
\frac{\partial \text{mse}(a, b)}{\partial a} &= -2E_{X,Y}[XY] + 2aE_X[X^2] + 2bE_X[X] = 0 \\
\frac{\partial \text{mse}(a, b)}{\partial b} &= -2E_Y[Y] + 2aE_X[X] + 2b = 0
\end{aligned}
$$

and rearranging yields the two simultaneous linear equations

$$
\begin{aligned}
E_X[X^2]a + E_X[X]b &= E_{X,Y}[XY] \\
E_X[X]a + b &= E_Y[Y].
\end{aligned}
$$

The solution is easily shown to be

$$
\begin{aligned}
a_{\text{opt}} &= \frac{E_{X,Y}[XY] - E_X[X]E_Y[Y]}{E_X[X^2] - E_X^2[X]} = \frac{\text{cov}(X, Y)}{\text{var}(X)} \\
b_{\text{opt}} &= E_Y[Y] - a_{\text{opt}}E_X[X] = E_Y[Y] - \frac{\text{cov}(X, Y)}{\text{var}(X)}E_X[X]
\end{aligned}
$$

so that the optimal linear prediction of Y given the outcome $X = x$ is

$$
\begin{aligned}
\hat{Y} &= a_{\text{opt}}x + b_{\text{opt}} \\
&= \frac{\text{cov}(X,Y)}{\text{var}(X)}x + E_Y[Y] - \frac{\text{cov}(X,Y)}{\text{var}(X)}E_X[X]
\end{aligned}
$$

or finally

$$
\hat{Y} = E_Y[Y] + \frac{\text{cov}(X,Y)}{\text{var}(X)}(x - E_X[X]). \tag{7.41}
$$

Note that we refer to $\hat{Y} = aX + b$ as a predict*or* but $\hat{Y} = ax + b$ as the predict*ion*, which is the *value* of the predictor. As expected, the prediction of Y based on $X = x$ depends on the covariance. In fact, if the covariance is zero, then $\hat{Y} = E_Y[Y]$, which is the best linear predictor of Y without knowledge of the outcome of X. In this case, X provides no information about Y. An example follows.

Example 7.13 – Predicting one random variable outcome from knowledge of second random variable outcome

Consider the joint PMF shown in Figure 7.9a as solid circles where all the outcomes are equally probable. Then, $\mathcal{S}_{X,Y} = \{(0,0),(1,1),(2,2),(2,3)\}$ and the marginals

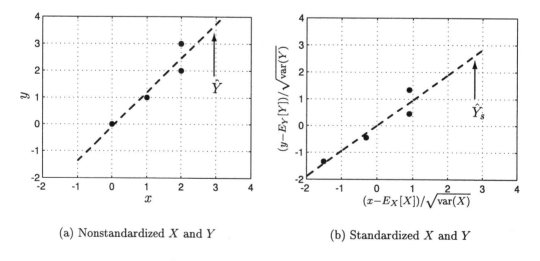

(a) Nonstandardized X and Y (b) Standardized X and Y

Figure 7.9: Joint PMF (shown as solid circles having equal probabilities) and best linear prediction of Y when $X = x$ is observed (shown as dashed line).

are found by summing along each direction to yield

$$p_X[i] = \begin{cases} \frac{1}{4} & i = 0 \\ \frac{1}{4} & i = 1 \\ \frac{1}{2} & i = 2 \end{cases}$$

$$p_Y[j] = \begin{cases} \frac{1}{4} & j = 0 \\ \frac{1}{4} & j = 1 \\ \frac{1}{4} & j = 2 \\ \frac{1}{4} & j = 3. \end{cases}$$

As a result, we have from the marginals that $E_X[X] = 5/4$, $E_Y[Y] = 3/2$, $E_X[X^2] = 9/4$, and $\text{var}(X) = E_X[X^2] - E_X^2[X] = 9/4 - (5/4)^2 = 11/16$. From the joint PMF we find that $E_{X,Y}[XY] = (0)(0)1/4 + (1)(1)1/4 + (2)(2)1/4 + (2)(3)1/4 = 11/4$, which results in $\text{cov}(X,Y) = E_{X,Y}[XY] - E_X[X]E_Y[Y] = 11/4 - (5/4)(3/2) = 7/8$. Substituting these values into (7.41) yields the best linear prediction of Y as

$$\begin{aligned} \hat{Y} &= \frac{3}{2} + \frac{7/8}{11/16}\left(x - \frac{5}{4}\right) \\ &= \frac{14}{11}x - \frac{1}{11} \end{aligned}$$

which is shown in Figure 7.9a as the dashed line. The line shown in Figure 7.9a is referred to as a *regression line* in statistics. What do you think would happen if the probability of $(2,3)$ were zero, and the remaining three points had probabilities of $1/3$?

$\Diamond$

The reader should be aware that we could also have predicted X from $Y = y$ by interchanging X and Y in (7.41). Also, we note that if $\text{cov}(X,Y) = 0$, then $\hat{Y} = E_Y[Y]$ or $X = x$ provides no information to help us predict Y. Clearly, this will be the case if X and Y are independent (see (7.37)) since independence of two random variables implies a covariance of zero. However, even if the covariance is zero, the random variables can still be dependent (see Figure 7.8) and so prediction should be possible. This apparent paradox is explained by the fact that in this case we must use a *nonlinear* predictor, not the simple linear function $aX + b$ (see Problem 8.27).

The optimal linear prediction of (7.41) can also be expressed in *standardized form*. A *standardized random variable* is defined to be one for which *the mean is zero and the variance is one*. An example would be a random variable that takes on the values ± 1 with equal probability. Any random variable can be standardized by subtracting the mean and dividing the result by the square root of the variance to form

$$X_s = \frac{X - E_X[X]}{\sqrt{\text{var}(X)}}$$

(see Problem 7.42). For example, if $X \sim \text{Pois}(\lambda)$, then $X_s = (X - \lambda)/\sqrt{\lambda}$, which is easily shown to have a mean of zero and a variance of one. We next seek the best linear prediction of the *standardized* Y based on a *standardized* $X = x$. To do so we define the standardized predictor based on a standardized $X_s = x_s$ as

$$\hat{Y}_s = \frac{\hat{Y} - E_Y[Y]}{\sqrt{\text{var}(Y)}}.$$

Then from (7.41), we have

$$\frac{\hat{Y} - E_Y[Y]}{\sqrt{\text{var}(Y)}} = \frac{\text{cov}(X,Y)}{\sqrt{\text{var}(Y)\text{var}(X)}} \frac{x - E_X[X]}{\sqrt{\text{var}(X)}}$$

and therefore

$$\hat{Y}_s = \frac{\text{cov}(X,Y)}{\sqrt{\text{var}(X)\text{var}(Y)}} x_s. \tag{7.42}$$

Example 7.14 – Previous example continued

For the previous example we have that

$$x_s = \frac{x - 5/4}{\sqrt{11/16}}$$

$$\hat{Y}_s = \frac{\hat{Y} - 3/2}{\sqrt{5/4}}$$

and

$$\frac{\text{cov}(X,Y)}{\sqrt{\text{var}(X)\text{var}(Y)}} = \frac{7/8}{\sqrt{(11/16)(5/4)}} \approx 0.94$$

so that

$$\hat{Y}_s = 0.94 x_s$$

and is displayed in Figure 7.9b.

$\diamondsuit$

The factor that scales x_s to produce $\hat{Y}_s$ is denoted by

$$\rho_{X,Y} = \frac{\text{cov}(X,Y)}{\sqrt{\text{var}(X)\text{var}(Y)}} \tag{7.43}$$

and is called the *correlation coefficient*. When X and Y have $\rho_{X,Y} \neq 0$, then X and Y are said to be *correlated*. If, however, the covariance is zero and hence $\rho_{X,Y} = 0$, then the random variables are said to be *uncorrelated*. Clearly, independent random variables are always uncorrelated, but not the other way around. Using the correlation coefficient allows us to express the best linear prediction in its standardized form as $\hat{Y}_s = \rho_{X,Y} x_s$. The correlation coefficient has an important property in that it is always less than one in magnitude. In the previous example, we had $\rho_{X,Y} \approx 0.94$.

Property 7.7 – Correlation coefficient is always less than or equal to one in magnitude or $|\rho_{X,Y}| \leq 1$.

Proof: The proof relies on the Cauchy-Schwarz inequality for random variables. This inequality is analogous to the usual one for the dot product of Euclidean vectors $\mathbf{v}$ and $\mathbf{w}$, which is

$$|\mathbf{v} \cdot \mathbf{w}| \leq ||\mathbf{v}|| \, ||\mathbf{w}||$$

where $||\mathbf{v}||$ denotes the length of the vector. Equality holds if and only if the vectors are collinear. Collinearity means that $\mathbf{w} = c\mathbf{v}$ for c a constant or the vectors point in the same direction. For random variables V and W the Cauchy-Schwarz inequality says that

$$|E_{V,W}[VW]| \leq \sqrt{E_V[V^2]}\sqrt{E_W[W^2]} \tag{7.44}$$

with equality if and only if $W = cV$ for c a constant. See Appendix 7A for a derivation. Thus letting $V = X - E_X[X]$ and $W = Y - E_Y[Y]$, we have

$$
\begin{aligned}
|\rho_{X,Y}| &= \frac{|\mathrm{cov}(X,Y)|}{\sqrt{\mathrm{var}(X)\mathrm{var}(Y)}} \\
&= \frac{|E_{V,W}[VW]|}{\sqrt{E_V[V^2]}\sqrt{E_W[W^2]}} \leq 1
\end{aligned}
$$

using (7.44). Equality will hold if and only if $W = cV$ or equivalently if $Y - E_Y[Y] = c(X - E_X[X])$, which is easily shown to imply that (see Problem 7.45)

$$\rho_{X,Y} = \begin{cases} 1 & \text{if } Y = aX + b \text{ with } a > 0 \\ -1 & \text{if } Y = aX + b \text{ with } a < 0 \end{cases}$$

for a and b constants.

$\square$

Note that when $\rho_{X,Y} = \pm 1$, Y can be perfectly predicted from X by using $Y = aX + b$. See also Figures 7.6a and 7.6b for examples of when $\rho_{X,Y} = +1$ and $\rho_{X,Y} = -1$, respectively.

⚠ **Correlation between random variables does not imply a causal relationship between the random variables.**

A frequent misapplication of probability is to assert that two quantities that are correlated ($\rho_{X,Y} \neq 0$) are such because one causes the other. To dispel this myth consider a survey in which all individuals older than 55 years of age in the U.S. are asked whether they have ever had prostate cancer and also their height in inches. Then, for each height in inches we compute the average number of individuals per

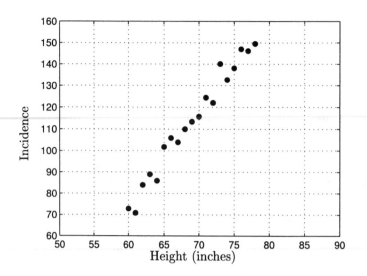

Figure 7.10: Incidence of prostate cancer per 1000 individuals older than age 55 versus height.

1000 who have had cancer. If we plot the average number, also called the incidence of cancer, versus height, a typical result would be as shown in Figure 7.10. This indicates a strong positive correlation of cancer with height. One might be tempted to conclude that growing taller causes prostate cancer. This is of course nonsense. What is actually shown is that segments of the population who are tall are associated with a higher incidence of cancer. This is because the portion of the population of individuals who are taller than the rest are predominately male. Females are not subject to prostate cancer, as they have no prostates! In summary, correlation between two variables only indicates an *association*, i.e., if one increases, then so does the other (if positively correlated). No physical or causal relationship need exist.

7.10   Joint Characteristic Functions

The characteristic function of a discrete random variable was introduced in Section 6.7. For two random variables we can define a *joint* characteristic function. For the random variables X and Y it is defined as

$$\phi_{X,Y}(\omega_X, \omega_Y) = E_{X,Y}[\exp[j(\omega_X X + \omega_Y Y)]]. \tag{7.45}$$

Assuming both random variables take on integer values, it is evaluated using (7.29) as

$$\phi_{X,Y}(\omega_X, \omega_Y) = \sum_{k=-\infty}^{\infty} \sum_{l=-\infty}^{\infty} p_{X,Y}[k,l] \exp[j(\omega_X k + \omega_Y l)]. \qquad (7.46)$$

It is seen to be the two-dimensional Fourier transform of the two-dimensional sequence $p_{X,Y}[k,l]$ (note the use of $+j$ as opposed to the more common $-j$ in the exponential). As in the case of a single random variable, the characteristic function can be used to find moments. In this case, the joint moments are given by the formula

$$E_{X,Y}[X^m Y^n] = \frac{1}{j^{m+n}} \frac{\partial^{m+n} \phi_{X,Y}(\omega_X, \omega_Y)}{\partial \omega_X^m \partial \omega_Y^n} \Bigg|_{\omega_X = \omega_Y = 0}. \qquad (7.47)$$

In particular, the first joint moment is found as

$$E_{X,Y}[XY] = -\frac{\partial^2 \phi_{X,Y}(\omega_X, \omega_Y)}{\partial \omega_X \partial \omega_Y} \Bigg|_{\omega_X = \omega_Y = 0}.$$

Another important application is to finding the PMF for the sum of two independent random variables. This application is based on the result that if X and Y are independent random variables, the joint characteristic function factors due to the property $E_{X,Y}[g(X)h(Y)] = E_X[g(X)]E_Y[h(Y)]$ (see (7.31)). Before deriving the PMF for the sum of two independent random variables, we prove the factorization result, and then give a theoretical application. The factorization of the characteristic function follows as

$$
\begin{aligned}
\phi_{X,Y}(\omega_X, \omega_Y) &= \sum_{k=-\infty}^{\infty} \sum_{l=-\infty}^{\infty} p_{X,Y}[k,l] \exp[j(\omega_X k + \omega_Y l)] \\
&= \sum_{k=-\infty}^{\infty} \sum_{l=-\infty}^{\infty} p_X[k] p_Y[l] \exp[j\omega_X k] \exp[j\omega_Y l] \quad \text{(joint PMF factors)} \\
&= \sum_{k=-\infty}^{\infty} p_X[k] \exp[j\omega_X k] \sum_{l=-\infty}^{\infty} p_Y[l] \exp[j\omega_Y l] \\
&= \phi_X(\omega_X) \phi_Y(\omega_Y). \quad \text{(definition of characteristic function} \qquad (7.48) \\
&\qquad\qquad\qquad \text{for single random variable).}
\end{aligned}
$$

The converse is also true—if the joint characteristic function factors, then X and Y are independent random variables. This can easily be shown to follow from the inverse Fourier transform relationship. As an application of the converse result, consider the tranformed random variables $W = g(X)$ and $Z = h(Y)$, where X and Y are independent. We prove that W and Z are independent as well, which is to say *functions of independent random variables are independent*. To do so we show that the joint characteristic function factors. The joint characteristic function of the transformed random variables is

$$\phi_{W,Z}(\omega_W, \omega_Z) = E_{W,Z}[\exp[j(\omega_W W + \omega_Z Z)]].$$

But we have that

$$
\begin{aligned}
\phi_{W,Z}(\omega_W, \omega_Z) &= E_{X,Y}[\exp[j(\omega_W g(X) + \omega_Z h(Y))]] && \text{(slight extension of (7.28))} \\
&= E_X[\exp(j\omega_W g(X))]E_Y[\exp(j\omega_Z h(Y))] && \text{(same argument as used to} \\
&&& \text{yield (7.31))} \\
&= E_W[\exp(j\omega_W W)]E_Z[\exp(j\omega_Z Z)] && \text{(from (6.5))} \\
&= \phi_W(\omega_W)\phi_Z(\omega_Z) && \text{(definition)}
\end{aligned}
$$

and hence W and Z are independent random variables. As a general result, we can now assert that *if X and Y are independent random variables, then so are $g(X)$ and $h(Y)$ for any functions g and h.*

Finally, consider the problem of determining the PMF for $Z = X + Y$, where X and Y are independent random variables. We have already solved this problem using the joint PMF approach with the final result given by (7.22). By using characteristic functions, we can simplify the derivation. The derivation proceeds as follows.

$$
\begin{aligned}
\phi_Z(\omega_Z) &= E_Z[\exp(j\omega_Z Z)] && \text{(definition)} \\
&= E_{X,Y}[\exp(j\omega_Z(X+Y))] && \text{(from (7.28) and (7.29))} \\
&= E_{X,Y}[\exp(j\omega_Z X)\exp(j\omega_Z Y)] \\
&= E_X[\exp(j\omega_Z X)]E_Y[\exp(j\omega_Z Y)] && \text{(from (7.31))} \\
&= \phi_X(\omega_Z)\phi_Y(\omega_Z).
\end{aligned}
$$

To find the PMF we take the inverse Fourier transform of $\phi_Z(\omega_Z)$, replacing ω_Z by the more usual notation ω, to yield

$$
\begin{aligned}
p_Z[k] &= \int_{-\pi}^{\pi} \phi_X(\omega)\phi_Y(\omega)\exp(-j\omega k)\frac{d\omega}{2\pi} \\
&= \sum_{i=-\infty}^{\infty} p_X[i]p_Y[k-i]
\end{aligned}
$$

which agrees with (7.22). The last result follows from the property that the Fourier transform of a convolution sum is the product of the Fourier transforms of the individual sequences.

7.11   Computer Simulation of Random Vectors

The method of generating realizations of a two-dimensional discrete random vector is nearly identical to the one-dimensional case. In fact, if X and Y are independent, then we generate a realization of X, say x_i, according to $p_X[x_i]$ and a realization of Y, say y_j, according to $p_Y[y_j]$ using the method of Chapter 5. Then we concatenate the realizations together to form the realization of the vector random variable as (x_i, y_j). Furthermore, independence reduces the problems of estimating a joint PMF, a joint CDF, etc. to that of the one-dimensional case. The joint PMF, for example, can be estimated by first estimating $p_X[x_i]$ as $\hat{p}_X[x_i]$, then estimating $p_Y[y_j]$ as $\hat{p}_Y[y_j]$, and finally forming the estimate of the joint PMF as $\hat{p}_{X,Y}[x_i, y_j] = \hat{p}_X[x_i]\hat{p}_Y[y_j]$.

When the random variables are not independent, we need to generate a realization of (X, Y) *simultaneously* since the value obtained for X is dependent on the value obtained for Y and vice versa. If both $\mathcal{S}_X$ and $\mathcal{S}_Y$ are finite, then a simple procedure is to consider each possible realization (x_i, y_j) as a single outcome with probability $p_{X,Y}[x_i, y_j]$. Then, we can apply the techniques of Section 5.9 directly. An example is given next.

Example 7.15 – Generating realizations of jointly distributed random variables

Assume a joint PMF as given in Table 7.7. A simple MATLAB program to generate

	$j = 0$	$j = 1$
$i = 0$	$\frac{1}{8}$	$\frac{1}{8}$
$i = 1$	$\frac{1}{4}$	$\frac{1}{2}$

Table 7.7: Joint PMF values for Example 7.15.

a set of M realizations of (X, Y) is given below.

```
for m=1:M
  u=rand(1,1);
  if u<=1/8
    x(m,1)=0;y(m,1)=0;
  elseif u>1/8&u<=1/4
    x(m,1)=0;y(m,1)=1;
  elseif u>1/4&u<=1/2
    x(m,1)=1;y(m,1)=0;
  else
    x(m,1)=1;y(m,1)=1;
  end
end
```

Once the realizations are available we can estimate the joint PMF and marginal

PMFs as

$$\hat{p}_{X,Y}[i,j] = \frac{\text{Number of outcomes equal to } (i,j)}{M} \qquad i = 0,1; j = 0,1$$
$$\hat{p}_X[i] = \hat{p}_{X,Y}[i,0] + \hat{p}_{X,Y}[i,1] \qquad i = 0,1$$
$$\hat{p}_Y[j] = \hat{p}_{X,Y}[0,j] + \hat{p}_{X,Y}[1,j] \qquad j = 0,1$$

and the joint moments are estimated as

$$\widehat{E_{X,Y}[X^k Y^l]} = \frac{1}{M} \sum_{m=1}^{M} x_m^k y_m^l$$

where (x_m, y_m) is the mth realization. Other quantities of interest are discussed in Problems 7.49 and 7.51.

7.12   Real-World Example – Assessing Health Risks

An increasingly common health problem in the United States is obesity. It has been found to be associated with many life-threatening illnesses, especially diabetes. One way to define what constitutes an obese person is via the body mass index (BMI) [CDC 2003]. It is computed as

$$\text{BMI} = \frac{703W}{H^2} \tag{7.49}$$

where W is the weight of the person in pounds and H is the person's height in inches. BMIs greater than 25 and less than 30 are considered to indicate an overweight person, and 30 and above an obese person [CDC 2003]. It is of great importance to be able to estimate the PMF of the BMI for a population of people. For example, in Chapter 4 we displayed a table of the joint probabilities of heights and weights for a hypothetical population of college students. For this population we would like to know the probability or percentage of obese persons. This percentage of the population would then be at risk for developing diabetes. To do so we could first determine the PMF of the BMI and then determine the probability of a BMI of 30 and above. From Table 4.1 or Figure 7.1 we have the joint PMF for the random vector (H, W). To find the PMF for the BMI we note that it is a function of H and W or in our previous notation, we wish to determine the PMF of $Z = g(X, Y)$, where Z denotes the BMI, X denotes the height, and Y denotes the weight. The solution follows immediately from (7.26). One slight modification that we must make in order to fit the data of Table 4.1 into our theoretical framework is to replace the height and weight intervals by their midpoint values. For example, in Table 4.1 the probability of observing a person with a height between $5'8''$ and $6'$ and a weight of between 130 and 160 lbs. is 0.06. We convert these intervals so that we can say that the probability of a person having a height of $5'10''$ and a weight of 145 lbs. is 0.06. Next to determine the PMF we first find the BMI for each height and weight using (7.49), rounding the result to the nearest integer. This is displayed in Table 7.8.

	W_1	W_2	W_3	W_4	W_5
	115	145	175	205	235
H_1 $5'2''$	21	27	32	37	43
H_2 $5'6''$	19	23	28	33	38
H_3 $5'10''$	16	21	25	29	34
H_4 $6'2''$	15	19	22	26	30
H_5 $6'6''$	13	17	20	24	27

Table 7.8: Body mass indexes for heights and weights of hypothetical college students.

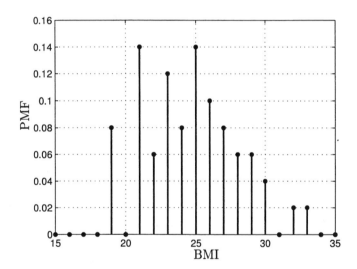

Figure 7.11: Probability mass function for body mass index of hypothetical college population.

Then, we determine the PMF by using (7.26). For example, for a BMI = 21, we require from Table 7.8 the entries $(H, W) = (5'2'', 115)$ and $(H, W) = (5'10'', 145)$. But from Table 4.1 we see that

$$P[H = 5'2'', W = 115] = 0.08$$
$$P[H = 5'10'', W = 145] = 0.06$$

and therefore $P[\text{BMI} = 21] = 0.14$. The other values of the PMF of the BMI are found similarly. This produces the PMF shown in Figure 7.12. It is seen that the probability of being obese as defined by the BMI (BMI $\geq$ 30) is 0.08. Stated another way 8% of the population of college students are obese and so are at risk for diabetes.

References

CDC, "Nutrition and Physical Activity," Center for Disease Control, http://www. cdc.gov/nccdphp/dnpa/bmi/bmi-adult-formula.htm, 2003.

Jackson, L.B., *Signals, Systems, and Transforms*, Addison-Wesley, Reading, MA, 1991.

Problems

7.1 (w) A chess piece is placed on a chessboard, which consists of an 8×8 array of 64 squares. Specify a numerical sample space $\mathcal{S}_{X,Y}$ for the location of the chess piece.

7.2 (w) Two coins are tossed in succession with a head being mapped into a $+1$ and a tail being mapped into a -1. If a random vector is defined as (X, Y) with X representing the mapping of the first toss and Y representing the mapping of the second toss, draw the mapping. Use Figure 7.2 as a guide. Also, what is $\mathcal{S}_{X,Y}$?

7.3 ($\ddot\smile$) (w) A woman has a penny, a nickel, and a dime in her pocket. If she chooses two coins from her pocket in succession, what is the sample space $\mathcal{S}$ of possible outcomes? If these outcomes are next mapped into the values of the coins, what is the numerical sample space $\mathcal{S}_{X,Y}$?

7.4 (w) If $\mathcal{S}_X = \{1,2\}$ and $\mathcal{S}_Y = \{3,4\}$, plot the points in the plane comprising $\mathcal{S}_{X,Y} = \mathcal{S}_X \times \mathcal{S}_Y$. What is the size of $\mathcal{S}_{X,Y}$?

7.5 (w) Two dice are tossed. The number of dots observed on the dice are added together to form the random variable X and also differenced to form Y. Determine the possible outcomes of the random vector (X, Y) and plot them in the plane. How many possible outcomes are there?

7.6 (f) A two-dimensional sequence is given by

$$p_{X,Y}[i,j] = c(1 - p_1)^i(1 - p_2)^j \qquad i = 1, 2, \ldots; j = 1, 2, \ldots$$

where $0 < p_1 < 1$, $0 < p_2 < 1$, and c is a constant. Find c to make $p_{X,Y}$ a valid joint PMF.

7.7 (f) Is

$$p_{X,Y}[i,j] = \left(\frac{1}{2}\right)^{i+j} \qquad i = 0, 1, \ldots; j = 0, 1, \ldots$$

a valid joint PMF?

7.8 (☺) (w) A single coin is tossed twice. A head outcome is mapped into a 1 and a tail outcome into a 0 to yield a numerical outcome. Next, a random vector (X, Y) is defined as

$$X = \text{outcome of first toss} + \text{outcome of second toss}$$
$$Y = \text{outcome of first toss} - \text{outcome of second toss}.$$

Find the joint PMF for (X, Y), assuming the outcomes (x_i, y_j) are equally likely.

7.9 (f) Find the joint PMF for the experiment described in Example 7.1. Assume each outcome in S is equally likely. How can you check your answer?

7.10 (☺) (f) The sample space for a random vector is $S_{X,Y} = \{(i,j) : i = 1, 2, 3, 4, 5; j = 1, 2, 3, 4\}$. If the outcomes are equally likely, find $P[(X,Y) \in A]$, where $A = \{(i,j) : 1 \le i \le 2; 3 \le j \le 4\}$.

7.11 (f) A joint PMF is given as $p_{X,Y}[i,j] = (1/2)^{i+j}$ for $i = 1, 2, \ldots; j = 1, 2, \ldots$. If $A = \{(i,j) : 1 \le i \le 3; j \ge 2\}$, find $P[A]$.

7.12 (f) The values of a joint PMF are given in Table 7.9. Determine the marginal PMFs.

	$j = 0$	$j = 1$	$j = 2$
$i = 0$	$\frac{1}{8}$	0	$\frac{1}{4}$
$i = 1$	0	$\frac{1}{8}$	$\frac{1}{4}$
$i = 2$	$\frac{1}{8}$	0	$\frac{1}{8}$

Table 7.9: Joint PMF values for Problem 7.12.

7.13 (☺) (f) If a joint PMF is given by

$$p_{X,Y}[i,j] = p^2(1-p)^{i+j-2} \qquad i = 1, 2, \ldots; j = 1, 2, \ldots$$

find the marginal PMFs.

7.14 (f) If a joint PMF is given by $p_{X,Y}[i,j] = 1/36$ for $i = 1, 2, 3, 4, 5, 6; j = 1, 2, 3, 4, 5, 6$, find the marginal PMFs.

7.15 (w) A joint PMF is given by

$$p_{X,Y}[i,j] = c \binom{10}{j} \left(\frac{1}{2}\right)^{10} \qquad i = 0,1; j = 0,1,\ldots,10$$

where c is some unknown constant. Find c so that the joint PMF is valid and then determine the marginal PMFs. Hint: Recall the binomial PMF.

7.16 (☺) (w) Find another set of values for the joint PMF that will yield the same marginal PMFs as given in Table 7.2.

7.17 (t) Prove Properties 7.3 and 7.4 for the joint CDF by relying on the standard properties of probabilities of events.

7.18 (w) Sketch the joint CDF for the joint PMF given in Table 7.2. Do this by shading each region in the x-y plane that has the same value.

7.19 (☺) (w) A joint PMF is given by

$$p_{X,Y}[i,j] = \begin{cases} \frac{1}{4} & (i,j) = (0,0) \\ \frac{1}{4} & (i,j) = (1,1) \\ \frac{1}{4} & (i,j) = (1,0) \\ \frac{1}{4} & (i,j) = (1,-1) \end{cases}$$

Are X and Y independent?

7.20 (t) Prove that if the random variables X and Y are independent, then the joint CDF factors as $F_{X,Y}(x,y) = F_X(x)F_Y(y)$.

7.21 (t) If a joint PMF is given by

$$p_{X,Y}[i,j] = \begin{cases} a & (i,j) = (0,0) \\ b & (i,j) = (0,1) \\ c & (i,j) = (1,0) \\ d & (i,j) = (1,1) \end{cases}$$

where of course we must have $a+b+c+d = 1$, show that a necessary condition for the random variables to be independent is $ad = bc$. This can be used to quickly assert that the random variables are not independent as for the case shown in Table 7.5.

7.22 (f) If $X \sim \text{Ber}(p_X)$ and $Y \sim \text{Ber}(p_Y)$, and X and Y are independent, what is the joint PMF?

7.23 ($\smile$) **(w)** If the joint PMF is given as

$$p_{X,Y}[i,j] = \binom{10}{i}\binom{11}{j}\left(\frac{1}{2}\right)^{21} \qquad i = 0, 1, \ldots, 10; j = 0, 1, \ldots, 11$$

are X and Y independent? What are the marginal PMFs?

7.24 (t) Assume that X and Y are discrete random variables that take on all integer values and are independent. Prove that the PMF of $Z = X - Y$ is given by

$$p_Z[l] = \sum_{k=-\infty}^{\infty} p_X[k]p_Y[k-l] \qquad l = \ldots, -1, 0, 1, \ldots$$

by following the same procedure as was used to derive (7.22). Note that the transformation from (X, Y) to (W, Z) is one-to-one. Next show that if X and Y take on nonnegative integer values only, then

$$p_Z[l] = \sum_{k=\max(0,l)}^{\infty} p_X[k]p_Y[k-l] \qquad l = \ldots, -1, 0, 1, \ldots \quad .$$

7.25 (f) Using the result of Problem 7.24 find the PMF for $Z = X - Y$ if $X \sim \text{Pois}(\lambda_X)$, $Y \sim \text{Pois}(\lambda_Y)$, and X and Y are independent. Hint: The result will be in the form of infinite sums.

7.26 (w) Find the PMF for $Z = \max(X, Y)$ if the joint PMF is given in Table 7.5.

7.27 ($\smile$) **(f)** If $X \sim \text{Ber}(1/2)$, $Y \sim \text{Ber}(1/2)$, and X and Y are independent, find the PMF for $Z = X + Y$. Why does the width of the PMF increase? Does the variance increase?

7.28 (t) Prove that $E_{X,Y}[g(X)] = E_X[g(X)]$. Do X and Y have to be independent?

7.29 (t) Prove that

$$E_{X,Y}[ag(X) + bh(Y)] = aE_X[g(X)] + bE_Y[h(Y)].$$

7.30 (t) Prove (7.31).

7.31 (t) Find a formula for $\text{var}(X - Y)$ similar to (7.33). What can you say about the relationship between $\text{var}(X + Y)$ and $\text{var}(X - Y)$ if X and Y are uncorrelated?

7.32 (f) Find the covariance for the joint PMF given in Table 7.4. How do you know the value that you obtained is correct?

7.33 ($\smile$) **(f)** Find the covariance for the joint PMF given in Table 7.5.

7.34 (t) Prove the following properties of the covariance:

$$\begin{aligned}
\operatorname{cov}(X,Y) &= E_{X,Y}[XY] - E_X[X]E_Y[Y] \\
\operatorname{cov}(X,X) &= \operatorname{var}(X) \\
\operatorname{cov}(Y,X) &= \operatorname{cov}(X,Y) \\
\operatorname{cov}(cX,Y) &= c\left[\operatorname{cov}(X,Y)\right] \\
\operatorname{cov}(X,cY) &= c\left[\operatorname{cov}(X,Y)\right] \\
\operatorname{cov}(X,X+Y) &= \operatorname{cov}(X,X) + \operatorname{cov}(X,Y) \\
\operatorname{cov}(X+Y,X) &= \operatorname{cov}(X,X) + \operatorname{cov}(Y,X)
\end{aligned}$$

for c a constant.

7.35 (t) If X and Y have a covariance of $\operatorname{cov}(X,Y)$, we can transform them to a new pair of random variables whose covariance is zero. To do so we let

$$\begin{aligned}
W &= X \\
Z &= aX + Y
\end{aligned}$$

where $a = -\operatorname{cov}(X,Y)/\operatorname{var}(X)$. Show that $\operatorname{cov}(W,Z) = 0$. This process is called *decorrelating the random variables*. See also Example 9.4 for another method.

7.36 (f) Apply the results of Problem 7.35 to the joint PMF given in Table 7.5. Verify by direct calculation that $\operatorname{cov}(W,Z) = 0$.

7.37 (⌣) (f) If the joint PMF is given as

$$p_{X,Y}[i,j] = \left(\frac{1}{2}\right)^{i+j} \qquad i = 1,2,\ldots; j = 1,2,\ldots$$

compute the covariance.

7.38 (⌣) (f) Determine the minimum mean square error for the joint PMF shown in Figure 7.9a. You will need to evaluate $E_{X,Y}[(Y - ((14/11)X - 1/11))^2]$.

7.39 (t,f) Prove that the minimum mean square error of the optimal linear predictor is given by

$$\operatorname{mse}_{\min} = E_{X,Y}[(Y - (a_{\mathrm{opt}}X + b_{\mathrm{opt}}))^2] = \operatorname{var}(Y)\left(1 - \rho_{X,Y}^2\right).$$

Use this formula to check your result for Problem 7.38.

7.40 (⌣) (w) In this problem we compare the prediction of a random variable with and without the knowledge of a second random variable outcome. Consider the joint PMF shown below. First determine the optimal linear prediction of Y

	$j = 0$	$j = 1$
$i = 0$	$\frac{1}{8}$	$\frac{1}{4}$
$i = 1$	$\frac{1}{4}$	$\frac{3}{8}$

Table 7.10: Joint PMF values for Problem 7.40.

without any knowledge of the outcome of X (see Section 6.6). Also, compute the minimum mean square error. Next determine the optimal linear prediction of Y based on the knowledge that $X = x$ and compute the minimum mean square error. Plot the predictions versus x in the plane. How do the minimum mean square errors compare?

7.41 (☺) **(w,c)** For the joint PMF of height and weight shown in Figure 7.1 determine the best linear prediction of weight based on a knowledge of height. You will need to use Table 4.1 as well as a computer to carry out this problem. Does your answer seem reasonable? Is your prediction of a person's weight if the height is 70 inches reasonable? How about if the height is 78 inches? Can you explain the difference?

7.42 **(f)** Prove that the transformed random variable

$$\frac{X - E_X[X]}{\sqrt{\text{var}(X)}}$$

has an expected value of 0 and a variance of 1.

7.43 (☺) **(w)** The linear prediction of one random variable based on the outcome of another becomes more difficult if noise is present. We model noise as the addition of an uncorrelated random variable. Specifically, assume that we wish to predict X based on observing $X + N$, where N represents the noise. If X and N are both zero mean random variables that are uncorrelated with each other, determine the correlation coefficient between $W = X$ and $Z = X + N$. How does it depend on the power in X, which is defined as $E_X[X^2]$, and the power in N, also defined as $E_N[N^2]$?

7.44 **(w)** Consider $\text{var}(X + Y)$, where X and Y are correlated random variables. How is the variance of a sum of random variables affected by the correlation between the random variables? Hint: Express the variance of the sum of the random variables using the correlation coefficient.

7.45 **(f)** Prove that if $Y = aX + b$, where a and b are constants, then $\rho_{X,Y} = 1$ if $a > 0$ and $\rho_{X,Y} = -1$ if $a < 0$.

7.46 (ツ) (w) If $X \sim \text{Ber}(1/2)$, $Y \sim \text{Ber}(1/2)$, and X and Y are independent, find the PMF for $Z = X + Y$. Use the characteristic function approach to do so. Compare your results to that of Problem 7.27.

7.47 (w) Using characteristic functions prove that the binomial PMF has the reproducing property. That is to say, if $X \sim \text{bin}(M_X, p)$, $Y \sim \text{bin}(M_Y, p)$, and X and Y are independent, then $Z = X + Y \sim \text{bin}(M_X + M_Y, p)$. Why does this make sense in light of the fact that a sequence of independent Bernoulli trials can be used to derive the binomial PMF?

7.48 (ツ) (c) Using the joint PMF shown in Table 7.7 generate realizations of the random vector (X, Y) and estimate its joint and marginal PMFs. Compare your estimated results to the true values.

7.49 (ツ) (c) For the joint PMF shown in Table 7.7 determine the correlation coefficient. Next use a computer simulation to generate realizations of the random vector (X, Y) and estimate the correlation coefficient as

$$\hat{\rho}_{X,Y} = \frac{\frac{1}{M} \sum_{m=1}^{M} x_m y_m - \bar{x}\bar{y}}{\sqrt{\left(\frac{1}{M} \sum_{m=1}^{M} x_m^2 - \bar{x}^2 \right) \left(\frac{1}{M} \sum_{m=1}^{M} y_m^2 - \bar{y}^2 \right)}}$$

where

$$\bar{x} = \frac{1}{M} \sum_{m=1}^{M} x_m$$

$$\bar{y} = \frac{1}{M} \sum_{m=1}^{M} y_m$$

and (x_m, y_m) is the mth realization.

7.50 (w,c) If $X \sim \text{geom}(p)$, $Y \sim \text{geom}(p)$, and X and Y are independent, show that the PMF of $Z = X + Y$ is given by

$$p_Z[k] = p^2(k-1)(1-p)^{k-2} \qquad k = 2, 3, \ldots.$$

To avoid errors use the discrete unit step sequence. Next, for $p = 1/2$ generate realizations of Z by first generating realizations of X, then generating realizations of Y and adding each pair of realizations together. Estimate the PMF of Z and compare it to the true PMF.

7.51 (w,c) Using the joint PMF given in Table 7.5 determine the covariance to show that it is nonzero and hence X and Y are correlated. Next use the procedure of Problem 7.35 to determine transformed random variables W and

Z that are uncorrelated. Verify that W and Z are uncorrelated by estimating the covariance as

$$\widehat{\mathrm{cov}(W, Z)} = \frac{1}{M} \sum_{m=1}^{M} w_m z_m - \bar{w}\bar{z}$$

where

$$\bar{w} = \frac{1}{M} \sum_{m=1}^{M} w_m$$

$$\bar{z} = \frac{1}{M} \sum_{m=1}^{M} z_m$$

and (w_m, z_m) is the mth realization. Be sure to generate the realizations of W and Z as $w_m = x_m$ and $z_m = ax_m + y_m$, where (x_m, y_m) is the mth realization of (X, Y).

Appendix 7A

Derivation of the Cauchy-Schwarz Inequality

The Cauchy-Schwarz inequality was given by

$$|E_{V,W}[VW]| \leq \sqrt{E_V[V^2]}\sqrt{E_W[W^2]} \qquad (7A.1)$$

with equality holding if and only if $W = cV$, for c a constant. To prove this, we first note that for all $\alpha \neq 0$ and $\beta \neq 0$

$$E_{V,W}[(\alpha V - \beta W)^2] \geq 0. \qquad (7A.2)$$

If we let

$$\begin{aligned} \alpha &= \sqrt{E_W[W^2]} \\ \beta &= \sqrt{E_V[V^2]} \end{aligned}$$

then we have that

$$\begin{aligned} E_{V,W}[(\sqrt{E_W[W^2]}V - \sqrt{E_V[V^2]}W)^2] &\geq 0 \\ E_{V,W}[E_W[W^2]V^2 - 2\sqrt{E_W[W^2]}\sqrt{E_V[V^2]}VW + E_V[V^2]W^2] &\geq 0 \\ E_W[W^2]E_V[V^2] - 2\sqrt{E_W[W^2]}\sqrt{E_V[V^2]}E_{V,W}[VW] + E_V[V^2]E_W[W^2] &\geq 0 \end{aligned}$$

since $E_{V,W}[g(W)] = E_W[g(W)]$, etc. , which results in

$$E_W[W^2]E_V[V^2] - \sqrt{E_W[W^2]}\sqrt{E_V[V^2]}E_{V,W}[VW] \geq 0.$$

Dividing by $E_W[W^2]E_V[V^2]$ produces

$$1 - \frac{E_{V,W}[VW]}{\sqrt{E_W[W^2]}\sqrt{E_V[V^2]}} \geq 0$$

or finally, upon rearranging terms we have that

$$\frac{E_{V,W}[VW]}{\sqrt{E_V[V^2]}\sqrt{E_W[W^2]}} \leq 1$$

or

$$E_{V,W}[VW] \leq \sqrt{E_V[V^2]}\sqrt{E_W[W^2]}.$$

By replacing the negative sign in (7A.2) by a positive sign and proceeding in an identical manner, we will obtain

$$-E_{V,W}[VW] \leq \sqrt{E_V[V^2]}\sqrt{E_W[W^2]}$$

and hence combining the two results yields the desired inequality. To determine when the equal sign will hold, we note that

$$E_{V,W}[(\alpha V - \beta W)^2] = \sum_{v_i}\sum_{w_j}(\alpha v_i - \beta w_j)^2 p_{V,W}[v_i, w_j]$$

which can only equal zero when $(\alpha v_i - \beta w_j)^2 = 0$ for all i and j since $p_{V,W}[v_i, w_j] > 0$. Thus, for equality to hold we must have

$$\alpha v_i = \beta w_j \qquad \text{all } i \text{ and } j$$

which is equivalent to requiring

$$\alpha V = \beta W$$

or finally dividing by β (asssumed not equal to zero), we obtain the condition for equality as

$$W = \frac{\alpha}{\beta}V = cV$$

for c a constant.

Chapter 8

Conditional Probability Mass Functions

8.1 Introduction

In Chapter 4 we discussed the concept of conditional probability. We recall that a conditional probability $P[A|B]$ is the probability of an event A, given that we know that some other event B has occurred. Except for the case when the two events are independent of each other, the knowledge that B has occurred will change the probability $P[A]$. In other words, $P[A|B]$ is our new probability in light of the additional knowledge. In many practical situations, two random mechanisms are at work and are described by events A and B. An example of such a compound experiment was given in Example 4.2. To compute probabilities for a compound experiment it is usually convenient to use a conditioning argument to simplify the reasoning. For example, say we choose one of two coins and toss it 4 times. We might inquire as to the probability of observing 2 or more heads. However, this probability will depend upon which coin was chosen, as for example in the situation where one coin is fair and the other coin is weighted. It is therefore convenient to define *conditional probability mass functions*, $p_X[k|\text{coin 1 chosen}]$ and $p_X[k|\text{coin 2 chosen}]$, since once we know which coin is chosen, we can easily specify the PMF. In particular, for this example the conditional PMF is a binomial one whose value of p depends upon which coin is chosen and with k denoting the number of heads (see (5.6)). Once the conditional PMFs are known, we have by the law of total probability (see (4.4)) that the probability of observing k heads for this experiment is given by the PMF

$$
\begin{aligned}
p_X[k] \quad = \quad & p_X[k|\text{coin 1 chosen}]P[\text{coin 1 chosen}] \\
& + p_X[k|\text{coin 2 chosen}]P[\text{coin 2 chosen}].
\end{aligned}
$$

Therefore, the desired probability of observing 2 or more heads is

$$P[X \geq 2] = \sum_{k=2}^{4} p_X[k]$$

$$= \sum_{k=2}^{4} (p_X[k|\text{coin 1 chosen}]P[\text{coin 1 chosen}]$$

$$+ p_X[k|\text{coin 2 chosen}]P[\text{coin 2 chosen}]).$$

The PMF that is required depends directly on the conditional PMFs (of which there are two). The use of conditional PMFs greatly simplifies our task in that *given* the event, i.e., the coin chosen, the PMF of the number of heads observed readily follows. Also, in many problems, including this one, it is actually the conditional PMFs that are specified in the description of the experimental procedure. It makes sense, therefore, to define a conditional PMF and study its properties. For the most part, the definitions and properties will mirror those of the conditional probabillity $P[A|B]$, where A and B are events defined on $S_{X,Y}$.

8.2 Summary

The utility of defining a conditional PMF is illustrated in Section 8.3. It is especially appropriate when the experiment is a compound one, in which the second part of the experiment depends upon the outcome of the first part. The definition of the conditional PMF is given in (8.7). It has the usual properties of a PMF, that of being between 0 and 1 and also summing to one. Its properties and relationships are summarized by Properties 8.1–8.5. The conditional PMF is related to the joint PMF and the marginal PMFs by these properties. They are also depicted in Figure 8.4 for easy reference. If the random variables are independent, then the conditional PMF reduces to the usual marginal PMF as shown in (8.22). For general probability calculations based on the conditional PMF one can use (8.23). In Section 8.5 it is shown how to use conditioning arguments to simplify the derivation of the PMF for $Z = g(X, Y)$. The PMF can be found using (8.24), which makes use of the conditional PMF. In particular, if X and Y are independent, the procedure is especially simplified with examples given in Section 8.5. The mean of the conditional PMF is defined by (8.30). It is computed by the usual procedures but uses the conditional PMF as the "averaging" PMF. It is next shown that the mean of the unconditional PMF can be found by averaging over the means of the conditional PMFs as given by (8.35). This simplifies the computation. Generation of realizations of random vectors (X, Y) can be simplified using conditioning arguments. An illustration and MATLAB code segment is given in Section 8.7. Finally, an application of conditioning to the modeling of human learning is described in Section 8.8. Utilizing the posterior PMF, which is a conditional PMF, one can demonstrate that "learning" takes place as the result of observing the outcomes of repeated experiments. The degree of learning is embodied in the posterior PMF.

8.3 Conditional Probability Mass Function

We continue with the introductory example to illustrate the utility of the conditional probability mass function. Summarizing the introductory problem, we have an experimental procedure in which we first choose a coin, either coin 1 or coin 2. Coin 1 has a probability of heads of p_1, while coin 2 has a probability of heads of p_2. Let X be the discrete random variable describing the outcome of the coin choice so that

$$X = \begin{cases} 1 & \text{if coin 1 is chosen} \\ 2 & \text{if coin 2 is chosen.} \end{cases}$$

Since $\mathcal{S}_X = \{1, 2\}$, we assign a PMF to X of

$$p_X[i] = \begin{cases} \alpha & i = 1 \\ 1 - \alpha & i = 2 \end{cases} \tag{8.1}$$

where $0 < \alpha < 1$. The second part of the experiment consists of tossing the chosen coin 4 times in succession. Call the outcome of the number of heads observed as Y and note that $\mathcal{S}_Y = \{0, 1, 2, 3, 4\}$. Hence, the overall set of outcomes of the compound experiment is $\mathcal{S}_{X,Y} = \mathcal{S}_X \times \mathcal{S}_Y$, which is shown in Figure 8.1. The overall

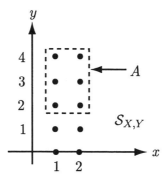

Figure 8.1: Mapping for coin toss example, x denotes the coin chosen while y denotes the number of heads observed.

outcome is described by the random vector (X, Y), where X is the coin chosen and Y is the number of heads observed for the 4 coin tosses. If we wish to determine the probability of 2 or more heads, then this is the probability of the set A shown in Figure 8.1. It is given mathematically as

$$P[A] = \sum_{\{(i,j):(i,j)\in A\}} p_{X,Y}[i,j]$$

$$= \sum_{i=1}^{2}\sum_{j=2}^{4} p_{X,Y}[i,j]. \tag{8.2}$$

Hence, we need only specify the joint PMF to determine the desired probability. To do so we make use of our definition of the joint PMF as well as our earlier concepts from conditional probability (see Chapter 4). Recall from Chapter 7 the definitions of the joint PMF and marginal PMF as

$$p_{X,Y}[i,j] = P[X = i, Y = j]$$
$$p_X[i] = P[X = i].$$

By using the definition of conditional probability for events we have

$$
\begin{aligned}
p_{X,Y}[i,j] &= P[X = i, Y = j] && \text{(definition of joint PMF)}\\
&= P[Y = j|X = i]P[X = i] && \text{(definition of conditional prob.)}\\
&= P[Y = j|X = i]p_X[i] && \text{(definition of marginal PMF).} \tag{8.3}
\end{aligned}
$$

From (8.1) we have $p_X[i]$ and from the experimental description we can determine $P[Y = j|X = i]$. When $X = 1$, we toss a coin with a probability of heads p_1, and when $X = 2$, we toss a coin with a probability of heads p_2. Also, we have previously shown that for a coin with a probability of heads p_i that is tossed 4 times, the number of heads observed has a binomial PMF. Thus, for $i = 1, 2$

$$P[Y = j|X = i] = \binom{4}{j} p_i^j (1 - p_i)^{4-j} \qquad j = 0, 1, 2, 3, 4. \tag{8.4}$$

Note that the probability depends on the outcome $X = i$ via p_i. Also, *for a given value* of $X = i$, the probability has all the usual properties of a PMF. These properties are

$$0 \leq P[Y = j|X = i] \leq 1$$

$$\sum_{j=0}^{4} P[Y = j|X = i] = 1.$$

It is therefore appropriate to define $P[Y = j|X = i]$ as a *conditional PMF*. We will denote it by

$$p_{Y|X}[j|i] = P[Y = j|X = i] \qquad j = 0, 1, 2, 3, 4.$$

Examples are plotted in Figure 8.2 for $p_1 = 1/4$ and $p_2 = 1/2$. Returning to our

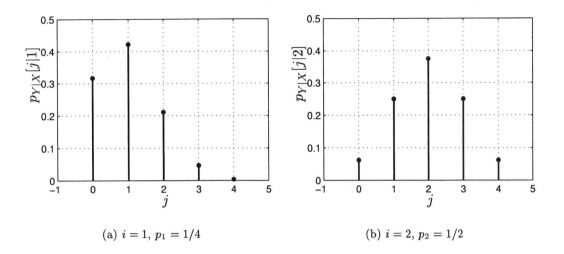

(a) $i = 1$, $p_1 = 1/4$ (b) $i = 2$, $p_2 = 1/2$

Figure 8.2: Conditional PMFs given by (8.4).

problem we can now determine the joint PMF. Using (8.3) we have

$$p_{X,Y}[i,j] = p_{Y|X}[j|i]p_X[i] \qquad (8.5)$$

and using (8.4) and (8.1) the joint PMF is

$$
\begin{aligned}
p_{X,Y}[i,j] &= \binom{4}{j} p_1^j (1-p_1)^{4-j} \alpha & i = 1; j = 0,1,2,3,4 \\
&= \binom{4}{j} p_2^j (1-p_2)^{4-j} (1-\alpha) & i = 2; j = 0,1,2,3,4.
\end{aligned}
$$

Finally the desired probability is from (8.2)

$$
\begin{aligned}
P[A] &= \sum_{j=2}^{4} p_{X,Y}[1,j] + \sum_{j=2}^{4} p_{X,Y}[2,j] \\
&= \sum_{j=2}^{4} \binom{4}{j} p_1^j (1-p_1)^{4-j} \alpha + \sum_{j=2}^{4} \binom{4}{j} p_2^j (1-p_2)^{4-j} (1-\alpha).
\end{aligned}
$$

As an example, if $p_1 = 1/4$ and $p_2 = 3/4$, we have for $\alpha = 1/2$, that $P[A] = 0.6055$, but if $\alpha = 1/8$, then $P[A] = 0.8633$. Can you explain this?

Note from (8.5) that the conditional PMF is also expressed as

$$p_{Y|X}[j|i] = \frac{p_{X,Y}[i,j]}{p_X[i]} \qquad (8.6)$$

and is only a renaming for the conditional probability of the event that $A_j = \{s : Y(s) = j\}$ given that $B_i = \{s : X(s) = i\}$. To make this connection we have

$$
\begin{aligned}
p_{Y|X}[j|i] &= P[Y = j | X = i] = \frac{P[X = i, Y = j]}{P[X = i]} \\
&= \frac{P[A_j \cap B_i]}{P[B_i]} \\
&= P[A_j | B_i]
\end{aligned}
$$

and hence $p_{Y|X}[j|i]$ is a conditional probability for the events A_j and B_i.

8.4   Joint, Conditional, and Marginal PMFs

As evidenced by (8.6), there are relationships between the joint, conditional, and marginal PMFs. In this section we describe these relationships. To do so we rewrite the definition of the conditional PMF in slightly more generality as

$$
p_{Y|X}[y_j|x_i] = \frac{p_{X,Y}[x_i, y_j]}{p_X[x_i]} \tag{8.7}
$$

for a sample space $\mathcal{S}_{X,Y}$ which may not consist solely of integer two-tuples. *It is always assumed that $p_X[x_i] \neq 0$.* Otherwise, the definition does not make any sense. The conditional PMF, although appearing to be a function of two variables, x_i and y_j, should be viewed as a *family* or set of PMFs. Each PMF in the family is a valid PMF *when x_i is considered to be a constant.* In the example of the previous section, we had $p_{Y|X}[j|1]$ and $p_{Y|X}[j|2]$. The family is therefore $\{p_{Y|X}[j|1], p_{Y|X}[j|2]\}$ and each member is a valid PMF, whose values depend on j. Hence, we would expect that (see Problem 8.4)

$$
\begin{aligned}
\sum_{j=-\infty}^{\infty} p_{Y|X}[j|1] &= 1 \\
\sum_{j=-\infty}^{\infty} p_{Y|X}[j|2] &= 1
\end{aligned}
$$

but *not* $\sum_{i=-\infty}^{\infty} p_{Y|X}[j|i] = 1$ (see also Problem 4.9). Before proceeding to list the relationships between the various PMFs, we give an example of the calculation of the conditional PMF based on (8.7).

Example 8.1 –   Two dice toss

Two dice are tossed with all outcomes assumed to be equally likely. The number of dots observed on each die are added together. What is the conditional PMF of the sum if it is known that the sum is even? We begin by letting Y be the sum and define $X = 1$ if the sum is even and $X = 0$ if the sum is odd. Thus, we wish to determine

$p_{Y|X}[j|1]$ and $p_{Y|X}[j|0]$ for all j. The sample space for Y is $\mathcal{S}_Y = \{2, 3, \ldots, 12\}$ as can be seen from Table 8.1, which lists the sum of the two dice outcomes as a function of the outcomes for each die. The boldfaced entries are the ones for which

	$j = 1$	$j = 2$	$j = 3$	$j = 4$	$j = 5$	$j = 6$
$i = 1$	**2**	3	**4**	5	**6**	7
$i = 2$	3	**4**	5	**6**	7	**8**
$i = 3$	**4**	5	**6**	7	**8**	9
$i = 4$	5	**6**	7	**8**	9	**10**
$i = 5$	**6**	7	**8**	9	**10**	11
$i = 6$	7	**8**	9	**10**	11	**12**

Table 8.1: The sum of the number of dots observed for two dice – boldface indicates an even sum.

the sum is even and therefore comprise the sample space for $p_{Y|X}[j|1]$. Note that each outcome (i, j) has an assumed probability of occurring of $1/36$. Now, using (8.7)

$$p_{Y|X}[j|1] = \frac{p_{X,Y}[1, j]}{p_X[1]} \qquad j = 2, 4, 6, 8, 10, 12 \qquad (8.8)$$

where $p_{X,Y}[1, j]$ is the probability of the sum being even and also equaling j. Since we assume in (8.8) that j is even (otherwise $p_{Y|X}[j|1] = 0$), we have that $p_{X,Y}[1, j] = p_Y[j]$ for $j = 2, 4, 6, 8, 10, 12$. Also, there are 18 even outcomes, which results in $p_X[1] = 1/2$. Thus, (8.8) becomes

$$
\begin{aligned}
p_{Y|X}[j|1] &= \frac{p_Y[j]}{1/2} \\
&= \frac{N_j(1/36)}{1/2} \\
&= \frac{1}{18} N_j
\end{aligned}
$$

where N_j is the number of outcomes in $\mathcal{S}_{X,Y}$ for which the sum is j. From Table 8.1 we can easily find N_j so that

$$p_{Y|X}[j|1] = \begin{cases} \frac{1}{18} & j = 2 \\ \frac{3}{18} & j = 4 \\ \frac{5}{18} & j = 6 \\ \frac{5}{18} & j = 8 \\ \frac{3}{18} & j = 10 \\ \frac{1}{18} & j = 12. \end{cases} \qquad (8.9)$$

Note that as expected $\sum_j p_{Y|X}[j|1] = 1$. The reader is asked to verify by a similar calculation that (see Problem 8.7)

$$p_{Y|X}[j|0] = \begin{cases} \frac{2}{18} & j = 3 \\ \frac{4}{18} & j = 5 \\ \frac{6}{18} & j = 7 \\ \frac{4}{18} & j = 9 \\ \frac{2}{18} & j = 11. \end{cases} \tag{8.10}$$

These conditional PMFs are shown in Figure 8.3. Also, note that $p_{Y|X}[j|0] \neq$

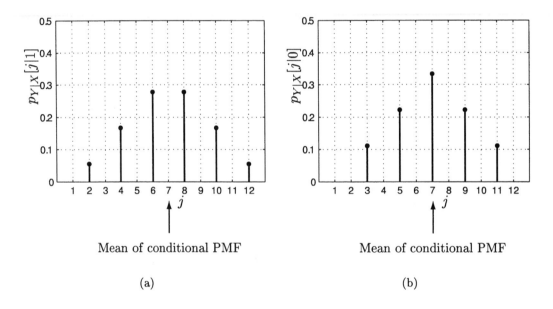

(a)   (b)

Figure 8.3: Conditional PMFs for Example 8.1.

$1 - p_{Y|X}[j|1]$. Each conditional PMF is generally different.

$\diamondsuit$

There are several relationships between the joint, marginal, and conditional PMFs. We now summarize these as properties.

Property 8.1 – Joint PMF yields conditional PMFs.
If the joint PMF $p_{X,Y}[x_i, y_j]$ is known, then the conditional PMFs are found as

$$p_{Y|X}[y_j|x_i] = \frac{p_{X,Y}[x_i, y_j]}{\sum_j p_{X,Y}[x_i, y_j]} \tag{8.11}$$

$$p_{X|Y}[x_i|y_j] = \frac{p_{X,Y}[x_i, y_j]}{\sum_i p_{X,Y}[x_i, y_j]}. \tag{8.12}$$

<u>Proof</u>: Since the marginal PMF $p_X[x_i]$ is found as $\sum_j p_{X,Y}[x_i, y_j]$, the denominator of (8.7) can be replaced by this to yield (8.11). The equation (8.12) is similarly proven.

□

Hence, we see that the conditional PMF is just the joint PMF with x_i fixed and then normalized by $\sum_j p_{X,Y}[x_i, y_j]$ so that it sums to one. In Figure 8.3a, the conditional PMF $p_{Y|X}[j|1]$ evaluated at $j = 8$ is just $p_{X,Y}[1, 8] = 5/36$ divided by the sum of the probabilities $p_{X,Y}[1, \cdot] = 18/36$, where "$\cdot$" indicates all possible values of j. This yields $p_{Y|X}[8|1] = 5/18$.

Property 8.2 – Conditional PMFs are related.

$$p_{X|Y}[x_i|y_j] = \frac{p_{Y|X}[y_j|x_i]p_X[x_i]}{p_Y[y_j]} \tag{8.13}$$

<u>Proof</u>: By interchanging X and Y in (8.7) we have

$$p_{X|Y}[x_i|y_j] = \frac{p_{Y,X}[y_j, x_i]}{p_Y[y_j]}$$

but

$$
\begin{aligned}
p_{Y,X}[y_j, x_i] &= P[Y = y_j, X = x_i] \\
&= P[X = x_i, Y = y_j] \quad \text{(since } A \cap B = B \cap A\text{)} \\
&= p_{X,Y}[x_i, y_j]
\end{aligned}
$$

and therefore

$$p_{X|Y}[x_i|y_j] = \frac{p_{X,Y}[x_i, y_j]}{p_Y[y_j]}. \tag{8.14}$$

Using $p_{X,Y}[x_i, y_j] = p_{Y|X}[y_j|x_i]p_X[x_i]$ from (8.7) in (8.14) yields the desired result (8.13).

□

Property 8.3 – Conditional PMF is expressible using Bayes' rule.

$$p_{Y|X}[y_j|x_i] = \frac{p_{X|Y}[x_i|y_j]p_Y[y_j]}{\sum_j p_{X|Y}[x_i|y_j]p_Y[y_j]} \tag{8.15}$$

<u>Proof</u>: From (8.11) we have that

$$p_{Y|X}[y_j|x_i] = \frac{p_{X,Y}[x_i, y_j]}{\sum_j p_{X,Y}[x_i, y_j]} \tag{8.16}$$

and using (8.14) we have

$$p_{X,Y}[x_i, y_j] = p_{X|Y}[x_i|y_j]p_Y[y_j] \tag{8.17}$$

which when substituted into (8.16) yields the desired result.

$\square$

Property 8.4 – Conditional PMF and its corresponding marginal PMF yields the joint PMF.

$$p_{X,Y}[x_i, y_j] = p_{Y|X}[y_j|x_i]p_X[x_i] \qquad (8.18)$$
$$p_{X,Y}[x_i, y_j] = p_{X|Y}[x_i|y_j]p_Y[y_j] \qquad (8.19)$$

Proof: (8.18) follows from definition of conditional PMF (8.7) and (8.19) is just (8.17).

$\square$

Property 8.5 – Conditional PMF and its corresponding marginal PMF yields the other marginal PMF.

$$p_Y[y_j] = \sum_i p_{Y|X}[y_j|x_i]p_X[x_i] \qquad (8.20)$$

Proof: This is just the law of total probability in disguise or equivalently just $p_Y[y_j] = \sum_i p_{X,Y}[x_i, y_j]$ (marginal PMF from joint PMF).

$\square$

These relationships are summarized in Figure 8.4. Notice that the joint PMF can be used to find all the marginals and conditional PMFs (see Figure 8.4a). The conditional PMF and its corresponding marginal PMF can be used to find the joint PMF (see Figure 8.4b). Finally, the conditional PMF and its corresponding marginal PMF can be used to find the other conditional PMF (see Figure 8.4c). As emphasized earlier, we cannot determine the joint PMF from the marginals. This is only possible if X and Y are independent random variables since in this case

$$p_{X,Y}[x_i, y_j] = p_X[x_i]p_Y[y_j]. \qquad (8.21)$$

In addition, *for independent random variables*, the use of (8.21) in (8.7) yields

$$p_{Y|X}[y_j|x_i] = \frac{p_X[x_i]p_Y[y_j]}{p_X[x_i]} = p_Y[y_j] \qquad (8.22)$$

or *the conditional PMF is the same as the unconditional PMF*. There is no change in the probabilities of Y whether or not X is observed. This is of course consistent with our previous definition of statistical independence.

Finally, for more general conditional probability calculations we sum the appropriate values of the conditional PMF to yield (see Problem 8.14)

$$P[Y \in A|X = x_i] = \sum_{\{j:y_j \in A\}} p_{Y|X}[y_j|x_i]. \qquad (8.23)$$

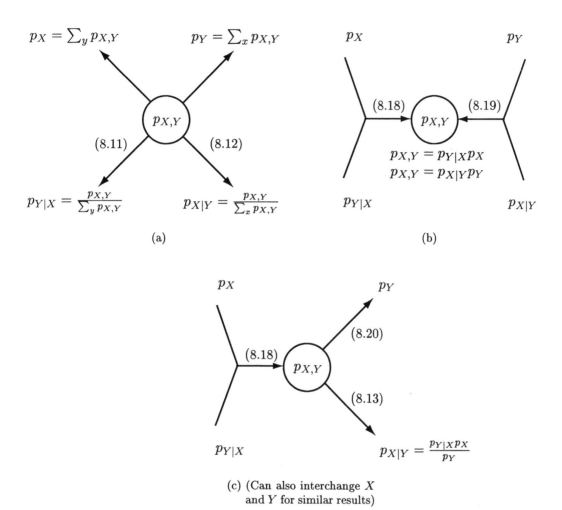

Figure 8.4: Conditional PMF relationships.

8.5 Simplifying Probability Calculations using Conditioning

As alluded to in the introduction, conditional PMFs can be used to simplify probability calculations. To illustrate the use of this approach we once again consider the determination of the PMF for $Z = X + Y$, where X and Y are independent discrete random variables that take on integer values. We have already seen that the solution is $p_Z = p_X \star p_Y$, where $\star$ denotes discrete convolution (see (7.22)). To solve this problem using conditional PMFs, we ask ourselves the question: Could I find the PMF of Z if X were known? If so, then we should be able to use conditioning arguments to first find the conditional PMF of Z given X, and then *uncondition* the result to yield the PMF of Z. Let us say that X is known and that $X = i$. As a result, we have that conditionally $Z = i + Y$, where i is just a constant. This is sometimes denoted by $Z|(X = i)$. But this is a transformation from one discrete random variable Y to another discrete random variable Z. We therefore wish to determine the PMF of a random variable that has been summed with a *constant*. It is not difficult to show that if a discrete random variable U has a PMF $p_U[j]$, then $U + i$ has the PMF $p_U[j - i]$ or the PMF is just shifted to the right by i units. Thus, the *conditional PMF* of Z evaluated at $Z = j$ is $p_{Z|X}[j|i] = p_{Y|X}[j - i|i]$. Now to find the unconditional PMF of Z we use (8.20) with an appropriate change of variables to yield

$$p_Z[j] = \sum_{i=-\infty}^{\infty} p_{Z|X}[j|i] p_X[i]$$

and since $p_{Z|X}[j|i] = p_{Y|X}[j - i|i]$, we have

$$p_Z[j] = \sum_{i=-\infty}^{\infty} p_{Y|X}[j - i|i] p_X[i].$$

But X and Y are independent so that $p_{Y|X} = p_Y$ and therefore we have the final result

$$p_Z[j] = \sum_{i=-\infty}^{\infty} p_Y[j - i] p_X[i]$$

which agrees with our earlier one. Another example follows.

Example 8.2 – PMF for $Z = \max(X, Y)$

Let X and Y be discrete random variables that take on integer values. Also, assume independence of the random variables X and Y and that the marginal PMFs of X and Y are known. To find the PMF of Z we use (8.20) or the law of total probability to yield

$$p_Z[k] = \sum_{i=-\infty}^{\infty} p_{Z|X}[k|i] p_X[i]. \tag{8.24}$$

Now p_X is known so that we only need to determine $p_{Z|X}$ for $X = i$. But given that $X = i$, we have that $Z = \max(i, Y)$ for which the PMF is easily found. We have thus reduced the original problem, which is to determine the PMF for the random variable obtained by transforming from (X, Y) to Z, to determining the PMF for a function of *only one random variable*. Letting $g(Y) = \max(i, Y)$ we see that the function appears as shown in Figure 8.5. Hence, using (5.9) for the PMF of a single transformed discrete random variable we have

$$p_{Z|X}[k|i] = \sum_{\{j : g(j) = k\}} p_{Y|X}[j|i].$$

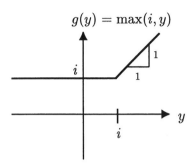

Figure 8.5: Plot of the function $g(y) = \max(i, y)$.

Solving for j in $g(j) = k$ (refer to Figure 8.5) yields no solution for $k < i$, the multiple solutions $j = \ldots, i - 1, i$ for $k = i$, and the single solution $j = k$ for $k = i + 1, i + 2, \ldots$. This produces

$$p_{Z|X}[k|i] = \begin{cases} 0 & k = \ldots, i - 2, i - 1 \\ \sum_{j=-\infty}^{i} p_{Y|X}[j|i] & k = i \\ p_{Y|X}[k|i] & k = i + 1, i + 2, \ldots \end{cases} \tag{8.25}$$

Using this in (8.24) produces

$$
\begin{aligned}
p_Z[k] &= \sum_{i=-\infty}^{k-1} p_{Z|X}[k|i] p_X[i] + p_{Z|X}[k|k] p_X[k] + \sum_{i=k+1}^{\infty} p_{Z|X}[k|i] p_X[i] \quad \text{(break up sum)} \\
&= \sum_{i=-\infty}^{k-1} p_{Y|X}[k|i] p_X[i] + \sum_{j=-\infty}^{k} p_{Y|X}[j|k] p_X[k] + 0 \quad \text{(use (8.25))} \\
&= \sum_{i=-\infty}^{k-1} p_Y[k] p_X[i] + \sum_{j=-\infty}^{k} p_Y[j] p_X[k] \quad \text{(since X and Y are independent)} \\
&= p_Y[k] \sum_{i=-\infty}^{k-1} p_X[i] + p_X[k] \sum_{j=-\infty}^{k} p_Y[j].
\end{aligned}
$$

Note that due to the independence assumption this final result can also be written as

$$p_Z[k] = \sum_{i=-\infty}^{k-1} p_{X,Y}[i, k] + \sum_{j=-\infty}^{k} p_{X,Y}[k, j]$$

so that the PMF of Z is obtained by summing all the points of the joint PMF shown in Figure 8.6 for $k = 2$, as an example. These point comprise the set $\{(x, y) : \max(x, y) = 2 \text{ and } x = i, y = j\}$. It is now clear that we could have solved this problem in a more direct fashion by making this observation. As in most problems, however, the solution is usually trivial once it is known!

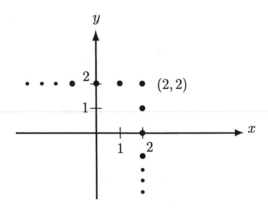

Figure 8.6: Points of joint PMF to be summed to find PMF of $Z = \max(X, Y)$ for $k = 2$.

$$\diamond$$

As we have seen, a general procedure for determining the PMF for $Z = g(X, Y)$ when X and Y are independent is as follows:

1. Fix $X = x_i$ and let $Z|(X = x_i) = g(x_i, Y)$

2. Find the PMF for $Z|X$ by using the techniques for a transformation of a single random variable Y into another random variable Z. The formula is from (5.9), where the PMFs are first converted to conditional PMFs

$$
\begin{aligned}
p_{Z|X}[z_k|x_i] &= \sum_{\{j:g(x_i,y_j)=z_k\}} p_{Y|X}[y_j|x_i] && \text{for each } x_i \\
&= \sum_{\{j:g(x_i,y_j)=z_k\}} p_Y[y_j] && \text{for each } x_i \quad \text{(due to independence)}.
\end{aligned}
$$

3. Uncondition the conditional PMF to yield the desired PMF

$$p_Z[z_k] = \sum_i p_{Z|X}[z_k|x_i]p_X[x_i].$$

In general, to compute probabilities of events it is advantageous to use a conditioning argument, whether or not X and Y are independent. Where previously we have used the formula

$$P[Y \in A] = \sum_{\{j:y_j \in A\}} p_Y[y_j]$$

to compute the probability, a conditioning approach would replace $p_Y[y_j]$ by $\sum_i p_{Y|X}[y_j|x_i]p_X[x_i]$ to yield

$$P[Y \in A] = \sum_{\{j:y_j \in A\}} \sum_i p_{Y|X}[y_j|x_i]p_X[x_i] \tag{8.26}$$

to determine the probability. Equivalently, we have that

$$P[Y \in A] = \sum_i \underbrace{\left[\underbrace{\sum_{\{j:y_j \in A\}} p_{Y|X}[y_j|x_i]}_{\text{conditioning}} \right] \underbrace{p_X[x_i]}_{\text{unconditioning}}}. \qquad (8.27)$$

In this form we recognize the conditional probability of (8.23), which is

$$P[Y \in A|X = x_i] = \sum_{\{j:y_j \in A\}} p_{Y|X}[y_j|x_i]$$

and the unconditional probability

$$P[Y \in A] = \sum_i P[Y \in A|X = x_i]p_X[x_i] \qquad (8.28)$$

with the latter being just a restatement of the law of total probability.

8.6 Mean of the Conditional PMF

Since the conditional PMF is a PMF, it exhibits all the usual properties. In particular, we can determine attributes such as the expected value of a random variable Y, when it is known that $X = x_i$. This expected value is the *mean of the conditional PMF* $p_{Y|X}$. Its definition is the usual one

$$\sum_j y_j p_{Y|X}[y_j|x_i] \qquad (8.29)$$

where we have replaced p_Y by $p_{Y|X}$. It should be emphasized that since the conditional PMF depends on x_i, so will its mean. Hence, the mean of the conditional PMF is a constant when we set x_i equal to a fixed value. We adopt the notation for the mean of the conditional PMF as $E_{Y|X}[Y|x_i]$. This notation includes the subscript "$Y|X$" to remind us that the averaging PMF is the conditional PMF $p_{Y|X}$. Also, the use of "$Y|x_i$" as the argument will remind us that the averaging PMF is the conditional PMF that is specified by $X = x_i$ in the family of conditional PMFs. The mean is therefore defined as

$$E_{Y|X}[Y|x_i] = \sum_j y_j p_{Y|X}[y_j|x_i]. \qquad (8.30)$$

Although we have previously asserted that the mean is a constant, here *it is to be regarded as a function of x_i*. An example of its calculation follows.

Example 8.3 – Mean of conditional PMF – continuation of Example 8.1

We now compute all the possible values of $E_{Y|X}[Y|x_i]$ for the problem described in Example 8.1. There $x_i = 1$ or $x_i = 0$ and the corresponding conditional PMFs are given by (8.9) and (8.10), respectively. The means of the conditional PMFs are therefore

$$E_{Y|X}[Y|1] = 2\left(\frac{1}{18}\right) + 4\left(\frac{3}{18}\right) + 6\left(\frac{5}{18}\right) + 8\left(\frac{5}{18}\right) + 10\left(\frac{3}{18}\right) + 12\left(\frac{1}{18}\right) = 7$$

$$E_{Y|X}[Y|0] = 3\left(\frac{2}{18}\right) + 5\left(\frac{4}{18}\right) + 7\left(\frac{6}{18}\right) + 9\left(\frac{4}{18}\right) + 11\left(\frac{2}{18}\right) = 7$$

and are shown in Figure 8.3. In this example the means of the conditional PMFs are the same, but will not be in general. We can expect that $g(x_i) = E_{Y|X}[Y|x_i]$ *will vary with* x_i.

$\Diamond$

We could also compute the variance of the conditional PMFs. This would be

$$\text{var}(Y|x_i) = \sum_j \left(y_j - E_{Y|X}[Y|x_i]\right)^2 p_{Y|X}[y_j|x_i]. \qquad (8.31)$$

The reader is asked to do this in Problem 8.22. (See also Problem 8.23 for an alternate expression for var($Y|x_i$).) Note from Figure 8.3 that we do not expect these to be the same.

 What is the "conditional expectation"?

The *function* $g(x_i) = E_{Y|X}[Y|x_i]$ is the mean of the conditional PMF $p_{Y|X}[y_j|x_i]$. Alternatively, it is known as the *conditional mean*. This terminology is widespread and so we will adhere to it, although we should keep in mind that it is meant to denote the usual mean of the *conditional* PMF. It is also of interest to determine the expectation of other quantities besides Y with respect to the conditional PMF. This is called the *conditional expectation* and is symbolized by $E_{Y|X}[g(Y)|x_i]$. The latter is called the conditional expectation of $g(Y)$. For example, if $g(Y) = Y^2$, then it becomes the conditional expectation of Y^2 or equivalently the *conditional second moment*. Lastly, the reader should be aware that the conditional mean is the optimal predictor of a random variable based on observation of a second random variable (see Problem 8.27).

We now give another example of the computation of the conditional mean.

Example 8.4 – Toss one of two dice.

There are two dice having different numbers of dots on their faces. Die 1 is the usual type of die with faces having $1, 2, 3, 4, 5,$ or 6 dots. Die 2 has been mislabled with its faces having $2, 3, 2, 3, 2,$ or 3 dots. A die is selected at random and tossed. Each face of the die is equally likely to occur. What is the expected number of dots observed for the tossed die? To solve this problem first observe that the outcomes will depend upon which die has been tossed. As a result, the *conditional expectation* of the number of dots will depend upon which die is initially chosen. We can view this problem as a conditional one by letting

$$X = \begin{cases} 1 & \text{if die 1 is chosen} \\ 2 & \text{if die 2 is chosen} \end{cases}$$

and Y is the number of dots observed. Thus, we wish to determine $E_{Y|X}[Y|1]$ and $E_{Y|X}[Y|2]$. But if die 1 is chosen, the conditional PMF is

$$p_{Y|X}[j|1] = \frac{1}{6} \qquad j = 1, 2, 3, 4, 5, 6 \tag{8.32}$$

and if die 2 is chosen

$$p_{Y|X}[j|2] = \frac{1}{2} \qquad j = 2, 3. \tag{8.33}$$

The latter conditional PMF is due to the fact that for die 2 half the sides show 2 dots and the other half of the sides show 3 dots. Using (8.30) with (8.32) and (8.33), we have that

$$E_{Y|X}[Y|1] = \sum_{j=1}^{6} j p_{Y|X}[j|1] = \frac{7}{2}$$

$$E_{Y|X}[Y|2] = \sum_{j=2}^{3} j p_{Y|X}[j|2] = \frac{5}{2}. \tag{8.34}$$

An example of typical outcomes for this experiment is shown in Figure 8.7. For 50 trials of the experiment Figure 8.7a displays the outcomes for which die 1 was chosen and Figure 8.7b displays the outcomes for which die 2 was chosen. It is interesting to note that the estimated mean for Figure 8.7a is 3.88 and for Figure 8.7b it is 2.58. Note that from (8.34) the theoretical conditional means are 3.5 and 2.5, respectively.

$\diamondsuit$

In the previous example, we have determined the conditional means, which are the means of the conditional PMFs. We also might wish to determine the *unconditional mean*, which is the mean of Y. This is the number of dots observed as a result of the overall experiment, without first conditioning on which die was chosen. In essence, we wish to determine $E_Y[Y]$. Intuitively, this is the average number of dots observed if we combined Figures 8.7a and 8.7b together (just overlay Figure 8.7b onto Figure 8.7a) and continued the experiment indefinitely. Hence, we wish to determine $E_Y[Y]$ for the following experiment:

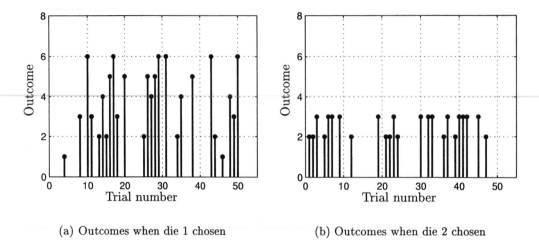

(a) Outcomes when die 1 chosen   (b) Outcomes when die 2 chosen

Figure 8.7: Computer simulated outcomes of randomly selected die toss experiment.

1. Choose die 1 or die 2 with probability of 1/2.

2. Toss the chosen die.

3. Count the number of dots on the face of tossed die and call this the outcome of
   the random variable Y.

A simple MATLAB program to simulate this experiment is given as

```
for m=1:M
  if rand(1,1)<0.5
    y(m,1)=PMFdata(1,[1 2 3 4 5 6]',[1/6 1/6 1/6 1/6 1/6 1/6]');
  else
    y(m,1)=PMFdata(1,[2 3]',[1/2 1/2]');
  end
end
```

where the subprogram `PMFdata.m` is listed in Appendix 6B. After the code is executed there is an array y, which is $M \times 1$, containing M realizations of Y. By
taking the sample mean of the elements in the array y, we will have estimated
$E_Y[Y]$. But we expect about half of the realizations to have used the fair die and
the other half to use the mislabled die. As a result, we might suppose that the
unconditional mean is just the average of the two conditional means. This would be
$(1/2)(7/2) + (1/2)(5/2) = 3$, which turns out to be the true result. This conjecture
is also strengthened by the results of Figure 8.7. By overlaying the plots we have
50 outcomes of the experiment for which the sample mean is 3.25. Let's see how to
verify this.

To determine the theoretical mean of Y, i.e., the unconditional mean, we will need $p_Y[j]$. But given the conditional PMF and the marginal PMF we know from Figure 8.4c that the joint PMF can be found. Hence, from (8.32) and (8.33) and $p_X[i] = 1/2$ for $i = 1, 2$, we have

$$
\begin{aligned}
p_{X,Y}[i,j] &= p_{Y|X}[j|i]p_X[i] \\
&= \begin{cases} \frac{1}{12} & i = 1; j = 1, 2, 3, 4, 5, 6 \\ \frac{1}{4} & i = 2; j = 2, 3. \end{cases}
\end{aligned}
$$

To find $p_Y[j]$ we use

$$
\begin{aligned}
p_Y[j] &= \sum_{i=1}^{2} p_{X,Y}[i,j] \\
&= \begin{cases} p_{X,Y}[1,j] = \frac{1}{12} & j = 1, 4, 5, 6 \\ p_{X,Y}[1,j] + p_{X,Y}[2,j] = \frac{1}{12} + \frac{1}{4} = \frac{1}{3} & j = 2, 3. \end{cases}
\end{aligned}
$$

Thus, the unconditional mean becomes

$$
\begin{aligned}
E_Y[Y] &= \sum_{j=1}^{6} j p_Y[j] \\
&= 1\left(\frac{1}{12}\right) + 2\left(\frac{1}{3}\right) + 3\left(\frac{1}{3}\right) + 4\left(\frac{1}{12}\right) + 5\left(\frac{1}{12}\right) + 6\left(\frac{1}{12}\right) \\
&= 3.
\end{aligned}
$$

This value is sometimes called the *unconditional expectation*. Note that for this example, we have upon using (8.34)

$$
E_Y[Y] = E_{Y|X}[Y|1]p_X[1] + E_{Y|X}[Y|2]p_X[2]
$$

or the *unconditional mean is the average of the conditional means*. This is true in general and is summarized by the relationship

$$
E_Y[Y] = \sum_i E_{Y|X}[Y|x_i]p_X[x_i]. \tag{8.35}
$$

To prove this relationship is straightforward. Starting with (8.35) we have

$$\sum_i E_{Y|X}[Y|x_i]p_X[x_i]$$

$$= \sum_i \left(\sum_j y_j p_{Y|X}[y_j|x_i] \right) p_X[x_i] \quad \text{(definition of conditional mean)}$$

$$= \sum_i \sum_j y_j \frac{p_{X,Y}[x_i, y_j]}{p_X[x_i]} p_X[x_i] \quad \text{(definition of conditional PMF)}$$

$$= \sum_j y_j \sum_i p_{X,Y}[x_i, y_j]$$

$$= \sum_j y_j p_Y[y_j] \quad \text{(marginal PMF from joint PMF)}$$

$$= E_Y[Y].$$

In (8.35) we can consider $g(x_i) = E_{Y|X}[Y|x_i]$ as the transformed outcome of the coin choice part of the experiment, where $X = x_i$ is the outcome of the coin choice. Since before we choose the coin to toss, we do not know which one it will be, we can consider $g(X)$ as a transformed *random variable* whose values are $g(x_i)$. By this way of viewing things, we can define a random variable as $g(X) = E_{Y|X}[Y|X]$ and therefore rewrite (8.35) as

$$E_Y[Y] = E_X[g(X)]$$

or explicitly we have that

$$E_Y[Y] = E_X[E_{Y|X}[Y|X]]. \tag{8.36}$$

In effect, we have computed the expectation of a random variable in two steps. Step 1 is to compute a conditional expectation $E_{Y|X}$ while step 2 is to undo the conditioning by averaging the result with respect to the PMF of X. An example is the previous coin tossing experiment. The utility in doing so is that the conditional PMFs were easily found and hence also the means of the conditional PMFs, and finally the averaging with respect to p_X is easily carried out to yield the desired result. We illustrate the use of (8.36) with another example.

Example 8.5 – Random number of coin tosses

An experiment is conducted in which a coin with a probability of heads p is tossed M times. However, M is a *random variable* with $M \sim \text{Pois}(\lambda)$. For example, if a realization of M is generated, say $M = 5$, then the coin is tossed 5 times in succession. We wish to determine the average number of heads observed. Conditionally on knowing the value of M, we have a binomial PMF for the number of heads Y. Hence, for $M = i$ we have upon using the binomial PMF (see (5.6))

$$p_{Y|M}[j|i] = \binom{i}{j} p^j (1-p)^{i-j} \qquad j = 0, 1, \ldots, i; i = 0, 1, \ldots.$$

Now using (8.36) and replacing X with M we have

$$E_Y[Y] = E_M[E_{Y|M}[Y|M]]$$

and for a binomial PMF we know that $E_{Y|M}[Y|i] = ip$ so that

$$E_Y[Y] = E_M[Mp] = pE_M[M].$$

But for a Poisson random variable $E_M[M] = \lambda$, which yields the final result

$$E_Y[Y] = \lambda p.$$

It can be shown more generally that $Y \sim \text{Pois}(\lambda p)$ (see Problem 8.26) so that our result for the mean of Y follows directly from knowledge of the mean of a Poisson random variable.

8.7 Computer Simulation Based on Conditioning

In Section 7.11 we discussed a simple method for generating realizations of jointly distributed discrete random variables (X, Y) using MATLAB. To do so we required the joint PMF. Using conditioning arguments, however, we can frequently simplify the procedure. Since $p_{X,Y}[x_i, y_j] = p_{Y|X}[y_j|x_i]p_X[x_i]$, a realization of (X, Y) can be obtained by first generating a realization of X according to its marginal PMF $p_X[x_i]$. Then, assuming that $X = x_i$ is obtained, we next generate a realization of Y according to the *conditional PMF* $p_{Y|X}[y_j|x_i]$. (Of course, if X and Y are independent, we replace the second step by the generation of Y according to $p_Y[y_j]$ since in this case $p_{Y|X}[y_j|x_i] = p_Y[y_j]$.) This is also advantageous when the problem description is formulated in terms of conditional PMFs, as in a compound experiment. To illustrate this approach with the one described previously we repeat Example 7.15.

Example 8.6 – Generating realizations of jointly distributed random variables – Example 7.15 (continued)

The joint PMF of Example 7.15 is shown in Figure 8.8, where the solid circles represent the sample points and the values of the joint PMF are shown to the right of the sample points. To use a conditioning approach we need to find p_X and $p_{Y|X}$. But from Figure 8.8, if we sum along the columns we obtain

$$p_X[i] = \begin{cases} \frac{1}{4} & i = 0 \\ \frac{3}{4} & i = 1 \end{cases}$$

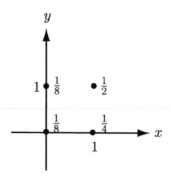

Figure 8.8: Joint PMF for Example 8.6.

and using the definition of the conditional PMF, we have

$$p_{Y|X}[j|0] = \frac{p_{X,Y}[0,j]}{p_X[0]}$$

$$= \begin{cases} \frac{1/8}{1/4} = \frac{1}{2} & j = 0 \\ \frac{1/8}{1/4} = \frac{1}{2} & j = 1 \end{cases}$$

and

$$p_{Y|X}[j|1] = \frac{p_{X,Y}[1,j]}{p_X[1]}$$

$$= \begin{cases} \frac{1/4}{3/4} = \frac{1}{3} & j = 0 \\ \frac{1/2}{3/4} = \frac{2}{3} & j = 1. \end{cases}$$

The MATLAB segment of code shown below generates M realizations of (X,Y) using this conditioning approach.

```
for m=1:M
  ux=rand(1,1);
  uy=rand(1,1);
  if ux<=1/4; % Refer to px[i]
    x(m,1)=0;
    if uy<=1/2 % Refer to py|x[j|0]
      y(m,1)=0;
    else
      y(m,1)=1;
    end
  else
    x(m,1)=1;  % Refer to px[i]
```

```
            if uy<=1/3 % Refer to py|x[j|1]
               y(m,1)=0;
            else
               y(m,1)=1;
            end
         end
      end
```

The reader is asked to test this program in Problem 8.29.

8.8 Real-World Example – Modeling Human Learning

A 2 year-old child who has learned to walk can perform tasks that not even the most sophisticated robots can match. For example, a 2 year-old child can easily maneuver her way to a favorite toy, pick it up, and start to play with it. Robots, powered by machine vision and mechanical grippers, have a hard time performing this supposedly simple task. It is not surprisingly, therefore, that one of the holy grails in cognitive science and also machine learning is to figure out how a child does this. If we were able to understand the thought processes that were used to successfully complete this task, then it is conceivable that a machine might be built to do the same thing. Many models of human learning employ a Bayesian framework [Tenenbaum 1999]. This approach appears to be fruitful in that using Bayesian modeling we are able to discriminate with more and more accuracy as we repeatedly perform an experiment and observe the outcome. This is analogous to a child attempting to pick up the toy, dropping it, picking it up again after having learned something about how to pick it up, dropping it, etc., until finally she is successful. Each time the experiment, attempting to pick up the toy, is repeated the child learns something or equivalently narrows down the number of possible strategies. In Bayesian analysis, as we will show next, the width of the PMF decreases as we observe more outcomes. This is in some sense saying that our uncertainty about the outcome of the experiment decreases as it is performed more times. Although not a perfect analogy, it does seem to possess some critical elements of the human learning process. Therefore, we illustrate this modeling with the simple example of coin tossing.

Suppose we wish to "learn" whether a coin is fair ($p = 1/2$) or is weighted ($p \neq 1/2$). One way to do this is to repeatedly toss the coin and count the number of heads observed. We would expect that our certainty about the conclusion, that the coin is fair or not, would increase as the number of trials increases. In the Bayesian model we quantify our knowledge about the value of p by assuming that p is a *random variable*. Our particular coin, however, has a fixed probability of heads. It is just that we do not know what it is and hence our *belief* about the value of p is embodied in an assumed PMF. This is a slightly different interpretation of probability than our previous relative frequency interpretation. To conform to our previous notation we let the probability of heads be denoted by the random variable Y and its values by y_j. Then, we determine its PMF. Our state of knowledge will be high if the PMF is highly concentrated about a particular value, as for example in Figure 8.9a. If, however, the PMF is spread out or "diffuse", our state of knowledge will be low, as for example in Figure 8.9b. Now let's say that we wish to learn the

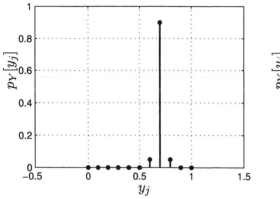
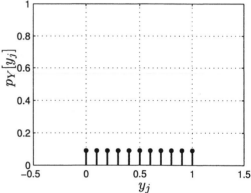

(a) Y = probability of heads – state of knowledge is high.

(b) Y = probability of heads – state of knowledge is low.

Figure 8.9: PMFs reflecting state of knowledge about coin's probability of heads.

value of the probability of heads. Before we toss the coin we have no idea what it is, and therefore it is reasonable to assume a PMF that is uniform, as, for example, the one shown in Figure 8.9b. Such a PMF is given by

$$p_Y[y_j] = \frac{1}{M+1} \qquad \text{for } y_j = 0, \tfrac{1}{M}, \tfrac{2}{M}, \ldots, \tfrac{M-1}{M}, 1 \qquad (8.37)$$

for some large M (in Figure 8.9b $M = 11$). This is also called the *prior PMF* since it summarizes our state of knowledge *before* the experiment is performed. Now we begin to toss the coin and examine our state of knowledge as the number of tosses increases. Let N be the number of coin tosses and X denote the number of heads observed in the N tosses. We know that the PMF of the number of heads is binomially distributed. However, to specify the PMF completely, we require knowledge of the probability of heads. Since this is unknown, we can only specify the PMF of X conditionally or if $Y = y_j$ is the probability of heads, then the conditional PMF of the number of heads for $X = i$ is

$$p_{X|Y}[i|y_j] = \binom{N}{i} y_j^i (1 - y_j)^{N-i} \qquad i = 0, 1, \ldots, N. \qquad (8.38)$$

Since we are actually interested in the probability of heads or the PMF of Y after observing the outcomes of N coin tosses, we need to determine the conditional PMF $p_{Y|X}[y_j|i]$. The latter is also called the *posterior PMF*, since it is to be determined *after* the experiment is peformed. The reader may wish to compare this terminology with that used in Chapter 4. The posterior PMF contains all the information about the probability of heads that results from our prior knowledge, summarized by p_Y, and our "data" knowledge, summarized by $p_{X|Y}$. The posterior PMF is given by Bayes' rule (8.15) with $x_i = i$ as

$$p_{Y|X}[y_j|i] = \frac{p_{X|Y}[i|y_j]p_Y[y_j]}{\sum_j p_{X|Y}[i|y_j]p_Y[y_j]}.$$

Using (8.37) and (8.38) we have

$$p_{Y|X}[y_j|i] = \frac{\binom{N}{i} y_j^i (1-y_j)^{N-i} \frac{1}{M+1}}{\sum_{j=0}^{M} \binom{N}{i} y_j^i (1-y_j)^{N-i} \frac{1}{M+1}} \qquad y_j = 0, 1/M, \ldots, 1; i = 0, 1, \ldots, N$$

or finally,

$$p_{Y|X}[y_j|i] = \frac{y_j^i (1-y_j)^{N-i}}{\sum_{j=0}^{M} y_j^i (1-y_j)^{N-i}} \qquad y_j = 0, 1/M, \ldots, 1; i = 0, 1, \ldots, N. \qquad (8.39)$$

Note that the posterior PMF depends on the number of heads observed, which is i. To understand what this PMF is saying about our state of knowledge, assume that we toss the coin $N = 10$ times and observe $i = 4$ heads. The posterior PMF is shown in Figure 8.10a. For $N = 20$, $i = 11$ and $N = 40$, $i = 19$, the posterior PMFs are shown in Figures 8.10b and 8.10c, respectively. Note that as the number

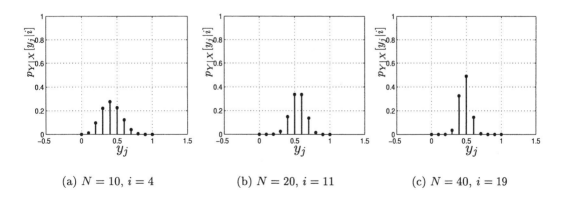

(a) $N = 10$, $i = 4$ \qquad (b) $N = 20$, $i = 11$ \qquad (c) $N = 40$, $i = 19$

Figure 8.10: Posterior PMFs for coin tossing analogy to human learning – coin appears to be fair. The y_j's are possible probability values for a head.

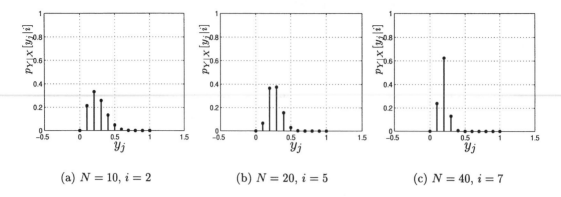

(a) $N = 10$, $i = 2$ (b) $N = 20$, $i = 5$ (c) $N = 40$, $i = 7$

Figure 8.11: Posterior PMFs for coin tossing analogy to human learning – coin appears to be weighted. The y_j's are possible probability values for a head.

of tosses increases the posterior PMF becomes narrower and centered about the value of 0.5. The Bayesian model has "learned" the value of p, with our confidence increasing as the number of trials increases. Note that for no trials (just set $N = 0$ and hence $i = 0$ in (8.39)) we have just the uniform prior PMF of Figure 8.9b. From our experiments we could now conclude with some certainty that the coin is fair. However, if the outcomes were $N = 10$, $i = 2$, and $N = 20$, $i = 5$, and $N = 40$, $i = 7$, then the posterior PMFs would appear as in Figure 8.11. We would then conclude that the coin is weighted and is biased against yielding a head, since the posterior PMF is concentrated about 0.2. See [Kay 1993] for futher descriptions of Bayesian approaches to estimation.

References

Kay, S., *Fundamentals of Statistical Signal Processing; Estimation Theory, Vol. I*, Prentice-Hall, Englewood Cliffs, NJ, 1993.

Tennebaum, J.B. "Bayesian modeling of human learning", in *Advances in Neural Information Processing Systems 11*, MIT Press, Cambridge, MA, 1999.

Problems

8.1 (w) A fair coin is tossed. If it comes up heads, then $X = 1$ and if it comes up tails, then $X = 0$. Next, a point is selected at random from the area A if $X = 1$ and from the area B if $X = 0$ as shown in Figure 8.12. Note that the area of the square is 4 and A and B both have areas of 3/2. If the point selected is in an upper quadrant, we set $Y = 1$ and if it is in a lower quadrant, we set $Y = 0$. Find the conditional PMF $p_{Y|X}[j|i]$ for all values of i and j. Next, compute $P[Y = 0]$.

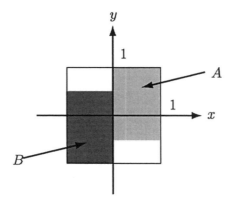

Figure 8.12: Areas for Problem 8.1.

8.2 ($\ddot{\smile}$) (w) A fair coin is tossed with the outcome mapped into $X = 1$ for a head and $X = 0$ for a tail. If it comes up heads, then a fair die is tossed. The outcome of the die is denoted by Y and is set equal to the number of dots observed. If the coin comes up tails, then we set $Y = 0$. Find the conditional PMF $p_{Y|X}[j|i]$ for all values of i and j. Next, compute $P[Y = 1]$.

8.3 (w) A fair coin is tossed 3 times in succession. All the outcomes (i.e., the 3-tuples) are equally likely. The random variables X and Y are defined as

$$X = \begin{cases} 0 & \text{if outcome of first toss is a tail} \\ 1 & \text{if outcome of first toss is a head} \end{cases}$$
$$Y = \text{number of heads observed for the three tosses}$$

Determine the conditional PMF $p_{Y|X}[j|i]$ for all i and j.

8.4 (t) Prove that $\sum_{j=-\infty}^{\infty} p_{Y|X}[y_j|x_i] = 1$ for all x_i.

8.5 ($\ddot{\smile}$) (w) Are the following functions valid conditional PMFs

    a. $p_{Y|X}[j|x_i] = (1 - x_i)^j x_i \qquad j = 1, 2, \ldots; x_i = 1/4, 1/2, 3/4$

    b. $p_{Y|X}[j|x_i] = \binom{N}{j} x_i^j (1 - x_i)^{N-j} \qquad j = 0, 1, \ldots, N; x_i = -1/2, 1/2$

    c. $p_{Y|X}[j|x_i] = c x_i^j \qquad j = 2, 3, \ldots; x_i = 2$ for c some constant?

8.6 ($\ddot{\smile}$) (f) If

$$p_{X,Y}[i,j] = \begin{cases} \frac{1}{6} & i = 0, j = 0 \\ \frac{1}{3} & i = 0, j = 1 \\ \frac{1}{3} & i = 1, j = 0 \\ \frac{1}{6} & i = 1, j = 1 \end{cases}$$

find $p_{Y|X}$ and $p_{X|Y}$.

8.7 (f) Verify the conditional PMF given in (8.10).

8.8 (☺) (f) For the sample space shown in Figure 8.1 determine $p_{Y|X}$ and $p_{X|Y}$ if all the outcomes are equally likely. Explain your results.

8.9 (w) Explain the need for the denominator term in (8.11) and (8.12).

8.10 (w) If $p_{Y|X}$ and p_Y are known, can you find $p_{X,Y}$?

8.11 (☺) (w) A box contains three types of replacement light bulbs. There is an equal proportion of each type. The types vary in their quality so that the probability that the light bulb *fails* at the jth use is given by

$$p_{Y|X}[j|1] = (0.99)^{j-1}0.01$$
$$p_{Y|X}[j|2] = (0.9)^{j-1}0.1$$
$$p_{Y|X}[j|3] = (0.8)^{j-1}0.2$$

for $j = 1, 2, \ldots$. Note that $p_{Y|X}[j|i]$ is the PMF of the bulb failing at the jth use if it is of type i. If a bulb is selected at random from the box, what is the probability that it will operate satisfactorily for at least 10 uses?

8.12 (f) A joint PMF $p_{X,Y}[i, j]$ has the values shown in Table 8.2. Determine the conditional PMF $p_{Y|X}$. Are the random variables independent?

	$j = 1$	$j = 2$	$j = 3$
$i = 1$	$\frac{1}{10}$	$\frac{1}{10}$	$\frac{2}{10}$
$i = 2$	$\frac{1}{20}$	$\frac{1}{20}$	$\frac{1}{10}$
$i = 3$	$\frac{3}{10}$	$\frac{1}{20}$	$\frac{1}{20}$

Table 8.2: Joint PMF for Problem 8.12.

8.13 (☺) (w) A random vector (X, Y) has a sample space shown in Figure 8.13 with the sample points depicted as solid circles. The four points are equally probable. Note that the points in Figure 8.13b are the corners of the square shown in Figure 8.13a after rotation by $+45°$. For both cases compute $p_{Y|X}$ and p_Y to determine if the random variables are independent.

8.14 (t) Use the properties of conditional probability and the definition of the conditional PMF to prove (8.23). Hint: Let $A = \cup_j \{s : Y(s) = y_j\}$ and note that the events $\{s : Y(s) = y_j\}$ are mutually exclusive.

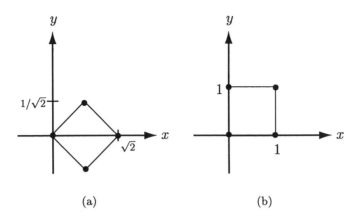

Figure 8.13: Joint PMFs – each point is equally probable.

8.15 (w) If X and Y are independent random variables, find the PMF of $Z = |X - Y|$. Assume that $\mathcal{S}_X = \{0, 1, \ldots\}$ and $\mathcal{S}_Y = \{0, 1, \ldots\}$. Hint: The answer is

$$p_Z[k] = \begin{cases} \sum_{i=0}^{\infty} p_X[i]p_Y[i] & k = 0 \\ \sum_{i=0}^{\infty} \left(p_Y[i]p_X[i+k] + p_X[i]p_Y[i+k]\right) & k = 1, 2, \ldots \end{cases}.$$

As an intermediate step show that

$$p_{Z|X}[k|i] = \begin{cases} p_Y[i] & k = 0 \\ p_Y[i+k] + p_Y[i-k] & k \neq 0. \end{cases}$$

8.16 (w) Two people agree to meet at a specified time. Person A will be late by i minutes with a probability $p_X[i] = (1/2)^{i+1}$ for $i = 0, 1, \ldots$, while person B will be late by j minutes with a probability of $p_Y[j] = (1/2)^{j+1}$ for $j = 0, 1, \ldots$. The persons arrive independently of each other. The first person to arrive will wait a maximum of 2 minutes for the second person to arrive. If the second person is more than 2 minutes late, the first person will leave. What is the probability that the two people will meet? Hint: Use the results of Problem 8.15.

8.17 (☺) (w) If X and Y are independent random variables, both of whose PMFs take on values $\{0, 1, \ldots\}$, find the PMF of $Z = \min(X, Y)$.

8.18 (w) If X and Y have the joint PMF

$$p_{X,Y}[i, j] = p_1 p_2 (1 - p_1)^i (1 - p_2)^j \qquad i = 0, 1, \ldots; j = 0, 1, \ldots$$

where $0 < p_1 < 1$, $0 < p_2 < 1$, find $P[Y > X]$ using a conditioning argument. In particular, make use of (8.23) and $P[Y > X|X = i] = P[Y > i|X = i]$.

8.19 (f) If X and Y have the joint PMF given in Problem 8.6, find $E_{Y|X}[Y|x_i]$.

8.20 (f) If X and Y have the joint PMF

$$p_{X,Y}[i,j] = \left(\frac{1}{2}\right)^{i+1} \exp(-\lambda)\frac{\lambda^j}{j!} \qquad i = 0,1,\ldots; j = 0,1,\ldots$$

find $E_{Y|X}[Y|i]$ for all i.

8.21 (☺) (f) Find the conditional mean of Y given X if the joint PMF is uniformly distributed over the points $S_{X,Y} = \{(0,0),(1,0),(1,1),(2,0),(2,1),(2,2)\}$.

8.22 (☺) (f) For the joint PMF given in Problem 8.21 determine $\text{var}(Y|x_i)$ for all x_i. Explain why your results appear to be reasonable.

8.23 (t) Prove that $\text{var}(Y|x_i) = E_{Y|X}[Y^2|x_i] - E_{Y|X}^2[Y|x_i]$ by using (8.31).

8.24 (f) Find $E_Y[Y]$ for the joint PMF given in Problem 8.21. Do this by using the definition of the expected value and also by using (8.36).

8.25 (t) Prove the extension of (8.36) which is

$$E_Y[g(Y)] = E_X\left[E_{Y|X}[g(Y)|X]\right]$$

where $h(X) = E_{Y|X}[g(Y)|X]$ is a function of the random variable X which takes on values

$$h(x_i) = E_{Y|X}[g(Y)|x_i] = \sum_j g(y_j)p_{Y|X}[y_j|x_i].$$

This says that $E_Y[g(Y)]$ can be computed using the formula

$$E_Y[g(Y)] = \sum_i \left[\sum_j g(y_j)p_{Y|X}[y_j|x_i]\right] p_X[x_i].$$

8.26 (t) In this problem we prove that if $M \sim \text{Pois}(\lambda)$ and Y conditioned on M is a binomial PMF with parameter p, then the unconditional PMF of Y is $\text{Pois}(\lambda p)$. This means that if

$$p_M[m] = \exp(-\lambda)\frac{\lambda^m}{m!} \qquad m = 0,1,\ldots$$

and

$$p_{Y|M}[j|m] = \binom{m}{j} p^j(1-p)^{m-j} \qquad j = 0,1,\ldots,m$$

then

$$p_Y[j] = \exp(-\lambda p)\frac{(\lambda p)^j}{j!} \qquad j = 0,1,\ldots.$$

To prove this you will need to derive the characteristic function of Y and show that it corresponds to a Pois(λp) random variable. Proceed as follows, making use of the results of Problem 8.25

$$\begin{aligned}
\phi_Y(\omega) &= E_Y[\exp(j\omega Y)] \\
&= E_M\left[E_{Y|M}[\exp(j\omega Y)|M]\right] \\
&= E_M\left[[p\exp(j\omega) + (1-p)]^M\right]
\end{aligned}$$

and complete the derivation.

8.27 (t) In Chapter 7 the optimal *linear* predictor of Y based on $X = x_i$ was found. The criterion of optimality was the minimum mean square error, where the mean square error was defined as $E_{X,Y}[(Y - (aX + b))^2]$. In this problem we prove that the best predictor, now allowing for *nonlinear predictors* as well, is given by the conditional mean $E_{Y|X}[Y|x_i]$. To prove this we let the predictor be $\hat{Y} = g(X)$ and minimize

$$\begin{aligned}
E_{X,Y}[(Y - g(X))^2] &= \sum_i \sum_j (y_j - g(x_i))^2 p_{X,Y}[x_i, y_j] \\
&= \sum_i \left[\sum_j (y_j - g(x_i))^2 p_{Y|X}[y_j|x_i]\right] p_X[x_i].
\end{aligned}$$

But since $p_X[x_i]$ is nonnegative and we can choose a different value of $g(x_i)$ for each x_i, we can equivalently minimize

$$\left[\sum_j (y_j - g(x_i))^2 p_{Y|X}[y_j|x_i]\right]$$

where we consider $g(x_i) = c$ as a constant. Prove that this is minimized for $g(x_i) = E_{Y|X}[Y|x_i]$. Hint: You may wish to review Section 6.6.

8.28 (⌣) (f) For random variables X and Y with the joint PMF

$$p_{X,Y}[i,j] = \begin{cases} \frac{1}{4} & (i,j) = (-1,0) \\ \frac{1}{8} & (i,j) = (0,-1) \\ \frac{3}{8} & (i,j) = (0,1) \\ \frac{1}{4} & (i,j) = (1,0) \end{cases}$$

we wish to predict Y based on our knowledge of the outcome of X. Find the optimal predictor using the results of Problem 8.27. Also, find the optimal *linear* predictor for this problem (see Section 7.9) and compare your results. Draw a picture of the sample space using solid circles to indicate the sample points in a plane and then plot the prediction for each outcome of $X = i$ for $i = -1, 0, 1$. Explain your results.

8.29 (c) Test out the MATLAB program given in Section 8.7 to generate realizations of the vector random variable (X, Y) whose joint PMF is given in Figure 8.8. Do so by estimating the joint PMF or $p_{X,Y}[i, j]$. You may wish to review Section 7.11.

8.30 (⌣) (w,c) For the joint PMF given in Figure 8.8 determine the conditional mean $E_{Y|X}[j|i]$ and then verify your results using a computer simulation. Note that you will have to separate the realizations (x_m, y_m) into two sets, one in which $x_m = 0$ and one in which $x_m = 1$, and then use the sample average of each set as your estimator.

8.31 (w,c) For the joint PMF given in Figure 8.8 determine $E_Y[Y]$. Then, verify (8.36) by using your results from Problem 8.30, and computing

$$\widehat{E_Y[Y]} = \widehat{E_{Y|X}[Y|0]}\hat{p}_X[0] + \widehat{E_{Y|X}[Y|1]}\hat{p}_X[1]$$

where $\widehat{E_{Y|X}[Y|0]}$ and $\widehat{E_{Y|X}[Y|1]}$ are the values obtained in Problem 8.30. Also, the PMF of X, which needs to be estimated, can be done so as described in Section 5.9.

8.32 (w,c) For the posterior PMF given by (8.39) plot the PMF for $i = N/2$, $M = 11$ and increasing N, say $N = 10, 30, 50, 70$. What happens as N becomes large? Explain your results. Hint: You will need a computer to evaluate and plot the posterior PMF.

Chapter 9

Discrete N-Dimensional Random Variables

9.1 Introduction

In this chapter we extend the results of Chapters 5–8 to N-dimensional random variables, which are represented as an $N \times 1$ random vector. Hence, our discussions will apply to the 2×1 random vector previously studied. In fact, most of the concepts introduced earlier are trivially extended so that we do not dwell on the conceptualization. The only exception is the introduction of the covariance *matrix*, which we have not seen before. We will introduce more general notation in combination with vector/matrix representations to allow the convenient manipulation of $N \times 1$ random vectors. This representation allows many results to be easily derived and is useful for the more advanced theory of probability that the reader may encounter later. Also, it lends itself to straightforward computer implementations, particularly if one uses MATLAB, which is a vector-based programming language. Since many of the methods and subsequent properties rely on linear and matrix algebra, a brief summary of relevant concepts is given in Appendix C.

9.2 Summary

The N-dimensional joint PMF is given by (9.1) and satisfies the usual properties of (9.3) and (9.4). The joint PMF of any subset of the N random variables is obtained by summing the joint PMF over the undesired ones. If the joint PMF factors as in (9.7), the random variables are independent and vice versa. The joint PMF of a transformed random vector is given by (9.9). In particular, if the transformed random variable is the sum of N independent random variables with the same PMF, then the PMF is most easily found from (9.14). The expected value of a random vector is defined by (9.15) and the expected value of a scalar function of a random vector is found via (9.16). As usual, the expectation operator is linear with a special case given by (9.17). The variance of a sum of N random variables is given by (9.20) or (9.21). If the random variables are uncorrelated, then this variance is the sum of the variances as per (9.22). The covariance matrix of a random vector is defined by (9.25). It has many important properties that are summarized in Properties 9.1– 5. Particularly useful results are the covariance matrix of a linearly transformed random vector given by (9.27) and the ability to decorrelate the elements of a random vector using a linear transformation as explained in the proof of Property 9.5. An example of this procedure is given in Example 9.4. The joint moments and characteristic function of an N-dimensional PMF are defined by (9.32) and (9.34), respectively. The joint moments are obtainable from the characteristic function by using (9.36). An important relationship is the factorization of the joint PMF into a product of conditional PMFs as given by (9.39). When the random variables exhibit the Markov property, then this factorization simplifies even further into the product of first-order conditional PMFs as given by (9.41). The estimates of the mean vector and the covariance matrix of a random vector are given by (9.44) and (9.46), respectively. Some MATLAB code for implementing these estimates is listed in Section 9.8. Finally, a real-world example of the use of transform coding to store/transmit image data is described in Section 9.9. It is based on decorrelation of random vectors and so makes direct use of the properties of the covariance matrix.

9.3   Random Vectors and Probability Mass Functions

Previously, we denoted a two-dimensional random vector by either of the equivalent notations (X, Y) or $[X\,Y]^T$. Since we now wish to extend our results to an $N \times 1$ random vector, we shall use $(X_1, X_2, \ldots, X_N)$ or $\mathbf{X} = [X_1\,X_2\ldots X_N]^T$. Note that a boldface character will always denote a vector or a matrix, in contrast to a scalar variable. Also, all vectors are assumed to be *column* vectors. A random vector is defined as a mapping from the original sample space S of the experiment to a numerical sample space, which we term $S_{X_1, X_2, \ldots, X_N}$. The latter is normally referred to as R^N, which is the N-dimensional Euclidean space. Hence, $\mathbf{X}$ takes on values in R^N so that

$$\mathbf{X}(s) = \begin{bmatrix} X_1(s) \\ X_2(s) \\ \vdots \\ X_N(s) \end{bmatrix}$$

will have values

$$\mathbf{x} = \begin{bmatrix} x_1 \\ x_2 \\ \vdots \\ x_N \end{bmatrix}$$

where $\mathbf{x}$ is a point in the N-dimensional Euclidean space R^N. A simple example is $S = \{$all lottery tickets$\}$ with $\mathbf{X}(s)$ representing the number printed on the ticket. Then, $X_1(s)$ is the first digit of the number, $X_2(s)$ is the second digit of the number, ..., and $X_N(s)$ is the Nth digit of the number.

We are, as usual, interested in the probability that $\mathbf{X}$ takes on its possible values. This probability is $P[X_1 = x_1, X_2 = x_2, \ldots, X_N = x_N]$ and it is defined as the joint PMF. The joint PMF is therefore defined as

$$p_{X_1,X_2,\ldots,X_N}[x_1, x_2, \ldots, x_N] = P[X_1 = x_1, X_2 = x_2, \ldots, X_N = x_N] \qquad (9.1)$$

or more succinctly using vector notation as

$$p_{\mathbf{X}}[\mathbf{x}] = P[\mathbf{X} = \mathbf{x}]. \qquad (9.2)$$

When $\mathbf{x}$ consists of integer values only, we will replace x_i by k_i. Then, the joint PMF will be $p_{X_1,X_2,\ldots,X_N}[k_1, k_2, \ldots, k_N]$ or more succinctly as $p_{\mathbf{X}}[\mathbf{k}]$, where $\mathbf{k} = [k_1 \, k_2 \ldots k_N]^T$. An example of an N-dimensional joint PMF, which is of considerable importance, is the multinomial PMF (see (4.19)). In our new notation the joint PMF is

$$p_{X_1,X_2,\ldots,X_N}[k_1, k_2, \ldots, k_N] = \binom{M}{k_1, k_2, \ldots, k_N} p_1^{k_1} p_2^{k_2} \ldots p_N^{k_N}$$

where $k_i \geq 0$ with $\sum_{i=1}^{N} k_i = M$, and $0 \leq p_i \leq 1$ for all i with $\sum_{i=1}^{N} p_i = 1$. That this is a valid joint PMF follows from its adherence to the usual properties

$$0 \leq p_{X_1,X_2,\ldots,X_N}[k_1, k_2, \ldots, k_N] \leq 1 \qquad (9.3)$$

$$\sum_{k_1} \sum_{k_2} \cdots \sum_{k_N} p_{X_1,X_2,\ldots,X_N}[k_1, k_2, \ldots, k_N] = 1. \qquad (9.4)$$

To prove (9.4) we need only use the multinomial expansion, which is (see Problem 9.3)

$$(a_1 + a_2 + \cdots + a_N)^M = \sum_{k_1} \sum_{k_2} \cdots \sum_{k_N} \binom{M}{k_1, k_2, \ldots, k_N} a_1^{k_1} a_2^{k_2} \ldots a_N^{k_N} \qquad (9.5)$$

where $\sum_{i=1}^{N} k_i = M$.

The marginal PMFs are obtained from the joint PMF by summing over the *other variables*. For example, if $p_{X_1}[x_1]$ is desired, then

$$p_{X_1}[x_1] = \sum_{\{x_2 : x_2 \in S_{X_2}\}} \sum_{\{x_3 : x_3 \in S_{X_3}\}} \cdots \sum_{\{x_N : x_N \in S_{X_N}\}} p_{X_1,X_2,\ldots,X_N}[x_1, x_2, \ldots, x_N] \quad (9.6)$$

and similarly for the other $N - 1$ marginals. This is because the right-hand side of (9.6) is

$$P[X_1 = x_1, X_2 \in S_{X_2}, X_3 \in S_{X_3}, \ldots, X_N \in S_{X_N}] = P[X_1 = x_1].$$

When the random vector is composed of more than two random variables, we can also obtain the joint PMF of any subset of the random variables. We do this by summing over the variables that we wish to eliminate. If, say, we wish to determine the joint PMF of X_1 and X_N, we have

$$p_{X_1,X_N}[x_1,x_N] = \sum_{x_2}\sum_{x_3}\cdots\sum_{x_{N-1}} p_{X_1,X_2,\ldots,X_N}[x_1,x_2,\ldots,x_N].$$

As in the case of $N = 2$ the marginal PMFs do not determine the joint PMF, unless of course the random variables are independent. In the N-dimensional case the random variables are defined to be independent if the joint PMF factors or if

$$p_{X_1,X_2,\ldots,X_N}[x_1,x_2,\ldots,x_N] = p_{X_1}[x_1]p_{X_2}[x_2]\ldots p_{X_N}[x_N]. \tag{9.7}$$

Hence, if (9.7) holds, the random variables are independent, and if the random variables are independent (9.7) holds. Unlike the case of $N = 2$, it is possible that the joint PMF may factor into two or more joint PMFs. Then, the subsets of random variables are said to be independent of each other. For example, if $N = 4$ and the joint PMF factors as $p_{X_1,X_2,X_3,X_4}[x_1,x_2,x_3,x_4] = p_{X_1,X_2}[x_1,x_2]p_{X_3,X_4}[x_3,x_4]$, then the random variables (X_1, X_2) are independent of the random variables (X_3, X_4). An example of the determination of a joint PMF follows.

Example 9.1 – Joint PMF for independent Bernoulli trials

Consider an experiment in which we toss a coin with a probability of heads p, N times in succession. We let $X_i = 1$ if the ith outcome is a head and $X_i = 0$ if it is a tail. Furthermore, *assume* that the trials are *independent*. As defined in Chapter 4, this means that the probability of the outcome on any trial is not affected by the outcomes of any of the other trials. Thus, the experiment is a sequence of independent Bernoulli trials. The sample space is N-dimensional and is given by $S_{X_1,X_2,\ldots,X_N} = \{(k_1,k_2,\ldots,k_N) : k_i = 0,1 \; ; i = 1,2,\ldots,N\}$, and since $p_{X_i}[k_i] = p^{k_i}(1-p)^{1-k_i}$, we have the joint PMF from (9.7)

$$
\begin{aligned}
p_{X_1,X_2,\ldots,X_N}[k_1,k_2,\ldots,k_N] &= \prod_{i=1}^{N} p_{X_i}[k_i] \\
&= \prod_{i=1}^{N} p^{k_i}(1-p)^{1-k_i} \\
&= p^{\sum_{i=1}^{N}k_i}(1-p)^{N-\sum_{i=1}^{N}k_i}.
\end{aligned}
\tag{9.8}
$$

$\diamond$

A joint cumulative distribution function (CDF) can be defined in the N-dimensional case as

$$F_{X_1,X_2,\ldots,X_N}(x_1,x_2,\ldots,x_N) = P[X_1 \le x_1, X_2 \le x_2, \ldots, X_N \le x_N].$$

It has the usual properties of being between 0 and 1, being monotonically increasing as any of the variables increases, and being "right continuous". Also,

$$
\begin{aligned}
F_{X_1, X_2, \ldots, X_N}(-\infty, -\infty, \ldots, -\infty) &= 0 \\
F_{X_1, X_2, \ldots, X_N}(+\infty, +\infty, \ldots, +\infty) &= 1.
\end{aligned}
$$

The marginal CDFs are easily found by letting the undesired variables be evaluated at $+\infty$. For example, to determine the marginal CDF for X_1, we have

$$
F_{X_1}[x_1] = F_{X_1, X_2, \ldots, X_N}(x_1, +\infty, +\infty, \ldots, +\infty).
$$

9.4 Transformations

Since $\mathbf{X}$ is an $N \times 1$ random vector, a transformation or mapping to a random vector $\mathbf{Y}$ can yield another $N \times 1$ random vector or an $M \times 1$ random vector with $M < N$. In the former case the formula for the joint PMF of $\mathbf{Y}$ is an extension of the usual one (see (7.12)). If the transformation is given as $\mathbf{y} = \mathbf{g}(\mathbf{x})$, where $\mathbf{g}$ represents an N-dimensional function or more explicitly

$$
\begin{aligned}
y_1 &= g_1(x_1, x_2, \ldots, x_N) \\
y_2 &= g_2(x_1, x_2, \ldots, x_N) \\
&\vdots \\
y_N &= g_N(x_1, x_2, \ldots, x_N)
\end{aligned}
$$

then

$$
p_{Y_1, Y_2, \ldots, Y_N}[y_1, y_2, \ldots, y_N] = \sum \sum \cdots \sum_{\substack{\{(x_1, \ldots, x_N): \\ g_1(x_1, \ldots, x_N) = y_1, \ldots, \\ g_N(x_1, \ldots, x_N) = y_N\}}} p_{X_1, X_2, \ldots, X_N}[x_1, x_2, \ldots, x_N]. \quad (9.9)
$$

In the case where the transformation is one-to-one, there is only one solution for $\mathbf{x}$ in the equation $\mathbf{y} = \mathbf{g}(\mathbf{x})$, which we denote symbolically by $\mathbf{x} = \mathbf{g}^{-1}(\mathbf{y})$. The transformed joint PMF becomes from (9.9) $p_{\mathbf{Y}}[\mathbf{y}] = p_{\mathbf{X}}[\mathbf{g}^{-1}(\mathbf{y})]$, using vector notation. A simple example of this is when the transformation is linear and so can be represented by $\mathbf{y} = \mathbf{A}\mathbf{x}$, where $\mathbf{A}$ is an $N \times N$ nonsingular matrix. Then, the solution is $\mathbf{x} = \mathbf{A}^{-1}\mathbf{y}$ and the transformed joint PMF becomes

$$
p_{\mathbf{Y}}[\mathbf{y}] = p_{\mathbf{X}}[\mathbf{A}^{-1}\mathbf{y}]. \quad (9.10)
$$

The other case, in which $\mathbf{Y}$ has dimension less than N, can be solved using the technique of auxiliary random variables. We add enough random variables to make the dimension of the transformed random vector equal to N, find the joint PMF via

(9.9), and finally sum the N-dimensional PMF over the auxiliary random variables. More specifically, if $\mathbf{Y}$ is $M \times 1$ with $M < N$, we define a new $N \times 1$ random vector

$$\mathbf{Z} = [Y_1 \ Y_2 \ldots Y_M \ Z_{M+1} = X_{M+1} \ Z_{M+2} = X_{M+2} \ldots Z_N = X_N]^T$$

so that the transformation becomes one-to-one, if possible. Once the joint PMF of $\mathbf{Z}$ is found, we can determine the joint PMF of $\mathbf{Y}$ as

$$p_{Y_1,Y_2,\ldots,Y_M}[y_1, y_2, \ldots, y_M] = \sum_{z_{M+1}} \sum_{z_{M+2}} \cdots \sum_{z_N} p_{Z_1,Z_2,\ldots,Z_N}[z_1, z_2, \ldots, z_N].$$

The determination of the PMF of a transformed random vector is in general *not an easy task*. Even to determine the possible values of $\mathbf{Y}$ can be quite difficult. An example follows that illustrates the work involved.

Example 9.2 – PMF for one-to-one transformation of N-dimensional random vector

In Example 9.1 $\mathbf{X}$ has the joint PMF given by (9.8). We define a transformed random vector as

$$
\begin{aligned}
Y_1 &= X_1 \\
Y_2 &= X_1 + X_2 \\
Y_3 &= X_1 + X_2 + X_3.
\end{aligned}
$$

This is a linear transformation that maps a 3×1 random vector $\mathbf{X}$ into another 3×1 random vector $\mathbf{Y}$. It can be represented by the 3×3 matrix

$$
\mathbf{A} = \begin{bmatrix} 1 & 0 & 0 \\ 1 & 1 & 0 \\ 1 & 1 & 1 \end{bmatrix}.
$$

Note that the transformed random variables are the sums of the outcomes of the first Bernoulli trial, the first and second Bernoulli trials, and finally the sum of the first three Bernoulli trials. As such the values of the transformed random variables must take on certain values. In particular, $y_1 \leq y_2 \leq y_3$ or the outcomes must increase as the index i increases. This is sometimes called a *counting process* and will be studied in more detail when we discuss random processes. Some typical realizations of the random vector $\mathbf{Y}$ are shown in Figure 9.1. To determine the sample space for $\mathbf{Y}$ we enumerate the possible values, making sure that the values in the vector increase or stay the same and that the increase is at most one unit from y_i to y_{i+1}. The sample space is composed of integer 3-tuples (l_1, l_2, l_3), which is given by

$$\mathcal{S}_{Y_1,Y_2,Y_3} = \{(0,0,0), (0,0,1), (0,1,1), (1,1,1), (0,1,2), (1,1,2), (1,2,2), (1,2,3)\}.$$
$$(9.11)$$

These are the values of $\mathbf{y}$ for which p_{Y_1,Y_2,Y_3} is nonzero and are seen to be integer-valued. Next, we need to solve for $\mathbf{x}$ according to (9.10). It is easily shown that the

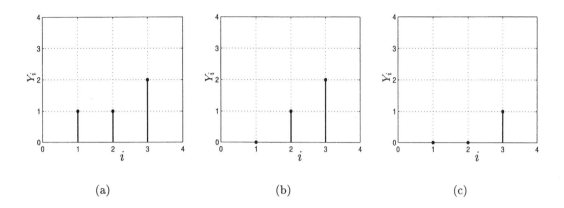

Figure 9.1: Typical realizations for sum of outcomes of independent Bernoulli trials.

linear transformation is one-to-one since $\mathbf{A}$ has an inverse (note that the determinant of $\mathbf{A}$ is nonzero since $\det(\mathbf{A}) = 1$, and so $\mathbf{A}$ has an inverse), which is

$$\mathbf{A}^{-1} = \begin{bmatrix} 1 & 0 & 0 \\ -1 & 1 & 0 \\ 0 & -1 & 1 \end{bmatrix}.$$

This says that $\mathbf{x} = \mathbf{A}^{-1}\mathbf{y}$ or $x_1 = y_1, x_2 = y_2 - y_1, x_3 = y_3 - y_2$. Thus, we can use (9.10) and then (9.8) to find the joint PMF of $\mathbf{Y}$, which becomes from (9.10)

$$p_{Y_1,Y_2,Y_3}[l_1, l_2, l_3] = p_{X_1,X_2,X_3}[l_1, l_2 - l_1, l_3 - l_2]$$

and since from (9.8)

$$p_{X_1,X_2,X_3}[k_1, k_2, k_3] = p^{k_1+k_2+k_3}(1-p)^{3-(k_1+k_2+k_3)}$$

we have that

$$p_{Y_1,Y_2,Y_3}[l_1, l_2, l_3] = p^{l_3}(1-p)^{3-l_3}. \tag{9.12}$$

Note that the joint PMF is nonzero only over the sample space $\mathcal{S}_{Y_1,Y_2,Y_3}$ given in (9.11).

$\diamondsuit$

 Always make sure PMF values sum to one.

The result of the previous example looks strange in that the joint PMF of $\mathbf{Y}$ does not depend on l_1 and l_2. A simple check that should always be made when working these types of problems is to verify that the PMF values sum to one. If not, then there is an error in the calculation. If they do sum to one, then there could still be an error but it is not likely. For the previous example, we have from (9.11) 1 outcome for which $l_3 = 0$, 3 outcomes for which $l_3 = 1$, 3 outcomes for which $l_3 = 2$, and 1 outcome for which $l_3 = 3$. If we sum the probabilities of these outcomes we have from (9.12)

$$1(1-p)^3 + 3p(1-p)^2 + 3p^2(1-p) + p^3 = 1$$

and hence we can assert with some confidence that the result is correct.

A transformation that is not one-to-one but that frequently is of interest is the sum of N independent discrete random variables. It is given by

$$Y = \sum_{i=1}^{N} X_i \qquad (9.13)$$

where the X_i's are independent random variables with integer values. For the case of $N = 2$ and integer-valued discrete random variables we saw in Section 7.6 that $p_Y = p_{X_1} \star p_{X_2}$, where $\star$ denotes discrete convolution. This is most easily evaluated using the characteristic functions and the inverse Fourier transform to yield

$$p_Y[k] = \int_{-\pi}^{\pi} \phi_{X_1}(\omega)\phi_{X_2}(\omega)\exp(-j\omega k)\frac{d\omega}{2\pi}.$$

For a sum of N independent random variables we have the similar result

$$p_Y[k] = \int_{-\pi}^{\pi} \prod_{i=1}^{N} \phi_{X_i}(\omega)\exp(-j\omega k)\frac{d\omega}{2\pi}$$

and if all the X_i's have the same PMF and hence the same characteristic function, this becomes

$$p_Y[k] = \int_{-\pi}^{\pi} \phi_X^N(\omega)\exp(-j\omega k)\frac{d\omega}{2\pi} \qquad (9.14)$$

where $\phi_X(\omega)$ is the common characteristic function. An example follows (see also Problem 9.9).

Example 9.3 – Binomial PMF derived as PMF of sum of independent Bernoulli random variables

We had previously derived the binomial PMF by examining the number of successes in N independent Bernoulli trials (see Section 4.6.2). We can rederive this result by using (9.14) with $X_i = 1$ for a success and $X_i = 0$ for a failure and determining the

PMF of $Y = \sum_{i=1}^{N} X_i$. The random variable Y will be the number of successes in N trials. The characteristic function of X is for a single Bernoulli trial

$$
\begin{aligned}
\phi_X(\omega) &= E_X[\exp(j\omega X)] \\
&= \exp(j\omega(1))p + \exp(j\omega(0))(1-p) \\
&= p\exp(j\omega) + (1-p).
\end{aligned}
$$

Now using (9.14) we have

$$
\begin{aligned}
p_Y[k] &= \int_{-\pi}^{\pi} [p\exp(j\omega) + (1-p)]^N \exp(-j\omega k)\frac{d\omega}{2\pi} \\
&= \int_{-\pi}^{\pi} \sum_{i=0}^{N} \binom{N}{i} [p\exp(j\omega)]^i (1-p)^{N-i} \exp(-j\omega k)\frac{d\omega}{2\pi} \\
&\qquad\qquad\qquad\qquad\qquad\qquad\qquad\qquad \text{(use binomial theorem)} \\
&= \sum_{i=0}^{N} \binom{N}{i} p^i (1-p)^{N-i} \int_{-\pi}^{\pi} \exp[j\omega(i-k)]\frac{d\omega}{2\pi}.
\end{aligned}
$$

But the integral can be shown to be 0 if $i \neq k$ and 1 if $i = k$ (see Problem 9.8). Using this result we have as the only term in the sum being nonzero the one for which $i = k$, and therefore

$$
p_Y[k] = \binom{N}{k} p^k (1-p)^{N-k} \qquad k = 0, 1, \ldots, N.
$$

The sum of N independent Bernoulli random variables has the PMF bin(N, p) in accordance with our earlier results.

$\diamondsuit$

9.5 Expected Values

The expected value of a random vector is defined as the vector of the expected values of the elements of the random vector. This is to say that we define

$$
E_{\mathbf{X}}[\mathbf{X}] = E_{X_1, X_2, \ldots, X_N}\left[\begin{bmatrix} X_1 \\ X_2 \\ \vdots \\ X_N \end{bmatrix}\right] = \begin{bmatrix} E_{X_1}[X_1] \\ E_{X_2}[X_2] \\ \vdots \\ E_{X_N}[X_N] \end{bmatrix}. \tag{9.15}
$$

We can view this definition as "passing" the expectation "through" the left bracket of the vector since $E_{X_1, X_2, \ldots, X_N}[X_i] = E_{X_i}[X_i]$.

A particular expectation of interest is that of a scalar function of $X_1, X_2, \ldots, X_N$, say $g(X_1, X_2, \ldots, X_N)$. Similar to previous results (see Section 7.7) this is determined by using

$$E_{X_1,X_2,\ldots,X_N}[g(X_1, X_2, \ldots, X_N)]$$

$$= \sum_{x_1} \sum_{x_2} \cdots \sum_{x_N} g(x_1, x_2, \ldots, x_N) p_{X_1,X_2,\ldots,X_N}[x_1, x_2, \ldots, x_N]. \tag{9.16}$$

As an example, if $g(X_1, X_2, \ldots, X_N) = \sum_{i=1}^{N} X_i$, then

$$E_{X_1,X_2,\ldots,X_N}\left[\sum_{i=1}^{N} X_i\right]$$

$$= \sum_{x_1} \sum_{x_2} \cdots \sum_{x_N} (x_1 + x_2 + \cdots + x_N) p_{X_1,X_2,\ldots,X_N}[x_1, x_2, \ldots, x_N]$$

$$= \sum_{x_1} \sum_{x_2} \cdots \sum_{x_N} x_1 p_{X_1,X_2,\ldots,X_N}[x_1, x_2, \ldots, x_N]$$

$$+ \sum_{x_1} \sum_{x_2} \cdots \sum_{x_N} x_2 p_{X_1,X_2,\ldots,X_N}[x_1, x_2, \ldots, x_N]$$

$$+ \cdots + \sum_{x_1} \sum_{x_2} \cdots \sum_{x_N} x_N p_{X_1,X_2,\ldots,X_N}[x_1, x_2, \ldots, x_N]$$

$$= E_{X_1}[X_1] + E_{X_2}[X_2] + \cdots + E_{X_N}[X_N].$$

By a slight modification we can also show that

$$E_{X_1,X_2,\ldots,X_N}\left[\sum_{i=1}^{N} a_i X_i\right] = \sum_{i=1}^{N} a_i E_{X_i}[X_i] \tag{9.17}$$

which says that the expectation is a linear operator. It is also possible to write (9.17) more succinctly by defining the $N \times 1$ vector $\mathbf{a} = [a_1\, a_2 \ldots a_N]^T$ to yield

$$E_{\mathbf{X}}[\mathbf{a}^T \mathbf{X}] = \mathbf{a}^T E_{\mathbf{X}}[\mathbf{X}]. \tag{9.18}$$

We next determine the variance of a sum of random variables. Previously it was shown that

$$\text{var}(X_1 + X_2) = \text{var}(X_1) + \text{var}(X_2) + 2\text{cov}(X_1, X_2). \tag{9.19}$$

Our goal is to extend this to $\text{var}(\sum_{i=1}^{N} X_i)$ for any N. To do so we proceed as follows.

$$\text{var}\left(\sum_{i=1}^{N} X_i\right) = E_{\mathbf{X}}\left[\left(\sum_{i=1}^{N} X_i - E_{\mathbf{X}}\left[\sum_{i=1}^{N} X_i\right]\right)^2\right]$$

$$= E_{\mathbf{X}}\left[\left(\sum_{i=1}^{N}(X_i - E_{X_i}[X_i])\right)^2\right] \quad (\text{since } E_{\mathbf{X}}[X_i] = E_{X_i}[X_i])$$

and by letting $U_i = X_i - E_{X_i}[X_i]$ we have

$$
\mathrm{var}\left(\sum_{i=1}^{N} X_i\right) = E_\mathbf{X}\left[\left(\sum_{i=1}^{N} U_i\right)^2\right]
$$
$$
= E_\mathbf{X}\left[\sum_{i=1}^{N}\sum_{j=1}^{N} U_i U_j\right]
$$
$$
= \sum_{i=1}^{N}\sum_{j=1}^{N} E_\mathbf{X}[U_i U_j].
$$

But

$$
\begin{aligned}
E_\mathbf{X}[U_i U_j] &= E_\mathbf{X}[(X_i - E_{X_i}[X_i])(X_j - E_{X_j}[X_j])] \\
&= E_{X_i X_j}[(X_i - E_{X_i}[X_i])(X_j - E_{X_j}[X_j])] \\
&= \mathrm{cov}(X_i, X_j)
\end{aligned}
$$

so that we have as our final result

$$
\mathrm{var}\left(\sum_{i=1}^{N} X_i\right) = \sum_{i=1}^{N}\sum_{j=1}^{N} \mathrm{cov}(X_i, X_j). \tag{9.20}
$$

Noting that since $\mathrm{cov}(X_i, X_i) = \mathrm{var}(X_i)$ and $\mathrm{cov}(X_j, X_i) = \mathrm{cov}(X_i, X_j)$, we have for $N = 2$ our previous result (9.19). Also, we can write (9.20) in the alternative form

$$
\mathrm{var}\left(\sum_{i=1}^{N} X_i\right) = \sum_{i=1}^{N} \mathrm{var}(X_i) + \underset{\{(i,j):i \neq j\}}{\sum_{i=1}^{N}\sum_{j=1}^{N}} \mathrm{cov}(X_i, X_j). \tag{9.21}
$$

As an immediate and important consequence, we see that if all the random variables are uncorrelated so that $\mathrm{cov}(X_i, X_j) = 0$ for $i \neq j$, then

$$
\mathrm{var}\left(\sum_{i=1}^{N} X_i\right) = \sum_{i=1}^{N} \mathrm{var}(X_i) \tag{9.22}
$$

which says that *the variance of a sum of uncorrelated random variables is the sum of the variances.*

We wish to explore (9.20) further since it embodies some important concepts that we have not yet touched upon. For clarity let $N = 2$. Then (9.20) becomes

$$
\mathrm{var}(X_1 + X_2) = \sum_{i=1}^{2}\sum_{j=1}^{2} \mathrm{cov}(X_i, X_j). \tag{9.23}
$$

If we define a 2×2 matrix $\mathbf{C}_X$ as

$$\mathbf{C}_X = \left[\begin{array}{cc} \text{var}(X_1) & \text{cov}(X_1, X_2) \\ \text{cov}(X_2, X_1) & \text{var}(X_2) \end{array} \right]$$

then we can rewrite (9.23) as

$$\text{var}(X_1 + X_2) = \left[\begin{array}{cc} 1 & 1 \end{array} \right] \mathbf{C}_X \left[\begin{array}{c} 1 \\ 1 \end{array} \right] \tag{9.24}$$

as is easily verified. The matrix $\mathbf{C}_X$ is called the *covariance matrix*. It is a matrix with the variances along the main diagonal and the covariances off the main diagonal. For $N = 3$ it is given by

$$\mathbf{C}_X = \left[\begin{array}{ccc} \text{var}(X_1) & \text{cov}(X_1, X_2) & \text{cov}(X_1, X_3) \\ \text{cov}(X_2, X_1) & \text{var}(X_2) & \text{cov}(X_2, X_3) \\ \text{cov}(X_3, X_1) & \text{cov}(X_3, X_2) & \text{var}(X_3) \end{array} \right]$$

and in general it becomes

$$\mathbf{C}_X = \left[\begin{array}{cccc} \text{var}(X_1) & \text{cov}(X_1, X_2) & \cdots & \text{cov}(X_1, X_N) \\ \text{cov}(X_2, X_1) & \text{var}(X_2) & \cdots & \text{cov}(X_2, X_N) \\ \vdots & \vdots & \ddots & \vdots \\ \text{cov}(X_N, X_1) & \text{cov}(X_N, X_2) & \cdots & \text{cov}(X_N, X_N) \end{array} \right]. \tag{9.25}$$

The covariance matrix has many important properties, which are discussed next.

Property 9.1 – Covariance matrix is symmetric, i.e., $\mathbf{C}_X^T = \mathbf{C}_X$.
<u>Proof:</u>

$$\text{cov}(X_j, X_i) = \text{cov}(X_i, X_j) \qquad \text{(Why?)}$$

$\square$

Property 9.2 – Covariance matrix is positive semidefinite.

Being positive semidefinite means that if $\mathbf{a}$ is the $N \times 1$ column vector $\mathbf{a} = [a_1 \, a_2 \ldots a_N]^T$, then $\mathbf{a}^T \mathbf{C}_X \mathbf{a} \geq 0$ for all $\mathbf{a}$. Note that $\mathbf{a}^T \mathbf{C}_X \mathbf{a}$ is a scalar and is referred to as a *quadratic form* (see Appendix C).
<u>Proof:</u> Consider the case of $N = 2$ since the extension is immediate. Let $U_i = X_i - E_{X_i}[X_i]$, which is zero mean, and therefore we have

$$\text{var}(a_1 X_1 + a_2 X_2)$$

$$
\begin{aligned}
&= \ \text{var}(a_1 U_1 + a_2 U_2) \quad \text{(since } a_1 X_1 + a_2 X_2 = a_1 U_1 + a_2 U_2 + c \text{ for } c \text{ a constant)}\\
&= \ E_{\mathbf{X}}[(a_1 U_1 + a_2 U_2)^2] \quad (E_{\mathbf{X}}[U_1] = E_{\mathbf{X}}[U_2] = 0)\\
&= \ a_1^2 E_{\mathbf{X}}[U_1^2] + a_2^2 E_{\mathbf{X}}[U_2^2] + a_1 a_2 E_{\mathbf{X}}[U_1 U_2] + a_2 a_1 E_{\mathbf{X}}[U_2 U_1] \quad \text{(linearity of } E_{\mathbf{X}})\\
&= \ a_1^2 \text{var}(X_1) + a_2^2 \text{var}(X_2) + a_1 a_2 \text{cov}(X_1, X_2) + a_2 a_1 \text{cov}(X_2, X_1)\\
&= \ \begin{bmatrix} a_1 & a_2 \end{bmatrix} \begin{bmatrix} \text{var}(X_1) & \text{cov}(X_1, X_2) \\ \text{cov}(X_2, X_1) & \text{var}(X_2) \end{bmatrix} \begin{bmatrix} a_1 \\ a_2 \end{bmatrix}\\
&= \ \mathbf{a}^T \mathbf{C}_X \mathbf{a}.
\end{aligned}
$$

Since $\text{var}(a_1 X_1 + a_2 X_2) \geq 0$ for all a_1 and a_2, it follows that $\mathbf{C}_X$ is positive semidefinite.

$\square$

Also, note that the covariance matrix of random variables that are not perfectly predictable by a linear predictor is *positive definite*. A positive definite covariance matrix is one for which $\mathbf{a}^T \mathbf{C}_X \mathbf{a} > 0$ for all $\mathbf{a} \neq \mathbf{0}$. If, however, perfect prediction is possible, as would be the case if for $N = 2$ we had $a_1 X_1 + a_2 X_2 + c = 0$, for c a constant and for some a_1 and a_2, or equivalently if $X_2 = -(a_1/a_2)X_1 - (c/a_2)$, then the covariance matrix is only positive *semi*definite. This is because $\text{var}(a_1 X_1 + a_2 X_2) = \mathbf{a}^T \mathbf{C}_X \mathbf{a} = 0$ in this case.

Finally, with the general result that (see Problem 9.14)

$$
\text{var} \left(\sum_{i=1}^{N} a_i X_i \right) = \mathbf{a}^T \mathbf{C}_X \mathbf{a} \tag{9.26}
$$

we have upon letting $\mathbf{a} = \mathbf{1} = [1\,1 \ldots 1]^T$ be an $N \times 1$ vector of ones that

$$
\text{var} \left(\sum_{i=1}^{N} X_i \right) = \mathbf{1}^T \mathbf{C}_X \mathbf{1}
$$

which is another way of writing (9.20) (the effect of premultiplying a matrix by $\mathbf{1}^T$ and postmultiplying by $\mathbf{1}$ is to sum all the elements in the matrix).

The fact that the covariance matrix is a symmetric positive semidefinite matrix is important in that it must exhibit all the properties of that type of matrix. For example, if a matrix is symmetric positive semidefinite, then it can be shown that its determinant is nonnegative. As a result, it follows that the correlation coefficient must have a magnitude less than or equal to one (see Problem 9.18). Some other properties of a covariance matrix follow.

Property 9.3 – Covariance matrix for uncorrelated random variables is a diagonal matrix.

Note that a diagonal matrix is one for which all the off-diagonal elements are zero.
<u>Proof</u>: Let $\text{cov}(X_i, X_j) = 0$ for $i \neq j$ in (9.25).

$\square$

Before listing the next property a new definition is needed. Similar to the definition that the expected value of a random vector is the vector of expected values of the elements, we define the expectation of a random matrix as the matrix of expected values of its elements. As an example, if $N = 2$ the definition is

$$E_{\mathbf{X}} \begin{bmatrix} g_{11}(\mathbf{X}) & g_{12}(\mathbf{X}) \\ g_{21}(\mathbf{X}) & g_{22}(\mathbf{X}) \end{bmatrix} = \begin{bmatrix} E_{\mathbf{X}}[g_{11}(\mathbf{X})] & E_{\mathbf{X}}[g_{12}(\mathbf{X})] \\ E_{\mathbf{X}}[g_{21}(\mathbf{X})] & E_{\mathbf{X}}[g_{22}(\mathbf{X})] \end{bmatrix}.$$

Property 9.4 – Covariance matrix of $\mathbf{Y} = \mathbf{AX}$, where $\mathbf{A}$ is an $M \times N$ matrix (with $M \leq N$), is easily determined.

The covariance matrix of $\mathbf{Y}$ is

$$\mathbf{C}_Y = \mathbf{A}\mathbf{C}_X\mathbf{A}^T. \tag{9.27}$$

Proof:

To prove this result without having to explicitly write out each element of the various matrices requires the use of matrix algebra. We therefore only sketch the proof and leave some details to the problems. The covariance matrix of $\mathbf{Y}$ can alternatively be defined by (see Problem 9.21)

$$\mathbf{C}_Y = E_{\mathbf{Y}}\left[(\mathbf{Y} - E_{\mathbf{Y}}[\mathbf{Y}])(\mathbf{Y} - E_{\mathbf{Y}}[\mathbf{Y}])^T\right].$$

Therefore,

$$\begin{aligned} \mathbf{C}_Y &= E_{\mathbf{X}}\left[(\mathbf{AX} - E_{\mathbf{X}}[\mathbf{AX}])(\mathbf{AX} - E_{\mathbf{X}}[\mathbf{AX}])^T\right] \\ &= E_{\mathbf{X}}\left[\mathbf{A}(\mathbf{X} - E_{\mathbf{X}}[\mathbf{X}])(\mathbf{A}(\mathbf{X} - E_{\mathbf{X}}[\mathbf{X}]))^T\right] \quad \text{(see Problem 9.22)} \\ &= \mathbf{A}E_{\mathbf{X}}\left[(\mathbf{X} - E_{\mathbf{X}}[\mathbf{X}])(\mathbf{X} - E_{\mathbf{X}}[\mathbf{X}])^T\right]\mathbf{A}^T \quad \text{(see Problem 9.23)} \\ &= \mathbf{A}\mathbf{C}_X\mathbf{A}^T. \end{aligned}$$

$\square$

This result subsumes many of our previous ones (try $\mathbf{A} = \mathbf{1}^T = [1\,1\ldots 1]$ and note that $\mathbf{C}_Y = \text{var}(Y)$ if $M = 1$, for example!).

Property 9.5 – Covariance matrix can always be diagonalized.

The importance of this property is that a diagonalized covariance matrix implies that the random variables are uncorrelated. Hence, by transforming a random vector of correlated random variable elements to one whose covariance matrix is diagonal, we can *decorrelate* the random variables. It is exceedingly fortunate that this transformation is a linear one and is easily found. In summary, if $\mathbf{X}$ has a covariance matrix $\mathbf{C}_X$, then we can find an $N \times N$ matrix $\mathbf{A}$ so that $\mathbf{Y} = \mathbf{AX}$ has the covariance matrix

$$\mathbf{C}_Y = \begin{bmatrix} \text{var}(Y_1) & 0 & \cdots & 0 \\ 0 & \text{var}(Y_2) & \cdots & 0 \\ \vdots & \vdots & \ddots & \vdots \\ 0 & 0 & \cdots & \text{var}(Y_N) \end{bmatrix}.$$

The matrix $\mathbf{A}$ is not unique (see Problem 7.35 for a particular method). One possible determination of $\mathbf{A}$ is contained within the proof given next.

Proof:

We only sketch the proof of this result since it relies heavily on linear and matrix algebra (see also Appendix C). More details are available in [Noble and Daniel 1977]. Since $\mathbf{C}_X$ is a symmetric matrix, it has a set of N orthonormal eigenvectors with corresponding real eigenvalues. Since $\mathbf{C}_X$ is also positive semidefinite, the eigenvalues are nonnegative. Hence, we can find $N \times 1$ eigenvectors $\{\mathbf{v}_1, \mathbf{v}_2, \ldots, \mathbf{v}_N\}$ so that

$$\mathbf{C}_X \mathbf{v}_i = \lambda_i \mathbf{v}_i \qquad i = 1, 2, \ldots, N$$

where $\mathbf{v}_i^T \mathbf{v}_j = 0$ for $i \neq j$ (orthogonality), $\mathbf{v}_i^T \mathbf{v}_i = 1$ (normalized to unit length), and $\lambda_i \geq 0$. We can arrange the $N \times 1$ column vectors $\mathbf{C}_X \mathbf{v}_i$ and also $\lambda_i \mathbf{v}_i$ into $N \times N$ matrices so that

$$\begin{bmatrix} \mathbf{C}_X \mathbf{v}_1 & \mathbf{C}_X \mathbf{v}_2 & \ldots & \mathbf{C}_X \mathbf{v}_N \end{bmatrix} = \begin{bmatrix} \lambda_1 \mathbf{v}_1 & \lambda_2 \mathbf{v}_2 & \ldots & \lambda_N \mathbf{v}_N \end{bmatrix}. \tag{9.28}$$

But it may be shown that for an $N \times N$ matrix $\mathbf{A}$ and $N \times 1$ vectors $\mathbf{b}_1, \mathbf{b}_2, \mathbf{d}_1, \mathbf{d}_2$, using $N = 2$ for simplicity (see Problem 9.24),

$$\begin{bmatrix} \mathbf{A}\mathbf{b}_1 & \mathbf{A}\mathbf{b}_2 \end{bmatrix} = \mathbf{A} \begin{bmatrix} \mathbf{b}_1 & \mathbf{b}_2 \end{bmatrix} \tag{9.29}$$

$$\begin{bmatrix} c_1 \mathbf{d}_1 & c_2 \mathbf{d}_2 \end{bmatrix} = \begin{bmatrix} \mathbf{d}_1 & \mathbf{d}_2 \end{bmatrix} \begin{bmatrix} c_1 & 0 \\ 0 & c_2 \end{bmatrix}. \tag{9.30}$$

Using these relationships (9.28) becomes

$$\mathbf{C}_X \underbrace{\begin{bmatrix} \mathbf{v}_1 & \mathbf{v}_2 & \ldots & \mathbf{v}_N \end{bmatrix}}_{\mathbf{V}} = \begin{bmatrix} \mathbf{v}_1 & \mathbf{v}_2 & \ldots & \mathbf{v}_N \end{bmatrix} \underbrace{\begin{bmatrix} \lambda_1 & 0 & \ldots & 0 \\ 0 & \lambda_2 & \ldots & 0 \\ \vdots & \vdots & \ddots & \vdots \\ 0 & 0 & \ldots & \lambda_N \end{bmatrix}}_{\mathbf{\Lambda}}$$

or

$$\mathbf{C}_X \mathbf{V} = \mathbf{V}\mathbf{\Lambda}.$$

(The matrix $\mathbf{V}$ is known as the *modal* matrix and is invertible.) Premultiplying both sides by $\mathbf{V}^{-1}$ produces

$$\mathbf{V}^{-1} \mathbf{C}_X \mathbf{V} = \mathbf{\Lambda}.$$

Next we use the property that the eigenvectors are orthonormal to assert that $\mathbf{V}^{-1} = \mathbf{V}^T$ (a property of orthogonal matrices), and therefore

$$\mathbf{V}^T \mathbf{C}_X \mathbf{V} = \mathbf{\Lambda} \tag{9.31}$$

Now recall from Property 9.4 that if $\mathbf{Y} = \mathbf{AX}$, then $\mathbf{C}_Y = \mathbf{AC}_X\mathbf{A}^T$. Thus, if we let $\mathbf{Y} = \mathbf{AX} = \mathbf{V}^T\mathbf{X}$, we will have

$$
\begin{aligned}
\mathbf{C}_Y &= \mathbf{V}^T\mathbf{C}_X\mathbf{V} &&\text{(from Property 9.4)} \\
&= \mathbf{\Lambda} &&\text{(from (9.31))}
\end{aligned}
$$

and the covariance matrix of $\mathbf{Y}$ will be diagonal with ith diagonal element $\text{var}(Y_i) = \lambda_i \geq 0$.

$\square$

This important result is used extensively in many disciplines. Later we will see that for some types of *continuous* random vectors, the use of this linear transformation will make the random variables not only uncorrelated but independent as well (see Example 12.14). An example follows.

Example 9.4 – Decorrelation of random variables

We consider a two-dimensional example whose joint PMF is given in Table 9.1. We

	$x_2 = -8$	$x_2 = 0$	$x_2 = 2$	$x_2 = 6$	$p_{X_1}[x_1]$
$x_1 = -8$	0	$\frac{1}{4}$	0	0	$\frac{1}{4}$
$x_1 = 0$	$\frac{1}{4}$	0	0	0	$\frac{1}{4}$
$x_1 = 2$	0	0	0	$\frac{1}{4}$	$\frac{1}{4}$
$x_1 = 6$	0	0	$\frac{1}{4}$	0	$\frac{1}{4}$
$p_{X_2}[x_2]$	$\frac{1}{4}$	$\frac{1}{4}$	$\frac{1}{4}$	$\frac{1}{4}$	

Table 9.1: Joint PMF values.

first determine the covariance matrix $\mathbf{C}_X$ and then $\mathbf{A}$ so that $\mathbf{Y} = \mathbf{AX}$ consists of uncorrelated random variables. From Table 9.1 we have that

$$
\begin{aligned}
E_{X_1}[X_1] &= E_{X_2}[X_2] = 0 \\
E_{X_1}[X_1^2] &= E_{X_2}[X_2^2] = 26 \\
E_{X_1X_2}[X_1X_2] &= 6
\end{aligned}
$$

and therefore we have that

$$
\begin{aligned}
\text{var}(X_1) &= \text{var}(X_2) = 26 \\
\text{cov}(X_1, X_2) &= 6
\end{aligned}
$$

yielding a covariance matrix

$$
\mathbf{C}_X = \begin{bmatrix} 26 & 6 \\ 6 & 26 \end{bmatrix}.
$$

9.6 Joint Moments and the Characteristic Function

The joint moments corresponding to an N-dimensional PMF are defined as

$$E_{X_1,X_2,\ldots,X_N}[X_1^{l_1} X_2^{l_2} \ldots X_N^{l_N}] = \sum_{x_1} \sum_{x_2} \cdots \sum_{x_N} x_1^{l_1} x_2^{l_2} \ldots x_N^{l_N} p_{X_1,X_2,\ldots,X_N}[x_1,x_2,\ldots,x_N].$$
(9.32)

As usual if the random variables are independent, the joint PMF factors and therefore

$$E_{X_1,X_2,\ldots,X_N}[X_1^{l_1} X_2^{l_2} \ldots X_N^{l_N}] = E_{X_1}[X_1^{l_1}] E_{X_2}[X_2^{l_2}] \ldots E_{X_N}[X_N^{l_N}].$$
(9.33)

The joint characteristic function is defined as

$$\phi_{X_1,X_2,\ldots,X_N}(\omega_1,\omega_2,\ldots,\omega_N) = E_{X_1,X_2,\ldots,X_N}\left[\exp[j(\omega_1 X_1 + \omega_2 X_2 + \cdots + \omega_N X_N)]\right]$$
(9.34)

and is evaluated as

$$\phi_{X_1,X_2,\ldots,X_N}(\omega_1,\omega_2,\ldots,\omega_N)$$

$$= \sum_{x_1} \sum_{x_2} \cdots \sum_{x_N} \exp[j(\omega_1 x_1 + \omega_2 x_2 + \cdots + \omega_N x_N)] p_{X_1,X_2,\ldots,X_N}[x_1,x_2,\ldots,x_N].$$

In particular, for independent random variables, we have (see Problem 9.28)

$$\phi_{X_1,X_2,\ldots,X_N}(\omega_1,\omega_2,\ldots,\omega_N) = \phi_{X_1}(\omega_1)\phi_{X_1}(\omega_2) \ldots \phi_{X_1}(\omega_N).$$

Also, if $\mathbf{X}$ takes on integer values, the joint PMF can be found from the joint characteristic function using the inverse Fourier transform or

$$p_{X_1,X_2,\ldots,X_N}[k_1,k_2,\ldots,k_N]$$

$$= \int_{-\pi}^{\pi} \int_{-\pi}^{\pi} \cdots \int_{-\pi}^{\pi} \phi_{X_1,X_2,\ldots,X_N}(\omega_1,\omega_2,\ldots,\omega_N)$$

$$\cdot \exp[-j(\omega_1 k_1 + \omega_2 k_2 + \cdots + \omega_N k_N)] \frac{d\omega_1}{2\pi} \frac{d\omega_2}{2\pi} \cdots \frac{d\omega_N}{2\pi}.$$
(9.35)

All the properties of the 2-dimensional characteristic function extend to the general case. Note that once $\phi_{X_1,X_2,\ldots,X_N}(\omega_1,\omega_2,\ldots,\omega_N)$ is known, the characteristic function for any subset of the X_i's is found by setting ω_i equal to zero for the ones not in the subset. For example, to find $p_{X_1,X_2}[x_1,x_2]$, we let $\omega_3 = \omega_4 = \cdots = \omega_N = 0$ in the joint characteristic function to yield

$$p_{X_1,X_2}[k_1,k_2] = \int_{-\pi}^{\pi} \int_{-\pi}^{\pi} \underbrace{\phi_{X_1,X_2,\ldots,X_N}(\omega_1,\omega_2,0,0,\ldots,0)}_{\phi_{X_1,X_2}(\omega_1,\omega_2)} \exp[-j(\omega_1 k_1 + \omega_2 k_2)] \frac{d\omega_1}{2\pi} \frac{d\omega_2}{2\pi}.$$

As seen previously, the joint moments can be obtained from the characteristic function. The general formula is

$$E_{X_1,X_2,\ldots,X_N}[X_1^{l_1}X_2^{l_2}\ldots X_N^{l_N}]$$

$$= \frac{1}{j^{l_1+l_2+\cdots+l_N}}\frac{\partial^{l_1+l_2+\cdots+l_N}}{\partial\omega_1^{l_1}\partial\omega_2^{l_2}\ldots\partial\omega_N^{l_N}}\phi_{X_1,X_2,\ldots,X_N}(\omega_1,\omega_2,\ldots,\omega_N)\bigg|_{\omega_1=\omega_2=\cdots=\omega_N=0}. \tag{9.36}$$

9.7   Conditional Probability Mass Functions

When we have an N-dimensional random vector, many different conditional PMFs can be defined. A straightforward extension of the conditional PMF $p_{Y|X}$ encountered in Chapter 8 is the conditional PMF of a single random variable conditioned on knowledge of the outcomes of all the other random variables. For example, it is of interest to study $p_{X_N|X_1,X_2,\ldots,X_{N-1}}$, whose definition is

$$p_{X_N|X_1,X_2,\ldots,X_{N-1}}[x_N|x_1,x_2,\ldots,x_{N-1}] = \frac{p_{X_1,X_2,\ldots,X_N}[x_1,x_2,\ldots,x_N]}{p_{X_1,X_2,\ldots,X_{N-1}}[x_1,x_2,\ldots,x_{N-1}]}. \tag{9.37}$$

Then by rearranging (9.37) we have upon omitting the arguments

$$p_{X_1,X_2,\ldots,X_N} = p_{X_N|X_1,X_2,\ldots,X_{N-1}}p_{X_1,X_2,\ldots,X_{N-1}}. \tag{9.38}$$

If we replace N by $N-1$ in (9.37), we have

$$p_{X_{N-1}|X_1,X_2,\ldots,X_{N-2}} = \frac{p_{X_1,X_2,\ldots,X_{N-1}}}{p_{X_1,X_2,\ldots,X_{N-2}}}$$

or

$$p_{X_1,X_2,\ldots,X_{N-1}} = p_{X_{N-1}|X_1,X_2,\ldots,X_{N-2}}p_{X_1,X_2,\ldots,X_{N-2}}.$$

Inserting this into (9.38) yields

$$p_{X_1,X_2,\ldots,X_N} = p_{X_N|X_1,X_2,\ldots,X_{N-1}}p_{X_{N-1}|X_1,X_2,\ldots,X_{N-2}}p_{X_1,X_2,\ldots,X_{N-2}}.$$

Continuing this process results in the *general chain rule* for joint PMFs (see also (4.10))

$$p_{X_1,X_2,\ldots,X_N} = p_{X_N|X_1,X_2,\ldots,X_{N-1}}p_{X_{N-1}|X_1,X_2,\ldots,X_{N-2}}\cdots p_{X_2|X_1}p_{X_1}. \tag{9.39}$$

A particularly useful special case of this relationship occurs when the conditional PMFs satisfies

$$p_{X_n|X_1,X_2,\ldots,X_{n-1}} = p_{X_n|X_{n-1}} \qquad \text{for } n = 3,4,\ldots,N \tag{9.40}$$

or X_n is independent of $X_1 \ldots, X_{n-2}$ *if X_{n-1} is known* for all $n \geq 3$. If we view n as a time index, then this says that the probability of the current random variable X_n is independent of the past outcomes once the most recent past outcome X_{n-1} is known. This is called the *Markov property*, which was described in Section 4.6.4. When the Markov property holds, we can rewrite (9.39) in the particularly simple form

$$p_{X_1,X_2,\ldots,X_N} = p_{X_N|X_{N-1}}p_{X_{N-1}|X_{N-2}} \cdots p_{X_2|X_1}p_{X_1} \tag{9.41}$$

which is a factorization of the N-dimensional joint PMF into a product of *first-order* conditional PMFs. It can be considered as the logical extension of the factorization of the N-dimensional joint PMF of independent random variables into the product of its marginals. As such it enjoys many useful properties, which are discussed in Chapter 22. A simple example of when (9.40) holds is for a "running" sum of independent random variables or $X_n = \sum_{i=1}^{n} U_i$, where the U_i's are independent. Then, we have

$$
\begin{aligned}
X_1 &= U_1 \\
X_2 &= U_1 + U_2 = X_1 + U_2 \\
X_3 &= U_1 + U_2 + U_3 = X_2 + U_3 \\
\vdots &= \vdots \\
X_N &= X_{N-1} + U_N.
\end{aligned}
$$

For example, X_2 is known, the PMF of $X_3 = X_2 + U_3$ depends only on U_3 and not on X_1. Also, it is seen from the definition of the random variables that U_3 and $U_1 = X_1$ are independent. Thus, once X_2 is known, X_3 (a function of U_3) is independent of X_1 (a function of U_1). As a result, $p_{X_3|X_2,X_1} = p_{X_3|X_2}$ and in general

$$p_{X_n|X_1,X_2,\ldots,X_{n-1}} = p_{X_n|X_{n-1}} \qquad \text{for } n = 3, 4, \ldots, N$$

or (9.40) is satisfied. It is said that "the PMF of X_n given the past samples depends only on the most recent past sample". To illustrate this we consider a particular running sum of independent random variables known as a random walk.

Example 9.5 – Random walk

Let U_i for $i = 1, 2, \ldots, N$ be independent random variables with the same PMF

$$p_U[k] = \begin{cases} 1-p & k = -1 \\ p & k = 1 \end{cases}$$

and define

$$X_n = \sum_{i=1}^{n} U_i.$$

At each "time" n the new random variable X_n changes from the old random variable X_{n-1} by ± 1 since $X_n = X_{n-1} + U_n$. The joint PMF is from (9.41)

$$p_{X_1,X_2,\ldots,X_N} = \prod_{n=1}^{N} p_{X_n|X_{n-1}} \tag{9.42}$$

where $p_{X_1|X_0}$ is defined as p_{X_1}. But $p_{X_n|X_{n-1}}$ can be found by noting that $X_n = X_{n-1} + U_n$ and therefore if $X_{n-1} = x_{n-1}$ we have that

$$
\begin{aligned}
p_{X_n|X_{n-1}}[x_n|x_{n-1}] &= p_{U_n|X_{n-1}}[x_n - x_{n-1}|x_{n-1}] &\text{(step 1 -- transform PMF)}\\
&= p_{U_n}[x_n - x_{n-1}] &\text{(step 2 -- independence)}\\
&= p_U[x_n - x_{n-1}] &(U_n\text{'s have same PMF).}
\end{aligned}
$$

Step 1 results from the transformed random variable $Y = X + c$, where c is a constant, having a PMF $p_Y[y_i] = p_X[y_i - c]$. Step 2 results from U_n being independent of $X_{n-1} = \sum_{i=1}^{n-1} U_i$ since all the U_i's are independent. Finally, we have from (9.42)

$$p_{X_1,X_2,\ldots,X_N}[x_1, x_2, \ldots, x_N] = \prod_{n=1}^{N} p_U[x_n - x_{n-1}]. \tag{9.43}$$

A realization of the random variables for $p = 1/2$ is shown in Figure 9.3. As justified by the character of the outcomes in Figure 9.3b, this *random process* is termed a *random walk*. We will say more about this later in Chapter 16. Note that the

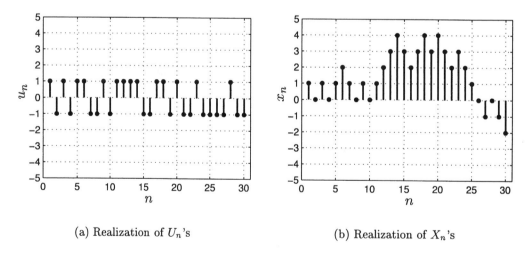

(a) Realization of U_n's   (b) Realization of X_n's

Figure 9.3: Typical realization of a random walk.

probability of the realization in Figure 9.3b is from (9.43)

$$p_{X_1,X_2,\ldots,X_{30}}[1, 0, \ldots, -2] = \prod_{n=1}^{30} p_U[x_n - x_{n-1}] = \prod_{n=1}^{30} \frac{1}{2} = \left(\frac{1}{2}\right)^{30}$$

since $p_U[-1] = p_U[1] = 1/2$.

9.8 Computer Simulation of Random Vectors

To generate a realization of a random vector we can use the direct method described in Section 7.11 or the conditional approach of Section 8.7. The latter uses the general chain rule (see (9.39)). We will not pursue this further as the extension to an $N \times 1$ random vector is obvious. Instead we concentrate on two important descriptors of a random vector, those being the mean vector given by (9.15) and the covariance matrix given by (9.25). We wish to see how to estimate these quantities. In practice, the N-dimensional PMF is usually quite difficult to estimate and so we settle for the estimation of the means and covariances. The mean vector is easily estimated by estimating each element by its sample mean as we have done in Section 6.8. Here we assume to have M realizations of the $N \times 1$ random vector $\mathbf{X}$, which we denote as $\{\mathbf{x}_1, \mathbf{x}_2, \ldots, \mathbf{x}_M\}$. The mean vector estimate becomes

$$\widehat{E_\mathbf{X}[\mathbf{X}]} = \frac{1}{M} \sum_{m=1}^{M} \mathbf{x}_m \tag{9.44}$$

which is the same as estimating the ith component of $E_\mathbf{X}[\mathbf{X}]$ by $(1/M)\sum_{m=1}^{M}[\mathbf{x}_m]_i$, where $[\boldsymbol{\xi}]_i$ denotes the ith component of the vector $\boldsymbol{\xi}$. To estimate the $N \times N$ covariance matrix we first recall that the vector/matrix definition is

$$\mathbf{C}_X = E_\mathbf{X}\left[(\mathbf{X} - E_\mathbf{X}[\mathbf{X}])(\mathbf{X} - E_\mathbf{X}[\mathbf{X}])^T\right].$$

This can also be shown to be equivalent to (see Problem 9.31)

$$\mathbf{C}_X = E_\mathbf{X}\left[\mathbf{X}\mathbf{X}^T\right] - (E_\mathbf{X}[\mathbf{X}])(E_\mathbf{X}[\mathbf{X}])^T. \tag{9.45}$$

We can now replace $E_\mathbf{X}[\mathbf{X}]$ by the estimate of (9.44). To estimate the $N \times N$ matrix

$$E_\mathbf{X}\left[\mathbf{X}\mathbf{X}^T\right]$$

we replace it by $(1/M)\sum_{m=1}^{M}\mathbf{x}_m\mathbf{x}_m^T$ since it is easily shown that the (i,j) element of $E_\mathbf{X}\left[\mathbf{X}\mathbf{X}^T\right]$ is

$$\left[E_\mathbf{X}[\mathbf{X}\mathbf{X}^T]\right]_{ij} = E_\mathbf{X}[X_iX_j] = E_{X_iX_j}[X_iX_j]$$

and

$$\left[\frac{1}{M}\sum_{m=1}^{M}\mathbf{x}_m\mathbf{x}_m^T\right]_{ij} = \frac{1}{M}\sum_{m=1}^{M}[\mathbf{x}_m]_i[\mathbf{x}_m]_j.$$

Thus we have that

$$\widehat{\mathbf{C}_X} = \frac{1}{M} \sum_{m=1}^{M} \mathbf{x}_m \mathbf{x}_m^T - \left(\frac{1}{M} \sum_{m=1}^{M} \mathbf{x}_m \right) \left(\frac{1}{M} \sum_{m=1}^{M} \mathbf{x}_m \right)^T$$

which can also be written as

$$\widehat{\mathbf{C}_X} = \frac{1}{M} \sum_{m=1}^{M} \left(\mathbf{x}_m - \widehat{E_\mathbf{X}[\mathbf{X}]} \right) \left(\mathbf{x}_m - \widehat{E_\mathbf{X}[\mathbf{X}]} \right)^T \qquad (9.46)$$

where $\widehat{E_\mathbf{X}[\mathbf{X}]}$ is given by (9.44). The latter form of the covariance matrix estimate is also more easily implemented. An example follows.

Example 9.6 – Decorrelation of random variables – continued

In Example 9.4 we showed that we could decorrelate the random variable components of a random vector by applying the appropriate linear transformation to the random vector. In particular, if the 2×1 random vector $\mathbf{X}$ whose joint PMF is given in Table 9.1 is transformed to a random vector $\mathbf{Y}$, where

$$\mathbf{Y} = \begin{bmatrix} \frac{1}{\sqrt{2}} & -\frac{1}{\sqrt{2}} \\ \frac{1}{\sqrt{2}} & \frac{1}{\sqrt{2}} \end{bmatrix} \mathbf{X}$$

then the covariance matrix for $\mathbf{X}$

$$\mathbf{C}_X = \begin{bmatrix} 26 & 6 \\ 6 & 26 \end{bmatrix}$$

becomes the diagonal covariance matrix for $\mathbf{Y}$

$$\mathbf{C}_Y = \begin{bmatrix} 20 & 0 \\ 0 & 32 \end{bmatrix}.$$

To check this we generate realizations of $\mathbf{X}$, as explained in Section 7.11 and then use the estimate of the covariance matrix given by (9.46). The results are for $M = 1000$ realizations

$$\widehat{\mathbf{C}_X} = \begin{bmatrix} 25.9080 & 6.1077 \\ 6.1077 & 25.8558 \end{bmatrix}$$

$$\widehat{\mathbf{C}_Y} = \begin{bmatrix} 19.7742 & 0.0261 \\ 0.0261 & 31.9896 \end{bmatrix}$$

and are near to the true covariance matrices. The entire MATLAB program is given next.

```
%  covexample.m
clear all % clears out all previous variables from workspace
rand('state',0); % sets random number generator to initial value
M=1000;
for m=1:M % generate realizations of X (see Section 7.11)
   u=rand(1,1);
   if u<=0.25
      x(1,m)=-8;x(2,m)=0;
   elseif u>0.25&u<=0.5
      x(1,m)=0;x(2,m)=-8;
   elseif u>0.5&u<=0.75
      x(1,m)=2;x(2,m)=6;
   else
      x(1,m)=6;x(2,m)=2;
   end
end
meanx=[0 0]'; % estimate mean vector of X
for m=1:M
   meanx=meanx+x(:,m)/M;
end
meanx
CX=zeros(2,2);
for m=1:M % estimate covariance matrix of X
   xbar(:,m)=x(:,m)-meanx;
   CX=CX+xbar(:,m)*xbar(:,m)'/M;
end
CX
A=[1/sqrt(2) -1/sqrt(2);1/sqrt(2) 1/sqrt(2)];
for m=1:M % transform random vector X
   y(:,m)=A*x(:,m);
end
meany=[0 0]'; %estimate mean vector or Y
for m=1:M
   meany=meany+y(:,m)/M;
end
meany
CY=zeros(2,2);
for m=1:M % estimate covariance matrix of Y
   ybar(:,m)=y(:,m)-meany;
   CY=CY+ybar(:,m)*ybar(:,m)'/M;
end
CY
```

◇

9.9   Real-World Example – Image Coding

The methods for digital storage and transmission of images is an important consideration in the modern digital age. One of the standard procedures used to convert an image to its digital representation is the JPEG encoding format [Sayood 1996]. It makes the observation that many images contain portions that do not change significantly in content. Such would be the case for the image of a house in which the color and texture of the siding, whether it be aluminum siding or clapboards, is relatively constant as the image is scanned in the horizontal direction. To store and transmit all this redundant information is costly and time consuming. Hence, it is desirable to reduce the image to its basic set of information. Consider a gray scale image for simplicity. Each pixel, which is a dot of a given intensity level, is modeled as a random variable. For the house image example, note that for the siding pixels, the random variables are heavily correlated. For example, if X_1 and X_2 denote neighboring pixels in the horizontal direction, then we would expect the correlation coefficient $\rho_{X_1,X_2} = 1$. If this is the case, then we know from Section 7.9 that $X_1 = X_2$, assuming zero mean random variables in our model. There is no economy in storing/transmitting the values $X_1 = x_1$ and $X_2 = x_2 = x_1$. We should just store/transmit $X_1 = x_1$ and when it is necessary to reconstruct the image let $\hat{X}_2 = X_1 = x_1$. In this case, there is no image degradation in doing so. If, however, $|\rho_{X_1,X_2}| < 1$, then there will be an error in the reconstructed X_2. If the correlation coefficient is close to ± 1, this error will be small. Even if it is not, for many images the errors introduced are perceptually unimportant. Human visual perception can tolerate gross errors before the image becomes unsatisfactory.

To apply this idea to image coding we will consider a simple yet illustrative example. The amount of correlation between random variables is quantified by the covariances. In particular, for multiple random variables this information is embodied in the covariance matrix. For example, if $N = 3$ a covariance matrix of

$$\mathbf{C}_X = \begin{bmatrix} 4 & 0 & 0 \\ 0 & 4 & 3.8 \\ 0 & 3.8 & 4 \end{bmatrix} \tag{9.47}$$

indicates that

$$\rho_{X_1,X_2} = \rho_{X_1,X_3} = 0$$

but

$$\rho_{X_2,X_3} = \frac{3.8}{\sqrt{4 \cdot 4}} = 0.95.$$

Clearly, then (X_1, X_2) or (X_1, X_3) contain most of the information. For more complicated covariance matrices these relationships are not so obvious. For example, if

$$\mathbf{C}_X = \begin{bmatrix} 4 & 1 & 5 \\ 1 & 4 & 5 \\ 5 & 5 & 10 \end{bmatrix} \tag{9.48}$$

it is not obvious that $X_3 = X_1 + X_2$ (assuming zero mean random variables). (This is verified by showing that $E[(X_3 - (X_1 + X_2))^2] = 0$ (see Problem 9.33)).

The technique of *transform coding* [Sayood 1996] used in the JPEG encoding scheme takes advantage of the correlation between random variables. The particular version we describe here can be shown to be an optimal approach [Kramer and Mathews 1956]. It is termed the *Karhunen-Loeve transform* and an approximate version is used in the JPEG encoding. Transform coding operates on a random vector **X** and proceeds as follows:

1. Transform the random variables into uncorrelated ones via a linear transformation $\mathbf{Y} = \mathbf{AX}$, where **A** is an invertible $N \times N$ matrix.

2. Discard the random variables whose variance is small relative to the others by setting the corresponding elements of **Y** equal to zero. This yields a new $N \times 1$ random vector $\hat{\mathbf{Y}}$. This vector would be stored or transmitted. (Of course, the zero vector elements would not require encoding, thereby effecting data compression. Their locations, though, would need to be specified.)

3. Transform back to $\hat{\mathbf{X}} = \mathbf{A}^{-1}\hat{\mathbf{Y}}$ to recover an approximation to the original random variables (if the values $\hat{\mathbf{Y}}$ were stored then this would occur upon retrieval or if they were transmitted, this would occur at the receiver).

By decorrelating the random variables first it becomes obvious which components can be discarded without significantly affecting the reconstructed vector. To accomplish the first step we have already determined that a suitable decorrelation matrix is $\mathbf{V}^T$, where **V** is the matrix of eigenvectors of $\mathbf{C}_X$. Thus, we have that

$$
\begin{aligned}
\mathbf{C}_Y &= \mathbf{A}\mathbf{C}_X\mathbf{A}^T \\
&= \mathbf{V}^T\mathbf{C}_X\mathbf{V} \\
&= \mathbf{\Lambda} = \begin{bmatrix} \text{var}(Y_1) & 0 & 0 \\ 0 & \text{var}(Y_2) & 0 \\ 0 & 0 & \text{var}(Y_3) \end{bmatrix}.
\end{aligned}
$$

We now carry out the transform coding procedure for the covariance matrix of (9.48). This is done numerically using MATLAB. The statement `[V Lambda]=eig(CX)` will produce the matrices **V** and $\mathbf{\Lambda}$, as

$$
\mathbf{V} = \begin{bmatrix} 0.4082 & -0.7071 & 0.5774 \\ 0.4082 & -0.7071 & 0.5774 \\ 0.8165 & 0 & -0.5774 \end{bmatrix}
$$

$$
\mathbf{\Lambda} = \begin{bmatrix} 15 & 0 & 0 \\ 0 & 3 & 0 \\ 0 & 0 & 0 \end{bmatrix}.
$$

Hence, $\text{var}(Y_3) = \lambda_3 = 0$ so that we discard it by setting $\hat{Y}_3 = 0$ and therefore

$$\hat{\mathbf{Y}} = \begin{bmatrix} Y_1 \\ Y_2 \\ 0 \end{bmatrix} = \underbrace{\begin{bmatrix} 1 & 0 & 0 \\ 0 & 1 & 0 \\ 0 & 0 & 0 \end{bmatrix}}_{\mathbf{B}} \underbrace{\begin{bmatrix} Y_1 \\ Y_2 \\ Y_3 \end{bmatrix}}_{\mathbf{Y}}.$$

The reconstructed random vector becomes with $\mathbf{A} = \mathbf{V}^T$

$$\begin{aligned} \hat{\mathbf{X}} = \mathbf{A}^{-1}\hat{\mathbf{Y}} &= \mathbf{V}\hat{\mathbf{Y}} \\ &= \mathbf{V}\mathbf{B}\mathbf{Y} \\ &= \mathbf{V}\mathbf{B}\mathbf{V}^T\mathbf{X} \end{aligned}$$

and since

$$\mathbf{V}\mathbf{B}\mathbf{V}^T = \begin{bmatrix} \frac{2}{3} & -\frac{1}{3} & \frac{1}{3} \\ -\frac{1}{3} & \frac{2}{3} & \frac{1}{3} \\ \frac{1}{3} & \frac{1}{3} & \frac{2}{3} \end{bmatrix}$$

we have that

$$\begin{aligned} \hat{\mathbf{X}} &= \begin{bmatrix} \frac{2}{3}X_1 - \frac{1}{3}X_2 + \frac{1}{3}X_3 \\ -\frac{1}{3}X_1 + \frac{2}{3}X_2 + \frac{1}{3}X_3 \\ \frac{1}{3}X_1 + \frac{1}{3}X_2 + \frac{2}{3}X_3 \end{bmatrix} \\ &= \begin{bmatrix} X_1 \\ X_2 \\ X_1 + X_2 \end{bmatrix} \quad (\text{using } X_3 = X_1 + X_2, \text{ see Problem 9.33}) \\ &= \begin{bmatrix} X_1 \\ X_2 \\ X_3 \end{bmatrix}. \end{aligned}$$

Here we see that the reconstructed vector $\hat{\mathbf{X}}$ is identical to the original one. Generally, however, there will be an error. For the covariance matrix of (9.47) there will be an error since X_2 and X_3 are not perfectly correlated. For that covariance matrix the eigenvector and eigenvalue matrices are

$$\mathbf{V} = \begin{bmatrix} 0 & 1 & 0 \\ 0.7071 & 0 & 0.7071 \\ 0.7071 & 0 & -0.7071 \end{bmatrix}$$

$$\mathbf{\Lambda} = \begin{bmatrix} 7.8 & 0 & 0 \\ 0 & 4 & 0 \\ 0 & 0 & 0.2 \end{bmatrix}$$

and it is seen that the decorrelated random variables all have a nonzero variance (recall that $\mathrm{var}(Y_i) = \lambda_i$). This indicates that no component of $\mathbf{Y}$ can be discarded without causing an error upon reconstruction. By discarding Y_3, which has the smallest variance, we will incur the least amount of error. Doing so produces the reconstructed random vector

$$\hat{\mathbf{X}} = \mathbf{VBV}^T\mathbf{X}$$
$$= \begin{bmatrix} 1 & 0 & 0 \\ 0 & \frac{1}{2} & \frac{1}{2} \\ 0 & \frac{1}{2} & \frac{1}{2} \end{bmatrix} \mathbf{X}$$

which becomes

$$\hat{\mathbf{X}} = \begin{bmatrix} X_1 \\ \dfrac{X_2 + X_3}{2} \\ \dfrac{X_2 + X_3}{2} \end{bmatrix}.$$

It is seen that the components X_2 and X_3 are replaced by their averages. This is due to the nearly unity correlation coefficient coefficient ($\rho_{X_2,X_3} = 0.95$) between these components. As an example, we generate 20 realizations of $\mathbf{X}$ as shown in Figure 9.4a, where the first realization is displayed in samples $1, 2, 3$; the second realization in samples $4, 5, 6$, etc. The reconstructed realizations are shown in Figure 9.4b.

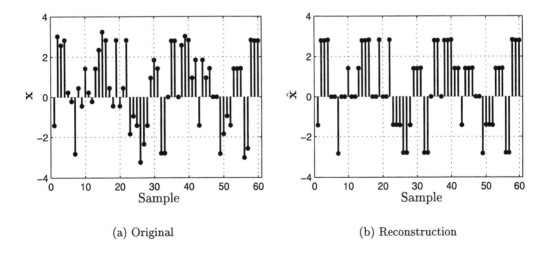

(a) Original   (b) Reconstruction

Figure 9.4: Realizations of original random vector $\{\mathbf{x}_1, \mathbf{x}_2, \ldots, \mathbf{x}_{20}\}$ and reconstructed random vectors $\{\hat{\mathbf{x}}_1, \hat{\mathbf{x}}_2, \ldots, \hat{\mathbf{x}}_{20}\}$. The displayed samples shown are components of $\mathbf{x}_1$, followed by components of $\mathbf{x}_2$, etc.

Finally, the error between the two is shown in Figure 9.5. Note that the total average

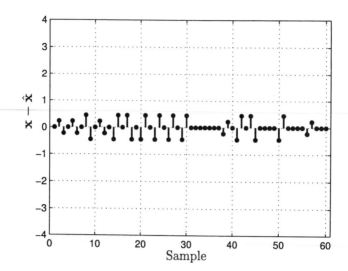

Figure 9.5: Error between original random vector realizations and reconstructed ones shown in Figure 9.4.

squared error or the total mean square error (MSE) is given by $\sum_{i=1}^{3} E_{\mathbf{X}}[(X_i - \hat{X}_i)^2]$ which is

$$
\begin{aligned}
\text{Total mse} &= E[(X_1 - \hat{X}_1)^2 + (X_2 - \hat{X}_2)^2 + (X_3 - \hat{X}_3)^2] \\
&= E[(X_2 - (X_2 + X_3)/2)^2] + E[(X_3 - (X_2 + X_3)/2)^2] \\
&= E[((X_2 - X_3)/2)^2] + E[((X_3 - X_2)/2)^2] \\
&= \frac{1}{2}E[(X_2 - X_3)^2] \\
&= \frac{1}{2}[\text{var}(X_2) + \text{var}(X_3) - 2\text{cov}(X_2, X_3)] \\
&= \frac{1}{2}[4 + 4 - 2(3.8)] = 0.2.
\end{aligned}
$$

This total MSE is estimated by taking the sum of the squares of the values in Figure 9.5 and dividing by 20, the number of vector realizations. Also, note what the total MSE would have been if $\rho_{X_2,X_3} = 1$.

Finally, to appreciate the error in terms of human vision perception, we can convert the realizations of $\mathbf{X}$ and $\hat{\mathbf{X}}$ into an image. This is shown in Figure 9.6. The grayscale bar shown at the right can be used to convert the various shades of gray into numerical values. Also, note that as expected (see $\mathbf{C}_X$ in (9.47)) X_1 is uncorrelated with X_2 and X_3, while X_2 and X_3 are heavily correlated in the upper image. In the lower image X_2 and X_3 have been replaced by their average.

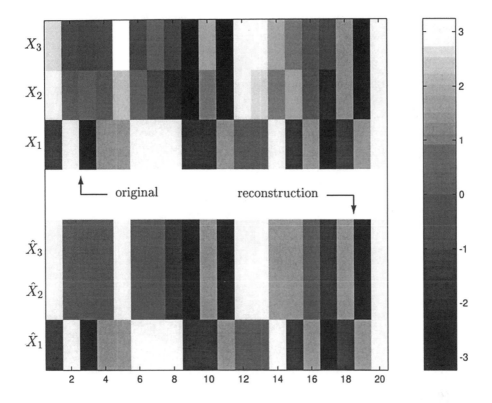

Figure 9.6: Realizations of original random vector and reconstructed random vectors displayed as gray-scale images. The upper image is the original and the lower image is the reconstructed image.

References

Kramer, H.P., Mathews, M.V., "A Linear Coding for Transmitting a Set of Correlated Signals," *IRE Trans. on Information Theory*, Vol. IT-2, pp. 41–46, 1956.

Noble, B., Daniel, J.W., *Applied Linear Algebra*, Prentice-Hall, Englewood Cliffs, NJ, 1977.

Sayood, K. *Introduction to Data Compression*, Morgan Kaufman, San Francisco, 1996.

Problems

9.1 (☺) (**w**) A retired person gets up in the morning and decides what to do that day. He will go fishing with probability 0.3, or he will visit his daughter with probability 0.2, or else he will stay home and tend to his garden. If the decision that he makes each day is independent of the decisions made on the other days, what is the probability that he will go fishing for 3 days, visit his daughter for 2 days, and garden for 2 days of the week?

9.2 (f,c) Compute the values of a multinomial PMF if $N = 3$, $M = 4$, $p_1 = 0.2$, and $p_2 = 0.4$ for all possible k_1, k_2, k_3. Do the sum of the values equal one? Hint: You will need a computer to do this.

9.3 (t) Prove the multinomial formula given by (9.5) for $N = 3$ by the following method. Use the binomial formula to yield

$$(a_1 + b)^M = \sum_{k_1=0}^{M} \frac{M!}{k_1!(M - k_1)!} a_1^{k_1} b^{M-k_1}.$$

Then let $b = a_2 + a_3$ so that upon using the binomial formula again we have

$$b^{M-k_1} = (a_2 + a_3)^{M-k_1} = \sum_{k_2=0}^{M-k_1} \frac{(M - k_1)!}{k_2!(M - k_1 - k_2)!} a_2^{k_2} a_3^{M-k_1-k_2}.$$

Finally, rearrange the sums and note that $k_3 = M - k_1 - k_2$ so that there is actually only a double sum in (9.5) for $N = 3$ due to this constraint.

9.4 (⌣) (f) Is the following function a valid PMF?

$$p_{X_1,X_2,X_3}[k_1, k_2, k_3] = \frac{1}{8} \left(\frac{1}{2}\right)^{k_1} \left(\frac{1}{4}\right)^{k_2} \qquad \begin{matrix} k_1 = 0, 1, \ldots \\ k_2 = 0, 1, \ldots \\ k_3 = -1, 0, 1. \end{matrix}$$

9.5 (w) For the joint PMF

$$p_{X_1,X_2,X_3}[k_1, k_2, k_3] = (1 - a)(1 - b)(1 - c)a^{k_1} b^{k_2} c^{k_3} \qquad \begin{matrix} k_1 = 0, 1, \ldots \\ k_2 = 0, 1, \ldots \\ k_3 = 0, 1, \ldots \end{matrix}$$

where $0 < a < 1$, $0 < b < 1$, and $0 < c < 1$, find the marginal PMFs p_{X_1}, p_{X_2} and p_{X_3}.

9.6 (⌣) (w) For the joint PMF given below are there any subsets of the random variables that are independent of each other?

$$p_{X_1,X_2,X_3}[k_1, k_2, k_3] = \binom{M}{k_1, k_2} p_1^{k_1} p_2^{k_2} (1 - p_3) p_3^{k_3} \qquad \begin{matrix} k_1 = 0, 1, \ldots, M \\ k_2 = M - k_1 \\ k_3 = 0, 1, \ldots \end{matrix}$$

where $0 < p_1 < 1$, $p_2 = 1 - p_1$, and $0 < p_3 < 1$.

9.7 (f) A random vector $\mathbf{X}$ with the joint PMF

$$p_{X_1,X_2,X_3}[k_1, k_2, k_3] = \exp[-(\lambda_1 + \lambda_2 + \lambda_3)]\frac{\lambda_1^{k_1} \lambda_2^{k_2} \lambda_3^{k_3}}{k_1! k_2! k_3!} \qquad \begin{array}{l} k_1 = 0, 1, \ldots \\ k_2 = 0, 1, \ldots \\ k_3 = 0, 1, \ldots \end{array}$$

is transformed according to $\mathbf{Y} = \mathbf{AX}$ where

$$\mathbf{A} = \begin{bmatrix} 1 & 0 & 0 \\ 1 & 1 & 0 \\ 1 & 1 & 1 \end{bmatrix}.$$

Find the joint PMF of $\mathbf{Y}$.

9.8 (t) Prove that

$$\int_{-\pi}^{\pi} \exp(j\omega k)\frac{d\omega}{2\pi} = \begin{cases} 0 & k \neq 0 \\ 1 & k = 0. \end{cases}$$

Hint: Expand $\exp(j\omega k)$ into its real and imaginary parts and note that $\int (g(\omega) + jh(\omega))d\omega = \int g(\omega)d\omega + j \int h(\omega)d\omega$.

9.9 (t) Prove that the sum of N independent Poisson random variables with $X_i \sim$ Pois(λ_i) for $i = 1, 2, \ldots, N$ is again Poisson distributed but with parameter $\lambda = \sum_{i=1}^{N} \lambda_i$. Hint: See Section 9.4.

9.10 (☺) (w) The components of a random vector $\mathbf{X} = [X_1 \, X_2 \ldots X_N]^T$ all have the same mean $E_X[X]$ and the same variance var(X). The "sample mean" random variable

$$\bar{X} = \frac{1}{N} \sum_{i=1}^{N} X_i$$

is formed. If the X_i's are independent, find the mean and variance of $\bar{X}$. What happens to the variance as $N \to \infty$? Does this tell you anything about the PMF of $\bar{X}$ as $N \to \infty$?

9.11 (w) Repeat Problem 9.10 if we know that each $X_i \sim$ Ber(p). How can this result be used to motivate the relative frequency interpretation of probability?

9.12 (f) If the covariance matrix of a 3×1 random vector $\mathbf{X}$ is

$$\mathbf{C}_X = \begin{bmatrix} 1 & 0 & 1 \\ 0 & 2 & 2 \\ 1 & 2 & 4 \end{bmatrix}$$

find the correlation coefficients ρ_{X_1,X_2}, ρ_{X_1,X_3}, and ρ_{X_2,X_3}.

9.13 (☺) (w) A 2×1 random vector is given by

$$\mathbf{X} = \begin{bmatrix} U \\ 2U \end{bmatrix}$$

where $\text{var}(U) = 1$. Find the covariance matrix for $\mathbf{X}$. Next find the correlation coefficient ρ_{X_1, X_2}. Finally, compute the determinant of the covariance matrix. Is the covariance matrix positive definite? Hint: A positive definite matrix must have a positive determinant.

9.14 (t) Prove (9.26) by noting that

$$\mathbf{a}^T \mathbf{C}_X \mathbf{a} = \sum_{i=1}^{N} \sum_{j=1}^{N} a_i a_j \text{cov}(X_i, X_j).$$

9.15 (f) For the covariance matrix given in Problem 9.12, find $\text{var}(X_1 + X_2 + X_3)$.

9.16 (t) Is it ever possible that $\text{var}(X_1 + X_2) = \text{var}(X_1)$ without X_2 being a constant?

9.17 (☺) (w) Which of the following matrices are not valid covariance matrices and why?

$$\text{a.} \begin{bmatrix} 1 & 2 \\ 2 & 1 \end{bmatrix} \quad \text{b.} \begin{bmatrix} -1 & 0 \\ 0 & -1 \end{bmatrix} \quad \text{c.} \begin{bmatrix} 2 & 1 \\ 1 & 2 \end{bmatrix} \quad \text{d.} \begin{bmatrix} 2 & 1 \\ 0 & 1 \end{bmatrix}$$

9.18 (f) A positive semidefinite matrix $\mathbf{A}$ must have $\det(\mathbf{A}) \geq 0$. Since a covariance matrix must be positive semidefinite, use this property to prove that the correlation coefficient satisfies $|\rho_{X_1, X_2}| \leq 1$. Hint: Consider a 2×2 covariance matrix.

9.19 (f) If a random vector $\mathbf{X}$ is transformed according to

$$\begin{aligned} Y_1 &= X_1 \\ Y_2 &= X_1 + X_2 \end{aligned}$$

and the mean of $\mathbf{X}$ is

$$E_{\mathbf{X}}[\mathbf{X}] = \begin{bmatrix} 3 \\ 4 \end{bmatrix}$$

find the mean of $\mathbf{Y} = [Y_1 \, Y_2]^T$.

9.20 (☺) (f) If the random vector $\mathbf{X}$ given in Problem 9.19 has a covariance matrix

$$\mathbf{C}_X = \begin{bmatrix} 2 & 1 \\ 1 & 2 \end{bmatrix}$$

find the covariance matrix for $\mathbf{Y} = [Y_1 \, Y_2]^T$.

9.21 (t) For $N = 2$ show that the covariance matrix may be defined as

$$\mathbf{C}_X = E_{\mathbf{X}}\left[(\mathbf{X} - E_{\mathbf{X}}[\mathbf{X}])(\mathbf{X} - E_{\mathbf{X}}[\mathbf{X}])^T\right].$$

Hint: Recall that the expected value of a matrix is the matrix of the expected values of its elements.

9.22 (t) In this problem you are asked to prove that if $\mathbf{Y} = \mathbf{AX}$, where both $\mathbf{X}$ and $\mathbf{Y}$ are $N \times 1$ random vectors and $\mathbf{A}$ is an $N \times N$ matrix, then $E_{\mathbf{Y}}[\mathbf{Y}] = \mathbf{A}E_{\mathbf{X}}[\mathbf{X}]$. If we let $[\mathbf{A}]_{ij}$ be the (i, j) element of $\mathbf{A}$, then you will need to prove that

$$[E_{\mathbf{Y}}[\mathbf{Y}]]_i = \sum_{j=1}^{N}[\mathbf{A}]_{ij}[E_{\mathbf{X}}[\mathbf{X}]]_j.$$

This is because if $\mathbf{b} = \mathbf{A}\mathbf{x}$, then $b_i = \sum_{j=1}^{N} a_{ij}x_j$, for $i = 1, 2, \ldots, N$ where b_i is the ith element of $\mathbf{b}$ and a_{ij} is the (i, j) element of $\mathbf{A}$.

9.23 (t) In this problem we prove that

$$E_{\mathbf{X}}[\mathbf{A}\mathbf{G}(\mathbf{X})\mathbf{A}^T] = \mathbf{A}E_{\mathbf{X}}[\mathbf{G}(\mathbf{X})]\mathbf{A}^T$$

where $\mathbf{A}$ is an $N \times N$ matrix and $\mathbf{G}(\mathbf{X})$ is an $N \times N$ matrix whose elements are all functions of $\mathbf{X}$. To do so we note that if $\mathbf{A}, \mathbf{B}, \mathbf{C}, \mathbf{D}$ are all $N \times N$ matrices then $\mathbf{D} = \mathbf{ABC}$ is an $N \times N$ matrix with (i, l) element

$$
\begin{aligned}
[\mathbf{D}]_{il} &= \sum_{k=1}^{N}[\mathbf{AB}]_{ik}[\mathbf{C}]_{kl} \\
&= \sum_{k=1}^{N}\left(\sum_{j=1}^{N}[\mathbf{A}]_{ij}[\mathbf{B}]_{jk}\right)[\mathbf{C}]_{kl} \\
&= \sum_{k=1}^{N}\sum_{j=1}^{N}[\mathbf{A}]_{ij}[\mathbf{B}]_{jk}[\mathbf{C}]_{kl}.
\end{aligned}
$$

Using this result and replacing $\mathbf{A}$ by itself, $\mathbf{B}$ by $\mathbf{G}(\mathbf{X})$, and $\mathbf{C}$ by $\mathbf{A}^T$ will allow the desired result to be proven.

9.24 (f) Prove (9.29) and (9.30) for the case of $N = 2$ by letting

$$
\mathbf{A} = \begin{bmatrix} a_{11} & a_{12} \\ a_{21} & a_{22} \end{bmatrix}
$$

$$
\mathbf{b}_1 = \begin{bmatrix} b_1^{(1)} \\ b_2^{(1)} \end{bmatrix} \qquad \mathbf{b}_2 = \begin{bmatrix} b_1^{(2)} \\ b_2^{(2)} \end{bmatrix}
$$

$$
\mathbf{d}_1 = \begin{bmatrix} d_1^{(1)} \\ d_2^{(1)} \end{bmatrix} \qquad \mathbf{d}_2 = \begin{bmatrix} d_1^{(2)} \\ d_2^{(2)} \end{bmatrix}
$$

and multiplying out all the matrices and vectors. Then, verify that the relationships are true by showing that the elements of the resultant $N \times N$ matrices are identical.

9.25 (c) Using MATLAB, find the eigenvectors and corresponding eigenvalues for the covariance matrix

$$\mathbf{C}_X = \begin{bmatrix} 26 & 6 \\ 6 & 26 \end{bmatrix}$$

To do so use the statement `[V Lambda]=eig(CX)`.

9.26 (⌣) (f,c) Find a linear transformation to decorrelate the random vector $\mathbf{X} = [X_1 \, X_2]^T$ that has the covariance matrix

$$\mathbf{C}_X = \begin{bmatrix} 10 & 6 \\ 6 & 20 \end{bmatrix}.$$

What are the variances of the decorrelated random variables?

9.27 (t) Prove that an orthogonal matrix, i.e., one that has the property $\mathbf{U}^T = \mathbf{U}^{-1}$, *rotates* a vector $\mathbf{x}$ to a new vector $\mathbf{y}$. Do this by letting $\mathbf{y} = \mathbf{U}\mathbf{x}$ and showing that the length of $\mathbf{y}$ is the same as the length of $\mathbf{x}$. The length of a vector is defined to be $||\mathbf{x}|| = \sqrt{\mathbf{x}^T\mathbf{x}} = \sqrt{x_1^2 + x_2^2 + \cdots + x_N^2}$.

9.28 (t) Prove that if the random variables $X_1, X_2, \ldots, X_N$ are independent, then the joint characteristic function factors as

$$\phi_{X_1, X_2, \ldots, X_N}(\omega_1, \omega_2, \ldots \omega_N) = \phi_{X_1}(\omega_1)\phi_{X_2}(\omega_2) \ldots \phi_{X_N}(\omega_N).$$

Alternatively, if the joint characteristic function factors, what does this say about the random variables and why?

9.29 (f) For the random walk described in Example 9.5 find the mean and the variance of X_n as a function of n if $p = 3/4$. What do they indicate about the probable outcomes of $X_1, X_2, \ldots, X_N$?

9.30 (c) For the random walk of Problem 9.29 simulate several realizations of the random vector $\mathbf{X} = [X_1 \, X_2 \ldots X_N]^T$ and plot these as x_n versus n for $n = 1, 2, \ldots, N = 50$. Does the appearance of the outcomes corroborate your results in Problem 9.29? Also, compare your results to those shown in Figure 9.3b.

9.31 (t) Prove the relationship given by (9.45) as follows. Consider the (i, j) element of $\mathbf{C}_X$, which is $\text{cov}(X_i, X_j) = E_{X_i, X_j}[X_i X_j] - E_{X_i}[X_i]E_{X_j}[X_j]$. Then, show that the latter is just the (i, j) element of the right-hand side of (9.45). Recall the definition of the expected value of a matrix/vector as the matrix/vector of expected values.

9.32 (c) A random vector is defined as $\mathbf{X} = [X_1 \, X_2 \ldots X_N]^T$, where each component is $X_i \sim \text{Ber}(1/2)$ and all the random variables are independent. Since the random variables are independent, the covariance matrix should be diagonal. Using MATLAB, generate realizations of $\mathbf{X}$ for $N = 10$ by using `x=floor(rand(10,1)+0.5)` to generate a single vector realization. Next generate multiple random vector realizations and use them to estimate the covariance matrix. Presumably the random numbers that MATLAB produces are "pseudo-independent" and hence "pseudo-uncorrelated". Does this appear to be the case? Hint: Use the MATLAB command `mesh(CXest)` to plot the estimated covariance matrix `CXest`.

9.33 (w) Prove that if X_1, X_2, X_3 are zero mean random variables, then $E[(X_3 - (X_1 + X_2))^2] = 0$ for the covariance matrix given by (9.48).

9.34 (t) In this problem we explain how to generate a computer realization of a random vector with a given covariance matrix. This procedure was used to produce the realizations shown in Figure 9.4a. For simplicity the desired $N \times 1$ random vector $\mathbf{X}$ is assumed to have a zero mean vector. The procedure is to first generate an $N \times 1$ random vector $\mathbf{U}$ whose elements are zero mean, uncorrelated random variables with unit variances so that its covariance matrix is $\mathbf{I}$. Then transform $\mathbf{U}$ according to $\mathbf{X} = \mathbf{BU}$, where $\mathbf{B}$ is an appropriate $N \times N$ matrix. The matrix $\mathbf{B}$ is obtained from the $N \times N$ matrix $\sqrt{\mathbf{\Lambda}}$ whose elements are obtained from the eigenvalue matrix $\mathbf{\Lambda}$ of $\mathbf{C}_X$ by taking the square root of the elements of $\mathbf{\Lambda}$, and $\mathbf{V}$, where $\mathbf{V}$ is the eigenvector matrix of $\mathbf{C}_X$, to form $\mathbf{B} = \mathbf{V}\sqrt{\mathbf{\Lambda}}$. Prove that the covariance matrix of $\mathbf{BU}$ will be $\mathbf{C}_X$.

9.35 (☺) (f) Using the results of Problem 9.34 find a matrix transformation $\mathbf{B}$ of $\mathbf{U} = [U_1 \, U_2]^T$, where $\mathbf{C}_U = \mathbf{I}$, so that $\mathbf{X} = \mathbf{BU}$ has the covariance matrix

$$\mathbf{C}_X = \begin{bmatrix} 4 & 1 \\ 1 & 4 \end{bmatrix}.$$

9.36 (☺) (c) Generate 30 realizations of a 2×1 random vector $\mathbf{X}$ that has a zero mean vector and the covariance matrix given in Problem 9.35. To do so use the results from Problem 9.35. For the random vector $\mathbf{U}$ assume that U_1 and U_2 are uncorrelated and have the same PMF

$$p_U[k] = \begin{cases} \frac{1}{2} & k = -1 \\ \frac{1}{2} & k = 1. \end{cases}$$

Note that the mean of $\mathbf{U}$ is zero and the covariance matrix of $\mathbf{U}$ is $\mathbf{I}$. Next estimate the covariance matrix $\mathbf{C}_X$ using your realizations and compare it to the true covariance matrix.

Chapter 10

Continuous Random Variables

10.1 Introduction

In Chapters 5–9 we discussed discrete random variables and the methods employed to describe them probabilistically. The principal assumption necessary in order to do so is that the sample space, which is the set of all possible outcomes, is finite or at most countably infinite. It followed then that a probability mass function (PMF) could be defined as the probability of each sample point and used to calculate the probability of all possible events (which are subsets of the sample space). Most physical measurements, however, do not produce a discrete set of values but rather a *continuum* of values such as the rainfall measurement data previously shown in Figures 1.1 and 1.2. Another example is the maximum temperature measured during the day, which might be anywhere between 20°F and 60°F. The number of possible temperatures in the interval $[20, 60]$ is infinite and *uncountable*. Therefore, we cannot assign a valid PMF to the temperature random variable. Of course, we could always choose to "round off" the measurement to the nearest degree so that the possible outcomes would then become $\{20, 21, \ldots, 60\}$. Then, many valid PMFs could be assigned. But this approach compromises the measurement precision and so is to be avoided if possible. What we are ultimately interested in is the probability of *any interval*, such as the probability of the temperature being in the interval $[20, 25]$ or $[55, 60]$ or the union of intervals $[20, 25] \cup [55, 60]$. To do so we must extend our previous approaches to be able to handle this new case. And if we later decide that less precision is warranted, such that the rounding of 20.6° to 21° is acceptable, we will still be able to determine the probability of observing 21°. To do so we can regard the rounded temperature of 21° as having arisen from all temperatures in the interval $A = [20.5, 21.5)$. Then, $P[\text{rounded temperature } = 21] = P[A]$, so that we have lost nothing by considering a continuum of outcomes (see Problem 10.2).

Chapters 10–14 discuss continuous random variables in a manner similar to Chapters 5–9 for discrete random variables. Since many of the concepts are the same, we will not belabor the discussion but will concentrate our efforts on the algebraic manipulations required to analyze continuous random variables. It may be of interest to note that discrete and continuous random variables can be subsumed under the topic of a general random variable. There exists the mathematical machinery to analyze both types of random variables simultaneously. This theory is called *measure theory* [Capinski, Kopp 2004]. It requires an advanced mathematical background and does not easily lend itself to intuitive interpretations. An alternative means of describing the general random variable that appeals more to engineers and scientists makes use of the Dirac delta function. This approach is discussed later in this chapter under the topic of *mixed random variables*.

In the course of our discussions we will revisit some of the concepts alluded to in Chapters 1 and 2. With the appropriate mathematical tools we will now be able to define these concepts. Hence, the reader may wish to review the relevant sections in those chapters.

10.2   Summary

The definition of a continuous random variable is given in Section 10.3 and illustrated in Figure 10.1. The probabilistic description of a continuous random variable is the probability density function (PDF) $p_X(x)$ with its interpretation as the probability per unit length. As such the probability of an interval is given by the area under the PDF (10.4). The properties of a PDF are that it is nonnegative and integrates to one, as summarized by Properties 10.1 and 10.2 in Section 10.4. Some important PDFs are given in Section 10.5, such as the uniform (10.6), the exponential (10.5), the Gaussian or normal (10.7), the Laplacian (10.8), the Cauchy (10.9), the Gamma (10.10), and the Rayleigh (10.14). Special cases of the Gamma PDF are the exponential, the chi-squared (10.12), and the Erlang (10.13). The cumulative distribution function (CDF) for a continuous random variable is defined the same as for the discrete random variable and is given by (10.16). The corresponding CDFs for the PDFs of Section 10.5 are given in Section 10.6. In particular, the CDF for the standard normal is denoted by $\Phi(x)$ and is related to the Q function by (10.17). The latter function cannot be evaluated in closed form but may be found numerically using the MATLAB subprogram Q.m listed in Appendix 10B. An approximation to the Q function is given by (10.23). The CDF is useful in that probabilities of intervals are easily found via (10.25) once the CDF is known. The transformation of a continuous random variable by a one-to-one function produces the PDF of (10.30). If the transformation is many-to-one, then (10.33) can be used to determine the PDF of the transformed random variable. Mixed random variables, ones that exhibit nonzero probabilities for some points but are continuous otherwise, are described in Section 10.8. They can be described by a PDF if we allow the use of the Dirac delta function or impulse. For a general mixed random variable the PDF is given by (10.36). To generate realizations of a continuous random variable on a digital computer one can use a transformation of a uniform random variable as summarized in Theorem 10.9.1. Examples are given in Section 10.9. Estimation of the PDF and CDF can be accomplished by using (10.38) and (10.39). Finally, an example of the application of the theory to the problem of speech clipping is given in Section 10.10.

10.3 Definition of a Continuous Random Variable

A *continuous random variable* X is defined as a mapping from the experimental sample space S to a numerical (or measurement) sample space $\mathcal{S}_X$, which is a subset of the real line R^1. In contrast to the sample space of a discrete random variable, $\mathcal{S}_X$ consists of an infinite *and uncountable* number of outcomes. As an example, consider an experiment in which a dart is thrown at the circular dartboard shown in Figure 10.1. The outcome of the dart-throwing experiment is a *point s_1* in the circle

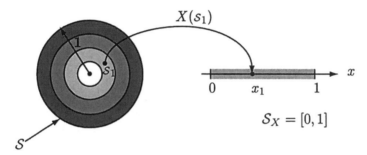

Figure 10.1: Mapping of the outcome of a thrown dart to the real line (example of continuous random variable).

of radius one. The distance from the bullseye (center of the dartboard) is measured and that value is assigned to the random variable as $X(s_1) = x_1$. Clearly then, the possible outcomes of the random variable are in the interval $[0, 1]$, which is an uncountably infinite set. We cannot assign a nonzero probability to each value of X and expect the sum of the probabilities to be one. One way out of this dilemma is to assign probabilities to intervals, as was done in Section 3.6. There we had a one-dimensional dartboard and we assigned a probability of the dart landing in an interval to be the *length* of the interval. Similarly, for our problem if each value of X is equally likely so that intervals of the same length are equally likely, we could assign

$$P[a \leq X \leq b] = b - a \qquad 0 \leq a \leq b \leq 1 \qquad (10.1)$$

for the probability of the dart landing in the interval $[a, b]$. This probability assignment satisfies the probability axioms given in Section 3.6 and so would suffice to calculate the probability of any interval or union of disjoint intervals (use Axiom 3 for disjoint intervals). But what would we do if the probability of all equal length intervals were not the same? For example, a champion dart thrower would be more likely to obtain a value near $x = 0$ than near $x = 1$. We therefore need a more general approach. For discrete random variables it was just as easy to assign PMFs that were not uniform as ones that were uniform. Our goal then is to extend this approach to encompass continuous random variables. We will do so by examining the approximation afforded by using the PMF to calculate interval probabilities for continuous random variables.

Consider first a possible approximation of (10.1) by a uniform PMF as

$$p_X[x_i] = \frac{1}{M} \qquad x_i = i\Delta x \text{ for } i = 1, 2, \ldots, M$$

where $\Delta x = 1/M$, so that $M\Delta x = 1$ as shown in Figure 10.2. Then to approximate

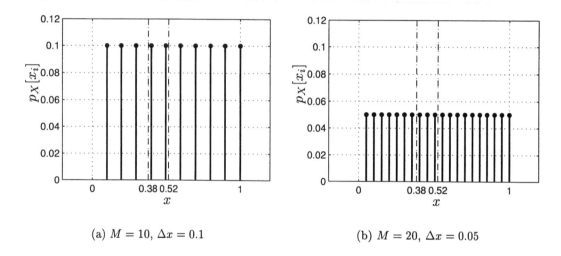

(a) $M = 10$, $\Delta x = 0.1$ (b) $M = 20$, $\Delta x = 0.05$

Figure 10.2: Approximating the probability of an interval for a continuous random variable by using a PMF.

the probability of the outcome of X in the interval $[a, b]$ we can use

$$P[a \le X \le b] = \sum_{\{i : a \le x_i \le b\}} \frac{1}{M}. \qquad (10.2)$$

For example, referring to Figure 10.2a, if $a = 0.38$ and $b = 0.52$, then there are two values of x_i that lie in that interval and therefore $P[0.38 \le X \le 0.52] = 2/M = 0.2$, even though we know that the true value from (10.1) is 0.14 . To improve the quality of our approximation we increase M to $M = 20$ as shown in Figure 10.2b. Then, we have three values of x_i that lie in the interval and therefore $P[0.38 \le X \le 0.52] = 3/M = 0.15$, which is closer to the true value. Clearly, if we let $M \to \infty$ or equivalently let $\Delta x \to 0$, our approximation will become exact. Considering again

(10.2) with $\Delta x = 1/M$, we have

$$P[a \leq X \leq b] = \sum_{\{i:a \leq x_i \leq b\}} 1 \cdot \Delta x$$

and defining $p_X(x) = 1$ for $0 < x < 1$ and zero otherwise, we can write this as

$$P[a \leq X \leq b] = \sum_{\{i:a \leq x_i \leq b\}} p_X(x_i) \Delta x. \tag{10.3}$$

Finally, letting $\Delta x \to 0$ to yield no error in the approximation, the sum in (10.3) becomes an integral and $p_X(x_i) \to p_X(x)$ so that

$$P[a \leq X \leq b] = \int_a^b p_X(x)dx \tag{10.4}$$

which gives the same result for the probability of an interval as (10.1). Note that $p_X(x)$ is defined to be 1 for *all* $0 < x < 1$. To interpret this new function $p_X(x)$ we have from (10.3) with $x_0 = k\Delta x$ for k an integer

$$P[x_0 - \Delta x/2 \leq X \leq x_0 + \Delta x/2]$$

$$
\begin{aligned}
&= \sum_{\{i:x_0 - \Delta x/2 \leq x_i \leq x_0 + \Delta x/2\}} p_X(x_i)\Delta x \\
&= \sum_{\{i:x_i = x_0\}} p_X(x_i)\Delta x \qquad \text{(only one value of } x_i \text{ within interval)} \\
&= p_X(x_0)\Delta x
\end{aligned}
$$

which yields

$$p_X(x_0) = \frac{P[x_0 - \Delta x/2 \leq X \leq x_0 + \Delta x/2]}{\Delta x}.$$

This is the probability of X being in the interval $[x_0 - \Delta x/2, x_0 + \Delta x/2]$ divided by the interval length Δx. Hence, $p_X(x_0)$ is the *probability per unit length* and is termed the *probability density function* (PDF). It can be used to find the probability of any interval by using (10.4). Equivalently, since the value of an integral may be interpreted as the area under a curve, the probability is found by determining the area under the PDF curve. This is shown in Figure 10.3. The PDF is denoted by $p_X(x)$, where we now use *parentheses* since the argument is no longer discrete but continuous. Also, for the same reason we omit the subscript i, which was used for the PMF argument. Hence, the PDF for a continuous random variable is the extension of the PMF that we sought. Before continuing we examine this example further.

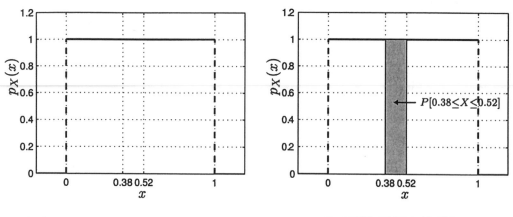

(a) Probability density function   (b) Probability shown as shaded area

Figure 10.3: Example of probability density function and how probability is found as the area under it.

Example 10.1 – PDF for a uniform random variable and the MATLAB command rand

The PDF given by

$$p_X(x) = \begin{cases} 1 & 0 < x < 1 \\ 0 & \text{otherwise} \end{cases}$$

is known as a *uniform* PDF. Equivalently, X is said to be a uniform random variable or we say that X is uniformly distributed on $(0,1)$. The shorthand notation is $X \sim \mathcal{U}(0,1)$. Observe that this is the continuous random variable for which MATLAB uses **rand** to produce a realization. Hence, in simulating a coin toss with a probability of heads of $p = 0.75$, we use (10.4) to obtain

$$
\begin{aligned}
P[a \leq X \leq b] &= \int_a^b p_X(x)dx \\
&= \int_a^b 1\,dx \\
&= b - a = 0.75
\end{aligned}
$$

and choose $a = 0$ and $b = 0.75$. The probability of obtaining an outcome in the interval $(0, 0.75]$ for a random variable $X \sim \mathcal{U}(0,1)$ is now seen to be 0.75. Hence, the code below can be used to generate the outcomes of a repeated coin tossing experiment with $p = 0.75$.

```
for i=1:M
u=rand(1,1);
```

```
            if u<=0.75
               x(i,1)=1; % head mapped into 1
            else
               x(i,1)=0; % tail mapped into 0
            end
      end
```

Could we have used any other values for a and b?

$\diamond$

Now returning to our dart thrower, we can acknowledge her superior dart-throwing ability by assigning a nonuniform PDF as shown in Figure 10.4. The probability of

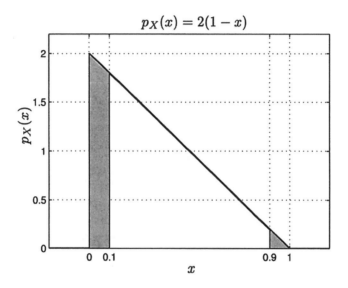

Figure 10.4: Nonuniform PDF.

throwing a dart within a circle of radius 0.1 or $X \in [0, 0.1]$ will be larger than for the region between the circles with radii 0.9 and 1 or $X \in [0.9, 1]$. Specifically, using (10.4)

$$P[0 \leq X \leq 0.1] = \int_0^{0.1} 2(1-x)dx = 2(x - x^2/2)\big|_0^{0.1} = 0.19$$

$$P[0.9 \leq X \leq 1] = \int_{0.9}^1 2(1-x)dx = 2(x - x^2/2)\big|_{0.9}^1 = 0.01.$$

Note that in this example $p_X(x) \geq 0$ for all x and also $\int_{-\infty}^{\infty} p_X(x)dx = 1$. These are properties that must be satisfied for a valid PDF. We will say more about these properties in the next section.

It may be helpful to consider a mass analogy to the PDF. An example is shown in Figure 10.5. It can be thought of as a slice of Jarlsberg cheese with length 2

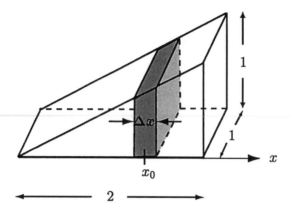

Figure 10.5: Jarlsberg cheese slice used for mass analogy to PDF.

meters, height of 1 meter, and depth of 1 meter, which might be purchased for a
New Year's Eve party (with a lot of guests!). If its mass is 1 kilogram (it is a new
"lite" cheese), then its overall density D is

$$D = \frac{\text{mass}}{\text{volume}} = \frac{M}{V} = \frac{1 \text{ kg}}{1 \text{ m}^3} = 1 \text{ kg/m}^3.$$

However, its *linear* density or mass per meter which is defined as $\Delta M / \Delta x$ will change
with x. If each guest is allowed to cut a wedge of cheese of length Δx as shown in
Figure 10.5, then clearly the hungriest guests should choose a wedge near $x = 2$ for
the greatest amount of cheese. To determine the linear density we compute $\Delta M / \Delta x$
versus x. To do so first note that $\Delta M = D \Delta V = \Delta V$ and $\Delta V = 1 \cdot$ (area of face),
where the face is seen to be trapezoidal. Thus,

$$\Delta V = \frac{1}{2} \Delta x \left(\frac{x_0 - \Delta x/2}{2} + \frac{x_0 + \Delta x/2}{2} \right) = \frac{1}{2} x_0 \Delta x.$$

Hence, $\Delta M / \Delta x = \Delta V / \Delta x = x_0/2$ and this is the same even as $\Delta x \to 0$. Thus,

$$\frac{dM}{dx} = \frac{1}{2} x \qquad 0 \le x \le 2$$

and to obtain the mass for any wedge from $x = a$ to $x = b$ we need only integrate
dM/dx to obtain the mass as a function of x. This yields

$$M([a, b]) = \int_a^b \frac{1}{2} x \, dx = \int_a^b m(x) \, dx$$

where $m(x) = x/2$ is the *linear mass density or the mass per unit length*. It is
perfectly analogous to the PDF which is the *probability per unit length*. Can you
find the total mass of cheese from $M([a, b])$? See also Problem 10.3.

10.4 The PDF and Its Properties

The PDF must have certain properties so that the probabilities obtained using (10.4) satisfy the axioms given in Section 3.4. Since the probability of an interval is given by

$$P[a \le X \le b] = \int_a^b p_X(x)dx$$

the PDF must have the following properties.

Property 10.1 – PDF must be nonnegative.

$$p_X(x) \ge 0 \qquad -\infty < x < \infty.$$

Proof: If $p_X(x) < 0$ on some small interval $[x_0 - \Delta x/2, x_0 + \Delta x/2]$, then

$$P[x_0 - \Delta x/2 \le X \le x_0 + \Delta x/2] = \int_{x_0 - \Delta x/2}^{x_0 + \Delta x/2} p_X(x)dx < 0$$

which violates Axiom 1 that $P[E] \ge 0$ for all events E.

$\square$

Property 10.2 – PDF must integrate to one.

$$\int_{-\infty}^{\infty} p_X(x)dx = 1$$

Proof:

$$1 = P[X \in \mathcal{S}_X] = P[-\infty < X < \infty] = \int_{-\infty}^{\infty} p_X(x)dx$$

$\square$

Hence, any nonnegative function that integrates to one can be considered as a PDF. An example follows.

Example 10.2 – Exponential PDF

Consider the function

$$p_X(x) = \begin{cases} \lambda \exp(-\lambda x) & x \ge 0 \\ 0 & x < 0 \end{cases} \qquad (10.5)$$

for $\lambda > 0$. This is called the *exponential PDF* and is shown in Figure 10.6. Note that it is discontinuous at $x = 0$. Hence, a PDF need not be continuous (see also Figure 10.3a for the uniform PDF which also has points of discontinuity). Also, for $\lambda > 1$, we have $p_X(0) = \lambda > 1$. In contrast to a PMF, the PDF can exceed one in

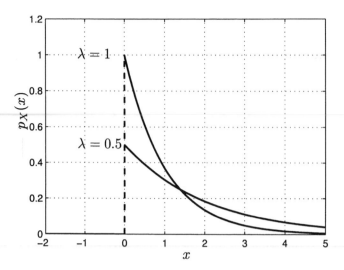

Figure 10.6: Exponential PDF.

value. It is the *area* under the PDF that cannot exceed one. As expected $p_X(x) \geq 0$ for $-\infty < x < \infty$ and

$$\int_{-\infty}^{\infty} p_X(x)dx = \int_{0}^{\infty} \lambda \exp(-\lambda x)dx$$
$$= -\exp(-\lambda x)|_{0}^{\infty} = 1$$

for $\lambda > 0$. This PDF is often used as a model for the lifetime of a product. For example, if X is the failure time in days of a lightbulb, then $P[X > 100]$ is the probability that the lightbulb will fail after 100 days or it will last for at least 100 days. This is found to be

$$P[X > 100] = \int_{100}^{\infty} \lambda \exp(-\lambda x)dx$$
$$= -\exp(-\lambda x)|_{100}^{\infty}$$
$$= \exp(-100\lambda)$$
$$= \begin{cases} 0.367 & \lambda = 0.01 \\ 0.904 & \lambda = 0.001 \,. \end{cases}$$

  The probability of a sample point is zero.

If X is a continuous random variable, then it was argued in Section 3.6 that the probability of a point is zero. This is consistent with our definition of a PDF. If the width of the interval shrinks to zero, then the area under the PDF also goes to zero. Hence, $P[X = x] = 0$. This is true whether or not $p_X(x)$ is continuous at the point of interest (as long as the discontinuity is a finite jump). In the previous example of an exponential PDF $P[X = 0] = 0$ even though $p_X(0)$ is discontinuous at $x = 0$. This means that we could, if desired, have defined the exponential PDF as

$$p_X(x) = \begin{cases} \lambda \exp(-\lambda x) & x > 0 \\ 0 & x \leq 0 \end{cases}$$

for which $p_X(0)$ is now defined to be 0. It makes no difference in our probability calculations whether we include $x = 0$ in the interval or not. Hence, we see that

$$\int_{0^-}^b p_X(x)dx = \int_{0^+}^b p_X(x)dx = \int_0^b p_X(x)dx$$

and in a similar manner if X is a continuous random variable, then

$$P[a \leq X \leq b] = P[a < X \leq b] = P[a \leq X < b] = P[a < X < b].$$

In summary, the value assigned to the PDF at a discontinuity is arbitrary since it does not affect any subsequent probability calculation involving a continuous random variable. However, for discontinuities other than step discontinuities (which are jumps of finite magnitude) we will see in Section 10.8 that we must be more careful.

10.5 Important PDFs

There are a multitude of PDFs in use in various scientific disciplines. The books by [Johnson, Kotz, and Balakrishnan 1994] contain a summary of many of these and should be consulted for further information. We now describe some of the more important PDFs.

10.5.1 Uniform

We have already encountered a special case of the uniform PDF in Figure 10.3. More generally it is defined as

$$p_X(x) = \begin{cases} \frac{1}{b-a} & a < x < b \\ 0 & \text{otherwise} \end{cases} \tag{10.6}$$

and examples are shown in Figure 10.7. It is given the shorthand notation $X \sim \mathcal{U}(a, b)$. If $a = 0$ and $b = 1$, then an outcome of a $\mathcal{U}(0, 1)$ random variable can be generated in MATLAB using **rand**.

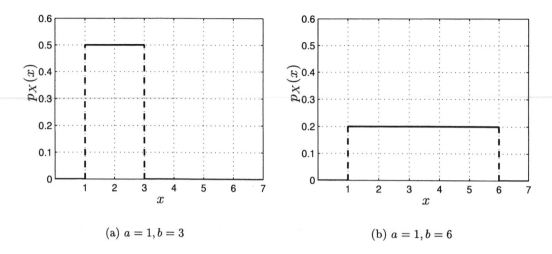

(a) $a = 1, b = 3$ (b) $a = 1, b = 6$

Figure 10.7: Examples of uniform PDF.

10.5.2 Exponential

This was previously defined in Example 10.2. The shorthand notation is $X \sim$ $\exp(\lambda)$.

10.5.3 Gaussian or Normal

This is the famous "bell-shaped" curve first introduced in Section 1.3. It is given by

$$p_X(x) = \frac{1}{\sqrt{2\pi\sigma^2}} \exp\left[-\frac{1}{2\sigma^2}(x - \mu)^2\right] \qquad -\infty < x < \infty \qquad (10.7)$$

where $\sigma^2 > 0$ and $-\infty < \mu < \infty$. Its application in practical problems is ubiquitous. It is shown to integrate to one in Problem 10.9. Some examples of this PDF as well as some outcomes for various values of the parameters (μ, σ^2) are shown in Figures 10.8 and 10.9. It is characterized by the two parameters μ and σ^2. The parameter μ indicates the center of the PDF which is seen in Figures 10.8a and 10.8c. It depicts the "average value" of the random variable as can be observed by examining Figures 10.8b and 10.8d. In Chapter 11 we will show that μ is actually the mean of X. The parameter σ^2 indicates the width of the PDF as is seen in Figures 10.9a and 10.9c. It is related to the variability of the outcomes as seen in Figures 10.9b and 10.9d. In Chapter 11 we will show that σ^2 is actually the variance of X. The PDF is called the *Gaussian* PDF after the famous German mathematician K.F. Gauss and also the *normal* PDF, since "normal" populations tend to exhibit this type of distribution. A *standard* normal PDF is one for which $\mu = 0$ and $\sigma^2 = 1$. The shorthand notation is $X \sim \mathcal{N}(\mu, \sigma^2)$. MATLAB generates a realization of a *standard normal* random variable using **randn**. This was used extensively in Chapter 2.

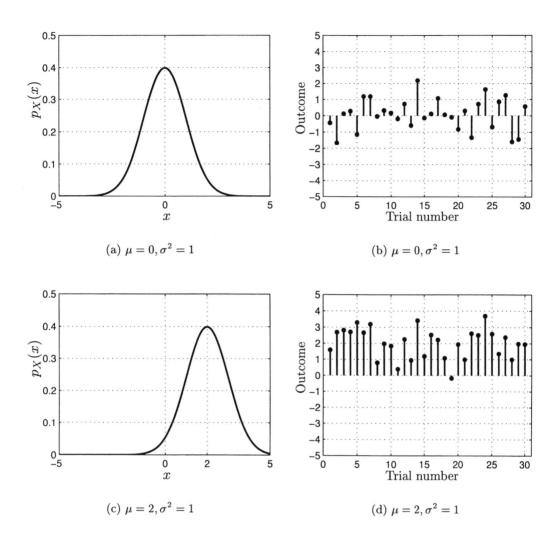

Figure 10.8: Examples of Gaussian PDF with different μ's.

To find the probability of the outcome of a Gaussian random variable lying within an interval requires numerical integration (see Problem 1.14) since the integral

$$\int_a^b \frac{1}{\sqrt{2\pi}} \exp(-(1/2)x^2)dx$$

cannot be evaluated analytically. A MATLAB subprogram will be provided and described shortly to do this. The Gaussian PDF is commonly used to model noise in a communication system (see Section 2.6), as well as for numerous other applications. We will see in Chapter 15 that the PDF arises quite naturally as the PDF of a large number of independent random variables that have been added together.

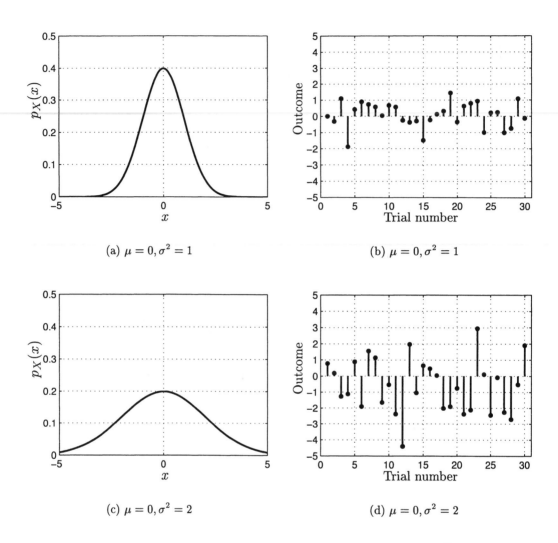

Figure 10.9: Examples of Gaussian PDF with different σ^2's.

10.5.4   Laplacian

This PDF is named after Laplace, the famous French mathematician. It is similar to the Gaussian except that it does not decrease as rapidly from its maximum value. Its PDF is

$$p_X(x) = \frac{1}{\sqrt{2\sigma^2}} \exp\left(-\sqrt{\frac{2}{\sigma^2}}|x|\right) \qquad -\infty < x < \infty \qquad (10.8)$$

where $\sigma^2 > 0$. Again the parameter σ^2 specifies the width of the PDF, and will be shown in Chapter 11 to be the variance of X. It is seen to be symmetric about $x = 0$. Some examples of the PDF and outcomes are shown in Figure 10.10. Note that for

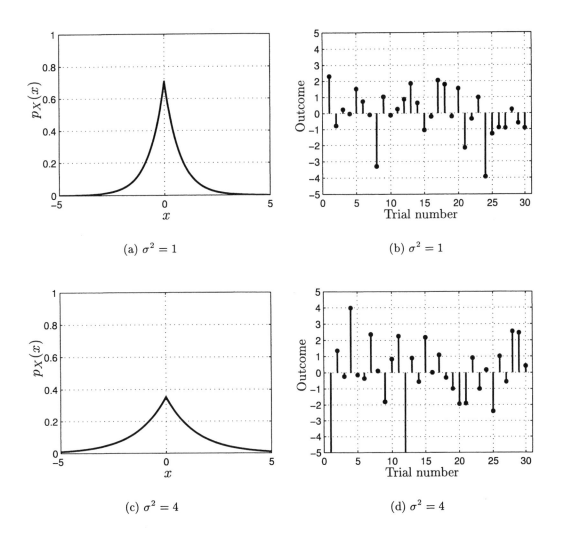

Figure 10.10: Examples of Laplacian PDF with different σ^2's.

the same σ^2 as the Gaussian PDF, the outcomes are larger as seen by comparing Figure 10.10b to Figure 10.9b. This is due to the larger probability in the "tails" of the PDF. The "tail" region of the PDF is that for which $|x|$ is large. The Laplacian PDF is easily integrated to find the probability of an interval. This PDF is used as a model for speech amplitudes [Rabiner and Schafer 1978].

10.5.5 Cauchy

The Cauchy PDF is named after another famous French mathematician and is defined as

$$p_X(x) = \frac{1}{\pi(1 + x^2)} \qquad -\infty < x < \infty. \tag{10.9}$$

It is shown in Figure 10.11 and is seen to be symmetric about $x = 0$. The Cauchy PDF can easily be integrated to find the probability of any interval. It arises as the PDF of the ratio of two independent $\mathcal{N}(0, 1)$ random variables (see Chapter 12).

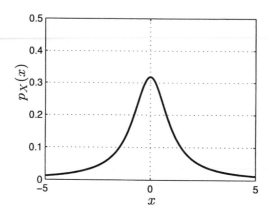

Figure 10.11: Cauchy PDF.

10.5.6   Gamma

The Gamma PDF is a very general PDF that is used for nonnegative random variables. It is given by

$$p_X(x) = \begin{cases} \frac{\lambda^\alpha}{\Gamma(\alpha)} x^{\alpha-1} \exp(-\lambda x) & x \geq 0 \\ 0 & x < 0 \end{cases} \qquad (10.10)$$

where $\lambda > 0$, $\alpha > 0$, and $\Gamma(z)$ is the *Gamma* function which is defined as

$$\Gamma(z) = \int_0^\infty t^{z-1} \exp(-t) dt. \qquad (10.11)$$

Clearly, the $\Gamma(\alpha)$ factor in (10.10) is the normalizing constant needed to ensure that the PDF integrates to one. Some examples of this PDF are shown in Figure 10.12. The shorthand notation is $X \sim \Gamma(\alpha, \lambda)$. Some useful properties of the Gamma function are as follows.

Property 10.3 − $\Gamma(z + 1) = z\Gamma(z)$
 Proof: See Problem 10.16.

$\Box$

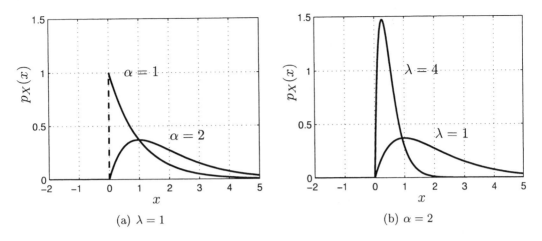

Figure 10.12: Examples of Gamma PDF.

Property 10.4 – $\Gamma(N) = (N-1)!$

<u>Proof:</u> Follows from Property 10.3 with $z = N - 1$ since

$$
\begin{aligned}
\Gamma(N) &= (N-1)\Gamma(N-1) \\
&= (N-1)(N-2)\Gamma(N-3) \qquad (\text{let } z = N-2 \text{ now}) \\
&= (N-1)(N-2)\ldots 1 = (N-1)!
\end{aligned}
$$

$\square$

Property 10.5 – $\Gamma(1/2) = \sqrt{\pi}$

<u>Proof:</u>

$$
\Gamma(1/2) = \int_0^\infty t^{-1/2} \exp(-t) dt
$$

(Note that near $t = 0$ the integrand becomes infinite but $t^{-1/2} \exp(-t) \approx t^{-1/2}$ which is integrable.) Now let $t = u^2/2$ and thus $dt = udu$ which yields

$$
\begin{aligned}
\Gamma(1/2) &= \int_0^\infty \frac{1}{\sqrt{u^2/2}} \exp(-u^2/2) u\,du \\
&= \int_0^\infty \sqrt{2} \exp(-u^2/2) du \\
&= \frac{\sqrt{2}}{2} \underbrace{\int_{-\infty}^\infty \exp(-u^2/2) du}_{=\sqrt{2\pi} \text{ why?}} \qquad (\text{integrand is symmetric about } u = 0) \\
&= \sqrt{\pi}.
\end{aligned}
$$

$\square$

The Gamma PDF reduces to many well known PDFs for appropriate choices of the parameters α and λ. Some of these are:

1. Exponential for $\alpha = 1$
   From (10.10) we have

   $$p_X(x) = \begin{cases} \frac{\lambda}{\Gamma(1)} \exp(-\lambda x) & x \geq 0 \\ 0 & x < 0. \end{cases}$$

   But $\Gamma(1) = 0! = 1$, which results from Property 10.4 so that we have the exponential PDF.

2. Chi-squared PDF with N degrees of freedom for $\alpha = N/2$ and $\lambda = 1/2$
   From (10.10) we have

   $$p_X(x) = \begin{cases} \frac{1}{2^{N/2}\Gamma(N/2)} x^{N/2-1} \exp(-x/2) & x \geq 0 \\ 0 & x < 0. \end{cases} \tag{10.12}$$

   This is called the *chi-squared PDF with N degrees of freedom* and is important in statistics. It can be shown to be the PDF for the sum of the squares of N independent random variables all with the same PDF $\mathcal{N}(0,1)$ (see Problem 12.44). The shorthand notation is $X \sim \chi_N^2$.

3. Erlang for $\alpha = N$
   From (10.10) we have

   $$p_X(x) = \begin{cases} \frac{\lambda^N}{\Gamma(N)} x^{N-1} \exp(-\lambda x) & x \geq 0 \\ 0 & x < 0 \end{cases}$$

   and since $\Gamma(N) = (N-1)!$ from Property 10.4, this becomes

   $$p_X(x) = \begin{cases} \frac{\lambda^N}{(N-1)!} x^{N-1} \exp(-\lambda x) & x \geq 0 \\ 0 & x < 0. \end{cases} \tag{10.13}$$

   This PDF arises as the PDF of a sum of N independent exponential random variables all with the same λ (see also Problem 10.17).

10.5.7   Rayleigh

The Rayleigh PDF is named after the famous British physicist Lord Rayleigh and is defined as

$$p_X(x) = \begin{cases} \frac{x}{\sigma^2} \exp\left(-\frac{1}{2}\frac{x^2}{\sigma^2}\right) & x \geq 0 \\ 0 & x < 0. \end{cases} \tag{10.14}$$

It is shown in Figure 10.13. The Rayleigh PDF is easily integrated to yield the

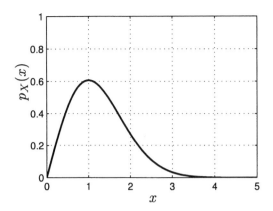

Figure 10.13: Rayleigh PDF with $\sigma^2 = 1$.

probability of any interval. It can be shown to arise as the PDF of the square root of the sum of the squares of two independent $\mathcal{N}(0,1)$ random variables (see Example 12.12).

Finally, note that many of these PDFs arise as the PDFs of transformed Gaussian random variables. Therefore, realizations of the random variable may be obtained by first generating multiple realizations of independent standard normal or $\mathcal{N}(0,1)$ random variables, and then performing the appropriate transformation. An alternative and more general approach to generating realizations of a random variable, once the PDF is known, is via the probability integral transformation to be discussed in Section 10.9.

10.6 Cumulative Distribution Functions

The cumulative distribution function (CDF) for a continuous random variable is defined exactly the same as for a discrete random variable. It is

$$F_X(x) = P[X \leq x] \qquad -\infty < x < \infty \tag{10.15}$$

and is evaluated using the PDF as

$$F_X(x) = \int_{-\infty}^{x} p_X(t)dt \qquad -\infty < x < \infty. \tag{10.16}$$

  Avoiding confusion in evaluating CDFs

It is important to note that in evaluating a definite integral such as in (10.16) it is best to replace the variable of integration with another symbol. This is because the upper limit depends on x which would conflict with the dummy variable of integration. We have chosen to use t but of course any other symbol that does not conflict with x can be used.

Some examples of the evaluation of the CDF are given next.

10.6.1 Uniform

Using (10.6) we have

$$F_X(x) = \begin{cases} 0 & x \le a \\ \int_a^x \frac{1}{b-a} dt & a < x < b \\ 1 & x \ge b \end{cases}$$

which is

$$F_X(x) = \begin{cases} 0 & x \le a \\ \frac{1}{b-a}(x-a) & a < x < b \\ 1 & x \ge b. \end{cases}$$

An example is shown in Figure 10.14 for $a = 1$ and $b = 2$.

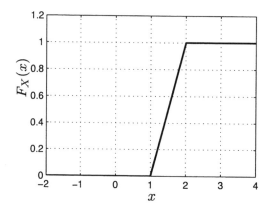

Figure 10.14: CDF for uniform random variable over interval $(1, 2)$.

10.6.2 Exponential

Using (10.5) we have

$$F_X(x) = \begin{cases} 0 & x < 0 \\ \int_0^x \lambda \exp(-\lambda t) dt & x \ge 0. \end{cases}$$

But

$$\int_0^x \lambda \exp(-\lambda t) dt = -\exp(-\lambda t)\big|_0^x = 1 - \exp(-\lambda x)$$

so that

$$F_X(x) = \begin{cases} 0 & x < 0 \\ 1 - \exp(-\lambda x) & x \geq 0 . \end{cases}$$

An example is shown in Figure 10.15 for $\lambda = 1$.

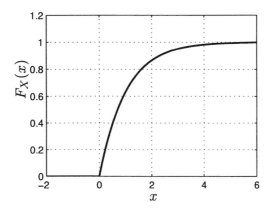

Figure 10.15: CDF for exponential random variable with $\lambda = 1$.

Note that for the uniform and exponential random variables the CDFs are continuous even though the PDFs are discontinuous. This property motivates an alternative definition of a continuous random variable as one for which *the CDF is continuous*. Recall that the CDF of a discrete random variable is always discontinuous, displaying multiple jumps.

10.6.3 Gaussian

Consider a standard normal PDF, which is a Gaussian PDF with $\mu = 0$ and $\sigma^2 = 1$. (If $\mu \neq 0$ and/or $\sigma^2 \neq 1$ the CDF is a simple modification as shown in Problem 10.22.) Then from (10.7) we have

$$F_X(x) = \int_{-\infty}^{x} \frac{1}{\sqrt{2\pi}} \exp\left(-\frac{1}{2}t^2\right) dt \quad -\infty < x < \infty.$$

This cannot be evaluated further but can be found numerically and is shown in Figure 10.16. The CDF for a standard normal is usually given the special symbol $\Phi(x)$ so that

$$\Phi(x) = \int_{-\infty}^{x} \frac{1}{\sqrt{2\pi}} \exp\left(-\frac{1}{2}t^2\right) dt \quad -\infty < x < \infty.$$

Hence, $\Phi(x)$ represents the area under the PDF to the *left* of the point x as seen in Figure 10.17a. It is sometimes more convenient, however, to have knowledge of the area to the *right* instead. This is called the *right-tail probability* of a standard

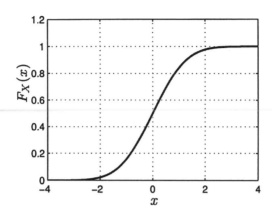

Figure 10.16: CDF for standard normal or Gaussian random variable.

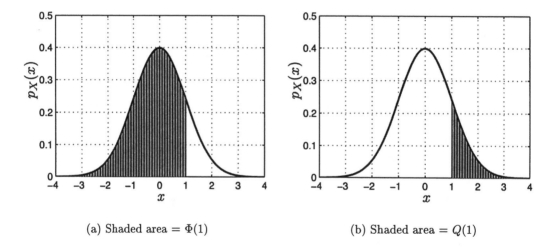

(a) Shaded area $= \Phi(1)$
(b) Shaded area $= Q(1)$

Figure 10.17: Definitions of $\Phi(x)$ and $Q(x)$ functions.

normal and is given the symbol $Q(x)$. It is termed the "Q" function and is defined as the area to the right of x, an example of which is shown in Figure 10.17b. By its definition we have

$$Q(x) = 1 - \Phi(x) \tag{10.17}$$

$$= \int_x^\infty \frac{1}{\sqrt{2\pi}} \exp\left(-\frac{1}{2}t^2\right) dt \quad -\infty < x < \infty \tag{10.18}$$

and is shown in Figure 10.18, plotted on a linear as well as a logarithmic vertical scale. Some of the properties of the Q function that are easily verified are (see

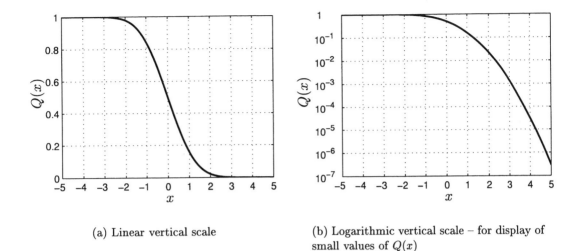

(a) Linear vertical scale

(b) Logarithmic vertical scale – for display of small values of $Q(x)$

Figure 10.18: $Q(x)$ function.

Problem 10.25)

$$Q(-\infty) = 1 \tag{10.19}$$
$$Q(\infty) = 0 \tag{10.20}$$
$$Q(0) = \frac{1}{2} \tag{10.21}$$
$$Q(-x) = 1 - Q(x). \tag{10.22}$$

Although the Q function cannot be evaluated analytically, it is related to the well known "error function". Thus, making use of the latter function a MATLAB subprogram `Q.m`, which is listed in Appendix 10B, can be used to evaluate it. An example follows.

Example 10.3 – Probability of error in a communication system

In Section 2.6 we analyzed the probability of error for a PSK digital communication system. The probability of error P_e was given by

$$P_e = P[A/2 + W \le 0]$$

where $W \sim \mathcal{N}(0,1)$. (In the MATLAB code we used `w=randn(1,1)` and hence the random variable representing the noise was a standard normal random variable.)

To explicitly evaluate P_e we have that

$$
\begin{aligned}
P_e &= P[A/2 + W \le 0] \\
&= 1 - P[A/2 + W > 0] \\
&= 1 - P[W > -A/2] \\
&= 1 - Q(-A/2) \qquad \text{(definition)} \\
&= Q(A/2) \qquad\qquad \text{(use (10.22))}.
\end{aligned}
$$

Hence, the true P_e shown in Figure 2.15 as the dashed line can be found by using the MATLAB subprogram Q.m, which is listed in Appendix 10B, for the argument $A/2$ (see Problem 10.26). It is also sometimes important to determine A to yield a given P_e. This is found as $A = 2Q^{-1}(P_e)$, where Q^{-1} is the inverse of the Q function. It is defined as the value of x necessary to yield a given value of $Q(x)$. It too cannot be expressed analytically but may be evaluated using the MATLAB subprogram Qinv.m, also listed in Appendix 10B.

$$\Diamond$$

The Q function can also be approximated for large values of x using [Abramowitz and Stegun 1965]

$$
Q(x) \approx \frac{1}{\sqrt{2\pi}x} \exp\left(-\frac{1}{2}x^2\right) \qquad x > 3. \tag{10.23}
$$

A comparison of the approximation to the true value is shown in Figure 10.19. If

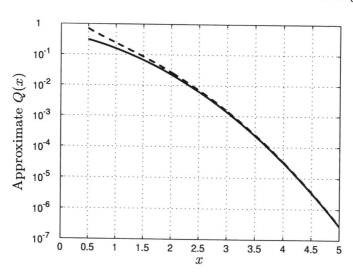

Figure 10.19: Approximation of Q function – true value is shown dashed.

$X \sim \mathcal{N}(\mu, \sigma^2)$, then the right-tail probability becomes

$$
P[X > x] = Q\left(\frac{x - \mu}{\sqrt{\sigma^2}}\right) \tag{10.24}
$$

(see Problem 10.24). Finally, note that the area under the standard normal Gaussian PDF is mostly contained in the interval $[-3, 3]$. As seen in Figure 10.19 $Q(3) \approx 0.001$, which means that the area to the right of $x = 3$ is only 0.001. Since the PDF is symmetric, the total area to the right of $x = 3$ and to the left of $x = -3$ is 0.002 or the area in the $[-3, 3]$ interval is 0.998. Hence, 99.8% of the probability lies within this interval. We would not expect to see a value greater than 3 in magnitude very often. This is borne out by an examination of Figure 10.8b. How many realizations would you expect to see in the interval $(1, \infty)$? Is this consistent with Figure 10.8b ?

As we have seen, the CDF for a continuous random variable has certain properties. For the most part they are the same as for a discrete random variable: the CDF is 0 at $x = -\infty$, 1 at $x = \infty$, and is monotonically increasing (or stays the same) between these limits. However, now it is continuous, having no jumps. The most important property for practical purposes is that which allows us to compute probabilities of intervals. This follows from the property

$$P[a \leq X \leq b] = P[a < X \leq b] = F_X(b) - F_X(a) \qquad (10.25)$$

which is easily proven (see Problem 10.35). It can be seen to be valid by referring to Figure 10.20. Using the CDF we *no longer have to integrate the PDF* to determine

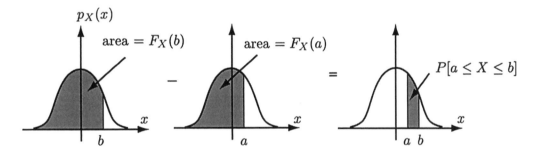

Figure 10.20: Illustration of use of CDF to find probability of interval.

probabilities of intervals. In effect, all the integration has been done for us in finding the CDF. Some examples follow.

Example 10.4 – Probability of interval for exponential PDF
Since $F_X(x) = 1 - \exp(-\lambda x)$ for $x \geq 0$, we have for $a > 0$ and $b > 0$

$$
\begin{aligned}
P[a \leq X \leq b] &= F_X(b) - F_X(a) \\
&= (1 - \exp(-\lambda b)) - (1 - \exp(-\lambda a)) \\
&= \exp(-\lambda a) - \exp(-\lambda b)
\end{aligned}
$$

which should be compared to

$$\int_a^b \lambda \exp(-\lambda x)\,dx.$$

◇

Since we obtained the CDF from the PDF, we might suppose that the PDF could be recovered from the CDF. For a discrete random variable this was the case since $p_X[x_i] = F_X(x_i^+) - F_X(x_i^-)$. For a continuous random variable we consider a small interval $[x_0 - \Delta x/2, x_0 + \Delta x/2]$ and evaluate its probability using (10.25) with

$$F_X(x) = \int_{-\infty}^{x} p_X(t)dt.$$

Then, we have

$$F_X(x_0 + \Delta x/2) - F_X(x_0 - \Delta x/2)$$

$$= \int_{-\infty}^{x_0+\Delta x/2} p_X(t)dt - \int_{-\infty}^{x_0-\Delta x/2} p_X(t)dt$$

$$= \int_{x_0-\Delta x/2}^{x_0+\Delta x/2} p_X(t)dt$$

$$\approx p_X(x_0) \int_{x_0-\Delta x/2}^{x_0+\Delta x/2} 1\, dt \quad (p_X(t) \approx \text{constant as } \Delta x \to 0)$$

$$= p_X(x_0)\Delta x$$

so that

$$p_X(x_0) \quad \approx \quad \frac{F_X(x_0 + \Delta x/2) - F_X(x_0 - \Delta x/2)}{\Delta x}$$

$$\rightarrow \quad \left.\frac{dF_X(x)}{dx}\right|_{x=x_0} \qquad \text{as } \Delta x \to 0.$$

Hence, we can obtain the PDF from the CDF by differentiation or

$$p_X(x) = \frac{dF_X(x)}{dx}. \tag{10.26}$$

This relationship is really just the *fundamental theorem of calculus* [Widder 1989]. Note the similarity to the discrete case in which $p_X[x_i] = F_X(x_i^+) - F_X(x_i^-)$. As an example, if $X \sim \exp(\lambda)$, then

$$F_X(x) = \begin{cases} 1 - \exp(-\lambda x) & x \geq 0 \\ 0 & x < 0. \end{cases}$$

For all x except $x = 0$ (at which the CDF does not have a derivative due to the change in slope as seen in Figure 10.15) we have

$$p_X(x) = \frac{dF_X(x)}{dx} \quad = \quad 0 \qquad\qquad x < 0$$

$$= \quad \lambda \exp(-\lambda x) \qquad x > 0$$

and as remarked earlier, $p_X(0)$ can be assigned any value.

10.7 Transformations

In discussing transformations for discrete random variables we noted that a transformation can be either one-to-one or many-to-one. For example, the function $g(x) = 2x$ is one-to-one while $g(x) = x^2$ is many-to-one (in this case two-to-one since $-x$ and $+x$ both map into x^2). The determination of the PDF of $Y = g(X)$ will depend upon which type of transformation we have. Initially, we will consider the one-to-one case, which is simpler. For the transformation of a discrete random variable we saw from (5.9) that the PMF of $Y = g(X)$ for any g could be found from the PMF of X using

$$p_Y[y_i] = \sum_{\{j:g(x_j)=y_i\}} p_X[x_j].$$

But if g is one-to-one we have only a single solution for $g(x_j) = y_i$, so that $x_j = g^{-1}(y_i)$ and therefore

$$p_Y[y_i] = p_X[g^{-1}(y_i)] \tag{10.27}$$

and we are done. For example, assume X takes on values $\{1, 2\}$ with a PMF $p_X[1]$ and $p_X[2]$ and we wish to determine the PMF of $Y = g(X) = 2X$, which is shown in Figure 10.21. Then from (10.27)

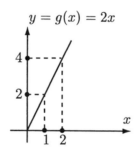

$$y = g(x) = 2x$$

Figure 10.21: Transformation of a discrete random variable.

$$p_Y[2] = p_X[g^{-1}(2)] = p_X[1]$$
$$p_Y[4] = p_X[g^{-1}(4)] = p_X[2].$$

Because we are now dealing with a PDF, which is a density function, and not a PMF, which is a probability function, the simple relationship of (10.27) is no longer valid. To see what happens instead, consider the problem of determining the PDF of $Y = 2X$, where $X \sim \mathcal{U}(1, 2)$. Clearly, $\mathcal{S}_X = \{x : 1 < x < 2\}$ and therefore $\mathcal{S}_Y = \{y : 2 < y < 4\}$ so that $p_Y(y)$ must be zero outside the interval $(2, 4)$. The results of a MATLAB computer simulation are shown in Figure 10.22. A total of 50 realizations were obtained for X and Y. The generated X outcomes are shown on the x-axis and the resultant Y outcomes obtained from $y = 2x$ are shown on the

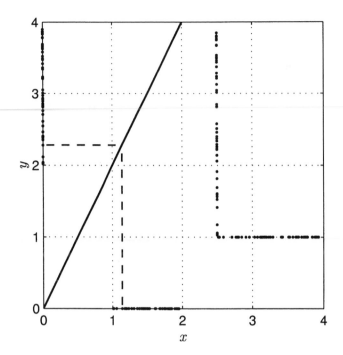

Figure 10.22: Computer generated realizations of X and $Y = 2X$ for $X \sim \mathcal{U}(1,2)$. A 50% expanded version of the realizations is shown to the right.

y-axis. Also, a 50% expanded version of the points is shown to the right. It is seen that the *density* of points on the y-axis is less than that on the x-axis. After some thought the reader will realize that this is the result of the scaling by a factor of 2 due to the transformation. Since the PDF is *probability per unit length*, we should expect $p_Y = p_X/2$ for $2 < y < 4$. To prove that this is so, we note that a small interval on the x-axis, say $[x_0 - \Delta x/2, x_0 + \Delta x/2]$, will map into $[2x_0 - \Delta x, 2x_0 + \Delta x]$ on the y-axis. However, the intervals are equivalent events and so their probabilities must be equal. It follows then that

$$\int_{x_0 - \Delta x/2}^{x_0 + \Delta x/2} p_X(x)dx = \int_{2x_0 - \Delta x}^{2x_0 + \Delta x} p_Y(y)dy$$

and as $\Delta x \to 0$, we have that $p_X(x) \to p_X(x_0)$ and $p_Y(y) \to p_Y(2x_0)$ in the small intervals so that

$$p_X(x_0)\Delta x = p_Y(2x_0)2\Delta x$$

or

$$p_Y(2x_0) = p_X(x_0)\frac{1}{2}.$$

As expected, the PDF of Y is scaled by $1/2$. If we now let $y_0 = 2x_0$, then this becomes

$$p_Y(y_0) = p_X(y_0/2)\frac{1}{2}$$

or for any arbitrary value of y

$$p_Y(y) = p_X(y/2)\frac{1}{2} \qquad 2 < y < 4. \tag{10.28}$$

This results in the final PDF using $p_X(x) = 1$ for $1 < x < 2$ as

$$p_Y(y) = \begin{cases} \frac{1}{2} & 2 < y < 4 \\ 0 & \text{otherwise} \end{cases} \tag{10.29}$$

and thus if $X \sim \mathcal{U}(1,2)$, then $Y = 2X \sim \mathcal{U}(2,4)$. The general result for the PDF of $Y = g(X)$ is given by

$$p_Y(y) = p_X(g^{-1}(y))\left|\frac{dg^{-1}(y)}{dy}\right|. \tag{10.30}$$

For our example, the use of (10.30) with $g(x) = 2x$ and therefore $g^{-1}(y) = y/2$ results in (10.29). The absolute value is needed to allow for the case when g is decreasing and hence g^{-1} is decreasing since otherwise the scaling term would be negative (see Problem 10.57). A formal derivation is given in Appendix 10A. Note the similarity of (10.30) to (10.27). The principal difference is the presence of the derivative or *Jacobian* factor $dg^{-1}(y)/dy$. It is needed to account for the change in scaling due to the mapping of a given length interval into an interval of a different length as illustrated in Figure 10.22. Some examples of the use of (10.30) follow.

Example 10.5 – PDF for linear (actually an affine) transformation

To determine the PDF of $Y = aX + b$, for a and b constants first assume that $S_X = \{x : -\infty < x < \infty\}$ and hence $S_Y = \{y : -\infty < y < \infty\}$. Here we have $g(x) = ax + b$ so that the inverse function g^{-1} is found by solving $y = ax + b$ for x. This yields $x = (y - b)/a$ so that

$$g^{-1}(y) = \frac{y - b}{a}$$

and from (10.30) the general result is

$$p_Y(y) = p_X\left(\frac{y - b}{a}\right)\left|\frac{1}{a}\right|. \tag{10.31}$$

As a further example, consider $X \sim \mathcal{N}(0,1)$ and the transformation $Y = \sqrt{\sigma^2}X + \mu$.

Then, letting $\sigma = \sqrt{\sigma^2} > 0$ we have

$$
\begin{aligned}
p_Y(y) &= p_X\left(\frac{y-\mu}{\sigma}\right)\left|\frac{1}{\sigma}\right| \\
&= p_X\left(\frac{y-\mu}{\sigma}\right)\frac{1}{\sigma} \\
&= \frac{1}{\sqrt{2\pi}}\exp\left[-\frac{1}{2}\left(\frac{y-\mu}{\sigma}\right)^2\right]\frac{1}{\sigma} \\
&= \frac{1}{\sqrt{2\pi\sigma^2}}\exp\left[-\frac{1}{2\sigma^2}(y-\mu)^2\right]
\end{aligned}
$$

and therefore $Y \sim \mathcal{N}(\mu, \sigma^2)$. *A linear transformation of a Gaussian random variable results in another Gaussian random variable whose Gaussian PDF has different values of the parameters.* Because of this property we can easily generate a realization of a $\mathcal{N}(\mu, \sigma^2)$ random variable using the MATLAB construction `y=sqrt(sigma2)*randn(1,1)+mu`, since `randn(1,1)` produces a realization of a standard normal random variable (see Problem 10.60).

Example 10.6 – PDF of $Y = \exp(X)$ for $X \sim \mathcal{N}(0, 1)$

Here we have that $S_Y = \{y : y > 0\}$. To find $g^{-1}(y)$ we let $y = \exp(x)$ and solve for x, which is $x = \ln(y)$. Thus, $g^{-1}(y) = \ln(y)$. From (10.30) it follows that

$$
p_Y(y) = p_X(\ln(y))\left|\frac{d\ln(y)}{dy}\right| = \begin{cases} p_X(\ln(y))\frac{1}{y} & y > 0 \\ 0 & y \le 0 \end{cases}
$$

or

$$
p_Y(y) = \begin{cases} \frac{1}{\sqrt{2\pi}y}\exp\left[-\frac{1}{2}(\ln(y))^2\right] & y > 0 \\ 0 & y \le 0. \end{cases}
$$

This PDF is called the *log-normal* PDF. It is frequently used as a model for a quantity that is measured in decibels (dB) and which has a normal PDF in dB quantities [Members of Technical Staff 1970].

  Always determine the possible values for Y before using (10.30).

A common error in determining the PDF of a transformed random variable is to forget that $p_Y(y)$ may be zero over some regions. In the previous example of $y = \exp(x)$, the mapping of $-\infty < x < \infty$ is into $y > 0$. Hence, the PDF of Y must be zero for $y \le 0$ since there are no values of X that produce a zero or negative value of Y. Nonsensical results occur if we attempt to insert values in $p_Y(y)$ for $y \leq 0$. To avoid this potential problem, we should first determine $\mathcal{S}_Y$ and then use (10.30) to find the PDF over the sample space.

When the transformation is not one-to-one, we will have multiple solutions for x in $y = g(x)$. An example is for $y = x^2$ for which the solutions are

$$
\begin{aligned}
x_1 &= -\sqrt{y} = g_1^{-1}(y) \\
x_2 &= +\sqrt{y} = g_2^{-1}(y).
\end{aligned}
$$

This is shown in Figure 10.23. In this case we use (10.30) but must add the PDFs

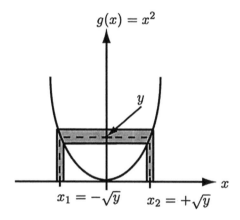

Figure 10.23: Solutions for x in $y = g(x) = x^2$.

(since both the x-intervals map into the same y-interval and the x-intervals are disjoint) to yield

$$
p_Y(y) = p_X(g_1^{-1}(y)) \left| \frac{dg_1^{-1}(y)}{dy} \right| + p_X(g_2^{-1}(y)) \left| \frac{dg_2^{-1}(y)}{dy} \right|. \tag{10.32}
$$

Example 10.7 – PDF of $Y = X^2$ for $X \sim \mathcal{N}(0,1)$

Since $-\infty < X < \infty$, we must have $Y \geq 0$. Next because $g_1^{-1}(y) = -\sqrt{y}$ and $g_2^{-1}(y) = \sqrt{y}$ we have from (10.32)

$$
p_Y(y) = \begin{cases} p_X(-\sqrt{y}) \left| -\frac{1}{2\sqrt{y}} \right| + p_X(\sqrt{y}) \left| \frac{1}{2\sqrt{y}} \right| & y \geq 0 \\ 0 & y < 0 \end{cases}
$$

which reduces to

$$p_Y(y) = \begin{cases} \left[\frac{1}{\sqrt{2\pi}}\exp(-y/2)\right]\frac{1}{2\sqrt{y}} + \left[\frac{1}{\sqrt{2\pi}}\exp(-y/2)\right]\frac{1}{2\sqrt{y}} & y \geq 0 \\ 0 & y < 0 \end{cases}$$

$$= \begin{cases} \frac{1}{\sqrt{2\pi y}}\exp(-y/2) & y \geq 0 \\ 0 & y < 0. \end{cases}$$

This is shown in Figure 10.24 and should be compared to Figure 2.10 in which this PDF was estimated (see also Problem 10.59). Note that the PDF is undefined at

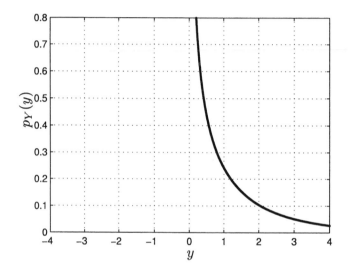

Figure 10.24: PDF for $Y = X^2$ for $X \sim \mathcal{N}(0,1)$.

$y = 0$ since $p_Y(0) \to \infty$, although the total area under the PDF is finite and of course is equal to 1. Also, $Y \sim \chi_1^2$ as can be seen by referring to (10.12) with $N = 1$.

$\Diamond$

In general, if $y = g(x)$ has solutions $x_i = g_i^{-1}(y)$ for $i = 1, 2, \ldots, M$, then

$$p_Y(y) = \sum_{i=1}^{M} p_X(g_i^{-1}(y)) \left| \frac{dg_i^{-1}(y)}{dy} \right|. \tag{10.33}$$

An alternative means of finding the PDF of a transformed random variable is to first find the CDF and then differentiate it (see (10.26)). We illustrate this approach by redoing the previous example.

Example 10.8 – CDF approach to determine PDF of $Y = X^2$ for $X \sim \mathcal{N}(0,1)$

First we determine the CDF of Y in terms of the CDF for X as

$$
\begin{aligned}
F_Y(y) &= P[Y \le y] \\
&= P[X^2 \le y] \\
&= P[-\sqrt{y} \le X \le \sqrt{y}] \\
&= F_X(\sqrt{y}) - F_X(-\sqrt{y}). \qquad \text{(from (10.25))}
\end{aligned}
$$

Then, differentiating we have

$$
\begin{aligned}
p_Y(y) &= \frac{dF_Y(y)}{dy} \\
&= \frac{d}{dy}[F_X(\sqrt{y}) - F_X(-\sqrt{y})] \\
&= p_X(\sqrt{y})\frac{d\sqrt{y}}{dy} - p_X(-\sqrt{y})\frac{d(-\sqrt{y})}{dy} \quad \text{(from (10.25) and chain rule of calculus)} \\
&= p_X(\sqrt{y})\frac{1}{2\sqrt{y}} + p_X(-\sqrt{y})\frac{1}{2\sqrt{y}} \\
&= \begin{cases} p_X(\sqrt{y})\frac{1}{\sqrt{y}} & y \ge 0 \\ 0 & y < 0 \end{cases} \quad \text{(since } p_X(-x) = p_X(x) \text{ for } X \sim \mathcal{N}(0,1)) \\
&= \begin{cases} \frac{1}{\sqrt{2\pi y}}\exp(-y/2) & y \ge 0 \\ 0 & y < 0. \end{cases}
\end{aligned}
$$

$\diamond$

10.8 Mixed Random Variables

We have so far described two types of random variables, the discrete random variable and the continuous random variable. The sample space for a discrete random variable consists of a countable (either finite or infinite) set of points while that for a continuous random variable has an infinite and uncountable set of points. The points in $\mathcal{S}_X$ for a discrete random variable have a nonzero probability while those for a continuous random variable have a zero probability. In some physical situations, however, we wish to assign a nonzero probability to some points but not others. As an example, consider an experiment in which a fair coin is tossed. If it comes up heads, we generate the outcome of a continuous random variable $X \sim \mathcal{N}(0,1)$ and if it comes up tails we set $X = 0$. Then, the possible outcomes are $-\infty < x < \infty$ and the probability of any point except $x = 0$ has a zero probability of occurring. However, the point $x = 0$ occurs with a probability of $1/2$ since the probability of a tail is $1/2$. A typical sequence of outcomes is shown in Figure 10.25. One could

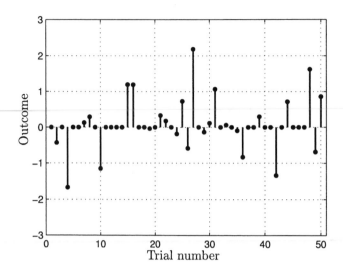

Figure 10.25: Sequence of outcomes for mixed random variable – $X = 0$ with nonzero probability.

define a random variable as

$$
\begin{aligned}
X &\sim \mathcal{N}(0,1) &&\text{if heads} \\
X &= 0 &&\text{if tails}
\end{aligned}
$$

which is neither a discrete nor a continuous random variable. To find its CDF we use the law of total probability to yield

$$
\begin{aligned}
F_X(x) &= P[X \le x] \\
&= P[X \le x|\text{heads}]P[\text{heads}] + P[X \le x|\text{tails}]P[\text{tails}] \\
&= \begin{cases} \Phi(x)\frac{1}{2} + 0(\frac{1}{2}) & x < 0 \\ \Phi(x)\frac{1}{2} + 1(\frac{1}{2}) & x \ge 0 \end{cases}
\end{aligned}
$$

which can be written more succinctly using the unit step function. The unit step function is defined as $u(x) = 1$ for $x \ge 0$ and $u(x) = 0$ for $x < 0$. With this definition the CDF becomes

$$
F_X(x) = \frac{1}{2}\Phi(x) + \frac{1}{2}u(x) \qquad -\infty < x < \infty.
$$

The CDF is shown in Figure 10.26. Note the jump at $x = 0$, indicative of the contribution of the discrete part of the random variable. The CDF is continuous for all $x \ne 0$ but has a jump at $x = 0$ of $1/2$. It corresponds to neither a discrete random variable, whose CDF consists *only* of jumps, nor a continuous random variable, whose CDF is continuous everywhere. Hence, it is called a *mixed random variable*.

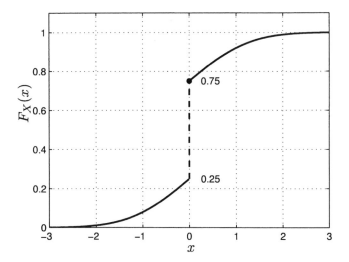

Figure 10.26: CDF for mixed random variable.

Its CDF is in general continuous except for a countable number of jumps (either finite or infinite). As usual it is right-continuous at the jump.

Strictly speaking, a mixed random variable does not have a PMF or a PDF. However, by the use of the Dirac delta function (also called an *impulse*), we can define a PDF which may then be used to find the probability of an interval via integration by using (10.4). To first find the PDF we attempt to differentiate the CDF

$$p_X(x) = \frac{d}{dx} \left[\frac{1}{2} \Phi(x) + \frac{1}{2} u(x) \right].$$

The difficulty encountered is that $u(x)$ is discontinuous at $x = 0$ and thus formally its derivative does not exist there. We can, however, *define* a derivative for the purposes of probability calculations as well as for conceptualization. To do so requires the introduction of the Dirac delta function $\delta(x)$ which is defined as (see also Appendix D)

$$\delta(x) = \frac{du(x)}{dx}.$$

The function $\delta(x)$ is usually thought of as a very narrow pulse with a very large amplitude which is centered at $x = 0$. It has the property that $\delta(t) = 0$ for all $t \neq 0$ but

$$\int_{-\epsilon}^{\epsilon} \delta(t) dt = 1$$

for ϵ a small positive number. Hence, the area under the narrow pulse is one. Using this definition we can now differentiate the CDF to find that

$$p_X(x) = \frac{1}{2} \frac{1}{\sqrt{2\pi}} \exp\left(-\frac{1}{2}x^2\right) + \frac{1}{2}\delta(x) \tag{10.34}$$

which is shown in Figure 10.27. This may be thought of as a *generalized PDF*. Note

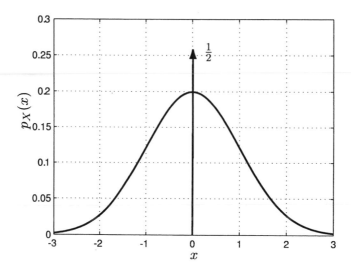

Figure 10.27: PDF for mixed random variable.

that it is the *strength*, which is defined as the area under the approximating narrow pulse, that is equal to 1/2. The amplitude is theoretically infinite. The CDF can be recovered using (10.16) and the result that

$$u(x) = \int_{-\infty}^{x^+} \delta(t)dt$$

where x^+ means that the integration interval is $(-\infty, x + \epsilon]$ for ϵ a small positive number. Thus, the impulse should be included in the integration interval if $x = 0$ so that $u(0) = 1$ according to the definition of the unit step function.

   When do we include the impulse in the integration interval?

For a mixed random variable the presence of impulses in the PDF requires a modification to (10.4). This is because an endpoint of the interval can have a nonzero probability. As a result, the probabilities $P[0 < X < 1]$ and $P[0 \leq X < 1]$ will be different if there is an impulse at $x = 0$. Specifically, consider the computation of $P[0 \leq X < 1]$ and note that the probability of $X = 0$ should be included. Therefore, if there is an impulse at $x = 0$, the area under the PDF should include the contribution of the impulse. Thus, the integration interval should be chosen as $[0^-, 1]$ so that

$$P[0 \leq X < 1] = \int_{0-}^{1} p_X(x)dx.$$

The more general modifications to (10.4) are

$$P[a \leq X \leq b] = \int_{a^-}^{b^+} p_X(x)dx$$

$$P[a < X \leq b] = \int_{a^+}^{b^+} p_X(x)dx$$

$$P[a \leq X < b] = \int_{a^-}^{b^-} p_X(x)dx$$

$$P[a < X < b] = \int_{a^+}^{b^-} p_X(x)dx$$

where x^- is a number slightly less than x and x^+ is a number slightly greater than x. Of course, if the PDF does not have any impulses at $x = a$ or $x = b$, then all the integrals above will be the same and, therefore there is no need to choose between them. See also Problem 10.51.

Continuing with our example, let's say we wish to determine $P[-2 \leq X \leq 2]$. Then, using (10.4) since the impulse does not occur at one of the interval endpoints, and our generalized PDF of (10.34) yields

$$
\begin{aligned}
P[-2 \leq X \leq 2] &= \int_{-2}^{2} p_X(x)dx \\
&= \int_{-2}^{2} \left[\frac{1}{2} \frac{1}{\sqrt{2\pi}} \exp\left(-\frac{1}{2}x^2\right) + \frac{1}{2}\delta(x) \right] dx \\
&= \frac{1}{2} \int_{-2}^{2} \frac{1}{\sqrt{2\pi}} \exp\left(-\frac{1}{2}x^2\right) dx + \frac{1}{2} \int_{-2}^{2} \delta(x)dx \\
&= \frac{1}{2} [Q(-2) - Q(2)] + \frac{1}{2} \\
&= \frac{1}{2} [1 - 2Q(2)] + \frac{1}{2} = 1 - Q(2).
\end{aligned}
$$

Alternatively, we could have obtained this result using $P[-2 \leq X \leq 2] = F_X(2) - F_X(-2)$ with $F_X(x) = (1/2)(1 - Q(x)) + (1/2)u(x)$.

Mixed random variables often arise as a result of a transformation of a continuous random variable. A final example follows.

Example 10.9 – PDF for amplitude-limited Rayleigh random variable

Consider a Rayleigh random variable whose PDF is given by (10.14) that is input to a device that limits its output. One might envision a physical quantity such as temperature and the device being a thermometer which can only read temperatures up to a maximum value. All temperatures above this maximum value are read as the maximum. Then the effect of the device can be represented by the transformation

$$y = g(x) = \begin{cases} x & 0 \le x < x_{\text{max}} \\ x_{\text{max}} & x \ge x_{\text{max}} \end{cases}$$

which is shown in Figure 10.28. The PDF of Y is zero for $y < 0$ since X can only

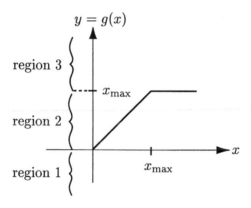

Figure 10.28: Amplitude limiter.

take on nonnegative values. For $0 \le y < x_{\text{max}}$ it is seen from Figure 10.28 that $g^{-1}(y) = y$. Finally, for $y \ge x_{\text{max}}$ we have from Figure 10.28 the infinite number of solutions $x \in [x_{\text{max}}, \infty)$. Thus, we have for region 1 or for $y < 0$ that $p_Y(y) = 0$. For region 2 or for $0 \le y < x_{\text{max}}$ where $g^{-1}(y) = y$, we have from (10.30)

$$\begin{aligned} p_Y(y) &= p_X(g^{-1}(y)) \left| \frac{dg^{-1}(y)}{dy} \right| \\ &= p_X(y). \end{aligned}$$

For region 3 which is $y \ge x_{\text{max}}$, we note that Y cannot exceed x_{max} and so $y = x_{\text{max}}$ is the only possible value for y in region 3. The probability of $Y = x_{\text{max}}$ is equal to the probability that $X \ge x_{\text{max}}$. In particular, it is

$$P[Y = x_{\text{max}}] = \int_{x_{\text{max}}}^{\infty} p_X(x)dx \tag{10.35}$$

since from Figure 10.28 the x-interval $[x_{\text{max}}, \infty)$ is mapped into the y-point given by $y = x_{\text{max}}$. Since the probability of Y at the point $y = x_{\text{max}}$ is nonzero, we represent its contribution to the PDF by using an impulse as

$$p_Y(y) = \left[\int_{x_{\text{max}}}^{\infty} p_X(x)dx \right] \delta(y - x_{\text{max}}) \qquad y = x_{\text{max}}.$$

In summary, the PDF of the transformed random variable is

$$
p_Y(y) = \begin{cases} 0 & y < 0 \\ p_X(y) & 0 \le y < x_{\max} \\ \left[\int_{x_{\max}}^{\infty} p_X(x)dx \right] \delta(y - x_{\max}) & y = x_{\max} \\ 0 & y > x_{\max} \; . \end{cases}
$$

It is seen to be the PDF of a mixed random variable in that it contains an impulse. Finally, for $x \ge 0$ the Rayleigh PDF is for $\sigma^2 = 1$

$$
p_X(x) = x \exp\left(-\frac{1}{2}x^2 \right)
$$

so that the PDF of Y becomes

$$
p_Y(y) = \begin{cases} 0 & y < 0 \\ y \exp\left(-\frac{1}{2}y^2\right) & 0 \le y < x_{\max} \\ \left[\int_{x_{\max}}^{\infty} x \exp\left(-\frac{1}{2}x^2\right) dx \right] \delta(y - x_{\max}) & y = x_{\max} \\ 0 & y > x_{\max} . \end{cases}
$$

$$
= \begin{cases} 0 & y < 0 \\ y \exp\left(-\frac{1}{2}y^2\right) & 0 \le y < x_{\max} \\ \exp\left(-\frac{1}{2}x_{\max}^2\right) \delta(y - x_{\max}) & y = x_{\max} \\ 0 & y > x_{\max}. \end{cases}
$$

This is plotted in Figure 10.29b.

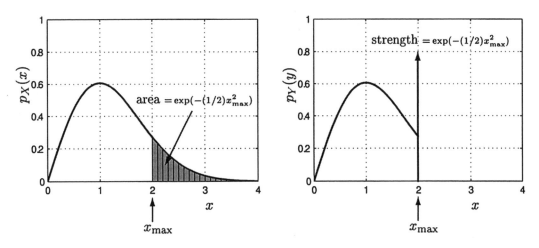

(a) PDF of X – continuous random variable   (b) PDF of $Y = g(X)$ – mixed random variable

Figure 10.29: PDFs before and after transformation of Figure 10.28.

In general, if a random variable X can take on a continuum of values as well as discrete values $\{x_1, x_2 \ldots\}$ with corresponding nonzero probabilities $\{p_1, p_2, \ldots\}$, then the PDF of the mixed random variable X can be written in the succinct form

$$p_X(x) = p_c(x) + \sum_{i=1}^{\infty} p_i \delta(x - x_i) \tag{10.36}$$

where $p_c(x)$ represents the contribution to the PDF of the continuous part (its integral must be < 1) and must satisfy $p_c(x) \geq 0$. To be a valid PDF we require that

$$\int_{-\infty}^{\infty} p_c(x) dx + \sum_{i=1}^{\infty} p_i = 1.$$

For solely discrete random variables we can use the generalized PDF

$$p_X(x) = \sum_{i=1}^{\infty} p_i \delta(x - x_i)$$

or equivalently the PMF

$$p_X[x_i] = p_i \qquad i = 1, 2, \ldots$$

to perform probability calculations.

10.9 Computer Simulation

In simulating the outcome of a discrete random variable X we saw in Figure 5.14 that first an outcome of a $U \sim \mathcal{U}(0, 1)$ random variable is generated and then mapped into a value of X. The mapping needed was the inverse of the CDF. This result is also valid for a continuous random variable so that $X = F_X^{-1}(U)$ is a random variable with CDF $F_X(x)$. Stated another way, we have that $U = F_X(X)$ or if a random variable is transformed according to its CDF, the transformed random variable $U \sim \mathcal{U}(0, 1)$. This latter transformation is termed *the probability integral transformation*. The transformation $X = F_X^{-1}(U)$ is called the *inverse probability integral transformation*. Before proving these results we give an example.

Example 10.10 – Probability integral transformation of exponential random variable

Since the exponential PDF is given for $\lambda = 1$ by

$$p_X(x) = \begin{cases} \exp(-x) & x \geq 0 \\ 0 & x < 0 \end{cases}$$

the CDF is from (10.16)

$$F_X(x) = \begin{cases} 0 & x \leq 0 \\ 1 - \exp(-x) & x > 0. \end{cases}$$

The probability integral transformation asserts that $Y = g(X) = F_X(X)$ has a $\mathcal{U}(0,1)$ PDF. Considering the transformation $g(x) = 1 - \exp(-x)$ for $x > 0$ and zero otherwise, we have that $y = 1 - \exp(-x)$ and, therefore the unique solution for x is $x = -\ln(1-y)$ for $0 < y < 1$ and zero otherwise. Hence,

$$g^{-1}(y) = \begin{cases} -\ln(1-y) & 0 < y < 1 \\ 0 & \text{otherwise} \end{cases}$$

and using (10.30), we have for $0 < y < 1$

$$\begin{aligned} p_Y(y) &= p_X(g^{-1}(y)) \left| \frac{dg^{-1}(y)}{dy} \right| \\ &= \exp\left[-(-\ln(1-y))\right] \left| \frac{1}{1-y} \right| \\ &= 1. \end{aligned}$$

Finally, then

$$p_Y(y) = \begin{cases} 1 & 0 < y < 1 \\ 0 & \text{otherwise} \end{cases}$$

which is the PDF of a $\mathcal{U}(0,1)$ random variable.

$\diamond$

To summarize our results we have the following theorem.

Theorem 10.9.1 (Inverse Probability Integral Transformation) *If a continuous random variable X is given as $X = F_X^{-1}(U)$, where $U \sim \mathcal{U}(0,1)$, then X has the PDF $p_X(x) = dF_X(x)/dx$.*

Proof:
Let $V = F_X^{-1}(U)$ and consider the CDF of V.

$$\begin{aligned} F_V(v) &= P[V \le v] = P[F_X^{-1}(U) \le v] \\ &= P[U \le F_X(v)] \qquad (F_X \text{ is monotonically increasing} - \text{see Problem 10.58}) \\ &= \int_0^{F_X(v)} p_U(u) du \\ &= \int_0^{F_X(v)} 1 \, du \\ &= F_X(v). \end{aligned}$$

Hence, the CDFs of V and X are equal and therefore the PDF of $V = F_X^{-1}(U)$ is $p_X(x)$.

$\triangle$

Another example follows.

Example 10.11 – Computer generation of outcome of Laplacian random variable

The Laplacian random variable has a PDF

$$p_X(x) = \frac{1}{\sqrt{2\sigma^2}} \exp\left[-\sqrt{\frac{2}{\sigma^2}}|x|\right] \qquad -\infty < x < \infty$$

and therefore its CDF is found as

$$F_X(x) = \int_{-\infty}^{x} \frac{1}{\sqrt{2\sigma^2}} \exp\left[-\sqrt{\frac{2}{\sigma^2}}|t|\right] dt.$$

For $x < 0$ we have

$$
\begin{aligned}
F_X(x) &= \int_{-\infty}^{x} \frac{1}{\sqrt{2\sigma^2}} \exp\left[\sqrt{\frac{2}{\sigma^2}}t\right] dt \\
&= \frac{1}{2} \exp\left[\sqrt{\frac{2}{\sigma^2}}t\right]\Bigg|_{-\infty}^{x} \\
&= \frac{1}{2} \exp\left[\sqrt{\frac{2}{\sigma^2}}x\right]
\end{aligned}
$$

and for $x \geq 0$ we have

$$
\begin{aligned}
F_X(x) &= \int_{-\infty}^{0} \frac{1}{\sqrt{2\sigma^2}} \exp\left[\sqrt{\frac{2}{\sigma^2}}t\right] dt + \int_{0}^{x} \frac{1}{\sqrt{2\sigma^2}} \exp\left[-\sqrt{\frac{2}{\sigma^2}}t\right] dt \\
&= \frac{1}{2} - \frac{1}{2} \exp\left[-\sqrt{\frac{2}{\sigma^2}}t\right]\Bigg|_{0}^{x} \qquad \text{(first integral is 1/2 since } p_X(-x) = p_X(x)) \\
&= 1 - \frac{1}{2} \exp\left[-\sqrt{\frac{2}{\sigma^2}}x\right].
\end{aligned}
$$

By letting $y = F_X(x)$, we have

$$
y = \begin{cases} \frac{1}{2} \exp\left[\sqrt{\frac{2}{\sigma^2}}x\right] & x < 0 \\ 1 - \frac{1}{2} \exp\left[-\sqrt{\frac{2}{\sigma^2}}x\right] & x \geq 0. \end{cases}
$$

We note that if $x < 0$, then $0 < y < 1/2$, and if $x \geq 0$, then $1/2 \leq y < 1$. Thus, solving for x to produce $F_X^{-1}(y)$ yields

$$
x = \begin{cases} \sqrt{\sigma^2/2}\,\ln(2y) & 0 < y < 1/2 \\ \sqrt{\sigma^2/2}\,\ln\left(\frac{1}{2(1-y)}\right) & 1/2 \leq y < 1. \end{cases}
$$

Finally to generate the outcome of a Laplacian random variable we can use

$$x = \begin{cases} \sqrt{\sigma^2/2}\ln(2u) & 0 < u < 1/2 \\ \sqrt{\sigma^2/2}\ln\left(\frac{1}{2(1-u)}\right) & 1/2 \le u < 1 \end{cases} \tag{10.37}$$

where u is a realization of a $\mathcal{U}(0,1)$ random variable. An example of the outcomes of a Laplacian random variable with $\sigma^2 = 1$ is shown in Figure 10.30a. In Figure 10.30b the true PDF (the solid curve) along with the estimated PDF (the bar plot) is shown based on $M = 1000$ outcomes. The estimate of the PDF was accomplished by

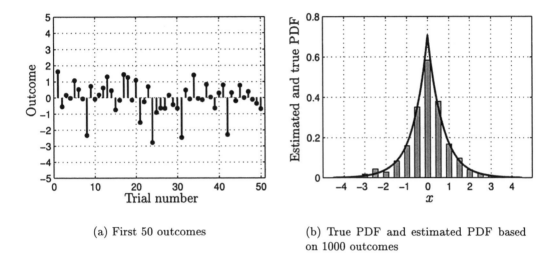

(a) First 50 outcomes

(b) True PDF and estimated PDF based on 1000 outcomes

Figure 10.30: Computer generation of Laplacian random variable outcomes using inverse probability integral transformation.

the procedure described in Example 2.1 (see Figure 2.7 for the code for a Gaussian PDF). We can now justify that procedure. Since from Section 10.3 we have

$$p_X(x_0) \approx \frac{P[x_0 - \Delta x/2 \le X \le x_0 + \Delta x/2]}{\Delta x}$$

and

$$P[x_0 - \Delta x/2 \le X \le x_0 + \Delta x/2] \approx \frac{\text{Number of outcomes in } [x_0 - \Delta x/2, x_0 + \Delta x/2]}{M}$$

we use as our PDF estimator

$$\hat{p}_X(x_0) = \frac{\text{Number of outcomes in } [x_0 - \Delta x/2, x_0 + \Delta x/2]}{M\Delta x}. \tag{10.38}$$

In Figure 10.30b we have chosen the *bins* or intervals to be $[-4.25, -3.75], [-3.75, -3.25],$ $\ldots, [3.75, 4.25]$ so that $\Delta x = 0.5$. We have therefore estimated $p_X(-4), p_X(-3.5), \ldots,$

$p_X(4)$. To estimate the PDF at more points we would have to decrease the *binwidth* or Δx. However, in doing so we cannot make it too small. This is because as the binwidth decreases, the probability of an outcome falling within the bin also decreases. As a result, fewer of the outcomes will occur within each bin, resulting in a poor estimate. The only way to remedy this situation is to increase the number of trials M. What do you suppose would happen if we wanted to estimate $p_X(5)$? The MATLAB code for producing the PDF estimate is given below.

```
% Assume outcomes are in x, which is M x 1 vector
M=1000;
bincenters=[-4:0.5:4]'; % set binwidth = 0.5
bins=length(bincenters);
h=zeros(bins,1);
for i=1:length(x)  % count outcomes in each bin
  for k=1:bins
    if x(i)>bincenters(k)-0.5/2. ...
      & x(i)<bincenters(k)+0.5/2
      h(k,1)=h(k,1)+1;
    end
  end
end
pxest=h/(M*0.5); % see (10.38)
```

The CDF can be estimated by using

$$\hat{F}_X(x) = \frac{\text{Number of outcomes} \leq x}{M} \qquad (10.39)$$

and is the same for either a discrete or a continuous random variable. See also Problems 10.60–62.

10.10   Real-World Example - Setting Clipping Levels for Speech Signals

In order to communicate speech over a transmission channel it is important to make sure that the equipment does not "clip" the speech signal. Commercial broadcast stations commonly use VU meters to monitor the power of the speech. If the power becomes too large, then the amplifier gains are manually decreased. Clipped speech sounds distorted and is objectionable. In other situations, the amplifier gains must be set automatically, as for example, in telephone speech transmission. This is necessary so that the speech, if transmitted in an analog form, is not distorted at the receiver, and if transmitted in a digital form is not clipped by an analog-to-digital convertor. To determine the highest amplitude of the speech signal that can be expected to occur a common model is to use a Laplacian PDF for the amplitudes [Rabiner and Schafer 1978]. Hence, most of the amplitudes are near zero but larger level ones are possible according to

$$p_X(x) = \frac{1}{\sqrt{2\sigma^2}} \exp\left[-\sqrt{\frac{2}{\sigma^2}}|x|\right] \qquad -\infty < x < \infty.$$

As seen in Figure 10.10, the width of the PDF increases as σ^2 increases. In effect, σ^2 measures the width of the PDF and is actually its variance (to be shown in Problem 11.34). The parameter σ^2 is also a measure of the speech power. In order to avoid excessive clipping we must be sure that an amplifier can accommodate a high level, even if it occurs rather infrequently. A design requirement might then be to transmit a speech signal without clipping 99% of the time. A model for a clipper is shown in Figure 10.31. As long as the input signal, i.e., x, remains in the interval

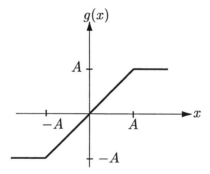

Figure 10.31: Clipper input-output characteristics.

$-A \leq x \leq A$, the output will be the same as the input and no clipping takes place. However, if $x > A$, the output will be limited to A and similarly if $x < -A$. Clipping will then occur whenever $|x| > A$. To satisfy the design requirement that clipping should not occur for 99% of the time, we should choose A (which is a characteristic of the amplifier or analog-to-digital convertor) so that $P_{\text{clip}} \leq 0.01$. But

$$P_{\text{clip}} = P[X > A \text{ or } X < -A]$$

and since the Laplacian PDF is symmetric about $x = 0$ this is just

$$
\begin{aligned}
P_{\text{clip}} &= 2P[X > A] = 2\int_A^\infty \frac{1}{\sqrt{2\sigma^2}} \exp\left[-\sqrt{\frac{2}{\sigma^2}}x\right] dx \\
&= 2\left[-\frac{1}{2}\exp\left[-\sqrt{\frac{2}{\sigma^2}}x\right]\Big|_A^\infty\right] \\
&= \exp\left[-\sqrt{\frac{2}{\sigma^2}}A\right].
\end{aligned}
\qquad (10.40)
$$

Hence, if this probability is to be no more than 0.01, we must have

$$\exp\left[-\sqrt{\frac{2}{\sigma^2}}A\right] \le 0.01$$

or solving for A produces the requirement that

$$A \ge \sqrt{\frac{\sigma^2}{2}}\ln\left(\frac{1}{0.01}\right). \qquad (10.41)$$

It is seen that as the speech power σ^2 increases, so must the clipping level A. If the clipping level is fixed, then speech with higher powers will be clipped more often. As an example, consider a speech signal with $\sigma^2 = 1$. The Laplacian model outcomes are shown in Figure 10.32 along with a clipping level of $A = 1$. According to (10.40)

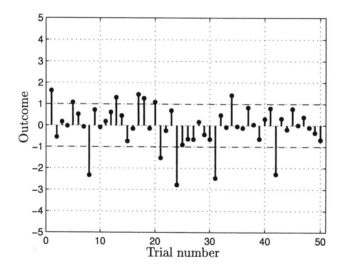

Figure 10.32: Outcomes of Laplacian random variable with $\sigma^2 = 1$ – model for speech amplitudes.

the probability of clipping is $\exp(-\sqrt{2}) = 0.2431$. Since there are 50 outcomes in Figure 10.32 we would expect about $50 \cdot 0.2431 \approx 12$ instances of clipping. From the figure we see that there are exactly 12. To meet the specification we should have that

$$A \ge \sqrt{1/2}\ln\left(\frac{1}{0.01}\right) = 3.25.$$

As seen from Figure 10.32 there are no instances of clipping for $A = 3.25$. In order to set the appropriate clipping level A, we need to know σ^2. In practice, this too must be estimated since different speakers have different volumes and even the same speaker will exhibit a different volume over time!

References

Abramowitz, M., I.A. Stegun, *Handbook of Mathematical Functions*, Dover, New York, 1965.

Capinski, M., P.E. Kopp, *Measure, Integral, and Probability*, Springer-Verlag, New York, 2004.

Johnson, N.L., S. Kotz, N. Balakrishnan, *Continuous Univariate Distributions, Vols. 1,2*, John Wiley & Sons, New York, 1994.

Members of Technical Staff, *Transmission Systems for Communications*, Western Electric Co., Inc., Winston-Salem, NC, 1970.

Rabiner, L.R., R.W. Schafer, *Digital Processing of Speech Signals*, Prentice-Hall, Englewood Cliffs, NJ, 1978.

Widder, D.A., *Advanced Calculus*, Dover, New York, 1989.

Problems

10.1 (w) Are the following random variables continuous or discrete?

    a. Temperature in degrees Fahrenheit

    b. Temperature rounded off to nearest $1°$

    c. Temperature rounded off to nearest $1/2°$

    d. Temperature rounded off to nearest $1/4°$

10.2 (☺) (w) The temperature in degrees Fahrenheit is modeled as a uniform random variable with $T \sim \mathcal{U}(20, 60)$. If T is rounded off to the nearest $1/2°$ to form $\hat{T}$, what is $P[\hat{T} = 30°]$? What can you say about the use of a PDF versus a PMF to describe the probabilistic outcome of a physical experiment?

10.3 (w) A wedge of cheese as shown in Figure 10.5 is sliced from $x = a$ to $x = b$. If $a = 0$ and $b = 0.2$, what is the mass of cheese in the wedge? How about if $a = 1.8$ and $b = 2$?

10.4 (☺) (w) Which of the following functions are valid PDFs? If a function is not a PDF, why not?

10.5 (f) Determine the value of c to make the following function a valid PDF

$$g(x) = \begin{cases} c(1 - |x/5|) & |x| < 5 \\ 0 & \text{otherwise.} \end{cases}$$

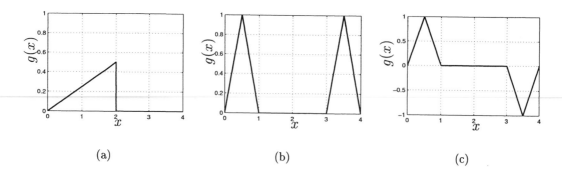

Figure 10.33: Possible PDFs for Problem 10.4.

10.6 (☺) (w) A Gaussian mixture PDF is defined as

$$p_X(x) = \alpha_1 \frac{1}{\sqrt{2\pi\sigma_1^2}} \exp\left(-\frac{1}{2\sigma_1^2}x^2\right) + \alpha_2 \frac{1}{\sqrt{2\pi\sigma_2^2}} \exp\left(-\frac{1}{2\sigma_2^2}x^2\right)$$

for $\sigma_1^2 \neq \sigma_2^2$. What are the possible values for α_1 and α_2 so that this is a valid PDF?

10.7 (w) Find the area under the curves given by the following functions:

$$g_1(x) = \begin{cases} x & 0 \leq x < 1 \\ 1+x & 1 \leq x \leq 2 \\ 0 & \text{otherwise} \end{cases}$$

$$g_2(x) = \begin{cases} x & 0 \leq x \leq 1 \\ 1+x & 1 < x \leq 2 \\ 0 & \text{otherwise} \end{cases}$$

and explain your results.

10.8 (w) A memory chip has a projected lifetime X in days that is modeled as $X \sim \exp(0.001)$. What is the probability that it will fail within one year?

10.9 (t) In this problem we prove that the Gaussian PDF integrates to one. First we let

$$I = \int_{-\infty}^{\infty} \frac{1}{\sqrt{2\pi}} \exp\left(-\frac{1}{2}x^2\right) dx$$

and write I^2 as the iterated integral

$$I^2 = \int_{-\infty}^{\infty} \int_{-\infty}^{\infty} \frac{1}{\sqrt{2\pi}} \exp\left(-\frac{1}{2}x^2\right) \frac{1}{\sqrt{2\pi}} \exp\left(-\frac{1}{2}y^2\right) dy dx.$$

Next, convert (x, y) into polar coordinates and evaluate the expression to prove that $I^2 = 1$. Finally, you can conclude that $I = 1$ (why?).

10.10 (f,c) If $X \sim \mathcal{N}(\mu, \sigma^2)$, find $P[X > \mu + a\sigma]$ for $a = 1, 2, 3$, where $\sigma = \sqrt{\sigma^2}$.

10.11 (t) The *median* of a PDF is defined as the point $x = \text{med}$ for which $P[X \leq \text{med}] = 1/2$. Prove that if $x \sim \mathcal{N}(\mu, \sigma^2)$, then $\text{med} = \mu$.

10.12 (☺) (w) A constant or DC current source that outputs 1 amp is connected to a resistor of nominal resistance of 1 ohm. If the resistance value can vary according to $R \sim \mathcal{N}(1, 0.1)$, what is the probability that the voltage across the resistor will be between 0.99 and 1.01 volts?

10.13 (w) An analog-to-digital convertor can convert voltages in the range $[-3, 3]$ volts to a digital number. Outside this range, it will "clip" a positive voltage at the highest positive level, i.e., $+3$, or a negative voltage at the most negative level, i.e., -3. If the input to the convertor is modeled as $X \sim \mathcal{N}(\mu, 1)$, how should μ be chosen to minimize the probability of clipping?

10.14 (☺) (f) Find $P[X > 3]$ for the two PDFs given by the Gaussian PDF with $\mu = 0, \sigma^2 = 1$ and the Laplacian PDF with $\sigma^2 = 1$. Which probability is larger and why? Plot both PDFs.

10.15 (f) Verify that the Cauchy PDF given in (10.9) integrates to one.

10.16 (t) Prove that $\Gamma(z + 1) = z\Gamma(z)$ by using integration by parts (see Appendix B and Problem 11.7).

10.17 (☺) (f) The arrival time in minutes of the Nth person at a ticket counter has a PDF that is Erlang with $\lambda = 0.1$. What is the probability that the first person will arrive within the first 5 minutes of the opening of the ticket counter? What is the probability that the first two persons will arrive within the first 5 minutes of opening?

10.18 (f) A person cuts off a wedge of cheese as shown in Figure 10.5 starting at $x = 0$ and ending at some value $x = x_0$. Determine the mass of the wedge as a function of the value x_0. Can you relate this to the CDF?

10.19 (☺) (f) Determine the CDF for the Cauchy PDF.

10.20 (f) If $X \sim \mathcal{N}(0, 1)$ find the probability that $|X| \leq a$, where $a = 1, 2, 3$. Also, plot the PDF and shade in the corresponding areas under the PDF.

10.21 (f,c) If $X \sim \mathcal{N}(0, 1)$, determine the number of outcomes out of 1000 that you would expect to occur within the interval $[1, 2]$. Next conduct a computer simulation to carry out this experiment. How many outcomes actually occur within this interval?

10.22 (☺) (w) If $X \sim \mathcal{N}(\mu, \sigma^2)$, find the CDF of X in terms of $\Phi(x)$.

10.23 (t) If a PDF is symmetric about $x = 0$ (also called an *even function*), prove that $F_X(-x) = 1 - F_X(x)$. Does this property hold for a Gaussian PDF with $\mu = 0$? Hint: See Figure 10.16.

10.24 (t) Prove that if $X \sim \mathcal{N}(\mu, \sigma^2)$, then

$$P[X > a] = Q\left(\frac{a - \mu}{\sigma}\right)$$

where $\sigma = \sqrt{\sigma^2}$.

10.25 (t) Prove the properties of the Q function given by (10.19)–(10.22).

10.26 (f) Plot the function $Q(A/2)$ versus A for $0 \leq A \leq 5$ to verify the true probability of error as shown in Figure 2.15.

10.27 (c) If $X \sim \mathcal{N}(0, 1)$, evaluate $P[X > 4]$ and then verify your results using a computer simulation. How easy do you think it would be to determine $P[X > 7]$ using a computer simulation? (See Section 11.10 for an alternative approach.)

10.28 (☺) (w) A survey is taken of the incomes of a large number of people in a city. It is determined that the income in dollars is distributed as $X \sim \mathcal{N}(50000, 10^8)$. What percentage of the people have incomes above \$70,000?

10.29 (w) In Chapter 1 an example was given of the length of time in minutes an office worker spends on the telephone in a given 10-minute period. The length of time T was given as $\mathcal{N}(7, 1)$ as shown in Figure 1.5. Determine the probability that a caller is on the telephone more than 8 minutes by finding $P[T > 8]$.

10.30 (☺) (w) A population of high school students in the eastern United States score X points on their SATs, where $X \sim \mathcal{N}(500, 4900)$. A similar population in the western United States score X points, where $X \sim \mathcal{N}(525, 3600)$. Which group is more likely to have scores above 700?

10.31 (f) Verify the numerical results given in (1.3).

10.32 (f) In Example 2.2 we asserted that $P[X > 2]$ for a standard normal random variable is 0.0228. Verify this result.

10.33 (☺) (w) Is the following function a valid CDF?

$$F_X(x) = \frac{1}{1 + \exp(-x)} \qquad -\infty < x < \infty.$$

10.34 (f) If $F_X(x) = (2/\pi)\arctan(x)$ for $0 \leq x < \infty$, determine $P[0 \leq X \leq 1]$.

10.35 (t) Prove that (10.25) is true.

10.36 (⌣) (w) Professor Staff always scales his test scores. He adds a number of points c to each score so that 50% of the class get a grade of C. A C is given if the score is between 70 and 80. If the scores have the distribution $\mathcal{N}(65, 38)$, what should c be? Hint: There are two possible solutions to this problem but the students will prefer only one of them.

10.37 (w) A Rhode Island weatherman says that he can accurately predict the temperature for the following day 95% of the time. He makes his prediction by saying that the temperature will be between T_1° Fahrenheit and T_2° Fahrenheit. If he knows that the actual temperature is a random variable with PDF $\mathcal{N}(50, 10)$, what should his prediction be for the next day?

10.38 (f) For the CDF given in Figure 10.14 find the PDF by differentiating. What happens at $x = 1$ and $x = 2$?

10.39 (f,c) If $Y = \exp(X)$, where $X \sim \mathcal{U}(0, 1)$, find the PDF of Y. Next generate realizations of X on a computer and transform them according to $\exp(X)$ to yield the realizations of Y. Plot the x's and y's in a similar manner to that shown in Figure 10.22 and discuss your results.

10.40 (⌣) (f) Find the PDF of $Y = X^4 + 1$ if $X \sim \exp(\lambda)$.

10.41 (w) Find the constants a and b so that $Y = aX + b$, where $X \sim \mathcal{U}(0, 1)$, yields $Y \sim \mathcal{U}(2, 6)$.

10.42 (f) If $Y = aX$, find the PDF of Y if the PDF of X is $p_X(x)$. Next, assume that $X \sim \exp(1)$ and find the PDFs of Y for $a > 1$ and $0 < a < 1$. Plot these PDFs and explain your results.

10.43 (⌣) (f) Find a general formula for the PDF of $Y = |X|$. Next, evaluate your formula if X is a standard normal random variable.

10.44 (f) If $X \sim \mathcal{N}(0, 1)$ is transformed according to $Y = \exp(X)$, determine $p_Y(y)$ by using the CDF approach. Compare your results to those given in Example 10.6. Hint: You will need Leibnitz's rule

$$\frac{d}{dy} \int_a^{g(y)} p(x)dx = p(g(y))\frac{dg(y)}{dy}.$$

10.45 (w) A random voltage X is input to a full wave rectifier that produces at its output the absolute value of the voltage. If X is a standard normal random variable, what is the probability that the output of the rectifier will exceed 2?

10.46 (☺) (f,c) If $Y = X^2$, where $X \sim \mathcal{U}(0,1)$, determine the PDF of Y. Next perform a computer simulation using the realizations of Y (obtained as $y_m = x_m^2$, where x_m is the mth realization of X) to estimate the PDF $p_Y(y)$. Do your theoretical results match the simulated results?

10.47 (w) If a discrete random variable X has a Ber(p) *PMF*, find the *PDF* of X using impulses. Next find the CDF of X by integrating the PDF.

10.48 (t) In this problem we point out that the use of impulses or Dirac delta functions serves mainly as a tool to allow sums to be written as integrals. For example, the sum

$$S = \sum_{i=1}^{N} a_i$$

can be written as the integral

$$S = \int_{-\infty}^{\infty} g(x)dx$$

if we define $g(x)$ *as*

$$g(x) = \sum_{i=1}^{N} a_i \delta(x - i).$$

Verify that this is true and show how it applies to computing probabilities of events of discrete random variables by using integration.

10.49 (f) Evaluate the expression

$$\int_{1}^{2^-} \left(\frac{1}{2}\delta(x - 2) + \frac{3}{8}\delta(x - 4) + \frac{1}{8}\delta(x - 3/2) \right) dx.$$

Could the integrand represent a PDF? If it does, what does this integral represent?

10.50 (w) Plot the PDF and CDF if

$$p_X(x) = \frac{1}{2} \exp(-x)u(x) + \frac{1}{4}\delta(x + 1) + \frac{1}{4}\delta(x - 1).$$

10.51 (☺) (w) For the PDF given in Problem 10.50 determine the following: $P[-2 \le X \le 2]$, $P[-1 \le X \le 1]$, $P[-1 < X \le 1]$, $P[-1 < X < 1]$, $P[-1 \le X < 1]$.

10.52 (f) Find and plot the PDF of the transformed random variable

$$Y = \begin{cases} 2X & 0 \le X < 1 \\ 2 & X \ge 1 \end{cases}$$

where $X \sim \exp(1)$.

10.53 (f) Find the PDF representation of the PMF of a bin$(3, 1/2)$ random variable. Plot the PMF and the PDF.

10.54 (☺) (f) Determine the function g so that $X = g(U)$, where $U \sim \mathcal{U}(0, 1)$, has a Rayleigh PDF with $\sigma^2 = 1$.

10.55 (f) Find a transformation so that $X = g(U)$, where $U \sim \mathcal{U}(0, 1)$, has the PDF shown in Figure 10.34.

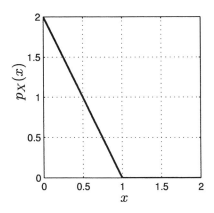

Figure 10.34: PDF for Problem 10.55

10.56 (c) Verify your results in Problem 10.55 by generating realizations of the random variable whose PDF is shown in Figure 10.34. Next estimate the PDF and compare it to the true PDF.

10.57 (t) A monotonically increasing function $g(x)$ is defined as one for which if $x_2 \geq x_1$, then $g(x_2) \geq g(x_1)$. A monotonically decreasing function is one for which if $x_2 \geq x_1$, then $g(x_2) \leq g(x_1)$. It can be shown that if $g(x)$ is differentiable, then a function is monotonically increasing (decreasing) if $dg(x)/dx \geq 0$ $(dg(x)/dx \leq 0)$ for all x. Which of the following functions are monotonically increasing or decreasing: $\exp(x)$, $\ln(x)$, and $1/x$?

10.58 (t) Explain why the values of x for which the inequality $x \geq x_0$ is true do not change if we take the logarithm of both sides to yield $\ln(x) \geq \ln(x_0)$. Would the inequality still hold if we inverted both sides or equivalently applied the function $g(x) = 1/x$ to both sides? Hint: See Problem 10.57.

10.59 (w) Compare the true PDF given in Figure 10.24 with the estimated PDF shown in Figure 2.10. Are they the same and if not, why not?

10.60 (c) Generate on the computer realizations of the random variable $X \sim \mathcal{N}(1, 4)$. Estimate the PDF and compare it to the true one.

10.61 (c) Determine the PDF of $Y = X^3$ if $X \sim \mathcal{U}(0,1)$. Next generate realizations of X on the computer, apply the transformation $g(x) = x^3$ to each realization to yield realizations of Y, and finally estimate the PDF of Y from these realizations. Does it agree with the true PDF?

10.62 (c) For the random variable Y described in Problem 10.61 determine the CDF. Then, generate realizations of Y, estimate the CDF, and compare it to the true one.

Appendix 10A

Derivation of PDF of a Transformed Continuous Random Variable

The proof uses the CDF approach as described in Section 10.7. It assumes that g is a one-to-one function. If $Y = g(X)$, where g is a one-to-one and monotonically *increasing* function, then there is a single solution for x in $y = g(x)$. Thus,

$$
\begin{aligned}
F_Y(y) &= P[g(X) \le y] \\
&= P[X \le g^{-1}(y)] \\
&= F_X(g^{-1}(y)).
\end{aligned}
$$

But $p_Y(y) = dF_Y(y)/dy$ so that

$$
\begin{aligned}
p_Y(y) &= \frac{d}{dy} F_X(g^{-1}(y)) \\
&= \frac{d\,F_X(x)}{dx}\bigg|_{x=g^{-1}(y)} \frac{dg^{-1}(y)}{dy} \quad \text{(chain rule of calculus)} \\
&= p_X(g^{-1}(y)) \frac{dg^{-1}(y)}{dy}.
\end{aligned}
$$

If $g(x)$ is one-to-one and monotonically *decreasing*, then

$$
\begin{aligned}
F_Y(y) &= P[g(X) \le y] \\
&= P[X \ge g^{-1}(y)] \\
&= 1 - P[X \le g^{-1}(y)] \quad \text{(since } P[X = g^{-1}(y)] = 0) \\
&= 1 - F_X(g^{-1}(y))
\end{aligned}
$$

and

$$\begin{aligned} p_Y(y) &= \frac{dF_Y(y)}{dy} = -\frac{d}{dy}F_X(g^{-1}(y)) \\ &= -p_X(g^{-1}(y))\frac{dg^{-1}(y)}{dy}. \end{aligned}$$

Note that if g is montonically decreasing, then g^{-1} is also montonically decreasing. Hence, $dg^{-1}(y)/dy$ will be negative. Thus, both cases can be subsumed by the formula

$$p_Y(y) = p_X(g^{-1}(y))\left|\frac{dg^{-1}(y)}{dy}\right|.$$

Appendix 10B

MATLAB Subprograms to Compute Q and Inverse Q Functions

```
%  Q.m
%
%  This program computes the right-tail probability
%  (complementary cumulative distribution function) for
%  a N(0,1) random variable.
%
%  Input Parameters:
%
%    x - Real column vector of x values
%
%  Output Parameters:
%
%    y - Real column vector of right-tail probabilities
%
%  Verification Test Case:
%
%  The input x=[0 1 2]'; should produce y=[0.5 0.1587 0.0228]'.
%
   function y=Q(x)
   y=0.5*erfc(x/sqrt(2));  % complementary error function
```

```
%  Qinv.m
```

```
%
% This program computes the inverse Q function or the value
% which is exceeded by a N(0,1) random variable with a
% probability of x.
%
% Input Parameters:
%
%    x - Real column vector of right-tail probabilities
%        (in interval [0,1])
%
% Output Parameters:
%
%    y - Real column vector of values of random variable
%
% Verification Test Case:
%
% The input x=[0.5 0.1587 0.0228]'; should produce
% y=[0 0.9998 1.9991]'.
%
    function y=Qinv(x)
    y=sqrt(2)*erfinv(1-2*x);  % inverse error function
```

Chapter 11

Expected Values for Continuous Random Variables

11.1 Introduction

We now define the expectation of a continuous random variable. In doing so we parallel the discussion of expected values for discrete random variables given in Chapter 6. Based on the probability density function (PDF) description of a continuous random variable, the expected value is defined and its properties explored. The discussion is conceptually much the same as before, only the particular method of evaluating the expected value is different. Hence, we will concentrate on the manipulations required to obtain the expected value.

11.2 Summary

The expected value $E[X]$ for a continuous random variable is motivated from the analogous definition for a discrete random variable in Section 11.3. Its definition is given by (11.3). An analogy with the center of mass of a wedge is also described. For the expected value to exist we must have $E[|X|] < \infty$ or the expected value of the absolute value of the random variable must be finite. The expected values for the common continuous random variables are given in Section 11.4 with a summary given in Table 11.1. The expected value of a function of a continuous random variable can be easily found using (11.10), eliminating the need to find the PDF of the transformed random variable. The expectation is shown to be linear in Example 11.2. For a mixed random variable the expectation is computed using (11.11). The variance is defined by (11.12) with some examples given in Section 11.6. It has the same properties as for a discrete random variable, some of which are given in (11.13), and is a nonlinear operation. The moments of a continuous random variable are defined as $E[X^n]$ and can be found either by using a direct integral evaluation as in Example 11.6 or by using the characteristic function (11.18). The characteristic function is the Fourier transform of the PDF as given by (11.17). Central moments, which are the moments about the mean, are related to the moments by (11.15). The second central moment is just the variance. Although the probability of an event cannot in general be determined from the mean and variance, the Chebyshev inequality of (11.21) provides a formula for bounding the probability. The mean and variance can be estimated using (11.22) and (11.23). Finally, an application of mean estimation to test highly reliable software is described in Section 11.10. It is based on importance sampling, which provides a means of estimating small probabilities with a reasonable number of Monte Carlo trials.

11.3   Determining the Expected Value

The expected value for a discrete random variable X was defined in Chapter 6 to be

$$E[X] = \sum_i x_i p_X[x_i] \tag{11.1}$$

where $p_X[x_i]$ is the probability mass function (PMF) of X and the sum is over all i for which the PMF $p_X[x_i]$ is nonzero. In the case of a continuous random variable, the sample space $\mathcal{S}_X$ is not countable and hence (11.1) can no longer be used. For example, if $X \sim \mathcal{U}(0,1)$, then X can take on any value in the interval $(0,1)$, which consists of an uncountable number of values. We might expect that the average value is $E[X] = 1/2$ since the probability of X being in any *equal length* interval in $(0,1)$ is the same. To verify this conjecture we employ the same strategy used previously, that of approximating a uniform PDF by a uniform *PMF*, using a fine partitioning of the interval $(0,1)$. Letting

$$p_X[x_i] = \frac{1}{M} \qquad x_i = i\Delta x$$

for $i = 1, 2, \ldots, M$ and with $\Delta x = 1/M$, we have from (11.1)

$$\begin{aligned}
E[X] &= \sum_{i=1}^{M} x_i p_X[x_i] = \sum_{i=1}^{M} (i\Delta x)\left(\frac{1}{M}\right) \\
&= \sum_{i=1}^{M} \frac{i}{M^2} = \frac{1}{M^2}\sum_{i=1}^{M} i.
\end{aligned} \tag{11.2}$$

But $\sum_{i=1}^{M} i = (M/2)(M+1)$ so that

$$E[X] = \frac{\frac{M}{2}(M+1)}{M^2} = \frac{1}{2} + \frac{1}{2M}$$

and as $M \to \infty$ or the partition of $(0, 1)$ becomes infinitely fine, we have $E[X] \to 1/2$, as expected. To extend these results to more general PDFs we first note from (11.2) that

$$
\begin{aligned}
E[X] &= \sum_{i=1}^{M} x_i P[x_i - \Delta x/2 \leq X \leq x_i + \Delta x/2] \\
&= \sum_{i=1}^{M} x_i \frac{P[x_i - \Delta x/2 \leq X \leq x_i + \Delta x/2]}{\Delta x} \Delta x.
\end{aligned}
$$

But

$$
\frac{P[x_i - \Delta x/2 \leq X \leq x_i + \Delta x/2]}{\Delta x} = \frac{1/M}{\Delta x} = 1
$$

and as $\Delta x \to 0$, this is the probability per unit length for all small intervals centered about x_i, which is the *PDF* evaluated at $x = x_i$. In this example, $p_X(x_i)$ does not depend on the interval center, which is x_i, so that the PDF is uniform or $p_X(x) = 1$ for $0 < x < 1$. Thus, as $\Delta x \to 0$

$$
E[X] \to \sum_{i=1}^{M} x_i p_X(x_i) \Delta x
$$

and this becomes the integral

$$
E[X] = \int_0^1 x p_X(x) dx
$$

where $p_X(x) = 1$ for $0 < x < 1$ and is zero otherwise. To confirm that this integral produces a result consistent with our earlier value of $E[X] = 1/2$, we have

$$
\begin{aligned}
E[X] &= \int_0^1 x p_X(x) dx \\
&= \int_0^1 x \cdot 1 dx = \frac{1}{2} x^2 \Big|_0^1 = \frac{1}{2}.
\end{aligned}
$$

In general, the expected value for a continuous random variable X is defined as

$$
E[X] = \int_{-\infty}^{\infty} x p_X(x) dx \tag{11.3}
$$

where $p_X(x)$ is the PDF of X. Another example follows.

Example 11.1 – Expected value for random variable with a nonuniform PDF

Consider the computation of the expected value for the PDF shown in Figure 11.1a. From the PDF and some typical outcomes shown in Figure 11.1b the expected value

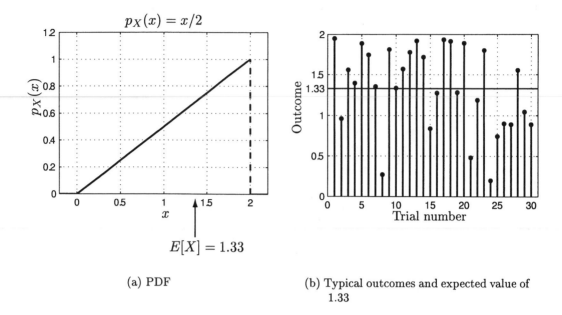

(a) PDF   (b) Typical outcomes and expected value of
                                         1.33

Figure 11.1: Example of nonuniform PDF and its mean.

should be between 1 and 2. Using (11.3) we have

$$E[X] = \int_0^2 x \left(\frac{1}{2}x\right) dx$$
$$= \frac{1}{6} x^3 \Big|_0^2 = \frac{4}{3}$$

which appears to be reasonable.

$\diamondsuit$

As an analogy to the expected value we can revisit our Jarlsberg cheese first described in Section 10.3, and which is shown in Figure 11.2. The integral

$$\text{CM} = \int_0^2 x m(x) dx \tag{11.4}$$

is the *center of mass*, assuming that the total mass or $\int_0^2 m(x)dx$, is one. Here, $m(x)$ is the linear mass density or mass per unit length. The center of mass is the point at which one could balance the cheese on the point of a pencil. Recall that the linear mass density is $m(x) = x/2$ for which CM = 4/3 from Example 11.1. To show that CM is the balance point we first note that $\int_0^2 m(x)dx = 1$ so that we can

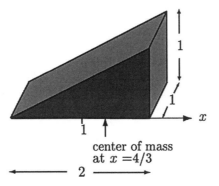

Figure 11.2: Center of mass (CM) analogy to average value.

write (11.4) as

$$\int_0^2 xm(x)dx - \text{CM} = 0$$

$$\int_0^2 xm(x)dx - \text{CM} \int_0^2 m(x)dx = 0$$

$$\int_0^2 \underbrace{(x - \text{CM})}_{\text{moment arm}} \underbrace{m(x)dx}_{\text{mass}} = 0.$$
$$\underbrace{\phantom{\int_0^2 (x - \text{CM}) m(x)dx}}_{\text{sum}}$$

Since the "sum" of the mass times moment arms is zero, the cheese is balanced at $x = \text{CM} = 4/3$.

By the same argument the expected value can also be found by solving

$$\int_{-\infty}^{\infty} (x - E[X])p_X(x)dx = 0 \tag{11.5}$$

for $E[X]$. If, however, the PDF is symmetric about some point $x = a$, which is to say that $p_X(a + u) = p_X(a - u)$ for $-\infty < u < \infty$, then (see Problem 11.2)

$$\int_{-\infty}^{\infty} (x - a)p_X(x)dx = 0 \tag{11.6}$$

and therefore $E[X] = a$. Such was the case for $X \sim \mathcal{U}(0, 1)$, whose PDF is symmetric about $a = 1/2$. Another example is the Gaussian PDF which is symmetric about $a = \mu$ as seen in Figures 10.8a and 10.8c. Hence, $E[X] = \mu$ for a Gaussian random variable (see also the next section for a direct derivation). In summary, if the PDF is symmetric about a point, then that point is $E[X]$. However, the PDF need not be symmetric about any point as in Example 11.1.

Not all PDFs have expected values.

Before computing the expected value of a random variable using (11.3) we must make sure that it exists (see similar discussion in Section 6.4 for discrete random variables). Not all integrals of the form $\int_{-\infty}^{\infty} x p_X(x) dx$ exist, even if $\int_{-\infty}^{\infty} p_X(x) dx = 1$. For example, if

$$p_X(x) = \begin{cases} \dfrac{1}{2x^{3/2}} & x \geq 1 \\ 0 & x < 1 \end{cases}$$

then

$$\int_1^\infty \frac{1}{2} x^{-3/2} dx = -\frac{1}{\sqrt{x}} \Big|_1^\infty = 1$$

but

$$\int_1^\infty x \frac{1}{2} x^{-3/2} dx = \sqrt{x} \Big|_1^\infty \to \infty.$$

A more subtle and somewhat surprising example is the Cauchy PDF. Recall that it is given by

$$p_X(x) = \frac{1}{\pi(1 + x^2)} \qquad -\infty < x < \infty.$$

Since the PDF is symmetric about $x = 0$, we would expect that $E[X] = 0$. However, if we are careful about our definition of expected value by correctly interpreting the region of integration in a limiting sense, we would have

$$E[X] = \lim_{L \to -\infty} \int_L^0 x p_X(x) dx + \lim_{U \to \infty} \int_0^U x p_X(x) dx.$$

But for a Cauchy PDF

$$\begin{aligned} E[X] &= \lim_{L \to -\infty} \int_L^0 x \frac{1}{\pi(1+x^2)} dx + \lim_{U \to \infty} \int_0^U x \frac{1}{\pi(1+x^2)} dx \\ &= \lim_{L \to -\infty} \frac{1}{2\pi} \ln(1+x^2) \Big|_L^0 + \lim_{U \to \infty} \frac{1}{2\pi} \ln(1+x^2) \Big|_0^U \\ &= \lim_{L \to -\infty} -\frac{1}{2\pi} \ln(1+L^2) + \lim_{U \to \infty} \frac{1}{2\pi} \ln(1+U^2) \\ &= -\infty + \infty = ? \end{aligned}$$

Hence, if the limits are taken independently, then the result is indeterminate. To make the expected value useful in practice the independent choice of limits (and not $L = U$) is necessary. The indeterminancy can be avoided, however, if we require "absolute convergence" or

$$\int_{-\infty}^{\infty} |x| p_X(x) dx < \infty.$$

Hence, $E[X]$ is defined to exist if $E[|X|] < \infty$. This surprising result can be "verified" by a computer simulation, the results of which are shown in Figure 11.3. In

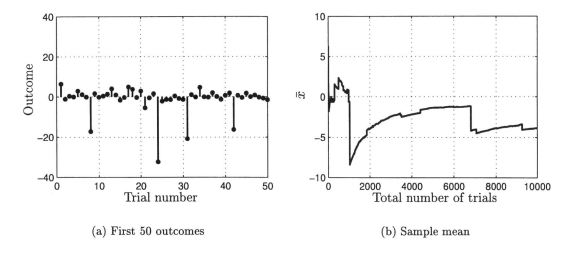

(a) First 50 outcomes   (b) Sample mean

Figure 11.3: Illustration of nonexistence of Cauchy PDF mean.

Figure 11.3a the first 50 outcomes of a total of 10,000 are shown. Because of the slow decay of the "tails" of the PDF or since the PDF decays only as $1/x^2$, very large outcomes are possible. As seen in Figure 11.3b the sample mean does not converge to zero as might be expected because of these infrequent but very large outcomes (see also Problem 12.41). See also Problem 11.3 on the simulation used in this example and Problems 11.4 and 11.9 on how to make the sample mean converge by truncating the PDF.

Finally, as for discrete random variables the expected value is the best guess of the outcome of the random variable. By "best" we mean that the use of $b = E[X]$ as our estimator. This estimator minimizes the mean square error, which is defined as mse $= E[(X - b)^2]$ (see Problem 11.5).

11.4 Expected Values for Important PDFs

We now determine the expected values for the important PDFs described in Chapter 10. Of course, the Cauchy PDF is omitted.

11.4.1 Uniform

If $X \sim \mathcal{U}(a, b)$, then it is easy to prove that $E[X] = (a + b)/2$ or the mean lies at the midpoint of the interval (see Problem 11.8).

11.4.2 Exponential

If $X \sim \exp(\lambda)$, then

$$
\begin{aligned}
E[X] &= \int_0^\infty x\lambda \exp(-\lambda x)dx \\
&= \left[-x \exp(-\lambda x) - \frac{1}{\lambda} \exp(-\lambda x) \right]\Big|_0^\infty = \frac{1}{\lambda}.
\end{aligned}
\tag{11.7}
$$

Recall that the exponential PDF spreads out as λ decreases (see Figure 10.6) and hence so does the mean.

11.4.3 Gaussian or Normal

If $X \sim \mathcal{N}(\mu, \sigma^2)$, then since the PDF is symmetric about the point $x = \mu$, we know that $E[X] = \mu$. A direct appeal to the definition of the expected value yields

$$
\begin{aligned}
E[X] &= \int_{-\infty}^\infty x \frac{1}{\sqrt{2\pi\sigma^2}} \exp\left[-\frac{1}{2\sigma^2}(x-\mu)^2 \right] dx \\
&= \int_{-\infty}^\infty (x-\mu) \frac{1}{\sqrt{2\pi\sigma^2}} \exp\left[-\frac{1}{2\sigma^2}(x-\mu)^2 \right] dx \\
&\quad + \int_{-\infty}^\infty \mu \frac{1}{\sqrt{2\pi\sigma^2}} \exp\left[-\frac{1}{2\sigma^2}(x-\mu)^2 \right] dx.
\end{aligned}
$$

Letting $u = x - \mu$ in the first integral we have

$$
E[X] = \underbrace{\int_{-\infty}^\infty u \frac{1}{\sqrt{2\pi\sigma^2}} \exp\left[-\frac{1}{2\sigma^2}u^2 \right] du}_{0} + \mu \underbrace{\int_{-\infty}^\infty \frac{1}{\sqrt{2\pi\sigma^2}} \exp\left[-\frac{1}{2\sigma^2}(x-\mu)^2 \right] dx}_{=1} = \mu.
$$

The first integral is zero since the integrand is an odd function $(g(-u) = -g(u)$, see also Problem 11.6) and the second integral is one since it is the total area under the Gaussian PDF.

11.4.4 Laplacian

The Laplacian PDF is given by

$$
p_X(x) = \frac{1}{\sqrt{2\sigma^2}} \exp\left[-\sqrt{\frac{2}{\sigma^2}}|x| \right] \qquad -\infty < x < \infty
\tag{11.8}
$$

and since it is symmetric about $x = 0$ (and the expected value *exists* – needed to avoid the situation of the Cauchy PDF), we must have $E[X] = 0$.

11.4.5 Gamma

If $X \sim \Gamma(\alpha, \lambda)$, then from (10.10)

$$E[X] = \int_0^\infty x \frac{\lambda^\alpha}{\Gamma(\alpha)} x^{\alpha-1} \exp(-\lambda x) dx.$$

To evaluate this integral we attempt to modify the integrand so that it becomes the PDF of a $\Gamma(\alpha', \lambda')$ random variable. Then, we can immediately equate the integral to one. Using this strategy

$$
\begin{aligned}
E[X] &= \frac{\lambda^\alpha}{\Gamma(\alpha)} \int_0^\infty \frac{\lambda^{\alpha+1}}{\Gamma(\alpha+1)} x^\alpha \exp(-\lambda x) dx \frac{\Gamma(\alpha+1)}{\lambda^{\alpha+1}} \\
&= \frac{\Gamma(\alpha+1)}{\lambda \Gamma(\alpha)} \qquad \text{(integrand is } \Gamma(\alpha+1, \lambda) \text{ PDF)} \\
&= \frac{\alpha \Gamma(\alpha)}{\lambda \Gamma(\alpha)} \qquad \text{(using Property 10.3)} \\
&= \frac{\alpha}{\lambda}.
\end{aligned}
$$

11.4.6 Rayleigh

It can be shown that $E[X] = \sqrt{(\pi \sigma^2)/2}$ (see Problem 11.16).

The reader should indicate on Figures 10.6–10.10, 10.12, and 10.13 where the mean occurs.

11.5 Expected Value for a Function of a Random Variable

If $Y = g(X)$, where X is a continuous random variable, then assuming that Y is also a continuous random variable with PDF $p_Y(y)$, we have by the definition of expected value of a continuous random variable

$$E[Y] = \int_{-\infty}^\infty y p_Y(y) dy. \tag{11.9}$$

Even if Y is a mixed random variable, its expected value is still given by (11.9), although in this case $p_Y(y)$ will contain impulses. Such would be the case if for example, $Y = \max(0, X)$ for X taking on values $-\infty < x < \infty$ (see Section 10.8). As in the case of a discrete random variable, it is not necessary to use (11.9) directly, which requires us to first determine $p_Y(y)$ from $p_X(x)$. Instead, we can use for $Y = g(X)$ the formula

$$E[g(X)] = \int_{-\infty}^\infty g(x) p_X(x) dx. \tag{11.10}$$

A partial proof of this formula is given in Appendix 11A. Some examples of its use follows.

Example 11.2 – Expectation of linear (affine) function
If $Y = aX + b$, then since $g(x) = ax + b$, we have from (11.10) that

$$
\begin{aligned}
E[g(X)] &= \int_{-\infty}^{\infty} (ax + b)p_X(x)dx \\
&= a\int_{-\infty}^{\infty} xp_X(x)dx + b\int_{-\infty}^{\infty} p_X(x)dx \\
&= aE[X] + b
\end{aligned}
$$

or equivalently

$$
E[aX + b] = aE[X] + b.
$$

It indicates how to easily change the expectation or mean of a random variable. For example, to increase the mean value by b just replace X by $X + b$. More generally, it is easily shown that

$$
E[a_1g_1(X) + a_2g_2(X)] = a_1E[g_1(X)] + a_2E[g_2(X)].
$$

This says that the expectation operator is *linear*.

$\Diamond$

Example 11.3 – Power of $\mathcal{N}(0,1)$ random variable
If $X \sim \mathcal{N}(0,1)$ and $Y = X^2$, consider $E[Y] = E[X^2]$. The quantity $E[X^2]$ is the average squared value of X and can be interpreted physically as a *power*. If X is a voltage across a 1 ohm resistor, then X^2 is the power and therefore $E[X^2]$ is the *average power*. Now according to (11.10)

$$
\begin{aligned}
E[X^2] &= \int_{-\infty}^{\infty} x^2 \frac{1}{\sqrt{2\pi}} \exp\left(-\frac{1}{2}x^2\right) dx \\
&= 2\int_{0}^{\infty} x^2 \frac{1}{\sqrt{2\pi}} \exp\left(-\frac{1}{2}x^2\right) dx \quad \text{(integrand is symmetric about } x = 0).
\end{aligned}
$$

To evaluate this integral we use integration by parts ($\int U dV = UV - \int V dU$, see also Problem 11.7) with $U = x$, $dU = dx$, $dV = (1/\sqrt{2\pi})x \exp[-(1/2)x^2]dx$ and therefore $V = -(1/\sqrt{2\pi})\exp[-(1/2)x^2]$ to yield

$$
\begin{aligned}
E[X^2] &= 2\left[-x\frac{1}{\sqrt{2\pi}}\exp\left(-\frac{1}{2}x^2\right)\Big|_0^{\infty} - \int_0^{\infty} -\frac{1}{\sqrt{2\pi}}\exp\left(-\frac{1}{2}x^2\right)dx\right] \\
&= 0 + 1 = 1.
\end{aligned}
$$

The first term is zero since

$$
\lim_{x\to\infty} x\exp\left(-\frac{1}{2}x^2\right) = \lim_{x\to\infty}\frac{x}{\exp\left(\frac{1}{2}x^2\right)} = \lim_{x\to\infty}\frac{1}{x\exp\left(\frac{1}{2}x^2\right)} = 0
$$

using L'Hospital's rule and the second term is evaluated using

$$\int_0^\infty \frac{1}{\sqrt{2\pi}} \exp\left(-\frac{1}{2}x^2\right) dx = \frac{1}{2} \qquad \text{(Why?)}.$$

Example 11.4 – Expected value of indicator random variable

An indicator function indicates whether a point is in a given set. For example, if the set is $A = [3, 4]$, then the indicator function is defined as

$$I_A(x) = \begin{cases} 1 & 3 \le x \le 4 \\ 0 & \text{otherwise} \end{cases}$$

and is shown in Figure 11.4. The subscript on I refers to the set of interest. The

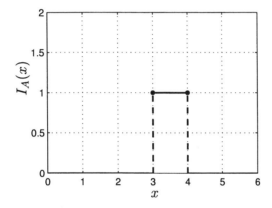

Figure 11.4: Example of indicator function for set $A = [3, 4]$.

indicator function may be thought of as a generalization of the unit step function since if $u(x) = 1$ for $x \ge 0$ and zero otherwise, we have that

$$I_{[0,\infty)}(x) = u(x).$$

Now if X is a random variable, then $I_A(X)$ is a transformed random variable that takes on values 1 and 0, depending upon whether the outcome of the experiment lies within the set A or not, respectively. (It is actually a Bernoulli random variable.) On the average, however, it has a value *between* 0 and 1, which from (11.10) is

$$
\begin{aligned}
E[I_A(x)] &= \int_{-\infty}^{\infty} I_A(x) p_X(x) dx \\
&= \int_{\{x:x\in A\}} 1 \cdot p_X(x) dx \qquad \text{(definition)} \\
&= \int_{\{x:x\in A\}} p_X(x) dx \\
&= P[A].
\end{aligned}
$$

Therefore, *the expected value of the indicator random variable is the probability of the set or event.* As an example of its utility, consider the estimation of $P[3 \leq X \leq 4]$. But this is just $E[I_A(X)]$ when $I_A(x)$ is given in Figure 11.4. To estimate the expected value of a transformed random variable we first generate the outcomes of X, say $x_1, x_2, \ldots, x_M$, then transform each one to the new random variable producing for $i = 1, 2, \ldots, M$

$$I_A(x_i) = \begin{cases} 1 & 3 \leq x_i \leq 4 \\ 0 & \text{otherwise} \end{cases}$$

and finally compute the sample mean for our estimate using

$$\widehat{E[I_A(X)]} = \frac{1}{M} \sum_{i=1}^{M} I_A(x_i).$$

However, since $P[A] = E[I_A(X)]$, we have as our estimate of the probability

$$\widehat{P[A]} = \frac{1}{M} \sum_{i=1}^{M} I_A(x_i).$$

But this is just what we have been using all along, since $\sum_{i=1}^{M} I_A(x_i)$ counts all the outcomes for which $3 \leq x \leq 4$. Thus, *the indicator function provides a means to connect the expected value with the probability.* This is a very useful for later theoretical work in probability.

$\Diamond$

Lastly, if the random variable is a mixed one with PDF

$$p_X(x) = p_c(x) + \sum_{i=1}^{\infty} p_i \delta(x - x_i)$$

where $p_c(x)$ is the continuous part of the PDF, then the expected value becomes

$$
\begin{aligned}
E[X] &= \int_{-\infty}^{\infty} x \left(p_c(x) + \sum_{i=1}^{\infty} p_i \delta(x - x_i) \right) dx \\
&= \int_{-\infty}^{\infty} x p_c(x) dx + \int_{-\infty}^{\infty} x \sum_{i=1}^{\infty} p_i \delta(x - x_i) dx \\
&= \int_{-\infty}^{\infty} x p_c(x) dx + \sum_{i=1}^{\infty} p_i \int_{-\infty}^{\infty} x \delta(x - x_i) dx \\
&= \int_{-\infty}^{\infty} x p_c(x) dx + \sum_{i=1}^{\infty} x_i p_i \quad\quad\quad (11.11)
\end{aligned}
$$

since $\int_{-\infty}^{\infty} g(x) \delta(x - x_i) dx = g(x_i)$ for $g(x)$ a function continuous at $x = x_i$. This is known as the *sifting* property of a Dirac delta function (see Appendix D). A

	Values	PDF	$E[X]$	$\text{var}(X)$	$\phi_X(\omega)$
Uniform	$a < x < b$	$\frac{1}{b-a}$	$\frac{1}{2}(a+b)$	$\frac{(b-a)^2}{12}$	$\frac{\exp(j\omega b) - \exp(j\omega a)}{j\omega(b-a)}$
Exponential	$x \geq 0$	$\lambda \exp(-\lambda x)$	$\frac{1}{\lambda}$	$\frac{1}{\lambda^2}$	$\frac{\lambda}{\lambda - j\omega}$
Gaussian	$-\infty < x < \infty$	$\frac{\exp[-(1/(2\sigma^2))(x-\mu)^2]}{\sqrt{2\pi\sigma^2}}$	μ	σ^2	$\exp[j\omega\mu - \sigma^2\omega^2/2]$
Laplacian	$-\infty < x < \infty$	$\frac{1}{\sqrt{2\sigma^2}}\exp(-\sqrt{2/\sigma^2}|x|)$	0	σ^2	$\frac{2/\sigma^2}{\omega^2 + 2/\sigma^2}$
Gamma	$x \geq 0$	$\frac{\lambda^\alpha}{\Gamma(\alpha)}x^{\alpha-1}\exp(-\lambda x)$	$\frac{\alpha}{\lambda}$	$\frac{\alpha}{\lambda^2}$	$\frac{1}{(1-j\omega/\lambda)^\alpha}$
Rayleigh	$x \geq 0$	$\frac{x}{\sigma^2}\exp[-x^2/(2\sigma^2)]$	$\sqrt{\frac{\pi\sigma^2}{2}}$	$(2-\pi/2)\sigma^2$	[Johnson et al 1994]

Table 11.1: Properties of continuous random variables.

summary of the means for the important PDFs is given in Table 11.1. Lastly, note that the expected value of a random variable can also be determined from the CDF as shown in Problem 11.28.

11.6 Variance and Moments of a Continuous Random Variable

The variance of a continuous random variable, as for a discrete random variable, measures the average squared deviation from the mean. It is defined as $\text{var}(X) = E[(X - E[X])^2]$ (exactly the same as for a discrete random variable). To evaluate the variance we use (11.10) to yield

$$\text{var}(X) = \int_{-\infty}^{\infty} (x - E[X])^2 p_X(x)dx. \tag{11.12}$$

As an example, consider a $\mathcal{N}(\mu, \sigma^2)$ random variable. In Figure 10.9 we saw that the width of the PDF increases as σ^2 increases. This is because the parameter σ^2 is actually the variance, as we now show. Using (11.12) and the definition of a Gaussian PDF

$$
\begin{aligned}
\text{var}(X) &= \int_{-\infty}^{\infty} (x - E[X])^2 \frac{1}{\sqrt{2\pi\sigma^2}} \exp\left[-\frac{1}{2\sigma^2}(x-\mu)^2\right] dx \\
&= \int_{-\infty}^{\infty} (x - \mu)^2 \frac{1}{\sqrt{2\pi\sigma^2}} \exp\left[-\frac{1}{2\sigma^2}(x-\mu)^2\right] dx \quad \text{(recall that } E[X] = \mu\text{)}.
\end{aligned}
$$

Letting $u = (x - \mu)/\sigma$ produces (recall that $\sigma = \sqrt{\sigma^2} > 0$)

$$
\begin{aligned}
\text{var}(X) &= \int_{-\infty}^{\infty} \sigma^2 u^2 \frac{1}{\sqrt{2\pi\sigma^2}} \exp\left[-\frac{1}{2\sigma^2}u^2\right] \sigma du \\
&= \sigma^2 \underbrace{\int_{-\infty}^{\infty} u^2 \frac{1}{\sqrt{2\pi}} \exp\left[-\frac{1}{2}u^2\right] du}_{=1} \qquad \text{(see Example 11.3)} \\
&= \sigma^2.
\end{aligned}
$$

Hence, we now know that a $\mathcal{N}(\mu, \sigma^2)$ random variable has a mean of μ and a variance of σ^2.

It is common to refer to the square-root of the variance as the *standard deviation*. For a $\mathcal{N}(\mu, \sigma^2)$ random variable it is given by σ. The standard deviation indicates how closely outcomes tend to cluster about the mean. (See Problem 11.29 for an alternative interpretation.) Again if the random variable is $\mathcal{N}(\mu, \sigma^2)$, then 68.2% of the outcomes will be within the interval $[\mu - \sigma, \mu + \sigma]$, 95.5% will be within $[\mu - 2\sigma, \mu + 2\sigma]$, and 99.8% will be within $[\mu - 3\sigma, \mu + 3\sigma]$. This is illustrated in Figure 11.5. Of course, other PDFs will have concentrations that are different for $E[X] \pm k\sqrt{\text{var}(X)}$. Another example follows.

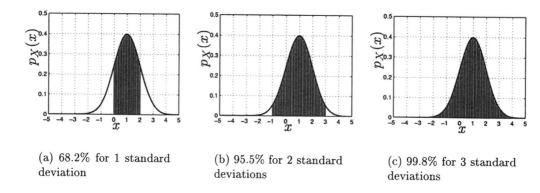

(a) 68.2% for 1 standard deviation

(b) 95.5% for 2 standard deviations

(c) 99.8% for 3 standard deviations

Figure 11.5: Percentage of outcomes of $\mathcal{N}(1, 1)$ random variable that are within $k = 1, 2,$ and 3 standard deviations from the mean. Shaded regions denote area within interval $\mu - k\sigma \le x \le \mu + k\sigma$.

Example 11.5 – Variance of a uniform random variable
If $X \sim \mathcal{U}(a, b)$, then

$$
\begin{aligned}
\text{var}(X) &= \int_{-\infty}^{\infty} (x - E[X])^2 p_X(x) dx \\
&= \int_{a}^{b} \left(x - \frac{1}{2}(a + b)\right)^2 \frac{1}{b - a} dx
\end{aligned}
$$

and letting $u = x - (a+b)/2$, we have

$$
\begin{aligned}
\text{var}(X) &= \frac{1}{b-a} \int_{-(b-a)/2}^{(b-a)/2} u^2 \, du \\
&= \frac{1}{b-a} \frac{1}{3} u^3 \Big|_{-(b-a)/2}^{(b-a)/2} \\
&= \frac{(b-a)^2}{12}.
\end{aligned}
$$

$\diamondsuit$

A summary of the variances for the important PDFs is given in Table 11.1. The variance of a continuous random variable enjoys the same properties as for a discrete random variable. Recall that an alternate form for variance computation is

$$
\text{var}(X) = E[X^2] - E^2[X]
$$

and if c is a constant then

$$
\begin{aligned}
\text{var}(c) &= 0 \\
\text{var}(X+c) &= \text{var}(X) \\
\text{var}(cX) &= c^2 \text{var}(X).
\end{aligned} \tag{11.13}
$$

Also, the variance is a nonlinear type of operation in that

$$
\text{var}(g_1(X) + g_2(X)) \neq \text{var}(g_1(X)) + \text{var}(g_2(X))
$$

(see Problem 11.32). Recall from the discussions for a discrete random variable that $E[X]$ and $E[X^2]$ are termed the *first and second moments*, respectively. In general, $E[X^n]$ is termed the nth moment and it is defined to exist if $E[|X|^n] < \infty$. If it is known that $E[X^s]$ exists, then it can be shown that $E[X^r]$ exists for $r < s$ (see Problem 6.23). This also says that *if $E[X^r]$ is known not to exist, then $E[X^s]$ cannot exist for $s > r$.* An example is the Cauchy PDF for which we saw that $E[X]$ does not exist and therefore all the higher order moments do not exist. In particular, the Cauchy PDF does not have a second-order moment and therefore its variance does not exist. We next give an example of the computation of all the moments of a PDF.

Example 11.6 – Moments of an exponential random variable

Using (11.10) we have for $X \sim \exp(\lambda)$ that

$$
E[X^n] = \int_0^\infty x^n \lambda \exp(-\lambda x) \, dx.
$$

To evalute this we first show how the nth moment can be written recursively in terms of the $(n-1)$st moment. Since we know that $E[X] = 1/\lambda$, we can then determine all the moments using the recursion. We can begin to evaluate the integral using integration by parts. This will yield the recursive formula for the moments. Letting $U = x^n$ and $dV = \lambda \exp(-\lambda x)dx$ so that $dU = nx^{n-1}dx$ and $V = -\exp(-\lambda x)$, we have

$$
\begin{aligned}
E[X^n] &= -x^n \exp(-\lambda x)\big|_0^\infty - \int_0^\infty -\exp(-\lambda x)nx^{n-1}dx \\
&= 0 + n \int_0^\infty x^{n-1} \exp(-\lambda x)dx \\
&= \frac{n}{\lambda} \int_0^\infty x^{n-1} \lambda \exp(-\lambda x)dx \\
&= \frac{n}{\lambda} E[X^{n-1}].
\end{aligned}
$$

Hence, the nth moment can be written in term of the $(n-1)$st moment. Since we know that $E[X] = 1/\lambda$, we have upon using the recursion that

$$
\begin{aligned}
E[X^2] &= \frac{2}{\lambda}E[X] = \frac{2}{\lambda}\frac{1}{\lambda} = \frac{2}{\lambda^2} \\
E[X^3] &= \frac{3}{\lambda}E[X^2] = \frac{3}{\lambda}\frac{2}{\lambda^2} = \frac{3 \cdot 2}{\lambda^3}
\end{aligned}
$$

etc.

and in general

$$
E[X^n] = \frac{n!}{\lambda^n}. \tag{11.14}
$$

The variance can be found to be $\text{var}(X) = 1/\lambda^2$ using these results.

$\diamondsuit$

In the next section we will see how to use characteristic functions to simplify the complicated integration process required for moment evaluation.

Lastly, it is sometimes important to be able to compute moments *about some point*. For example, the variance is the second moment about the point $E[X]$. In general, the nth *central moment* about the point $E[X]$ is defined as $E[(X - E[X])^n]$. The relationship between the moments and the central moments is of interest. For $n = 2$ the central moment is related to the moments by the usual formula $E[(X - E[X])^2] = E[X^2] - E^2[X]$. More generally, this relationship is found using the binomial theorem as follows.

$$
\begin{aligned}
E[(X - E[X])^n] &= E\left[\sum_{k=0}^n \binom{n}{k} X^k (-E[X])^{n-k}\right] \\
&= \sum_{k=0}^n \binom{n}{k} E[X^k](-E[X])^{n-k} \quad \text{(linearity of expectation operator)}
\end{aligned}
$$

or finally we have that

$$E[(X - E[X])^n] = \sum_{k=0}^{n} (-1)^{n-k} \binom{n}{k} (E[X])^{n-k} E[X^k]. \qquad (11.15)$$

11.7 Characteristic Functions

As first introduced for discrete random variables, the characteristic function is a valuable tool for the calculation of moments. It is defined as

$$\phi_X(\omega) = E[\exp(j\omega X)] \qquad (11.16)$$

and always exists (even though the moments of a PDF may not). For a continuous random variable it is evaluated using (11.10) for the real and imaginary parts of $E[\exp(j\omega X)]$, which are $E[\cos(\omega X)]$ and $E[\sin(\omega X)]$. This results in

$$\phi_X(\omega) = \int_{-\infty}^{\infty} \exp(j\omega x) p_X(x) dx$$

or in more familiar form as

$$\phi_X(\omega) = \int_{-\infty}^{\infty} p_X(x) \exp(j\omega x) dx. \qquad (11.17)$$

The characteristic function is seen to be the Fourier transform of the PDF, although with a $+j$ in the definition as opposed to the more common $-j$. Once the characteristic function has been found, the moments are given as

$$E[X^n] = \frac{1}{j^n} \left. \frac{d^n \phi_X(\omega)}{d\omega^n} \right|_{\omega=0}. \qquad (11.18)$$

An example follows.

Example 11.7 – Moments of the exponential PDF

Using the definition of the exponential PDF (see (10.5)) we have

$$\begin{aligned}
\phi_X(\omega) &= \int_0^{\infty} \lambda \exp(-\lambda x) \exp(j\omega x) dx \\
&= \int_0^{\infty} \lambda \exp[-(\lambda - j\omega)x] dx \\
&= \lambda \left. \frac{\exp[-(\lambda - j\omega)x]}{-(\lambda - j\omega)} \right|_0^{\infty} \\
&= -\frac{\lambda}{\lambda - j\omega} \left(\exp[-(\lambda - j\omega)\infty] - 1 \right).
\end{aligned}$$

But $\exp[-(\lambda - j\omega)x] \to 0$ as $x \to \infty$ since $\lambda > 0$ and hence we have

$$\phi_X(\omega) = \frac{\lambda}{\lambda - j\omega}. \tag{11.19}$$

To find the moments using (11.18) we need to differentiate the characteristic function n times. Proceeding to do so

$$
\begin{aligned}
\frac{d\phi_X(\omega)}{d\omega} &= \frac{d}{d\omega}\lambda(\lambda - j\omega)^{-1} \\
&= \lambda(-1)(\lambda - j\omega)^{-2}(-j) \\
\frac{d^2\phi_X(\omega)}{d\omega^2} &= \lambda(-1)(-2)(\lambda - j\omega)^{-3}(-j)^2 \\
&\;\;\vdots \\
\frac{d^n\phi_X(\omega)}{d\omega^n} &= \lambda(-1)(-2)\ldots(-n)(\lambda - j\omega)^{-n-1}(-j)^n \\
&= \lambda j^n n!(\lambda - j\omega)^{-n-1}
\end{aligned}
$$

and therefore

$$
\begin{aligned}
E[X^n] &= \frac{1}{j^n}\left.\frac{d^n\phi_X(\omega)}{d\omega^n}\right|_{\omega=0} \\
&= \lambda n!\,(\lambda - j\omega)^{-n-1}\big|_{\omega=0} \\
&= \frac{n!}{\lambda^n}
\end{aligned}
$$

which agrees with our earlier results (see (11.14)).

$\Diamond$

 Moment formula only valid if moments exist

Just because a PDF has a characteristic function, and all do, does not mean that (11.18) can be applied. For example, the Cauchy PDF has the characteristic function (see Problem 11.40)

$$\phi_X(\omega) = \exp(-|\omega|)$$

(although the derivative does not exist at $\omega = 0$). However, as we have already seen, the mean does not exist and hence all higher order moments also do not exist. Thus, no moments exist at all for the Cauchy PDF.

The characteristic function has nearly the same properties as for a discrete random variable, namely

1. The characteristic function always exists.

2. The PDF can be recovered from the characteristic function by the inverse Fourier transform, which in this case is

$$p_X(x) = \int_{-\infty}^{\infty} \phi_X(\omega) \exp(-j\omega x) \frac{d\omega}{2\pi}. \tag{11.20}$$

3. Convergence of a sequence of characteristic functions $\phi_X^{(n)}(\omega)$ for $n = 1, 2, \ldots$ to a given characteristic function $\phi(\omega)$ guarantees that the corresponding sequence of PDFs $p_X^{(n)}(x)$ for $n = 1, 2, \ldots$ converges to $p(x)$, where from (11.20)

$$p(x) = \int_{-\infty}^{\infty} \phi(\omega) \exp(-j\omega x) \frac{d\omega}{2\pi}.$$

(See Problem 11.42 for an example.) This property is also essential for proving the central limit theorem described in Chapter 15.

A slight difference from the characteristic function of a discrete random variable is that now $\phi_X(\omega)$ is *not* periodic in ω. It does, however, have the usual properties of the continuous-time Fourier transform [Jackson 1991]. A summary of the characteristic functions for the important PDFs is given in Table 11.1.

11.8 Probability, Moments, and the Chebyshev Inequality

The mean and variance of a random variable indicate the average value and variability of the outcomes of a repeated experiment. As such, they summarize important information about the PDF. However, they are not sufficient to determine probabilities of events. For example, the PDFs

$$p_X(x) = \frac{1}{\sqrt{2\pi}} \exp\left(-\frac{1}{2}x^2\right) \qquad \text{(Gaussian)}$$

$$p_X(x) = \frac{1}{\sqrt{2}} \exp\left(-\sqrt{2}|x|\right) \qquad \text{(Laplacian)}$$

both have $E[X] = 0$ (due to symmetry about $x = 0$) and $\text{var}(X) = 1$. Yet, the probability of a given interval can be very different. Although the relationship between the mean and variance, and the probability of an event is not a direct one, we can still obtain some information about the probabilities based on the mean and variance. In particular, it is possible to *bound* the probability or to be able to assert that

$$P[|X - E[X]| > \gamma] \le B$$

where B is a number less than one. This is especially useful if we only wish to make sure the probability is below a certain value, without explicitly having to find the probability. For example, if the probability of a speech signal of mean 0 and variance 1 exceeding a given magnitude γ (see Section 10.10) is to be no more than 1%, then we would be satisfied if we could determine a γ so that

$$P[|X - E[X]| > \gamma] \leq 0.01.$$

We now show that the probability for the event $|X - E[X]| > \gamma$ can be bounded if we know the mean and variance. Computation of the probability is not required and therefore *the PDF does not need to be known.* Estimating the mean and variance is much easier than the entire PDF (see Section 11.9). The inequality to be developed is called the *Chebyshev inequality.* Using the definition of the variance we have

$$
\begin{aligned}
\text{var}(X) &= \int_{-\infty}^{\infty} (x - E[X])^2 p_X(x)dx \\
&= \int_{\{x:|x-E[X]|>\gamma\}} (x - E[X])^2 p_X(x)dx + \int_{\{x:|x-E[X]|\leq\gamma\}} (x - E[X])^2 p_X(x)dx \\
&\geq \int_{\{x:|x-E[X]|>\gamma\}} (x - E[X])^2 p_X(x)dx \quad \text{(omitted integral is nonnegative)} \\
&\geq \int_{\{x:|x-E[X]|>\gamma\}} \gamma^2 p_X(x)dx \quad \text{(since for each } x, |x - E[X]| > \gamma) \\
&= \gamma^2 \int_{\{x:|x-E[X]|>\gamma\}} p_X(x)dx \\
&= \gamma^2 P[|X - E[X]| > \gamma]
\end{aligned}
$$

so that we have the Chebyshev inequality

$$P[|X - E[X]| > \gamma] \leq \frac{\text{var}(X)}{\gamma^2}. \tag{11.21}$$

Hence, the probability that a random variable deviates from its mean by more than γ (in either direction) is less than or equal to $\text{var}(X)/\gamma^2$. This agrees with our intuition in that the probability of an outcome departing from the mean must become smaller as the width of the PDF decreases or equivalently as the variance decreases. An example follows.

Example 11.8 – Bounds for different PDFs

Assuming $E[X] = 0$ and $\text{var}(X) = 1$, we have from (11.21)

$$P[|X| > \gamma] \leq \frac{1}{\gamma^2}.$$

If $\gamma = 3$, then we have that $P[|X| > 3] \leq 1/9 \approx 0.11$. This is a rather "loose" bound in that if $X \sim \mathcal{N}(0,1)$, then the actual value of this probability is $P[|X| >$

3] $= 2Q(3) = 0.0027$. Hence, the actual probability is indeed less than or equal to the bound of 0.11, but quite a bit less. In the case of a Laplacian random variable with mean 0 and variance 1, the bound is the same but the actual value is now

$$
\begin{aligned}
P[|X| > 3] &= \int_{-\infty}^{-3} \frac{1}{\sqrt{2}} \exp\left(-\sqrt{2}|x|\right) dx + \int_{3}^{\infty} \frac{1}{\sqrt{2}} \exp\left(-\sqrt{2}|x|\right) dx \\
&= 2 \int_{3}^{\infty} \frac{1}{\sqrt{2}} \exp\left(-\sqrt{2}x\right) dx \quad \text{(PDF is symmetric about } x = 0\text{)} \\
&= -\exp\left(-\sqrt{2}x\right)\Big|_{3}^{\infty} \\
&= \exp\left(-3\sqrt{2}\right) = 0.0144.
\end{aligned}
$$

Once again the bound is seen to be correct but provides a gross overestimation of the probability. A graph of the Chebyshev bound as well as the actual probabilities of $P[|X| > \gamma]$ versus γ is shown in Figure 11.6. The reader may also wish to consider

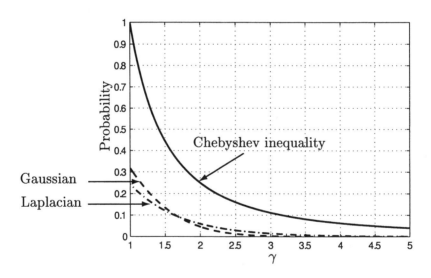

Figure 11.6: Probabilities $P[|X| > \gamma]$ for Gaussian and Laplacian random variables with zero mean and unity variance compared to Chebyshev inequality.

what would happen if we used the Chebyshev inequality to bound $P[|X| > 0.5]$ if $X \sim \mathcal{N}(0,1)$.

11.9 Estimating the Mean and Variance

The mean and variance of a continuous random variable are estimated in exactly the same way as for a discrete random variable (see Section 6.8). Assuming that we have the M outcomes $\{x_1, x_2, \ldots, x_M\}$ of a random variable X the mean estimate is

$$\widehat{E[X]} = \frac{1}{M} \sum_{i=1}^{M} x_i \tag{11.22}$$

and the variance estimate is

$$\begin{aligned}
\mathrm{var}(X) &= \widehat{E[X^2]} - \left(\widehat{E[X]}\right)^2 \\
&= \frac{1}{M} \sum_{i=1}^{M} x_i^2 - \left(\frac{1}{M} \sum_{i=1}^{M} x_i\right)^2.
\end{aligned} \tag{11.23}$$

An example of the use of (11.22) was given in Example 2.6 for a $\mathcal{N}(0,1)$ random variable. Some practice with the estimation of the mean and variance is provided in Problem 11.46.

11.10   Real-World Example – Critical Software Testing Using Importance Sampling

Computer software is a critical component of nearly every device used today. The failure of such software can range from being an annoyance, as in the outage of a cellular telephone, to being a catastrophe, as in the breakdown of the control system for a nuclear power plant. Testing of software is of course a prerequisite for reliable operation, but some events, although potentially catastrophic, will (hopefully) occur only rarely. Therefore, the question naturally arises as to how to test software that is designed to only fail once every 10^7 hours (≈ 1400 years). In other words, although a theoretical analysis might predict such a low failure rate, there is no way to test the software by running it and waiting for a failure. A technique that is often used in other fields to test a system is to "stress" the system to induce more frequent failures, say by a factor of 10^5, then estimate the probability of failure per hour, and finally readjust the probability for the increased stress factor. An analogous approach can be used for highly reliable software if we can induce a higher failure rate and then readjust our failure probability estimate by the increased factor. A proposed method to do this is to stress the software to cause the probability of a failure to increase [Hecht and Hecht 2000]. Conceivably we could do this by inputting data to the software that is suspected to cause failures but at a much higher rate than is normally encountered in practice. This means that if T is the time to failure, then we would like to replace the PDF of T so that $P[T > \gamma]$ increases by a significant factor. Then, after estimating this probability by exercising the software we could adjust the estimate back to the original unstressed value. This probabilitic approach is called *importance sampling* [Rubinstein 1981].

As an example of the use of importance sampling, assume that X is a continuous random variable and we wish to estimate $P[X > \gamma]$. As usual, we could generate realizations of X, count the number that exceed γ, and then divide this by the total number of realizations. But what if the probability sought is 10^{-7}? Then we would need about 10^9 realizations to do this. As a specific example, suppose that $X \sim \mathcal{N}(0, 1)$, although in practice we would not have knowledge of the PDF at our disposal, and that we wish to estimate $P[X > 5]$ based on observed realization values. The true probability is known to be $Q(5) = 2.86 \times 10^{-7}$. The importance sampling approach first recognizes that the desired probability is given by

$$\mathcal{I} = \int_5^\infty \frac{1}{\sqrt{2\pi}} \exp\left(-\frac{1}{2}x^2\right) dx$$

and is equivalent to

$$\mathcal{I} = \int_5^\infty \frac{\frac{1}{\sqrt{2\pi}} \exp\left(-\frac{1}{2}x^2\right)}{p_{X'}(x)} p_{X'}(x) dx$$

where $p_{X'}(x)$ is a more suitable PDF. By "more suitable" we mean that its probability of $X' > 5$ is larger, and therefore, generating realizations based on it will produce more occurrences of the desired event. One possibility is $X' \sim \exp(1)$ or $p_{X'}(x) = \exp(-x)u(x)$ for which $P[X > 5] = \exp(-5) = 0.0067$. Using this new PDF we have the desired probability

$$\mathcal{I} = \int_5^\infty \frac{\frac{1}{\sqrt{2\pi}} \exp\left(-\frac{1}{2}x^2\right)}{\exp(-x)} \exp(-x) dx$$

or using the indicator function, this can be written as

$$\mathcal{I} = \int_0^\infty \underbrace{I_{(5,\infty)}(x) \frac{1}{\sqrt{2\pi}} \exp\left(-\frac{1}{2}x^2 + x\right)}_{g(x)} p_{X'}(x) dx.$$

Now the desired probability can be interpreted as $E[g(X')]$, where $X' \sim \exp(1)$. To estimate it using a Monte Carlo computer simulation we first generate M realizations of an $\exp(1)$ random variable and then use as our estimate

$$\begin{aligned}
\hat{\mathcal{I}} &= \frac{1}{M} \sum_{i=1}^M g(x_i) \\
&= \frac{1}{M} \sum_{i=1}^M I_{(5,\infty)}(x_i) \underbrace{\frac{1}{\sqrt{2\pi}} \exp\left(-\frac{1}{2}x_i^2 + x_i\right)}_{\substack{\text{weight with value} \ll 1 \\ \text{for } x_i \gg 5}}.
\end{aligned} \qquad (11.24)$$

The advantage of the importance sampling approach is that the realizations whose values exceed 5, which are the ones contributing to the sum, are *much more proba-ble*. In fact, as we have noted $P[X' > 5] = 0.0067$ and therefore with $N = 10,000$

realizations we would expect about 67 realizations to contribute to the sum. Contrast this with a $\mathcal{N}(0,1)$ random variable for which we would expect $NQ(5) = (10^4)(2.86 \times 10^{-7}) \approx 0$ realizations to exceed 5. The new PDF $p_{X'}$ is called the *importance function* and hence the generation of realizations from this PDF, which is also called *sampling from the PDF*, is termed *importance sampling*. As seen from (11.24), its success requires a weighting factor that downweights the counting of threshold exceedances.

In software testing the portions of software that are critical to the operation of the overall system would be exercised more often than in normal operation, thus effectively replacing the operational PDF or p_X by the importance function PDF or $p_{X'}$. The ratio of these two would be needed as seen in (11.24) to adjust the weight for each incidence of a failure. *This ratio would also need to be estimated in practice.* In this way a good estimate of the probability of failure could be obtained by exercising the software a reasonable number of times with different inputs. Otherwise, the critical software might not exhibit a failure a sufficient number of times to estimate its probability.

As a numerical example, if $X' \sim \exp(1)$, we can generate realizations using the inverse probability transformation method (see Section 10.9) via $X' = -\ln(1-U)$, where $U \sim \mathcal{U}(0,1)$. A MATLAB computer program to estimate $\mathcal{I}$ is given below.

```
rand('state',0) % sets random number generator to
                % initial value
M=10000;gamma=5;% change M for different estimates
u=rand(M,1);    % generates M U(0,1) realizations
x=-log(1-u);    % generates M exp(1) realizations
k=0;
for i=1:M       % computes estimate of P[X>gamma]
   if x(i)>gamma
      k=k+1;
      y(k,1)=(1/sqrt(2*pi))*exp(-0.5*x(i)^2+x(i)); % computes weights
                                                   % for estimate
   end
end
Qest=sum(y)/M  % final estimate of P[X>gamma]
```

The results are summarized in Table 11.2 for different values of M, along with the true value of $Q(5)$. Also shown are the number of times γ was exceeded. Without the use of importance sampling the number of exceedances would be expected to be $MQ(5) \approx 0$ in all cases.

M	Estimated $P[X > 5]$	True $P[X > 5]$	Exceedances
10^3	1.11×10^{-7}	2.86×10^{-7}	4
10^4	2.96×10^{-7}	2.86×10^{-7}	66
10^5	2.51×10^{-7}	2.86×10^{-7}	630
10^6	2.87×10^{-7}	2.86×10^{-7}	6751

Table 11.2: Importance sampling approach to estimation of small probabilities.

References

Hecht, M., H. Hecht, "Use of Importance Sampling and Related Techniques to Measure Very High Reliability Software," 2000 IEEE Aerospace Conference Proc., Vol. 4, pp. 533–546.

Jackson, L.B., *Signals, Systems, and Transforms*, Addison-Wesley, Reading, MA, 1991.

Johnson, N.L., S. Kotz, N. Balakrishnan, *Continuous Univariate Distributions, Vol. 1*, see pp. 456–459 for moments, John Wiley & Sons, New York, 1994.

Parzen, E., *Modern Probability Theory and its Applications*, John Wiley & Sons, New York, 1960.

Rubinstein, R.Y., *Simulation and the Monte Carlo Method*, John Wiley & Sons, New York, 1981.

Problems

11.1 (⌣) **(f)** The block shown in Figure 11.7 has a mass of 1 kg. Find the center of mass for the block, which is the point along the x-axis where the block could be balanced (in practice the point would also be situated in the depth direction at $1/2$).

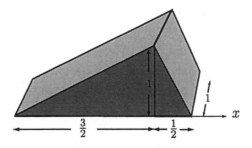

Figure 11.7: Block for Problem 11.1.

11.2 (t) Prove that if the PDF is symmetric about a point $x = a$, which is to say that it satisfies $p_X(a+u) = p_X(a-u)$ for all $-\infty < u < \infty$, then the mean will be a. Hint: Write the integral $\int_{-\infty}^{\infty} x p_X(x)dx$ as $\int_{-\infty}^{a} x p_X(x)dx + \int_{a}^{\infty} x p_X(x)dx$ and then let $u = x - a$ in the first integral and $u = a - x$ in the second integral.

11.3 (c) Generate and plot 50 realizations of a Cauchy random variable. Do so by using the inverse probability integral transformation method. You should be able to show that $X = \tan(\pi(U - 1/2))$, where $U \sim \mathcal{U}(0,1)$, will generate the Cauchy realizations.

11.4 (c) In this problem we show via a computer simulation that the mean of a *truncated Cauchy* PDF exists and is equal to zero. A truncated Cauchy random variable is one in which the realizations of a Cauchy PDF are set to $x = x_{\max}$ if $x > x_{\max}$ and $x = -x_{\max}$ if $x < -x_{\max}$. Generate realizations of this random variable with $x_{\max} = 50$ and plot the sample mean versus the number of realizations. What does the sample mean converge to?

11.5 (t) Prove that the best prediction of the outcome of a continuous random variable is its mean. Best is to be interpreted as the value that minimizes the mean square error $\mathrm{mse}(b) = E[(X - b)^2]$.

11.6 (t) An even function is one for which $g(-x) = g(x)$, as for example $\cos(x)$. An odd function is one for which $g(-x) = -g(x)$, as for example $\sin(x)$. First prove that $\int_{-\infty}^{\infty} g(x)dx = 2\int_{0}^{\infty} g(x)dx$ if $g(x)$ is even and that $\int_{-\infty}^{\infty} g(x)dx = 0$ if $g(x)$ is odd. Next, prove that if $p_X(x)$ is even, then $E[X] = 0$ and also that $\int_{0}^{\infty} p_X(x)dx = 1/2$.

11.7 (f) Many integrals encountered in probability can be evaluated using *integration by parts*. This useful formula is

$$\int U\,dV = UV - \int V\,dU$$

where U and V are functions of x. As an example, if we wish to evaluate $\int x \exp(ax)dx$, we let $U = x$ and $dV = \exp(ax)dx$. The function U is easily differentiated to yield $dU = dx$ and the differential dV is easily integrated to yield $V = (1/a)\exp(ax)$. Continue the derivation to determine the integral of the function $x \exp(ax)$.

11.8 (f) Find the mean for a uniform PDF. Do so by first using the definition and then rederive it using the results of Problem 11.2.

11.9 (t) Consider a continuous random variable that can take on values $x_{\min} \leq x \leq x_{\max}$. Prove that the expected value of this random variable must satisfy $x_{\min} \leq E[X] \leq x_{\max}$. Hint: Use the fact that if $M_1 \leq g(x) \leq M_2$, then $M_1 a \leq \int_{a}^{b} g(x)dx \leq M_2 b$.

11.10 (☺) **(w)** The signal-to-noise ratio (SNR) of a random variable quantifies the accuracy of a measurement of a physical quantity. It is defined as $E^2[X]/\text{var}(X)$ and is seen to increase as the mean, which represents the true value, increases and also as the variance, which represents the power of the measurement error, i.e., $X - E[X]$, decreases. For example, if $X \sim \mathcal{N}(\mu, \sigma^2)$, then SNR $= \mu^2/\sigma^2$. Determine the SNR if the measurement is $X = A + U$, where A is the true value and U is the measurement error with $U \sim \mathcal{U}(-1/2, 1/2)$. For an SNR of 1000 what should A be?

11.11 (☺) **(w)** A toaster oven has a failure time that has an exponential PDF. If the mean time to failure is 1000 hours, what is the probability that it will not fail for at least 2000 hours?

11.12 (w) A bus always arrives late. On the average it is 10 minutes late. If the lateness time is an exponential random variable, determine the probability that the bus will be less than 1 minute late.

11.13 (w) In Section 1.3 we described the amount of time an office worker spends on the phone in a 10-minute period. From Figure 1.5 what is the average amount of time he spends on the phone?

11.14 (☺) **(f)** Determine the mean of a χ_N^2 PDF. See Chapter 10 for the definition of this PDF.

11.15 (f) Determine the mean of an Erlang PDF using the definition of expected value. See Chapter 10 for the definition of this PDF.

11.16 (f) Determine the mean of a Rayleigh PDF using the definition of expected value. See Chapter 10 for the definition of this PDF.

11.17 (w) The *mode* of a PDF is the value of x for which the PDF is maximum. It can be thought of as the most probable value of a random variable (actually most probable small interval). Find the mode for a Gaussian PDF and a Rayleigh PDF. How do they relate to the mean?

11.18 (f) Indicate on the PDFs shown in Figures 10.7–10.13 the location of the mean value.

11.19 (☺) **(w)** A dart is thrown at a circular dartboard. If the distance from the bullseye is a Rayleigh random variable with a mean value of 10, what is the probability that the dart will land within 1 unit of the bullseye?

11.20 (f) For the random variables described in Problems 2.8–2.11 what are the means? Note that the uniform random variable is $\mathcal{U}(0, 1)$ and the Gaussian random variable is $\mathcal{N}(0, 1)$.

11.21 (⌣) (w) In Problem 2.14 it was asked whether the mean of $\sqrt{U}$, where $U \sim$ $\mathcal{U}(0,1)$, is equal to $\sqrt{\text{mean of } U}$. There we relied on a computer simulation to answer the question. Now prove or disprove this equivalence.

11.22 (⌣) (w) A sinusoidal oscillator outputs a waveform $s(t) = \cos(2\pi F_0 t + \phi)$, where t indicates time, F_0 is the frequency in Hz, and ϕ is a phase angle that varies depending upon when the oscillator is turned on. If the phase is modeled as a random variable with $\phi \sim \mathcal{U}(0, 2\pi)$, determine the average value of $s(t)$ for a given $t = t_0$. Also, determine the average power, which is defined as $E[s^2(t)]$ for a given $t = t_0$. Does this make sense? Explain your results.

11.23 (f) Determine $E[X^2]$ for a $\mathcal{N}(\mu, \sigma^2)$ random variable.

11.24 (f) Determine $E[(2X + 1)^2]$ for a $\mathcal{N}(\mu, \sigma^2)$ random variable.

11.25 (f) Determine the mean and variance for the indicator random variable $I_A(X)$ as a function of $P[A]$.

11.26 (⌣) (w) A half-wave rectifier passes a zero or positive voltage undisturbed but blocks any negative voltage by outputting a zero voltage. If a noise sample with PDF $\mathcal{N}(0, \sigma^2)$ is input to a half-wave rectifier, what is the average power at the output? Explain your result.

11.27 (⌣) (w) A mixed PDF is given as

$$p_X(x) = \frac{1}{2}\delta(x) + \frac{1}{\sqrt{2\pi\sigma^2}} \exp\left(-\frac{1}{2\sigma^2}x^2\right) u(x).$$

What is $E[X^2]$ for this PDF? Can this PDF be interpreted physically? Hint: See Problem 11.26.

11.28 (t) In this problem we derive an alternative formula for the mean of a non-negative random variable. A more general formula exists for random variables that can take on both positive and negative values [Parzen 1960]. If X can only take on values $x \geq 0$, then

$$E[X] = \int_0^\infty (1 - F_X(x))\, dx.$$

First verify that this formula holds for $X \sim \exp(\lambda)$. To prove that the formula is true in general, we use integration by parts (see Problem 11.7) as follows.

$$E[X] = \int_0^\infty (1 - F_X(x))\, dx$$
$$= \int_0^\infty \underbrace{\int_x^\infty p_X(t)dt}_{U}\, \underbrace{dx}_{dV}.$$

Finish the proof by using $\lim_{x \to \infty} x \int_x^\infty p_X(t)dt = 0$, which must be true if the expected value exists (see if this holds for $X \sim \exp(\lambda)$).

11.29 (t) The standard deviation σ of a Gaussian PDF can be interpreted as the distance from the mean at which the PDF curve goes through an inflection point. This means that at the points $x = \mu \pm \sigma$ the second derivative of $p_X(x)$ is zero. The curve then changes from being concave (shaped like a ∩) to being convex (shaped like a ∪). Show that the second derivative is zero at these points.

11.30 (⌣) (w) The office worker described in Section 1.3 will spend an average of 7 minutes on the phone in any 10-minute interval. However, the probability that he will spend *exactly* 7 minutes on the phone is zero since the length of this interval is zero. If we wish to assert that he will spend between $T_{\min}$ and $T_{\max}$ minutes on the phone 95% of the time, what should $T_{\min}$ and $T_{\max}$ be? Hint: There are multiple solutions – choose any convenient one.

11.31 (w) A group of students is found to weigh an average of 150 lbs. with a standard deviation of 30 lbs. If we assume a normal population (in the probabilistic sense!) of students, what is the range of weights for which approximately 99.8% of the students will lie? Hint: There are multiple solutions – choose any convenient one.

11.32 (w) Provide a counterexample to disprove that $\text{var}(g_1(X) + g_2(X)) = \text{var}(g_1(X)) + \text{var}(g_2(X))$ in general.

11.33 (w) The SNR of a random variable was defined in Problem 11.10. Determine the SNR for exponential random variable and explain why it doesn't increase as the mean increases. Compare your results to a $\mathcal{N}(\mu, \sigma^2)$ random variable and explain.

11.34 (f) Verify the mean and variance for a Laplacian random variable given in Table 11.1.

11.35 (⌣) (f) Determine $E[X^3]$ if $X \sim \mathcal{N}(\mu, \sigma^2)$. Next find the third *central* moment.

11.36 (f) An example of a Gaussian mixture PDF is

$$p_X(x) = \frac{1}{2}\frac{1}{\sqrt{2\pi}}\exp\left[-\frac{1}{2}(x-1)^2\right] + \frac{1}{2}\frac{1}{\sqrt{2\pi}}\exp\left[-\frac{1}{2}(x+1)^2\right].$$

Determine its mean and variance.

11.37 (t) Prove that if a PDF is symmetric about $x = 0$, then all its odd-order moments are zero.

11.38 (☺) **(f)** For a Laplacian PDF with $\sigma^2 = 2$ determine all the moments. Hint: Let

$$\frac{1}{\omega^2 + 1} = \frac{1}{2j} \left(\frac{1}{\omega - j} - \frac{1}{\omega + j} \right).$$

11.39 (f) If $X \sim \mathcal{N}(0, \sigma^2)$, determine $E[X^2]$ using the characteristic function approach.

11.40 (t) To determine the characteristic function of a Cauchy random variable we must evaluate the integral

$$\int_{-\infty}^{\infty} \frac{1}{\pi(1 + x^2)} \exp(j\omega x) dx.$$

A result from Fourier transform theory called the *duality theorem* asserts that the Fourier transform and inverse Fourier transform are nearly the same if we replace x by ω and ω by x. As an example, for a Laplacian PDF with $\sigma^2 = 2$ we have from Table 11.1 that

$$\int_{-\infty}^{\infty} p_X(x) \exp(j\omega x) dx = \int_{-\infty}^{\infty} \frac{1}{2} \exp(-|x|) \exp(j\omega x) dx = \frac{1}{1 + \omega^2}.$$

The inverse Fourier transform relationship is therefore

$$\int_{-\infty}^{\infty} \frac{1}{1 + \omega^2} \exp(-j\omega x) \frac{d\omega}{2\pi} = \frac{1}{2} \exp(-|x|).$$

Use the latter integral, with appropriate modifications (note that x and ω are just variables which we can redefine as desired), to obtain the characteristic function of a Cauchy random variable.

11.41 (f) If the characteristic function of a random variable is

$$\phi_X(\omega) = \left(\frac{\sin \omega}{\omega} \right)^2$$

find the PDF. Hint: Recall that when we convolve two functions together the Fourier transform of the new function is the product of the individual Fourier transforms. Also, see Table 11.1 for the characteristic function of a $\mathcal{U}(-1, 1)$ random variable.

11.42 (☺) **(w)** If $X^{(n)} \sim \mathcal{N}(\mu, 1/n)$, determine the PDF of the limiting random variable X as $n \to \infty$. Use characteristic functions to do so.

11.43 (f) Find the mean and variance of a χ_N^2 random variable using the characteristic function.

11.44 (☺) (f) The probability that a random variable deviates from its mean by an amount γ in either direction is to be less than or equal to $1/2$. What should γ be?

11.45 (f) Determine the probability that $|X| > \gamma$ if $X \sim \mathcal{U}[-a, a]$. Next compare these results to the Chebyshev bound for $a = 2$.

11.46 (☺) (c) Estimate the mean and variance of a Rayleigh random variable with $\sigma^2 = 1$ using a computer simulation. Compare your estimated results to the theoretical values.

11.47 (c) Use the importance sampling method described in Section 11.10 to determine $Q(7)$. If you were to generate M realizations of a $\mathcal{N}(0, 1)$ random variable and count the number that exceed $\gamma = 7$ as is usually done to estimate a right-tail probability, what would M have to be (in terms of order of magnitude)?

Appendix 11A

Partial Proof of Expected Value of Function of Continuous Random Variable

For simplicity assume that $Y = g(X)$ is a continuous random variable with PDF $p_Y(y)$ (having no impulses). Also, assume that $y = g(x)$ is monotonically increasing so that it has a single solution to the equation $y = g(x)$ for all y as shown in Figure 11A.1. Then

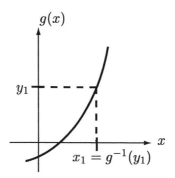

Figure 11A.1: Monotonically increasing function used to derive $E[g(X)]$.

$$\begin{aligned} E[Y] &= \int_{-\infty}^{\infty} y p_Y(y) dy \\ &= \int_{-\infty}^{\infty} y p_X(g^{-1}(y)) \left| \frac{dg^{-1}(y)}{dy} \right| dy \qquad \text{(from (10.30).} \end{aligned}$$

Next change variables from y to x using $x = g^{-1}(y)$. Since we have assumed that $g(x)$ is monotonically increasing, the limits for y of $\pm\infty$ also become $\pm\infty$ for x.

Then, since $x = g^{-1}(y)$, we have that $y p_X(g^{-1}(y))$ becomes $g(x) p_X(x)$ and

$$\left| \frac{dg^{-1}(y)}{dy} \right| dy \;=\; \frac{dg^{-1}(y)}{dy} dy \qquad (g \text{ is monotonically increasing,}$$

$$\text{implies } g^{-1} \text{ is monotonically increasing,}$$

$$\text{implies derivative is positive)}$$

$$=\; \frac{dx}{dy} dy = dx$$

from which (11.10) follows. The more general result for nonmonotonic functions follows along these lines.

Chapter 12

Multiple Continuous Random Variables

12.1 Introduction

In Chapter 7 we discussed multiple discrete random variables. We now proceed to parallel that discussion for *multiple continuous random variables*. We will consider in this chapter only the case of two random variables, also called *bivariate random variables*, with the extension to any number of continuous random variables to be presented in Chapter 14. In describing bivariate discrete random variables, we used the example of height and weight of a college student. Figure 7.1 displayed the probabilities of a student having a height in a given interval and a weight in a given interval. For example, the probability of having a height in the interval $[5'8'', 6']$ and a weight in the interval $[160, 190]$ lbs. is 0.14 as listed in Table 4.1 and as seen in Figure 7.1 for the values of $H = 70$ inches and $W = 175$ lbs. For physical measurements such as height and weight, however, we would expect to observe a continuum of values. As such, height and weight are more appropriately modeled by multiple *continuous* random variables. For example, we might have a population of college students, all of whose heights and weights lie in the intervals $60 \leq H \leq 80$ inches and $100 \leq W \leq 250$ lbs. Therefore, the *continuous* random variables (H, W) would take on values in the sample space

$$\mathcal{S}_{H,W} = \{(h, w) : 60 \leq h \leq 80, 100 \leq w \leq 250\}$$

which is a subset of the plane, i.e., R^2. We might wish to determine probabilities such as $P[61 \leq H \leq 67.5, 98.5 \leq W \leq 154]$, which cannot be found from Figure 7.1. In order to compute such a probability we will define a *joint PDF* for the continuous random variables H and W. It will be a two-dimensional function of h and w. In the case of a single random variable we needed to integrate to find the area under the PDF as the desired probability. Now integration of the joint PDF, which is a function of two variables, will produce the probability. However, we will now be determining the *volume* under the joint PDF. All our concepts for a single continuous random variable will extend to the case of two random variables. Computationally, however, we will encounter more difficulty since two-dimensional integrals, also known as double integrals, will need to be evaluated. Hence, the reader should be acquainted with double integrals and their evaluation using iterated integrals.

12.2   Summary

The concept of jointly distributed continuous random variables is introduced in Section 12.3. Given the joint PDF the probability of any event defined on the plane is given by (12.2). The standard bivariate Gaussian PDF is given by (12.3) and is plotted in Figure 12.9. The concept of constant PDF contours is also illustrated in Figure 12.9. The marginal PDF is found from the joint PDF using (12.4). The joint CDF is defined by (12.6) and is evaluated using (12.7). Its properties are listed in P12.1–P12.6. To obtain the joint PDF from the joint CDF we use (12.9). Independence of jointly distributed random variables is defined by (12.10) and can be verified by the factorization of either the PDF as in (12.11) or the CDF as in (12.12). Section 12.6 addresses the problem of determining the PDF of a function of two random variables—see (12.13), and that of determining the joint PDF of a function which maps two random variables into two new random variables. See (12.18) for a linear transformation and (12.22) for a nonlinear transformation. The general bivariate Gaussian PDF is defined in (12.24) and some useful properties are discussed in Section 12.7. In particular, Theorem 12.7.1 indicates that a linear transformation of a bivariate Gaussian random vector produces another bivariate Gaussian random vector, although with different means and covariances. Example 12.14 indicates how a bivariate Gaussian random vector may be transformed to one with independent components. Also, a formula for computation of the expected value of a function of two random variables is given as (12.28). Section 12.9 discusses prediction of a random variable from the observation of a second random variable while Section 12.10 summarizes the joint characteristic function and its properties. In particular, the use of (12.47) allows the determination of the PDF of the sum of two continuous and independent random variables. It is used to prove that two independent Gaussian random variables that are added together produce another Gaussian random variable in Example 12.15. Section 12.11 shows how to simulate on a computer a random vector with any desired mean vector and covariance matrix by using the Cholesky decomposition of the covariance matrix—see (12.53). If the desired random vector is bivariate Gaussian, then the procedure provides a general method for generating Gaussian random vectors on a computer. Finally, an application to optical character recognition is described in Section 12.12.

12.3 Jointly Distributed Random Variables

We consider two continuous random variables that will be denoted by X and Y. As alluded to in the introduction, they represent the functions that map an outcome s of an experiment to a point in the plane. Hence, we have that

$$\begin{bmatrix} X(s) \\ Y(s) \end{bmatrix} = \begin{bmatrix} x \\ y \end{bmatrix}$$

for all $s \in \mathcal{S}$. An example is shown in Figure 12.1 in which the outcome of a dart toss s, which is a point within a unit radius circular dartboard, is mapped into a point in the plane, which is within the unit circle. The random variables X and Y

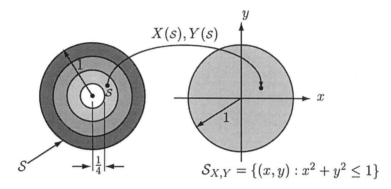

Figure 12.1: Mapping of the outcome of a thrown dart to the plane (example of jointly continuous random variables).

are said to be *jointly distributed continuous random variables*. As before, we will denote the random variables as (X, Y) or $[X \, Y]^T$, in either case referring to them as a random vector. Note that a different mapping would result if we chose to represent the point in $\mathcal{S}_{X,Y}$ in polar coordinates (r, θ). Then we would have

$$\mathcal{S}_{R,\Theta} = \{(r, \theta) : 0 \le r \le 1, 0 \le \theta < 2\pi\}.$$

This is a different random vector but is of course related to (X, Y). Depending upon the shape of the mapped region in the plane, it may be more convenient to use either rectangular coordinates or polar coordinates for probability calculations (see also Problem 12.1).

Typical outcomes of the random variables are shown in Figure 12.2 as points in $\mathcal{S}_{X,Y}$ for two different players. In Figure 12.2a 100 outcomes for a novice dart player are shown while those for a champion dart player are displayed in Figure 12.2b. We might be interested in the probability that $\sqrt{X^2 + Y^2} \le 1/4$, which is the event that a bullseye is attained. Now our event of interest is a two-dimensional region as opposed to a one-dimensional interval for a single continuous random variable. In

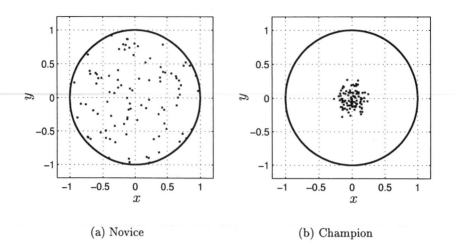

(a) Novice (b) Champion

Figure 12.2: Typical outcomes for novice and champion dart player.

the case of the novice dart player the dart is equally likely to land anywhere in the unit circle and hence the probability is

$$P[\text{bullseye}] = \frac{\text{Area of bullseye}}{\text{Total area of dartboard}}$$

$$= \frac{\pi(1/4)^2}{\pi(1)^2} = \frac{1}{16}.$$

However, for a champion dart player we see from Figure 12.2b that the probability of a bullseye is much higher. How should we compute this probability? For the novice dart player we can interpret the probability calculation geometrically as shown in Figure 12.3 as the *volume* of the inner cylinder since

$$P[\text{bullseye}] = \pi(1/4)^2 \times \frac{1}{\pi}$$

$$= \underbrace{\text{Area of bullseye}}_{\text{Area of event}} \times \underbrace{\frac{1}{\pi}}_{\text{Height}}.$$

If we define a function

$$p_{X,Y}(x,y) = \begin{cases} \frac{1}{\pi} & x^2 + y^2 \le 1 \\ 0 & \text{otherwise} \end{cases} \tag{12.1}$$

then this volume is also given by

$$P[A] = \iint_A p_{X,Y}(x,y)\,dx\,dy \tag{12.2}$$

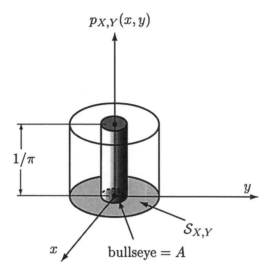

Figure 12.3: Geometric interpretation of bullseye probability calculation for novice dart thrower.

since then

$$
\begin{aligned}
P[A] &= \iint_{\{(x,y):x^2+y^2\leq(1/4)^2\}} \frac{1}{\pi} dx\, dy \\
&= \frac{1}{\pi} \iint_{\{(x,y):x^2+y^2\leq(1/4)^2\}} dx\, dy \\
&= \frac{1}{\pi} \times \text{Area of } A = \frac{1}{\pi}\pi\left(\frac{1}{4}\right)^2 = \frac{1}{16}.
\end{aligned}
$$

In analogy with the definition of the PDF for a single random variable X, we define $p_{X,Y}(x,y)$ as the *joint PDF* of X and Y. For this example, it is given by (12.1) and is used to evaluate the probability that (X,Y) lies in a given region A by (12.2). The region A can be any subset of the plane. Note that in using (12.2) we are determining the *volume* under $p_{X,Y}$, hence the need for a double integral. Another example follows.

Example 12.1 – Pyramid-like joint PDF

A joint PDF is given by

$$
p_{X,Y}(x,y) = \begin{cases} 4(1-|2x-1|)(1-|2y-1|) & 0 \leq x \leq 1, 0 \leq y \leq 1 \\ 0 & \text{otherwise.} \end{cases}
$$

We wish to first verify that the PDF integrates to one. Then, we consider the evaluation of $P[1/4 \leq X \leq 3/4, 1/4 \leq Y \leq 3/4]$. A three-dimensional plot of the PDF is shown in Figure 12.4 and appears pyramid-like. Since it is often difficult to visualize the PDF in 3-D, it is helpful to plot the contours of the PDF as shown

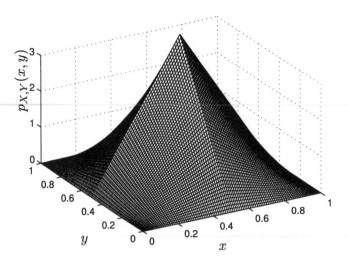

Figure 12.4: Three-dimensional plot of joint PDF.

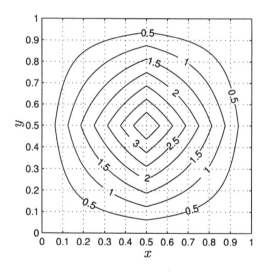

Figure 12.5: Contour plot of joint PDF.

in Figure 12.5.   As seen in the contour plot (also called a topographical map) the innermost contour consists of all values of (x, y) for which $p_{X,Y}(x, y) = 3.5$. This contour is obtained by slicing the solid shown in Figure 12.4 with a plane parallel to the x-y plane and at a height of 3.5 and similarly for the other contours. These contours are called *contours of constant PDF*.

To verify that $p_{X,Y}$ is indeed a valid joint PDF, we need to show that the volume under the PDF is equal to one. Since the sample space is $\mathcal{S}_{X,Y} = \{(x, y) : 0 \le x \le$

$1, 0 \le y \le 1\}$ we have that

$$P[\mathcal{S}_{X,Y}] = \int_0^1 \int_0^1 4(1 - |2x - 1|)(1 - |2y - 1|)dx\, dy$$

$$= \int_0^1 2(1 - |2x - 1|)dx \int_0^1 2(1 - |2y - 1|)dy.$$

The two definite integrals are seen to be identical and hence we need only evaluate one of these. But each integral is the area under the function shown in Figure 12.6a which is easily found to be 1. Hence, $P[\mathcal{S}_{X,Y}] = 1 \cdot 1 = 1$, verifying that $p_{X,Y}$ is a

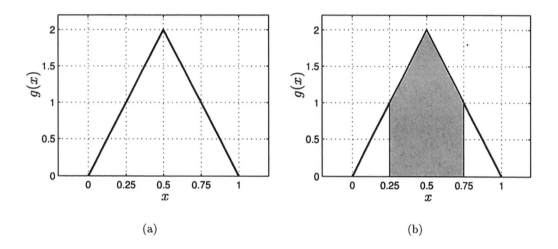

(a)   (b)

Figure 12.6: Plot of function $g(x) = 2(1 - |2x - 1|)$.

valid PDF. Next to find $P[1/4 \le X \le 3/4, 1/4 \le Y \le 3/4]$ we use (12.2) to yield

$$P[A] = \int_{1/4}^{3/4} \int_{1/4}^{3/4} 4(1 - |2x - 1|)(1 - |2y - 1|)dx\, dy.$$

By the same argument as before we have

$$P[A] = \left[\int_{1/4}^{3/4} 2(1 - |2x - 1|)dx \right]^2$$

and referring to Figure 12.6b, we have that each unshaded triangle has an area of $(1/2)(1/4)(1) = 1/8$ and so

$$P[A] = \left[1 - \frac{1}{8} - \frac{1}{8} \right]^2 = \left(\frac{6}{8} \right)^2 = \frac{9}{16}.$$

In summary, a joint PDF has the expected properties of being a nonnegative two-dimensional function that integrates to one over R^2.

$\diamond$

For the previous example the double integral was easily evaluated since

1. The integrand $p_{X,Y}(x,y)$ was separable (we will see shortly that this property will hold when the random variables are independent).

2. The integration region in the x-y plane was rectangular.

More generally this will not be the case. Consider, for example, the computation of $P[Y \leq X]$. We need to integrate $p_{X,Y}$ over the shaded region shown in Figure 12.7. To do so we first integrate in the y direction for a fixed x, shown as the darkly

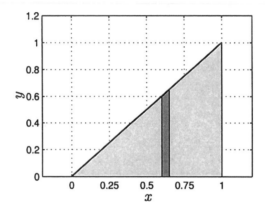

Figure 12.7: Integration region to determine $P[Y \leq X]$.

shaded region. Since $0 \leq y \leq x$ for a fixed x, we have the limits of 0 to x for the integration over y and the limits of 0 to 1 for the final integration over x. This results in

$$
\begin{aligned}
P[Y \leq X] &= \int_0^1 \int_0^x p_{X,Y}(x,y) dy\, dx \\
&= \int_0^1 \int_0^x 4(1 - |2x - 1|)(1 - |2y - 1|) dy\, dx.
\end{aligned}
$$

Although the integration can be carried out, it is tedious. In this illustration the joint PDF is separable but the integration region is not rectangular.

 Zero probability events are more complex in two dimensions.

Recall that for a single continuous random variable the probability of X attaining any value is zero. This is because the area under the PDF is zero for any zero length *interval.* Similarly, for jointly continuous random variables X and Y the probability of any event defined on the x-y plane will be zero *if the region of the event in the plane has zero area.* Then, the volume under the joint PDF will be zero. Some examples of these zero probability events are shown in Figure 12.8.

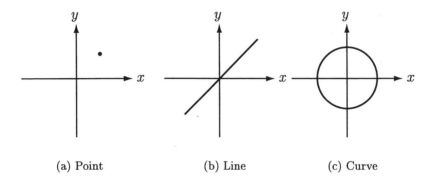

(a) Point   (b) Line   (c) Curve

Figure 12.8: Examples of zero probability events for jointly distributed continuous random variables X and Y. All regions in the x-y plane have zero area.

An important joint PDF is the *standard bivariate Gaussian or normal PDF*, which is defined as

$$p_{X,Y}(x,y) = \frac{1}{2\pi\sqrt{1-\rho^2}} \exp\left[-\frac{1}{2(1-\rho^2)}(x^2 - 2\rho xy + y^2)\right] \quad \begin{array}{c} -\infty < x < \infty \\ -\infty < y < \infty \end{array}$$
(12.3)

where ρ is a parameter that takes on values $-1 < \rho < 1$. (The use of the term *standard* is because as is shown later the means of X and Y are 0 and the variances are 1.) The joint PDF is shown in Figure 12.9 for various values of ρ. We will see shortly that ρ is actually the correlation coefficient $\rho_{X,Y}$ first introduced in Section 7.9. The contours of constant PDF shown in Figures 12.9b,d,f are given by the values of (x,y) for which

$$x^2 - 2\rho xy + y^2 = r^2$$

where r is a constant. This is because for these values of (x,y) the joint PDF takes on the fixed value

$$p_{X,Y}(x,y) = \frac{1}{2\pi\sqrt{1-\rho^2}} \exp\left[-\frac{1}{2(1-\rho^2)}r^2\right].$$

If $\rho = 0$, these contours are circular as seen in Figure 12.9d and otherwise they are elliptical. Note that our use of r^2, which implies that $x^2 - 2\rho xy + y^2 > 0$, is valid

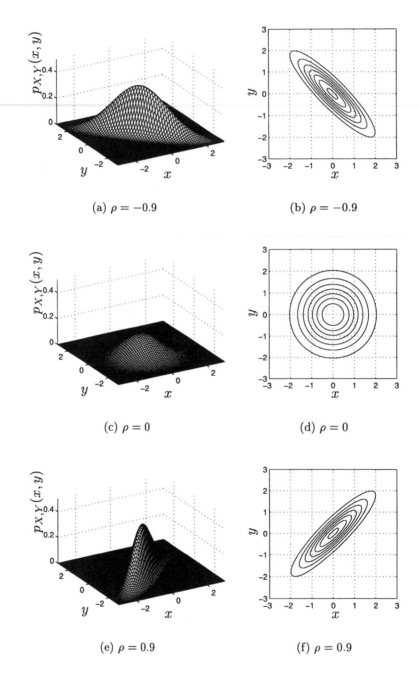

(a) $\rho = -0.9$ (b) $\rho = -0.9$

(c) $\rho = 0$ (d) $\rho = 0$

(e) $\rho = 0.9$ (f) $\rho = 0.9$

Figure 12.9: Three-dimensional and constant PDF contour plots of standard bivariate Gaussian PDF.

since in vector/matrix notation

$$x^2 - 2\rho xy + y^2 = \begin{bmatrix} x \\ y \end{bmatrix}^T \begin{bmatrix} 1 & -\rho \\ -\rho & 1 \end{bmatrix} \begin{bmatrix} x \\ y \end{bmatrix}$$

which is a *quadratic form*. Because $-1 < \rho < 1$, the matrix is positive definite (its principal minors are all positive—see Appendix C) and hence the quadratic form is positive. We will frequently use the standard bivariate Gaussian PDF and its generalizations as examples to illustrate other concepts. This is because its mathematical tractability lends itself to easy algebraic manipulations.

12.4 Marginal PDFs and the Joint CDF

The marginal PDF $p_X(x)$ of jointly distributed continuous random variables X and Y is the usual PDF which yields the probability of $a \le X \le b$ when integrated over the interval $[a, b]$. To determine $p_X(x)$ if we are given the joint PDF $p_{X,Y}(x, y)$, we consider the event

$$A = \{(x, y) : a \le x \le b, -\infty < y < \infty\}$$

whose probability must be the same as

$$A_X = \{x : a \le x \le b\}.$$

Thus, using (12.2)

$$
\begin{aligned}
P[a \le X \le b] &= P[A_X] = P[A] \\
&= \iint_A p_{X,Y}(x, y) dx\, dy \\
&= \int_{-\infty}^{\infty} \int_a^b p_{X,Y}(x, y) dx\, dy \\
&= \int_a^b \underbrace{\int_{-\infty}^{\infty} p_{X,Y}(x, y) dy}_{p_X(x)} dx.
\end{aligned}
$$

Clearly then, we must have that

$$p_X(x) = \int_{-\infty}^{\infty} p_{X,Y}(x, y) dy \tag{12.4}$$

as the marginal PDF for X. This operation is shown in Figure 12.10. In effect, we "sum" the probabilites of all the y values associated with the desired x, much the same as summing along a row to determine the marginal PMF $p_X[x_i]$ from the joint PMF $p_{X,Y}[x_i, y_j]$. The marginal PDF can also be viewed as the limit as $\Delta x \to 0$ of

$$
\begin{aligned}
p_X(x_0) &= \frac{P[x_0 - \Delta x/2 \le X \le x_0 + \Delta x/2, -\infty < Y < \infty]}{\Delta x} \\
&= \frac{\int_{x_0 - \Delta x/2}^{x_0 + \Delta x/2} \int_{-\infty}^{\infty} p_{X,Y}(x, y) dy\, dx}{\Delta x}
\end{aligned}
$$

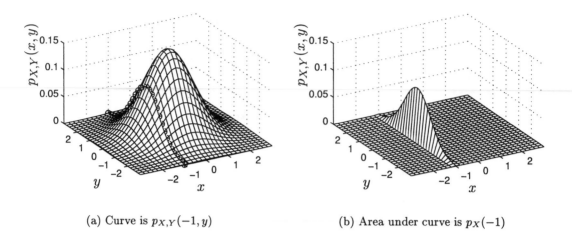

(a) Curve is $p_{X,Y}(-1, y)$ (b) Area under curve is $p_X(-1)$

Figure 12.10: Obtaining the marginal PDF of X from the joint PDF of (X, Y).

for a small Δx. An example follows.

Example 12.2 – Marginal PDFs for Standard Bivariate Gaussian PDF
From (12.3) and (12.4) we have that

$$p_X(x) = \int_{-\infty}^{\infty} \frac{1}{2\pi\sqrt{1-\rho^2}} \exp\left[-\frac{1}{2(1-\rho^2)}(x^2 - 2\rho xy + y^2)\right] dy. \qquad (12.5)$$

To carry out the integration we convert the integrand to one we recognize, i.e., a Gaussian, for which the integral over $(-\infty, \infty)$ is known. The trick here is to "complete the square" in y as follows:

$$\begin{aligned}
Q &= y^2 - 2\rho xy + x^2 \\
&= y^2 - 2\rho xy + \rho^2 x^2 + x^2 - \rho^2 x^2 \\
&= (y - \rho x)^2 + (1 - \rho^2)x^2.
\end{aligned}$$

Substituting into (12.5) produces

$$\begin{aligned}
p_X(x) &= \exp(-(1/2)x^2) \int_{-\infty}^{\infty} \frac{1}{2\pi\sqrt{1-\rho^2}} \exp\left[-\frac{1}{2(1-\rho^2)}(y - \rho x)^2\right] dy \\
&= \frac{1}{\sqrt{2\pi}} \exp(-(1/2)x^2) \underbrace{\int_{-\infty}^{\infty} \frac{1}{\sqrt{2\pi\sigma^2}} \exp\left[-\frac{1}{2\sigma^2}(y - \mu)^2\right] dy}_{=1}
\end{aligned}$$

where $\mu = \rho x$ and $\sigma^2 = 1 - \rho^2$, so that we have

$$p_X(x) = \frac{1}{\sqrt{2\pi}} \exp\left(-\frac{1}{2}x^2\right)$$

or $X \sim \mathcal{N}(0,1)$. Hence, the marginal PDF for X is a standard Gaussian PDF. By reversing the roles of X and Y, we will also find that $Y \sim \mathcal{N}(0,1)$. Note that since the marginal PDFs are standard Gaussian PDFs, the corresponding bivariate Gaussian PDF is also referred to as a standard one.

$\diamond$

In the previous example we saw that the marginals could be found from the joint PDF. However, in general the reverse process is not possible—given the marginal PDFs we cannot determine the joint PDF. For example, knowing that $X \sim \mathcal{N}(0,1)$ and $Y \sim \mathcal{N}(0,1)$ does not allow us to determine ρ, which characterizes the joint PDF. Furthermore, the marginal PDFs are the same for any ρ in the interval $(-1,1)$. This is just a restatement of the conclusion that we arrived at for joint and marginal PMFs. In that case there were many possible two-dimensional sets of numbers, i.e., specified by a joint PMF, that could sum to the same one-dimensional set, i.e., specified by a marginal PMF.

We next define the joint CDF for continuous random variables (X, Y). It is given by

$$F_{X,Y}(x,y) = P[X \leq x, Y \leq y]. \tag{12.6}$$

From (12.2) it is evaluated using

$$F_{X,Y}(x,y) = \int_{-\infty}^{y} \int_{-\infty}^{x} p_{X,Y}(t, u) dt \, du. \tag{12.7}$$

Some examples follow.

Example 12.3 – Joint CDF for an exponential joint PDF
If (X, Y) have the joint PDF

$$p_{X,Y}(x,y) = \begin{cases} \exp[-(x+y)] & x \geq 0, y \geq 0 \\ 0 & \text{otherwise} \end{cases}$$

then for $x \geq 0$, $y \geq 0$

$$\begin{aligned}
F_{X,Y}(x,y) &= \int_0^y \int_0^x \exp[-(t+u)] dt \, du \\
&= \int_0^y \exp(-u) \underbrace{\int_0^x \exp(-t) dt}_{1-\exp(-x)} du \\
&= \int_0^y [1 - \exp(-x)] \exp(-u) du \\
&= [1 - \exp(-x)] \int_0^y \exp(-u) du
\end{aligned}$$

so that

$$F_{X,Y}(x,y) = \begin{cases} [1 - \exp(-x)][1 - \exp(-y)] & x \geq 0, y \geq 0 \\ 0 & \text{otherwise.} \end{cases} \tag{12.8}$$

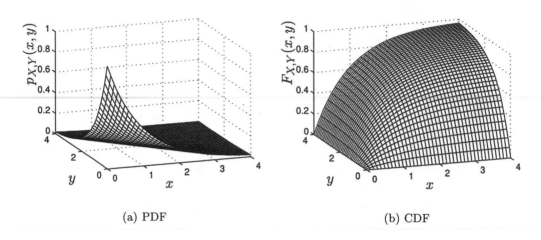

(a) PDF   (b) CDF

Figure 12.11: Joint exponential PDF and CDF.

The joint CDF is shown in Figure 12.11 along with the joint PDF. Once the joint CDF is obtained the probability for any rectangular region is easily found.

$\diamondsuit$

Example 12.4 – Probability from CDF for exponential random variables

Consider the rectangular region $A = \{(x,y) : 1 < x \le 2, 2 < y \le 3\}$. Then referring

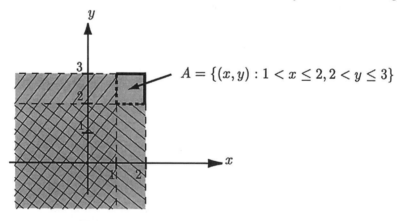

Figure 12.12: Evaluation of probability of rectangular region A using joint CDF.

to Figure 12.12 we determine the probability of A by determining the probability of the shaded region, then subtracting out the probability of each cross-hatched region (one running from south-east to north-west and the other running from south-west to north-east), and finally adding back in the probability of the double cross-hatched region (which has been subtracted out twice). This results in

$$
\begin{aligned}
P[A] \quad &= \quad P[-\infty < X \le 2, -\infty < Y \le 3] - P[-\infty < X \le 2, -\infty < Y \le 2] \\
&\quad -P[-\infty < X \le 1, -\infty < Y \le 3] + P[-\infty < X \le 1, -\infty < Y \le 2] \\
&= \quad F_{X,Y}[2,3] - F_{X,Y}[2,2] - F_{X,Y}[1,3] + F_{X,Y}[1,2].
\end{aligned}
$$

For the joint CDF given by (12.8) this becomes

$$
\begin{aligned}
P[A] \quad &= \quad [1 - \exp(-2)][1 - \exp(-3)] - [1 - \exp(-2)]^2 \\
&\quad -[1 - \exp(-1)][1 - \exp(-3)] + [1 - \exp(-1)][1 - \exp(-2)].
\end{aligned}
$$

Upon simplication we have the result

$$
P[A] = [\exp(-1) - \exp(-2)][\exp(-2) - \exp(-3)]
$$

which can also be verified by a direct evaluation as

$$
P[A] = \int_2^3 \int_1^2 \exp[-(x+y)]dx\,dy.
$$

We see that the advantage here is that no integration is required. However, the event A must be a rectangular region.

$\diamond$

The joint PDF can be recovered from the joint CDF by partial differentiation as

$$
p_{X,Y}(x,y) = \frac{\partial^2 F_{X,Y}(x,y)}{\partial x \partial y} \tag{12.9}
$$

which is the two-dimensional version of the fundamental theorem of calculus. As an example we continue the previous one.

Example 12.5 – Obtaining the joint PDF from the joint CDF for exponential random variables

Continuing with the previous example we have from (12.8) that

$$
p_{X,Y}(x,y) = \begin{cases} \frac{\partial^2 [1-\exp(-x)][1-\exp(-y)]}{\partial x \partial y} & x \ge 0, y \ge 0 \\ 0 & \text{otherwise.} \end{cases}
$$

For $x > 0, y > 0$

$$
\begin{aligned}
p_{X,Y}(x,y) \quad &= \quad \frac{\partial}{\partial x} \frac{\partial [1 - \exp(-x)][1 - \exp(-y)]}{\partial y} \\
&= \quad \frac{\partial [1 - \exp(-x)]}{\partial x} \frac{\partial [1 - \exp(-y)]}{\partial y} \\
&= \quad \exp(-x)\exp(-y) = \exp[-(x+y)].
\end{aligned}
$$

$\diamond$

Finally, the properties of the joint CDF are for the most part identical to those for the CDF (see Section 7.4 for the properties of the joint CDF for discrete random variables). They are (see Figure 12.11b for an illustration):

P12.1 $F_{X,Y}(-\infty, -\infty) = 0$

P12.2 $F_{X,Y}(+\infty, +\infty) = 1$

P12.3 $F_{X,Y}(x, \infty) = F_X(x)$

P12.4 $F_{X,Y}(\infty, y) = F_Y(y)$

P12.5 $F_{X,Y}(x, y)$ is monotonically increasing, which means that if $x_2 \geq x_1$ and $y_2 \geq y_1$, then $F_{X,Y}(x_2, y_2) \geq F_{X,Y}(x_1, y_1)$.

P12.6 $F_{X,Y}(x, y)$ is continuous with no jumps (assuming that X and Y are jointly continuous random variables). This property is different from the case of jointly discrete random variables.

12.5   Independence of Multiple Random Variables

The definition of independence of two continuous random variables is the same as for discrete random variables. Two continuous random variables X and Y are defined to be independent if for *all* events $A \in R$ and $B \in R$

$$P[X \in A, Y \in B] = P[X \in A]P[Y \in B]. \tag{12.10}$$

Using the definition of conditional probability this is equivalent to

$$
\begin{aligned}
P[Y \in B | X \in A] &= \frac{P[X \in A, Y \in B]}{P[X \in A]} \\
&= P[Y \in B]
\end{aligned}
$$

and similarly $P[X \in A | Y \in B] = P[X \in A]$. It can be shown that X and Y are independent if and only if the joint PDF factors as (see Problem 12.20)

$$p_{X,Y}(x, y) = p_X(x)p_Y(y). \tag{12.11}$$

Alternatively, X and Y are independent if and only if (see Problem 12.21)

$$F_{X,Y}(x, y) = F_X(x)F_Y(y). \tag{12.12}$$

An example follows.

Example 12.6 – Independence of exponential random variables

From Example 12.3 we have for the joint PDF

$$p_{X,Y}(x, y) = \begin{cases} \exp[-(x+y)] & x \geq 0, y \geq 0 \\ 0 & \text{otherwise.} \end{cases}$$

Recalling that the unit step function $u(x)$ is defined as $u(x) = 1$ for $x \geq 0$ and $u(x) = 0$ for $x < 0$, we have

$$p_{X,Y}(x,y) = \exp[-(x+y)]u(x)u(y)$$

since $u(x)u(y) = 1$ if and only if $u(x) = 1$ *and* $u(y) = 1$, which will be true for $x \geq 0, y \geq 0$. Hence, we have

$$p_{X,Y}(x,y) = \underbrace{\exp(-x)u(x)}_{p_X(x)}\underbrace{\exp(-y)u(y)}_{p_Y(y)}.$$

To assert independence we need only factor $p_{X,Y}(x,y)$ as $g(x)h(y)$, where g and h are nonnegative functions. However, to assert that $g(x)$ is actually $p_X(x)$ and $h(y)$ is actually $p_Y(y)$, each function, g and h, must integrate to one. For example, we could have factored $p_{X,Y}(x,y)$ into $(1/2)\exp(-x)u(x)$ and $2\exp(-y)u(y)$, but then we could not claim that $p_X(x) = (1/2)\exp(-x)u(x)$ since it does not integrate to one. Note also that the joint CDF given in Example 12.3 is also factorable as given in (12.8) and in general, factorization of the CDF is also necessary and sufficient to assert independence.

$\Diamond$

 Assessing independence – careful with domain of PDF

The joint PDF given by

$$p_{X,Y}(x,y) = \begin{cases} 2\exp[-(x+y)] & x \geq 0, y \geq 0, \text{ and } y < x \\ 0 & \text{otherwise} \end{cases}$$

is *not* factorable, although it is very similar to our previous example. The reason is that the region in the x-y plane where $p_{X,Y}(x,y) \neq 0$ cannot be written as $u(x)u(y)$ or for that matter as any $g(x)h(y)$. See Figure 12.13.

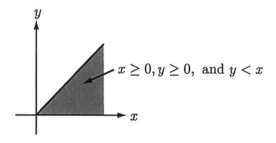

Figure 12.13: Nonfactorable region in x-y plane.

Example 12.7 – Standard bivariate Gaussian PDF

From (12.3) we see that $p_{X,Y}(x, y)$ is only factorable if $\rho = 0$. From Figure 12.9d this corresponds to the case of circular PDF contours. Specifically, for $\rho = 0$, we have

$$
\begin{aligned}
p_{X,Y}(x, y) &= \frac{1}{2\pi} \exp\left[-\frac{1}{2}(x^2 + y^2)\right] \qquad -\infty < x < \infty, -\infty < y < \infty \\
&= \underbrace{\frac{1}{\sqrt{2\pi}} \exp\left[-\frac{1}{2}x^2\right]}_{p_X(x)} \underbrace{\frac{1}{\sqrt{2\pi}} \exp\left[-\frac{1}{2}y^2\right]}_{p_Y(y)}.
\end{aligned}
$$

Hence, we observe that if $\rho = 0$, then X and Y are independent. Furthermore, each marginal PDF is a standard Gaussian (normal) PDF, but as shown in Example 12.2 this holds regardless of the value of ρ.

$\Diamond$

Finally, note that if we can assume that X and Y are independent, then knowledge of $p_X(x)$ and $p_Y(y)$ is sufficient to determine the joint PDF according to (12.11). In practice, the independence assumption greatly simplifies the problem of joint PDF estimation as we need only to estimate the two one-dimensional PDFs $p_X(x)$ and $p_Y(y)$.

12.6   Transformations

We will consider two types of transformations. The first one maps two continuous random variables into a single continuous random variable as $Z = g(X, Y)$, and the second one maps two continuous random variables into two new continuous random variables as $W = g(X, Y)$ and $Z = h(X, Y)$. The first type of transformation $Z = g(X, Y)$ is now discussed. The approach is to find the CDF of Z and then differentiate it to obtain the PDF. The CDF of Z is given as

$$
\begin{aligned}
F_Z(z) &= P[Z \le z] && \text{(definition of CDF)} \\
&= P[g(X, Y) \le z] && \text{(definition of } Z\text{)} \\
&= \iint_{\{(x,y):g(x,y)\le z\}} p_{X,Y}(x, y) dx\, dy && \text{(from (12.2)).} \qquad (12.13)
\end{aligned}
$$

We see that it is necessary to integrate the joint PDF over the region in the plane where $g(x, y) \le z$. Depending upon the form of g, this may be a simple task or unfortunately a very complicated one. A simple example follows. It is the continuous version of (7.22), which yields the PMF for the sum of two independent discrete random variables.

Example 12.8 – Sum of independent $\mathcal{U}(0,1)$ random variables

In Section 2.3 we inquired as to the distribution of the outcomes of an experiment that added U_1, a number chosen at random from 0 to 1, to U_2, another number chosen at random from 0 to 1. A histogram of the outcomes of a computer simulation indicated that there is a higher probability of the sum being near 1, as opposed to being near 0 or 2. We now know that $U_1 \sim \mathcal{U}(0,1)$, $U_2 \sim \mathcal{U}(0,1)$. Also, in the experiment of Section 2.3 the two numbers were chosen independently of each other. Hence, we can determine the probabilities of the sum random variable if we first find the CDF of $X = U_1 + U_2$, where U_1 and U_2 are independent, and then differentiate it to find the PDF of X. We will use (12.13) and replace x, y, z, and $g(x, y)$ by u_1, u_2, x, and $g(u_1, u_2)$, respectively. Then

$$F_X(x) = \iint_{\{(u_1, u_2):u_1 + u_2 \leq x\}} p_{U_1, U_2}(u_1, u_2) du_1 \, du_2.$$

To determine the possible values of X, we note that both U_1 and U_2 take on values in $(0, 1)$ and so $0 < X < 2$. In evaluating the CDF we need two different intervals for x as shown in Figure 12.14. Since U_1 and U_2 are independent, we have $p_{U_1, U_2} = p_{U_1} p_{U_2}$

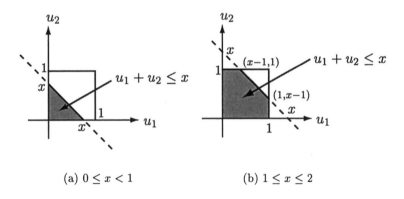

(a) $0 \leq x < 1$ (b) $1 \leq x \leq 2$

Figure 12.14: Shaded areas are regions of integration used to find CDF.

and therefore $p_{U_1, U_2}(u_1, u_2) = 1$ for $0 < u_1 < 1$ and $0 < u_2 < 1$, which results in

$$F_X(x) = \iint_{\{(u_1, u_2):u_1 + u_2 \leq x\}} 1 \, du_1 \, du_2 = \text{shaded area in Figure 12.14.}$$

Hence, the CDF is given by

$$F_X(x) = \begin{cases} 0 & x < 0 \\ \frac{1}{2}x^2 & 0 \leq x < 1 \\ 1 - \frac{1}{2}(2 - x)^2 & 1 \leq x \leq 2 \\ 1 & x > 2. \end{cases}$$

and the PDF is finally

$$p_X(x) = \frac{dF_X(x)}{dx}$$

$$= \begin{cases} 0 & x < 0 \\ x & 0 \le x < 1 \\ 2-x & 1 \le x \le 2 \\ 0 & x > 2. \end{cases}$$

This PDF is shown in Figure 12.15. This is in agreement with our computer results

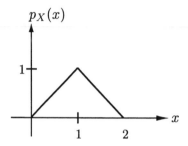

Figure 12.15: PDF for the sum of two independent $\mathcal{U}(0,1)$ random variables.

shown in Figure 2.2. The highest probability is at $x = 1$, which concurs with our computer generated results of Section 2.3. Also, note that $p_X(x) = p_{U_1}(x) \star p_{U_2}(x)$, where $\star$ denotes integral *convolution* (see Problem 12.28).

More generally, we can derive a useful formula for the PDF of the sum of two independent continuous random variables. According to (12.13), we first need to determine the region in the plane for which $x+y \le z$. This inequality can be written as $y \le z - x$, where z is to be regarded for the present as a *constant*. To integrate $p_{X,Y}(x,y)$ over this region, which is shown in Figure 12.16 as the shaded region, we can use an iterated integral. Thus,

$$\begin{aligned} F_Z(z) &= \int_{-\infty}^{\infty} \int_{-\infty}^{z-x} p_{X,Y}(x,y)dy\,dx \\ &= \int_{-\infty}^{\infty} \int_{-\infty}^{z-x} p_X(x)p_Y(y)dy\,dx && \text{(independence)} \\ &= \int_{-\infty}^{\infty} p_X(x) \int_{-\infty}^{z-x} p_Y(y)dy\,dx \\ &= \int_{-\infty}^{\infty} p_X(x)F_Y(z-x)dx && \text{(definition of CDF).} \end{aligned}$$

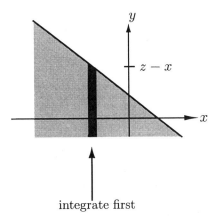

integrate first

Figure 12.16: Iterated integral evaluation – shaded region is $y \leq z - x$. Integrate first in y direction for a fixed x and then integrate over $-\infty < x < \infty$.

If we now differentiate the CDF, we have

$$
\begin{aligned}
p_Z(z) &= \frac{d}{dz} \int_{-\infty}^{\infty} p_X(x) F_Y(z - x)\, dx \\
&= \int_{-\infty}^{\infty} p_X(x) \frac{d}{dz} F_Y(z - x)\, dx && \text{(assume interchange is valid)} \\
&= \int_{-\infty}^{\infty} p_X(x) \left. \frac{d}{du} F_Y(u) \right|_{u=z-x} \frac{du}{dz}\, dx && \text{(chain rule with } u = z - x\text{)}
\end{aligned}
$$

so that finally we have our formula

$$
p_Z(z) = \int_{-\infty}^{\infty} p_X(x) p_Y(z - x)\, dx. \tag{12.14}
$$

This is the analogous result to (7.22). It is recognized as a convolution integral, which we can express more succinctly as $p_Z = p_X \star p_Y$, and thus may be more easily evaluated by using characteristic functions. The latter approach is explored in Section 12.10.

A second approach to obtaining the PDF of $g(X, Y)$ is to let $W = X$, $Z = g(X, Y)$, find the joint PDF of W and Z, i.e., $p_{W,Z}(w, z)$, and finally integrate out W to yield the desired PDF for Z. This method was encountered previously in Chapter 7, where it was used for discrete random variables, and was termed the method of auxiliary random variables. To implement it now requires us to determine the joint PDF of two new random variables that result from having transformed two random variables. This is the second type of transformation we were interested in. Hence, we now consider the more general transformation

$$
\begin{aligned}
W &= g(X, Y) \\
Z &= h(X, Y).
\end{aligned}
$$

The final result will be a formula relating the joint PDF of (W, Z) to that of the given joint PDF of (X, Y). It will be a generalization of the single random variable transformation formula

$$p_Y(y) = p_X(g^{-1}(y)) \left| \frac{dg^{-1}(y)}{dy} \right| \tag{12.15}$$

for $Y = g(X)$.

To understand what is involved, consider as an example the transformation

$$\begin{bmatrix} X_1 \\ X_2 \end{bmatrix} = \begin{bmatrix} U_1 \\ (U_1 + U_2)/2 \end{bmatrix} \tag{12.16}$$

where $U_1 \sim \mathcal{U}(0, 1)$, $U_2 \sim \mathcal{U}(0, 1)$, and U_1 and U_2 are independent. In Figure 2.13 we plotted realizations of $[U_1 \, U_2]^T$ and $[X_1 \, X_2]^T$. Note that the original joint PDF p_{U_1, U_2} is nonzero on the unit square while the transformed PDF is nonzero on a parallelogram. In either case the PDFs appear to be uniformly distributed. Similar observations about the region for which the PDF of the transformed random variable is nonzero were made in the one-dimensional case for $Y = g(X)$, where $X \sim \mathcal{U}(0, 1)$, in Figure 10.22. In general, a linear transformation will change the support area of the joint PDF, which is the region in the plane where the PDF is nonzero. In Figure 2.13 it is seen that the area of the square is 1 while that for the parallelogram is $1/2$. It can furthermore be shown that if we have the linear transformation (see Problem 12.29)

$$\begin{bmatrix} W \\ Z \end{bmatrix} = \underbrace{\begin{bmatrix} a & b \\ c & d \end{bmatrix}}_{\mathbf{G}} \begin{bmatrix} X \\ Y \end{bmatrix} \tag{12.17}$$

then

$$\begin{aligned} \frac{\text{Area in } w\text{-}z \text{ plane}}{\text{Area in } x\text{-}y \text{ plane}} &= |\det(\mathbf{G})| \\ &= |ad - bc|. \end{aligned}$$

It is always assumed that $\mathbf{G}$ is invertible so that $\det(\mathbf{G}) \neq 0$. In the previous example of (12.16) for which in our new notation we have $W = X$ and $Z = (X + Y)/2$, the linear transformation matrix is

$$\mathbf{G} = \begin{bmatrix} 1 & 0 \\ \frac{1}{2} & \frac{1}{2} \end{bmatrix}$$

and it is seen that $|\det(\mathbf{G})| = 1/2$. Thus, the PDF support region is decreased by a factor of 2. We therefore expect the joint PDF of $[X \, (X + Y)/2]^T$ to be uniform with a height of 2 (as opposed to a height of 1 for the original joint PDF). Hence, the transformed PDF should have a factor of $1/|\det(\mathbf{G})|$ to make it integrate to one.

This amplification factor, which is $1/|\det(\mathbf{G})| = |\det(\mathbf{G}^{-1})|$ must be included in the expression for the transformed joint PDF. Also, we have that $[x\,y]^T = \mathbf{G}^{-1}[w\,z]^T$. Hence, it should not be surprising that for the linear transformation of (12.17) we have the formula for the transformed joint PDF

$$p_{W,Z}(w,z) = p_{X,Y}\left(\mathbf{G}^{-1}\begin{bmatrix} w \\ z \end{bmatrix}\right)|\det(\mathbf{G}^{-1})|. \tag{12.18}$$

An example follows.

Example 12.9 – Linear transformation for standard bivariate Gaussian PDF

Assume that (X,Y) has the PDF of (12.3) and consider the linear transformation

$$\begin{bmatrix} W \\ Z \end{bmatrix} = \underbrace{\begin{bmatrix} \sigma_W & 0 \\ 0 & \sigma_Z \end{bmatrix}}_{\mathbf{G}}\begin{bmatrix} X \\ Y \end{bmatrix}.$$

Then,

$$\mathbf{G}^{-1} = \begin{bmatrix} 1/\sigma_W & 0 \\ 0 & 1/\sigma_Z \end{bmatrix}$$

and

$$\mathbf{G}^{-1}\begin{bmatrix} w \\ z \end{bmatrix} = \begin{bmatrix} w/\sigma_W \\ z/\sigma_Z \end{bmatrix}$$

$$\det(\mathbf{G}^{-1}) = \frac{1}{\sigma_W \sigma_Z}$$

so that from (12.3) and (12.18)

$$p_{W,Z}(w,z)$$

$$= \frac{1}{2\pi\sqrt{1-\rho^2}}\exp\left[-\frac{1}{2(1-\rho^2)}\left((w/\sigma_W)^2 - 2\rho wz/(\sigma_W\sigma_Z) + (z/\sigma_Z)^2\right)\right]\frac{1}{\sigma_W\sigma_Z}$$

$$= \frac{1}{2\pi\sqrt{(1-\rho^2)\sigma_W^2\sigma_Z^2}}$$

$$\cdot \exp\left[-\frac{1}{2(1-\rho^2)}\left(\left(\frac{w}{\sigma_W}\right)^2 - 2\rho\left(\frac{w}{\sigma_W}\right)\left(\frac{z}{\sigma_Z}\right) + \left(\frac{z}{\sigma_Z}\right)^2\right)\right]. \tag{12.19}$$

Note that since $-\infty < x < \infty$, $-\infty < y < \infty$, we have that the region of support for $p_{W,Z}$ is $-\infty < w < \infty$, $-\infty < z < \infty$. Also, the joint PDF can be written in vector/matrix form as (see Problem 12.31)

$$p_{W,Z}(w,z) = \frac{1}{2\pi\det^{1/2}(\mathbf{C})}\exp\left(-\frac{1}{2}\begin{bmatrix} w \\ z \end{bmatrix}^T\mathbf{C}^{-1}\begin{bmatrix} w \\ z \end{bmatrix}\right) \tag{12.20}$$

where

$$\mathbf{C} = \begin{bmatrix} \sigma_W^2 & \rho \sigma_W \sigma_Z \\ \rho \sigma_Z \sigma_W & \sigma_Z^2 \end{bmatrix}. \tag{12.21}$$

The matrix $\mathbf{C}$ will be shown later to be the *covariance matrix* of W and Z (see Section 9.5 for the definition of the covariance matrix, which is also valid for continuous random variables).

$\diamond$

For nonlinear transformations a result similar to (12.18) is obtained. This is because a two-dimensional nonlinear function can be linearized about a point by replacing the usual tangent or derivative approximation for a one-dimensional function by a tangent plane approximation (see Problem 12.32). Hence, if the transformation is given by

$$\begin{aligned} W &= g(X, Y) \\ Z &= h(X, Y) \end{aligned}$$

then a given point in the w-z plane is obtained via $w = g(x, y)$, $z = h(x, y)$. Assume that the latter set of equations has a single solution for all (w, z), say

$$\begin{aligned} x &= g^{-1}(w, z) \\ y &= h^{-1}(w, z). \end{aligned}$$

Then it can be shown that

$$p_{W,Z}(w, z) = p_{X,Y}(\underbrace{g^{-1}(w, z)}_{x}, \underbrace{h^{-1}(w, z)}_{y}) \left| \det \left(\frac{\partial(x, y)}{\partial(w, z)} \right) \right| \tag{12.22}$$

where

$$\frac{\partial(x, y)}{\partial(w, z)} = \begin{bmatrix} \frac{\partial x}{\partial w} & \frac{\partial x}{\partial z} \\ \frac{\partial y}{\partial w} & \frac{\partial y}{\partial z} \end{bmatrix} \tag{12.23}$$

is called the Jacobian matrix of the inverse transformation from $[w \ z]^T$ to $[x \ y]^T$ and is sometimes referred to as $\mathbf{J}^{-1}$. It represents the compensation for the amplification/reduction of the areas due to the transformation. For a linear transformation $\mathbf{G}$ it is given by $\mathbf{J} = \mathbf{G}$ (see also (12.15) for a single random variable). We now illustrate the use of this formula.

Example 12.10 – Affine transformation for standard bivariate Gaussian PDF

Let (X, Y) have a standard bivariate Gaussian PDF and consider the affine transformation

$$\begin{bmatrix} W \\ Z \end{bmatrix} = \begin{bmatrix} \sigma_W & 0 \\ 0 & \sigma_Z \end{bmatrix} \begin{bmatrix} X \\ Y \end{bmatrix} + \begin{bmatrix} \mu_W \\ \mu_Z \end{bmatrix}.$$

Then using (12.22) we first solve for (x, y) as

$$x = \frac{w - \mu_W}{\sigma_W}$$

$$y = \frac{z - \mu_Z}{\sigma_Z}.$$

The inverse Jacobian matrix becomes

$$\frac{\partial(x, y)}{\partial(w, z)} = \begin{bmatrix} 1/\sigma_W & 0 \\ 0 & 1/\sigma_Z \end{bmatrix}$$

and therefore, since

$$p_{X,Y}(x, y) = \frac{1}{2\pi\sqrt{1 - \rho^2}} \exp\left[-\frac{1}{2(1 - \rho^2)}(x^2 - 2\rho xy + y^2)\right]$$

we have from (12.22)

$$p_{W,Z}(w, z) = \frac{1}{2\pi\sqrt{1 - \rho^2}}$$

$$\cdot \exp\left[-\frac{1}{2(1 - \rho^2)}\left(\left(\frac{w - \mu_W}{\sigma_W}\right)^2 - 2\rho\left(\frac{w - \mu_W}{\sigma_W}\right)\left(\frac{z - \mu_Z}{\sigma_Z}\right) + \left(\frac{z - \mu_Z}{\sigma_Z}\right)^2\right)\right]\frac{1}{\sigma_W\sigma_Z}$$

or finally

$$p_{W,Z}(w, z) = \frac{1}{2\pi\sqrt{(1 - \rho^2)\sigma_W^2\sigma_Z^2}}$$

$$\cdot \exp\left[-\frac{1}{2(1 - \rho^2)}\left(\left(\frac{w - \mu_W}{\sigma_W}\right)^2 - 2\rho\left(\frac{w - \mu_W}{\sigma_W}\right)\left(\frac{z - \mu_Z}{\sigma_Z}\right) + \left(\frac{z - \mu_Z}{\sigma_Z}\right)^2\right)\right].$$

$$(12.24)$$

This is called the *bivariate Gaussian PDF*. If $\mu_W = \mu_Z = 0$ and $\sigma_W = \sigma_Z = 1$, then it reverts back to the usual *standard* bivariate Gaussian PDF. If $\mu_W = \mu_Z = 0$, we have the joint PDF in Example 12.9. An example of the PDF is shown in Figure 12.17.

$\Diamond$

The bivariate Gaussian PDF can also be written more compactly in vector/matrix form as

$$p_{W,Z}(w, z) = \frac{1}{2\pi \det^{1/2}(\mathbf{C})} \exp\left(-\frac{1}{2}\begin{bmatrix} w - \mu_W \\ z - \mu_Z \end{bmatrix}^T \mathbf{C}^{-1} \begin{bmatrix} w - \mu_W \\ z - \mu_Z \end{bmatrix}\right) \quad (12.25)$$

where $\mathbf{C}$ is the covariance matrix given by (12.21). It can also be shown that the marginal PDFs are $W \sim \mathcal{N}(\mu_W, \sigma_W^2)$ and $Z \sim \mathcal{N}(\mu_Z, \sigma_Z^2)$ (see Problem 12.36).

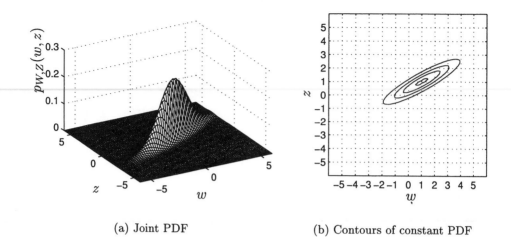

(a) Joint PDF   (b) Contours of constant PDF

Figure 12.17: Example of bivariate Gaussian PDF with $\mu_W = 1, \mu_Z = 1, \sigma_W^2 = 3, \sigma_Z^2 = 1$, and $\rho = 0.9$.

Hence, *the marginal PDFs of the bivariate Gaussian PDF are obtained by inspection* (see Problem 12.37).

Example 12.11 – Transformation of independent Gaussian random variables to a Cauchy random variable

Let $X \sim \mathcal{N}(0,1)$, $Y \sim \mathcal{N}(0,1)$, and X and Y be independent. Then consider the transformation $W = X, Z = Y/X$. To determine $\mathcal{S}_{W,Z}$ note that $w = x$ so that $-\infty < w < \infty$ and since $z = y/x$ with $-\infty < x < \infty, -\infty < y < \infty$, we have $-\infty < z < \infty$. Hence, $\mathcal{S}_{W,Z}$ is the entire plane. To find the joint PDF we first solve for (x,y) as $x = w$ and $y = xz = wz$. The inverse Jacobian matrix is

$$\frac{\partial(x,y)}{\partial(w,z)} = \begin{bmatrix} 1 & 0 \\ z & w \end{bmatrix}$$

so that $|\det(\partial(x,y)/\partial(w,z))| = |w|$. Using (12.22), we have

$$
\begin{aligned}
p_{W,Z}(w,z) &= \frac{1}{2\pi} \exp\left[-\frac{1}{2}(x^2 + y^2)\right]\Bigg|_{x=w, y=wz} |w| \\
&= \frac{1}{2\pi} \exp\left[-\frac{1}{2}(w^2 + w^2 z^2)\right] |w| \\
&= \frac{1}{2\pi} \exp\left[-\frac{1}{2}(1 + z^2)w^2\right] |w|.
\end{aligned}
$$

It is of interest to determine the marginal PDFs. Clearly, the marginal of $W = X$ is just the original PDF $\mathcal{N}(0,1)$. The marginal PDF for Z, which is the ratio of two independent $\mathcal{N}(0,1)$ random variables, is found from (12.4) as

$$
\begin{aligned}
p_Z(z) &= \int_{-\infty}^{\infty} p_{W,Z}(w,z)\,dw \\
&= \int_{-\infty}^{\infty} \frac{1}{2\pi} \exp\left[-\frac{1}{2}(1+z^2)w^2\right]|w|\,dw \\
&= \frac{1}{\pi}\int_0^{\infty} w \exp\left[-\frac{1}{2}(1+z^2)w^2\right]dw \quad \text{(integrand is even function)} \\
&= \frac{1}{\pi}\left.\frac{\exp[-(1/2)(1+z^2)w^2]}{-(1+z^2)}\right|_0^{\infty} \\
&= \frac{1}{\pi(1+z^2)} \qquad -\infty < z < \infty
\end{aligned}
$$

which is recognized as the Cauchy PDF. Hence, the PDF of Y/X, where X and Y are independent standard Gaussian random variables is Cauchy. We have implicitly used the method of *auxiliary random variables* to derive this result. Finally, the observation that the denominator of Y/X is a standard Gaussian random variable, with significant probability of being near zero, may help to explain why the outcomes of a Cauchy random variable are as large as they are. See Figure 11.3.

$\diamond$

The next example is of great importance in many fields of science and engineering.

Example 12.12 – Magnitude and angle of jointly Gaussian distributed random variables

Let $X \sim \mathcal{N}(0, \sigma^2)$, $Y \sim \mathcal{N}(0, \sigma^2)$, and X and Y be independent random variables. Then, it is desired to find the joint PDF when X and Y, considered as Cartesian coordinates, are converted to polar coordinates via

$$
\begin{aligned}
R &= \sqrt{X^2 + Y^2} \qquad R \geq 0 \\
\Theta &= \arctan\frac{Y}{X} \qquad 0 \leq \Theta < 2\pi.
\end{aligned}
\tag{12.26}
$$

It is common in many engineering disciplines, for example, in radar, sonar, and communications, to transmit a sinusoidal signal and to process the received signal by a digital computer. The received signal will be given by $s(t) = A\cos(2\pi F_0 t) + B\sin(2\pi F_0 t)$ for a transmit frequency of F_0 Hz. However, because the received signal is due to the sum of multiple reflections from an aircraft, as in the radar example, the values of A and B are generally not known. Consequently, they are modeled as continuous random variables with marginal PDFs $A \sim \mathcal{N}(0, \sigma^2)$, $B \sim \mathcal{N}(0, \sigma^2)$, and where A and B are independent. Since the received signal can equivalently be written in terms of a single sinusoid as (see Problem 12.42)

$$
s(t) = \sqrt{A^2 + B^2}\cos(2\pi F_0 t - \arctan(B/A))
$$

the amplitude is a random variable as is the phase angle. Thus, the transformation of (12.26) is of interest in order to determine the joint PDF of the sinusoid's amplitude and phase. This motivates our interest in this particular transformation.

We first solve for (x, y) as $x = r \cos\theta$, $y = r \sin\theta$. Then using (12.22) and replacing w by r and z by θ we have the inverse Jacobian

$$\frac{\partial(x, y)}{\partial(r, \theta)} = \begin{bmatrix} \cos\theta & -r\sin\theta \\ \sin\theta & r\cos\theta \end{bmatrix}$$

and thus

$$\det\left(\frac{\partial(x, y)}{\partial(r, \theta)}\right) = r \geq 0.$$

Since

$$p_{X,Y}(x, y) = p_X(x)p_Y(y) = \frac{1}{2\pi\sigma^2} \exp\left[-\frac{1}{2\sigma^2}(x^2 + y^2)\right]$$

we have upon using (12.22)

$$
\begin{aligned}
p_{R,\Theta}(r, \theta) &= \frac{1}{2\pi\sigma^2} \exp\left[-\frac{1}{2\sigma^2}r^2\right] r \qquad r \geq 0, \, 0 \leq \theta < 2\pi \\
&= \underbrace{\frac{r}{\sigma^2} \exp\left[-\frac{1}{2\sigma^2}r^2\right]}_{p_R(r)} \underbrace{\frac{1}{2\pi}}_{p_\Theta(\theta)} \qquad r \geq 0, \, 0 \leq \theta < 2\pi.
\end{aligned}
$$

Here we see that R has a Rayleigh PDF with parameter σ^2, Θ has a uniform PDF, and R and Θ are independent random variables.

◇

12.7   Expected Values

The expected value of two jointly distributed continuous random variables X and Y, or equivalently the random vector $[X\,Y]^T$, is defined as the vector of the expected values. That is to say

$$E_{X,Y}\left[\begin{bmatrix} X \\ Y \end{bmatrix}\right] = \begin{bmatrix} E_X[X] \\ E_Y[Y] \end{bmatrix}.$$

Of course this is equivalent to the vector of the expected values of the marginal PDFs. As an example, for the bivariate Gaussian PDF as given by (12.24) with W, Z replaced by X, Y, the marginals are $\mathcal{N}(\mu_X, \sigma_X^2)$ and $\mathcal{N}(\mu_Y, \sigma_Y^2)$ and hence the expected value or equivalently the mean of the random vector is

$$E_{X,Y}\left[\begin{bmatrix} X \\ Y \end{bmatrix}\right] = \begin{bmatrix} \mu_X \\ \mu_Y \end{bmatrix}$$

as shown in Figure 12.17 for $\mu_X = \mu_Y = 1$.

We frequently require the expected value of a function of two jointly distributed random variables or of $Z = g(X, Y)$. By definition this is

$$E[Z] = \int_{-\infty}^{\infty} z p_Z(z) dz.$$

But as in the case for jointly distributed *discrete* random variables we can avoid the determination of $p_Z(z)$ by employing instead the formula

$$E[Z] = E[g(X, Y)] = \int_{-\infty}^{\infty} \int_{-\infty}^{\infty} g(x, y) p_{X,Y}(x, y) dx \, dy. \qquad (12.27)$$

To remind us that the averaging PDF is $p_{X,Y}(x, y)$ we usually write this as

$$E_{X,Y}[g(X, Y)] = \int_{-\infty}^{\infty} \int_{-\infty}^{\infty} g(x, y) p_{X,Y}(x, y) dx \, dy. \qquad (12.28)$$

If the function g depends on only one of the variables, say X, then we have

$$\begin{aligned}
E_{X,Y}[g(X)] &= \int_{-\infty}^{\infty} \int_{-\infty}^{\infty} g(x) p_{X,Y}(x, y) dx \, dy \\
&= \int_{-\infty}^{\infty} g(x) \underbrace{\int_{-\infty}^{\infty} p_{X,Y}(x, y) dy}_{p_X(x)} \, dx \\
&= E_X[g(X)].
\end{aligned}$$

As in the case of discrete random variables (see Section 7.7), the expectation has the following properties:

1. Linearity

$$E_{X,Y}[aX + bY] = aE_X[X] + bE_Y[Y]$$

and more generally

$$E_{X,Y}[ag(X, Y) + bh(X, Y)] = aE_{X,Y}[g(X, Y)] + bE_{X,Y}[h(X, Y)].$$

2. Factorization for independent random variables

If X and Y are independent random variables

$$E_{X,Y}[XY] = E_X[X]E_Y[Y] \qquad (12.29)$$

and more generally

$$E_{X,Y}[g(X)h(Y)] = E_X[g(X)]E_Y[h(Y)]. \qquad (12.30)$$

Also, in determining the variance for a sum of random variables we have

$$\text{var}(X+Y) = \text{var}(X) + \text{var}(Y) + 2\text{cov}(X,Y) \qquad (12.31)$$

where $\text{cov}(X,Y) = E_{X,Y}[(X - E_X[X])(Y - E_Y[Y])]$. If X and Y are independent, then by (12.30)

$$
\begin{aligned}
\text{cov}(X,Y) &= E_{X,Y}[(X - E_X[X])(Y - E_Y[Y])] \\
&= E_X[(X - E_X[X]]E_Y[(Y - E_Y[Y])] \\
&= 0.
\end{aligned}
$$

The covariance can also be computed as

$$\text{cov}(X,Y) = E_{X,Y}[XY] - E_X[X]E_Y[Y] \qquad (12.32)$$

where

$$E_{X,Y}[XY] = \int_{-\infty}^{\infty}\int_{-\infty}^{\infty} xy p_{X,Y}(x,y)\,dx\,dy. \qquad (12.33)$$

An example follows.

Example 12.13 – Covariance for standard bivariate Gaussian PDF
For the standard bivariate Gaussian PDF of (12.3) we now determine $\text{cov}(X,Y)$. We have already seen that the marginal PDFs are $X \sim \mathcal{N}(0,1)$ and $Y \sim \mathcal{N}(0,1)$ so that $E_X[X] = E_Y[Y] = 0$. From (12.32) we have that $\text{cov}(X,Y) = E_{X,Y}[XY]$ and using (12.33) and (12.3)

$$\text{cov}(X,Y) = \int_{-\infty}^{\infty}\int_{-\infty}^{\infty} xy \frac{1}{2\pi\sqrt{1-\rho^2}} \exp\left[-\frac{1}{2(1-\rho^2)}(x^2 - 2\rho xy + y^2)\right] dx\,dy.$$

To evaluate this double integral we use iterated integrals and complete the square in the exponent of the exponential as was previously done in Example 12.2. This results in

$$Q = y^2 - 2\rho xy + x^2 = (y - \rho x)^2 + (1 - \rho^2)x^2$$

and produces

$$\text{cov}(X,Y)$$

$$
\begin{aligned}
&= \int_{-\infty}^{\infty}\int_{-\infty}^{\infty} xy \frac{1}{2\pi\sqrt{1-\rho^2}} \exp\left[-\frac{1}{2(1-\rho^2)}(y - \rho x)^2\right] \exp\left[-\frac{1}{2}x^2\right] dx\,dy \\
&= \int_{-\infty}^{\infty} x \frac{1}{\sqrt{2\pi}} \exp\left[-\frac{1}{2}x^2\right] \int_{-\infty}^{\infty} y \frac{1}{\sqrt{2\pi(1-\rho^2)}} \exp\left[-\frac{1}{2(1-\rho^2)}(y - \rho x)^2\right] dy\,dx.
\end{aligned}
$$

The inner integral over y is just $E_Y[Y] = \int_{-\infty}^{\infty} y p_Y(y)dy$, where $Y \sim \mathcal{N}(\rho x, 1 - \rho^2)$. Thus, $E_Y[Y] = \rho x$ so that

$$
\begin{aligned}
\text{cov}(X,Y) &= \int_{-\infty}^{\infty} \rho x^2 \frac{1}{\sqrt{2\pi}} \exp\left[-\frac{1}{2}x^2\right] dx \\
&= \rho E_X[X^2]
\end{aligned}
$$

where $X \sim \mathcal{N}(0,1)$. But $E_X[X^2] = \text{var}(X) + E_X^2[X] = 1 + 0^2 = 1$ and therefore we have finally that

$$\text{cov}(X,Y) = \rho.$$

$\diamondsuit$

With the result of the previous example we can now determine the correlation coefficient between X and Y for the standard bivariate Gaussian PDF. Since the marginal PDFs are $X \sim \mathcal{N}(0,1)$ and $Y \sim \mathcal{N}(0,1)$, the correlation coefficient between X and Y is

$$
\begin{aligned}
\rho_{X,Y} &= \frac{\text{cov}(X,Y)}{\sqrt{\text{var}(X)\text{var}(Y)}} \\
&= \frac{\rho}{\sqrt{1 \cdot 1}} \\
&= \rho.
\end{aligned}
$$

We have therefore established that *in the standard bivariate Gaussian PDF, the parameter ρ is the correlation coefficient*. This explains the orientation of the constant PDF contours shown in Figure 12.9. Also, we can now assert that if the correlation coefficient between X and Y is zero, i.e., $\rho = 0$, *and X and Y are jointly Gaussian distributed (i.e., a standard bivariate Gaussian PDF)*, then

$$
\begin{aligned}
p_{X,Y}(x,y) &= \frac{1}{2\pi\sqrt{1-\rho^2}} \exp\left[-\frac{1}{2(1-\rho^2)}(x^2 - 2\rho xy + y^2)\right] \\
&= \frac{1}{2\pi} \exp\left[-\frac{1}{2}(x^2 + y^2)\right] \\
&= \underbrace{\frac{1}{\sqrt{2\pi}} \exp\left[-\frac{1}{2}x^2\right]}_{p_X(x)} \underbrace{\frac{1}{\sqrt{2\pi}} \exp\left[-\frac{1}{2}y^2\right]}_{p_Y(y)}
\end{aligned}
$$

and *X and Y are independent.* This also holds for the general bivariate Gaussian PDF in which the marginal PDFs are $X \sim \mathcal{N}(\mu_X, \sigma_X^2)$ and $Y \sim \mathcal{N}(\mu_Y, \sigma_Y^2)$. This result provides a *partial converse* to the theorem that if X and Y are independent, then the random variables are uncorrelated, *but only for this particular joint PDF.*

Finally, since $\rho = \rho_{X,Y}$ we have from (12.21) upon replacing W by X and Z by Y, that

$$
\begin{aligned}
\mathbf{C} &= \begin{bmatrix} \sigma_X^2 & \rho\sigma_X\sigma_Y \\ \rho\sigma_Y\sigma_X & \sigma_Y^2 \end{bmatrix} \\
&= \begin{bmatrix} \sigma_X^2 & \rho_{X,Y}\sigma_X\sigma_Y \\ \rho_{X,Y}\sigma_Y\sigma_X & \sigma_Y^2 \end{bmatrix} \\
&= \begin{bmatrix} \text{var}(X) & \text{cov}(X,Y) \\ \text{cov}(Y,X) & \text{var}(Y) \end{bmatrix}
\end{aligned}
\tag{12.34}
$$

is the covariance matrix. We have now established that $\mathbf{C}$ as given by (12.34) is actually the covariance matrix. Hence, the general bivariate Gaussian PDF is given in succinct form as (see (12.24))

$$p_{X,Y}(x,y) = \frac{1}{2\pi \det^{1/2}(\mathbf{C})} \exp\left(-\frac{1}{2}\begin{bmatrix} x - \mu_X \\ y - \mu_Y \end{bmatrix}^T \mathbf{C}^{-1} \begin{bmatrix} x - \mu_X \\ y - \mu_Y \end{bmatrix}\right) \qquad (12.35)$$

where $\mathbf{C}$ is given by (12.34) and is the covariance matrix (see Section also 9.5)

$$\mathbf{C} = \begin{bmatrix} \text{var}(X) & \text{cov}(X,Y) \\ \text{cov}(Y,X) & \text{var}(Y) \end{bmatrix}. \qquad (12.36)$$

As previously mentioned, an extremely important property of the bivariate Gaussian PDF is that uncorrelated random variables implies independent random variables. Hence, if the covariance matrix in (12.36) is *diagonal*, then X and Y are independent. We have shown in Chapter 9 that it is always possible to diagonalize a covariance matrix by transforming the random vector using a linear transformation. Specifically, if the random vector $[X\,Y]^T$ is transformed to a new random vector $\mathbf{V}^T[X\,Y]^T$, where $\mathbf{V}$ is the modal matrix for the covariance matrix $\mathbf{C}$, then the transformed random vector will have a diagonal covariance matrix. Hence, the transformed random vector will have uncorrelated components. If furthermore, the transformed random vector also has a bivariate Gaussian PDF, then *its component random variables will be independent.* It is indeed fortunate that this is true—a *linearly transformed bivariate Gaussian random vector produces another bivariate Gaussian random vector*, as we now show. To do so it is more convenient to use a vector/matrix representation of the PDF. Let the linear transformation be

$$\begin{bmatrix} W \\ Z \end{bmatrix} = \mathbf{G} \begin{bmatrix} X \\ Y \end{bmatrix}$$

where $\mathbf{G}$ is an invertible 2×2 matrix. Assume for simplicity that $\mu_X = \mu_Y = 0$. Then, from (12.35)

$$p_{X,Y}(x,y) = \frac{1}{2\pi \det^{1/2}(\mathbf{C})} \exp\left(-\frac{1}{2}\begin{bmatrix} x \\ y \end{bmatrix}^T \mathbf{C}^{-1} \begin{bmatrix} x \\ y \end{bmatrix}\right)$$

and using (12.18)

$$\begin{aligned} p_{W,Z}(w,z) &= p_{X,Y}\left(\mathbf{G}^{-1}\begin{bmatrix} w \\ z \end{bmatrix}\right) |\det(\mathbf{G}^{-1})| \\ &= \frac{1}{2\pi \det^{1/2}(\mathbf{C})} \exp\left(-\frac{1}{2}\begin{bmatrix} w \\ z \end{bmatrix}^T \mathbf{G}^{-1^T}\mathbf{C}^{-1}\mathbf{G}^{-1}\begin{bmatrix} w \\ z \end{bmatrix}\right) |\det(\mathbf{G}^{-1})|. \end{aligned}$$

But it can be shown that (see Section C.3 of Appendix C for matrix inverse and determinant formulas)

$$\mathbf{G}^{-1^T}\mathbf{C}^{-1}\mathbf{G}^{-1} = \mathbf{G}^{T^{-1}}\mathbf{C}^{-1}\mathbf{G}^{-1} = (\mathbf{G}\mathbf{C}\mathbf{G}^T)^{-1}$$

and

$$
\begin{aligned}
\left|\det(\mathbf{G}^{-1})\right| &= \frac{1}{|\det(\mathbf{G})|} \\
&= \frac{1}{(\det(\mathbf{G})\det(\mathbf{G}))^{1/2}} \\
&= \frac{1}{(\det(\mathbf{G})\det(\mathbf{G}^T))^{1/2}}
\end{aligned}
$$

so that

$$
\begin{aligned}
\frac{\left|\det(\mathbf{G}^{-1})\right|}{\det^{1/2}(\mathbf{C})} &= \frac{1}{\det^{1/2}(\mathbf{C})(\det(\mathbf{G})\det(\mathbf{G}^T))^{1/2}} \\
&= \frac{1}{(\det(\mathbf{C})\det(\mathbf{G})\det(\mathbf{G}^T))^{1/2}} \\
&= \frac{1}{(\det(\mathbf{G})\det(\mathbf{C})\det(\mathbf{G}^T))^{1/2}} \\
&= \frac{1}{\det^{1/2}(\mathbf{G}\mathbf{C}\mathbf{G}^T)}.
\end{aligned}
$$

Thus, we have finally that the PDF of the linearly transformed random vector is

$$p_{W,Z}(w,z) = \frac{1}{2\pi\det^{1/2}(\mathbf{G}\mathbf{C}\mathbf{G}^T)}\exp\left(-\frac{1}{2}\begin{bmatrix} w \\ z \end{bmatrix}^T(\mathbf{G}\mathbf{C}\mathbf{G}^T)^{-1}\begin{bmatrix} w \\ z \end{bmatrix}\right)$$

which is recognized as a bivariate Gaussian PDF with zero means and a covariance matrix $\mathbf{G}\mathbf{C}\mathbf{G}^T$. This also agrees with Property 9.4. We summarize our results in a theorem.

Theorem 12.7.1 (Linear transformation of Gaussian random variables)
If (X,Y) has the bivariate Gaussian PDF

$$p_{X,Y}(x,y) = \frac{1}{2\pi\det^{1/2}(\mathbf{C})}\exp\left(-\frac{1}{2}\begin{bmatrix} x - \mu_X \\ y - \mu_Y \end{bmatrix}^T\mathbf{C}^{-1}\begin{bmatrix} x - \mu_X \\ y - \mu_Y \end{bmatrix}\right) \qquad (12.37)$$

and the random vector is linearly transformed as

$$\begin{bmatrix} W \\ Z \end{bmatrix} = \mathbf{G}\begin{bmatrix} X \\ Y \end{bmatrix}$$

where **G** *is invertible, then*

$$p_{W,Z}(w,z) = \frac{1}{2\pi \det^{1/2}(\mathbf{GCG}^T)} \exp\left(-\frac{1}{2}\begin{bmatrix} w - \mu_W \\ z - \mu_Z \end{bmatrix}^T (\mathbf{GCG}^T)^{-1} \begin{bmatrix} w - \mu_W \\ z - \mu_Z \end{bmatrix}\right)$$

where

$$\begin{bmatrix} \mu_W \\ \mu_Z \end{bmatrix} = \mathbf{G}\begin{bmatrix} \mu_X \\ \mu_Y \end{bmatrix}$$

is the transformed mean vector.

The bivariate Gaussian PDF with mean vector $\boldsymbol{\mu}$ and covariance matrix $\mathbf{C}$ is denoted by $\mathcal{N}(\boldsymbol{\mu}, \mathbf{C})$. Hence, the theorem may be paraphrased as follows—if $[X\,Y]^T \sim \mathcal{N}(\boldsymbol{\mu}, \mathbf{C})$, then $\mathbf{G}[X\,Y]^T \sim \mathcal{N}(\mathbf{G}\boldsymbol{\mu}, \mathbf{GCG}^T)$. An example, which uses results from Example 9.4, is given next.

Example 12.14 – Transforming correlated Gaussian random variables to independent Gaussian random variables

Let $\mu_X = \mu_Y = 0$ and

$$\mathbf{C} = \begin{bmatrix} 26 & 6 \\ 6 & 26 \end{bmatrix}$$

in (12.37). The joint PDF and its constant PDF contours are shown in Figure 12.18. Now transform X and Y according to

$$\begin{bmatrix} W \\ Z \end{bmatrix} = \mathbf{G}\begin{bmatrix} X \\ Y \end{bmatrix}$$

where **G** is the transpose of the modal matrix **V**, which is given in Example 9.4. Therefore

$$\mathbf{G} = \mathbf{V}^T = \begin{bmatrix} \frac{1}{\sqrt{2}} & -\frac{1}{\sqrt{2}} \\ \frac{1}{\sqrt{2}} & \frac{1}{\sqrt{2}} \end{bmatrix}$$

so that

$$\begin{aligned} W &= \frac{1}{\sqrt{2}}X - \frac{1}{\sqrt{2}}Y \\ Z &= \frac{1}{\sqrt{2}}X + \frac{1}{\sqrt{2}}Y. \end{aligned}$$

We have that

$$\mathbf{GCG}^T = \mathbf{V}^T\mathbf{CV} = \begin{bmatrix} 20 & 0 \\ 0 & 32 \end{bmatrix}$$

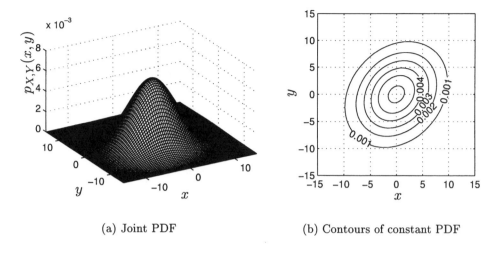

(a) Joint PDF

(b) Contours of constant PDF

Figure 12.18: Example of joint PDF for correlated Gaussian random variables.

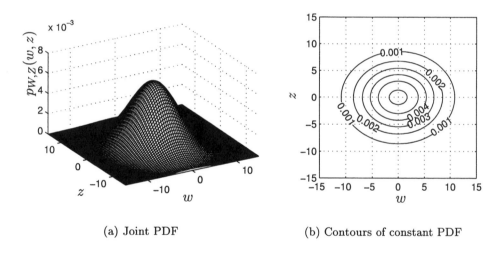

(a) Joint PDF

(b) Contours of constant PDF

Figure 12.19: Example of joint PDF for transformed correlated Gaussian random variables. The random variables are now uncorrelated and hence independent.

and thus $\det(\mathbf{GCG}^T) = 20 \cdot 32$. The transformed PDF is from Theorem 12.7.1

$$
\begin{aligned}
p_{W,Z}(w,z) &= \frac{1}{2\pi\sqrt{20 \cdot 32}} \exp\left(-\frac{1}{2}\begin{bmatrix} w \\ z \end{bmatrix}^T \begin{bmatrix} 1/20 & 0 \\ 0 & 1/32 \end{bmatrix} \begin{bmatrix} w \\ z \end{bmatrix}\right) \\
&= \frac{1}{\sqrt{2\pi \cdot 20}} \exp\left[-\frac{1}{2}\frac{w^2}{20}\right] \cdot \frac{1}{\sqrt{2\pi \cdot 32}} \exp\left[-\frac{1}{2}\frac{z^2}{32}\right]
\end{aligned}
$$

which is the factorization of the joint PDF of W and Z into the marginal PDFs $W \sim \mathcal{N}(0, 20)$ and $Z \sim \mathcal{N}(0, 32)$. Hence, W and Z are now *independent* random variables, each with a marginal Gaussian PDF. The joint PDF $p_{W,Z}$ is shown in Figure 12.19. Note the rotation of the contour plots in Figures 12.18b and 12.19b. This rotation was asserted in Example 9.4 (see also Problem 12.48).

$\diamond$

12.8   Joint Moments

For jointly distributed continuous random variables the k-lth joint moments are defined as $E_{X,Y}[X^k Y^l]$. They are evaluated as

$$E_{X,Y}[X^k Y^l] = \int_{-\infty}^{\infty} \int_{-\infty}^{\infty} x^k y^l p_{X,Y}(x, y) dx \, dy. \tag{12.38}$$

An example for $k = l = 1$ and for a standard bivariate Gaussian PDF of $E_{X,Y}[XY]$ was given Example 12.13. The k-lth joint *central* moments are defined as $E_{X,Y}[(X - E_X[X])^k (Y - E_Y[Y])^l]$ and are evaluated as

$$E_{X,Y}[(X - E_X[X])^k (Y - E_Y[Y])^l] = \int_{-\infty}^{\infty} \int_{-\infty}^{\infty} (x - E_X[X])^k (y - E_Y[Y])^l p_{X,Y}(x, y) dx \, dy.$$
$$\tag{12.39}$$

Of course, the most important case is for $k = l = 1$ for which we have the $\text{cov}(X, Y)$. For independent random variables the joint moments factor as

$$E_{X,Y}[X^k Y^l] = E_X[X^k] E_Y[Y^l]$$

and similarly for the joint central moments.

12.9   Prediction of Random Variable Outcome

In Section 7.9 we described the prediction of the outcome of a discrete random variable based on the observed outcome of another discrete random variable. We now examine the prediction problem for jointly distributed continuous random variables, and in particular, for the case of a bivariate Gaussian PDF. First we plot a scatter diagram of the outcomes of the random vector $[X \, Y]^T$ in the x-y plane. Shown in Figure 12.20 is the result for a random vector with a zero mean and a covariance matrix

$$\mathbf{C} = \begin{bmatrix} 1 & 0.9 \\ 0.9 & 1 \end{bmatrix}. \tag{12.40}$$

Note that the correlation coefficient is given by

$$\begin{aligned} \rho_{X,Y} &= \frac{\text{cov}(X, Y)}{\sqrt{\text{var}(X)\text{var}(Y)}} \\ &= \frac{0.9}{\sqrt{1 \cdot 1}} = 0.9. \end{aligned}$$

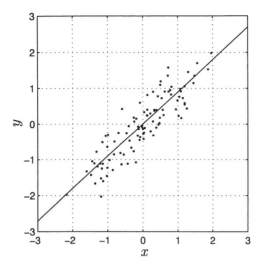

Figure 12.20: 100 outcomes of bivariate Gaussian random vector with zero means and covariance matrix given by (12.40). The best prediction of Y when $X = x$ is observed is given by the line.

It is seen from Figure 12.20 that knowledge of X should allow us to predict the outcome of Y with some accuracy. To do so we adopt as our error criterion the minimum mean square error (MSE) and use a linear predictor or $\hat{Y} = aX + b$. From Section 7.9 the best linear prediction when $X = x$ is observed is

$$\hat{Y} = E_Y[Y] + \frac{\text{cov}(X, Y)}{\text{var}(X)}(x - E_X[X]). \tag{12.41}$$

For this example the best linear prediction is

$$\hat{Y} = 0 + \frac{0.9}{1}(x - 0) = 0.9x \tag{12.42}$$

and is shown as the line in Figure 12.20. Note that the error $\epsilon = Y - 0.9X$ is also a random variable and can be shown to have the PDF $\epsilon \sim \mathcal{N}(0, 0.19)$ (see Problem 12.49). Finally, note that the predictor, which was constrained to be linear (actually affine but the use of the term linear is commonplace), cannot be improved upon by resorting to a *nonlinear* predictor. This is because it can be shown that *the optimal predictor, among all predictors, is linear if (X, Y) has a bivariate Gaussian PDF* (see Section 13.6). Hence, in this case the prediction of (12.42) is optimal among all linear *and* nonlinear predictors.

12.10   Joint Characteristic Functions

The joint characteristic function for two jointly continuous random variables X and Y is defined as

$$\phi_{X,Y}(\omega_X, \omega_Y) = E_{X,Y}[\exp[j(\omega_X X + \omega_Y Y)]]. \tag{12.43}$$

It is evaluated as

$$\phi_{X,Y}(\omega_X, \omega_Y) = \int_{-\infty}^{\infty} \int_{-\infty}^{\infty} p_{X,Y}(x,y) \exp[j(\omega_X x + \omega_Y y)] \, dx \, dy \tag{12.44}$$

and is seen to be the two-dimensional Fourier transform of the PDF (with a $+j$ instead of the more common $-j$ in the exponential). As in the case of discrete random variables, the joint moments can be found from the characteristic function using the formula

$$E_{X,Y}[X^k Y^l] = \frac{1}{j^{k+l}} \left. \frac{\partial^{k+l} \phi_{X,Y}(\omega_X, \omega_Y)}{\partial \omega_X^k \partial \omega_Y^l} \right|_{\omega_X = \omega_Y = 0}. \tag{12.45}$$

Another important application is in determining the PDF for the sum of two independent continuous random variables. As shown in Section 7.10 for discrete random variables and also true for jointly continuous random variables, if X and Y are independent, then the characteristic function of the sum $Z = X + Y$ is

$$\phi_Z(\omega) = \phi_X(\omega)\phi_Y(\omega). \tag{12.46}$$

If we were to take the inverse Fourier transform of both sides of (12.46), then the PDF of $X + Y$ would result. Hence, the procedure to determine the PDF of $X + Y$, where X and Y are independent random variables, is

1. Find the characteristic function $\phi_X(\omega)$ by evaluating the Fourier transform $\int_{-\infty}^{\infty} p_X(x) \exp(j\omega x) dx$ and similarly for $\phi_Y(\omega)$.

2. Multiply $\phi_X(\omega)$ and $\phi_Y(\omega)$ together to form $\phi_X(\omega)\phi_Y(\omega)$.

3. Finally, find the inverse Fourier transform to yield the PDF for the sum $Z = X + Y$ as

$$p_Z(z) = \int_{-\infty}^{\infty} \phi_X(\omega)\phi_Y(\omega) \exp(-j\omega z) \frac{d\omega}{2\pi}. \tag{12.47}$$

Alternatively, one could convolve the PDFs of X and Y using the convolution *integral* of (12.14) to yield the PDF of Z. However, the convolution approach is seldom easier. An example follows.

Example 12.15 – PDF for sum of independent Gaussian random variables
If $X \sim \mathcal{N}(\mu_X, \sigma_X^2)$ and $Y \sim \mathcal{N}(\mu_Y, \sigma_Y^2)$ and X and Y are independent, we wish to determine the PDF of $Z = X + Y$. A convolution approach is explored in

Problem 12.51. Here we use (12.47) to accomplish the same task. First we need the characteristic function of a Gaussian PDF. From Table 11.1 if $X \sim \mathcal{N}(\mu, \sigma^2)$, then

$$\phi_X(\omega) = \exp\left(j\omega\mu - \frac{1}{2}\sigma^2\omega^2\right).$$

Thus, the characteristic function for $X + Y$ is

$$\begin{aligned}
\phi_{X+Y}(\omega) &= \exp\left(j\omega\mu_X - \frac{1}{2}\sigma_X^2\omega^2\right)\exp\left(j\omega\mu_Y - \frac{1}{2}\sigma_Y^2\omega^2\right) \\
&= \exp\left(j\omega(\mu_X + \mu_Y) - \frac{1}{2}(\sigma_X^2 + \sigma_Y^2)\omega^2\right).
\end{aligned}$$

Since this is again the characteristic function of a Gaussian random variable, albeit with different parameters, we have that $X+Y \sim \mathcal{N}(\mu_X+\mu_Y, \sigma_X^2+\sigma_Y^2)$. (Recognizing that the characteristic function is that of a known PDF allows us to avoid inverting the characteristic function according to (12.47).) Hence, the PDF of the sum of independent Gaussian random variables is again a Gaussian random variable whose mean is $\mu = \mu_X + \mu_Y$ and whose variance is $\sigma^2 = \sigma_X^2 + \sigma_Y^2$. The Gaussian PDF is therefore called a *reproducing PDF*. By the same argument it follows that *the sum of any number of independent Gaussian random variables is again a Gaussian random variable with mean equal to the sum of the means and variance equal to the sum of the variances*. In Problem 12.53 it is shown that the Gamma PDF is also a reproducing PDF.

$\Diamond$

The result of the previous example could also be obtained by appealing to Theorem 12.7.1. If we let

$$\begin{bmatrix} W \\ Z \end{bmatrix} = \begin{bmatrix} 1 & 0 \\ 1 & 1 \end{bmatrix}\begin{bmatrix} X \\ Y \end{bmatrix}$$

then by Theorem 12.7.1, W and $Z = X+Y$ are bivariate Gaussian distributed. Also, we know that the marginals of a bivariate Gaussian PDF are Gaussian PDFs and therefore the PDF of $Z = X + Y$ is Gaussian. Its mean is $\mu_X + \mu_Y$ and its variance is $\sigma_X^2 + \sigma_Y^2$, the latter because X and Y are independent and hence uncorrelated.

12.11 Computer Simulation of Jointly Continuous Random Variables

For an arbitrary joint PDF the generation of continuous random variables is most easily done using ideas from conditional PDF theory. In Chapter 13 we will see how this is done. Here we will consider only the generation of a bivariate Gaussian random vector. The approach is based on the following properties:

1. Any affine transformation of two jointly Gaussian random variables results in two new jointly Gaussian random variables. A special case, the linear transformation, was proven in Section 12.7 and the general result summarized in Theorem 12.7.1. We will now consider the affine transformation

$$\begin{bmatrix} W \\ Z \end{bmatrix} = \mathbf{G} \begin{bmatrix} X \\ Y \end{bmatrix} + \begin{bmatrix} a \\ b \end{bmatrix}. \tag{12.48}$$

2. The mean vector and covariance matrix of $[W\ Z]^T$ transform according to

$$E\left[\begin{bmatrix} W \\ Z \end{bmatrix}\right] = \mathbf{G} E\left[\begin{bmatrix} X \\ Y \end{bmatrix}\right] + \begin{bmatrix} a \\ b \end{bmatrix} \quad \text{(see Problem 9.22)} \tag{12.49}$$

$$\mathbf{C}_{W,Z} = \mathbf{G}\mathbf{C}_{X,Y}\mathbf{G}^T \quad \text{(see (Theorem 12.7.1))} \tag{12.50}$$

where we now use subscripts on the covariance matrices to indicate the random variables.

The approach to be described next assumes that X and Y are standard Gaussian and independent random variables whose realizations are easily generated. In MATLAB the command `randn(1,1)` can be used. Otherwise, if only $\mathcal{U}(0,1)$ random variables are available, one can use the Box-Mueller transform to obtain X and Y (see Problem 12.54). Then, to obtain any bivariate Gaussian random variables (W, Z) with a given mean $[\mu_W\ \mu_Z]^T$ and covariance matrix $\mathbf{C}_{W,Z}$, we use (12.48) with a suitable $\mathbf{G}$ and $[a\ b]^T$ so that

$$E\left[\begin{bmatrix} W \\ Z \end{bmatrix}\right] = \begin{bmatrix} \mu_W \\ \mu_Z \end{bmatrix}$$

$$\mathbf{C}_{W,Z} = \begin{bmatrix} \sigma_W^2 & \rho\sigma_W\sigma_Z \\ \rho\sigma_Z\sigma_W & \sigma_Z^2 \end{bmatrix}. \tag{12.51}$$

Since it is assumed that X and Y are zero mean, from (12.49) we choose $a = \mu_W$ and $b = \mu_Z$. Also, since X and Y are assumed independent, hence uncorrelated, and with unit variances, we have

$$\mathbf{C}_{X,Y} = \begin{bmatrix} 1 & 0 \\ 0 & 1 \end{bmatrix} = \mathbf{I}.$$

It follows from (12.50) that $\mathbf{C}_{W,Z} = \mathbf{G}\mathbf{G}^T$. To find $\mathbf{G}$ if we are given $\mathbf{C}_{W,Z}$, we could use an eigendecomposition approach based on the relationship $\mathbf{V}^T\mathbf{C}_{W,Z}\mathbf{V} = \mathbf{\Lambda}$ (see Problem 12.55). Instead, we next explore an alternative approach which is somewhat easier to implement in practice. Let $\mathbf{G}$ be a *lower triangular* matrix

$$\mathbf{G} = \begin{bmatrix} a & 0 \\ b & c \end{bmatrix}.$$

Then, we have that

$$\mathbf{G}\mathbf{G}^T = \begin{bmatrix} a & 0 \\ b & c \end{bmatrix}\begin{bmatrix} a & b \\ 0 & c \end{bmatrix} = \begin{bmatrix} a^2 & ab \\ ab & b^2 + c^2 \end{bmatrix}. \tag{12.52}$$

The numerical procedure of decomposing a covariance matrix into a product such as $\mathbf{G}\mathbf{G}^T$, where $\mathbf{G}$ is lower triangular, is called the *Cholesky decomposition* [Golub and Van Loan 1983]. Here we can do so almost by inspection. We need only equate the elements of $\mathbf{C}_{W,Z}$ in (12.51) to those of $\mathbf{G}\mathbf{G}^T$ as given in (12.52). Doing so produces the result

$$\begin{aligned} a &= \sigma_W \\ b &= \rho\sigma_Z \\ c &= \sigma_Z\sqrt{1-\rho^2}. \end{aligned}$$

Hence, we have that

$$\mathbf{G} = \begin{bmatrix} \sigma_W & 0 \\ \rho\sigma_Z & \sigma_Z\sqrt{1-\rho^2} \end{bmatrix}.$$

In summary, to generate a realization of a bivariate Gaussian random vector we first generate two independent standard Gaussian random variables X and Y and then transform according to

$$\begin{bmatrix} W \\ Z \end{bmatrix} = \begin{bmatrix} \sigma_W & 0 \\ \rho\sigma_Z & \sigma_Z\sqrt{1-\rho^2} \end{bmatrix}\begin{bmatrix} X \\ Y \end{bmatrix} + \begin{bmatrix} \mu_W \\ \mu_Z \end{bmatrix}. \tag{12.53}$$

As an example, we let $\mu_W = \mu_Z = 1$, $\sigma_W = \sigma_Z = 1$, and $\rho = 0.9$. The constant PDF contours as well as 500 realizations of $[W\ Z]^T$ are shown in Figure 12.21. To verify that the mean vector and covariance matrix are correct, we can estimate these quantities using (9.44) and (9.46) which are

$$\widehat{E_{W,Z}}\left[\begin{bmatrix} W \\ Z \end{bmatrix}\right] = \frac{1}{M}\sum_{m=1}^{M}\begin{bmatrix} w_m \\ z_m \end{bmatrix}$$

$$\widehat{\mathbf{C}_{W,Z}} = \frac{1}{M}\sum_{m=1}^{M}\left(\begin{bmatrix} w_m \\ z_m \end{bmatrix} - \widehat{E_{W,Z}}\left[\begin{bmatrix} W \\ Z \end{bmatrix}\right]\right)\left(\begin{bmatrix} w_m \\ z_m \end{bmatrix} - \widehat{E_{W,Z}}\left[\begin{bmatrix} W \\ Z \end{bmatrix}\right]\right)^T$$

where $[w_m\ z_m]^T$ is the mth realization of $[W\ Z]^T$. The results and the true values for $M = 2000$ are

$$\widehat{E_{W,Z}}\left[\begin{bmatrix} W \\ Z \end{bmatrix}\right] = \begin{bmatrix} 1.0326 \\ 1.0252 \end{bmatrix} \quad E_{W,Z}\left[\begin{bmatrix} W \\ Z \end{bmatrix}\right] = \begin{bmatrix} 1 \\ 1 \end{bmatrix}$$

$$\widehat{\mathbf{C}_{W,Z}} = \begin{bmatrix} 0.9958 & 0.9077 \\ 0.9077 & 1.0166 \end{bmatrix} \quad \mathbf{C}_{W,Z} = \begin{bmatrix} 1 & 0.9 \\ 0.9 & 1 \end{bmatrix}.$$

The MATLAB code used to generate the realizations and to estimate the mean vector and covariance matrix is given next.

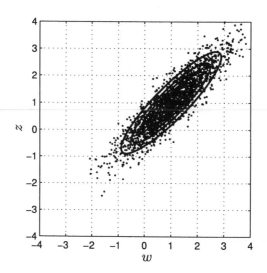

Figure 12.21: 500 outcomes of bivariate Gaussian random vector with mean $[1\ 1]^T$ and covariance matrix given by (12.40).

```
randn('state',0) % set random number generator to initial value
G=[1 0;0.9 sqrt(1-0.9^2)]; % define G matrix
M=2000; % set number of realizations
for m=1:M
  x=randn(1,1);y=randn(1,1); % generate realizations of two
                             % independent N(0,1) random variables
  wz=G*[x y]'+[1 1]'; % transform to desired mean and covariance
  WZ(:,m)=wz; % save realizations in 2 x M array
end
Wmeanest=mean(WZ(1,:)); % estimate mean of W
Zmeanest=mean(WZ(2,:)); % estimate mean of Z
WZbar(1,:)=WZ(1,:)-Wmeanest; % subtract out mean of W
WZbar(2,:)=WZ(2,:)-Zmeanest; % subtract out mean of Z
Cest=[0 0;0 0];
for m=1:M
  Cest=Cest+(WZbar(:,m)*WZbar(:,m)')/M; % compute estimate of
                                        % covariance matrix
end
Wmeanest % write out estimate of mean of W
Zmeanest % write out estimate of mean of Z
Cest % write out estimate of covariance matrix
```

12.12 Real-World Example – Optical Character Recognition

An important use of computers is to be able to scan a document and automatically read the characters. For example, bank checks are routinely scanned to ascertain the account numbers, which are usually printed on the bottom. Also, scanners are used to take a page of alphabetic characters and convert the text to a computer file that can later be edited in a computer. In this section we briefly describe how this might be done. A more comprehensive description can be found in [Trier, Jain, and Taxt 1996]. To simplify the discussion we consider recognition of the digits $0, 1, 2, \ldots, 9$ that have been generated by a printer (as opposed to handwritten, the recognition of which is much more complex due to the potential variations of the characters). An example of these characters is shown in Figure 12.22. They were obtained by printing the characters from a computer to a laser printer and then scanning them back into a computer. Note that each digit consists of an 80×80

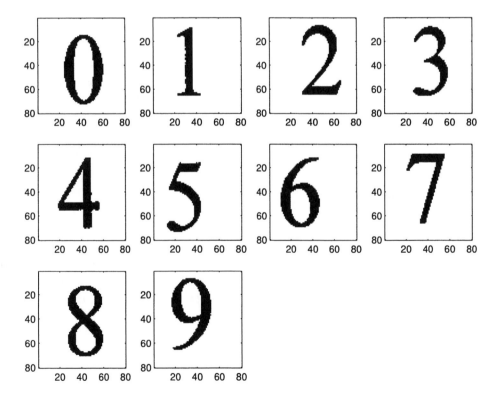

Figure 12.22: Scanned digits for optical character recognition.

array of pixels and each pixel is either black or white. This is termed a *binary* image. A magnified version of the digit "1" is shown in Figure 12.23, where the "pixelation" is clearly evident. Also, some of the black pixels have been omitted due to errors in the scanning process. In order for a computer to be able to recognize and decode

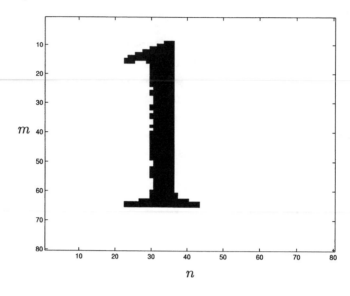

Figure 12.23: Magnified version of the digit "1".

the digits it is necessary to reduce each 80×80 image to a number or set of numbers that characterize the digit. These numbers are called the *features* and they compose a *feature vector* that must be different for each digit. This will allow a computer to distinguish between the digits and be less susceptible to noise effects as is evident in the "1" image. For our example, we will choose only two features, although in practice many more are used. A typical feature based on the geometric character of the digit images is the *geometric moments*. It attempts to measure the distribution of the black pixels and is completely analogous to our usual joint moments. (Recall our motivation of the expected value using the idea of the center of mass of an object in Section 11.3.) Let $g[m,n]$ denote the pixel value at location $[m,n]$ in the image, where $m = 1, 2, \ldots, 80$, $n = 1, 2, \ldots, 80$ and either $g[m,n] = 1$ for a black pixel or $g[m,n] = 0$ for a white pixel. Note from Figure 12.23 that the indices for the $[m,n]$ pixel are specified in matrix format, where m indicates the row and n indicates the column. The geometric moments are defined as

$$\mu'[k,l] = \frac{\sum_{m=1}^{80}\sum_{n=1}^{80} m^k n^l g[m,n]}{\sum_{m=1}^{80}\sum_{n=1}^{80} g[m,n]}. \tag{12.54}$$

If we were to define

$$p[m,n] = \frac{g[m,n]}{\sum_{m=1}^{80}\sum_{n=1}^{80} g[m,n]} \qquad m = 1, 2, \ldots, 80; n = 1, 2, \ldots, 80$$

then $p[m,n]$ would have the properties of a joint PMF, in that it is nonnegative and sums to one. A somewhat better feature is obtained by using the *central* geometric moments which will yield the same number even as the digit is translated in the horizontal and vertical directions. This may be seen to be of value by referring to Figure 12.22, in which the center of the digits do not all lie at the same location. Using central geometric moments alleviates having to center each digit. The definition is

$$\mu[k,l] = \frac{\sum_{m=1}^{80} \sum_{n=1}^{80} (m - \bar{m})^k (n - \bar{n})^l g[m,n]}{\sum_{m=1}^{80} \sum_{n=1}^{80} g[m,n]} \tag{12.55}$$

where

$$\bar{m} = \mu'[1,0] = \frac{\sum_{m=1}^{80} \sum_{n=1}^{80} m g[m,n]}{\sum_{m=1}^{80} \sum_{n=1}^{80} g[m,n]}$$

$$\bar{n} = \mu'[0,1] = \frac{\sum_{m=1}^{80} \sum_{n=1}^{80} n g[m,n]}{\sum_{m=1}^{80} \sum_{n=1}^{80} g[m,n]}$$

The coordinate pair $(\bar{m}, \bar{n})$ is the center of mass of the character and is completely analogous to the mean of the "joint PDF" $p[m,n]$.

To demonstrate the procedure by which optical character recognition is accomplished we will add noise to the characters. To simulate a "dropout", in which a black pixel becomes a white one (see Figure 12.23 for an example), we change each black pixel to a white one with a probability of 0.4, and make no change with probability of 0.6. To simulate spurious scanning marks we change each white pixel to a black one with probability of 0.1, and make no change with probability of 0.9. An example of the corrupted digits is shown in Figure 12.24. As a feature vector we will use the pair $(\mu[1,1], \mu[2,2])$. For the digits "1" and "8", 50 realizations of the feature vector are shown in Figure 12.25a. The black square indicates the center of mass $(\bar{m}, \bar{n})$ for each digit's feature vector. Note that we could distinguish between the two characters without error if we recognize an outcome as belonging to a "1" if we are below the line boundary shown and as a "8" otherwise. However, for the digits "1" and "3" there is an overlap region where the outcomes could belong to either character as seen in Figure 12.25b. For these digits we could not separate the digits without a large error. The latter is more typically the case and can only be resolved by using a larger dimension feature vector. The interested reader should consult [Duda, Hart, and Stork 2001] for a further discussion of pattern recognition (also called *pattern classification*). Also, note that the digits "3" and "8" would produce outcomes that would overlap greatly. Can you explain why? You might consider some typical scanned digits as shown in Figure 12.26 that have been designed to make recognition easier!

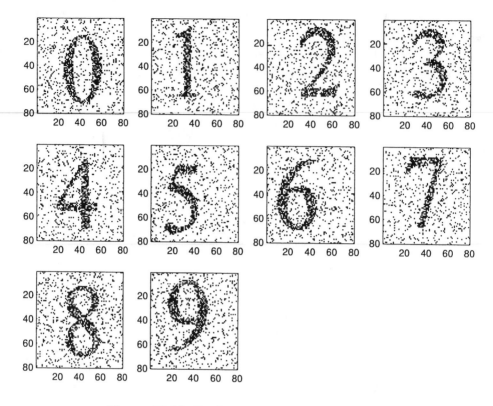

Figure 12.24: Realization of corrupted digits.

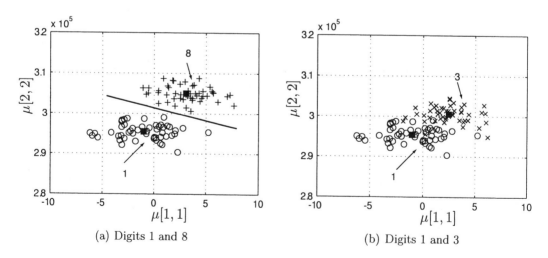

(a) Digits 1 and 8

(b) Digits 1 and 3

Figure 12.25: 50 realizations of feature vector for two competing digits.

Figure 12.26: Some scanned digits typically used in optical character recognition. They were scanned into a computer, which accounts for the obvious errors.

References

Duda, R.O., P.E. Hart, D.G. Stork, *Pattern Classification, 2nd Ed.*, John Wiley & Sons, New York, 2001.

Golub, G.H., C.F. Van Loan, *Matrix Computations*, Johns Hopkins University Press, Baltimore, MD, 1983.

Trier, O.D., A.K. Jain, T. Taxt, "Feature Extraction Methods for Character Recognition – A Survey," *Pattern Recognition*, Vol. 29, No. 4, pp. 641–662, 1996, Pergamon, Elsevier Science.

Problems

12.1 (☺) (w) For the dartboard shown in Figure 12.1 determine the probability that the novice dart player will land his dart in the outermost ring, which has radii $3/4 \leq r \leq 1$. Do this by using geometrical arguments and also using double integrals. Hint: For the latter approach convert to polar coordinates (r, θ) and remember to use $dx\,dy = r\,dr\,d\theta$.

12.2 (c) Reproduce Figure 12.2a by letting $X \sim \mathcal{U}(-1,1)$ and $Y \sim \mathcal{U}(-1,1)$, where X and Y are independent. Omit any realizations of (X,Y) for which $\sqrt{X^2 + Y^2} > 1$. Explain why this produces a uniform distribution of points in the unit circle. See also Problem 13.23 for a more formal justification of this procedure.

12.3 (☺) (w) For the novice dart player is $P[0 \leq R \leq 0.5] = 0.5$ (R is the distance from the center of the dartboard)? Explain your results.

12.4 (w) Find the volume of a cylinder of height h and whose base has radius r by using a double integral evaluation.

12.5 (☺) (c) In this problem we estimate π using probability arguments. Let $X \sim \mathcal{U}(-1,1)$ and $Y \sim \mathcal{U}(-1,1)$ for X and Y independent. First relate $P[X^2+Y^2 \leq 1]$ to the value of π. Then generate realizations of X and Y and use them to estimate π.

12.6 (f) For the joint PDF

$$p_{X,Y}(x,y) = \begin{cases} \frac{1}{\pi} & x^2 + y^2 \leq 1 \\ 0 & \text{otherwise} \end{cases}$$

find $P[|X| \leq 1/2]$. Hint: You will need

$$\int \sqrt{1 - x^2}\, dx = \frac{1}{2}x\sqrt{1 - x^2} + \frac{1}{2}\arcsin(x).$$

12.7 (☺) (f) If a joint PDF is given by

$$p_{X,Y}(x,y) = \begin{cases} \frac{c}{\sqrt{xy}} & 0 \leq x \leq 1, 0 \leq y \leq 1 \\ 0 & \text{otherwise} \end{cases}$$

find c.

12.8 (w) A point is chosen at random from the sample space $\mathcal{S} = \{(x,y) : 0 \leq x \leq 1, 0 \leq y \leq 1\}$. Find $P[Y \leq X]$.

12.9 (f) For the joint PDF $p_{X,Y}(x,y) = \exp[-(x + y)]u(x)u(y)$, find $P[Y \leq X]$.

12.10 (☺) (w,c) Two persons play a game in which the first person thinks of a number from 0 to 1, while the second person tries to guess player one's number. The second player claims that he is telepathic and knows what number the first player has chosen. In reality the second player just chooses a number at random. If player one also thinks of a number at random, what is the probability that player two will choose a number whose difference from player one's number is less than 0.1? Add credibility to your solution by simulating the game and estimating the desired probability.

12.11 (☺) (f) If (X,Y) has a standard bivariate Gaussian PDF, find $P[X^2 + Y^2 = 10]$.

12.12 (f,c) Plot the values of (x,y) for which $x^2 - 2\rho xy + y^2 = 1$ for $\rho = -0.9$, $\rho = 0$, and $\rho = 0.9$. Hint: Solve for y in terms of x.

12.13 (w,c) Plot the standard bivariate PDF in three dimensions for $\rho = 0.9$. Next examine your plot if $\rho \to 1$ and determine what happens. As $\rho \to 1$, can you predict Y based on $X = x$?

12.14 (f) If $p_{X,Y}(x,y) = \exp[-(x + y)]u(x)u(y)$, determine the marginal PDFs.

12.15 (☺) (f) If

$$p_{X,Y}(x,y) = \begin{cases} 2 & 0 < x < 1, 0 < y < x \\ 0 & \text{otherwise} \end{cases}$$

find the marginal PDFs.

12.16 (t) Assuming that $(x, y) \neq (0, 0)$, prove that $x^2 - 2\rho xy + y^2 > 0$ for $-1 < \rho < 1$.

12.17 (f) If $p_X(x) = (1/2) \exp[-(1/2)x]u(x)$ and $p_Y(y) = (1/4) \exp[-(1/4)y]u(y)$, find the joint PDF of X and Y.

12.18 (⌣) (f) Determine the joint CDF if X and Y are independent with

$$p_X(x) = \begin{cases} \frac{1}{2} & 0 < x < 2 \\ 0 & \text{otherwise} \end{cases}$$

$$p_Y(y) = \begin{cases} \frac{1}{4} & 0 < y < 4 \\ 0 & \text{otherwise.} \end{cases}$$

12.19 (f) Determine the joint CDF corresponding to the joint PDF

$$p_{X,Y}(x, y) = \begin{cases} xy \exp\left[-\frac{1}{2}(x^2 + y^2)\right] & x \geq 0, y \geq 0 \\ 0 & \text{otherwise.} \end{cases}$$

Next verify Properties 12.1–12.6 for the CDF.

12.20 (t) Prove that (12.10) is true if (12.11) is true and vice versa. Hint: Let $A = \{a \leq x \leq b\}$ and $B = \{y : c \leq y \leq d\}$ for the first part and let $A = \{x : x_0 - \Delta x/2 \leq x \leq x_0 + \Delta x/2\}$ and $B = \{y : y_0 - \Delta y/2 \leq y \leq y_0 + \Delta y/2\}$ with x_0 and y_0 arbitrary for the second part.

12.21 (t) Prove that (12.11) and (12.12) are equivalent.

12.22 (w) Two independent speech signals are added together. If each one has a Laplacian PDF with parameter σ^2, what is the power of the resultant signal?

12.23 (⌣) (w) Lightbulbs fail with a time to failure modeled as an exponential random variable with a mean time to failure of 1000 hours. If two lightbulbs are used to illuminate a room, what is the probability that both bulbs will fail before 2000 hours? Assume that the failure time of one bulb does not affect the failure time of the other bulb.

12.24 (f) If a joint PDF is given as $p_{X,Y}(x, y) = 6 \exp[-(2x + 3y)]u(x)u(y)$, what is the probability of $A = \{(x, y) : 0 < x < 2, 0 < y < 1\}$? Are the two random variables independent?

12.25 (⌣) (w) A joint PDF is uniform over the region $\{(x, y) : 0 \leq y < x, 0 \leq x < 1\}$ and zero elsewhere. Are X and Y independent?

12.26 (⌣) (w) The temperature in Antarctica is modeled as a random variable $X \sim \mathcal{N}(20, 1500)$ degrees Fahrenheit, while that in Ecuador is modeled also as a random variable with $Y \sim \mathcal{N}(100, 100)$ degrees Fahrenheit. What is the probability that it will be hotter in Antarctica than in Ecuador? Assume the random variables are independent.

12.27 (w,c) In Section 2.3 we discussed the outcomes resulting from adding together two random variables uniform on $(0,1)$. We claimed that the probability of 500 outcomes in the interval $[0, 0.5]$ and 500 outcomes in the interval $[1.5, 2]$ resulting from a total of 1000 outcomes is

$$\binom{1000}{500}\left(\frac{1}{8}\right)^{1000} \approx 2.2 \times 10^{-604}.$$

Can you now justify this result? What assumptions are implicit in its calculation? Hint: For each trial consider the 3 possible outcomes $(0, 0.5)$, $[0.5, 1.5)$, and $[1.5, 2)$. Also, see Problem 3.48 on how to evaluate expressions with large factorials.

12.28 (f) Find the PDF of $X = U_1 + U_2$, where $U_1 \sim \mathcal{U}(0,1)$, $U_2 \sim \mathcal{U}(0,1)$, and U_1, U_2 are independent. Use a convolution integral to do this.

12.29 (w) In this problem we show that the ratio of areas for the linear transformation

$$\begin{bmatrix} w \\ z \end{bmatrix} = \underbrace{\begin{bmatrix} a & b \\ c & d \end{bmatrix}}_{\mathbf{G}} \underbrace{\begin{bmatrix} x \\ y \end{bmatrix}}_{\boldsymbol{\xi}}$$

is $|\det(\mathbf{G})|$. To do so let $\boldsymbol{\xi} = [x\, y]^T$ take on values in the region $\{(x,y) : 0 \leq x \leq 1, 0 \leq y \leq 1\}$ as shown by the shaded area in Figure 12.27. Then, consider a point in the unit square to be represented as $\boldsymbol{\xi} = \alpha \mathbf{e}_1 + \beta \mathbf{e}_2$, where $0 \leq \alpha \leq 1$, $0 \leq \beta \leq 1$, $\mathbf{e}_1 = [1\, 0]^T$, and $\mathbf{e}_2 = [0\, 1]^T$. The transformed vector is

$$\begin{aligned} \mathbf{G}\boldsymbol{\xi} &= \mathbf{G}(\alpha \mathbf{e}_1 + \beta \mathbf{e}_2) \\ &= \alpha \mathbf{G}\mathbf{e}_1 + \beta \mathbf{G}\mathbf{e}_2 \\ &= \alpha \begin{bmatrix} a \\ c \end{bmatrix} + \beta \begin{bmatrix} b \\ d \end{bmatrix}. \end{aligned}$$

It is seen that the natural basis vectors $\mathbf{e}_1, \mathbf{e}_2$ map into the vectors $[a\, c]^T$, $[b\, d]^T$, which appear as shown in Figure 12.27. The region in the w-z plane that results from mapping the unit square is shown as shaded. The area of the parallelogram can be found from Figure 12.28 as BH. Determine the ratio of areas to show that

$$\frac{\text{Area in } w\text{-}z \text{ plane}}{\text{Area in } x\text{-}y \text{ plane}} = ad - bc = \det(\mathbf{G}).$$

The absolute value is needed since if for example $a < 0, b < 0$, the parallelogram will be in the second quadrant and its determinant will be negative. The absolute value sign takes care of all the possible cases.

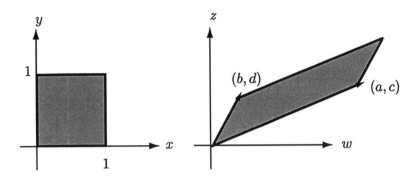

Figure 12.27: Mapping of areas for linear transformation.

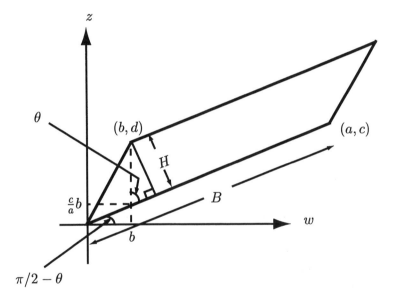

Figure 12.28: Geometry to determine area of parallelogram.

12.30 (☺) **(w,c)** The champion dart player described in Section 12.3 is able to land his dart at a point (x, y) according to the joint PDF

$$\begin{bmatrix} X \\ Y \end{bmatrix} \sim \mathcal{N}\left(\begin{bmatrix} 0 \\ 0 \end{bmatrix}, \begin{bmatrix} 1/64 & 0 \\ 0 & 1/64 \end{bmatrix} \right)$$

with some outcomes shown in Figure 12.2b. Determine the probability of a bullseye. Next simulate the game and plot the outcomes. Finally estimate the probability of a bullseye using the results of your computer simulation.

12.31 (t) Show that (12.19) can be written as (12.20).

12.32 (t) Consider the nonlinear transformation $w = g(x,y), z = h(x,y)$. Use a tangent approximation to both functions about the point (x_0, y_0) to express $[w\, z]^T$ as an approximate affine function of $[x\, y]^T$, and use matrix/vector notation. For example,

$$w = g(x,y) \approx g(x_0, y_0) + \left.\frac{\partial g}{\partial x}\right|_{\substack{x=x_0 \\ y=y_0}} (x - x_0) + \left.\frac{\partial g}{\partial y}\right|_{\substack{x=x_0 \\ y=y_0}} (y - y_0)$$

and similarly for $z = h(x,y)$. Compare the matrix to the Jacobian matrix of (12.23).

12.33 (f) If a joint PDF is given as $p_{X,Y}(x,y) = (1/4)^2 \exp[-\frac{1}{2}(|x| + |y|)]$ for $-\infty < x < \infty, -\infty < y < \infty$, find the joint PDF of

$$\begin{bmatrix} W \\ Z \end{bmatrix} = \begin{bmatrix} 2 & 2 \\ 2 & 1 \end{bmatrix} \begin{bmatrix} X \\ Y \end{bmatrix}.$$

12.34 (f) If a joint PDF is given as $p_{X,Y}(x,y) = \exp[-(x+y)]u(x)u(y)$, find the joint PDF of $W = XY, Z = Y/X$.

12.35 (w,c) Consider the nonlinear transformation

$$\begin{aligned} W &= X^2 + 5Y^2 \\ Z &= -5X^2 + Y^2. \end{aligned}$$

Write a computer program to plot in the x-y plane the points (x_i, y_j) for $x_i = 0.95 + (i-1)/100$ for $i = 1, 2, \ldots, 11$ and $y_j = 1.95 + (j-1)/100$ for $j = 1, 2, \ldots, 11$. Next transform all these points into the w-z plane using the given nonlinear transformation. What kind of figure do you see? Next calculate the area of the figure (you can use a rough approximation based on the computer generated figure output) and finally take the ratio of the areas of the figures in the two planes. Does this ratio agree with the Jacobian factor

$$\left| \det\left(\frac{\partial(w,z)}{\partial(x,y)} \right) \right|$$

when evaluated at $x = 1, y = 2$?

12.36 ($\smile$) (f) Find the marginal PDFs of the joint PDF given in (12.25).

12.37 (f) Determine the marginal PDFs for the joint PDF given by

$$\begin{bmatrix} X \\ Y \end{bmatrix} \sim \mathcal{N}\left(\begin{bmatrix} 1 \\ 2 \end{bmatrix}, \begin{bmatrix} 3 & 4 \\ 4 & 6 \end{bmatrix} \right).$$

12.38 (☺) (f) If X and Y have the joint PDF

$$\begin{bmatrix} X \\ Y \end{bmatrix} \sim \mathcal{N}\left(\begin{bmatrix} 1 \\ 2 \end{bmatrix}, \begin{bmatrix} 2 & -1 \\ -1 & 2 \end{bmatrix} \right)$$

find the joint PDF of the transformed random vector

$$\begin{bmatrix} W \\ Z \end{bmatrix} = \begin{bmatrix} 1 & 1 \\ 2 & 3 \end{bmatrix} \begin{bmatrix} X \\ Y \end{bmatrix}.$$

12.39 (t) Prove that the PDF of $Z = Y/X$, where X and Y are independent, is given by

$$p_Z(z) = \int_{-\infty}^{\infty} p_X(x) p_Y(xz) |x| dx.$$

12.40 (t) Prove that the PDF of $Z = XY$, where X and Y are independent is given by

$$p_Z(z) = \int_{-\infty}^{\infty} p_X(x) p_Y(z/x) \frac{1}{|x|} dx.$$

12.41 (c) Generate outcomes of a Cauchy random variable using Y/X, where $X \sim \mathcal{N}(0,1)$, $Y \sim \mathcal{N}(0,1)$ and X and Y are independent. Can you explain what happens when the Cauchy outcome becomes very large in magnitude?

12.42 (t) Prove that $s(t) = A\cos(2\pi F_0 t) + B\sin(2\pi F_0 t)$ can be written as $s(t) = \sqrt{A^2 + B^2}\cos(2\pi F_0 t - \arctan(B/A))$. Hint: Convert (A, B) to polar coordinates.

12.43 (☺) (w) A particle is subject to a force in a random force field. If the velocity of the particle is modeled in the x and y directions as $V_x \sim \mathcal{N}(0,10)$ and $V_y \sim \mathcal{N}(0,10)$ meters/sec, and V_x and V_y are assumed to be independent, how far will the particle move on the average in 1 second?

12.44 (f) Prove that if X and Y are independent standard Gaussian random variables, then $X^2 + Y^2$ will have a χ_2^2 PDF.

12.45 (☺) (w,f) Two independent random variables X and Y have zero means and variances of 1. If they are linearly transformed as $W = X + Y, Z = X - Y$, find the covariance between the transformed random variables. Are W and Z uncorrelated? Are W and Z independent?

12.46 (f) If

$$\begin{bmatrix} X \\ Y \end{bmatrix} \sim \mathcal{N}\left(\begin{bmatrix} 1 \\ 1 \end{bmatrix}, \begin{bmatrix} 2 & 1 \\ 1 & 2 \end{bmatrix} \right)$$

determine the mean of $X + Y$ and the variance of $X + Y$.

12.47 (☺) (w) The random vector $[X\,Y]^T$ has a covariance matrix

$$\mathbf{C} = \begin{bmatrix} 2 & 1 \\ 1 & 2 \end{bmatrix}.$$

Find a 2×2 matrix $\mathbf{G}$ so that $\mathbf{G}[X\,Y]^T$ is a random vector with uncorrelated components.

12.48 (t) Prove that if a random vector has a covariance matrix

$$\mathbf{C} = \begin{bmatrix} a & b \\ b & a \end{bmatrix}$$

then the matrix

$$\mathbf{G} = \begin{bmatrix} \frac{1}{\sqrt{2}} & -\frac{1}{\sqrt{2}} \\ \frac{1}{\sqrt{2}} & \frac{1}{\sqrt{2}} \end{bmatrix}$$

can always be used to diagonalize it. Show that the effect of this matrix transformation is to rotate the point (x, y) by $45°$ and relate this back to the contours of a standard bivariate Gaussian PDF.

12.49 (f) Find the MMSE estimator of Y based on observing $X = x$ if (X, Y) has the joint PDF

$$p_{X,Y}(x, y) = \frac{1}{2\pi\sqrt{0.19}} \exp\left[-\frac{1}{2(0.19)}(x^2 - 1.8xy + y^2)\right].$$

Also, find the PDF of the error $Y - \hat{Y} = Y - (aX + b)$, where a, b are the optimal values. Hint: See Theorem 12.7.1.

12.50 (w,c) A random signal voltage $V \sim \mathcal{N}(1, 1)$ is corrupted by an independent noise sample N, where $N \sim \mathcal{N}(0, 2)$, so that $V + N$ is observed. It is desired to estimate the signal voltage as accurately as possible using a linear MMSE estimator. Assuming that V and N are independent, find this estimator. Then plot the constant PDF contours for the random vector $(V+N, V)$ and indicate the estimated values on the plot.

12.51 (f) Using a convolution integral prove that if X and Y are independent standard Gaussian random variables, then $X + Y \sim \mathcal{N}(0, 2)$.

12.52 (☺) (f) If

$$\begin{bmatrix} X \\ Y \end{bmatrix} \sim \mathcal{N}\left(\begin{bmatrix} 0 \\ 0 \end{bmatrix}, \begin{bmatrix} 2 & 0 \\ 0 & 2 \end{bmatrix}\right).$$

find $P[X + Y > 2]$.

12.53 (t) Prove that if $X \sim \Gamma(\alpha_X, \lambda)$ and $Y \sim \Gamma(\alpha_Y, \lambda)$ and X and Y are independent, then $X + Y \sim \Gamma(\alpha_X + \alpha_Y, \lambda)$.

12.54 (f) To generate two independent standard Gaussian random variables on a computer one can use the *Box-Mueller transform*

$$
\begin{aligned}
X &= \sqrt{-2 \ln U_1} \cos(2\pi U_2) \\
Y &= \sqrt{-2 \ln U_1} \sin(2\pi U_2)
\end{aligned}
$$

where U_1, U_2 are both uniform on $(0, 1)$ and independent of each other. Prove that this result is true. Hint: To find the inverse transformation use a polar coordinate transformation.

12.55 (t) Prove that by using the eigendecomposition of a covariance matrix or $\mathbf{V}^T \mathbf{C} \mathbf{V} = \mathbf{\Lambda}$ that one can factor $\mathbf{C}$ as $\mathbf{C} = \mathbf{G}\mathbf{G}^T$, where $\mathbf{G} = \mathbf{V}\sqrt{\mathbf{\Lambda}}$, and $\sqrt{\mathbf{\Lambda}}$ is defined as the matrix obtained by taking the positive square roots of all the elements. Recall that $\mathbf{\Lambda}$ is a diagonal matrix with positive elements on the main diagonal. Next find $\mathbf{G}$ for the covariance matrix

$$
\mathbf{C} = \begin{bmatrix} 26 & 6 \\ 6 & 26 \end{bmatrix}
$$

and verify that $\mathbf{G}\mathbf{G}^T$ does indeed produce $\mathbf{C}$.

12.56 (c) Simulate on the computer realizations of the random vector

$$
\begin{bmatrix} W \\ Z \end{bmatrix} \sim \mathcal{N} \left(\begin{bmatrix} 1 \\ 1 \end{bmatrix}, \begin{bmatrix} 1 & -0.9 \\ -0.9 & 1 \end{bmatrix} \right).
$$

Plot these realizations as well as the contours of constant PDF on the same graph.

Chapter 13

Conditional Probability Density Functions

13.1 Introduction

A discussion of conditional probability mass functions (PMFs) was given in Chapter 8. The motivation was that many problems are stated in a conditional format so that the solution must naturally accommodate this conditional structure. In addition, the use of conditioning is useful for simplifying probability calculations when two random variables are statistically dependent. In this chapter we formulate the analogous approach for probability density functions (PDFs). A potential stumbling block is that the usual conditioning event $X = x$ has probability zero for a continuous random variable. As a result the conditional PMF cannot be extended in a straightforward manner. We will see, however, that using care, a conditional PDF can be defined and will prove to be useful.

13.2 Summary

The conditional PDF is defined in (13.3) and can be used to find conditional probabilities using (13.4). The conditional PDF for a standard bivariate Gaussian PDF is given by (13.5) and is seen to retain its Gaussian form. The joint, conditional, and marginal PDFs are related to each other as summarized by Properties 13.1–13.5. A conditional CDF is defined by (13.6) and is evaluated using (13.7). The use of conditioning can simplify probability calculations as described in Section 13.5. A version of the law of total probability is given by (13.12) and is evaluated using (13.13). An optimal predictor for the outcome of a random variable based on the outcome of a second random variable is given by the mean of the conditional PDF as defined by (13.14). An example is given for the bivariate Gaussian PDF in which the predictor becomes linear (actually affine). To generate realizations of two jointly distributed continuous random variables the procedure based on conditioning and described in Section 13.7 can be used. Lastly, an application to determining mortality rates for retirement planning is described in Section 13.8.

13.3   Conditional PDF

Recall that for two jointly *discrete* random variables X and Y, the conditional PMF is defined as

$$p_{Y|X}[y_j|x_i] = \frac{p_{X,Y}[x_i,y_j]}{p_X[x_i]} \qquad j = 1,2,\ldots . \tag{13.1}$$

This formula gives the probability of the event $Y = y_j$ for $j = 1,2,\ldots$ once we have observed that $X = x_i$. Since $X = x_i$ has occurred, the only joint events with a nonzero probability are $\{(x,y) : x = x_i, y = y_1, y_2, \ldots\}$. As a result we divide the joint probability $p_{X,Y}[x_i,y_j] = P[X = x_i, Y = y_j]$ by the probability of the reduced sample space, which is $p_X[x_i] = P[X = x_i, Y = y_1] + P[X = x_i, Y = y_2] + \cdots = \sum_{j=1}^{\infty} p_{X,Y}[x_i,y_j]$. This division assures us that

$$
\begin{aligned}
\sum_{j=1}^{\infty} p_{Y|X}[y_j|x_i] &= \sum_{j=1}^{\infty} \frac{p_{X,Y}[x_i,y_j]}{p_X[x_i]} \\
&= \frac{\sum_{j=1}^{\infty} p_{X,Y}[x_i,y_j]}{p_X[x_i]} \\
&= \frac{\sum_{j=1}^{\infty} p_{X,Y}[x_i,y_j]}{\sum_{j=1}^{\infty} p_{X,Y}[x_i,y_j]} = 1.
\end{aligned}
$$

In the case of *continuous* random variables X and Y a problem arises in defining a *conditional PDF*. If we observe $X = x$, then since $P[X = x] = 0$, the use of a formula like (13.1) is no longer valid due to the division by zero. Recall that our original definition of the conditional probability is

$$P[A|B] = \frac{P[A \cap B]}{P[B]}$$

which is undefined if $P[B] = 0$. How should we then proceed to extend (13.1) for continuous random variables?

We will motivate a viable approach using the example of the circular dartboard described in Section 12.3. In particular, we consider a revised version of the dart throwing contest. Referring to Figure 12.2 the champion dart player realizes that the novice presents little challenge. To make the game more interesting the champion proposes the following modification. If the novice player's dart lands outside the region $|x| \leq \Delta x/2$, then the novice player gets to go again. He continues until his dart lands within the region $|x| \leq \Delta x/2$ as shown cross-hatched in Figure 13.1a. The novice dart player even gets to pick the value of Δx. Hence, he reasons that it

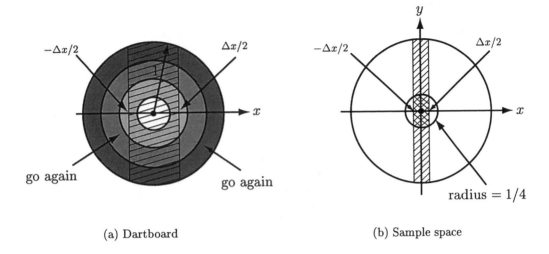

(a) Dartboard   (b) Sample space

Figure 13.1: Revised dart throwing game. Only dart outcomes in the cross-hatched region are counted.

should be small to exclude regions of the dart board that are outside the bullseye circle. As a result, he chooses a Δx as shown in Figure 13.1b, which allows him to continue throwing darts until one lands within the cross-hatched region. The champion, however, has taken a course in probability and so is not worried. In fact, in Problem 12.30 the probability of the champion's dart landing in the bullseye area was shown to be 0.8646. To find the probability of the novice player obtaining a bullseye, we recall that his dart is equally likely to land anywhere on the dartboard. Hence, using conditional probability we have that

$$P[\text{bullseye}| - \Delta x/2 \leq X \leq \Delta x/2] = \frac{P[\text{bullseye}, -\Delta x/2 \leq X \leq \Delta x/2]}{P[-\Delta x/2 \leq X \leq \Delta x/2]}.$$

Since $\Delta x/2$ is small, we can assume that it is much less than $1/4$ as shown in Figure 13.1b. Therefore, we have that the cross-hatched regions can be approximated by rectangles and so

$$P[\text{bullseye}| - \Delta x/2 \leq X \leq \Delta x/2]$$

$$= \frac{P[\text{double cross-hatched region}]}{P[\text{double cross-hatched region}] + P[\text{single cross-hatched region}]}$$

$$= \frac{\Delta x(1/2)/\pi}{\Delta x(2)/\pi} \quad \text{(probability = rectangle area/dartboard area)}$$

$$= \quad 0.25 < 0.865. \tag{13.2}$$

Hence, the revised strategy will still allow the champion to have a higher probability of winning for any Δx, no matter how small it is chosen. Even though $P[X = 0] = 0$, the conditional probability is well defined even as $\Delta x \to 0$ (but not equal 0). Some typical outcomes of this game are shown in Figure 13.2, where it is assumed the novice player has chosen $\Delta x/2 = 0.2$. In Figure 13.2a are shown the outcomes of X

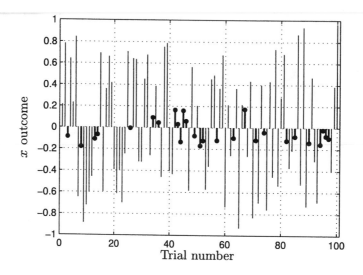

(a) All x outcomes – those with $|x| \le 0.2$ are shown as dark lines

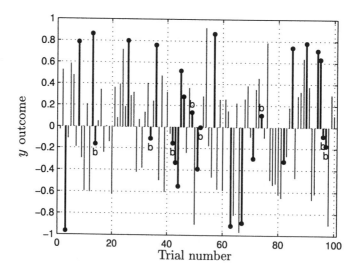

(b) Dark lines are y outcomes for which $|x| \le 0.2$, b indicates a bullseye ($\sqrt{x^2 + y^2} \le 1/4$) for the outcomes with $|x| \le 0.2$

Figure 13.2: Revised dart throwing game outcomes.

for the novice player. Only those for which $|x| \leq \Delta x/2 = 0.2$, which are shown as the darker lines, are kept. In Figure 13.2b the outcomes of Y are shown with the kept outcomes shown as the dark lines. Those outcomes that resulted in a bullseye are shown with a "b" over them. Note that there were 27 out of 100 outcomes that had $|x|$ values less than or equal to 0.2 (see Figure 13.2a), and of these, 8 outcomes resulted in a bullseye (see Figure 13.2b). Hence, the estimated probability of landing in either the single or double cross-hatched region of Figure 13.1b is $27/100 = 0.27$, while the theoretical probability is approximately $\Delta x(2)/\pi = 0.4(2)/\pi = 0.254$. Also, the estimated conditional probability of a bullseye is from Figure 13.2b, $8/27 = 0.30$ while from (13.2) the theoretical probability is approximately equal to 0.25. (The approximations are due to the use of rectangular approximations to the cross-hatched regions, which only become exact as $\Delta x \to 0$.) We will use the same strategy to define a conditional PDF. Let $A = \{(x, y) : \sqrt{x^2 + y^2} \leq 1/4\}$, which is the bullseye region. Then

$$P[A | |X| \leq \Delta x/2]$$

$$= \frac{P[A, |X| \leq \Delta x/2]}{P[|X| \leq \Delta x/2]} \qquad \text{(definition of cond. prob.)}$$

$$= \frac{P[\{(x, y) : |x| \leq \Delta x/2, |y| \leq \sqrt{1/16 - x^2}\}]}{P[\{(x, y) : |x| \leq \Delta x/2, |y| \leq 1\}]} \qquad \left(\frac{\text{double cross-hatched area}}{\text{cross-hatched area}}\right)$$

$$= \frac{P[\{(x, y) : |x| \leq \Delta x/2, |y| \leq \sqrt{1/16 - x^2}\}]}{P[\{x : |x| \leq \Delta x/2\}]}$$

$$= \frac{\int_{-\Delta x/2}^{\Delta x/2} \int_{-\sqrt{1/16 - x^2}}^{\sqrt{1/16 - x^2}} p_{X,Y}(x, y) dy \, dx}{\int_{-\Delta x/2}^{\Delta x/2} p_X(x) dx}.$$

As $\Delta x \to 0$, we can write

$$P[A | |X| \leq \Delta x/2]$$

$$\approx \frac{\int_{-\Delta x/2}^{\Delta x/2} \int_{-1/4}^{1/4} p_{X,Y}(x, y) dy \, dx}{\int_{-\Delta x/2}^{\Delta x/2} p_X(x) dx} \qquad \text{(since } \sqrt{1/16 - x^2} \approx 1/4 \text{ for } |x| \leq \Delta x/2)$$

$$\approx \frac{\int_{-1/4}^{1/4} p_{X,Y}(0, y) \Delta x dy}{p_X(0) \Delta x} \qquad \text{(since } p_{X,Y}(x, y) \approx p_{X,Y}(0, y) \text{ for } |x| \leq \Delta x/2)$$

$$= \int_{-1/4}^{1/4} \frac{p_{X,Y}(0, y)}{p_X(0)} dy.$$

We now define $p_{X,Y}/p_X$ as the conditional PDF

$$p_{Y|X}(y|x) = \frac{p_{X,Y}(x, y)}{p_X(x)}. \qquad (13.3)$$

Note that it is well defined as long as $p_X(x) \neq 0$. Thus, as $\Delta x \to 0$

$$P[A|\,|X| \leq \Delta x/2] = \int_{-1/4}^{1/4} p_{Y|X}(y|0)dy.$$

More generally, the conditional PDF allows us to compute probabilities as (see Problem 13.6)

$$P[a \leq Y \leq b|x - \Delta x/2 \leq X \leq x + \Delta x/2] = \int_a^b p_{Y|X}(y|x)dy.$$

This probability is usually written as

$$P[a \leq Y \leq b|X = x]$$

but the conditioning event should be understood to be $\{x : x - \Delta x/2 \leq X \leq x + \Delta x/2\}$ *for Δx small.* Finally, with this understanding, we have that

$$P[a \leq Y \leq b|X = x] = \int_a^b p_{Y|X}(y|x)dy \tag{13.4}$$

where $p_{Y|X}$ is defined by (13.3) and is termed the *conditional PDF*. The conditional PDF $p_{Y|X}(y|x)$ is the probability per unit length of Y when $X = x$ (actually $x - \Delta x/2 \leq X \leq x + \Delta x/2$) is observed. Since it is found using (13.3), it is seen to be a function of y *and* x. It should be thought of as a *family* of PDFs with y as the independent variable, and with a different PDF for each value x. An example follows.

Example 13.1 – Standard bivariate Gaussian PDF

Assume that (X, Y) have the joint PDF

$$p_{X,Y}(x,y) = \frac{1}{2\pi\sqrt{1-\rho^2}} \exp\left[-\frac{1}{2(1-\rho^2)}(x^2 - 2\rho xy + y^2)\right] \qquad \begin{array}{l} -\infty < x < \infty \\ -\infty < y < \infty \end{array}$$

and note that the marginal PDF for X is given by

$$p_X(x) = \frac{1}{\sqrt{2\pi}} \exp\left[-\frac{1}{2}x^2\right].$$

The conditional PDF is found from (13.3) as

$$p_{Y|X}(y|x) = \frac{\frac{1}{2\pi\sqrt{1-\rho^2}} \exp\left[-\frac{1}{2(1-\rho^2)}(x^2 - 2\rho xy + y^2)\right]}{\frac{1}{\sqrt{2\pi}} \exp\left[-\frac{1}{2}x^2\right]}$$

$$= \frac{1}{\sqrt{2\pi(1-\rho^2)}} \exp\left(-\frac{1}{2}Q\right)$$

where

$$Q = \frac{x^2 - 2\rho xy + y^2}{1 - \rho^2} - \frac{(1 - \rho^2)x^2}{1 - \rho^2}$$

$$= \frac{y^2 - 2\rho xy + \rho^2 x^2}{1 - \rho^2}$$

$$= \frac{(y - \rho x)^2}{1 - \rho^2}.$$

As a result we have that the conditional PDF is

$$p_{Y|X}(y|x) = \frac{1}{\sqrt{2\pi(1 - \rho^2)}} \exp\left[-\frac{1}{2(1 - \rho^2)}(y - \rho x)^2\right] \qquad (13.5)$$

and is seen to be Gaussian. This result, although of great importance, is not true in general. The form of the PDF usually changes from $p_Y(y)$ to $p_{Y|X}(y|x)$. We will denote this conditional PDF in shorthand notation as $Y|(X = x) \sim \mathcal{N}(\rho x, 1 - \rho^2)$. As expected, the conditional PDF depends on x, and in particular the mean of the conditional PDF is a function of x. It is a valid PDF in that *for each x value*, it is nonnegative and integrates to 1 over $-\infty < y < \infty$. These properties are true in general. In effect, the conditional PDF depends on the outcome of X so that we use a different PDF for each X outcome. For example, if $\rho = 0.9$ and we observe $X = -1$, then to compute $P[-1 \leq Y \leq -0.8|X = -1]$ and $P[-0.1 \leq Y \leq 0.1|X = -1]$, we first observe from (13.5) that $Y|(X = -1) \sim \mathcal{N}(-0.9, 0.19)$. Then

$$P[-1 \leq Y \leq -0.8|X = -1] = Q\left(\frac{-1 - (-0.9)}{\sqrt{0.19}}\right) - Q\left(\frac{-0.8 - (-0.9)}{\sqrt{0.19}}\right) = 0.1815$$

$$P[-0.1 \leq Y \leq 0.1|X = -1] = Q\left(\frac{-0.1 - (-0.9)}{\sqrt{0.19}}\right) - Q\left(\frac{0.1 - (-0.9)}{\sqrt{0.19}}\right) = 0.0223.$$

Can you explain the difference between these values? (See Figure 13.3b where the dark lines indicate $y = 0$ and $y = -0.9$.) In Figure 13.3b the cross-section of the joint PDF is shown. Once the cross-section is normalized so that it integrates to one, it becomes the conditional PDF $p_{Y|X}(y| - 1)$. This is easily verified since

$$p_{Y|X}(y| - 1) = \frac{p_{X,Y}(-1, y)}{p_X(-1)}$$

$$= \frac{p_{X,Y}(-1, y)}{\int_{-\infty}^{\infty} p_{X,Y}(-1, y)dy}.$$

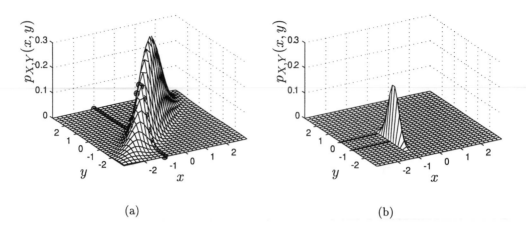

(a)   (b)

Figure 13.3: Standard bivariate Gaussian PDF and its cross-section at $x = -1$. The *normalized* cross-section is the conditional PDF.

13.4   Joint, Conditional, and Marginal PDFs

The relationships between the joint, conditional, and marginal PMFs as described in Section 8.4 also hold for the corresponding PDFs. Hence, we just summarize the properties and leave the proofs to the reader (see Problem 13.11).

Property 13.1 –   Joint PDF yields conditional PDFs.

$$p_{Y|X}(y|x) = \frac{p_{X,Y}(x,y)}{\int_{-\infty}^{\infty} p_{X,Y}(x,y)dy}$$

$$p_{X|Y}(x|y) = \frac{p_{X,Y}(x,y)}{\int_{-\infty}^{\infty} p_{X,Y}(x,y)dx}$$

$\square$

Property 13.2 –   Conditional PDFs are related.

$$p_{X|Y}(x|y) = \frac{p_{Y|X}(y|x)p_X(x)}{p_Y(y)}$$

$\square$

Property 13.3 – **Conditional PDF is expressible using Bayes' rule.**

$$p_{Y|X}(y|x) = \frac{p_{X|Y}(x|y)p_Y(y)}{\int_{-\infty}^{\infty} p_{X|Y}(x|y)p_Y(y)dy}$$

$\square$

Property 13.4 – **Conditional PDF and its corresponding marginal PDF yields the joint PDF**

$$p_{X,Y}(x,y) = p_{Y|X}(y|x)p_X(x) = p_{X|Y}(x|y)p_Y(y)$$

$\square$

Property 13.5 – **Conditional PDF and its corresponding marginal PDF yields the other marginal PDF**

$$p_Y(y) = \int_{-\infty}^{\infty} p_{Y|X}(y|x)p_X(x)dx$$

$\square$

A conditional CDF can also be defined. Based on (13.4) we have upon letting $a = -\infty$ and $b = y$

$$P[Y \leq y|X = x] = \int_{-\infty}^{y} p_{Y|X}(t|x)dt.$$

As a result the conditional CDF is defined as

$$F_{Y|X}(y|x) = P[Y \leq y|X = x] \tag{13.6}$$

and is evaluated using

$$F_{Y|X}(y|x) = \int_{-\infty}^{y} p_{Y|X}(t|x)dt. \tag{13.7}$$

As an example, if $Y|(X = x) \sim \mathcal{N}(\rho x, 1 - \rho^2)$ as was shown in Example 13.1, we have that

$$F_{Y|X}(y|x) = 1 - Q\left(\frac{y - \rho x}{\sqrt{1 - \rho^2}}\right). \tag{13.8}$$

Finally, as previously mentioned in Chapter 12 two continuous random variables X and Y are independent if and only if the joint PDF factors as $p_{X,Y}(x,y) = p_X(x)p_Y(y)$ or equivalently if the joint CDF factors as $F_{X,Y}(x,y) = F_X(x)F_Y(y)$.

This is consistent with our definition of the conditional PDF since if X and Y are independent

$$
\begin{aligned}
p_{Y|X}(y|x) &= \frac{p_{X,Y}(x,y)}{p_X(x)} \\
&= \frac{p_X(x)p_Y(y)}{p_X(x)} = p_Y(y)
\end{aligned}
\tag{13.9}
$$

(and similarly $p_{X|Y} = p_X$). Hence, the conditional PDF no longer depends on the observed value of X, i.e., x. This means that the knowledge that $X = x$ has occurred does not affect the PDF of Y (and thus does not affect the probability of events defined on $\mathcal{S}_Y$). Similarly, from (13.7), if X and Y are independent

$$
\begin{aligned}
F_{Y|X}(y|x) &= \int_{-\infty}^{y} p_{Y|X}(t|x)dt \\
&= \int_{-\infty}^{y} p_Y(t)dt \quad \text{(from (13.9))} \\
&= F_Y(y).
\end{aligned}
$$

An example would be if $\rho = 0$ for the standard bivariate Gaussian PDF. Then since $Y|(X = x) \sim \mathcal{N}(\rho x, 1 - \rho^2) = \mathcal{N}(0,1)$, we have that $p_{Y|X}(y|x) = p_Y(y)$. Also, from (13.8)

$$
\begin{aligned}
F_{Y|X}(y|x) &= 1 - Q\left(\frac{y - \rho x}{\sqrt{1 - \rho^2}}\right) \\
&= 1 - Q(y) = F_Y(y).
\end{aligned}
$$

Another example follows.

Example 13.2 – Lifetime PDF of spare lightbulb

A professor uses the overhead projector for his class. The time to failure of a new bulb X has the exponential PDF $p_X(x) = \lambda \exp(-\lambda x)u(x)$, where x is in hours. A new spare bulb also has a time to failure Y that is modeled as an exponential PDF. However, the time to failure of the spare bulb depends upon how long the spare bulb sits unused. Assuming the spare bulb is activated as soon as the original bulb fails, the time to activation is given by X. As a result, the *expected* time to failure of the spare bulb is decreased as

$$
\frac{1}{\lambda_Y} = \frac{1}{\lambda(1 + \alpha x)}
$$

where $0 < \alpha < 1$ is some factor that indicates the degradation of the unused bulb with storage time. The expected time to failure of the spare bulb decreases as the

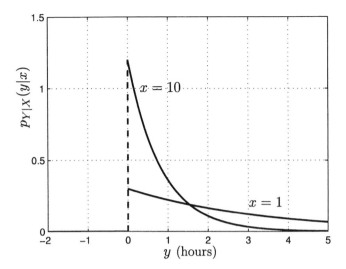

Figure 13.4: Conditional PDF for lifetime of spare bulb. Dependence is on time to failure x of original bulb.

original bulb is used longer (and hence the spare bulb must sit unused longer). Thus, we model the time to failure of the spare bulb as

$$
\begin{aligned}
p_{Y|X}(y|x) &= \lambda_Y \exp(-\lambda_Y y) u(y) \\
&= \lambda(1 + \alpha x) \exp\left[-\lambda(1 + \alpha x)y\right] u(y).
\end{aligned}
$$

This conditional PDF is shown in Figure 13.4 for $1/\lambda = 5$ hours and $\alpha = 0.5$. We now wish to determine the *unconditional PDF* of the time to failure of the spare bulb which is $p_Y(y)$. It is expected that the probability of failure of the spare bulb will increase than if the spare bulb were used rightaway or for $x = 0$. Note that if $x = 0$, then $p_{Y|X} = p_X$, which says that the spare bulb will fail with the same PDF as the original bulb. Using Property 13.5 we have

$$
\begin{aligned}
p_Y(y) &= \int_{-\infty}^{\infty} p_{Y|X}(y|x) p_X(x) dx \\
&= \int_0^{\infty} \lambda(1 + \alpha x) \exp\left[-\lambda(1 + \alpha x)y\right] \lambda \exp(-\lambda x) dx \\
&= \lambda^2 \exp(-\lambda y) \int_0^{\infty} (1 + \alpha x) \exp\left[-\lambda(\alpha y + 1)x\right] dx \\
&= \lambda^2 \exp(-\lambda y) \left[\int_0^{\infty} \exp(ax) dx + \alpha \int_0^{\infty} x \exp(ax) dx\right] \quad (\text{let } a = -\lambda(1 + \alpha y)) \\
&= \lambda^2 \exp(-\lambda y) \left[\left.\frac{\exp(ax)}{a}\right|_0^{\infty} + \alpha \left(\frac{1}{a} x \exp(ax) - \frac{1}{a^2} \exp(ax)\right)\Big|_0^{\infty}\right] \\
&= \lambda^2 \exp(-\lambda y) \left[-\frac{1}{a} + \frac{\alpha}{a^2}\right] \\
&= \lambda^2 \exp(-\lambda y) \left[\frac{1}{\lambda(\alpha y + 1)} + \frac{\alpha}{[\lambda(\alpha y + 1)]^2}\right].
\end{aligned}
$$

or finally

$$p_Y(y) = \lambda^2 \exp(-\lambda y) \left[\frac{1}{\lambda(\alpha y + 1)} + \frac{\alpha}{[\lambda(\alpha y + 1)]^2} \right] u(y).$$

This is shown in Figure 13.5 for $1/\lambda = 5$ hours and $\alpha = 0.5$ along with the PDF $p_X(x)$ of the time to failure of the original bulb. As expected the probability of the

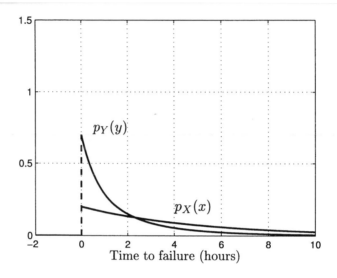

Figure 13.5: PDFs for time to failure of original bulb X and spare bulb Y.

spare bulb failing before 2 hours is greatly increased.

$\diamondsuit$

Finally, note that the conditional PDF is obtained by differentiating the conditional CDF. From (13.7) we have

$$p_{Y|X}(y|x) = \frac{\partial F_{Y|X}(y|x)}{\partial y}. \tag{13.10}$$

13.5   Simplifying Probability Calculations Using Conditioning

Following Section 8.6 we can easily find the PDF of $Z = g(X, Y)$ if X and Y are independent by using conditioning. We shall not repeat the argument other than to summarize the results and give an example. The procedure is

1. Fix $X = x$ and let $Z|(X = x) = g(x, Y)$.

2. Find the PDF of $Z|(X = x)$ using the standard approach for a transformation of a single random variable from Y to Z.

3. Uncondition the conditional PDF to yield the desired PDF $p_Z(z)$.

Example 13.3 – PDF for ratio of independent random variables
Consider the function $Z = Y/X$ where X and Y are independent random variables.
In Problem 12.39 we asserted that

$$p_Z(z) = \int_{-\infty}^{\infty} p_X(x) p_Y(xz) |x| dx.$$

We now derive this using the aforementioned approach. First recall that if $Z = aY$
for a a constant, then $p_Z(z) = p_Y(z/a)/|a|$ (see Example 10.5). Now we have that

$$Z|(X = x) = \frac{Y}{X}\Big|(X = x) = \frac{Y}{x}\Big|(X = x)$$

so that with $a = 1/x$ and noting the independence of X and Y, we have

$$p_{Z|X}(z|x) = p_{Y|X}(zx)|x| = p_Y(zx)|x| \tag{13.11}$$

and thus

$$
\begin{aligned}
p_Z(z) &= \int_{-\infty}^{\infty} p_{Z,X}(z, x) dx && \text{(marginal PDF from joint PDF)} \\
&= \int_{-\infty}^{\infty} p_{Z|X}(z|x) p_X(x) dx && \text{(definition of conditional PDF)} \\
&= \int_{-\infty}^{\infty} p_Y(zx)|x| p_X(x) dx && \text{(from (13.11))} \\
&= \int_{-\infty}^{\infty} p_X(x) p_Y(xz)|x| dx.
\end{aligned}
$$

Note that without the independence assumption, we could *not* assert that $p_{Y|X} = p_Y$
in (13.11).

$\diamondsuit$

In general to compute probabilities of events it is advantageous to use conditioning
arguments whether or not X and Y are independent. The analogous result to (8.28)
is (see Problem 13.15)

$$P[Y \in A] = \int_{-\infty}^{\infty} P[Y \in A | X = x] p_X(x) dx. \tag{13.12}$$

This is another form of the theorem of *total probability*. It can also be written as

$$P[Y \in A] = \int_{-\infty}^{\infty} \left[\int_A p_{Y|X}(y|x) dy \right] p_X(x) dx \tag{13.13}$$

where we have used (13.4) and replaced $\{y : a \le y \le b\}$ by the more general set
A. The formula of (13.13) is analogous to (8.27) for discrete random variables. An
example follows.

Example 13.4 – Probability of error for a digital communication system

Consider the PSK communication system shown in Figure 2.14. The probability of error was shown in Section 10.6 to be

$$P_e = P[W \leq -A/2] = Q(A/2)$$

since the noise sample $W \sim \mathcal{N}(0,1)$. In a wireless communication system such as is used in cellular telephone, the received amplitude A varies with time due to multipath propagation [Rappaport 2002]. As a result, it is usually modeled as a Rayleigh random variable whose PDF is

$$p_A(a) = \begin{cases} \frac{a}{\sigma_A^2} \exp\left(-\frac{1}{2\sigma_A^2}a^2\right) & a \geq 0 \\ 0 & a < 0. \end{cases}$$

We wish to determine the probability of error if A is a Rayleigh random variable. Thus, we need to evaluate $P[W + A/2 \leq 0]$ if $W \sim \mathcal{N}(0,1)$, A is a Rayleigh random variable, and we assume W and A are independent. A straightforward approach is to first find the PDF of $Z = W + A/2$, and then to integrate $p_Z(z)$ from $-\infty$ to 0. Alternatively, it is simpler to use (13.12) as follows.

$$
\begin{aligned}
P_e &= P[W \leq -A/2] \\
&= \int_{-\infty}^{\infty} P[W \leq -A/2 | A = a] p_A(a) da \quad \text{(from (13.12))} \\
&= \int_{-\infty}^{\infty} P[W \leq -a/2 | A = a] p_A(a) da \quad \text{(since } A = a \text{ has occurred)}.
\end{aligned}
$$

But since W and A are independent, $P[W \leq -a/2 | A = a] = P[W \leq -a/2]$ and thus

$$P_e = \int_{-\infty}^{\infty} P[W \leq -a/2] p_A(a) da.$$

Using $P[W \leq -a/2] = Q(a/2)$ we have

$$P_e = \int_{0}^{\infty} Q(a/2) \frac{a}{\sigma_A^2} \exp\left(-\frac{1}{2\sigma_A^2}a^2\right) da.$$

Unfortunately, this is not easily evaluated in closed form.

$\diamondsuit$

13.6 Mean of Conditional PDF

For a conditional PDF the mean is given by the usual mean definition except that the PDF now depends on x. We therefore have the definition

$$E_{Y|X}[Y|x] = \int_{-\infty}^{\infty} y p_{Y|X}(y|x) dy \tag{13.14}$$

which is analogous to (8.29) for discrete random variables. We also expect and it follows that (see Problem 13.19 and also the discussion in Section 8.6)

$$E_X[E_{Y|X}[Y|X]] = E_Y[Y] \tag{13.15}$$

where $E_{Y|X}[Y|X]$ is given by (13.14) except that the value x is now replaced by the random variable X. Therefore, $E_{Y|X}[Y|X]$ is viewed as a function of the random variable X. As an example, we saw that for the bivariate Gaussian PDF $Y|(X = x) \sim \mathcal{N}(\rho x, 1 - \rho^2)$. Hence, $E_{Y|X}[Y|x] = \rho x$, but regarding the mean of the conditional PDF as a function of the random variable X we have that $E_{Y|X}[Y|X] = \rho X$. To see that (13.15) holds for this example

$$E_X[E_{Y|X}[Y|X]] = E_X[\rho X] = \rho E_X[X] = 0$$

since the marginal PDF of X for the standard bivariate Gaussian PDF was shown to be $\mathcal{N}(0,1)$. Also, since $Y \sim \mathcal{N}(0,1)$ for the standard bivariate Gaussian PDF, $E_Y[Y] = 0$, and we see that (13.15) is satisfied.

The mean of the conditional PDF arises in optimal prediction, where it is proven that the minimum mean square error (MMSE) prediction of Y given $X = x$ has been observed is $E_{Y|X}[Y|x]$ (see Problem 13.17). This is optimal over all predictors, linear and *nonlinear*. For the standard Gaussian PDF, however, the optimal prediction turns out to be linear since $E_{Y|X}[Y|x] = \rho x$. More generally, it can be shown that if X and Y are jointly Gaussian with PDF given by (12.35), then

$$
\begin{aligned}
E_{Y|X}[Y|x] &= E_Y[Y] + \frac{\text{cov}(X,Y)}{\text{var}(X)}(x - E_X[X]) \\
&= \mu_Y + \frac{\rho \sigma_X \sigma_Y}{\sigma_X^2}(x - \mu_X) \\
&= \mu_Y + \frac{\rho \sigma_Y}{\sigma_X}(x - \mu_X).
\end{aligned}
$$

(See also Problem 13.20.)

13.7 Computer Simulation of Jointly Continuous Random Variables

In a manner similar to that described inn Section 8.7 we can generate realizations of a continuous random vector (X, Y) using the relationship

$$p_{X,Y}(x,y) = p_{Y|X}(y|x)p_X(x).$$

(Of course, if X and Y are independent, we can generate X based on $p_X(x)$ and Y based on $p_Y(y)$). Consider as an example the standard bivariate Gaussian PDF. We know that $Y|(X = x) \sim \mathcal{N}(\rho x, 1 - \rho^2)$ and $X \sim \mathcal{N}(0,1)$. Hence, we can generate realizations of (X, Y) as follows.

Step 1. Generate $X = x$ according to $\mathcal{N}(0,1)$.

Step 2. Generate $Y|(X = x)$ according to $\mathcal{N}(\rho x, 1 - \rho^2)$.

This procedure is conceptually simpler than what we implemented in Section 12.11 and much more general. There we used (12.53). Referring to (12.53), if we let $\mu_W = \mu_Z = 0$, $\sigma_W^2 = \sigma_Z^2 = 1$ and make the replacements of W, Z, X, Y with X, Y, U, V, we have

$$\left[\begin{array}{c} X \\ Y \end{array} \right] = \left[\begin{array}{cc} 1 & 0 \\ \rho & \sqrt{1 - \rho^2} \end{array} \right] \left[\begin{array}{c} U \\ V \end{array} \right] \tag{13.16}$$

where $U \sim \mathcal{N}(0,1)$, $V \sim \mathcal{N}(0,1)$, and U and V are independent. The transformation of (13.16) can be used to generate realizations of a standard bivariate Gaussian random vector. It is interesting to note that in this special case the two procedures for generating bivariate Gaussian random vectors lead to the identical algorithm. Can you see why from (13.16)?

As an example of the conditional PDF approach, if we let $\rho = 0.9$, we have the plot shown in Figure 13.6. It should be compared with Figure 12.21 (note that in Figure 12.21 the means of X and Y are 1). The MATLAB code used to generate

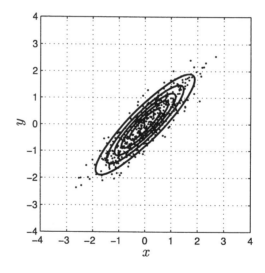

Figure 13.6: 500 outcomes of standard bivariate Gaussian random vector with $\rho = 0.9$ generated using conditional PDF approach.

realizations of a standard bivariate Gaussian random vector using conditioning is given below.

```
randn('state',0) % set random number generator to initial value
rho=0.9;
M=500; % set number of realizations to generate
for m=1:M
  x(m,1)=randn(1,1); % generate realization of N(0,1) random
                     % variable (Step 1)
```

```
ygx(m,1)=rho*x(m)+sqrt(1-rho^2)*randn(1,1); % generate
                                              % Y|(X=x) (Step 2)
end
```

13.8   Real-World Example – Retirement Planning

Professor Staff, who teaches too many courses a semester, plans to retire at age 65. He will have accumulated a total of $500,000 in a retirement account and wishes to use the money to live on during his retirement years. He assumes that his money will earn enough to offset the decrease in value due to inflation. Hence, if he lives to age 75 he could spend $50,000 a year and if he lives to age 85, then he could spend only $25,000 a year. How much should he figure on spending per year?

Besides the many courses Professor Staff has taught in history, English, mathematics, and computer science, he has also taught a course in probability. He therefore reasons that if he spends s dollars a year and lives for Y years during his retirement, then the probability that $500,000 - sY < 0$ should be small. Here sY is the total money spent during his retirement. In other words, he desires

$$P[500,000 - sY < 0] = 0.5. \tag{13.17}$$

He chooses 0.5 for the probability of outliving his retirement fund. This acknowledges the fact that choosing a lower probability will lead to an overly conservative approach and a small amount of expendable funds per year as we will see shortly. Equivalently, he requires that

$$P\left[Y > \frac{500,000}{s}\right] = 0.5. \tag{13.18}$$

As an example, if he spends $s = 50,000$ per year, then the probability he lives more than $500,000/s = 10$ years should be 0.5.

It should now be obvious that (13.18) is actually the right-tail probability or *complementary CDF* of the years lived in retirement. This type of information is of great interest not only to retirees but also to insurance companies who pay annuities. An annuity is a payment that an insurance company pays annually to an investor for the remainder of his/her life. The amount of the payment depends upon how much the investor originally invests, the age of the investor, and the insurance company's belief that the investor will live for so many years. To quantify answers to questions concerning years of life remaining, the *mortality rate*, which is the distribution of years lived past a given age is required. If Y is a continuous random variable that denotes the years lived past age $X = x$, then the mortality rate can be described by first defining the *conditional CDF*

$$F_{Y|X}(y|x) = P[Y \leq y|X = x].$$

For example, the probability that a person will live at least 10 more years if he is currently 65 years old is given by

$$P[Y > 10 | X = 65] = 1 - F_{Y|X}(10|65)$$

which is the complementary CDF or the *right-tail probability* of the conditional PDF $p_{Y|X}(y|x)$. It has been shown that for Canadian citizens the conditional CDF is well modeled by [Milevsky and Robinson 2000]

$$F_{Y|X}(y|x) = 1 - \exp\left[\exp\left(\frac{x-m}{l}\right)\left(1 - \exp\left(\frac{y}{l}\right)\right)\right] \qquad y \geq 0 \qquad (13.19)$$

where $m = 81.95, l = 10.6$ for males and $m = 87.8, l = 9.5$ for females. As an example, if $F_{Y|X}(y|x) = 0.5$, then you have a 50% chance of living more than y years if you are currently x years old. In other words, 50% of the population who are x years old will live more than y years and 50% will live less than y years. The number of years y is the *median* number of years to live. (Recall that the median is the value at which the probability of being less than or equal to this value is 0.5.) From (13.19) this will be true when

$$0.5 = \exp\left[\exp\left(\frac{x-m}{l}\right)\left(1 - \exp\left(\frac{y}{l}\right)\right)\right]$$

which results in the remaining number of years lived by 50% of the population who are currently x years old as

$$y = l \ln\left[1 - \exp\left(-\left(\frac{x-m}{l}\right)\right)\ln 0.5\right]. \qquad (13.20)$$

This is plotted in Figure 13.7a versus the current age x for males and females. In Figure 13.7a the median number of years left is shown while in Figure 13.7b the median life expectancy (which is $x + y$) is given.

Returning to Professor Staff, he can now determine how much money he can afford to spend each year. Since the probability of outliving one's retirement funds is a conditional probability based on current age, we rewrite (13.18) as

$$P\left[Y > \frac{500,000}{s}\,\middle|\, X = x\right] = P_L$$

where we allow the probability to be denoted in general by P_L. Since he will retire at age $x = 65$, we have from (13.19) that he will live more than y years with a probability of P_L given by

$$P_L = \exp\left[\exp\left(\frac{65-m}{l}\right)\left(1 - \exp\left(\frac{y}{l}\right)\right)\right]. \qquad (13.21)$$

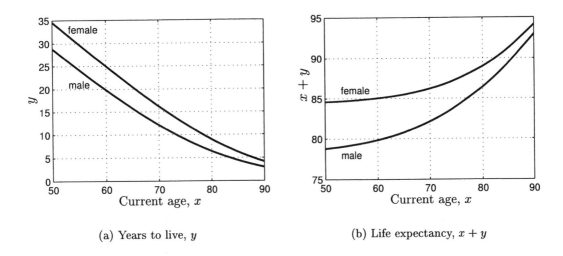

(a) Years to live, y

(b) Life expectancy, $x + y$

Figure 13.7: Mortality rates.

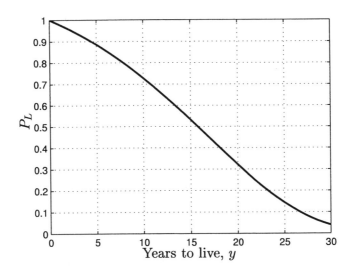

Figure 13.8: Probability P_L of exceeding y years in retirement for male who retires at age 65.

Assuming Professor Staff is a male, we use $m = 81.95, l = 10.6$ in (13.21) to produce a plot P_L versus y as shown in Figure 13.8. If the professor is overly conservative, he may want to assure himself that the probability of outliving his retirement fund is only about 0.1. Then, he should plan on living another 27 years, which means that his yearly expenses should not exceed $\$500,000/27 = \$18,500$. If he is less conservative and chooses a probability of 0.5, then he can plan on living about 15 years. Then his yearly expenses should not exceed $\$500,000/15 \approx \$33,000$.

References

Milevsky, M.A., C. Robinson, "Self-Annuitization and Ruin in Retirement," *North American Actuarial Journal*, Vol. 4, pp. 113–129, 2000.

Rappaport, T.S., *Wireless Communications, Principles and Practice*, Prentice-Hall, Upper Saddle River, NJ, 2002.

Problems

13.1 (w,c) In this problem we simulate on a computer the dartboard outcomes of the novice player for the game shown in Figure 13.1a. To do so, generate two independent $\mathcal{U}(-1, 1)$ random variables to serve as the x and y outcomes. Keep only the outcomes (x, y) for which $\sqrt{x^2 + y^2} \leq 1$ (see Problem 13.23 for why this produces a uniform joint PDF within the unit circle). Then, of the kept outcomes retain only the ones for which $\Delta x/2 \leq 0.2$ (see Figure 13.2a). Finally, estimate the probability that the novice player obtains a bullseye and compare it to the theoretical value. Note that the theoretical value of 0.25 as given by (13.2) is actually an approximation based on the areas in Figure 13.1b being rectangular.

13.2 ($\smile$) (w) Determine if the proposed conditional PDF

$$p_{Y|X}(y|x) = \begin{cases} c\exp(-y/x) & y \geq 0, x > 0 \\ 0 & \text{otherwise} \end{cases}$$

is a valid conditional PDF for some c. If so, find the required value of c.

13.3 (w) Is the proposed conditional PDF

$$p_{Y|X}(y|x) = \frac{1}{\sqrt{2\pi}} \exp\left[-\frac{1}{2}(y - x)^2\right] \qquad -\infty < y < \infty, -\infty < x < \infty$$

valid? If so, and if $X \sim \mathcal{N}(0, 1)$, design an experiment that will produce the random variables X and Y.

13.4 ($\smile$) (f) If

$$p_{X,Y}(x, y) = \begin{cases} 2\exp[-(x + y)] & 0 \leq y \leq x, x \geq 0 \\ 0 & \text{otherwise} \end{cases}$$

find $p_{Y|X}(y|x)$.

13.5 (w) Plot the joint PDF

$$p_{X,Y}(x, y) = \begin{cases} 2x & 0 < x < 1, 0 < y < 1 \\ 0 & \text{otherwise.} \end{cases}$$

Next determine by inspection the conditional PDF $p_{Y|X}(y|x)$. Recall that the conditional PDF is just the normalized cross-section of the joint PDF.

13.6 (t) In this problem we show that

$$\lim_{\Delta x \to 0} P[a \leq Y \leq b | x - \Delta x/2 \leq X \leq x + \Delta x/2] = \int_a^b p_{Y|X}(y|x) dy.$$

To do so first show that

$$\lim_{\Delta x \to 0} P[a \leq Y \leq b | x - \Delta x/2 \leq X \leq x + \Delta x/2]$$

$$= \int_a^b \lim_{\Delta x \to 0} \frac{\int_{x-\Delta x/2}^{x+\Delta x/2} p_{X,Y}(x,y) dx}{\int_{x-\Delta x/2}^{x+\Delta x/2} p_X(x) dx} dy.$$

13.7 (f) Determine $P[Y > \frac{1}{2} | X = 0]$ if the joint PDF is given as

$$p_{X,Y}(x,y) = \begin{cases} 2x & 0 < x < 1, 0 < y < 1 \\ 0 & \text{otherwise.} \end{cases}$$

13.8 (⌣) (f) If $X \sim \mathcal{U}(0,1)$ and $Y|(X = x) \sim \mathcal{U}(0,x)$, find the joint PDF for X and Y and also the marginal PDF for Y.

13.9 (f,t) For the standard bivariate Gaussian PDF find the conditional PDFs $p_{Y|X}$ and $p_{X|Y}$ and compare them. Explain your results. Are your results true in general?

13.10 (⌣) (f) If the joint PDF $p_{X,Y}$ is uniform over the region $0 < y < x$ and $0 < x < 1$ and zero otherwise, find the conditional PDFs $p_{Y|X}$ and $p_{X|Y}$.

13.11 (t) Prove Properties P13.1–13.5.

13.12 (f) Determine the PDF of Y/X if $X \sim \mathcal{N}(0,1)$, $Y \sim \mathcal{N}(0,1)$ and X and Y are independent. Do so by using the conditioning approach.

13.13 (t) Prove that the PDF of $X + Y$, where X and Y are independent is given as a convolution integral (see (12.14)). Do so by using the conditioning approach.

13.14 (⌣) (w) A game of darts is played using the linear dartboard shown in Figure 3.8. If two novice players throw darts at the board and each one's dart is equally likely to land anywhere in the interval $(-1/2, 1/2)$, prove that the probability of player 2 winning is 1/2. Hint: Let X_1 and X_2 be the outcomes and use $Y = |X_2| - |X_1|$ and $X = X_1$ in (13.12).

13.15 (t) Prove (13.12) by starting with (13.4).

13.16 (☺) (w) A resistor is chosen from a bin of 10 ohm resistors whose distribution satisfies $R \sim \mathcal{N}(10, 0.25)$. A $i = 1$ amp current source is applied to the resistor and the subsequent voltage V is measured with a voltmeter. The voltmeter has an error E that is modeled as $E \sim \mathcal{N}(0, 1)$. Find the probability that $V > 10$ volts if an 11 ohm resistor is chosen. Note that $V = iR + E$. What assumption do you need to make about the dependence between R and E?

13.17 (t) In this problem we prove that the minimum mean square error estimate of Y based on $X = x$ is given by $E_{Y|X}[Y|x]$. First let the estimate be denoted by $\hat{Y}(x)$ since it will depend in general on the outcome of X. Then note that the mean square error is

$$
\begin{aligned}
\text{mse} &= E_{X,Y}[(Y - \hat{Y}(X))^2] \\
&= \int_{-\infty}^{\infty} \int_{-\infty}^{\infty} (y - \hat{Y}(x))^2 p_{X,Y}(x, y)\, dx\, dy \\
&= \int_{-\infty}^{\infty} \int_{-\infty}^{\infty} (y - \hat{Y}(x))^2 p_{Y|X}(y|x) p_X(x)\, dx\, dy \\
&= \int_{-\infty}^{\infty} \underbrace{\left[\int_{-\infty}^{\infty} (y - \hat{Y}(x))^2 p_{Y|X}(y|x)\, dy \right]}_{J(\hat{Y}(x))} p_X(x)\, dx.
\end{aligned}
$$

Now we can minimize $J(\hat{Y}(x))$ for each value of x since $p_X(x) \geq 0$. Complete the derivation by differentiating $J(\hat{Y}(x))$ and setting the result equal to zero. Consider $\hat{Y}(x)$ as a constant (since x is assumed fixed inside the inner integral) in doing so. Finally justify all the steps in the derivation.

13.18 (f) For the joint PDF given in Problem 13.10 find the minimum mean square error estimate of Y given $X = x$. Plot the region in the x-y plane for which the joint PDF is nonzero and also the estimated value of Y versus x.

13.19 (t) Prove (13.15).

13.20 (w,c) If a bivariate Gaussian PDF has a mean vector $[\mu_X \ \mu_Y]^T = [1 \ 2]^T$ and a covariance matrix

$$
\mathbf{C} = \begin{bmatrix} 2 & 1 \\ 1 & 2 \end{bmatrix}
$$

plot the contours of constant PDF. Next find the minimum mean square error prediction of Y given $X = x$ and plot it on top of the contour plot. Explain the significance of the plot.

13.21 (☺) (w) A random variable X has a Laplacian PDF with variance σ^2. If the variance is chosen according to $\sigma^2 \sim \mathcal{U}(0, 1)$, what is average variance of the random variable?

13.22 (c) In this problem we use a computer simulation to illustrate the known result that $E_{Y|X}[Y|x] = \rho x$ for (X,Y) distributed according to a standard bivariate Gaussian PDF. Using (13.16) generate $M = 10,000$ realizations of a standard bivariate Gaussian random vector with $\rho = 0.9$. Then let $A = \{x : x_0 - \Delta x/2 \leq x \leq x_0 + \Delta x/2\}$ and discard the realizations for which x is not in A. Finally, estimate the mean of the conditional PDF by taking the sample mean of the remaining realizations. Choose $\Delta x/2 = 0.1$ and $x_0 = 1$ and compare the theoretical value of $E_{Y|X}[Y|x]$ to the estimated value based on your computer simulation.

13.23 (t) We now prove that the procedure described in Problem 13.1 will produce a random vector (X,Y) that is uniformly distributed within the unit circle. First consider the polar equivalent of (X,Y), which is (R, Θ), so that the conditional CDF is given by

$$P[R \leq r, \Theta \leq \theta | R \leq 1] \qquad 0 \leq r \leq 1, 0 \leq \theta < 2\pi.$$

But this is equal to

$$\frac{P[R \leq r, R \leq 1, \Theta \leq \theta]}{P[R \leq 1]} = \frac{P[R \leq r, \Theta \leq \theta]}{P[R \leq 1]}.$$

(Why?) Next show that

$$P[R \leq r, \Theta \leq \theta | R \leq 1] = \frac{\theta r^2}{2\pi}$$

and differentiate with respect to r and then θ to find the joint PDF $p_{R,\Theta}(r, \theta)$ (which is actually a conditional joint PDF due to the conditioning on the value of R being $r \leq 1$). Finally, transform this PDF back to that of (X,Y) to verify that it is uniform within the unit circle. Hint: You will need the result

$$\det\left(\frac{\partial(r,\theta)}{\partial(x,y)}\right) = \frac{1}{\det\left(\frac{\partial(x,y)}{\partial(r,\theta)}\right)}.$$

13.24 (☺) (f,c) For the conditional CDF of years left to live given current age, which is given by (13.19), find the conditional PDF. Plot the conditional PDF for a Canadian male who is currently 50 years old and also for one who is 75 years old. Next find the average life span for each of these individuals. Hint: You will need to use a computer evaluation of the integral for the last part.

13.25 (t) Verify that the conditional CDF given by (13.19) is a valid CDF.

Chapter 14

Continuous N-Dimensional Random Variables

14.1 Introduction

This chapter extends the results of Chapters 10–13 for one and two continuous random variables to N continuous random variables. Our discussion will mirror Chapter 9 quite closely, the difference being the consideration of continuous rather than discrete random variables. Therefore, the descriptions will be brief and will serve mainly to extend the usual definitions for one and two jointly distributed continuous random variables to an N-dimensional random vector. One new concept that is introduced is the orthogonality principle approach to prediction of the outcome of a random variable based on the outcomes of several other random variables. This concept will be useful later when we discuss prediction of random processes in Chapter 18.

14.2 Summary

The probability of an event defined on an N-dimensional sample space is given by (14.1). The most important example of an N-dimensional PDF is the multivariate Gaussian PDF, which is given by (14.2). If the components of the multivariate Gaussian random vector are uncorrelated, then they are also independent as shown in Example 14.2. Transformations of random vectors yield the transformed PDF given by (14.5). In particular, linear tranformations of Gaussian random vectors preserve the Gaussian nature but change the mean vector and covariance matrix as discussed in Example 14.3. Expected values are described in Section 14.5 with the mean and variance of a linear combination of random variables given by (14.8) and (14.10), respectively. The sample mean random variable is introduced in Example 14.4. The joint moment is defined by (14.13) and the joint characteristic function by (14.15). Joint moments can be found from the characteristic function using (14.17). The PDF for a sum of independent and identically distributed random variables is conveniently determined using (14.22). The prediction of the outcome of a random variable based on a linear combination of the outcomes of other random variables is given by (14.24). The linear prediction coefficients are found by solving the set of simultaneous linear equations in (14.27). The orthogonality principle is summarized by (14.29) and illustrated in Figure 14.3. Section 14.9 describes the computer generation of a multivariate Gaussian random vector. Finally, section 14.10 applies the results of this chapter to the real-world problem of signal detection with the optimal detector given by (14.33).

14.3 Random Vectors and PDFs

An N-dimensional random vector will be denoted by either $(X_1, X_2, \ldots, X_N)$ or $\mathbf{X} = [X_1 \, X_2 \ldots X_N]^T$. It is defined as a mapping from the original sample space of the experiment to a numerical sample space $\mathcal{S}_{X_1, X_2, \ldots, X_N} = R^N$. Hence, $\mathbf{X}$ takes on values in the N-dimensional Euclidean space R^N so that

$$\mathbf{X}(s) = \begin{bmatrix} X_1(s) \\ X_2(s) \\ \vdots \\ X_N(s) \end{bmatrix}$$

will have values

$$\mathbf{x} = \begin{bmatrix} x_1 \\ x_2 \\ \vdots \\ x_N \end{bmatrix}$$

where $\mathbf{x}$ is a point in R^N. The number of possible values is uncountably infinite. As an example, we might observe the temperature on each of N successive days. Then, the elements of the random vector would be $X_1(s) =$ temperature on day 1, $X_2(s) =$ temperature on day 2, ..., $X_N(s) =$ temperature on day N, and each temperature measurement would take on a continuum of values.

To compute probabilities of events defined on $\mathcal{S}_{X_1, X_2, \ldots, X_N}$ we will define the N-dimensional joint PDF (or more succinctly just the PDF) as

$$p_{X_1, X_2, \ldots, X_N}(x_1, x_2, \ldots, x_N)$$

and sometimes use the more compact notation $p_{\mathbf{X}}(\mathbf{x})$. The usual properties of a joint PDF must be valid

$$p_{X_1, X_2, \ldots, X_N}(x_1, x_2, \ldots, x_N) \geq 0$$

$$\int_{-\infty}^{\infty} \int_{-\infty}^{\infty} \cdots \int_{-\infty}^{\infty} p_{X_1, X_2, \ldots, X_N}(x_1, x_2, \ldots, x_N) dx_1 \, dx_2 \ldots dx_N = 1.$$

Then the probability of an event A defined on R^N is given by

$$P[A] = \int \int \cdots \int_A p_{X_1, X_2, \ldots, X_N}(x_1, x_2, \ldots, x_N) dx_1\, dx_2 \ldots dx_N. \qquad (14.1)$$

The most important example of an N-dimensional joint PDF is the *multivariate Gaussian* PDF. This PDF is the extension of the bivariate Gaussian PDF described at length in Chapter 12 (see (12.35)). It is given in vector/matrix form as

$$p_{\mathbf{X}}(\mathbf{x}) = \frac{1}{(2\pi)^{N/2} \det^{1/2}(\mathbf{C})} \exp\left[-\frac{1}{2}(\mathbf{x} - \boldsymbol{\mu})^T \mathbf{C}^{-1}(\mathbf{x} - \boldsymbol{\mu})\right] \qquad (14.2)$$

where $\boldsymbol{\mu} = [\mu_1\, \mu_2 \ldots \mu_N]^T$ is the $N \times 1$ mean vector so that

$$E_{\mathbf{X}}[\mathbf{X}] = \begin{bmatrix} E_{X_1}[X_1] \\ E_{X_2}[X_2] \\ \vdots \\ E_{X_N}[X_N] \end{bmatrix} = \boldsymbol{\mu}$$

and $\mathbf{C}$ is the $N \times N$ covariance matrix defined as

$$\mathbf{C} = \begin{bmatrix} \text{var}(X_1) & \text{cov}(X_1, X_2) & \cdots & \text{cov}(X_1, X_N) \\ \text{cov}(X_2, X_1) & \text{var}(X_2) & \cdots & \text{cov}(X_2, X_N) \\ \vdots & \vdots & \ddots & \vdots \\ \text{cov}(X_N, X_1) & \text{cov}(X_N, X_2) & \cdots & \text{var}(X_N) \end{bmatrix}.$$

Note that $\mathbf{C}$ is assumed to be positive definite and so it is invertible and has $\det(\mathbf{C}) > 0$ (see Appendix C). If the random variables have the multivariate Gaussian PDF, they are said to be *jointly Gaussian distributed*. Note that the covariance matrix can also be written as (see (Problem 9.21))

$$\mathbf{C} = E_{\mathbf{X}}\left[(\mathbf{X} - \boldsymbol{\mu})(\mathbf{X} - \boldsymbol{\mu})^T\right].$$

To denote a multivariate Gaussian PDF we will use the notation $\mathcal{N}(\boldsymbol{\mu}, \mathbf{C})$. Clearly, for $N = 2$ we have the bivariate Gaussian PDF. Evaluation of the probability of an event using (14.1) is in general quite difficult. Progress can, however, be made when A is a simple geometric region in R^N and $\mathbf{C}$ is a diagonal matrix. An example follows.

Example 14.1 – Probability of a point lying within a sphere

Assume $N = 3$ and let $\mathbf{X} \sim \mathcal{N}(\mathbf{0}, \sigma^2 \mathbf{I})$. We will determine the probability that an outcome falls within a sphere of radius R. The event is then given by $A = \{(x_1, x_2, x_3) : x_1^2 + x_2^2 + x_3^2 \leq R^2\}$. This event might represent the probability that a particle with mass m and random velocity components V_x, V_y, V_z has a kinetic energy $\mathcal{E} = (1/2)m(V_x^2 + V_y^2 + V_z^2)$ less than a given amount. This modeling is used in the kinetic theory of gases [Resnick and Halliday 1966] and is known as the Maxwellian distribution. From (14.2) we have with $\boldsymbol{\mu} = \mathbf{0}$, $\mathbf{C} = \sigma^2 \mathbf{I}$, and $N = 3$

$$
\begin{aligned}
P[A] &= \int \int \int_A \frac{1}{(2\pi)^{3/2} \det^{1/2}(\sigma^2 \mathbf{I})} \exp\left[-\frac{1}{2}\mathbf{x}^T (\sigma^2 \mathbf{I})^{-1} \mathbf{x}\right] dx_1\, dx_2\, dx_3 \\
&= \int \int \int_A \frac{1}{(2\pi\sigma^2)^{3/2}} \exp\left[-\frac{1}{2\sigma^2}(x_1^2 + x_2^2 + x_3^2)\right] dx_1\, dx_2\, dx_3
\end{aligned}
$$

since $\det(\sigma^2 \mathbf{I}) = (\sigma^2)^3$ and $(\sigma^2 \mathbf{I})^{-1} = (1/\sigma^2)\mathbf{I}$. Next we notice that the region of integration is the inside of a sphere. As a result of this and the observation that the integrand only depends on the squared-distance of the point from the origin, a reasonable approach is to convert the Cartesian coordinates to spherical coordinates. Doing so produces the inverse transformation

$$
\begin{aligned}
x_1 &= r\cos\theta\sin\phi \\
x_2 &= r\sin\theta\sin\phi \\
x_3 &= r\cos\phi
\end{aligned}
$$

where $r \geq 0$, $0 \leq \theta < 2\pi$, $0 \leq \phi \leq \pi$. We must be sure to include in the integral over r, θ, ϕ the absolute value of the Jacobian determinant of the inverse transformation which is $r^2 \sin\phi$ (see Problem 14.5). Thus,

$$
\begin{aligned}
P[A] &= \int_0^R \int_0^\pi \int_0^{2\pi} \frac{1}{(2\pi\sigma^2)^{3/2}} \exp\left(-\frac{1}{2\sigma^2}r^2\right) r^2 \sin\phi\, d\theta\, d\phi\, dr \\
&= \int_0^R \int_0^\pi \frac{1}{(2\pi\sigma^2)^{3/2}} r^2 \exp\left(-\frac{1}{2\sigma^2}r^2\right) 2\pi \sin\phi\, d\phi\, dr \\
&= \int_0^R \frac{1}{(2\pi\sigma^2)^{3/2}} r^2 \exp\left(-\frac{1}{2\sigma^2}r^2\right) 2\pi \underbrace{(-\cos\phi)|_0^\pi}_{2} dr \\
&= \int_0^R \frac{4\pi}{(2\pi\sigma^2)^{3/2}} r^2 \exp\left(-\frac{1}{2\sigma^2}r^2\right) dr \\
&= \sqrt{\frac{2}{\pi\sigma^2}} \int_0^R \frac{r^2}{\sigma^2} \exp\left(-\frac{1}{2\sigma^2}r^2\right) dr.
\end{aligned}
$$

To evaluate the integral

$$
I = \int_0^R \frac{r^2}{\sigma^2} \exp\left(-\frac{1}{2\sigma^2}r^2\right) dr
$$

we use integration by parts (see Problem 11.7) with $U = r$ and hence $dU = dr$ and

$dV = (r/\sigma^2)\exp[-r^2/(2\sigma^2)]dr$ so that $V = -\exp[-r^2/(2\sigma^2)]$. Then

$$
\begin{aligned}
I &= -r\exp\left[-\frac{1}{2}r^2/\sigma^2\right]\Big|_0^R + \int_0^R \exp\left[-\frac{1}{2}r^2/\sigma^2\right]dr \\
&= -R\exp\left[-\frac{1}{2}R^2/\sigma^2\right] + \sqrt{2\pi\sigma^2}\int_0^R \frac{1}{\sqrt{2\pi\sigma^2}}\exp\left[-\frac{1}{2}r^2/\sigma^2\right]dr \\
&= -R\exp\left[-\frac{1}{2}R^2/\sigma^2\right] + \sqrt{2\pi\sigma^2}\left[Q(0) - Q(R/\sigma)\right].
\end{aligned}
$$

Finally, we have that

$$
\begin{aligned}
P[A] &= \sqrt{\frac{2}{\pi\sigma^2}}\left[-R\exp\left[-\frac{1}{2}R^2/\sigma^2\right] + \sqrt{2\pi\sigma^2}\left[Q(0) - Q(R/\sigma)\right]\right] \\
&= 1 - 2Q(R/\sigma) - \sqrt{\frac{2}{\pi\sigma^2}}R\exp\left(-\frac{1}{2}R^2/\sigma^2\right).
\end{aligned}
$$

$\diamond$

The marginal PDFs are found by integrating out the other variables. For example, if $p_{X_1}(x_1)$ is desired, then

$$
p_{X_1}(x_1) = \int_{-\infty}^{\infty}\int_{-\infty}^{\infty}\cdots\int_{-\infty}^{\infty} p_{X_1,X_2,\ldots,X_N}(x_1,x_2,\ldots,x_N)dx_2\,dx_3\ldots dx_N.
$$

As an example, for the multivariate Gaussian PDF it can be shown that $X_i \sim \mathcal{N}(\mu_i, \sigma_i^2)$, where $\sigma_i^2 = \mathrm{var}(X_i)$ (see Problem 14.16). Also, the lower dimensional joint PDFs are similarly found. To determine $p_{X_1,X_N}(x_1, x_N)$ for example, we use

$$
p_{X_1,X_N}(x_1,x_N) = \int_{-\infty}^{\infty}\int_{-\infty}^{\infty}\cdots\int_{-\infty}^{\infty} p_{X_1,X_2,\ldots,X_N}(x_1,x_2,\ldots,x_N)dx_2\,dx_3\ldots dx_{N-1}.
$$

The random variables are defined to be independent if the joint PDF factors into the product of the marginal PDFs as

$$
p_{X_1,X_2,\ldots,X_N}(x_1,x_2,\ldots,x_N) = p_{X_1}(x_1)p_{X_2}(x_2)\ldots p_{X_N}(x_N). \tag{14.3}
$$

An example follows.

Example 14.2 – Condition for independence of multivariate Gaussian random variables

If the covariance matrix for a multivariate Gaussian PDF is diagonal, then the random variables are not only uncorrelated but also independent as we now show. Assume that

$$
\mathbf{C} = \mathrm{diag}\left(\sigma_1^2, \sigma_2^2, \ldots, \sigma_N^2\right)
$$

then it follows that

$$\det(\mathbf{C}) = \prod_{i=1}^{N} \sigma_i^2$$

$$\mathbf{C}^{-1} = \operatorname{diag}\left(\frac{1}{\sigma_1^2}, \frac{1}{\sigma_2^2}, \ldots, \frac{1}{\sigma_N^2}\right).$$

Using these results in (14.2) produces

$$
\begin{aligned}
p_{\mathbf{X}}(\mathbf{x}) &= \frac{1}{(2\pi)^{N/2}\left(\prod_{i=1}^{N}\sigma_i^2\right)^{1/2}} \exp\left[-\frac{1}{2}(\mathbf{x}-\boldsymbol{\mu})^T \operatorname{diag}\left(\frac{1}{\sigma_1^2}, \frac{1}{\sigma_2^2}, \ldots, \frac{1}{\sigma_N^2}\right)(\mathbf{x}-\boldsymbol{\mu})\right] \\
&= \frac{1}{\prod_{i=1}^{N}\sqrt{2\pi\sigma_i^2}} \exp\left[-\frac{1}{2}\sum_{i=1}^{N}(x_i-\mu_i)^2/\sigma_i^2\right] \\
&= \prod_{i=1}^{N}\frac{1}{\sqrt{2\pi\sigma_i^2}} \exp\left[-\frac{1}{2\sigma_i^2}(x_i-\mu_i)^2\right] \\
&= \prod_{i=1}^{N} p_{X_i}(x_i)
\end{aligned}
$$

where $X_i \sim \mathcal{N}(\mu_i, \sigma_i^2)$. *Hence, if a random vector has a multivariate Gaussian PDF and the covariance matrix is diagonal, which means that the random variables are uncorrelated, then the random variables are also independent.*

$\Diamond$

⚠ **Uncorrelated implies independence only for multivariate Gaussian PDF even if marginal PDFs are Gaussian!**

Consider the counterexample of a PDF for the random vector (X, Y) given by

$$
\begin{aligned}
p_{X,Y}(x,y) &= \frac{1}{2}\frac{1}{2\pi\sqrt{1-\rho^2}} \exp\left[-\frac{1}{2(1-\rho^2)}(x^2-2\rho xy+y^2)\right] \\
&\quad + \frac{1}{2}\frac{1}{2\pi\sqrt{1-\rho^2}} \exp\left[-\frac{1}{2(1-\rho^2)}(x^2+2\rho xy+y^2)\right] \quad (14.4)
\end{aligned}
$$

for $0 < \rho < 1$. This PDF is shown in Figure 14.1 for $\rho = 0.9$. Clearly, the random variables are not independent. Yet, it can be shown that $X \sim \mathcal{N}(0,1)$, $Y \sim \mathcal{N}(0,1)$, and X and Y are uncorrelated (see Problem 14.7). The difference here is that the joint PDF is not a bivariate Gaussian PDF.

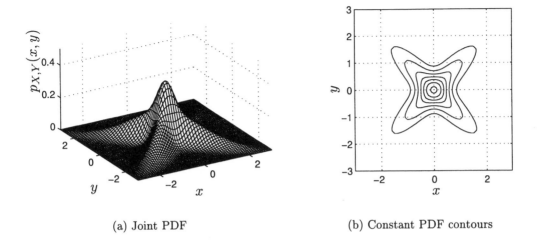

(a) Joint PDF   (b) Constant PDF contours

Figure 14.1: Uncorrelated but not independent random variables with Gaussian marginal PDFs.

A joint cumulative distribution function (CDF) can be defined in the N-dimensional case as

$$F_{X_1,X_2,\ldots,X_N}(x_1,x_2,\ldots,x_N) = P[X_1 \le x_1, X_2 \le x_2, \ldots, X_N \le x_N].$$

It has the usual properties of being between 0 and 1, being monotonically increasing as any of the variables increases, and being "right continuous". Also,

$$\begin{aligned} F_{X_1,X_2,\ldots,X_N}(-\infty,-\infty,\ldots,-\infty) &= 0 \\ F_{X_1,X_2,\ldots,X_N}(+\infty,+\infty,\ldots,+\infty) &= 1. \end{aligned}$$

The marginal CDFs are easily found by letting the undesired variables be evaluated at $+\infty$. For example, to determine the marginal CDF for X_1, we have

$$F_{X_1}(x_1) = F_{X_1,X_2,\ldots,X_N}(x_1,+\infty,+\infty,\ldots,+\infty).$$

14.4   Transformations

We consider the transformation from $\mathbf{X}$ to $\mathbf{Y}$ where

$$\begin{aligned} Y_1 &= g_1(X_1,X_2,\ldots,X_N) \\ Y_2 &= g_2(X_1,X_2,\ldots,X_N) \\ &\vdots \\ Y_N &= g_N(X_1,X_2,\ldots,X_N) \end{aligned}$$

and the transformation is one-to-one. Hence **Y** is a continuous random vector having a joint PDF (due to the one-to-one property). If we wish to find the PDF of a subset of the Y_i's, then we need only first find the PDF of **Y** and then integrate out the undesired variables. The extension of (12.22) for obtaining the joint PDF of two transformed random variables is

$$p_{Y_1,Y_2,\ldots,Y_N}(y_1, y_2, \ldots, y_N)$$

$$= p_{X_1,X_2,\ldots,X_N}(\underbrace{g_1^{-1}(\mathbf{y})}_{x_1}, \underbrace{g_2^{-1}(\mathbf{y})}_{x_2}, \ldots, \underbrace{g_N^{-1}(\mathbf{y})}_{x_N}) \left| \det\left(\frac{\partial(x_1, x_2, \ldots, x_N)}{\partial(y_1, y_2, \ldots, y_N)}\right)\right| \quad (14.5)$$

where

$$\frac{\partial(x_1, x_2, \ldots, x_N)}{\partial(y_1, y_2, \ldots, y_N)} = \begin{bmatrix} \frac{\partial x_1}{\partial y_1} & \frac{\partial x_1}{\partial y_2} & \cdots & \frac{\partial x_1}{\partial y_N} \\ \frac{\partial x_2}{\partial y_1} & \frac{\partial x_2}{\partial y_2} & \cdots & \frac{\partial x_2}{\partial y_N} \\ \vdots & \vdots & \ddots & \vdots \\ \frac{\partial x_N}{\partial y_1} & \frac{\partial x_N}{\partial y_2} & \cdots & \frac{\partial x_N}{\partial y_N} \end{bmatrix}$$

is the inverse Jacobian matrix. An example follows.

Example 14.3 – Linear transformation of multivariate Gaussian random vector

If $\mathbf{X} \sim \mathcal{N}(\boldsymbol{\mu}, \mathbf{C})$ and $\mathbf{Y} = \mathbf{GX}$, where **G** is an invertible $N \times N$ matrix, then we have from $\mathbf{y} = \mathbf{Gx}$ that

$$\mathbf{x} = \mathbf{G}^{-1}\mathbf{y}$$
$$\frac{\partial \mathbf{x}}{\partial \mathbf{y}} = \mathbf{G}^{-1}.$$

Hence, using (14.5) and (14.2)

$$\begin{aligned} p_{\mathbf{Y}}(\mathbf{y}) &= p_{\mathbf{X}}(\mathbf{G}^{-1}\mathbf{y}) \left|\det(\mathbf{G}^{-1})\right| \\ &= \frac{1}{(2\pi)^{N/2}\det^{1/2}(\mathbf{C})} \exp\left[-\frac{1}{2}(\mathbf{G}^{-1}\mathbf{y} - \boldsymbol{\mu})^T \mathbf{C}^{-1}(\mathbf{G}^{-1}\mathbf{y} - \boldsymbol{\mu})\right] \frac{1}{|\det(\mathbf{G})|} \\ &= \frac{1}{(2\pi)^{N/2}\det^{1/2}(\mathbf{GCG}^T)} \exp\left[-\frac{1}{2}(\mathbf{y} - \mathbf{G}\boldsymbol{\mu})^T (\mathbf{GCG}^T)^{-1}(\mathbf{y} - \mathbf{G}\boldsymbol{\mu})\right] \end{aligned}$$

(see Section 12.7 for details of matrix manipulations) so that $\mathbf{Y} \sim \mathcal{N}(\mathbf{G}\boldsymbol{\mu}, \mathbf{GCG}^T)$. This result is the extension of Theorem 12.7 from 2 to N jointly Gaussian random variables. See also Problems 14.8 and 14.15 for the case where **G** is $M \times N$ with $M < N$. It is shown there that the same result holds.

$\diamondsuit$

14.5 Expected Values

The expected value of a random vector is defined as the vector of the expected values of the elements of the random vector. This says that we define

$$E_{\mathbf{X}}[\mathbf{X}] = E_{X_1,X_2,\ldots,X_N}\left[\begin{bmatrix} X_1 \\ X_2 \\ \vdots \\ X_N \end{bmatrix}\right] = \begin{bmatrix} E_{X_1}[X_1] \\ E_{X_2}[X_2] \\ \vdots \\ E_{X_N}[X_N] \end{bmatrix}. \tag{14.6}$$

We can view this definition as "passing" the expectation "through" the left bracket of the vector since $E_{X_1,X_2,\ldots,X_N}[X_i] = E_{X_i}[X_i]$. A particular expectation of interest is that of a scalar function of $X_1, X_2, \ldots, X_N$, say $g(X_1, X_2, \ldots, X_N)$. Similar to previous results (see (12.28)) this is determined using

$$E_{X_1,X_2,\ldots,X_N}[g(X_1, X_2, \ldots, X_N)]$$

$$= \int_{-\infty}^{\infty} \int_{-\infty}^{\infty} \cdots \int_{-\infty}^{\infty} g(x_1, x_2, \ldots, x_N) p_{X_1,X_2,\ldots,X_N}(x_1, x_2, \ldots, x_N) dx_1 \, dx_2 \ldots dx_N. \tag{14.7}$$

Some specific results of interest are the linearity of the expectation operator or

$$E_{X_1,X_2,\ldots,X_N}\left[\sum_{i=1}^{N} a_i X_i\right] = \sum_{i=1}^{N} a_i E_{X_i}[X_i] \tag{14.8}$$

and in particular if $a_i = 1$ for all i, then we have

$$E_{X_1,X_2,\ldots,X_N}\left[\sum_{i=1}^{N} X_i\right] = \sum_{i=1}^{N} E_{X_i}[X_i]. \tag{14.9}$$

The variance of a linear combination of random variables is given by

$$\text{var}\left(\sum_{i=1}^{N} a_i X_i\right) = \sum_{i=1}^{N} \sum_{j=1}^{N} a_i a_j \text{cov}(X_i, X_j) = \mathbf{a}^T \mathbf{C}_X \mathbf{a} \tag{14.10}$$

where $\mathbf{C}_X$ is the covariance matrix of $\mathbf{X}$ and $\mathbf{a} = [a_1 \, a_2 \ldots a_N]^T$. The derivation of (14.10) is identical to that given in the proof of Property 9.2 for discrete random variables. If the random variables are *uncorrelated* so that the covariance matrix is diagonal or

$$\mathbf{C}_X = \text{diag}(\text{var}(X_1), \text{var}(X_2) \ldots, \text{var}(X_N))$$

then (see Problem 14.10)

$$\text{var}\left(\sum_{i=1}^{N} a_i X_i\right) = \sum_{i=1}^{N} a_i^2 \text{var}(X_i). \tag{14.11}$$

If furthermore, $a_i = 1$ for all i, then

$$\text{var}\left(\sum_{i=1}^{N} X_i\right) = \sum_{i=1}^{N} \text{var}(X_i). \tag{14.12}$$

An example follows.

Example 14.4 – Sample mean of independent and identically distributed random variables

Assume that $X_1, X_2, \ldots, X_N$ are independent random variables and each random variable has the same marginal PDF. When random variables have the same marginal PDF, they are said to be *identically distributed*. Hence, we are assuming that the random variables are *independent and identically distributed (IID)*. As a consequence of being identically distributed, $E_{X_i}[X_i] = \mu$ and $\text{var}(X_i) = \sigma^2$ for all i. It is of interest to examine the mean and variance of the *random variable* that we obtain by averaging the X_i's together. This averaged random variable is

$$\bar{X} = \frac{1}{N} \sum_{i=1}^{N} X_i$$

and is called the *sample mean random variable*. We have previously encountered the sample mean when referring to an average of a set of *outcomes* of a repeated experiment, which produced a number. Now, however, $\bar{X}$ is a function of the random variables $X_1, X_2, \ldots, X_N$ and so is a random variable itself. As such we may consider its probabilistic properties such as its mean and variance. The mean is from (14.8) with $a_i = 1/N$

$$E_{X_1, X_2, \ldots, X_N}[\bar{X}] = \frac{1}{N} \sum_{i=1}^{N} E_{X_i}[X_i] = \mu$$

and the variance is from (14.11) with $a_i = 1/N$ (since X_i's are independent and hence uncorrelated)

$$\begin{aligned}
\text{var}(\bar{X}) &= \sum_{i=1}^{N} \frac{1}{N^2} \text{var}(X_i) \\
&= \frac{1}{N^2} \sum_{i=1}^{N} \sigma^2 \\
&= \frac{\sigma^2}{N}.
\end{aligned}$$

Note that on the average the sample mean random variable will yield the value μ, which is the expected value of each X_i. Also as $N \to \infty$, $\text{var}(\bar{X}) \to 0$, so that the PDF of $\bar{X}$ will become more and more concentrated about μ. In effect, as $N \to \infty$, we have that $\bar{X} \to \mu$. This says that the sample mean random variable will converge to the true expected value of X_i. An example is shown in Figure 14.2 in which the marginal PDF of each X_i is $\mathcal{N}(2,1)$. In the next chapter we will prove that $\bar{X}$ does indeed converge to $E_{X_i}[X_i] = \mu$.

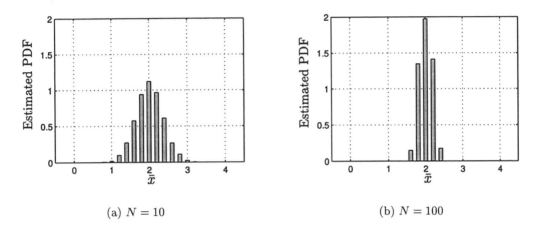

(a) $N = 10$ (b) $N = 100$

Figure 14.2: Estimated PDF for sample mean random variable, $\bar{X}$.

$\Diamond$

14.6 Joint Moments and the Characteristic Function

The joint moments corresponding to an N-dimensional PDF are defined as

$$E_{X_1, X_2, \ldots, X_N}[X_1^{l_1} X_2^{l_2} \ldots X_N^{l_N}]$$

$$= \int_{-\infty}^{\infty} \int_{-\infty}^{\infty} \cdots \int_{-\infty}^{\infty} x_1^{l_1} x_2^{l_2} \ldots x_N^{l_N} p_{X_1, X_2, \ldots, X_N}(x_1, x_2, \ldots, x_N) dx_1\, dx_2 \ldots dx_N.$$

$$(14.13)$$

As usual, if the random variables are *independent*, the joint PDF factors and therefore

$$E_{X_1, X_2, \ldots, X_N}[X_1^{l_1} X_2^{l_2} \ldots X_N^{l_N}] = E_{X_1}[X_1^{l_1}] E_{X_2}[X_2^{l_2}] \ldots E_{X_N}[X_N^{l_N}]. \qquad (14.14)$$

The joint characteristic function is defined as

$$\phi_{X_1, X_2, \ldots, X_N}(\omega_1, \omega_2, \ldots, \omega_N) = E_{X_1, X_2, \ldots, X_N}\left[\exp[j(\omega_1 X_1 + \omega_2 X_2 + \cdots + \omega_N X_N)]\right]$$

$$(14.15)$$

and is evaluated as

$$\phi_{X_1, X_2, \ldots, X_N}(\omega_1, \omega_2, \ldots, \omega_N)$$

$$= \int_{-\infty}^{\infty} \int_{-\infty}^{\infty} \cdots \int_{-\infty}^{\infty} \exp[j(\omega_1 x_1 + \omega_2 x_2 + \cdots + \omega_N x_N)]$$

$$p_{X_1, X_2, \dots, X_N}(x_1, x_2, \dots, x_N) dx_1 \, dx_2 \dots dx_N.$$

In particular, for independent random variables, we have (see Problem 14.13)

$$\phi_{X_1, X_2, \dots, X_N}(\omega_1, \omega_2, \dots, \omega_N) = \phi_{X_1}(\omega_1) \phi_{X_2}(\omega_2) \dots \phi_{X_N}(\omega_N).$$

Also, the joint PDF can be found from the joint characteristic function using the inverse Fourier transform as

$$p_{X_1, X_2, \dots, X_N}(x_1, x_2, \dots, x_N)$$

$$= \int_{-\infty}^{\infty} \int_{-\infty}^{\infty} \cdots \int_{-\infty}^{\infty} \phi_{X_1, X_2, \dots, X_N}(\omega_1, \omega_2, \dots, \omega_N)$$

$$\cdot \exp[-j(\omega_1 x_1 + \omega_2 x_2 + \cdots + \omega_N x_N)] \frac{d\omega_1}{2\pi} \frac{d\omega_2}{2\pi} \cdots \frac{d\omega_N}{2\pi}. \qquad (14.16)$$

All the properties of the 2-dimensional characteristic function extend to the general case. Note that once $\phi_{X_1, X_2, \dots, X_N}(\omega_1, \omega_2, \dots, \omega_N)$ is known, the characteristic function for any subset of the X_i's is found by setting ω_i equal to zero for the ones not in the subset. For example, to find $p_{X_1, X_2}(x_1, x_2)$, we can let $\omega_3 = \omega_4 = \cdots = \omega_N = 0$ in the joint characteristic function to yield (see Problem 14.14)

$$p_{X_1, X_2}(x_1, x_2) = \int_{-\infty}^{\infty} \int_{-\infty}^{\infty} \underbrace{\phi_{X_1, X_2, \dots, X_N}(\omega_1, \omega_2, 0, \dots, 0)}_{\phi_{X_1, X_2}(\omega_1, \omega_2)} \exp[-j(\omega_1 x_1 + \omega_2 x_2)] \frac{d\omega_1}{2\pi} \frac{d\omega_2}{2\pi}.$$

As seen previously, the joint moments can be obtained from the characteristic function. The general formula is

$$E_{X_1, X_2, \dots, X_N}[X_1^{l_1} X_2^{l_2} \dots X_N^{l_N}]$$

$$= \frac{1}{j^{l_1 + l_2 + \cdots + l_N}} \frac{\partial^{l_1 + l_2 + \cdots + l_N}}{\partial \omega_1^{l_1} \partial \omega_2^{l_2} \dots \partial \omega_N^{l_N}} \phi_{X_1, X_2, \dots, X_N}(\omega_1, \omega_2, \dots, \omega_N) \Big|_{\omega_1 = \omega_2 = \cdots = \omega_N = 0}. \qquad (14.17)$$

An example follows.

Example 14.5 – Second-order joint moments for multivariate Gaussian PDF

In this example we derive the second-order moments $E_{X_i X_j}[X_i X_j]$ if $\mathbf{X} \sim \mathcal{N}(\mathbf{0}, \mathbf{C})$. The characteristic function can be shown to be [Muirhead 1982]

$$\phi_{\mathbf{X}}(\boldsymbol{\omega}) = \exp\left(-\frac{1}{2} \boldsymbol{\omega}^T \mathbf{C} \boldsymbol{\omega}\right)$$

where $\boldsymbol{\omega} = [\omega_1 \, \omega_2 \ldots \omega_N]^T$. We first let

$$Q(\boldsymbol{\omega}) = \boldsymbol{\omega}^T \mathbf{C} \boldsymbol{\omega} = \sum_{m=1}^{N} \sum_{n=1}^{N} \omega_m \omega_n [\mathbf{C}]_{mn} \qquad (14.18)$$

and note that it is a quadratic form (see Appendix C). Also, we let $[\mathbf{C}]_{mn} = c_{mn}$ to simplify the notation. Then from (14.17) with $l_i = l_j = 1$ and the other l's equal to zero, we have

$$E_{X_i, X_j}[X_i X_j] = \frac{1}{j^2} \frac{\partial^2}{\partial \omega_i \partial \omega_j} \exp\left(-\frac{1}{2} Q(\boldsymbol{\omega})\right)\bigg|_{\boldsymbol{\omega}=0}.$$

Carrying out the partial differentiation produces

$$\frac{\partial \exp[-(1/2)Q(\boldsymbol{\omega})]}{\partial \omega_i} = -\frac{1}{2} \frac{\partial Q(\boldsymbol{\omega})}{\partial \omega_i} \exp\left(-\frac{1}{2} Q(\boldsymbol{\omega})\right)$$

$$\frac{\partial^2 \exp[-(1/2)Q(\boldsymbol{\omega})]}{\partial \omega_i \partial \omega_j} = \frac{1}{4} \frac{\partial Q(\boldsymbol{\omega})}{\partial \omega_i} \frac{\partial Q(\boldsymbol{\omega})}{\partial \omega_j} \exp\left(-\frac{1}{2} Q(\boldsymbol{\omega})\right)$$

$$-\frac{1}{2} \frac{\partial^2 Q(\boldsymbol{\omega})}{\partial \omega_i \partial \omega_j} \exp\left(-\frac{1}{2} Q(\boldsymbol{\omega})\right). \qquad (14.19)$$

But

$$\frac{\partial Q(\boldsymbol{\omega})}{\partial \omega_i}\bigg|_{\boldsymbol{\omega}=0} = \sum_{m=1}^{N} \sum_{n=1}^{N} \frac{\partial \omega_m \omega_n}{\partial \omega_i} c_{mn}\bigg|_{\boldsymbol{\omega}=0} \qquad \text{(from (14.18))}$$

$$= \sum_{m=1}^{N} \sum_{n=1}^{N} \left[\omega_m \frac{\partial \omega_n}{\partial \omega_i} c_{mn} + \omega_n \frac{\partial \omega_m}{\partial \omega_i} c_{mn} \right]\bigg|_{\boldsymbol{\omega}=0}$$

$$= 0 \qquad (14.20)$$

and also

$$\frac{\partial^2 Q(\boldsymbol{\omega})}{\partial \omega_i \partial \omega_j}\bigg|_{\boldsymbol{\omega}=0} = \sum_{m=1}^{N} \sum_{n=1}^{N} \frac{\partial^2 \omega_m \omega_n}{\partial \omega_i \partial \omega_j} c_{mn}\bigg|_{\boldsymbol{\omega}=0}. \qquad (14.21)$$

But

$$\frac{\partial^2 \omega_m \omega_n}{\partial \omega_i} = \omega_m \frac{\partial \omega_n}{\partial \omega_i} + \omega_n \frac{\partial \omega_m}{\partial \omega_i}$$

$$= \omega_m \delta_{ni} + \omega_n \delta_{mi}$$

where δ_{ij} is the Kronecker delta, which is defined to be 1 if $i = j$ and 0 otherwise. Hence

$$\frac{\partial^2 \omega_m \omega_n}{\partial \omega_i \partial \omega_j} = \delta_{mj} \delta_{ni} + \delta_{nj} \delta_{mi}$$

and $\delta_{mj}\delta_{ni}$ equals 1 if $(m, n) = (j, i)$ and equals 0 otherwise, and $\delta_{nj}\delta_{mi}$ equals 1 if $(m, n) = (i, j)$ and equals 0 otherwise. Thus,

$$\left.\frac{\partial^2 Q(\boldsymbol{\omega})}{\partial \omega_i \partial \omega_j}\right|_{\boldsymbol{\omega}=0} = c_{ji} + c_{ij} \quad \text{(from (14.21))}$$

$$= 2c_{ij} \quad \text{(recall that } \mathbf{C}^T = \mathbf{C}\text{).}$$

Finally, we have the expected result from (14.19) and (14.20) that

$$E_{X_i, X_j}[X_i X_j] = \frac{1}{j^2}\left[-\frac{1}{2}\frac{\partial^2 Q(\boldsymbol{\omega})}{\partial \omega_i \partial \omega_j}\exp\left(-\frac{1}{2}Q(\boldsymbol{\omega})\right)\right]\Bigg|_{\boldsymbol{\omega}=0}$$

$$= \frac{1}{j^2}\left(-\frac{1}{2}\right)(2c_{ij}) = c_{ij} = [\mathbf{C}]_{ij}.$$

$\diamondsuit$

Lastly, we extend the characteristic function approach to determining the PDF for a sum of IID random variables. Letting $Y = \sum_{i=1}^{N} X_i$, the characteristic function of Y is defined by

$$\phi_Y(\omega) = E_Y[\exp(j\omega Y)]$$

and is evaluated using (14.7) with $g(X_1, X_2, \ldots, X_N) = \exp[j\omega \sum_{i=1}^{N} X_i]$ (the real and imaginary parts are evaluated as separate integrals) as

$$\phi_Y(\omega) = E_{X_1, X_2, \ldots, X_N}\left[\exp\left(j\omega \sum_{i=1}^{N} X_i\right)\right]$$

$$= E_{X_1, X_2, \ldots, X_N}\left[\prod_{i=1}^{N} \exp(j\omega X_i)\right].$$

Now using the fact that the X_i's are IID, we have that

$$\phi_Y(\omega) = \prod_{i=1}^{N} E_{X_i}[\exp(j\omega X_i)] \quad \text{(independence)}$$

$$= \prod_{i=1}^{N} \phi_{X_i}(\omega)$$

$$= [\phi_X(\omega)]^N \quad \text{(identically distributed)}$$

where $\phi_X(\omega)$ is the common characteristic function of the random variables. To finally obtain the PDF of the sum random variable we use an inverse Fourier transform to yield

$$p_Y(y) = \int_{-\infty}^{\infty} [\phi_X(\omega)]^N \exp(-j\omega y)\frac{d\omega}{2\pi}. \tag{14.22}$$

This formula will form the basis for the exploration of the PDF of a sum of IID random variables in Chapter 15. See Problems 14.17 and 14.18 for some examples of its use.

14.7 Conditional PDFs

The discussion of Section 9.7 of the definitions and properties of the conditional PMF also hold for the conditional PDF. To accommodate continuous random variables we need only replace the PMF notation of the "bracket" with that of the PDF notation of the "parenthesis." Hence, we do not pursue this topic further.

14.8 Prediction of a Random Variable Outcome

We have seen in Section 7.9 that the optimal linear prediction of the outcome of Y when $X = x$ is observed to occur is

$$\hat{Y} = E_Y[Y] + \frac{\text{cov}(X, Y)}{\text{var}(X)}(x - E_X[X]). \qquad (14.23)$$

If (X, Y) has a bivariate Gaussian PDF, then the linear predictor is also the optimal predictor, amongst all linear *and* nonlinear predictors. We now extend these results to the prediction of a random variable after having observed the outcomes of *several* other random variables. In doing so the *orthogonality principle* will be introduced. Our discussions will assume only zero mean random variables, although the results are easily modified to yield the prediction for a nonzero mean random variable. To do so note that (14.23) can also be written as

$$\hat{Y} - E_Y[Y] = \frac{\text{cov}(X, Y)}{\text{var}(X)}(x - E_X[X]).$$

But if X and Y had been zero mean, then we would have obtained

$$\hat{Y} = \frac{\text{cov}(X, Y)}{\text{var}(X)}x.$$

It is clear that the modification from the zero mean case to the nonzero mean case is to replace each x_i by $x_i - E_{X_i}[X_i]$ and also $\hat{Y}$ by $\hat{Y} - E_Y[Y]$.

Now consider the $p + 1$ continuous random variables $\{X_1, X_2, \ldots, X_p, X_{p+1}\}$ and say we wish to predict X_{p+1} based on the knowledge of the outcomes of $X_1, X_2, \ldots, X_p$. Letting $X_1 = x_1, X_2 = x_2 \ldots, X_p = x_p$ be those outcomes, we consider the linear prediction

$$\hat{X}_{p+1} = \sum_{i=1}^{p} a_i x_i \qquad (14.24)$$

where the a_i's are the *linear prediction coefficients*, which are to be determined. The optimal coefficients are chosen to minimize the mean square error (MSE)

$$\text{mse} = E_{X_1, X_2, \ldots, X_{p+1}}[(X_{p+1} - \hat{X}_{p+1})^2]$$

or written more explicitly as

$$\text{mse} = E_{X_1, X_2, \ldots, X_{p+1}} \left[\left(X_{p+1} - \sum_{i=1}^{p} a_i X_i \right)^2 \right]. \tag{14.25}$$

We have used $\sum_{i=1}^{p} a_i X_i$, which is a random variable, as the predictor in order that the error measure be the *average* over all predictions. If we now differentiate the MSE with respect to a_1 we obtain

$$\frac{\partial E_{X_1, X_2, \ldots, X_{p+1}}[(X_{p+1} - \sum_{i=1}^{p} a_i X_i)^2]}{\partial a_1}$$

$$= E_{X_1, X_2, \ldots, X_{p+1}} \left[\frac{\partial}{\partial a_1} (X_{p+1} - \sum_{i=1}^{p} a_i X_i)^2 \right] \quad \begin{array}{l} \text{(interchange integration} \\ \text{and differentiation)} \end{array}$$

$$= E_{X_1, X_2, \ldots, X_{p+1}} \left[-2(X_{p+1} - \sum_{i=1}^{p} a_i X_i) X_1 \right] = 0. \tag{14.26}$$

This produces

$$E_{X_1, X_2, \ldots, X_{p+1}}[X_1 X_{p+1}] = E_{X_1, X_2, \ldots, X_{p+1}} \left[\sum_{i=1}^{p} a_i X_1 X_i \right]$$

or

$$E_{X_1, X_{p+1}}[X_1 X_{p+1}] = \sum_{i=1}^{p} a_i E_{X_1, X_i}[X_1 X_i].$$

Letting $c_{ij} = E_{X_i, X_j}[X_i X_j]$ denote the covariance (since the X_i's are zero mean) we have the equation

$$\sum_{i=1}^{p} c_{1i} a_i = c_{1, p+1}.$$

If we differentiate with respect to the other coefficients, similar equations are obtained. In all, there will be p simultaneous linear equations given by

$$\sum_{i=1}^{p} c_{ki} a_i = c_{k, p+1} \qquad k = 1, 2, \ldots, p$$

that need to be solved to yield the a_i's. These equations can be written in vector/matrix form as

$$\underbrace{\begin{bmatrix} c_{11} & c_{12} & \cdots & c_{1p} \\ c_{21} & c_{22} & \cdots & c_{2p} \\ \vdots & \vdots & \ddots & \vdots \\ c_{p1} & c_{p2} & \cdots & c_{pp} \end{bmatrix}}_{\mathbf{C}} \begin{bmatrix} a_1 \\ a_2 \\ \vdots \\ a_p \end{bmatrix} = \underbrace{\begin{bmatrix} c_{1, p+1} \\ c_{2, p+1} \\ \vdots \\ c_{p, p+1} \end{bmatrix}}_{\mathbf{c}}. \tag{14.27}$$

We note that $\mathbf{C}$ is the covariance matrix of the random vector $[X_1 \, X_2 \ldots X_p]^T$ and $\mathbf{c}$ is the vector of covariances between X_{p+1} and each X_i used in the predictor. The linear prediction coefficients are found by solving these linear equations. An example follows.

Example 14.6 – Linear prediction based on two random variable outcomes

Consider the prediction of X_3 based on the outcomes of X_1 and X_2 so that $\hat{X}_3 = a_1 x_1 + a_2 x_2$, where $p = 2$. If we know the covariance matrix of $\mathbf{X} = [X_1 \, X_2 \, X_3]^T$ say $\mathbf{C}_X$, then all the c_{ij}'s needed for (14.27) are known. Hence, suppose that

$$\mathbf{C}_X = \begin{bmatrix} c_{11} & c_{12} & c_{13} \\ c_{21} & c_{22} & c_{23} \\ c_{31} & c_{32} & c_{33} \end{bmatrix} = \begin{bmatrix} 1 & 2/3 & 1/3 \\ 2/3 & 1 & 2/3 \\ 1/3 & 2/3 & 1 \end{bmatrix}.$$

Thus, X_3 is correlated with X_2 with a correlation coefficient of $2/3$ and X_3 is correlated with X_1 but with a smaller correlation coefficient of $1/3$. Using (14.27) with $p = 2$ we must solve

$$\begin{bmatrix} 1 & 2/3 \\ 2/3 & 1 \end{bmatrix} \begin{bmatrix} a_1 \\ a_2 \end{bmatrix} = \begin{bmatrix} 1/3 \\ 2/3 \end{bmatrix}.$$

By inverting the covariance matrix we have the solution

$$\begin{bmatrix} a_{1_{\text{opt}}} \\ a_{2_{\text{opt}}} \end{bmatrix} = \frac{1}{1 - (2/3)^2} \begin{bmatrix} 1 & -2/3 \\ -2/3 & 1 \end{bmatrix} \begin{bmatrix} 1/3 \\ 2/3 \end{bmatrix}$$

$$= \begin{bmatrix} -\frac{1}{5} \\ \frac{4}{5} \end{bmatrix}.$$

Due to the larger correlation of X_3 with X_2, the prediction coefficient a_2 is larger. Note that if the covariance matrix is $\mathbf{C}_X = \sigma^2 \mathbf{I}$, then $c_{13} = c_{23} = 0$ and $a_{1_{\text{opt}}} = a_{2_{\text{opt}}} = 0$. This results in $\hat{X}_3 = 0$ or more generally for random variables with nonzero means, $\hat{X}_3 = E_{X_3}[X_3]$, as one might expect. See also Problem 14.24 to see how to determine the minimum value of the MSE.

$\diamondsuit$

As another simple example, observe what happens if $p = 1$ so that we wish to predict X_2 based on the outcome of X_1. In this case we have that $\hat{X}_2 = a_1 x_1$ and from (14.27), the solution for a_1 is $a_{1_{\text{opt}}} = c_{12}/c_{11} = \text{cov}(X_1, X_2)/\text{var}(X_1)$. Hence, $\hat{X}_2 = [\text{cov}(X_1, X_2)/\text{var}(X_1)]x_1$ and we recover our previous results for the bivariate case (see (14.23) and let $E_X[X] = E_Y[Y] = 0$) by replacing X_1 with X, x_1 with x, and X_2 with Y.

An interesting and quite useful interpretation of the linear prediction procedure can be made by reexamining (14.26). To simplify the discussion let $p = 2$ so that the equations to be solved are

$$E_{X_1,X_2,X_3}[(X_3 - a_1X_1 - a_2X_2)X_1] = 0$$
$$E_{X_1,X_2,X_3}[(X_3 - a_1X_1 - a_2X_2)X_2] = 0. \qquad (14.28)$$

Let the predictor error be denoted by ϵ, which is explicitly $\epsilon = X_3 - a_1X_1 - a_2X_2$. Then (14.28) becomes

$$E_{X_1,X_2,X_3}[\epsilon X_1] = 0$$
$$E_{X_1,X_2,X_3}[\epsilon X_2] = 0 \qquad (14.29)$$

which says that the optimal prediction coefficients a_1, a_2 are found by making the predictor error uncorrelated with the random variables used to predict X_3. Presumably if this were not the case, then some correlation would remain between the error and X_1, X_2, and this correlation could be exploited to reduce the error further (see Problem 14.23).

A geometric interpretation of (14.29) becomes apparent by considering X_1, X_2, and X_3 as vectors in a Euclidean space as depicted in Figure 14.3a. Since $\hat{X}_3 =$

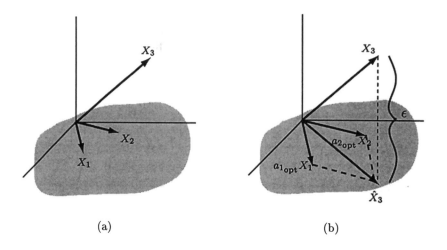

| (a) | (b) |

Figure 14.3: Geometrical interpretation of linear prediction.

$a_1X_1 + a_2X_2$, $\hat{X}_3$ can be any vector in the shaded region, which is the X_1-X_2 plane, depending upon the choice of a_1 and a_2. To minimize the error we should choose $\hat{X}_3$ as the *orthogonal projection* onto the plane as shown in Figure 14.3b. But this is equivalent to making the error vector ϵ orthogonal to any vector in the plane. In particular, then we have the requirement that

$$\epsilon \perp X_1$$
$$\epsilon \perp X_2 \qquad (14.30)$$

where $\perp$ denotes orthogonality. To relate these conditions back to those of (14.29) we define two zero mean random variables X and Y to be orthogonal if $E_{X,Y}[XY] = 0$. Hence, we have that (14.30) is equivalent to

$$E_{X_1,X_2,X_3}[\epsilon X_1] = 0$$
$$E_{X_1,X_2,X_3}[\epsilon X_2] = 0$$

or just the condition given by (14.29). (Since ϵ depends on (X_1, X_2, X_3), the expectation reflects this dependence.) This is called the *orthogonality principle*. It asserts that *to minimize the MSE the error "vector" should be orthogonal to each of the "data vectors" used to predict the desired "vector"*. The "vectors" X and Y are defined to be orthogonal if $E_{X,Y}[XY] = 0$, which is equivalent to being uncorrelated since we have assumed zero mean random variables. See also Problem 14.22 for the one-dimensional case of the orthogonality principle.

14.9 Computer Simulation of Gaussian Random Vectors

The method described in Section 12.11 for generating a bivariate Gaussian random vector is easily extended to the N-dimensional case. To generate a realization of $\mathbf{X} \sim \mathcal{N}(\boldsymbol{\mu}, \mathbf{C})$ we proceed as follows:

1. Perform a Cholesky decomposition of $\mathbf{C}$ to yield the $N \times N$ nonsingular matrix $\mathbf{G}$, where $\mathbf{C} = \mathbf{G}\mathbf{G}^T$.

2. Generate a realization $\mathbf{u}$ of an $N \times 1$ random vector $\mathbf{U}$ whose PDF is $\mathcal{N}(\mathbf{0}, \mathbf{I})$.

3. Form the realization of $\mathbf{X}$ as $\mathbf{x} = \mathbf{G}\mathbf{u} + \boldsymbol{\mu}$.

As an example, if $\boldsymbol{\mu} = \mathbf{0}$ and

$$C = \begin{bmatrix} 1 & 2/3 & 1/3 \\ 2/3 & 1 & 2/3 \\ 1/3 & 2/3 & 1 \end{bmatrix} \tag{14.31}$$

then

$$G = \begin{bmatrix} 1 & 0 & 0 \\ 0.6667 & 0.7454 & 0 \\ 0.3333 & 0.5963 & 0.7303 \end{bmatrix}.$$

We plot 100 realizations of $\mathbf{X}$ in Figure 14.4. The MATLAB code is given next.

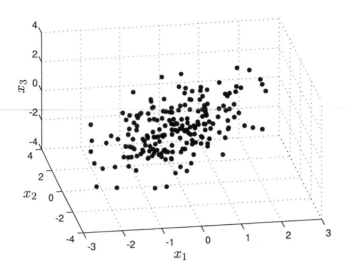

Figure 14.4: Realizations of 3×1 multivariate Gaussian random vector.

```
C=[1 2/3 1/3;2/3 1 2/3;1/3 2/3 1];
G=chol(C)'; % perform Cholesky decomposition
            % MATLAB produces C=A'*A so G=A'
M=200;
for m=1:M % generate realizations of x
  u=[randn(1,1) randn(1,1) randn(1,1)]';
  x(:,m)=G*u; % realizations stored as columns of 3 x 200 matrix
end
```

14.10   Real-World Example – Signal Detection

An important problem in sonar and radar is to be able to determine when an object, such as a submarine in sonar or an aircraft in radar, is present. To make this decision a pulse is transmitted into the water (sonar) or air (radar) and one looks to see if a reflected pulse from the object is returned. Typically, a digital computer is used to sample the received waveform in time and store the samples in memory for further processing. We will denote the received samples as $X_1, X_2, \ldots, X_N$. If there is no reflection, indicating no object is present, the received samples are due to noise only. If, however, there is a reflected pulse, also called an *echo*, the received samples will consist of a signal added to the noise. A standard model for the received samples is to assume that $X_i = W_i$, where $W_i \sim \mathcal{N}(0, \sigma^2)$ for noise only present and $X_i = s_i + W_i$ for a signal plus noise present. The noise samples W_i are usually also assumed to be independent and hence they are IID. With this modeling we can formulate the signal detection problem as the problem of deciding between the following two *hypotheses*

$$\mathcal{H}_W \quad : \quad X_i = W_i \qquad i = 1, 2, \ldots, N$$
$$\mathcal{H}_{s+W} \quad : \quad X_i = s_i + W_i \qquad i = 1, 2, \ldots, N.$$

It can be shown that a good decision procedure is to choose the hypothesis for which the received data samples have the highest probability of occurring. In other words, if the received data is more probable when $\mathcal{H}_{s+W}$ is true than when $\mathcal{H}_W$ is true, we say that a signal is present. Otherwise, we decide that noise only is present. To implement this approach we let $p_{\mathbf{X}}(\mathbf{x}; \mathcal{H}_W)$ be the PDF when noise only is present and $p_{\mathbf{X}}(\mathbf{x}; \mathcal{H}_{s+W})$ be the PDF when a signal plus noise is present. Then we decide a signal is present if

$$p_{\mathbf{X}}(\mathbf{x}; \mathcal{H}_{s+W}) > p_{\mathbf{X}}(\mathbf{x}; \mathcal{H}_W). \tag{14.32}$$

But from the modeling we have that $\mathbf{X} = \mathbf{W} \sim \mathcal{N}(\mathbf{0}, \sigma^2 \mathbf{I})$ for no signal present and $\mathbf{X} = \mathbf{s} + \mathbf{W} \sim \mathcal{N}(\mathbf{s}, \sigma^2 \mathbf{I})$ when a signal is present. Here we have defined the signal vector as $\mathbf{s} = [s_1 \, s_2 \ldots s_N]^T$. Hence, (14.32) becomes from (14.2)

$$\frac{1}{(2\pi\sigma^2)^{\frac{N}{2}}} \exp\left[-\frac{1}{2\sigma^2}(\mathbf{x} - \mathbf{s})^T(\mathbf{x} - \mathbf{s})\right] > \frac{1}{(2\pi\sigma^2)^{\frac{N}{2}}} \exp\left[-\frac{1}{2\sigma^2}\mathbf{x}^T\mathbf{x}\right]$$

An equivalent inequality is

$$-(\mathbf{x} - \mathbf{s})^T(\mathbf{x} - \mathbf{s}) > -\mathbf{x}^T\mathbf{x}$$

since the constant $1/(2\pi\sigma^2)^{N/2}$ is positive and the exponential function increases with its argument. Expanding the terms we have

$$-\mathbf{x}^T\mathbf{x} + \mathbf{x}^T\mathbf{s} + \mathbf{s}^T\mathbf{x} - \mathbf{s}^T\mathbf{s} > -\mathbf{x}^T\mathbf{x}$$

and since $\mathbf{s}^T\mathbf{x} = \mathbf{x}^T\mathbf{s}$ we have

$$\mathbf{x}^T\mathbf{s} > \frac{1}{2}\mathbf{s}^T\mathbf{s}$$

or finally we decide a signal is present if

$$\sum_{i=1}^{N} x_i s_i > \frac{1}{2}\sum_{i=1}^{N} s_i^2. \tag{14.33}$$

This detector is called a *replica correlator* [Kay 1998] since it correlates the data $x_1, x_2, \ldots, x_N$ with a replica of the signal $s_1, s_2, \ldots, s_N$. The quantity on the right-hand-side of (14.33) is called the *threshold*. If the value of $\sum_{i=1}^{N} x_i s_i$ exceeds the threshold, the signal is declared as being present.

As an example, assume that the signal is a "DC level" pulse or $s_i = A$ for $i = 1, 2, \ldots, N$ and that $A > 0$. Then (14.33) reduces to

$$A\sum_{i=1}^{N} x_i > \frac{1}{2}NA^2$$

and since $A > 0$, we decide a signal is present if

$$\frac{1}{N} \sum_{i=1}^{N} x_i > \frac{A}{2}.$$

Hence, the sample mean is compared to a threshold of $A/2$. To see how this detector performs we choose $A = 0.5$ and $\sigma^2 = 1$. The received data samples are shown in Figure 14.5a for the case of noise only and in Figure 14.5b for the case of a signal plus noise. A total of 100 received data samples are shown. Note that the noise samples

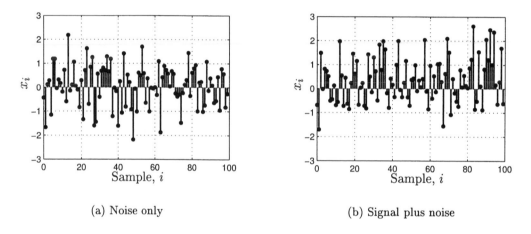

(a) Noise only   (b) Signal plus noise

Figure 14.5: Received data samples. Signal is $s_i = A = 0.5$ and noise consists of IID standard Gaussian random variables.

generated are different for each figure. The value of the sample mean $(1/N) \sum_{i=1}^{N} x_i$ is shown in Figure 14.6 versus the number of data samples N used in the averaging. For example, if $N = 10$, then the value shown is $(1/10) \sum_{i=1}^{10} x_i$, where x_i is found from the first 10 samples of Figure 14.5. To more easily observe the results they have been plotted as a continuous curve by connecting the points with straight lines. Also, the threshold of $A/2 = 0.25$ is shown as the dashed line. It is seen that as the number of data samples averaged increases, the sample mean converges to the mean of X_i (see also Example 14.4). When noise only is present, this becomes $E_X[X] = 0$ and when a signal is present, it becomes $E_X[X] = A = 0.5$. Thus by comparing the sample mean to the threshold of $A/2 = 0.25$ we should be able to decide if a signal is present or not most of the time (see also Problem 14.26).

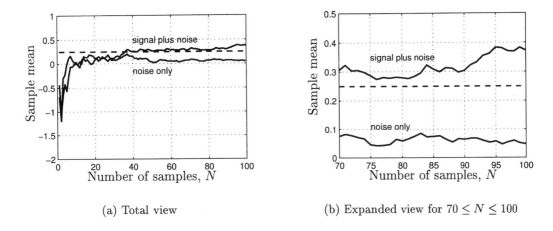

(a) Total view   (b) Expanded view for $70 \leq N \leq 100$

Figure 14.6: Value of sample mean versus the number of data samples averaged.

References

Kay, S., *Fundamentals of Statistical Signal Processing: Detection Theory*, Prentice-Hall, Englewood Cliffs, NJ, 1998.

Muirhead, R.J., *Aspects of Multivariate Statistical Theory*, John Wiley & Sons, New York, 1982.

Resnick, R., D. Halliday, *Physics, Part I*, John Wiley & Sons, New York, 1966.

Problems

14.1 (☺) (w,f) If $Y = X_1 + X_2 + X_3$, where $\mathbf{X} \sim \mathcal{N}(\boldsymbol{\mu}, \mathbf{C})$ and

$$\boldsymbol{\mu} = \begin{bmatrix} 1 \\ 2 \\ 3 \end{bmatrix}$$

$$\mathbf{C} = \begin{bmatrix} 1 & 1/2 & 1/4 \\ 1/2 & 1 & 1/2 \\ 1/4 & 1/2 & 1 \end{bmatrix}$$

find the mean and variance of Y.

14.2 (w,c) If $[X_1\ X_2]^T \sim \mathcal{N}(\mathbf{0}, \sigma^2 \mathbf{I})$, find $P[X_1^2 + X_2^2 > R^2]$. Next, let $\sigma^2 = 1$ and $R = 1$ and lend credence to your result by performing a computer simulation to estimate the probability.

14.3 (f) Find the PDF of $Y = X_1^2 + X_2^2 + X_3^2$ if $\mathbf{X} \sim \mathcal{N}(\mathbf{0}, \mathbf{I})$. Hint: Use the results of Example 14.1. Note that you should obtain the PDF for a χ_3^2 random variable.

14.4 (w) An airline has flights that depart according to schedule 95% of the time. This means that they depart late 1/2 hour or more 5% of the time due to mechanical problems, traffic delays, etc. (for less than 1/2 hour the plane is considered to be "on time"). The amount of time that the plane is late is modeled as an $\exp(\lambda)$ random variable. If a person takes a plane that makes two stops at intermediate destinations, what is the probability that he will be more than 1 1/2 hours late? Hint: You will need the PDF for a sum of independent exponential random variables.

14.5 (f) Consider the transformation from spherical to Cartesian coordinates. Show that the Jacobian has a determinant whose absolute value is equal to $r^2 \sin \phi$.

14.6 (☺) (w) A large group of college students have weights that can be modeled as a $\mathcal{N}(150, 30)$ random variable. If 4 students are selected at random, what is the probability that they will all weigh more than 150 lbs?

14.7 (t) Prove that the joint PDF given by (14.4) has $\mathcal{N}(0, 1)$ marginal PDFs and that the random variables are uncorrelated. Hint: Use the known properties of the standard bivariate Gaussian PDF.

14.8 (t) Assume that $\mathbf{X} \sim \mathcal{N}(\mathbf{0}, \mathbf{C})$ for $\mathbf{X}$ an $N \times 1$ random vector and that $\mathbf{Y} = \mathbf{G}\mathbf{X}$, where $\mathbf{G}$ is an $M \times N$ matrix with $M < N$. If the characteristic function of $\mathbf{X}$ is $\phi_{\mathbf{X}}(\boldsymbol{\omega}) = \exp\left(-\frac{1}{2}\boldsymbol{\omega}^T \mathbf{C}\boldsymbol{\omega}\right)$, find the characteristic function of $\mathbf{Y}$. Use the following

$$\phi_{\mathbf{Y}}(\boldsymbol{\omega}) = E_{\mathbf{Y}}[\exp(j\boldsymbol{\omega}^T \mathbf{Y})] = E_{\mathbf{X}}[\exp(j\boldsymbol{\omega}^T \mathbf{G}\mathbf{X})] = E_{\mathbf{X}}[\exp(j(\mathbf{G}^T\boldsymbol{\omega})^T \mathbf{X})].$$

Based on your results conclude that $\mathbf{Y} \sim \mathcal{N}(\mathbf{0}, \mathbf{G}\mathbf{C}\mathbf{G}^T)$.

14.9 (☺) (f) If $Y = X_1 + X_2 + X_3$, where $\mathbf{X} \sim \mathcal{N}(\mathbf{0}, \mathbf{C})$ and $\mathbf{C} = \text{diag}(\sigma_1^2, \sigma_2^2, \sigma_3^2)$, find the PDF of Y. Hint: See Problem 14.8.

14.10 (f) Show that if $\mathbf{C}_X$ is a diagonal matrix, then $\mathbf{a}^T \mathbf{C}_X \mathbf{a} = \sum_{i=1}^{N} a_i^2 \text{var}(X_i)$.

14.11 (c) Simulate a single realization of a random vector composed of IID random variables with PDF $X_i \sim \mathcal{N}(1, 2)$ for $i = 1, 2, \ldots, N$. Do this by repeating an experiment that successively generates $X \sim \mathcal{N}(1, 2)$. Then, find the outcome of the sample mean random variable and discuss what happens as N becomes large.

14.12 (☺) (w,c) An $N \times 1$ random vector $\mathbf{X}$ has $E_{X_i}[X_i] = \mu$ and $\text{var}(X_i) = i\sigma^2$ for $i = 1, 2, \ldots, N$. The components of $\mathbf{X}$ are independent. Does the sample mean random variable converge to μ as N becomes large? Carry out a computer simulation for this problem and explain your results.

14.13 (t) Prove that if $X_1, X_2, \ldots, X_N$ are independent random variables, then $\phi_{X_1, X_2, \ldots, X_N}(\omega_1, \omega_2, \ldots, \omega_N) = \prod_{i=1}^{N} \phi_{X_i}(\omega_i)$.

14.14 (t) Prove that $\phi_{X_1, X_2, \ldots, X_N}(\omega_1, \omega_2, 0, 0 \ldots, 0) = \phi_{X_1, X_2}(\omega_1, \omega_2)$.

14.15 (t) If $\mathbf{X} \sim \mathcal{N}(\boldsymbol{\mu}, \mathbf{C})$ with $\mathbf{X}$ an $N \times 1$ random vector, prove that the characteristic function is

$$\phi_{\mathbf{X}}(\boldsymbol{\omega}) = \exp\left(j\boldsymbol{\omega}^T \boldsymbol{\mu} - \frac{1}{2}\boldsymbol{\omega}^T \mathbf{C}\boldsymbol{\omega} \right).$$

To do so note that the characteristic function of a random vector distributed according to $\mathcal{N}(\mathbf{0}, \mathbf{C})$ is $\exp\left(-\frac{1}{2}\boldsymbol{\omega}^T \mathbf{C}\boldsymbol{\omega}\right)$. With these results show that the PDF of $\mathbf{Y} = \mathbf{G}\mathbf{X}$ for $\mathbf{G}$ an $M \times N$ matrix with $M < N$ is $\mathcal{N}(\mathbf{G}\boldsymbol{\mu}, \mathbf{G}\mathbf{C}\mathbf{G}^T)$.

14.16 (t) Prove that if $\mathbf{X} \sim \mathcal{N}(\boldsymbol{\mu}, \mathbf{C})$ for $\mathbf{X}$ an $N \times 1$ random vector, then the marginal PDFs are $X_i \sim \mathcal{N}(\mu_i, \sigma_i^2)$. Hint: Examine the PDF of $Y = \mathbf{e}_i^T \mathbf{X}$, where $\mathbf{e}_i$ is the $N \times 1$ vector whose elements are all zeros except for the ith element, which is a one. Also, make use of the results of Problem 14.15.

14.17 (f) Prove that if $X_i \sim \mathcal{N}(0, 1)$ for $i = 1, 2 \ldots, N$ and the X_i's are IID, then $\sum_{i=1}^{N} X_i^2 \sim \chi_N^2$. To do so first find the characteristic function of X_i^2. Hint: You will need the result that

$$\int_{-\infty}^{\infty} \frac{1}{\sqrt{2\pi c}} \exp\left(-\frac{1}{2}\frac{x^2}{c} \right) dx = 1$$

for c a *complex number*. Also, see Table 11.1.

14.18 (t) Prove that if $X_i \sim \exp(\lambda)$ and the X_i's are IID, then $\sum_{i=1}^{N} X_i$ has an Erlang PDF. Hint: See Table 11.1.

14.19 (☺) (w,c) Find the mean and variance of the random variable

$$Y = \sum_{i=1}^{12} (U_i - 1/2)$$

where $U_i \sim \mathcal{U}(0, 1)$ and the U_i's are IID. Estimate the PDF of Y using a computer simulation and compare it to a standard Gaussian PDF. See Section 15.5 for a theoretical justification of your results.

14.20 (w) Three different voltmeters measure the voltage of a 100 volt source. The measurements can be modeled as random variables with

$$V_1 \sim \mathcal{N}(100, 1)$$
$$V_2 \sim \mathcal{N}(100, 10)$$
$$V_3 \sim \mathcal{N}(100, 5).$$

Is it better to average the results or just use the most accurate voltmeter?

14.21 (☺) (f) If a 3×1 random vector has mean zero and covariance matrix

$$\mathbf{C}_X = \begin{bmatrix} 3 & 2 & 1 \\ 2 & 3 & 2 \\ 1 & 2 & 3 \end{bmatrix}$$

find the optimal prediction of X_3 given that we have observed $X_1 = 1$ and $X_2 = 2$.

14.22 (t) Consider the prediction of the random variable Y based on observing that $X = x$. Assuming (X, Y) is a zero mean random vector, we propose using the linear prediction $\hat{Y} = ax$. Determine the optimal value of a (being the value that minimizes the MSE) by using the orthogonality principle. Explain your results by drawing a diagram.

14.23 (f) If a 3×1 random vector $\mathbf{X}$ has a zero mean and covariance matrix

$$\mathbf{C}_X = \begin{bmatrix} 1 & \rho & \rho^2 \\ \rho & 1 & \rho \\ \rho^2 & \rho & 1 \end{bmatrix}$$

determine the optimal linear prediction of X_3 based on the observed outcomes of X_1 and X_2. Why is $a_{1_{\text{opt}}} = 0$? Hint: Consider the covariance between $\epsilon = X_3 - \rho X_2$, which is the predictor error for X_3 based on observing only X_2, and X_1.

14.24 (☺) (t,f) Explain why the minimum MSE of the predictor $\hat{X}_3 = a_{1_{\text{opt}}} X_1 + a_{2_{\text{opt}}} X_2$ is

$$
\begin{aligned}
\text{mse}_{\text{min}} &= E_{X_1, X_2, X_3}\left[(X_3 - a_{1_{\text{opt}}} X_1 - a_{2_{\text{opt}}} X_2)^2\right] \\
&= E_{X_1, X_2, X_3}\left[(X_3 - a_{1_{\text{opt}}} X_1 - a_{2_{\text{opt}}} X_2) X_3\right] \\
&= c_{33} - a_{1_{\text{opt}}} c_{13} - a_{2_{\text{opt}}} c_{23}.
\end{aligned}
$$

Next use this result to find the minimum MSE for Example 14.6.

14.25 (☺) (c) Use a computer simulation to generate realizations of the random vector **X** described in Example 14.6. Then, predict X_3 based on the outcomes of X_1 and X_2 and plot the true realizations and the predictions. Finally, estimate the average predictor error and compare your results to the theoretical minimum MSE obtained in Problem 14.24.

14.26 (w) For the signal detection example described in Section 14.9 prove that the probability of saying a signal is present when indeed there is one goes to 1 as $A \to \infty$.

14.27 (c) Generate on a computer 1000 realizations of the two different random variables $X_W \sim \mathcal{N}(0,1)$ and $X_{s+W} \sim \mathcal{N}(0.5,1)$. Next plot the outcomes of the sample mean random variable versus N, the number of successive samples averaged, or $\bar{x}_N = (1/N)\sum_{i=1}^{N} x_i$. What can you say about the sample means as N becomes large? Explain what this has to do with signal detection.

Chapter 15

Probability and Moment Approximations Using Limit Theorems

15.1 Introduction

So far we have described the methods for determining the exact probability of events using probability mass functions (PMFs) for discrete random variables and probability density functions (PDFs) for continuous random variables. Also of importance were the methods to determine the moments of these random variables. The procedures employed were all based on knowledge of the PMF/PDF and the implementation of its summation/integration. In many practical situations the PMF/PDF may be unknown or the summation/integration may not be easily carried out. It would be of great utility, therefore, to be able to approximate the desired quantities using much simpler methods. For random variables that are the sum of a *large* number of independent and identically distributed random variables this can be done. In this chapter we focus our discussions on two very powerful theorems in probability—*the law of large numbers* and *the central limit theorem*. The first theorem asserts that the sample mean random variable, which is the average of IID random variables and which was introduced in Chapter 14, converges to the *expected value*, a number, of each random variable in the average. The law of large numbers is also known colloquially as the *law of averages*. Another reason for its importance is that it provides a justification for the relative frequency interpretation of probability. The second theorem asserts that a properly normalized sum of IID random variables converges to a Gaussian *random variable*.

The theorems are actually the simplest forms of much more general results. For example, the theorems can be formulated to handle sums of nonidentically distributed random variables [Rao 1973] and dependent random variables [Brockwell and Davis 1987].

15.2   Summary

The Bernoulli law of large number is introduced in Section 15.4 as a prelude to the more general law of large numbers. The latter is summarized in Theorem 15.4.1 and asserts that the sample mean random variable of IID random variables will converge to the expected value of a single random variable. The central limit theorem is described in Section 15.5 where it is demonstrated that the repeated convolution of PDFs produces a Gaussian PDF. For continuous random variables the central limit theorem, which asserts that the sum of a large number of IID random variables has a Gaussian PDF, is summarized in Theorem 15.5.1. The precise statement is given by (15.6). For the sum of a large number of IID discrete random variables it is the CDF that converges to a Gaussian CDF. Theorem 15.5.2 is the central limit theorem for discrete random variables. The precise statement is given by (15.9). The concept of confidence intervals is introduced in Section 15.6. A 95% confidence interval for the sample mean estimate of the parameter p of a Ber(p) random variable is given by (15.14). It is then applied to the real-world problem of opinion polling.

15.3   Convergence and Approximation of a Sum

Since we will be dealing with the sum of a large number of random variables, it is worthwhile first to review some concepts of convergence. In particular, we need to understand the role that convergence plays in approximating the behavior of a sum of terms. As an illustrative example, consider the determination of the value of the sum

$$s_N = \sum_{i=1}^{N} \frac{(1+a^i)}{N}$$

for some large value of N. We have purposedly chosen a sum that may be evaluated in closed form to allow a comparison to its approximation. The exact value can be found as

$$
\begin{aligned}
s_N &= \frac{1}{N} \sum_{i=1}^{N} 1 + \frac{1}{N} \sum_{i=1}^{N} a^i \\
&= 1 + \frac{1}{N} \frac{a - a^{N+1}}{1-a}.
\end{aligned}
$$

Examples of s_N versus N are shown in Figure 15.1. The values of s_N have been connected by straight lines for easier viewing. It should be clear that as $N \to \infty$, $s_N \to 1$ if $|a| < 1$. This means that if N is sufficiently large, then s_N will differ from 1 by a very small amount. This small amount, which is the error in the approximation of s_N by 1, is given by

$$|s_N - 1| = \frac{1}{N} \frac{a - a^{N+1}}{1-a}$$

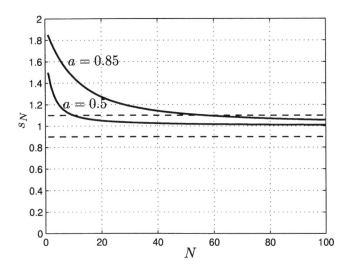

Figure 15.1: Convergence of sum to 1.

and will depend on a as well as N. For example, if we wish to claim that the error is less than 0.1, then N would have to be 10 for $a = 0.5$ but N would need to be 57 for $a = 0.85$, as seen in Figure 15.1. Thus, in general the error of the approximation will depend upon the particular sequence (value of a here). We can assert, without actually knowing the value of a as long as $|a| < 1$ and hence the sum converges, that s_N will eventually become close to 1. The error can be quite large for a fixed value of N (consider what would happen if $a = 0.999$). Such are the advantages (sum will be close to 1 for *all* $|a| < 1$) and disadvantages (how large does N have to be?) of limit theorems. We next describe the law of large numbers.

15.4 Law of Large Numbers

When we began our study of probability, we argued that if a fair coin is tossed N times in succession, then the relative frequency of heads, i.e., the number of heads observed divided by the number of coin tosses, should be close to 1/2. This was why we intuitively accepted the assignment of a probability of 1/2 to the event that the outcome of a fair coin toss would be a head. If we continue to toss the coin, then as $N \rightarrow \infty$, we expect the relative frequency to approach 1/2. We can now prove that this is indeed the case under certain assumptions. First we model the repeated coin toss experiment as a sequence of N Bernoulli subexperiments (see also Section 4.6.2). The result of the ith subexperiment is denoted by the discrete random variable X_i, where

$$X_i = \begin{cases} 1 & \text{if heads} \\ 0 & \text{if tails.} \end{cases}$$

We can then model the overall experimental output by the random vector $\mathbf{X} = [X_1 \, X_2 \ldots X_N]^T$. We next assume that the discrete random variables X_i are IID with marginal PMF

$$p_X[k] = \begin{cases} \frac{1}{2} & k = 0 \\ \frac{1}{2} & k = 1 \end{cases}$$

or the experiment is a sequence of *independent* and identical Bernoulli subexperiments. Finally, the relative frequency is given by the sample mean random variable

$$\bar{X}_N = \frac{1}{N} \sum_{i=1}^{N} X_i \tag{15.1}$$

which was introduced in Chapter 14, although there it was used for the average of *continuous* random variables. We subscript the sample mean random variable by N to remind us that N coin toss outcomes are used in its computation. Now consider what happens to the mean and variance of $\bar{X}_N$ as $N \to \infty$. The mean is

$$
\begin{aligned}
E_{\mathbf{X}}[\bar{X}_N] &= \frac{1}{N} \sum_{i=1}^{N} E_{\mathbf{X}}[X_i] \\
&= \frac{1}{N} \sum_{i=1}^{N} E_{X_i}[X_i] \\
&= \frac{1}{N} \sum_{i=1}^{N} \frac{1}{2} \\
&= \frac{1}{2} \quad \text{for all } N.
\end{aligned}
$$

The variance is

$$
\begin{aligned}
\text{var}(\bar{X}_N) &= \text{var}\left(\frac{1}{N} \sum_{i=1}^{N} X_i \right) \\
&= \frac{1}{N^2} \sum_{i=1}^{N} \text{var}(X_i) \quad (X_i\text{'s are independent} \Rightarrow \text{uncorrelated}) \\
&= \frac{\text{var}(X_i)}{N} \quad\quad (X_i\text{'s are identically distributed} \\
&\quad\quad\quad\quad\quad\quad\quad\quad \Rightarrow \text{have same variance}).
\end{aligned}
$$

But for a Bernoulli random variable, $X_i \sim \text{Ber}(p)$, the variance is $\text{var}(X_i) = p(1-p)$. Since $p = 1/2$ for a fair coin,

$$
\begin{aligned}
\text{var}(\bar{X}_N) &= \frac{p(1-p)}{N} \\
&= \frac{1}{4N} \to 0 \quad \text{as } N \to \infty.
\end{aligned}
$$

Therefore the width of the PMF of $\bar{X}_N$ must decrease as N increases and eventually go to zero. Since the variance is defined as

$$\text{var}(\bar{X}_N) = E_{\mathbf{X}}\left[\left(\bar{X}_N - E_{\mathbf{X}}[\bar{X}_N]\right)^2\right]$$

we must have that as $N \rightarrow \infty$, $\bar{X}_N \rightarrow E_{\mathbf{X}}[\bar{X}_N] = 1/2$. In effect the random variable $\bar{X}_N$ becomes not random at all but a constant. It is called a *degenerate* random variable. To further verify that the PMF becomes concentrated about its mean, which is $1/2$, we note that the sum of N IID Bernoulli random variables is a binomial random variable. Thus,

$$S_N = \sum_{i=1}^{N} X_i \sim \text{bin}\left(N, \frac{1}{2}\right)$$

and therefore the PMF is

$$p_{S_N}[k] = \binom{N}{k}\left(\frac{1}{2}\right)^N \qquad k = 0, 1, \ldots, N.$$

To find the PMF of $\bar{X}_N$ we let $\bar{X}_N = (1/N)\sum_{i=1}^{N} X_i = S_N/N$ and note that $\bar{X}_N$ can take on values $u_k = k/N$ for $k = 0, 1, \ldots, N$. Therefore, using the formula for the transformation of a discrete random variable, the PMF becomes

$$p_{\bar{X}_N}[u_k] = \binom{N}{Nu_k}\left(\frac{1}{2}\right)^N \qquad u_k = k/N; \; k = 0, 1, \ldots, N \qquad (15.2)$$

which is plotted in Figure 15.2 for various values of N. Because as N increases $\bar{X}_N$

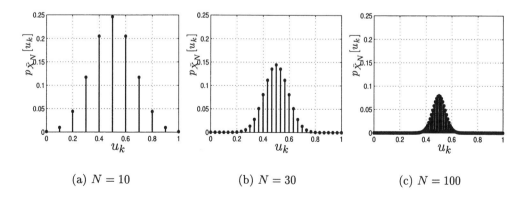

(a) $N = 10$ (b) $N = 30$ (c) $N = 100$

Figure 15.2: PMF for sample mean random variable of N IID Bernoulli random variables with $p = 1/2$. It models the relative frequency of heads obtained for N fair coin tosses.

takes on values more densely in the interval $[0, 1]$, we do not obtain a PMF with all its mass concentrated at 0.5, as we might expect. Nonetheless, the probability that the sample mean random variable will be concentrated about $1/2$ increases. As an example, the probability of being within the interval $[0.45, 0.55]$ is 0.2461 for $N = 10$, 0.4153 for $N = 30$, and 0.7287 for $N = 100$, as can be verified by summing the values of the PMF over this interval. Usually it is better to plot the CDF since as $N \to \infty$, it can be shown to converge to the unit step beginning at $u = 0.5$ (see Problem 15.1). Also, it is interesting to note that the PMF appears Gaussian, although it changes in amplitude and width for each N. This is an observation that we will focus on later when we discuss the central limit theorem. The preceding results say that for large enough N the sample mean random variable will *always* yield a *number*, which in this case is $1/2$. By "always" we mean that every time we perform a repeated Bernoulli experiment consisting of N independent and fair coin tosses, we will obtain a sample mean of $1/2$, *for N large enough.* As an example, we have plotted in Figure 15.3 five realizations of the sample mean random variable or $\bar{x}_N$ versus N. The values of $\bar{x}_N$ have been connected by straight lines for easier viewing.   We see that

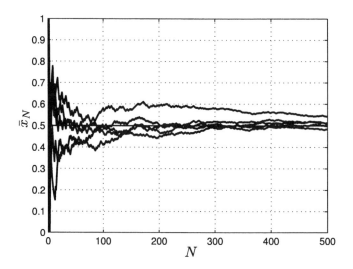

Figure 15.3: Realizations of sample mean random variable of N IID Bernoulli random variables with $p = 1/2$ as N increases.

$$\bar{X}_N \to \frac{1}{2} = E_X[X]. \tag{15.3}$$

This is called the *Bernoulli law of large numbers,* and is known to the layman as *the law of averages.* More generally for a Bernoulli subexperiment with probability p, we have that

$$\bar{X}_N \to p = E_X[X].$$

The sample mean random variable converges to the expected value of a single random variable. Note that since $\bar{X}_N$ is the relative frequency of heads and p is the probability of heads, we have shown that *the probability of a head in a single coin toss can be interpreted as the value obtained as the relative frequency of heads in a large number of independent and identical coin tosses.* This observation also justifies our use of the sample mean random variable as an estimator of a moment since

$$\widehat{E_X[X]} = \frac{1}{N} \sum_{i=1}^{N} X_i \to E_X[X] \quad \text{as } N \to \infty$$

and more generally, justifies our use of $(1/N) \sum_{i=1}^{N} X_i^n$ as an estimate of the nth moment $E[X^n]$ (see also Problem 15.6).

A more general law of large numbers is summarized in the following theorem. It is valid for the sample mean of IID random variables, either discrete, continuous, or mixed.

Theorem 15.4.1 (Law of Large Numbers) *If $X_1, X_2, \ldots, X_N$ are IID random variables with mean $E_X[X]$ and $\text{var}(X) = \sigma^2 < \infty$, then $\lim_{N \to \infty} \bar{X}_N = E_X[X]$.*

Proof:
Consider the probability of the sample mean random variable deviating from the expected value by more than ϵ, where ϵ is a small positive number. This probability is given by

$$P\left[|\bar{X}_N - E_X[X]| > \epsilon\right] = P\left[|\bar{X}_N - E_{\mathbf{X}}[\bar{X}_N]| > \epsilon\right].$$

Since $\text{var}(\bar{X}_N) = \sigma^2/N$, we have upon using Chebyshev's inequality (see Section 11.8)

$$P\left[|\bar{X}_N - E_X[X]| > \epsilon\right] \leq \frac{\text{var}(\bar{X}_N)}{\epsilon^2} = \frac{\sigma^2}{N\epsilon^2}$$

and taking the limit of both sides yields

$$\lim_{N \to \infty} P\left[|\bar{X}_N - E_X[X]| > \epsilon\right] \leq \lim_{N \to \infty} \frac{\sigma^2}{N\epsilon^2} = 0.$$

Since a probability must be greater than or equal to zero, we have finally that

$$\lim_{N \to \infty} P\left[|\bar{X}_N - E_X[X]| > \epsilon\right] = 0 \tag{15.4}$$

which is the mathematical statement that the sample mean random variable converges to the expected value of a single random variable.

$\square$

The limit in (15.4) says that for large enough N, the *probability of the error* in the approximation of $\bar{X}_N$ by $E_X[X]$ exceeding ϵ (which can be chosen as small as desired) will be exceedingly small. It is said that $\bar{X}_N \to E_X[X]$ *in probability* [Grimmett and Stirzaker 2001].

⚠️ **Convergence in probability does not mean all realizations will converge.**

Referring to Figure 15.3 it is seen that for all realizations except the top one, the error is small. The statement of (15.4) does allow some realizations to have an error greater than ϵ for a given large N. However, the probability of this happening becomes very small but not zero as N increases. For all practical purposes, then, we can ignore this occurrence. Hence, convergence in probability is somewhat different than what one may be familiar with in dealing with convergence of *deterministic sequences*. For deterministic sequences, all sequences (since there is only one) will have an error less than ϵ for all $N \geq N_\epsilon$, where N_ϵ will depend on ϵ (see Figure 15.1). The interested reader should consult [Grimmett and Stirzaker 2001] for further details. See also Problem 15.8 for an example.

We conclude our discussion with an example and some further comments.

Example 15.1 – Sample mean for IID Gaussian random variables
Recall from the real-world example in Chapter 14 that when a signal is present we have

$$X_{s+W_i} \sim \mathcal{N}(A, \sigma^2) \quad i = 1, 2, \ldots, N.$$

Since the random variables are IID, we have by the law of large numbers that

$$\bar{X}_N \to E_X[X] = A.$$

Thus, the upper curve shown in Figure 14.6 must approach $A = 0.5$ (with high probability) as $N \to \infty$.

◇

In applying the law of large numbers *we do not need to know the marginal PDF.* If in the previous example, we had $X_{s+W_i} \sim \mathcal{U}(0, 2A)$, then we also conclude that $\bar{X}_N \to A$. As long as the random variables are IID with mean A and a finite variance, $\bar{X}_N \to A$ (although the error in the approximation will depend upon the marginal PDF—see Problem 15.3).

15.5   Central Limit Theorem

By the law of large numbers the PMF/PDF of the sample mean random variable decreases in width until all the probability is concentrated about the mean. The theorem, however, does not say much about the PMF/PDF itself. However, by considering a slightly modified sample mean random variable, we can make some more definitive assertions about its probability distribution. To illustrate the necessity of doing so we consider the PDF of a continuous random variable that is the sum of N continuous IID random variables. A particularly illustrative example is for $X_i \sim \mathcal{U}(-1/2, 1/2)$.

Example 15.2 – PDF for sum of IID $\mathcal{U}(-1/2, 1/2)$ random variables

Consider the sum

$$S_N = \sum_{i=1}^{N} X_i$$

where the X_i's are IID random variables with $X_i \sim \mathcal{U}(-1/2, 1/2)$. If $N = 2$, then $S_2 = X_1 + X_2$ and the PDF of S_2 is easily found using a convolution integral as described in Section 12.6. Therefore,

$$p_{S_2}(x) = p_X(x) \star p_X(x) = \int_{-\infty}^{\infty} p_X(u) p_X(x - u) du$$

where $\star$ denotes convolution. The evaluation of the convolution integral is most easily done by plotting $p_X(u)$ and $p_X(x - u)$ versus u as shown in Figure 15.4a. This is necessary to determine the regions over which the product of $p_X(u)$ and $p_X(x - u)$ is nonzero and so contributes to the integral. The reader should be able to show, based upon Figure 15.4a, that the PDF of S_2 is that shown in Figure 15.4b. More generally, we have from (14.22) that

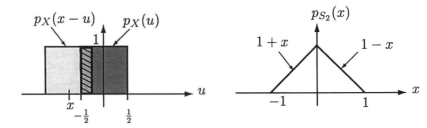

(a) Cross-hatched region con-   (b) Result of convolution
tributes to integral

Figure 15.4: Determining the PDF for the sum of two independent uniform random variables using a convolution integral evaluation.

$$
\begin{aligned}
p_{S_N}(x) &= \int_{-\infty}^{\infty} \phi_X^N(\omega) \exp(-j\omega x) \frac{d\omega}{2\pi} \\
&= \underbrace{p_X(x) \star p_X(x) \star \cdots \star p_X(x)}_{(N-1) \text{ convolutions}}.
\end{aligned}
$$

Hence to find $p_{S_3}(x)$ we must convolve $p_{S_2}(x)$ with $p_X(x)$ to yield $p_X(x) \star p_X(x) \star p_X(x)$ since $p_{S_2}(x) = p_X(x) \star p_X(x)$. This is

$$p_{S_3}(x) = \int_{-\infty}^{\infty} p_{S_2}(u) p_X(x - u) du$$

but since $p_X(-x) = p_X(x)$, we can express this in the more convenient form as

$$p_{S_3}(x) = \int_{-\infty}^{\infty} p_{S_2}(u)p_X(u-x)du.$$

The integrand may be determined by plotting $p_{S_2}(u)$ and the right-shifted version $p_X(u-x)$ and multiplying these two functions. The different regions that must be considered are shown in Figure 15.5. Hence, referring to Figure 15.5 we have

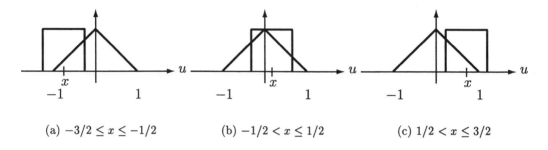

(a) $-3/2 \leq x \leq -1/2$ (b) $-1/2 < x \leq 1/2$ (c) $1/2 < x \leq 3/2$

Figure 15.5: Determination of limits for convolution integral.

$$
\begin{aligned}
p_{S_3}(x) &= \int_{-1}^{x+1/2} p_{S_2}(u) \cdot 1 du \\
&= \frac{1}{2}x^2 + \frac{3}{2}x + \frac{9}{8} & -\frac{3}{2} \leq x \leq -\frac{1}{2} \\
p_{S_3}(x) &= \int_{x-1/2}^{x+/2} p_{S_2}(u) \cdot 1 du \\
&= -x^2 + \frac{3}{4} & -\frac{1}{2} < x \leq \frac{1}{2} \\
p_{S_3}(x) &= \int_{x-1/2}^{1} p_{S_2}(u) \cdot 1 du \\
&= \frac{1}{2}x^2 - \frac{3}{2}x + \frac{9}{8} & \frac{1}{2} < x \leq \frac{3}{2}
\end{aligned}
$$

and $p_{S_3}(x) = 0$ otherwise. This is plotted in Figure 15.6 versus the PDF of a $\mathcal{N}(0, 3/12)$ random variable. Note the close agreement. We have chosen the mean and variance of the Gaussian approximation to *match* that of $p_{S_3}(x)$ (recall that $\text{var}(X) = (b-a)^2/12$ for $X \sim \mathcal{U}(a,b)$ and hence $\text{var}(X_i) = 1/12$). If we continue the convolution process, the mean will remain at zero but the variance of S_N will be $N/12$.

$\diamondsuit$

A MATLAB program that implements a repeated convolution for a PDF that is nonzero over the interval $(0,1)$ is given in Appendix 15A. It can be used to verify

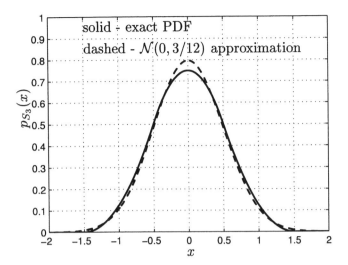

Figure 15.6: PDF for sum of 3 IID $\mathcal{U}(-1/2, 1/2)$ random variables and Gaussian approximation.

analytical results and also to try out other PDFs. An example of its use is shown in Figure 15.7 for the repeated convolution of a $\mathcal{U}(0,1)$ PDF. Note that as N increases the PDF moves to the right since $E[S_N] = NE_X[X] = N/2$ and the variance also increases since $\text{var}(S_N) = N\text{var}(X) = N/12$. Because of this behavior it is not possible to state that the PDF converges to any PDF. To circumvent this problem it is necessary to $normalize$ the sum so that its mean and variance are fixed as N increases. It is convenient, therefore, to have the mean fixed at 0 and the variance fixed at 1, resulting in a $standardized\ sum$. Recall from Section 7.9 that this is easily accomplished by forming

$$\frac{S_N - E[S_N]}{\sqrt{\text{var}(S_N)}} = \frac{S_N - NE_X[X]}{\sqrt{N\text{var}(X)}}. \tag{15.5}$$

By doing so, we can now assert that this standardized random variable will converge to a $\mathcal{N}(0,1)$ random variable. An example is shown in Figure 15.8 for $X_i \sim \mathcal{U}(0,1)$ and for $N = 2, 3, 4$. This is the famous $central\ limit\ theorem$, which says that the $PDF\ of\ the\ standardized\ sum\ of\ a\ large\ number\ of\ continuous\ IID\ random\ variables$ $will\ converge\ to\ a\ Gaussian\ PDF$. Its great importance is that in many practical situations one can model a random variable as having arisen from the contributions of many small and similar physical effects. By making the IID assumption we can assert that the PDF is Gaussian. There is no need to know the PDF of each random variable or even if it is known, to determine the exact PDF of the sum, which may not be possible. Some application areas are:

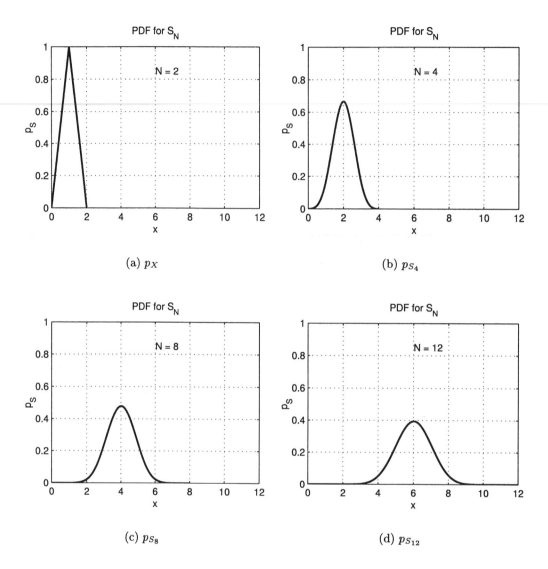

Figure 15.7: PDF of sum of N IID $\mathcal{U}(0,1)$ random variables. The plots were obtained using `clt_demo.m` listed in Appendix 15A.

1. Polling (see Section 15.6) [Weisburg, Krosnick, Bowen 1996]

2. Noise characterization [Middleton 1960]

3. Scattering effects modeling [Urick 1975]

4. Kinetic theory of gases [Reif 1965]

5. Economic modeling [Harvey 1989]

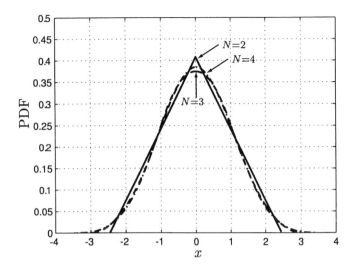

Figure 15.8: PDF of standardized sum of N IID $\mathcal{U}(0,1)$ random variables.

and many more.

We now state the theorem for continuous random variables.

Theorem 15.5.1 (Central limit theorem for continuous random variables)
If $X_1, X_2, \ldots, X_N$ are continuous IID random variables, each with mean $E_X[X]$ and variance $\mathrm{var}(X)$, and $S_N = \sum_{i=1}^{N} X_i$, then as $N \to \infty$

$$\frac{S_N - E[S_N]}{\sqrt{\mathrm{var}(S_N)}} = \frac{\sum_{i=1}^{N} X_i - N E_X[X]}{\sqrt{N \mathrm{var}(X)}} \to \mathcal{N}(0,1). \tag{15.6}$$

Equivalently, the CDF of the standardized sum converges to $\Phi(x)$ or

$$P\left[\frac{S_N - E[S_N]}{\sqrt{\mathrm{var}(S_N)}} \le x\right] \to \int_{-\infty}^{x} \frac{1}{\sqrt{2\pi}} \exp\left(-\frac{1}{2}t^2\right) dt = \Phi(x). \tag{15.7}$$

The proof is given in Appendix 15B and is based on the properties of characteristic functions and the continuity theorem. An example follows.

Example 15.3 – PDF of sum of squares of independent $\mathcal{N}(0,1)$ random variables

Let $X_i \sim \mathcal{N}(0,1)$ for $i = 1, 2, \ldots, N$ and assume that the X_i's are independent. We wish to determine the approximate PDF of $Y_N = \sum_{i=1}^{N} X_i^2$ as N becomes large. Note that the exact PDF for Y_N is a χ_N^2 PDF so that we will equivalently find an approximation to the PDF of the standardized χ_N^2 random variable. To apply the central limit theorem we first note that since the X_i's are IID so are the X_i^2's (why?). Then as $N \to \infty$ we have from (15.6)

$$\frac{\sum_{i=1}^{N} X_i^2 - N E_X[X^2]}{\sqrt{N \mathrm{var}(X^2)}} \to \mathcal{N}(0,1).$$

But $X^2 \sim \chi_1^2$ so that $E_X[X^2] = 1$ and $\text{var}(X^2) = 2$ (see Section 10.5.6 and Table 11.1 for a $\chi_N^2 = \Gamma(N/2, 1/2)$ PDF) and therefore

$$\frac{\sum_{i=1}^{N} X_i^2 - N}{\sqrt{2N}} \to \mathcal{N}(0, 1).$$

Noting that for finite N this result can be viewed as an approximation, we can use the approximate result

$$Y_N = \sum_{i=1}^{N} X_i^2 \sim \mathcal{N}(N, 2N)$$

in making probability calculations. The error in the approximation is shown in Figure 15.9, where the approximate PDF (shown as the solid curve) of Y_N, which is a $\mathcal{N}(N, 2N)$, is compared to the exact PDF, which is a χ_N^2 (shown as the dashed curve). It is seen that the approximation becomes better as N increases.

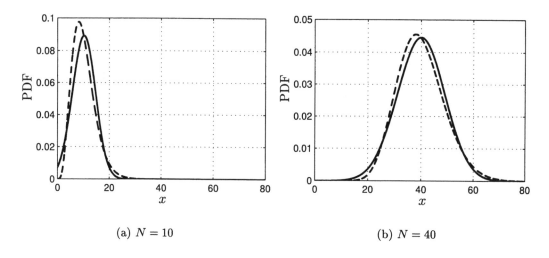

(a) $N = 10$ (b) $N = 40$

Figure 15.9: χ_N^2 PDF (dashed curve) and Gaussian PDF approximation of $\mathcal{N}(N, 2N)$ (solid curve).

◇

For the previous example it can be shown directly that the characteristic function of the standardized χ_N^2 random variable converges to that of the standardized Gaussian random variable, and hence so do their PDFs by the continuity theorem (see Section 11.7 for third property of characteristic function and also Problem 15.17). We next give an example that quantifies the numerical error of the central limit theorem approximation.

Example 15.4 – Central limit theorem and computation of probabilities—numerical results

Recall that the Erlang PDF is the PDF of the sum of N IID exponential random variables, where $X_i \sim \exp(\lambda)$ for $i = 1, 2, \ldots, N$ (see Section 10.5.6). Hence, letting $Y_N = \sum_{i=1}^{N} X_i$ the Erlang PDF is

$$p_{Y_N}(y) = \begin{cases} \frac{\lambda^N}{(N-1)!} y^{N-1} \exp(-\lambda y) & y \geq 0 \\ 0 & y < 0. \end{cases} \tag{15.8}$$

Its mean is N/λ and its variance is N/λ^2 since the mean and variance of an $\exp(\lambda)$ random variable is $1/\lambda$ and $1/\lambda^2$, respectively. If we wish to determine $P[Y_N > 10]$, then from (15.8) we can find the exact value for $\lambda = 1$ as

$$P[Y_N > 10] = \int_{10}^{\infty} \frac{1}{(N-1)!} y^{N-1} \exp(-y) \, dy.$$

But using

$$\int y^n \exp(-y) \, dy = -n! \exp(-y) \sum_{k=0}^{n} \frac{y^k}{k!}$$

[Gradshteyn and Ryzhik 1994], we have

$$\begin{aligned} P[Y_N > 10] &= \frac{1}{(N-1)!} \left[-(N-1)! \exp(-y) \sum_{k=0}^{N-1} \frac{y^k}{k!} \bigg|_{10}^{\infty} \right] \\ &= \exp(-10) \sum_{k=0}^{N-1} \frac{10^k}{k!}. \end{aligned}$$

A central limit theorem approximation would yield $Y_N \sim \mathcal{N}(N/\lambda, N/\lambda^2) = \mathcal{N}(N, N)$ so that

$$\hat{P}[Y_N > 10] = Q\left(\frac{10 - N}{\sqrt{N}}\right)$$

where the $\hat{P}$ denotes the approximation of P. The true and approximate values for this probability are shown in Figure 15.10. The probability values have been connected by straight lines for easier viewing.

◇

For the sum of IID *discrete* random variables the situation changes markedly. Consider the sum of N IID $\text{Ber}(p)$ random variables. We already know that the PMF is binomial so that

$$p_{S_N}[k] = \binom{N}{k} p^k (1 - p)^{N-k} \qquad k = 0, 1, \ldots, N - 1.$$

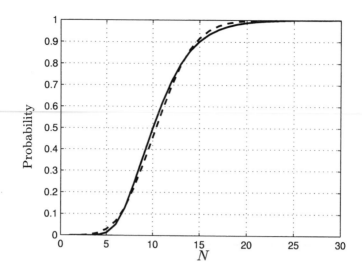

Figure 15.10: Exact and approximate calculation of probability that $Y_N > 10$ for Y_N an Erlang PDF. Exact value shown as dashed curve and Gaussian approximation as solid curve.

Hence, this example will allow us to compare the true PMF against any approximation. For reasons already explained we need to consider the PMF of the standardized sum or

$$\frac{S_N - E[S_N]}{\sqrt{\operatorname{var}(S_N)}} = \frac{S_N - Np}{\sqrt{Np(1-p)}}.$$

The PMF of the standardized binomial random variable PMF with $p = 1/2$ is shown in Figure 15.11 for various values on N. Note that it does not converge to any given

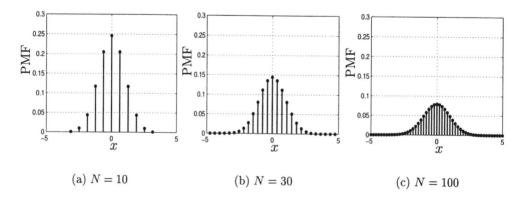

(a) $N = 10$ (b) $N = 30$ (c) $N = 100$

Figure 15.11: PMF for standardized binomial random variable with $p = 1/2$.

PMF, although the "envelope", whose amplitude decreases as N increases, appears

to be Gaussian. The lack of convergence is because the sample space or values that the standardized random variable can take on changes with N. The possible values are

$$x_k = \frac{k - Np}{\sqrt{Np(1-p)}} = \frac{k - N/2}{\sqrt{N/4}} \qquad k = 0, 1, \ldots, N$$

which become more dense as N increases. However, what does converge is the CDF as shown in Figure 15.12. Now as $N \to \infty$ we can assert that the CDF converges,

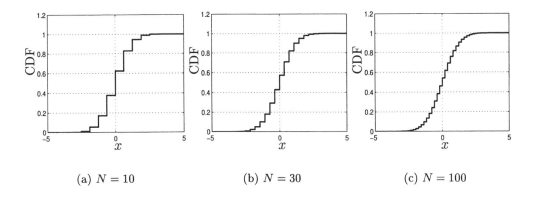

(a) $N = 10$ (b) $N = 30$ (c) $N = 100$

Figure 15.12: CDF for standardized binomial random variable with $p = 1/2$.

and furthermore it converges to the CDF of a $\mathcal{N}(0,1)$ random variable. Hence, the central limit theorem for discrete random variables is stated in terms of its CDF. It says that as $N \to \infty$

$$P\left[\frac{S_N - E[S_N]}{\sqrt{\operatorname{var}(S_N)}} \leq x\right] \to \int_{-\infty}^{x} \frac{1}{\sqrt{2\pi}} \exp\left(-\frac{1}{2}t^2\right) dt = \Phi(x)$$

and is also known as the *DeMoivre-Laplace theorem*. We summarize the central limit theorem for discrete random variables next.

Theorem 15.5.2 (Central limit theorem for discrete random variables)
If $X_1, X_2, \ldots, X_N$ are IID discrete random variables, each with mean $E_X[X]$ and variance $\operatorname{var}(X)$, and $S_N = \sum_{i=1}^{N} X_i$, then as $N \to \infty$

$$P\left[\frac{S_N - E[S_N]}{\sqrt{\operatorname{var}(S_N)}} \leq x\right] \to \int_{-\infty}^{x} \frac{1}{\sqrt{2\pi}} \exp\left(-\frac{1}{2}t^2\right) dt = \Phi(x) \qquad (15.9)$$

An example follows.

Example 15.5 – Computation of binomial probability
Assume that $Y_N \sim \operatorname{bin}(N, 1/2)$, which may be viewed as the PMF for the number of heads obtained in N fair coin tosses, and consider the probability $P[k_1 \leq Y_N \leq k_2]$.

Then the exact probability is

$$P[k_1 \leq Y_N \leq k_2] = \sum_{k=k_1}^{k_2} \binom{N}{k} \left(\frac{1}{2}\right)^N. \tag{15.10}$$

A central limit theorem approximation yields

$$
\begin{aligned}
P[k_1 \leq Y_N \leq k_2] &= P\left[\frac{k_1 - N/2}{\sqrt{N/4}} \leq \frac{Y_N - N/2}{\sqrt{N/4}} \leq \frac{k_2 - N/2}{\sqrt{N/4}}\right] \\
&\approx \Phi\left(\frac{k_2 - N/2}{\sqrt{N/4}}\right) - \Phi\left(\frac{k_1 - N/2}{\sqrt{N/4}}\right) \quad \text{(from (10.25))}
\end{aligned}
$$

since

$$\frac{Y_N - Np}{\sqrt{Np(1-p)}} = \frac{Y_N - N/2}{\sqrt{N/4}}$$

is the standardized random variable for $p = 1/2$. For example, if we wish to compute the probability of between 490 and 510 heads out of $N = 1000$ tosses, then

$$
\begin{aligned}
P[490 \leq Y_N \leq 510] &\approx \Phi\left(\frac{510 - 500}{\sqrt{250}}\right) - \Phi\left(\frac{490 - 500}{\sqrt{250}}\right) \\
&= 1 - Q\left(\frac{10}{\sqrt{250}}\right) - \left(1 - Q\left(\frac{-10}{\sqrt{250}}\right)\right) \\
&= 1 - 2Q\left(\frac{10}{\sqrt{250}}\right) = 0.4729.
\end{aligned}
$$

The exact value, however, is from (15.10)

$$P[490 \leq Y_N \leq 510] = \sum_{k=490}^{510} \binom{N}{k} \left(\frac{1}{2}\right)^N = 0.4933 \tag{15.11}$$

(see Problem 15.24 on how this was computed). A slightly better approximation using the central limit theorem can be obtained by replacing $P[490 \leq Y \leq 510]$ with $P[489.5 \leq Y \leq 510.5]$, which will more closely approximate the discrete random variable CDF by the continuous Gaussian CDF. This is because the binomial CDF has jumps at the integers as can be seen by referring to Figure 15.12. By taking a slighter larger interval to be used with the Gaussian approximation, the area under the Gaussian CDF more closely approximates these jumps at the endpoints of the interval. With this approximation we have

$$
\begin{aligned}
P[489.5 \leq Y \leq 510.5] &\approx Q\left(\frac{489.5 - 500}{\sqrt{250}}\right) - Q\left(\frac{510.5 - 500}{\sqrt{250}}\right) \\
&= 0.4934
\end{aligned}
$$

which is quite close to the true value!

$\diamondsuit$

15.6 Real-World Example – Opinion Polling

A frequent news topic of interest is the opinion of people on a major issue. For example, during the year of a presidential election in the United States, we hear almost on a daily basis the percentage of people who would vote for candidate A, with the remaining percentage voting for candidate B. It may be reported that 75% of the population would vote for candidate A and 25% would vote for candidate B. Upon reflection, it does not seem reasonable that a news organization would contact the entire population of the United States, almost 294,000,000 people, to determine their voter preferences. And indeed it is unreasonable! A more typical number of people contacted is only about 1000. How then can the news organization report that 75% of the population would vote for candidate A? The answer lies in the polling *error* – the results are actually stated as 75% with *a margin of error of* ±3%. Hence, it is not claimed that exactly 75% of the population would vote for candidate A, but between 72% and 78% would vote for candidate A. Even so, this seems like a lot of information to be gleaned from a very small sample of the population.

An analogous problem may help to unravel the mystery. Let's say we have a coin with an unknown probability of heads p. We wish to estimate p by tossing the coin N times. As we have already discussed, the law of large numbers asserts that we can determine p without error if we toss the coin an infinite number of times and use as our estimate the relative frequency of heads. However, in practice we are limited to only N coin tosses. How much will our estimate be in error? Or more precisely, how much can the true value deviate from our estimate? We know that the number of heads observed in N independent coin tosses can be anywhere from 0 to N. Hence, our estimate of p for $N = 1000$ can take on the possible values

$$\hat{p} = 0, \frac{1}{1000}, \frac{2}{1000}, \ldots, 1.$$

Of course, most of these estimates are not very probable. The probability that the estimate will take on these values is

$$P[\hat{p} = k/1000] = \binom{1000}{k} p^k (1-p)^{1000-k} \quad k = 0, 1, \ldots, 1000$$

which is shown in Figure 15.13 for $p = 0.75$. The probabilities for $\hat{p}$ outside the interval shown are approximately zero. Note that the maximum probability is for the true value $p = 0.75$. To assess the error in the estimate of p we can determine the *interval* over which say 95% of the $\hat{p}$'s will lie. The interval is chosen to be centered about $\hat{p} = 0.75$. In Figure 15.13 it is shown as the interval contained within the dashed vertical lines and is found by solving

$$\underbrace{\sum_{k=k_1}^{k_2} \binom{1000}{k} (0.75)^k (0.25)^{1000-k}}_{P[k \text{ heads}]} = 0.95 \qquad (15.12)$$

yielding $k_1 = 724$ and $k_2 = 776$, which results in $\hat{p}_1 = 0.724$ and $\hat{p}_2 = 0.776$. Hence,

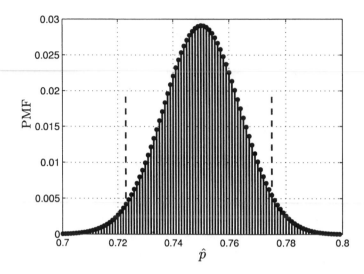

Figure 15.13: PMF for estimate of p for a binomial random variable. Also, shown as the dashed vertical lines are the boundaries of the interval within which 95% of the estimates will lie.

for $p = 0.75$ we see that 95% of the time (if we kept repeating the 1000 coin toss experiment), the value of $\hat{p}$ would be in the interval $[0.724, 0.776]$. We can assert that we are 95% confident that for $p = 0.75$

$$p - 0.026 \leq \hat{p} \leq p + 0.026$$

or

$$-p + 0.026 \geq -\hat{p} \geq -p - 0.026$$

or finally

$$\hat{p} - 0.026 \leq p \leq \hat{p} + 0.026.$$

The interval $[\hat{p} - 0.026, \hat{p} + 0.026]$ is called the 95% *confidence interval.* It is a *random interval* that covers the true value of $p = 0.75$ for 95% of the time. As an example a MATLAB simulation is shown in Figure 15.14. For each of 50 trials the estimate of p is shown by the dot while the confidence interval is indicated by a vertical line. Note that only 3 of the intervals fail to cover the true value of $p = 0.75$. With 50 trials and a probability of 0.95 we expect 2.5 intervals not to cover the true value.

Instead of having to compute k_1 and k_2 using (15.12), it is easier in practice to use the central limit theorem. Since $\hat{p} = \sum_{i=1}^{N} (X_i/N)$, with $X_i \sim \text{Ber}(p)$, is a sum of IID random variables we can assert from Theorem 15.5.2 that

$$P\left[-b \leq \frac{\hat{p} - E[\hat{p}]}{\sqrt{\text{var}(\hat{p})}} \leq b\right] \approx \Phi(b) - \Phi(-b).$$

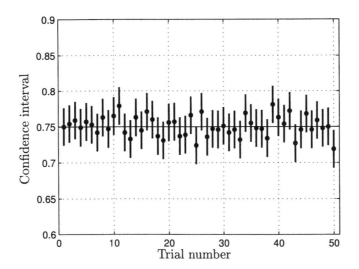

Figure 15.14: 95% confidence interval for estimate of $p = 0.75$ for a binomial random variable. The estimates are shown as dots.

Noting that $X_i \sim \text{Ber}(p)$, $E[\hat{p}] = E[\sum_{i=1}^{N} X_i/N] = Np/N = p$ and $\text{var}(\hat{p}) = \text{var}(\sum_{i=1}^{N} X_i/N) = Np(1-p)/N^2 = p(1-p)/N$, we have

$$P\left[-b \leq \frac{\hat{p} - p}{\sqrt{p(1-p)/N}} \leq b\right] \approx \Phi(b) - \Phi(-b).$$

For a 95% confidence interval or $\Phi(b) - \Phi(-b) = 0.95$, we have $b = 1.96$, as may be easily verified. Hence, we can use the approximation

$$-1.96 \leq \frac{\hat{p} - p}{\sqrt{p(1-p)/N}} \leq 1.96$$

which after the same manipulation as before yields the confidence interval

$$\hat{p} - 1.96\sqrt{\frac{p(1-p)}{N}} \leq p \leq \hat{p} + 1.96\sqrt{\frac{p(1-p)}{N}}. \tag{15.13}$$

The only difficulty in applying this result is that we don't know the value of p, which arose from the variance of $\hat{p}$. To circumvent this there are two approaches. We can replace p by its estimate to yield the confidence interval

$$\hat{p} - 1.96\sqrt{\frac{\hat{p}(1-\hat{p})}{N}} \leq p \leq \hat{p} + 1.96\sqrt{\frac{\hat{p}(1-\hat{p})}{N}}. \tag{15.14}$$

A more conservative approach is to note that $p(1-p)$ is maximum for $p = 1/2$. Using this number yields a larger interval than necessary. However, it allows us to determine before the experiment is performed and the value of $\hat{p}$ revealed, the length of the confidence interval. This is useful in planning how large N must be in order to have a confidence interval not exceeding a given length (see Problem 15.25). If we adopt the latter approach then the confidence interval becomes

$$\hat{p} \pm 1.96 \sqrt{\frac{p(1-p)}{N}} = \hat{p} \pm 1.96 \sqrt{\frac{1/4}{N}} \approx \hat{p} \pm \frac{1}{\sqrt{N}}.$$

In summary, if we toss a coin with a probability p of heads N times, then the interval $[\hat{p} - 1/\sqrt{N}, \hat{p} + 1/\sqrt{N}]$ will contain the true value of p more than 95% of the time. It is said that the error in our estimate of p is $\pm 1/\sqrt{N}$.

Finally, returning to our polling problem we ask N people if they will vote for candidate A. The probability that a person chosen at random will say "yes" is p, because the proportion of people in the population who will vote for candidate A is p. We liken this to tossing a single coin and noting if it comes up a head (vote "yes") or a tail (vote "no"). Then we continue to record the responses of N people (continue to toss the coin N times). Assume, for example, 750 people out of 1000 say "yes". Then $\hat{p} = 750/1000 = 0.75$ and the *margin of error* is $\pm 1/\sqrt{N} \approx 3\%$. Hence, we report the results as 75% of the population would vote for candidate A with a margin of error of 3%. (Probabilistically speaking, if we continue to poll groups of 1000 voters, estimating p for each group, then about 95 out of 100 groups would cover the true value of $100p\%$ by their estimated interval $[100\hat{p} - 3, 100\hat{p} + 3]\%$.) We needn't poll 294,000,000 people since we assume that the *percentage of the 1000 people polled who would vote for candidate A is representative of the percentage of the entire population.* Is this true? Certainly not if the 1000 people were all relatives of candidate A. Pollsters make their living by ensuring that their sample (1000 people polled) is a representative cross-section of the entire United States population.

References

Brockwell, P.J., R.A. Davis, *Time Series: Theory and Methods*, Springer-Verlag, New York, 1987.

Gradshteyn, I.S., I.M. Ryzhik, *Tables of Integrals, Series, and Products*, Fifth Ed., Academic Press, New York, 1994.

Grimmett, G., D. Stirzaker, *Probability and Random Processes*, Third Ed., Oxford University Press, New York, 2001.

Harvey, A.C., *Forecasting, Structural Time Series Models and the Kalman Filter*, Cambridge University Press, New York, 1989.

Middleton, D., *An Introduction to Statistical Communication Theory*, McGraw-Hill, New York, 1960.

Rao, C.R., *Linear Statistical Inference and Its Applications*, John Wiley & Sons, New York, 1973.

Reif, F., *Fundamentals of Statistical and Thermal Physics*, McGraw-Hill, New York, 1965.

Urick, R.J., *Principles of Underwater Sound*, McGraw-Hill, New York, 1975.

Weisburg, H.F., J.A. Krosnick, B.D. Bowen, *An Introduction to Survey Research, Polling and Data Analysis*, Sage Pubs., CA, 1996.

Problems

15.1 (f) For the PMF given by (15.2) plot the CDF for $N = 10$, $N = 30$, and $N = 100$. What function does the CDF appear to converge to?

15.2 (c) If $X_i \sim \mathcal{N}(1,1)$ for $i = 1, 2 \ldots, N$ are IID random variables, plot a realization of the sample mean random variable versus N. Should the realization converge and if so to what value?

15.3 (w,c) Let $X_{1_i} \sim \mathcal{U}(0,2)$ for $i = 1, 2 \ldots, N$ be IID random variables and let $X_{2_i} \sim \mathcal{N}(1,4)$ for $i = 1, 2 \ldots, N$ be another set of IID random variables. If the sample mean random variable is formed for each set of IID random variables, which one should converge faster? Implement a computer simulation to check your results.

15.4 ($\cdot\cdot$) (w) Consider the weighted sum of N IID random variables $Y_N = \sum_{i=1}^{N} \alpha_i X_i$. If $E_X[X] = 0$ and $\text{var}(X) = 1$, under what conditions will the sum converge to a number? Can you give an example, other than $\alpha_i = 1/N$, of a set of α_i's which will result in convergence?

15.5 (w) A random walk is defined as $X_N = X_{N-1} + U_N$ for $N = 2, 3, \ldots$ and $X_1 = U_1$, where the U_i's are IID random variables with $P[U_i = -1] = P[U_i = +1] = 1/2$. Will X_N converge to anything as $N \to \infty$?

15.6 (w) To estimate the second moment of a random variable it is proposed to use $(1/N) \sum_{i=1}^{N} X_i^2$. Under what conditions will the estimate converge to the true value?

15.7 ($\cdot\cdot$) (w) If X_i for $i = 1, 2 \ldots, N$ are IID random variables, will the random variable $(1/\sqrt{N}) \sum_{i=1}^{N} X_i$ converge to a number?

15.8 (t,c) In this problem we attempt to demonstrate that convergence in probability is different than standard convergence of a sequence of real numbers.

Consider the sequence of random variables

$$Y_N = \frac{X_N}{\sqrt{N}} + u\left(\frac{X_N}{\sqrt{N}} - 0.1\right)$$

where the X_N's are IID, each with PDF $X_N \sim \mathcal{N}(0,1)$ and $u(x)$ is the unit step function. Prove that $P[|Y_N| > \epsilon] \to 0$ as $N \to \infty$ by using the law of total probability as

$$
\begin{aligned}
P[|Y_N| > \epsilon] &= P[|Y_N| > \epsilon | X_N/\sqrt{N} > 0.1]P[X_N/\sqrt{N} > 0.1] \\
&\quad + P[|Y_N| > \epsilon | X_N/\sqrt{N} \le 0.1]P[X_N/\sqrt{N} \le 0.1].
\end{aligned}
$$

This says that $Y_N \to 0$ in probability. Next simulate this sequence on the computer for $N = 1, 2, \ldots, 200$ to generate 4 realizations of $\{Y_1, Y_2, \ldots, Y_{200}\}$. Examine whether for a given N all realizations lie within the "convergence band" of $[-0.2, 0.2]$. Next generate an additional 6 realizations and overlay all 10 realizations. What can you say about the convergence of any one realization?

15.9 (w) There are 1000 resistors in a bin labeled 10 ohms. Due to manufacturing tolerances, however, the resistance of the resistors are somewhat different. Assume that the resistance can be modeled as a random variable with a mean of 10 ohms and a variance of 2 ohms. If 100 resistors are chosen from the bin and connected in series (so the resistances add together), what is the approximate probability that the total resistance will exceed 1030 ohms?

15.10 (w) Consider a sequence of random variables $X_1, X_1, X_2, X_2, X_3, X_3, \ldots$, where $X_1, X_2, X_3 \ldots$ are IID random variables. Does the law of large numbers hold? How about the central limit theorem?

15.11 (w) Consider an Erlang random variable with parameter N. If N increases, does the PDF become Gaussian? Hint: Compare the characteristic functions of the exponential random variable and the $\Gamma(N, \lambda)$ random variable in Table 11.1.

15.12 (f) Find the approximate PDF of $Y = \sum_{i=1}^{100} X_i^2$, if the X_i's are IID with $X_i \sim \mathcal{N}(-4, 8)$.

15.13 (⌣) (f) Find the approximate PDF of $Y = \sum_{i=1}^{1000} X_i$, if the X_i's are IID with $X_i \sim \mathcal{U}(1, 3)$.

15.14 (f) Find the approximate probability that $Y = \sum_{i=1}^{10} X_i$ will exceed 7, if the X_i's are IID with the PDF

$$
p_X(x) = \begin{cases} 2x & 0 < x < 1 \\ 0 & \text{otherwise.} \end{cases}
$$

15.15 (c) Modify the computer program `clt_demo.m` listed in Appendix 15A to display the repeated convolution of the PDF

$$p_X(x) = \begin{cases} \frac{\pi}{2}\sin(\pi x) & 0 < x < 1 \\ 0 & \text{otherwise.} \end{cases}$$

and examine the results.

15.16 (c) Use the computer program `clt_demo.m` listed in Appendix 15A to display the repeated convolution of the PDF $\mathcal{U}(0, 1)$. Next modify the program to display the repeated convolution of the PDF

$$p_X(x) = \begin{cases} |2 - 4x| & 0 < x < 1 \\ 0 & \text{otherwise.} \end{cases}$$

Which PDF results in a faster convergence to a Gaussian PDF and why?

15.17 (t) In this problem we prove that the PDF of a standardized χ_N^2 random variable converges to a Gaussian PDF as $N \to \infty$. To do so let $Y_N \sim \chi_N^2$ and show that the characteristic function is

$$\phi_{Y_N}(\omega) = \frac{1}{(1 - 2j\omega)^{N/2}}$$

by using Table 11.1. Next define the standardized random variable

$$Z_N = \frac{Y_N - E[Y_N]}{\sqrt{\text{var}(Y_N)}}$$

and note that the mean and variance of a χ_N^2 random variable is N and $2N$, respectively. Show the characteristic function of Z_N is

$$\phi_{Z_N}(\omega) = \frac{\exp(-j\omega\sqrt{N/2})}{(1 - j\omega\sqrt{2/N})^{N/2}}.$$

Finally, take the natural logarithm of $\phi_{Z_N}(\omega)$ and note that for a complex variable x with $|x| \ll 1$, we have that $\ln(1 - x) \approx -x - x^2/2$. You should be able to show that as $N \to \infty$, $\ln\phi_{Z_N}(\omega) \to -\omega^2/2$.

15.18 (w) A particle undergoes collisions with other particles. Each collision causes its horizontal velocity to change according to a $\mathcal{N}(0, 0.1)$ cm/sec random variable. After 100 independent collisions what is the probability that the particle's velocity will exceed 5 cm/sec if it is initially at rest? Is this result exact or approximate?

15.19 (⌣) (f) The sample mean random variable of N IID random variables with $X_i \sim \mathcal{U}(0, 1)$ will converge to 1/2. How many random variables need to be averaged before we can assert that the approximate probability of an error of not more than 0.01 in magnitude is 0.99?

15.20 (☺) **(w)** An orange grove produces oranges whose weights are uniformly distributed between 3 and 7 ozs. If a truck can hold 4000 lbs. of oranges, what is the approximate probability that it can carry 15,000 oranges?

15.21 (w) A sleeping pill is effective for 75% of the population. If in a hospital 160 patients are given a sleeping pill, what is the approximate probability that 125 or more of them will sleep better?

15.22 (☺) **(w)** For which PDF will a sum of IID random variables when added together have a PDF that converges to a Gaussian PDF the fastest?

15.23 (☺) **(w)** A coin is tossed 1000 times, producing 750 heads. Is this a fair coin?

15.24 (f,c) To compute the probability of (15.11) we can use the following approach to compute each term in the summation. Each term can be written as

$$p_{Y_N}[k] = \binom{N}{k}\left(\frac{1}{2}\right)^N = \frac{N(N-1)\cdots(N-k+1)}{1(2)(3)\cdots(k)}\left(\frac{1}{2}\right)^N.$$

Taking the natural logarithm produces

$$\ln p_{Y_N}[k] = \sum_{i=N-k+1}^{N}\ln(i) - \sum_{i=1}^{k}\ln(i) - N\ln(2)$$

which is easily done on a computer. Next, exponentiate to find $p_{Y_N}[k]$ and add each of the terms together to finally implement the summation. Carry this out to verify the result given in (15.11). What happens if you try to compute each term directly?

15.25 (f) In a poll of candidate preferences for two candidates, we wish to report that the margin of error is only ±1%. What is the maximum number of people whom we will need to poll?

15.26 (☺) **(w)** A clinical trial is performed to determine if a particular drug is effective. A group of 100 people is split into two equal groups at random. The drug is administered to group 1 while group 2 is given a placebo. As a result of the study, 40 people in group 1 show a marked improvement while only 30 people in group 2 do so. Is the drug effective? Hint: Find the confidence intervals (using (15.14)) for the percentage of the people in each group who show an improvement.

Appendix 15A

MATLAB Program to Compute Repeated Convolution of PDFs

```
%  This program demonstrates the central limit theorem.  It determines
%  the PDF for the sum S_N of N IID random variables. Each marginal PDF
%  is assumed to be nonzero over the interval (0,1). The repeated
%  convolution integral is implemented using a discrete convolution. The
%  plots of the PDF of S_N as N increases are shown successively
%  (press carriage return for next plot).
%
%  clt_demo.m
clear all
delu=0.005;
u=[0:delu:1-delu]'; % p_X defined on interval [0,1]
p_X=ones(length(u),1); % try p_X=abs(2-4*u) for really strange PDF
x=[u;u+1]; % increase abcissa values since repeated
          % convolution increases nonzero width of output
p_S=zeros(length(x),1);
N=12; % number of random variables summed
for j=1:length(x) % start discrete convolution approximation
                % to continuous convolution
  for i=1:length(u)
    if j-i>0&j-i<=length(p_X)
      p_S(j)=p_S(j)+p_X(i)*p_X(j-i)*delu;
    end
  end
end
plot(x,p_S) % plot results for N=2
grid
```

```
axis([0 N 0 1]) % set axes lengths for plotting
xlabel('x')
ylabel('p_S')
title('PDF for S_N')
text(0.75*N,0.85,'N = 2') % label plot with the
                          % number of convolutions
for n=3:N
  pause
  x=[x;u+n-1]; % increase abcissa values since
               % repeated convolution increases
               % nonzero width of output
  p_S=[p_S;zeros(length(u),1)];
  g=zeros(length(p_S),1);
  for j=1:length(x) % start discrete convolution
  for i=1:length(u)
    if j-i>0
      g(j,1)=g(j,1)+p_X(i)*p_S(j-i)*delu;
    end
  end
end
p_S=g; % plot results for N=3,4,...,12
plot(x,p_S)
grid
axis([0 N 0 1])
xlabel('x')
ylabel('p_S')
title('PDF for S_N')
text(0.75*N,0.85,['N = ' num2str(n)])
end
```

Appendix 15B

Proof of Central Limit Theorem

In this appendix we prove the central limit theorem for continuous random variables. Consider the characteristic function of the standardized continuous random variable

$$Z_N = \frac{S_N - NE_X[X]}{\sqrt{N\mathrm{var}(X)}}$$

where $S_N = \sum_{i=1}^{N} X_i$ and the X_i's are IID. By definition of Z_N the characteristic function becomes

$$
\begin{aligned}
\phi_{Z_N}(\omega) &= E_{Z_N}[\exp(j\omega Z_N)] \\
&= E_{\mathbf{X}}\left[\exp\left(j\omega\frac{\sum_{i=1}^{N} X_i - NE_X[X]}{\sqrt{N\mathrm{var}(X)}}\right)\right] \\
&= E_{\mathbf{X}}\left[\prod_{i=1}^{N}\exp\left(j\omega\frac{X_i - E_X[X]}{\sqrt{N\mathrm{var}(X)}}\right)\right] \\
&= \prod_{i=1}^{N} E_{X_i}\left[\exp\left(j\omega\frac{X_i - E_X[X]}{\sqrt{N\mathrm{var}(X)}}\right)\right] \quad \text{(independence of } X_i\text{'s)} \\
&= \left[E_X\left[\exp\left(j\omega\frac{X - E_X[X]}{\sqrt{N\mathrm{var}(X)}}\right)\right]\right]^N \quad \text{(identically distributed } X_i\text{'s).}
\end{aligned}
$$

But for a complex variable ξ we can write its exponential as a Taylor series yielding

$$\exp(\xi) = \sum_{k=0}^{\infty}\frac{\xi^k}{k!} \quad \text{(see Problem 5.22).}$$

Thus,

$$E_X\left[\exp\left(j\omega\frac{X - E_X[X]}{\sqrt{N\mathrm{var}(X)}}\right)\right]$$

$$= E_X \left[\sum_{k=0}^{\infty} \frac{(j\omega)^k}{k!} \left(\frac{X - E_X[X]}{\sqrt{N\mathrm{var}(X)}} \right)^k \right]$$

$$= \sum_{k=0}^{\infty} \frac{(j\omega)^k}{k!} E_X \left[\left(\frac{X - E_X[X]}{\sqrt{N\mathrm{var}(X)}} \right)^k \right] \quad \text{(assume interchange valid)}$$

$$= 1 + j\omega E_X \left[\frac{X - E_X[X]}{\sqrt{N\mathrm{var}(X)}} \right] + \frac{1}{2}(j\omega)^2 E_X \left[\left(\frac{X - E_X[X]}{\sqrt{N\mathrm{var}(X)}} \right)^2 \right] + E_X[R(X)]$$

where $R(X)$ is the third-order and higher terms of the Taylor expansion. But

$$E_X \left[\frac{X - E_X[X]}{\sqrt{N\mathrm{var}(X)}} \right] = \frac{E_X[X] - E_X[X]}{\sqrt{N\mathrm{var}(X)}} = 0$$

$$E_X \left[\left(\frac{X - E_X[X]}{\sqrt{N\mathrm{var}(X)}} \right)^2 \right] = \frac{E_X[(X - E_X[X])^2]}{N\mathrm{var}(X)} = \frac{1}{N}$$

and so

$$\phi_{Z_N}(\omega) = \left[1 - \frac{\omega^2}{2N} + E_X[R(X)] \right]^N.$$

The terms comprising $R(X)$ are

$$R(X) = \frac{(j\omega)^3}{3!} E_X \left[\left(\frac{X - E_X[X]}{\sqrt{N\mathrm{var}(X)}} \right)^3 \right] + \cdots$$

$$= \frac{1}{N^{3/2}} \frac{(j\omega)^3}{3!} E_X \left[\left(\frac{X - E_X[X]}{\sqrt{\mathrm{var}(X)}} \right)^3 \right] + \cdots$$

which can be shown to be small, due to the division of the successive terms by $N^{3/2}, N^2, \ldots$, relative to the $-\omega^2/(2N)$ term. Hence as $N \to \infty$, they do not contribute to $\phi_{Z_N}(\omega)$ and therefore

$$\phi_{Z_N}(\omega) \to \left(1 - \frac{\omega^2}{2N} \right)^N$$

$$\to \exp\left(-\frac{1}{2}\omega^2 \right) = \phi_Z(\omega) \quad \text{(see Problem 5.15)}$$

where $Z \sim \mathcal{N}(0,1)$. Since the characteristic function of Z_N converges to the characteristic function of Z, we have by the continuity theorem (see Section 11.7) that the PDF of Z_N must converge to the PDF of Z. Therefore, we have finally that as $N \to \infty$

$$p_{Z_N}(z) \to p_Z(z) = \frac{1}{\sqrt{2\pi}} \exp\left(-\frac{1}{2}z^2 \right).$$

Chapter 16

Basic Random Processes

16.1 Introduction

So far we have studied the probabilistic description of a *finite* number of random variables. This is useful for random phenomena that have definite beginning and end times. Many physical phenomena, however, are more appropriately modeled as ongoing in time. Such is the case for the annual summer rainfall in Rhode Island as shown in Figure 1.1 and repeated for convenience in Figure 16.1. This physical

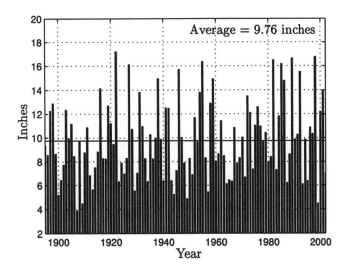

Figure 16.1: Annual summer rainfall in Rhode Island from 1895 to 2002.

process has been ongoing for all time and will undoubtedly continue into the future. It is only our limited ability to measure the rainfall over several lifetimes that has produced the data shown in Figure 16.1. It therefore seems more reasonable to attempt to study the probabilistic characteristics of the annual summer rainfall in

Rhode Island for *all time*. To do so let $X[n]$ be a random variable that denotes the annual summer rainfall for year n. Then, we will be interested in the behavior of the *infinite* tuple of random variables $(\ldots, X[-1], X[0], X[1], \ldots)$, where the corresponding year for $n = 0$ can be chosen for convenience (maybe according to the Christian or Hebrew calendars, as examples). Note that we cannot employ our previous probabilistic methods directly since the number of random variables is not finite or N-dimensional.

Given our interest in the annual summer rainfall, what types of questions are pertinent? A meterologist might wish to determine if the rainfall totals are increasing with time. Hence, he may question if the *average* rainfall is really constant. If it is not constant with time, then our estimate of the average, obtained by taking the sample mean of the values shown in Figure 16.1, is meaningless. As an example, we would also have obtained an average of 9.76 inches if the rainfall totals were increasing linearly with time, starting at 7.76 inches and ending at 11.76 inches. The meterologist might argue that due to global warming the rainfall totals should be increasing. We will return to this question in Section 16.8. Another question might be to assess the probability that the following year the rainfall will be 12 inches or more if we know the entire past history of rainfall totals. This is the problem of prediction, which is a fundamental problem in many scientific disciplines.

A second example of a random process, which is of intense interest, is a man-made one: the Dow-Jones industrial average (DJIA) for stocks. At the end of each trading day the average of the prices of a representative group of stocks is computed to give an indication of the health of the U.S. stock market. Its usefulness is that this value also gives an indication of the overall health of the U.S. economy. Some recent weekly values are shown in Figure 16.2. The overall trend beginning at week 10 is upward until about week 60, at which point it fluctuates up and down. Some questions of interest are whether the index will go back up again after week 92 and to what degree is it possible to predict the movement of the stock market, of which the DJIA is an indicator. The financial industry and in fact the health of the U.S. economy depends in a large degree upon the answers to these questions! In the remaining chapters we will describe the theory and application of random processes. As always, the theory will serve as a foundation upon which we will be able to analyze random processes. In any practical situation, however, the ideal theoretical analysis must be tempered with the constraints and additional complexities of the real world.

16.2   Summary

A random process is defined in Section 16.3. Four different types of random processes are described in Section 16.4. They are classified according to whether they are defined for all time or only for uniformly spaced time samples, and also according to their possible values as being discrete or continuous. Figure 16.5 illustrates the various types. A stationary random process is one for which its probabilistic

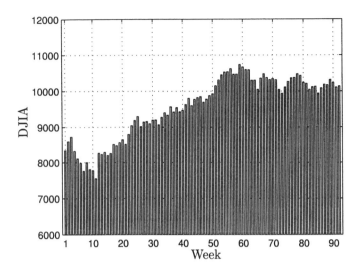

Figure 16.2: Dow-Jones industrial average at the end of each week from January 8, 2003 to September 29, 2004 [DowJones.com 2004].

description does not change with the chosen time origin, which is expressed mathematically by (16.3). An IID random process is stationary as shown in Example 16.3. The concept of a random process having stationary and independent increments is described in Section 16.5 with an illustration given in Example 16.5. Some more examples of random processes are given in Section 16.6. The most useful moments of a random process, the mean sequence and the covariance sequence, are defined by (16.5) and (16.7), respectively. Finally, in Section 16.8 an application of the estimation of the mean sequence to predicting average rainfall totals is described. The least squares estimator of the slope and intercept of a straight line is found using (16.9) and is commonly used in data analysis problems.

16.3 What Is a Random Process?

To define the concept of a random process we will begin by considering our usual example of a coin tossing experiment. Assume that at some start time we toss a coin and then repeat this subexperiment at one second intervals for all time. Letting n denote the time in seconds, we therefore generate successive outcomes at times $n = 0, 1, \ldots$. The experiment continues indefinitely. Since there are two possible outcomes for each coin toss and we will assume that the tosses are independent, we have an infinite sequence of Bernoulli trials. This is termed a *Bernoulli random process* and extends the finite Bernoulli set of random variables first introduced in Section 4.6.2, in which a finite number of trials were carried out. As usual, we let the probability of a head ($X = 1$) be p and the prob-

ability of a tail ($X = 0$) be $1 - p$ for each trial. With this setup, a random
process can be defined as a mapping from the original experimental sample space
$S = \{(H, H, T, \ldots), (H, T, H, \ldots), (T, T, H, \ldots), \ldots\}$ to the numerical sample space
$S_X = \{(1, 1, 0, \ldots), (1, 0, 1, \ldots), (0, 0, 1, \ldots), \ldots\}$. Note that each simple event or element of S is an infinite sequence of H's and T's which is then mapped into an
infinite sequence of 1's and 0's, which is the corresponding simple event in S_X. One
may picture a random process as being generated by the "random process generator" shown in Figure 16.3. The random process is composed of the infinite (but

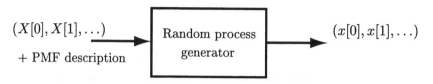

Figure 16.3: A conceptual random process generator. The input is an infinite sequence of random variables with their probabilistic description and the output is an
infinite sequence of numbers.

countable) "vector" of random variables $(X[0], X[1], \ldots)$, each of which is a Bernoulli
random variable, and each outcome of the random process is given by the infinite
sequence of numerical values $(x[0], x[1], \ldots)$. As usual, uppercase letters are used for
the random variables and lowercase letters for the values they take on. Some typical
outcomes of the Bernoulli random process are shown in Figure 16.4. They were
generated in MATLAB using x=floor(rand(31,1)+0.5) for each outcome. Each
sequence in Figure 16.4 is called an *outcome* or by its synonyms of *realization* or
sample sequence. We will prefer the use of the term "realization". Each realization
is an infinite sequence of numbers. Hence, the random process is a mapping from S,
which is a set of infinite sequential experimental outcomes, to S_X, which is a set of
infinite sequences of 1's and 0's or realizations. The total number of realizations is
not countable (see Problem 16.3). The set of all realizations is sometimes referred
to as the *ensemble of realizations*. Just as for the case of a single random variable,
which is a mapping from S to S_X and therefore is represented as the set function
$X(s)$, a similar notation is used for random processes. Now, however, we will use
$X[n, s]$ to represent the mapping from an element of S to a realization $x[n]$. In
Figure 16.4 we see the result of the mapping for $s = s_1$, which is $X[n, s_1] = x_1[n]$,
as well as others. It is important to note that if we fix n at $n = 18$, for example,
then $X[18, s]$ is a *random variable* that has a Bernoulli PMF. Three of its outcomes
are shown highlighted in Figure 16.4 with dashed boxes. Hence, all the methods
developed for a single random variable are applicable. Likewise, if we fix two samples at $n = 20$ and $n = 22$, then $X[20, s]$ and $X[22, s]$ becomes a bivariate random
vector. Again all our previous methods for two-dimensional random vectors apply.
    To summarize, a random process is defined to be an infinite sequence of random

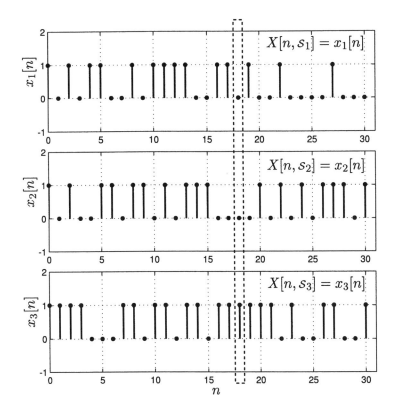

Figure 16.4: Typical outcomes of Bernoulli random process with $p = 0.5$. The realization starts at $n = 0$ and continues indefinitely. The dashed box indicates the realizations of the *random variable* $X[18, s]$.

variables $(X(0), X(1), \ldots)$, with one random variable for each time instant, and each realization of the random process takes on a value that is represented as an infinite sequence of numbers or $(x[0], x[1], \ldots)$. We will denote the random process more succinctly by $X[n]$ and the realization by $x[n]$ but *it is understood that the n denotes the values $n = 0, 1, \ldots$*. If we wish to indicate the random process at a *fixed time instant*, then we will use $n = n_0$ or $n = n_1$, etc. so that $X[n_0]$ is the random process at $n = n_0$ (which is just a random variable) and its realization at that time is $x[n_0]$ (which is a number). Finally, we have used the $[\cdot]$ notation to remind us that $X[n]$ is defined only for *discrete integer* times. This type of random process is known as a *discrete-time* random process. In the next section the *continuous-time* random process will be discussed. Before continuing, however, we look at a typical probability calculation for a random process.

Example 16.1 – Bernoulli random process

For the infinite coin tossing example, we might ask for the probability of the first

5 tosses coming up all heads. Thus, we wish to evaluate

$$P[X[0] = 1, X[1] = 1, X[2] = 1, X[3] = 1, X[4] = 1, X[5] = 0 \text{ or } 1, X[6] = 0 \text{ or } 1, \ldots].$$

It would seem that since we don't care what the outcomes of $X[n]$ for $n = 5, 6, \ldots$ are, then the probability expression could be replaced by

$$P[X[0] = 1, X[1] = 1, X[2] = 1, X[3] = 1, X[4] = 1]$$

and indeed this is the case, although it is not so easy to prove [Billingsley 1986]. Then, by using the assumption of independence of a Bernoulli random process we have

$$P[X[0] = 1, X[1] = 1, X[2] = 1, X[3] = 1, X[4] = 1] = \prod_{n=0}^{4} P[X[n] = 1] = p^5.$$

A related question is to determine the probability that we will *ever* observe 5 ones in a row. Intuitively, we expect this probability to be 1, but how do we prove this? It is not easy! Such is the difficulty encountered when we make the leap from a random vector, having a finite number of random variables, to a random process, having an infinite number of random variables.

◇

16.4   Types of Random Processes

The previous example of an infinite number of coin tosses produced a random process $X[n]$ for $n = 0, 1, \ldots$. In some cases, however, we wish to think of the random process as having started sometime in the infinite past. If $X[n]$ is defined for $n = \ldots, -1, 0, 1, \ldots$ or equivalently $-\infty < n < \infty$, where it is assumed that n is an integer, then $X[n]$ is called an *infinite* random process. In contrast, the previous example is referred to as a *semi-infinite* random process. Another categorization of random processes involves whether the times at which the random variables are defined and the values that they take on are either discrete or continuous. The infinite coin toss example is a *discrete-time* random process, since it is defined for $n = 0, 1, \ldots$, and is a *discrete-valued* random process, since it takes on values 0 and 1 only. It is referred to as a *discrete-time/discrete valued* (DTDV) random process. Other types of random processes are discrete-time/continuous-valued (DTCV), continuous-time/discrete-valued (CTDV), and continuous-time/continuous-valued (CTCV). A realization of each type is shown in Figure 16.5. In Figure 16.5a a realization of the Bernoulli random process, as previously described, is shown while in Figure 16.5b a realization of a Gaussian random process with $Y[n] \sim \mathcal{N}(0, 1)$ is shown. The Bernoulli random process is defined for $n = 0, 1, \ldots$ (semi-infinite) while the Gaussian random process is defined for $-\infty < n < \infty$ and n an integer (infinite).

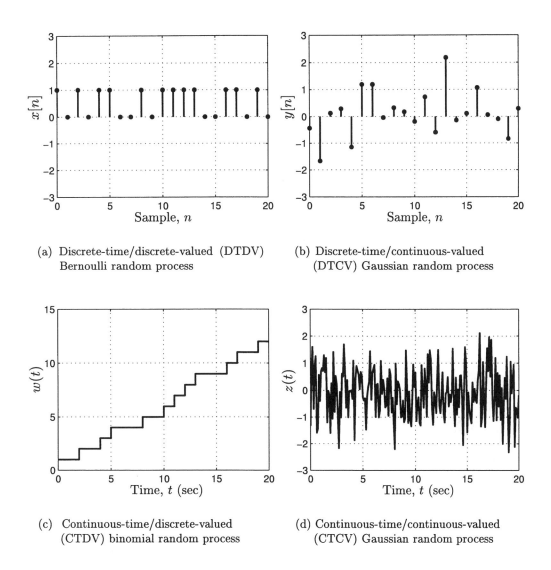

(a) Discrete-time/discrete-valued (DTDV)
    Bernoulli random process

(b) Discrete-time/continuous-valued
    (DTCV) Gaussian random process

(c) Continuous-time/discrete-valued
    (CTDV) binomial random process

(d) Continuous-time/continuous-valued
    (CTCV) Gaussian random process

Figure 16.5: Typical realizations of different types of random processes.

Both these random processes are discrete-time with the first one taking on only the values 0 and 1 and the second one taking on all real values. In Figure 16.5c is shown a random process, also known as a continuous-time binomial random process, which is defined as $W(t) = \sum_{n=0}^{[t]} X[n]$, where $X[n]$ is a Bernoulli random process and $[t]$ denotes the largest integer less than or equal to t. This process effectively counts the number of successes or ones of the Bernoulli random process (compare Figure 16.5c with Figure 16.5a). It is defined for all time; hence, it is a continuous-time random process, and it takes on only integer values in the range $\{0, 1, \ldots\}$; hence, it is discrete-valued. Finally, in Figure 16.5d is shown a realization of another

Gaussian process but with $Z(t) \sim \mathcal{N}(0,1)$ for all time t. This is a continuous-time random process that takes on all real values; hence, it is continuous-valued. We will generally use a discrete-time random process, with either discrete or continuous values, to introduce new concepts. This is because a continuous-time random process introduces a host of mathematical subtleties which in many cases are beyond the scope of this text. When possible, however, we will quote the analogous results for continuous-time random processes. Note finally that a realization of $X[n]$, which is $x[n]$, is also called a *sample sequence*, while a realization of $X(t)$, which is $x(t)$, is also called a *sample function*. We will, however, reserve the use of the word *sample* to refer to a *time sample* of the random process. Hence, a time sample will refer to either the random variable $X[n_0]$ $(X(t_0))$ or the realization $x[n_0]$ $(x(t_0))$ of the random process, with the meaning determined by the context of the discussion. We next revisit the random walk of Example 9.5.

Example 16.2 – Random walk (continued from Example 9.5)
Recall that

$$X_n = \sum_{i=1}^{n} U_i \qquad n = 1, 2, \ldots$$

where

$$p_U[k] = \begin{cases} \frac{1}{2} & k = -1 \\ \frac{1}{2} & k = 1 \end{cases} \qquad (16.1)$$

and the U_i's are IID. The random walk is a random process so that rewriting the definition in our new notation, we have

$$X[n] = \sum_{i=0}^{n} U[i] \qquad n = 0, 1, \ldots$$

where the $U[i]$'s are IID random variables having the PMF of (16.1). We also assume that the random walk starts at time $n = 0$. The $U[i]$'s comprise the random variables of a Bernoulli random process but with values of ± 1, instead of the usual 0 and 1. As such, we can view the $U[i]$'s as comprising a Bernoulli random process $U[n]$ for $n = 0, 1, \ldots$. Realizations of $U[n]$ and $X[n]$ are shown in Figure 16.6. One question that comes to mind is the behavior of the random walk for large n. For example, we might be interested in the PDF of $X[n]$ for large n. Relying on the central limit theorem (see Chapter 15), we can assert that the PDF is Gaussian, and therefore we need only determine the mean and variance. This easily follows from the definition of the random walk as

$$E[X[n]] = \sum_{i=0}^{n} E[U[i]] = (n+1)E[U[0]] = 0$$

$$\text{var}(X[n]) = \sum_{i=0}^{n} \text{var}(U[i]) = (n+1)\text{var}(U[0]) = n+1$$

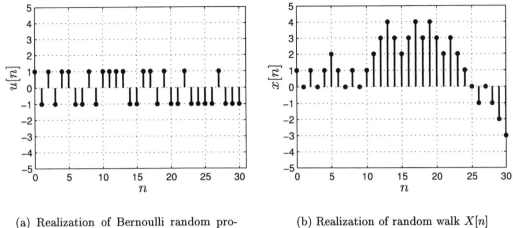

(a) Realization of Bernoulli random process $U[n]$

(b) Realization of random walk $X[n]$

Figure 16.6: Typical realization of a random walk.

since $E[U[i]] = 0$ and $\text{var}(U[i]) = 1$. (Note that since the $U[i]$'s are identically distributed, they all have the same mean and variance. We have arbitrarily chosen $U[0]$ in the expression for the mean and variance of a single sample.) Hence, for large n we have approximately that $X[n] \sim \mathcal{N}(0, n+1)$. Does this appear to explain the behavior of $x[n]$ shown in Figure 16.6b?

16.5 The Important Property of Stationarity

The simplest type of random process is an IID random process. The Bernoulli random process is an example of this. Each random variable $X[n_0]$ is independent of all the others and each random variable has the same marginal PMF. As such, the joint PMF of any finite number of samples can immediately be written as

$$p_{X[n_1],X[n_2],\ldots,X[n_N]}[x_1, x_2, \ldots, x_N] = \prod_{i=1}^{N} p_{X[n_i]}[x_i] \tag{16.2}$$

and used for probability calculations. For example, for a Bernoulli random process with values $0, 1$ the probability of the first 10 samples being $1, 0, 1, 0, 1, 0, 1, 0, 1, 0$ is $p^5(1-p)^5$. Note that we are able to specify the joint PMF for *any* finite number of sample times. This is sometimes referred to as being able to specify the *finite dimensional distribution* (FDD). It is the most complete probabilistic description that we can manage for a random process and reduces the analysis of a random process to the analysis of a *finite* but arbitrary set of random variables.

A generalization of the IID random process is a random process for which the FDD does not change with the time origin. This is to say that the PMF or PDF of the samples $\{X[n_1], X[n_2], \ldots, X[n_N]\}$ is the same as for $\{X[n_1 + n_0], X[n_2 + n_0], \ldots, X[n_N + n_0]\}$, where n_0 is an arbitrary integer. Alternatively, the set of samples can be shifted in time, with each one being shifted the same amount, without affecting the joint PMF or joint PDF. Mathematically, for the FDD not to change with the time origin, we must have that

$$p_{X[n_1+n_0],X[n_2+n_0],\ldots,X[n_N+n_0]} = p_{X[n_1],X[n_2],\ldots,X[n_N]} \qquad (16.3)$$

for all n_0, and for any arbitrary choice of N and $n_1, n_2, \ldots, n_N$. Such a random process is said to be *stationary*. It is implicit from (16.3) that all joint and marginal PMFs or PDFs must have probabilities that do not depend on the time origin. For example, by letting $N = 1$ in (16.3) we have that $p_{X[n_1+n_0]} = p_{X[n_1]}$ and setting $n_1 = 0$, we have that $p_{X[n_0]} = p_{X[0]}$ for all n_0. This says that the marginal PMF or PDF is the same for every sample in a stationary random process. We next prove that an IID random process is stationary.

Example 16.3 – IID random process is stationary.

To prove that the IID random process is a special case of a stationary random process we must show that (16.3) is satisfied. This follows from

$$p_{X[n_1+n_0],X[n_2+n_0],\ldots,X[n_N+n_0]} \;=\; \prod_{i=1}^{N} p_{X[n_i+n_0]} \qquad \text{(by independence)}$$

$$=\; \prod_{i=1}^{N} p_{X[n_i]} \qquad \text{(by identically distributed)}$$

$$=\; p_{X[n_1],X[n_2],\ldots,X[n_N]} \qquad \text{(by independence)}.$$

$\Diamond$

If a random process is stationary, then all its joint moments and more generally all expected values of functions of the random process, must also be stationary since

$$E_{X[n_1+n_0],\ldots,X[n_N+n_0]}[\cdot] = E_{X[n_1],\ldots,X[n_N]}[\cdot]$$

which follows from (16.3). Examples then of random processes that are not stationary are ones whose means and/or variances change in time, which implies that the marginal PMF or PDF change with time. In Figure 16.7 we show typical realizations of random processes whose mean in Figure 16.7a and whose variance in Figure 16.7b change with time. They were generated using the MATLAB code:

```
randn('state',0)
N=51;
x=randn(N,1)+0.1*[0:N-1]'; % for Figure 16.7a
y=sqrt(0.95.^[0:50]').*randn(N,1); % for Figure 16.7b
```

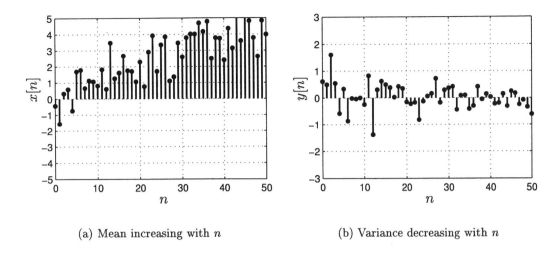

(a) Mean increasing with n (b) Variance decreasing with n

Figure 16.7: Random processes that are not stationary.

In Figure 16.7a the true mean increases linearly from 0 to 5 while in Figure 16.7b the variance decreases exponentially as 0.95^n. It is clear then that the samples all have different moments and therefore $p_{X[n_1+n_0]} \neq p_{X[n_1]}$ which violates the condition for stationarity.

⚠️ **It is impossible to determine if a random process is stationary from a single realization.**

A realization of a random process is a *single outcome* of the random process. This is analogous to observing a single outcome of a coin toss. We cannot determine if the coin is fair by observing that the outcome was a head. What is required are multiple realizations of the coin tossing experiment. So it is with random processes. In Figure 16.7b, although we generated the realization using a variance that decreased with time, and hence the random process is not stationary, the realization shown *could have been generated with a constant variance*. Then, the values of the realization near $n = 50$ just happen to be smaller than the ones near $n = 0$, which is possible, although maybe not very probable. To better discern whether a random process is stationary we require *multiple realizations*.

Another example of a random process that is not stationary follows.

Example 16.4 – Sum random process

A sum random process is a slight generalization of the random walk process of

Example 16.2. As before, $X[n] = \sum_{i=0}^{n} U[i]$, where the $U[i]$'s are IID but for the general sum process, the $U[i]$'s can have any, although the same, PMF or PDF. Thus, the sum random process is not stationary since

$$E[X[n]] = (n+1)E_U[U[0]]$$
$$\mathrm{var}(X[n]) = (n+1)\mathrm{var}(U[0])$$

both of which change with n. Hence, it violates the condition for stationarity.

$$\diamond$$

A random process that is not stationary is said to be *nonstationary*. In light of the fact that an IID random process lends itself to simple probability calculations, it is advantageous, if possible, to transform a nonstationary random process into a stationary one (see Problem 16.12 on transforming the random processes of Figure 16.7 into stationary ones). As an example, for the sum random process this can be done by "reversing" the summing operation. Specifically, we difference the random process. Then $X[n] - X[n-1] = U[n]$ for $n \geq 0$, where we define $X[-1] = 0$. This is an IID random process. The differences or *increment* random variables $U[n]$ are independent and identically distributed. More generally, for the sum random process any two *increments* of the form

$$X[n_2] - X[n_1] = \sum_{i=n_1+1}^{n_2} U[i]$$

$$X[n_4] - X[n_3] = \sum_{i=n_3+1}^{n_4} U[i]$$

are independent if $n_4 > n_3 \geq n_2 > n_1$. Thus, nonoverlapping increments for a sum random process are independent. (Recall that functions of independent random variables are themselves independent.) If furthermore, $n_4 - n_3 = n_2 - n_1$, then they also have the same PMF or PDF since they are composed of the same number of IID random variables. It is then said that for the sum random process, the increments are independent and stationary (equivalent to being identically distributed) or that it has *stationary independent increments*. The reader may wish to ponder whether a random process can have independent but nonstationary increments (see Problem 16.13). Many random processes (an example of which follows) that we will encounter have this property and it allows us to more easily analyze the probabilistic behavior.

Example 16.5 – Binomial counting random process

Consider the repeated coin tossing experiment where we are interested in the number of heads that occurs. Letting $U[n]$ be a Bernoulli random process with $U[n] = 1$ with probability p and $U[n] = 0$ with probability $1 - p$, the number of heads is given by the *binomial counting* or sum process

$$X[n] = \sum_{i=0}^{n} U[i] \qquad n = 0, 1, \ldots$$

or equivalently

$$X[n] = \begin{cases} U[0] & n = 0 \\ X[n-1] + U[n] & n \geq 1. \end{cases}$$

A typical realization is shown in Figure 16.8. The random process has stationary

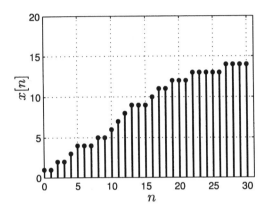

Figure 16.8: Typical realization of binomial counting random process with $p = 0.5$.

and independent increments since the changes over two nonoverlapping intervals are composed of different sets of identically distributed $U[i]$'s. We can use this property to more easily determine probabilities of events. For example, to determine $p_{X[1],X[2]}[1,2] = P[X[1] = 1, X[2] = 2]$, we can note that the event $X[1] = 1, X[2] = 2$ is equivalent to the event $Y_1 = X[1] - X[-1] = 1$, $Y_2 = X[2] - X[1] = 1$, where $X[-1]$ is defined to be identically zero. But Y_1 and Y_2 are nonoverlapping increments (but of unequal length), making them independent random variables. Thus,

$$\begin{aligned} P[X[1] = 1, X[2] = 2] &= P[Y_1 = 1, Y_2 = 1] = P[Y_1 = 1]P[Y_2 = 1] \\ &= P[\underbrace{U[0] + U[1]}_{\text{bin}(2,p)} = 1]P[U[2] = 1] \\ &= \binom{2}{1} p^1 (1-p)^1 \cdot p \\ &= 2p^2(1-p). \end{aligned}$$

$\diamondsuit$

16.6   Some More Examples

We continue our discussion by examining some random processes of practical interest.

Example 16.6 − White Gaussian noise

A common model for physical noise, such as resistor noise due to electron motion fluctuations in an electric field, is termed *white Gaussian noise* (WGN). It is assumed that the noise has been sampled in time to yield a DTCV random process $X[n]$. The WGN random process is defined to be an IID one whose marginal PDF is Gaussian so that $X[n] \sim \mathcal{N}(0, \sigma^2)$ for $-\infty < n < \infty$. Each random variable $X[n_0]$ has a mean of zero, consistent with our notion of a noise process, and the same variance or because the mean is zero, the same power $E[X^2[n_0]]$. A typical realization is shown in Figure 16.5b for $\sigma^2 = 1$. The WGN random process is stationary since it is an IID random process. Its joint PDF is

$$
\begin{aligned}
p_{X[n_1],X[n_2],\ldots,X[n_N]}(x_1, x_2, \ldots, x_N) &= \prod_{i=1}^{N} p_{X[n_i]}(x_i) \\
&= \prod_{i=1}^{N} \frac{1}{\sqrt{2\pi\sigma^2}} \exp\left(-\frac{1}{2\sigma^2}x_i^2\right) \\
&= \frac{1}{(2\pi\sigma^2)^{N/2}} \exp\left(-\frac{1}{2\sigma^2}\sum_{i=1}^{N} x_i^2\right). \quad (16.4)
\end{aligned}
$$

Note that the joint PDF is $\mathcal{N}(\mathbf{0}, \sigma^2\mathbf{I})$, which is a special form of the multivariate Gaussian PDF (see Problem 16.15). The terminology of "white" derives from the property that such a random process may be synthesized from a sum of different frequency random sinusoids each having the same power, much the same as white light is composed of equal contributions of each visible wavelength of light. We will justify this property in Chapter 17 when we discuss the power spectral density.

Example 16.7 − Moving average random process

The moving average (MA) random process is a DTCV random process defined as

$$X[n] = \tfrac{1}{2}(U[n] + U[n-1]) \qquad -\infty < n < \infty$$

where $U[n]$ is a WGN random process with variance σ_U^2. (To avoid confusion with the variance of other random variables we will sometimes use a subscript on σ^2, in this case σ_U^2, to refer to the variance of the $U[n_0]$ random variable.) The terminology of moving average refers to the averaging of the current random variable $U[n]$ with the previous random variable $U[n-1]$ to form the current moving average random variable. Also, this averaging "moves" in time, as for example,

$$
\begin{aligned}
X[0] &= \tfrac{1}{2}(U[0] + U[-1]) \\
X[1] &= \tfrac{1}{2}(U[1] + U[0]) \\
X[2] &= \tfrac{1}{2}(U[2] + U[1])
\end{aligned}
$$

etc.

A typical realization of $X[n]$ is shown in Figure 16.9 and should be compared to the realization of $U[n]$ shown in Figure 16.5b. It is seen that the moving average random process is "smoother" than the WGN random process, from which it was obtained. Further smoothing is possible by averaging more WGN samples together (see Problem 16.17). The MATLAB code shown below was used to generate the realization.

```
randn('state',0)
u=randn(21,1);
for i=1:21
  if i==1
    x(i,1)=0.5*(u(1)+randn(1,1)); % needed to initialize sequence
  else
    x(i,1)=0.5*(u(i)+u(i-1));
  end
end
```

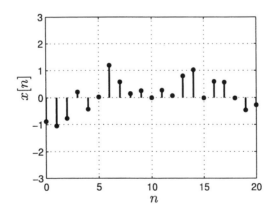

Figure 16.9: Typical realization of moving average random process. The realization of the $U[n]$ random process is shown in Figure 16.5b.

The joint PDF of $X[n]$ can be determined by observing that it is a linearly transformed version of $U[n]$. As an example, to determine the joint PDF of the random vector $[X[0]\,X[1]]^T$, we have from the definition of the MA random process

$$
\begin{bmatrix} X[0] \\ X[1] \end{bmatrix} = \begin{bmatrix} \frac{1}{2} & \frac{1}{2} & 0 \\ 0 & \frac{1}{2} & \frac{1}{2} \end{bmatrix} \begin{bmatrix} U[-1] \\ U[0] \\ U[1] \end{bmatrix}
$$

or in matrix/vector notation $\mathbf{X} = \mathbf{GU}$. Now recalling that $\mathbf{U}$ is a Gaussian random vector (see (16.4)) and that a linear transformation of a Gaussian random vector produces another Gaussian random vector, we have from Example 14.3 that

$$
\mathbf{X} \sim \mathcal{N}(\mathbf{G}E[\mathbf{U}], \mathbf{G}\mathbf{C}_U\mathbf{G}^T).
$$

Explicitly, since each sample of $U[n]$ is zero mean with variance σ_U^2 and all samples are independent, we have that $E[\mathbf{U}] = \mathbf{0}$ and $\mathbf{C}_U = \sigma_U^2\mathbf{I}$. This results in

$$
\mathbf{X} = \begin{bmatrix} X[0] \\ X[1] \end{bmatrix} \sim \mathcal{N}(\mathbf{0}, \sigma_U^2\mathbf{G}\mathbf{G}^T)
$$

where

$$
\mathbf{G}\mathbf{G}^T = \begin{bmatrix} \frac{1}{2} & \frac{1}{4} \\ \frac{1}{4} & \frac{1}{2} \end{bmatrix}.
$$

It can furthermore be shown that the MA random process is stationary (see Example 20.2 and Property 20.2).

◇

Example 16.8 – Randomly phased sinusoid (or sine wave)

Consider the DTCV random process given as

$$
X[n] = \cos(2\pi(0.1)n + \Theta) \qquad -\infty < n < \infty
$$

where $\Theta \sim \mathcal{U}(0, 2\pi)$. Some typical realizations are shown in Figure 16.10. The MATLAB statements n=[0:31]' and x=cos(2*pi*0.1*n+2*pi*rand(1,1)) can be used to generate each realization. This random process is frequently used to model an analog sinusoid whose phase is unknown and that has been sampled by an analog-to-digital convertor. It is nearly a deterministic signal, except for the phase uncertainty, and is therefore perfectly predictable. This is to say that once we observe two successive samples, then all the remaining ones are known (see Problem 16.20). This is in contrast to the WGN random process, for which regardless of how many samples we observe, we cannot predict any of the remaining ones due to the independence of the samples. Because of the predictability of the randomly phased sinusoidal process, the joint PDF can only be represented using impulsive functions. As an example, you might try to find the PDF of (X, Y) if (X, Y) has the bivariate Gaussian

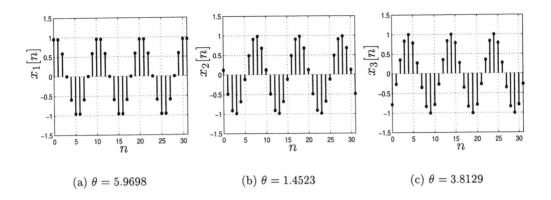

(a) $\theta = 5.9698$ (b) $\theta = 1.4523$ (c) $\theta = 3.8129$

Figure 16.10: Typical realizations for randomly phased sinusoid.

PDF with $\rho = 1$. We will not pursue this further. However, we can determine the marginal PDF $p_{X[n]}$. To do so we use the transformation formula of (10.30), where the Y random variable is $X[n_0]$ (considering the random process at a fixed time) and the X random variable is Θ. The transformation is shown in Figure 16.11 for $n_0 = 0$. Note that there are two solutions for any given $x[n_0] = y$ (except for the

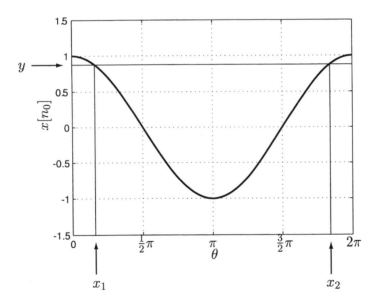

Figure 16.11: Function transforming Θ into $X[n_0]$ for the value $n_0 = 0$, where $X[n_0] = \cos(2\pi(0.1)n_0 + \Theta)$.

point at $\theta = \pi$, which has probability zero). We denote the solutions as $\theta = x_1, x_2$. Using our previous notation of $y = g(x)$ for a transformation of a single random variable we have that

$$y = \cos(2\pi(0.1)n_0 + x)$$

so that the solutions are

$$
\begin{aligned}
x_1 &= \arccos(y) - 2\pi(0.1)n_0 = g_1^{-1}(y) \\
x_2 &= 2\pi - [\arccos(y) - 2\pi(0.1)n_0] = g_2^{-1}(y)
\end{aligned}
$$

for $-1 < y < 1$ and thus $0 < \arccos(y) < \pi$. Using $d\arccos(y)/dy = 1/\sqrt{1-y^2}$, we have

$$
\begin{aligned}
p_Y(y) &= p_X(g_1^{-1}(y)) \left| \frac{dg_1^{-1}(y)}{dy} \right| + p_X(g_2^{-1}(y)) \left| \frac{dg_2^{-1}(y)}{dy} \right| \\
&= \frac{1}{2\pi} \left| \frac{1}{\sqrt{1-y^2}} \right| + \frac{1}{2\pi} \left| -\frac{1}{\sqrt{1-y^2}} \right| \\
&= \frac{1}{\pi\sqrt{1-y^2}}.
\end{aligned}
$$

Finally, in our original notation we have the marginal PDF for $X[n]$ for any n

$$
p_{X[n]}(x) = \begin{cases} \frac{1}{\pi\sqrt{1-x^2}} & -1 < x < 1 \\ 0 & \text{otherwise.} \end{cases}
$$

This PDF is shown in Figure 16.12. Note that the values of $X[n]$ that are most probable are near $x = \pm 1$. Can you explain why? (Hint: Determine the values of θ for which $0.9 < \cos\theta < 1$ and also $0 < \cos\theta < 0.1$ in Figure 16.11.)

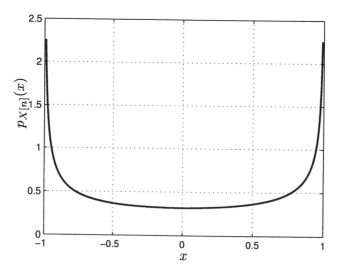

Figure 16.12: Marginal PDF for randomly phased sinusoid.

◇

16.7 Joint Moments

The first and second moments or equivalently the mean and variance of a random process at a given sample time are of great practical importance since they are easily determined. Also, the covariance between two samples of the random process at two different times is easily found. At worst, the first and second moments can always be estimated in practice. This is in contrast to the joint PMF or joint PDF, which in practice may be difficult to determine. Hence, we next define and give some examples of the mean, variance, and covariance sequences for a DTCV random process. The mean sequence is defined as

$$\mu_X[n] = E[X[n]] \qquad -\infty < n < \infty \tag{16.5}$$

while the variance sequence is defined as

$$\sigma_X^2[n] = \text{var}(X[n]) \qquad -\infty < n < \infty \tag{16.6}$$

and finally the covariance sequence is defined as

$$
\begin{aligned}
c_X[n_1, n_2] &= \text{cov}(X[n_1], X[n_2]) \\
&= E[(X[n_1] - \mu_X[n_1])(X[n_2] - \mu_X[n_2])] \qquad
\begin{array}{l} -\infty < n_1 < \infty \\ -\infty < n_2 < \infty. \end{array}
\end{aligned} \tag{16.7}
$$

The expectations for the mean and variance are taken with respect to the PMF or PDF $p_{X[n]}$ for a particular value of n. Similarly, the expectation needed for the evaluation of the covariance is with respect to the joint PMF or PDF $p_{X[n_1], X[n_2]}$ for particular values of n_1 and n_2. Since the required PMF or PDF should be clear from the context, we henceforth do not subscript the expectation operator as we have done so previously. Note that the usual symmetry property of the covariance holds, which results in $c_X[n_2, n_1] = c_X[n_1, n_2]$. Also, it follows from the definition of the covariance sequence that $c_X[n, n] = \sigma_X^2[n]$. The actual evaluation of the moments proceeds exactly the same as for random variables.

If the random process is a continuous-time one, then the corresponding definitions are

$$
\begin{aligned}
\mu_X(t) &= E[X(t)] \\
\sigma_X^2(t) &= \text{var}(X(t)) \\
c_X(t_1, t_2) &= E[(X(t_1) - \mu_X(t_1))(X(t_2) - \mu_X(t_2))].
\end{aligned}
$$

These are called the *mean function*, *variance function*, and *covariance function*, respectively. We next examine the moments for the examples of the previous section. Noting that the variance is just the covariance sequence evaluated at $n_1 = n_2 = n$, we need only determine the mean and covariance sequences.

Example 16.9 – White Gaussian noise

Since $X[n] \sim \mathcal{N}(0, \sigma^2)$ for all n, we have that

$$\mu_X[n] = 0 \qquad -\infty < n < \infty$$
$$\sigma_X^2[n] = \sigma^2 \qquad -\infty < n < \infty.$$

The covariance sequence for $n_1 \neq n_2$ must be zero since the random variables are all independent. Recalling that the covariance between $X[n]$ and itself is just the variance, we have that

$$c_X[n_1, n_2] = \begin{cases} 0 & n_1 \neq n_2 \\ \sigma^2 & n_1 = n_2. \end{cases}$$

This can be written in more succinct form by using the discrete delta function as

$$c_X[n_1, n_2] = \sigma^2 \delta[n_2 - n_1].$$

In summary, for a WGN random process we have that $\mu_X[n] = 0$ for all n and $c_X[n_1, n_2] = \sigma^2 \delta[n_2 - n_1]$.

$\diamondsuit$

Example 16.10 – Moving average random process

The mean sequence is

$$\mu_X[n] = E[X[n]] = E[\tfrac{1}{2}(U[n] + U[n-1])] = 0 \qquad -\infty < n < \infty$$

since $U[n]$ is white Gaussian noise, which has a zero mean for all n. To find the covariance sequence using $X[n] = (U[n] + U[n-1])/2$, we have

$$
\begin{aligned}
c_X[n_1, n_2] &= E[(X[n_1] - \mu_X[n_1])(X[n_2] - \mu_X[n_2])] \\
&= E[X[n_1]X[n_2]] \\
&= \frac{1}{4}E[(U[n_1] + U[n_1 - 1])(U[n_2] + U[n_2 - 1])] \\
&= \frac{1}{4}\left(E[U[n_1]U[n_2]] + E[U[n_1]U[n_2 - 1]] \right. \\
&\qquad \left. + E[U[n_1 - 1]U[n_2]] + E[U[n_1 - 1]U[n_2 - 1]]\right).
\end{aligned}
$$

But $E[U[k]U[l]] = \sigma_U^2 \delta[l - k]$ since $U[n]$ is WGN, and as a result

$$
\begin{aligned}
c_X[n_1, n_2] &= \frac{1}{4}\left(\sigma_U^2 \delta[n_2 - n_1] + \sigma_U^2 \delta[n_2 - 1 - n_1] + \sigma_U^2 \delta[n_2 - n_1 + 1] + \sigma_U^2 \delta[n_2 - n_1]\right) \\
&= \frac{\sigma_U^2}{2}\delta[n_2 - n_1] + \frac{\sigma_U^2}{4}\delta[n_2 - n_1 - 1] + \frac{\sigma_U^2}{4}\delta[n_2 - n_1 + 1].
\end{aligned}
$$

This is plotted in Figure 16.13 versus $\Delta n = n_2 - n_1$. It is seen that the covariance sequence is zero unless the two samples are at most one unit apart or $\Delta n = n_2 - n_1 =$

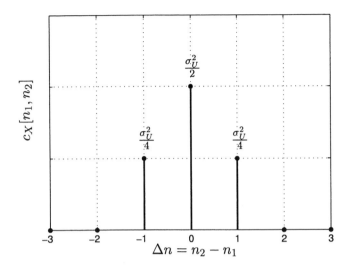

Figure 16.13: Covariance sequence for moving average random process.

± 1. Note that the covariance between any two samples spaced one unit apart is the same. Thus, for example, $X[1]$ and $X[2]$ have the covariance $c_X[1,2] = \sigma_U^2/4$, as do $X[9]$ and $X[10]$ since $c_X[9,10] = \sigma_U^2/4$, and as do $X[-3]$ and $X[-2]$ since $c_X[-3,-2] = \sigma_U^2/4$ (see Figure 16.13). Any samples that are spaced *more than* one unit apart are *uncorrelated*. This is because for $|n_2 - n_1| > 1$, $X[n_1]$ and $X[n_2]$ are independent, being composed of two sets of *different* WGN samples (recall that functions of independent random variables are independent). In summary, we have that

$$\mu_X[n] = 0$$

$$c_X[n_1, n_2] = \begin{cases} \frac{\sigma_U^2}{2} & n_1 = n_2 \\ \frac{\sigma_U^2}{4} & |n_2 - n_1| = 1 \\ 0 & |n_2 - n_1| > 1. \end{cases}$$

and the variance is $c_X[n,n] = \sigma_U^2/2$ for all n. Also, note from Figure 16.13 that the covariance sequence is symmetric about $\Delta n = 0$.

Example 16.11 – Randomly phased sinusoid

Recalling that the phase is uniformly distributed on $(0, 2\pi)$ we have that the mean sequence is

$$\begin{aligned}
\mu_X[n] &= E[X[n]] = E[\cos(2\pi(0.1)n + \Theta)] \\
&= \int_0^{2\pi} \cos(2\pi(0.1)n + \theta)\frac{1}{2\pi}d\theta \qquad \text{(use (11.10))} \\
&= \left.\frac{1}{2\pi}\sin(2\pi(0.1)n + \theta)\right|_0^{2\pi} = 0
\end{aligned}$$

for all n. Noting that the mean sequence is zero, the covariance sequence becomes

$$\begin{aligned}
c_X[n_1, n_2] &= E[X[n_1]X[n_2]] \\
&= \int_0^{2\pi} [\cos(2\pi(0.1)n_1 + \theta)\cos(2\pi(0.1)n_2 + \theta)]\frac{1}{2\pi}d\theta \\
&= \int_0^{2\pi} \left[\frac{1}{2}\cos[2\pi(0.1)(n_2 - n_1)] + \frac{1}{2}\cos[2\pi(0.1)(n_1 + n_2) + 2\theta]\right]\frac{1}{2\pi}d\theta \\
&= \left.\frac{1}{2}\cos[2\pi(0.1)(n_2 - n_1)] + \frac{1}{8\pi}\sin[2\pi(0.1)(n_1 + n_2) + 2\theta]\right|_0^{2\pi} \\
&= \frac{1}{2}\cos[2\pi(0.1)(n_2 - n_1)].
\end{aligned}$$

Once again the covariance sequence depends only on the spacing between the two

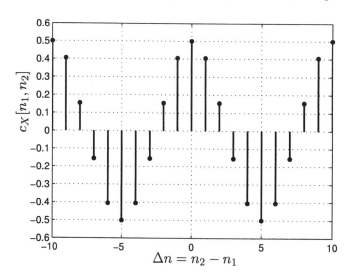

Figure 16.14: Covariance sequence for randomly phased sinusoid.

samples or on $n_2 - n_1$. The covariance sequence is shown in Figure 16.14. The reader should note the symmetry of the covariance sequence about $\Delta n = 0$. Also, the variance follows as $\sigma_X^2[n] = c_X[n, n] = 1/2$ for all n. It is interesting to observe that in this example the fact that the mean sequence is zero makes intuitive sense.

To see this we have plotted 50 realizations of the random process in an overlaid fashion in Figure 16.15. This representation is called a *scatter diagram*. Also is

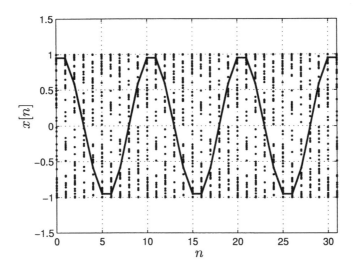

Figure 16.15: Fifty realizations of randomly phased sinusoid plotted in an overlaid format with one realization shown with its points connected by straight lines.

plotted the first realization with the values connected by straight lines for easier viewing. The difference in the realizations is due to the different values of phase realized. It is seen that for a given time instant the values are nearly symmetric about zero, as is predicted by the PDF shown in Figure 16.12 and that the majority of the values are near ± 1, again in agreement with the PDF. The MATLAB code used to generate Figure 16.15 (but omitting the solid curve) is given below.

```
clear all
rand('state',0)
n=[0:31]';
nreal=50;
for i=1:nreal
   x(:,i)=cos(2*pi*0.1*n+2*pi*rand(1,1));
end
plot(n,x(:,1),'.')
grid
hold on
for i=2:nreal
   plot(n,x(:,i),'.')
end
axis([0 31 -1.5 1.5])
```

$\Diamond$

In these three examples the covariance sequence only depends on $|n_2 - n_1|$. This is not always the case, as is illustrated in Problem 16.26. Also, another counterexample is the random process whose realization is shown in Figure 16.7b. This random process has $\text{var}(X[n]) = c_X[n, n]$ which is not a function of $n_2 - n_1 = n - n = 0$ since otherwise its variance would be a constant for all n.

16.8   Real-World Example – Statistical Data Analysis

It was mentioned in the introduction that some meterologists argue that the annual summer rainfall totals are increasing due to global warming. Referring to Figure 16.1 this supposition asserts that if $X[n]$ is the annual summer rainfall total for year n, then $\mu_X[n_2] > \mu_X[n_1]$ for $n_2 > n_1$. One way to attempt to confirm or dispute this supposition is to assume that $\mu_X[n] = an + b$ and then determine if $a > 0$, as would be the case if the mean were increasing. From the data shown in Figure 16.1 we can estimate a. To do so we let the year 1895, which is the beginning of our data set, be indexed as $n = 0$ and note that $an + b$ when plotted versus n is a straight line. We estimate a by fitting a straight line to the data set using a *least squares* procedure [Kay 1993]. The least squares estimate chooses as estimates of a and b the values that minimize the *least squares error*

$$J(a, b) = \sum_{n=0}^{N-1} (x[n] - (an + b))^2 \tag{16.8}$$

where $N = 108$ for our data set. This approach can be shown to be an optimal one under the condition that the random process is actually given by $X[n] = an + b + U[n]$, where $U[n]$ is a WGN random process [Kay 1993]. Note that if we did not suspect that the mean rainfall totals were changing, then we might assume that $\mu_X[n] = b$ and the least squares estimate of b would result from minimizing

$$J(b) = \sum_{n=0}^{N-1} (x[n] - b)^2.$$

If we differentiate $J(b)$ with respect to b, set the derivative equal to zero, and solve for b, we obtain (see Problem 16.32)

$$\hat{b} = \frac{1}{N} \sum_{n=0}^{N-1} x[n]$$

or $\hat{b} = \bar{x}$, where $\bar{x}$ is the sample mean, which for our data set is 9.76. Now, however, we obtain the least squares estimates of a and b by differentiating (16.8) with respect to b and a to yield

$$\frac{\partial J}{\partial b} = -2 \sum_{n=0}^{N-1} (x[n] - an - b) = 0$$

$$\frac{\partial J}{\partial a} = -2 \sum_{n=0}^{N-1} (x[n] - an - b)n = 0.$$

This results in two simultaneous linear equations

$$bN + a \sum_{n=0}^{N-1} n = \sum_{n=0}^{N-1} x[n]$$

$$b \sum_{n=0}^{N-1} n + a \sum_{n=0}^{N-1} n^2 = \sum_{n=0}^{N-1} nx[n].$$

In vector/matrix form this is

$$\begin{bmatrix} N & \sum_{n=0}^{N-1} n \\ \sum_{n=0}^{N-1} n & \sum_{n=0}^{N-1} n^2 \end{bmatrix} \begin{bmatrix} b \\ a \end{bmatrix} = \begin{bmatrix} \sum_{n=0}^{N-1} x[n] \\ \sum_{n=0}^{N-1} nx[n] \end{bmatrix} \tag{16.9}$$

which is easily solved to yield the estimates $\hat{b}$ and $\hat{a}$. For the data of Figure 16.1 the estimates are $\hat{a} = 0.0173$ and $\hat{b} = 8.8336$. The data along with the estimated mean sequence $\hat{\mu}_X[n] = 0.0173n + 8.8336$ are shown in Figure 16.16. Note that the

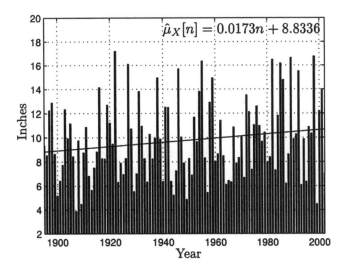

Figure 16.16: Annual summer rainfall in Rhode Island and the estimated mean sequence, $\hat{\mu}_X[n] = 0.0173n + 8.8336$, where $n = 0$ corresponds to the year 1895.

mean indeed appears to be increasing with time. The least squares error sequence, which is defined as $e[n] = x[n] - (\hat{a}n + \hat{b})$, is shown in Figure 16.17. It is sometimes referred to as the *fitting error*. Note that the error can be quite large. In fact, we

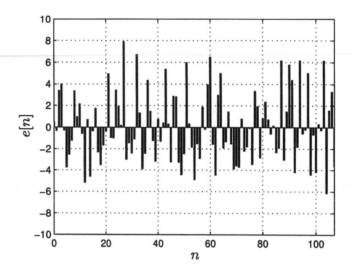

Figure 16.17: Least squares error sequence for annual summer rainfall in Rhode Island fitted with a straight line.

have that $(1/N) \sum_{n=0}^{N-1} e^2[n] = 10.05$.

Now the real question is whether the estimated mean increase in rainfall is significant. The increase is $\hat{a} = 0.0173$ *per year* for a total increase of about 1.85 inches over the course of 108 years. Is it possible that the true mean rainfall has not changed, or that it is really $\mu_X[n] = b$ with the true value of a being zero? In effect, is the value of $\hat{a} = 0.0173$ only due to estimation error? One way to answer this question is to hypothesize that $a = 0$ and then determine the probability density function of $\hat{a}$ as obtained from (16.9). This can be done analytically by assuming $X[n] = b + U[n]$, where $U[n]$ is white Gaussian noise (see Problem 16.33). However, we can gain some quick insight into the problem by resorting to a computer simulation. To do so we assume that the true model for the rainfall data is $X[n] = b + U[n] = 9.76 + U[n]$, where $U[n]$ is white Gaussian noise with variance σ^2. Since we do not know the value of σ^2, we estimate it by using the results shown in Figure 16.17. The least squares error sequence $e[n]$, which is the original data with its estimated mean sequence subtracted, should then be an estimate of $U[n]$. Therefore, we use $\hat{\sigma}^2 = (1/N) \sum_{n=0}^{N-1} e^2[n] = 10.05$ in our simulation. In summary, we generate 20 realizations of the random process $X[n] = 9.76 + U[n]$, where $U[n]$ is WGN with $\sigma^2 = 10.05$. Then, we use (16.9) to estimate a and b and finally we plot our mean sequence estimate, which is $\hat{\mu}_X[n] = \hat{a}n + \hat{b}$ for each realization. Using the MATLAB code shown at the end of this section, the results are shown in Figure 16.18. It is seen that even though the true value of a is zero, the estimated value will take on nonzero values with a high probability. Since some of the lines are decreasing, some

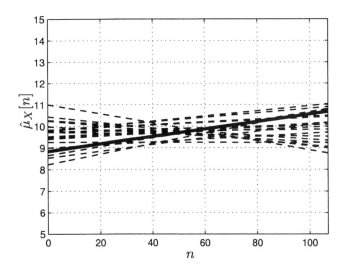

Figure 16.18: Twenty realizations of the estimated mean sequence $\hat{\mu}_X[n] = \hat{a}n + \hat{b}$ based on the random process $X[n] = 9.76 + U[n]$ with $U[n]$ being WGN with $\sigma^2 = 10.05$. The realizations are shown as dashed lines. The estimated mean sequence from Figure 16.16 is shown as the solid line.

of the estimated values of a are even negative. Hence, we would be hard pressed to say that the mean rainfall totals are indeed increasing. Such is the quandry that scientists must deal with on an everyday basis. The only way out of this dilemma is to accumulate more data so that hopefully our estimate of a will be more accurate (see also Problem 16.34).

```
clear all
randn('state',0)
years=[1895:2002]';
N=length(years);
n=[0:N-1]';
A=[N sum(n);sum(n) sum(n.^2)]; % precompute matrix (see (16.9))
B=inv(A); % invert matrix
for i=1:20
  xn=9.76+sqrt(10.05)*randn(N,1); % generate realizations
  baest=B*[sum(xn);sum(n.*xn)]; % estimate a and b using (16.9)
  aest=baest(2);best=baest(1);
  meanest(:,i)=aest*n+best; % determine mean sequence estimate
end
figure % plot mean sequence estimates and overlay
plot(n,meanest(:,1))
grid
xlabel('n')
```

```
ylabel('Estimated mean')
axis([0 107 5 15])
hold on
for i=2:20
  plot(n,meanest(:,i))
end
```

References

Billingsley, P., *Probability and Measure*, John Wiley & Sons, New York, 1986.

DowJones.com, "DowJones Averages," http://averages.dowjones.com/jsp/uiHistoricalIndexRep.jsp, 2004.

Kay, S., *Fundamentals of Statistical Signal Processing: Estimation Theory*, Prentice-Hall, Englewood Cliffs, NJ, 1993.

Problems

16.1 (⌣) (w) Describe a random process that you are likely to encounter in the following situations:

a. listening to the daily weather forecast

b. paying the monthly telephone bill

c. leaving for work in the morning

Why is each process a random one?

16.2 (w) A single die is tossed repeatedly. What are S and S_X? Also, can you determine the joint PMF for any N sample times?

16.3 (t) An infinite sequence of 0's and 1's, denoted as $b_1, b_2, \ldots$, can be used to represent any number x in the interval $[0, 1]$ using the binary representation formula

$$x = \sum_{i=1}^{\infty} b_i 2^{-i}.$$

For example, we can represent $3/4$ as $0.b_1 b_2 \ldots = 0.11000\ldots$ and $1/16$ as $0.b_1 b_2 \ldots = 0.0001000\ldots$. Find the representations for $7/8$ and $5/8$. Is the total number of infinite sequences of 0's and 1's countable?

16.4 (⌣) (w) For a Bernoulli random process determine the probability that we will observe an alternating sequence of 1's and 0's for the first 100 samples with the first sample being a 1. What is the probability that we will observe an alternating sequence of 1's and 0's for all n?

16.5 (w) Classify the following random processes as either DTDV, DTCV, CTDV, or CTCV:

a. temperature in Rhode Island

b. outcomes for continued spins of a roulette wheel

c. daily weight of person

d. number of cars stopped at an intersection

16.6 (c) Simulate a realization of the random walk process described in Example 16.2 on a computer. What happens as n becomes large?

16.7 ($\because$) (c,f) A *biased* random walk process is defined as $X[n] = \sum_{i=0}^{n} U[i]$, where $U[i]$ is a Bernoulli random process with

$$p_U[k] = \begin{cases} \frac{1}{4} & k = -1 \\ \frac{3}{4} & k = 1. \end{cases}$$

What is $E[X[n]]$ and $\text{var}(X[n])$ as a function of n? Next, simulate on a computer a realization of this random process. What happens as $n \to \infty$ and why?

16.8 (w) A random process $X[n]$ is stationary. If it is known that $E[X[10]] = 10$ and $\text{var}(X[10]) = 1$, then determine $E[X[100]]$ and $\text{var}(X[100])$.

16.9 ($\because$) (f) The IID random process $X[n]$ has the marginal PDF $p_X(x) = \exp(-x)u(x)$. What is the probability that $X[0], X[1], X[2]$ will all be greater than 1?

16.10 (w) If an IID random process $X[n]$ is transformed to the random process $Y[n] = X^2[n]$, is the transformed random process also IID?

16.11 (w) A Bernoulli random process $X[n]$ that takes on values 0 or 1, each with probability of $p = 1/2$, is transformed using $Y[n] = (-1)^n X[n]$. Is the random process $Y[n]$ IID?

16.12 (w,f) A nonstationary random process is defined as $X[n] = a^{|n|}U[n]$, where $0 < a < 1$ and $U[n]$ is WGN with variance σ_U^2. Find the mean and covariance sequences of $X[n]$. Can you transform the $X[n]$ random process to make it stationary?

16.13 ($\because$) (w) Consider the random process $X[n] = \sum_{i=0}^{n} U[i]$, which is defined for $n \geq 0$. The $U[n]$ random process consists of independent Gaussian random variables with marginal PDF $U[n] \sim \mathcal{N}(0, (1/2)^n)$. Are the increments independent? Are the increments stationary?

16.14 (c) Plot 50 realizations of a WGN random process $X[n]$ with $\sigma^2 = 1$ for $n = 0, 1, \ldots, 49$ using a scatter diagram (see Figure 16.15 for an example). Use the MATLAB commands `plot(x,y,'.')` and `hold on` to plot each realization as dots and to overlay the realizations on the same graph, respectively. For a fixed n can you explain the observed distribution of the dots?

16.15 (f) Prove that

$$\frac{1}{(2\pi)^{N/2} \det^{1/2}(\mathbf{C})} \exp\left(-\tfrac{1}{2}\mathbf{x}^T \mathbf{C}^{-1} \mathbf{x}\right)$$

where $\mathbf{x} = [x_1 \, x_2 \ldots x_N]^T$ and $\mathbf{C} = \sigma^2 \mathbf{I}$ for $\mathbf{I}$ an $N \times N$ identity matrix, reduces to (16.4).

16.16 ($\smile$) (f) A "white" uniform random process is defined to be an IID random process with $X[n] \sim \mathcal{U}(-\sqrt{3}, \sqrt{3})$ for all n. Determine the mean and covariance sequences for this random process and compare them to those of the WGN random process. Explain your results.

16.17 (w) A moving average random process can be defined more generally as one for which N samples of WGN are averaged, instead of only $N = 2$ samples as in Example 16.7. It is given by $X[n] = (1/N) \sum_{i=0}^{N-1} U[n - i]$ for all n, where $U[n]$ is a WGN random process with variance σ_U^2. Determine the correlation coefficient for $X[0]$ and $X[1]$. What happens as N increases?

16.18 ($\smile$) (f) For the moving average random process defined in Example 16.7 determine $P[X[n] > 3]$ and compare it to $P[U[n] > 3]$. Explain the difference in terms of "smoothing". Assume that $\sigma_U^2 = 1$.

16.19 (c) For the randomly phased sinusoid defined in Example 16.8 determine the mean sequence using a computer simulation.

16.20 (t) For the randomly phased sinusoid of Example 16.8 assume that the realization $x[n] = \cos(2\pi(0.1)n + 0)$ is generated. Prove that if we observe only the samples $x[0] = 1$ and $x[1] = \cos(2\pi(0.1)) = 0.8090$, then all the future samples can be found by using the recursive formula $x[n] = 2\cos(2\pi(0.1))x[n - 1] - x[n - 2]$ for $n \geq 2$. Could you also find the past samples or $x[n]$ for $n \leq -1$? See also Problem 18.25 for prediction of a sinusoidal random process.

16.21 (c) Verify the PDF of the randomly phased sinusoid given in Figure 16.12 by using a computer simulation.

16.22 ($\smile$) (f,c) A continuous-time random process known as the random *amplitude* sinusoid is defined as $X(t) = A\cos(2\pi t)$ for $-\infty < t < \infty$ and $A \sim \mathcal{N}(0, 1)$. Find the mean and covariance functions. Then, plot some realizations of $X(t)$ in an overlaid fashion.

16.23 (f) A random process is the sum of WGN and a deterministic sinusoid and is given as $X[n] = U[n] + \sin(2\pi f_0 n)$ for all n, where $U[n]$ is WGN with variance σ_U^2. Determine the mean and covariance sequences.

16.24 ($\ddot{\smile}$) (w) A random process is IID with samples $X[n] \sim \mathcal{N}(\mu, 1)$. It is desired to remove the mean of the random process by forming the new random process $Y[n] = X[n] - X[n-1]$. First determine the mean sequence of $Y[n]$. Next find $\text{cov}(Y[0], Y[1])$. Is $Y[n]$ an IID random process with a zero mean sequence?

16.25 (f) If a random process is defined as $X[n] = h[0]U[n] + h[1]U[n-1]$, where $h[0]$ and $h[1]$ are constants and $U[n]$ is WGN with variance σ_U^2, find the covariance for $X[0]$ and $X[1]$. Repeat for $X[9]$ and $X[10]$. How do they compare?

16.26 ($\ddot{\smile}$) (f) If a sum random process is defined as $X[n] = \sum_{i=0}^{n} U[i]$ for $n \geq 0$, where $E[U[i]] = 0$ and $\text{var}(U[i]) = \sigma_U^2$ for $i \geq 0$ and the $U[i]$ are IID, find the mean and covariance sequences of $X[n]$.

16.27 ($\ddot{\smile}$) (c) For the MA random process defined in Example 16.7 find $c_X[1,1]$, $c_X[1,2]$ and $c_X[1,3]$ if $\sigma_U^2 = 1$. Next simulate on a computer $M = 10,000$ realizations of the random process $X[n]$ for $n = 0, 1, \ldots, 10$. Estimate the previous covariance sequence samples using $\hat{c}_X[n_1, n_2] = (1/M) \sum_{i=1}^{M} x_i[n_1] x_i[n_2]$, where $x_i[n]$ is the ith realization of $X[n]$. Note that since $X[n]$ is zero mean, $c_X[n_1, n_2] = E[X[n_1]X[n_2]]$.

16.28 (w) For the randomly phased sinusoid described in Example 16.11 determine the minimum mean square estimate of $X[10]$ based on observing $x[0]$. How accurate do you think this prediction will be?

16.29 (f) For a random process $X[n]$ the mean sequence $\mu_X[n]$ and covariance sequence $c_X[n_1, n_2]$ are known. It is desired to predict k samples into the future. If $x[n_0]$ is observed, find the minimum mean square estimate of $X[n_0 + k]$. Next assume that $\mu_X[n] = \cos(2\pi f_0 n)$ and $c_X[n_1, n_2] = 0.9^{|n_2 - n_1|}$ and evaluate the estimate. Finally, what happens to your prediction as $k \to \infty$ and why?

16.30 (f) A random process is defined as $X[n] = As[n]$ for all n, where $A \sim \mathcal{N}(0, 1)$ and $s[n]$ is a deterministic signal. Find the mean and covariance sequences.

16.31 ($\ddot{\smile}$) (f) A random process is defined as $X[n] = AU[n]$ for all n, where $A \sim \mathcal{N}(0, \sigma_A^2)$ and $U[n]$ is WGN with variance σ_U^2, and A is independent of $U[n]$ for all n. Find the mean and covariance sequences. What type of random process is $X[n]$?

16.32 (f) Verify that by differentiating $\sum_{n=0}^{N-1} (x[n] - b)^2$ with respect to b, setting the derivative equal to zero, and solving for b, we obtain the sample mean.

16.33 (t) In this problem we show how to obtain the variance of $\hat{a}$ as obtained by solving (16.9). The variance of $\hat{a}$ is derived under the assumption that $X[n] = b + U[n]$, where $U[n]$ is WGN with variance σ^2. This says that we assume the true value of a is zero. The steps are as follows:

a. Let

$$
\mathbf{H} = \begin{bmatrix} 1 & 0 \\ 1 & 1 \\ 1 & 2 \\ \vdots & \vdots \\ 1 & N-1 \end{bmatrix}
\qquad
\mathbf{X} = \begin{bmatrix} X[0] \\ X[1] \\ X[2] \\ \vdots \\ X[N-1] \end{bmatrix}
$$

where $\mathbf{H}$ is an $N \times 2$ matrix and $\mathbf{X}$ is an $N \times 1$ random vector. Now show that that the equations of (16.9) can be written as

$$
\mathbf{H}^T \mathbf{H} \begin{bmatrix} b \\ a \end{bmatrix} = \mathbf{H}^T \mathbf{X}.
$$

b. The solution for b and a can now be written symbolically as

$$
\begin{bmatrix} \hat{b} \\ \hat{a} \end{bmatrix} = \underbrace{(\mathbf{H}^T \mathbf{H})^{-1} \mathbf{H}^T}_{\mathbf{G}} \mathbf{X}
$$

Since $\mathbf{X}$ is a Gaussian random vector, show that $[\hat{b}\,\hat{a}]^T$ is also a Gaussian random vector with mean $[b\,0]^T$ and covariance matrix $\sigma^2 (\mathbf{H}^T \mathbf{H})^{-1}$.

c. As a result we can assert that the marginal PDF of $\hat{a}$ is Gaussian with mean zero and variance equal to the $(2,2)$ element of $\sigma^2 (\mathbf{H}^T \mathbf{H})^{-1}$. Show then that $\hat{a} \sim \mathcal{N}(0, \text{var}(\hat{a}))$, where

$$
\text{var}(\hat{a}) = \frac{\sigma^2}{\sum_{n=0}^{N-1} n^2 - \frac{1}{N}\left(\sum_{n=0}^{N-1} n\right)^2}.
$$

Next assume that $\sigma^2 = 10.05$, $N = 108$ and find the probability that $\hat{a} > 0.0173$. Can we assert that the estimated mean sequence shown in Figure 16.16 is not just due to estimation error?

16.34 (☺) (f) Using the results of Problem 16.33 determine the required value of N so that the probability that $\hat{a} > 0.0173$ is less than 10^{-6}.

Chapter 17

Wide Sense Stationary Random Processes

17.1 Introduction

Having introduced the concept of a random process in the previous chapter, we now wish to explore an important subclass of stationary random processes. This is motivated by the very restrictive nature of the stationarity condition, which although mathematically expedient, is almost never satisfied in practice. A somewhat weaker type of stationarity is based on requiring the mean to be a constant in time and the covariance sequence to depend only on the separation in time between the two samples. We have already encountered these types of random processes in Examples 16.9–16.11. Such a random process is said to be stationary in the wide sense or *wide sense stationary* (WSS). It is also termed a *weakly stationary* random process to distinguish it from a stationary process, which is said to be *strictly stationary*. We will use the former terminology to refer to such a process as a WSS random process. In addition, as we will see in Chapter 19, if the random process is Gaussian, then wide sense stationarity implies stationarity. For this reason alone, it makes sense to explore WSS random processes since the use of Gaussian random processes for modeling is ubiquitous.

Once we have discussed the concept of a WSS random process, we will be able to define an extremely important measure of the WSS random process—the *power spectral density* (PSD). This function extends the idea of analyzing the behavior of a deterministic signal by decomposing it into a sum of sinusoids of different frequencies to that of a random process. The difference now is that the amplitudes and phases of the sinusoids will be random variables and so it will be convenient to quantify the *average power* of the various sinusoids. This description of a random phenomenon is important in nearly every scientific field that is concerned with the analysis of time series data such as systems control [Box and Jenkins 1970], signal processing [Schwartz and Shaw 1975], economics [Harvey 1989], geophysics [Robinson 1967], vibration testing [McConnell 1995], financial analysis [Taylor 1986], and others. As an example, in Figure 17.1 the Wolfer sunspot data [Tong 1990] is shown, with the data points connected by straight lines for easier viewing. It measures the average number of sunspots visually observed through a telescope each year. The importance of the sunspot number is that as it increases, an increase in solar flares occurs. This has the effect of disrupting all radio communications as the solar flare particles reach the earth. Clearly from the data we see a periodic type property. The estimated PSD of this data set is shown in Figure 17.2. We see that the distribution of power versus frequency is highest at a frequency of about 0.09 cycles per year. This means that the random process exhibits a large periodic component with a period of about $1/0.09 \approx 11$ years per cycle, as is also evident from Figure 17.1. This is a powerful prediction tool and therefore is of great interest. How the PSD is actually estimated will be discussed in this chapter, but before doing so, we will need to lay some groundwork.

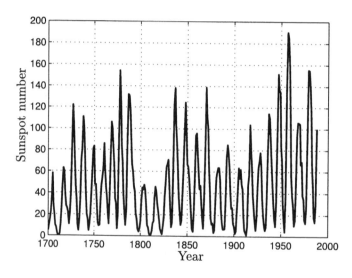

Figure 17.1: Annual number of sunspots – Wolfer sunspot data.

17.2   Summary

A less restrictive form of stationarity, termed wide sense stationarity, is defined by (17.4) and (17.5). The conditions require the mean to be the same for all n and the covariance sequence to depend only on the time difference between the samples. A random process that is stationary is also wide sense stationary as shown in Section 17.3. The autocorrelation sequence is defined by (17.9) with n being arbitrary. It is the covariance between two samples separated by k units for a zero mean WSS random process. Some of its properties are summarized by Properties 17.1–17.4. Under certain conditions the mean of a WSS random process can be found by using

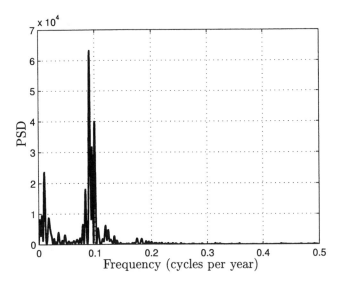

Figure 17.2: Estimated power spectral density for Wolfer sunspot data of Figure 17.1. The sample mean has been computed and removed from the data prior to estimation of the PSD.

the temporal average of (17.25). Such a process is said to be ergodic in the mean. For this to be true the variance of the temporal average given by (17.28) must converge to zero as the number of samples averaged becomes large. The power spectral density (PSD) of a WSS random process is defined by (17.30) and can be evaluated more simply using (17.34). The latter relationship says that the PSD is the Fourier transform of the autocorrelation sequence. It measures the amount of average power per unit frequency or the distribution of average power with frequency. Some of its properties are summarized in Properties 17.7–17.12. From a finite segment of a realization of the random process the autocorrelation sequence can be estimated using (17.43) and the PSD can be estimated by using the averaged periodogram estimate of (17.44) and (17.45). The analogous definitions for a continuous-time WSS random process are given in Section 17.8. Also, an important example is described that relates sampled continuous-time white Gaussian noise to discrete-time white Gaussian noise. Finally, an application of the use of PSDs to random vibration testing is given in Section 17.9.

17.3 Definition of WSS Random Process

Consider a discrete-time random process $X[n]$, which is defined for $-\infty < n < \infty$ with n an integer. Previously, we defined the mean and covariance sequences of

$X[n]$ to be

$$\mu_X[n] \;=\; E[X[n]] \quad -\infty < n < \infty \tag{17.1}$$

$$c_X[n_1, n_2] \;=\; E[(X[n_1] - \mu_X[n_1])(X[n_2] - \mu_X[n_2])] \quad \begin{matrix} -\infty < n_1 < \infty \\ -\infty < n_2 < \infty \end{matrix} \tag{17.2}$$

where n_1, n_2 are integers. Having knowledge of these sequences allows us to assess important characteristics of the random process such as the mean level and the correlation between samples. In fact, based on only this information we are able to predict $X[n_2]$ based on observing $X[n_1] = x[n_1]$ as

$$\hat{X}[n_2] = \mu_X[n_2] + \frac{c_X[n_1, n_2]}{c_X[n_1, n_1]}(x[n_1] - \mu_X[n_1]) \tag{17.3}$$

which is just the usual linear prediction formula of (7.41) with x replaced by $x[n_1]$ and Y replaced by $X[n_2]$, and which makes use of the mean and covariance sequences defined in (17.1) and (17.2), respectively. However, since in general the mean and covariance change with time, i.e., they are nonstationary, it would be exceedingly difficult to estimate them in practice. To extend the practical utility we would like the mean not to depend on time and the covariance only to depend on the separation between samples or on $|n_2 - n_1|$. This will allow us to estimate these quantities as described later. Thus, we are led to a weaker form of stationarity known as *wide sense stationarity*. A random process is defined to be WSS if

$$\mu_X[n] \;=\; \mu \quad \text{(a constant)} \quad -\infty < n < \infty \tag{17.4}$$

$$c_X[n_1, n_2] \;=\; g(|n_2 - n_1|) \quad -\infty < n_1 < \infty, -\infty < n_2 < \infty \tag{17.5}$$

for some function g. Note that since

$$c_X[n_1, n_2] = E[X[n_1]X[n_2]] - E[X[n_1]]E[X[n_2]]$$

these conditions are equivalent to requiring that $X[n]$ satisfy

$$E[X[n]] \;=\; \mu \quad -\infty < n < \infty$$

$$E[X[n_1]X[n_2]] \;=\; h(|n_2 - n_1|) \quad -\infty < n_1 < \infty, -\infty < n_2 < \infty$$

for some function h. *The mean should not depend on time and the average value of the product of two samples should depend only upon the time interval between the samples.* Some examples of WSS random processes have already been given in Examples 16.9–16.11. For the MA process of Example 16.10 we showed that

$$\mu_X[n] \;=\; 0 \quad -\infty < n < \infty$$

$$c_X[n_1, n_2] \;=\; \begin{cases} \frac{1}{2}\sigma_U^2 & |n_2 - n_1| = 0 \\ \frac{1}{4}\sigma_U^2 & |n_2 - n_1| = 1 \\ 0 & |n_2 - n_1| > 1. \end{cases}$$

It is seen that every random variable $X[n]$ for $-\infty < n < \infty$ has a mean of zero and the covariance for two samples depends only on the time interval between the samples, which is $|n_2 - n_1|$. Also, this implies that the variance does not depend on time since $\text{var}(X[n]) = c_X[n,n] = \sigma_U^2/2$ for $-\infty < n < \infty$. In contrast to this behavior consider the random processes for which typical realizations are shown in Figure 16.7. In Figure 16.7a the mean changes with time (with the variance being constant) and in Figure 16.7b the variance changes with time (with the mean being constant). Clearly, these random processes are not WSS.

A WSS random process is a special case of a stationary random process. To see this recall that if $X[n]$ is stationary, then from (16.3) with $N = 1$ and $n_1 = n$, we have

$$p_{X[n+n_0]} = p_{X[n]} \qquad \text{for all } n \text{ and for all } n_0.$$

As a consequence, if we let $n = 0$, then

$$p_{X[n_0]} = p_{X[0]} \qquad \text{for all } n_0$$

and since the PDF does not depend on the particular time n_0, the mean must not depend on time. Thus,

$$\mu_X[n] = \mu \qquad -\infty < n < \infty. \tag{17.6}$$

Next, using (16.3) with $N = 2$, we have

$$p_{X[n_1+n_0],X[n_2+n_0]} = p_{X[n_1],X[n_2]} \qquad \text{for all } n_1, n_2 \text{ and } n_0. \tag{17.7}$$

Now if $n_0 = -n_1$ we have from (17.7)

$$p_{X[0],X[n_2-n_1]} = p_{X[n_1],X[n_2]}$$

and if $n_0 = -n_2$, we have

$$p_{X[n_1-n_2],X[0]} = p_{X[n_1],X[n_2]}.$$

This results in

$$
\begin{aligned}
p_{X[n_1],X[n_2]} &= p_{X[0],X[n_2-n_1]} \\
p_{X[n_1],X[n_2]} &= p_{X[n_1-n_2],X[0]}
\end{aligned}
$$

which leads to

$$
\begin{aligned}
E[X[n_1]X[n_2]] &= E[X[0]X[n_2 - n_1]] \\
E[X[n_1]X[n_2]] &= E[X[n_1 - n_2]X[0]] = E[X[0]X[n_1 - n_2]].
\end{aligned}
$$

Finally, these two conditions combine to give

$$E[X[n_1]X[n_2]] = E[X[0]X[|n_2 - n_1|]] \tag{17.8}$$

which along with the mean being constant with time yields the second condition for wide sense stationarity of (17.5) that

$$c_X[n_1, n_2] = E[X[n_1]X[n_2]] - E[X[n_1]]E[X[n_2]] = E[X[0]X[|n_2 - n_1|]] - \mu^2.$$

This proves the assertion that a stationary random process is WSS but the converse is not generally true (see Problem 17.5).

17.4 Autocorrelation Sequence

If $X[n]$ is WSS, then as we have seen $E[X[n_1]X[n_2]]$ depends only on the separation in time between the samples. We can therefore define a new joint moment by letting $n_1 = n$ and $n_2 = n + k$ to yield

$$r_X[k] = E[X[n]X[n + k]] \tag{17.9}$$

which is called the *autocorrelation sequence* (ACS). It depends only on the time difference between samples which is $|n_2 - n_1| = |(n + k) - n| = |k|$ so that *the value of n used in the definition is arbitrary*. It is termed the *auto*correlation sequence (ACS) since it measures the correlation between two samples of the *same* random process. Later we will have occasion to define correlation between two different random processes (see Section 19.3). Note that the time interval between samples is also called the *lag*. An example of the computation of the ACS is given next.

Example 17.1 – A Differencer

Define a random process as $X[n] = U[n] - U[n - 1]$, where $U[n]$ is an IID random process with mean μ and variance σ_U^2. A realization of this random process for which $U[n]$ is a Gaussian random variable for all n is shown in Figure 17.3. Although $U[n]$ was chosen here to be a sequence of Gaussian random variables for the sake of displaying the realization in Figure 17.3, the ACS to be found will be the same regardless of the PDF of $U[n]$. This is because it relies on *only the first two moments of $U[n]$ and not its PDF*. The ACS is found as

$$
\begin{aligned}
r_X[k] &= E[X[n]X[n + k]] \\
&= E[(U[n] - U[n - 1])(U[n + k] - U[n + k - 1])] \\
&= E[U[n]U[n + k]] - E[U[n]U[n + k - 1]] \\
&\quad - E[U[n - 1]U[n + k]] + E[U[n - 1]U[n + k - 1]].
\end{aligned}
$$

But for $n_1 \neq n_2$

$$
\begin{aligned}
E[U[n_1]U[n_2]] &= E[U[n_1]]E[U[n_2]] \quad \text{(independence)} \\
&= \mu^2
\end{aligned}
$$

and for $n_1 = n_2 = n$

$$E[U[n_1]U[n_2]] = E[U^2[n]] = E[U^2[0]] = \sigma_U^2 + \mu^2 \quad \text{(identically distributed)}.$$

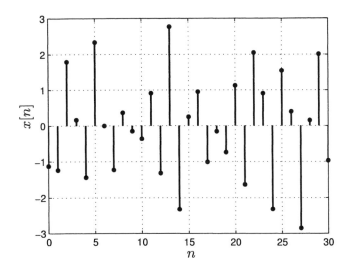

Figure 17.3: Typical realization of a differenced IID Gaussian random process with $U[n] \sim \mathcal{N}(1,1)$.

Combining these results we have that

$$E[U[n_1]U[n_2]] = \mu^2 + \sigma_U^2 \delta[n_2 - n_1]$$

and therefore the ACS becomes

$$r_X[k] = 2\sigma_U^2 \delta[k] - \sigma_U^2 \delta[k-1] - \sigma_U^2 \delta[k+1]. \qquad (17.10)$$

This is shown in Figure 17.4. Several observations can be made. The only nonzero correlation is between adjacent samples and this correlation is negative. This accounts for the observation that the realization shown in Figure 17.3 exhibits many adjacent samples that are opposite in sign. Some other observations are that $r_X[0] > 0$, $|r_X[k]| \leq r_X[0]$ for all k, and finally $r_X[-k] = r_X[k]$. In words, the ACS has a maximum at $k = 0$, which is positive, and is a symmetric sequence about $k = 0$ (also called an *even sequence*). These properties hold in general as we now prove.

$\diamond$

Property 17.1 – ACS is positive for the zero lag or $r_X[0] > 0$.
  Proof:

$$r_X[k] = E[X[n]X[n+k]] \qquad \text{(definition)}$$

so that with $k = 0$ we have $r_X[0] = E[X^2[n]] > 0$.

$\square$

Note that $r_X[0]$ is the *average power* of the random process at all sample times n. One can view $X[n]$ as the voltage across a 1 ohm resistor and hence $x^2[n]/1$

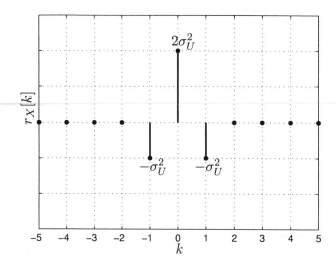

Figure 17.4: Autocorrelation sequence for differenced random process.

is the power for any particular realization of $X[n]$ at time n. The *average* power $E[X^2[n]] = r_X[0]$ does not change with time.

Property 17.2 – ACS is an even sequence or $r_X[-k] = r_X[k]$.
Proof:

$$\begin{aligned} r_X[k] &= E[X[n]X[n+k]] \qquad \text{(definition)} \\ r_X[-k] &= E[X[n]X[n-k]] \end{aligned}$$

and letting $n = m + k$ since the choice of n in the definition of the ACS is arbitrary, we have

$$\begin{aligned} r_X[-k] &= E[X[m+k]X[m]] \\ &= E[X[m]X[m+k]] \\ &= E[X[n]X[n+k]] \qquad \text{(ACS not dependent on } n) \\ &= r_X[k]. \end{aligned}$$

□

Property 17.3 – Maximum absolute value of ACS is at $k = 0$ or $|r_X[k]| \le r_X[0]$.

Note that it is possible for some values of $r_X[k]$ for $k \ne 0$ to also equal $r_X[0]$. As an example, for the randomly phased sinusoid of Example 16.11 we had $c_X[n_1, n_2] = \frac{1}{2}\cos[2\pi(0.1)(n_2 - n_1)]$ with a mean of zero. Thus, $r_X[k] = \frac{1}{2}\cos[2\pi(0.1)k]$ and therefore $r_X[10] = r_X[0]$. Hence, the property says that no value of the ACS can *exceed* $r_X[0]$, although there may be multiple values of the ACS that are equal to $r_X[0]$.

Proof: The proof is based on the Cauchy-Schwarz inequality, which from Appendix 7A is

$$|E_{V,W}[VW]| \le \sqrt{E_V[V^2]}\sqrt{E_W[W^2]}$$

with equality holding if and only if $W = cV$ for c a constant. Letting $V = X[n]$ and $W = X[n+k]$, we have

$$|E[X[n]X[n+k]]| \le \sqrt{E[X^2[n]]}\sqrt{E[X^2[n+k]]}$$

from which it follows that

$$|r_X[k]| \le \sqrt{r_X[0]}\sqrt{r_X[0]} = |r_X[0]| = r_X[0] \qquad (\text{since } r_X[0] > 0).$$

Note that equality holds if and only if $X[n+k] = cX[n]$ for all n. This implies perfect predictability of a sample based on the realization of another sample spaced k units ahead or behind in time (see Problem 17.10 for an example involving periodic random processes).

$\square$

Property 17.4 – ACS measures the predictability of a random process.
The correlation coefficient for two samples of a zero mean WSS random process is

$$\rho_{X[n],X[n+k]} = \frac{r_X[k]}{r_X[0]} \tag{17.11}$$

For a nonzero mean the expression is easily modified (see Problem 17.11).
Proof: Recall that the correlation coefficient for two random variables V and W is defined as

$$\rho_{V,W} = \frac{\text{cov}(V,W)}{\sqrt{\text{var}(V)\text{var}(W)}}.$$

Assuming that V and W are zero mean, this becomes

$$\rho_{V,W} = \frac{E_{V,W}[VW]}{\sqrt{E_V[V^2]E_W[W^2]}}$$

and letting $V = X[n]$ and $W = X[n+k]$, we have

$$\begin{aligned}
\rho_{X[n],X[n+k]} &= \frac{E[X[n]X[n+k]]}{\sqrt{E[X^2[n]]E[X^2[n+k]]}} \\
&= \frac{r_X[k]}{\sqrt{r_X[0]r_X[0]}} \\
&= \frac{r_X[k]}{|r_X[0]|} \\
&= \frac{r_X[k]}{r_X[0]} \qquad (\text{from Property 17.1}).
\end{aligned}$$

$\square$

As an example, for the differencer of Example 17.1 we have from Figure 17.4

$$\rho_{X[n],X[n+k]} = \begin{cases} 1 & k = 0 \\ -\frac{1}{2} & k = \pm 1 \\ 0 & \text{otherwise.} \end{cases}$$

As mentioned previously, the adjacent samples are negatively correlated and the magnitude of the correlation coefficient is now seen to be $1/2$.

We next give some more examples of the computation of the ACS.

Example 17.2 – White noise

White noise is defined as *a WSS random process with zero mean, identical variance* σ^2, *and uncorrelated samples*. It is a more general case of the white noise random process first described in Example 16.9. There we assumed the stronger condition of zero mean *IID* samples (hence they must have the same variance due to the identically distributed assumption and also be uncorrelated due to the independence assumption). In addition, it was assumed there that each sample had a *Gaussian PDF*. Note, however, that the definition given above for white noise does not specify a particular PDF. To find the ACS we note that from the definition of the white noise random process

$$
\begin{aligned}
r_X[k] &= E[X[n]X[n+k]] \\
&= E[X[n]]E[X[n+k]] = 0 \quad k \neq 0 \quad \text{(uncorrelated and} \\
&\qquad\qquad\qquad\qquad\qquad\qquad\qquad \text{zero mean samples)} \\
&= E[X^2[n]] = \sigma^2 \quad k = 0 \quad \text{(equal variance samples).}
\end{aligned}
$$

Therefore, we have that

$$r_X[k] = \sigma^2 \delta[k]. \tag{17.12}$$

Could you predict $X[1]$ from a realization of $X[0]$?

$\Diamond$

As an aside, for WSS random processes, we can find the covariance sequence from the ACS and the mean since

$$
\begin{aligned}
c_X[n_1, n_2] &= E[X[n_1]X[n_2]] - \mu_X[n_1]\mu_X[n_2] \\
&= r_X[n_2 - n_1] - \mu^2. \tag{17.13}
\end{aligned}
$$

Another property of the ACS that is evident from (17.13) concerns the behavior of the ACS as $k \to \infty$. Letting $n_1 = n$ and $n_2 = n + k$, we have that

$$r_X[k] = c_X[n, n+k] + \mu^2. \tag{17.14}$$

If two samples becomes uncorrelated or $c_X[n, n+k] \to 0$ as $k \to \infty$, then we see that $r_X[k] \to \mu^2$ as $k \to \infty$. Thus, as another property of the ACS we have the following.

Property 17.5 – ACS approaches μ^2 as $k \to \infty$
This assumes that the samples become uncorrelated for large lags, which is usually the case.

$\square$

If the mean is zero, then from (17.14)

$$r_X[k] = c_X[n, n+k] \tag{17.15}$$

and the ACS approaches zero as the lag increases. We continue with some more examples.

$\diamond$

Example 17.3 – MA random process
This random process was shown in Example 16.10 to have a zero mean and a covariance sequence

$$c_X[n_1, n_2] = \begin{cases} \frac{\sigma_U^2}{2} & n_1 = n_2 \\ \frac{\sigma_U^2}{4} & |n_2 - n_1| = 1 \\ 0 & \text{otherwise.} \end{cases} \tag{17.16}$$

Since the covariance sequence depends only on $|n_2 - n_1|$, $X[n]$ is WSS from (17.15). Specifically, the ACS follows from (17.15) and (17.16) with $k = n_2 - n_1$ as

$$r_X[k] = \begin{cases} \frac{\sigma_U^2}{2} & k = 0 \\ \frac{\sigma_U^2}{4} & k = \pm 1 \\ 0 & \text{otherwise.} \end{cases}$$

See Figure 16.13 for a plot of the ACS (replace Δn with k.) Could you predict $X[1]$ from a realization of $X[0]$?

$\diamond$

Example 17.4 – Randomly phased sinusoid
This random process was shown in Example 16.11 to have a zero mean and a covariance sequence $c_X[n_1, n_2] = \frac{1}{2} \cos[2\pi(0.1)(n_2 - n_1)]$. Since the covariance sequence depends only on $|n_2 - n_1|$, $X[n]$ is WSS. Hence, from (17.15) we have that

$$r_X[k] = \frac{1}{2} \cos[2\pi(0.1)k].$$

See Figure 16.14 for a plot of the ACS (replace Δn with k.) Could you predict $X[1]$ from a realization of $X[0]$?

In determining predictability of a WSS random process, it is convenient to consider the linear predictor, which depends only on the first two moments. Then, the MMSE linear prediction of $X[n_0 + k]$ given $x[n_0]$ is from (17.3) and (17.13) with $n_1 = n_0$ and $n_2 = n_0 + k$

$$\hat{X}[n_0 + k] = \mu + \frac{r_X[k] - \mu^2}{r_X[0] - \mu^2}(x[n_0] - \mu) \quad \text{for all } k \text{ and } n_0.$$

For a zero mean random process this becomes

$$\begin{aligned} \hat{X}[n_0 + k] &= \frac{r_X[k]}{r_X[0]}x[n_0] \\ &= \rho_{X[n_0],X[n_0+k]}x[n_0] \quad \text{for all } k \text{ and } n_0. \end{aligned}$$

One last example is the autoregressive random process which we will use to illustrate several new concepts for WSS random processes.

Example 17.5 – Autoregressive random process

An autoregressive (AR) random process $X[n]$ is defined to be a WSS random process with a zero mean that evolves according to the recursive difference equation

$$X[n] = aX[n-1] + U[n] \qquad -\infty < n < \infty \tag{17.17}$$

where $|a| < 1$ and $U[n]$ is WGN. The WGN random process $U[n]$ (see Example 16.6), has a zero mean and variance σ_U^2 for all n and its samples are all independent with a Gaussian PDF. The name *autoregressive* is due to the *regression* of $X[n]$ upon $X[n-1]$, which is another sample of the same random process, hence, the prefix *auto*. The evolution of $X[n]$ proceeds, for example, as

$$\begin{aligned} &\vdots \\ X[0] &= aX[-1] + U[0] \\ X[1] &= aX[0] + U[1] \\ X[2] &= aX[1] + U[2] \\ &\vdots \end{aligned}$$

Note that $X[n]$ depends only upon the present and past values of $U[n]$ since for example

$$\begin{aligned} X[2] &= aX[1] + U[2] = a(aX[0] + U[1]) + U[2] = a^2 X[0] + aU[1] + U[2] \\ &= a^2(aX[-1] + U[0]) + aU[1] + U[2] = a^3 X[-1] + a^2 U[0] + aU[1] + U[2] \\ &\vdots \\ &= \sum_{k=0}^{\infty} a^k U[2-k] \tag{17.18} \end{aligned}$$

where the term involving $a^k U[2 - k]$ decays to zero as $k \to \infty$ since $|a| < 1$. We see that $X[2]$ depends only on $\{U[2], U[1], \ldots\}$ and it is therefore uncorrelated with $\{U[3], U[4], \ldots\}$. More generally, it can be shown that (see also Problem 19.6)

$$E[X[n]U[n + k]] = 0 \qquad k \geq 1. \tag{17.19}$$

It is seen from (17.18) that in order for the recursion to be stable and hence $X[n]$ to be WSS it is required that $|a| < 1$. The AR random process can be used to model a wide variety of physical random processes with various ACSs, depending upon the choice of the parameters a and σ_U^2. Some typical realizations of the AR random process for different values of a are shown in Figure 17.5. The WGN random process $U[n]$ has been chosen to have a variance $\sigma_U^2 = 1 - a^2$. We will soon see that this choice of variance results in $r_X[0] = 1$ for both AR processes shown in Figure 17.5. The MATLAB code used to generate the realizations shown is given below.

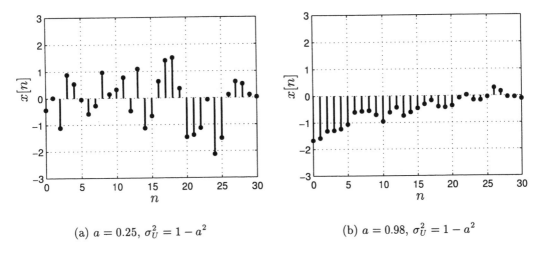

(a) $a = 0.25$, $\sigma_U^2 = 1 - a^2$ (b) $a = 0.98$, $\sigma_U^2 = 1 - a^2$

Figure 17.5: Typical realizations of autoregressive random process with different parameters.

```
clear all
randn('state',0)
a1=0.25;a2=0.98;
varu1=1-a1^2;varu2=1-a2^2;
varx1=varu1/(1-a1^2);varx2=varu2/(1-a2^2); % this is r_X[0]
x1(1,1)=sqrt(varx1)*randn(1,1); % set initial condition X[-1]
                                % see Problems 17.17, 17.18
x2(1,1)=sqrt(varx2)*randn(1,1);
for n=2:31
    x1(n,1)=a1*x1(n-1)+sqrt(varu1)*randn(1,1);
    x2(n,1)=a2*x2(n-1)+sqrt(varu2)*randn(1,1);
end
```

We next derive the ACS. In Chapter 18 we will see how to alternatively obtain the ACS using results from linear systems theory. Using (17.17) we have for $k \geq 1$

$$
\begin{aligned}
r_X[k] &= E[X[n]X[n+k]] \\
&= E[X[n](aX[n+k-1] + U[n+k])] \\
&= aE[X[n]X[n+k-1]] \qquad \text{(using (17.19))} \\
&= ar_X[k-1].
\end{aligned}
$$
(17.20)

The solution of this recursive linear difference equation is readily seen to be $r_X[k] = ca^k$, for c any constant and for $k \geq 1$. For $k = 1$ we have that $r_X[1] = ca$ and so from (17.20) $r_X[1] = ar_X[0]$, which implies $c = r_X[0]$. In Problem 17.15 it is shown that

$$
r_X[0] = \frac{\sigma_U^2}{1 - a^2}
$$

so that for all $k \geq 0$, $r_X[k] = r_X[0]a^k$ becomes

$$
r_X[k] = \frac{\sigma_U^2}{1 - a^2}a^k.
$$

Finally, noting that $r_X[-k] = r_X[k]$ from Property 17.2, we obtain the ACS as

$$
r_X[k] = \frac{\sigma_U^2}{1 - a^2}a^{|k|} \qquad -\infty < k < \infty.
$$
(17.21)

(See also Problem 17.16 for an alternative derivation of the ACS.) The ACS is plotted in Figure 17.6 for $a = 0.25$ and $a = 0.98$ and $\sigma_U^2 = 1 - a^2$. For both values of a the value of σ_U^2 has been chosen to ensure that $r_X[0] = 1$. Note that for $a = 0.25$ the ACS dies off very rapidly which means that the random process samples quickly become uncorrelated as the separation between them increases. This is consistent with the typical realization shown in Figure 17.5a. For $a = 0.98$ the ACS decays very slowly, indicating a strong positive correlation between samples, and again being consistent with the typical realization shown in Figure 17.5b. In either case the samples become uncorrelated as $k \to \infty$ since $|a| < 1$ and therefore, $r_X[k] \to 0$ as $k \to \infty$ in accordance with Property 17.5. However, the random process with the slower decaying ACS is more predictable.

$\diamond$

One last property that is necessary for a sequence to be a valid ACS is the property of positive definiteness. As its name implies, it is related to the positive definite property of the covariance matrix. As an example, consider the random vector $\mathbf{X} = [X[0]\, X[1]]^T$. Then we know from the proof of Property 9.2 (covariance matrix is positive semidefinite) that if $Y = a_0X[0] + a_1X[1]$ cannot be made equal to a constant by any choice of a_0 and a_1, then

$$
\text{var}(Y) = \underbrace{\left[\begin{array}{cc} a_0 & a_1 \end{array}\right]}_{\mathbf{a}^T} \underbrace{\left[\begin{array}{cc} \text{cov}(X[0], X[0]) & \text{cov}(X[0], X[1]) \\ \text{cov}(X[1], X[0]) & \text{cov}(X[1], X[1]) \end{array}\right]}_{\mathbf{C}_X} \underbrace{\left[\begin{array}{c} a_0 \\ a_1 \end{array}\right]}_{\mathbf{a}} > 0.
$$

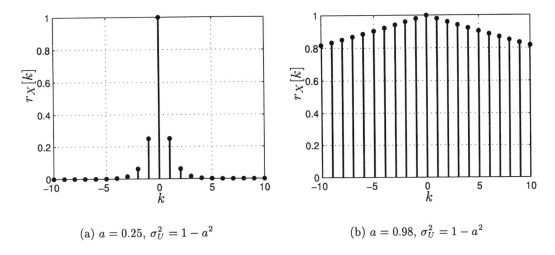

(a) $a = 0.25$, $\sigma_U^2 = 1 - a^2$ (b) $a = 0.98$, $\sigma_U^2 = 1 - a^2$

Figure 17.6: The autocorrelation sequence for autoregressive random processes with different parameters.

Since this holds for all $\mathbf{a} \neq \mathbf{0}$, the covariance matrix $\mathbf{C}_X$ is by definition positive definite (see Appendix C). (If it were possible to choose a_0 and a_1 so that $Y = c$, for c a constant, then $X[1]$ would be perfectly predictable from $X[0]$ as $X[1] = -(a_0/a_1)X[0] + (c/a_1)$. Therefore, we could have $\text{var}(Y) = \mathbf{a}^T \mathbf{C}_X \mathbf{a} = 0$, and $\mathbf{C}_X$ would only be positive *semi*definite.) Now if $X[n]$ is a zero mean WSS random process

$$\text{cov}(X[n_1], X[n_2]) = E(X[n_1]X[n_2]) = r_X[n_2 - n_1]$$

and the covariance matrix becomes

$$\mathbf{C}_X = \begin{bmatrix} r_X[0] & r_X[1] \\ r_X[-1] & r_X[0] \end{bmatrix} = \underbrace{\begin{bmatrix} r_X[0] & r_X[1] \\ r_X[1] & r_X[0] \end{bmatrix}}_{\mathbf{R}_X}.$$

Therefore, the covariance matrix, which we now denote by $\mathbf{R}_X$ and which is called the *autocorrelation matrix*, must be positive definite. This implies that all the *principal minors* (see Appendix C) are positive. For the 2×2 case this means that

$$\begin{aligned} r_X[0] &> 0 \\ r_X^2[0] - r_X^2[1] &> 0 \end{aligned} \tag{17.22}$$

with the first condition being consistent with Property 17.1 and the second condition producing $r_X[0] > |r_X[1]|$. The latter condition is nearly consistent with Property 17.3 with the slight difference, that $|r_X[1]|$ may equal $r_X[0]$ being excluded. This is because we assumed that $X[1]$ was not perfectly predictable from knowledge of $X[0]$. If we allow perfect predictability, then the autocorrelation matrix is only positive semidefinite and the $>$ sign in the second equation of (17.22) would be replaced with $\geq$. In general the $N \times N$ autocorrelation matrix $\mathbf{R}_X$ is given as the covariance matrix of the *zero mean* random vector $\mathbf{X} = [X[0]\, X[1] \ldots X[N-1]]^T$ as

$$\mathbf{R}_X = \begin{bmatrix} r_X[0] & r_X[1] & r_X[2] & \cdots & r_X[N-1] \\ r_X[1] & r_X[0] & r_X[1] & \cdots & r_X[N-2] \\ \vdots & \vdots & \vdots & \ddots & \vdots \\ r_X[N-1] & r_X[N-2] & r_X[N-3] & \cdots & r_X[0] \end{bmatrix}. \tag{17.23}$$

For a sequence to be a valid ACS the $N \times N$ autocorrelation matrix must be positive semidefinite for all $N = 1, 2, \ldots$ and positive definite if we exclude the possibility of perfect predictability [Brockwell and Davis 1987]. This imposes a large number of constraints on $r_X[k]$ and hence not all sequences satisfying Properties 17.1–17.3 are valid ACSs (see also Problem 17.19). In summary, for our last property of the ACS we have the following.

Property 17.6 – ACS is a positive semidefinite sequence.

Mathematically, this means that $r_X[k]$ must satisfy

$$\mathbf{a}^T \mathbf{R}_X \mathbf{a} \geq 0$$

for all $\mathbf{a} = [a_0\, a_1 \ldots a_{N-1}]^T$ and where $\mathbf{R}_X$ is the $N \times N$ autocorrelation matrix given by (17.23). This must hold for all $N \geq 1$.

$\square$

17.5   Ergodicity and Temporal Averages

When a random process is WSS, its mean does not depend on time. Hence, the random variables $\ldots, X[-1], X[0], X[1], \ldots$ all have the same mean. Then, at least as far as the mean is concerned, when we observe a realization of a random process, it is as if we are observing multiple realizations of the same random variable. This suggests that we may be able to determine the value of the mean from a single infinite length realization. To pursue this idea further we plot three realizations of an IID random process whose marginal PDF is Gaussian with mean $\mu_X[n] = \mu = 1$ and a variance $\sigma_X^2[n] = \sigma^2 = 1$ in Figure 17.7. If we let $x_i[18]$ denote the ith realization at time $n = 18$, then by definition of $E[X[18]]$

$$\lim_{M \to \infty} \frac{1}{M} \sum_{m=1}^{M} x_m[18] = E[X[18]] = \mu_X[18] = \mu = 1. \tag{17.24}$$

This is because as we observe all realizations of the random variable $X[18]$ they will conform to the Gaussian PDF (recall that $X[n] \sim \mathcal{N}(1,1)$). In fact, the original definition of expected value was based on the relationship given in (17.24). This

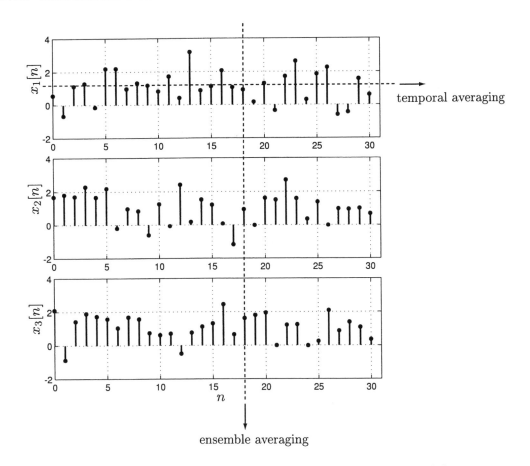

Figure 17.7: Several realizations of WSS random process with $\mu_X[n] = \mu = 1$. Vertical dashed line indicates "ensemble averaging" while horizontal dashed line indicates "temporal averaging."

type of averaging is called "averaging down the ensemble" and consequently is just a restatement of our usual notion of the expected value of a random variable. However, if we are given only a single realization such as $x_1[n]$, then it seems reasonable that

$$\hat{\mu}_N = \frac{1}{N} \sum_{n=0}^{N-1} x_1[n]$$

should also converge to μ as $N \to \infty$. This type of averaging is called "temporal averaging" since we are averaging the samples in time. If it is true that the temporal average converges to μ, then we can state that

$$\lim_{N \to \infty} \frac{1}{N} \sum_{n=0}^{N-1} x_1[n] = \mu = E[X[18]] = \lim_{M \to \infty} \frac{1}{M} \sum_{m=1}^{M} x_m[18]$$

and it is said that *temporal averaging is equivalent to ensemble averaging* or that the random process is *ergodic in the mean*. This property is of great practical importance since it assures us that by averaging enough samples of the realization, we can determine the mean of the random process. For the case of an IID random process ergodicity holds due to the law of large numbers (see Chapter 15). Recall that if $X_1, X_2, \ldots, X_N$ are IID random variables with mean μ and variance σ^2, then the sample mean random variable has the property that

$$\frac{1}{N} \sum_{i=1}^{N} X_i \to E[X] = \mu \qquad \text{as } N \to \infty.$$

Hence, if $X[n]$ is an IID random process, the conditions required for the law of large numbers to hold are satisfied, and we can immediately conclude that

$$\hat{\mu}_N = \frac{1}{N} \sum_{n=0}^{N-1} X[n] \to \mu. \tag{17.25}$$

Now the assumptions required for a random process to be IID are overly restrictive for (17.25) to hold. More generally, if $X[n]$ is a WSS random process, then since $E[X[n]] = \mu$, it follows that $E[\hat{\mu}_N] = (1/N) \sum_{n=0}^{N-1} E[X[n]] = \mu$. Therefore, the only further condition required for ergodicity in the mean is that

$$\lim_{N \to \infty} \text{var}(\hat{\mu}_N) = 0.$$

In the case of the IID random process it is easily shown that $\text{var}(\hat{\mu}_N) = \sigma^2/N \to 0$ as $N \to \infty$ and the condition is satisfied. More generally, however, the random process samples are correlated so that evaluation of this variance is slightly more complicated. We illustrate this computation next.

Example 17.6 – General MA random process

Consider the general MA random process given as $X[n] = (U[n] + U[n-1])/2$, where $E[U[n]] = \mu$ and $\text{var}(U[n]) = \sigma_U^2$ for $-\infty < n < \infty$ and the $U[n]$'s are all uncorrelated. This is similar to the MA process of Example 16.10 but is more general in that the mean of $U[n]$ is not necessarily zero, the samples of $U[n]$ are only uncorrelated, and hence, not necessarily independent, and the PDF of each sample need not be Gaussian. The general MA process $X[n]$ is easily shown to be WSS and to have a mean sequence $\mu_X[n] = \mu$ (see Problem 17.20). To determine if it is ergodic in the mean we must compute the $\text{var}(\hat{\mu}_N)$ and show that it converges to zero as $N \to \infty$. Now

$$\text{var}(\hat{\mu}_N) = \text{var}\left(\frac{1}{N} \sum_{n=0}^{N-1} X[n] \right).$$

Since the $X[n]$'s are now correlated, we use (9.26), where $\mathbf{a} = [a_0\, a_1 \ldots a_{N-1}]^T$ with $a_n = 1/N$, to yield

$$\operatorname{var}(\hat{\mu}_N) = \operatorname{var}\left(\sum_{n=0}^{N-1} a_n X[n]\right) = \mathbf{a}^T \mathbf{C}_X \mathbf{a}. \tag{17.26}$$

The covariance matrix has (i,j) element

$$[\mathbf{C}_X]_{ij} = E[(X[i]-E[X[i]])(X[j]-E[X[j]])] \qquad i = 0,1,\ldots,N-1; j = 0,1,\ldots,N-1.$$

But

$$\begin{aligned} X[n] - E[X[n]] &= \frac{1}{2}(U[n] + U[n-1]) - \frac{1}{2}(\mu + \mu) \\ &= \frac{1}{2}[(U[n]-\mu) + (U[n-1]-\mu)] \\ &= \frac{1}{2}[\bar{U}[n] + \bar{U}[n-1]] \end{aligned}$$

where $\bar{U}[n]$ is a *zero mean* random variable for each value of n. Thus,

$$\begin{aligned} [\mathbf{C}_X]_{ij} &= \frac{1}{4}E[(\bar{U}[i] + \bar{U}[i-1])(\bar{U}[j] + \bar{U}[j-1])] \\ &= \frac{1}{4}\left(E[\bar{U}[i]\bar{U}[j]] + E[\bar{U}[i]\bar{U}[j-1]] + E[\bar{U}[i-1]\bar{U}[j]] + E[\bar{U}[i-1]\bar{U}[j-1]]\right) \end{aligned}$$

and since $E[\bar{U}[n_1]\bar{U}[n_2]] = \operatorname{cov}(U[n_1], U[n_2]) = \sigma_U^2 \delta[n_2 - n_1]$ (all the $U[n]$'s are uncorrelated), we have

$$[\mathbf{C}_X]_{ij} = \frac{1}{4}\left(\sigma_U^2 \delta[j-i] + \sigma_U^2 \delta[j-1-i] + \sigma_U^2 \delta[j-i+1] + \sigma_U^2 \delta[j-i]\right).$$

Finally, we have the required covariance matrix

$$[\mathbf{C}_X]_{ij} = \begin{cases} \frac{1}{2}\sigma_U^2 & i = j \\ \frac{1}{4}\sigma_U^2 & |i-j| = 1 \\ 0 & \text{otherwise.} \end{cases} \tag{17.27}$$

Using this in (17.26) produces

$$\operatorname{var}(\hat{\mu}_N)$$

$$= \mathbf{a}^T \mathbf{C}_X \mathbf{a}$$

$$= \begin{bmatrix} \frac{1}{N} & \frac{1}{N} & \cdots & \frac{1}{N} \end{bmatrix} \begin{bmatrix} \frac{\sigma_U^2}{2} & \frac{\sigma_U^2}{4} & 0 & 0 & \cdots & 0 & 0 & 0 \\ \frac{\sigma_U^2}{4} & \frac{\sigma_U^2}{2} & \frac{\sigma_U^2}{4} & 0 & \cdots & 0 & 0 & 0 \\ \vdots & \vdots & \vdots & \vdots & \vdots & \vdots & \vdots & \vdots \\ 0 & 0 & 0 & 0 & \cdots & \frac{\sigma_U^2}{4} & \frac{\sigma_U^2}{2} & \frac{\sigma_U^2}{4} \\ 0 & 0 & 0 & 0 & \cdots & 0 & \frac{\sigma_U^2}{4} & \frac{\sigma_U^2}{2} \end{bmatrix} \begin{bmatrix} \frac{1}{N} \\ \frac{1}{N} \\ \vdots \\ \frac{1}{N} \end{bmatrix}$$

$$= \frac{1}{N^2} \sum_{i=0}^{N-1} \frac{\sigma_U^2}{2} + \frac{1}{N^2} \sum_{i=0}^{N-2} \frac{\sigma_U^2}{4} + \frac{1}{N^2} \sum_{i=1}^{N-1} \frac{\sigma_U^2}{4}$$

$$= \frac{\sigma_U^2}{2N} + \frac{\sigma_U^2}{4} \frac{N-1}{N^2} + \frac{\sigma_U^2}{4} \frac{N-1}{N^2} \to 0 \qquad \text{as } N \to \infty.$$

Finally, we see that the general MA random process is ergodic in the mean.

$\diamond$

In general, it can be shown that for a WSS random process to be ergodic in the mean, the variance of the sample mean

$$\text{var}(\hat{\mu}_N) = \frac{1}{N} \sum_{k=-(N-1)}^{N-1} \left(1 - \frac{|k|}{N}\right) (r_X[k] - \mu^2) \tag{17.28}$$

must converge to zero as $N \to \infty$ (see Problem 17.23 for the derivation of (17.28)). For this to occur, the covariance sequence $r_X[k] - \mu^2$ must decay to zero at a fast enough rate as $k \to \infty$, which is to say that as the samples are spaced further and further apart, they must eventually become uncorrelated. A little reflection on the part of the reader will reveal that ergodicity requires a single realization of the random process to display the behavior of the entire ensemble of realizations. If not, ergodicity will not hold. Consider the following simple nonergodic random process.

Example 17.7 – Random DC level

Define a random process as $X[n] = A$ for $-\infty < n < \infty$, where $A \sim \mathcal{N}(0,1)$. Some realizations are shown in Figure 17.8. This random process is WSS since

$$\begin{aligned} \mu_X[n] &= E[X[n]] = E[A] = 0 = \mu & -\infty < n < \infty & \quad \text{(not dependent on } n) \\ r_X[k] &= E[X[n]X[n+k]] = E[A^2] = 1 & & \quad \text{(not dependent on } n). \end{aligned}$$

However, it should be clear that $\hat{\mu}_N$ will not converge to $\mu = 0$. Referring to the realization $x_1[n]$ in Figure 17.8, the sample mean will produce -0.43 no matter how large N becomes. In addition, it can be shown that $\text{var}(\hat{\mu}_N) = 1$ (see Problem 17.24). Each realization is *not* representative of the *ensemble of realizations*.

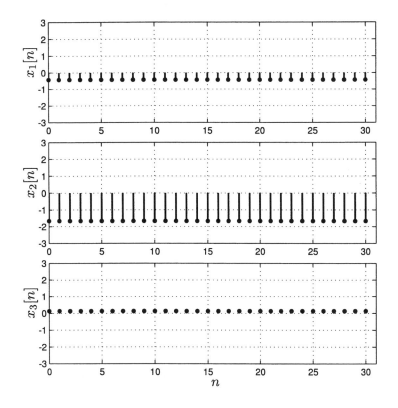

Figure 17.8: Several realizations of the random DC level process.

17.6   The Power Spectral Density

The ACS measures the correlation between samples of a WSS random process. For example, the AR random process was shown to have the ACS

$$r_X[k] = \frac{\sigma_U^2}{1 - a^2} a^{|k|}$$

which for $a = 0.25$ and $a = 0.98$ is shown in Figure 17.6, along with some typical realizations in Figure 17.5. Note that when the ACS dies out rapidly (see Figure 17.6a), the realization is more rapidly varying in time (see Figure 17.5a). In contrast, when the ACS decays slowly (see Figure 17.6b), the realization varies slowly (see Figure 17.5b). It would seem that the ACS is related to the rate of change of the random process. For deterministic signals the rate of change is usually measured by examining a discrete-time Fourier transform [Jackson 1991]. Signals with high frequency content exhibit rapid fluctutations in time while signals with only low frequency content exhibit slow variations in time. For WSS random processes we will be interested in the *power* at the various frequencies. In particular, we will introduce the measure known as the *power spectral density* (PSD) and show that it quantifies the distribution of power with frequency. Before doing so, however, we consider the following deterministically motivated measure of power with frequency based on the discrete-time Fourier transform

$$\hat{P}_X(f) = \frac{1}{N} \left| \sum_{n=0}^{N-1} X[n] \exp(-j2\pi f n) \right|^2 . \tag{17.29}$$

This is a normalized version of the magnitude-squared discrete-time Fourier transform of the random process over the time interval $0 \leq n \leq N-1$. It is called the *periodogram* since its original purpose was to find periodicities in random data sets [Schuster 1898]. In (17.29) f denotes the discrete-time frequency, which is assumed to be in the range $-1/2 \leq f \leq 1/2$ for reasons that will be elucidated later. The $1/N$ factor is required to normalize $\hat{P}_X(f)$ to be interpretable as a power spectral *density* or power per unit frequency. The use of a "hat" is meant to convey the notion that this quantity is an estimator. As we now show, the periodogram is not a suitable measure of the distribution of power with frequency, although it would be for some deterministic signals (such as periodic discrete-time signals with period N). As an example, we plot $\hat{P}_X(f)$ in Figure 17.9 for the realizations given in Figure 17.5. We

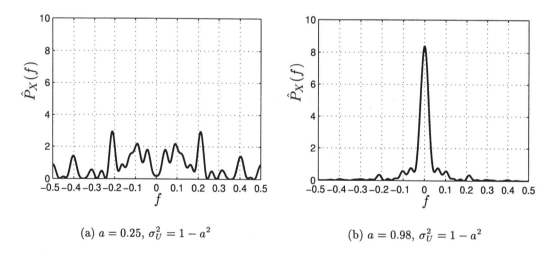

(a) $a = 0.25$, $\sigma_U^2 = 1 - a^2$ (b) $a = 0.98$, $\sigma_U^2 = 1 - a^2$

Figure 17.9: Periodogram for autoregressive random process with different parameters. The realizations shown in Figure 17.5 were used to generate these estimates.

see that the periodogram in Figure 17.9a exhibits many random fluctuations. Other realizations will also produce similar seemingly random curves. However, it does seem to produce a reasonable result—for the periodogram in Figure 17.9a there is more high frequency power than for the periodogram in Figure 17.9b. The reason for the random nature of the plot is that (17.29) is a function of N random variables and hence is a random variable itself for each frequency. As such, it exhibits the variability of a random process for which the usual dependence on time is replaced by frequency. What we would actually like is an *average* measure of the power distribution with frequency, suggesting the need for an expected value. Also, to ensure that we capture the entire random process behavior, an infinite length realization is required. We are therefore led to the following more suitable definition of the PSD

$$P_X(f) = \lim_{M \to \infty} \frac{1}{2M+1} E\left[\left| \sum_{n=-M}^{M} X[n] \exp(-j2\pi fn) \right|^2 \right]. \qquad (17.30)$$

The function $P_X(f)$ is called the *power spectral density* (PSD) and when integrated provides a measure of the average power within a band of frequencies. It is completely analogous to the PDF in that to find the *average power* of the random process in the frequency band $f_1 \le f \le f_2$ we should find the area under the PSD curve.

 Fourier analysis of a random process yields no phase information.

In our definition of the PSD we are using the magnitude-squared of the Fourier transform. It is obvious then, that the PSD does not tell us anything about the phases of the Fourier transform of the random process. This is in contrast to a Fourier transform of a deterministic signal. There the inverse Fourier transform can be viewed as a decomposition of the signal into sinusoids of different frequencies with deterministic amplitudes and phases. For a random process a similar decomposition called the *spectral representation theorem* [Brockwell and Davis 1987] yields sinusoids of different frequencies with *random amplitudes and random phases*. The PSD is essentially the *expected value of the power of the random sinusoidal amplitudes* per unit of frequency. No phase information is retained and therefore no phase information can be extracted from knowledge of the PSD.

We next give an example of the computation of a PSD.

Example 17.8 – White noise

Assume that $X[n]$ is white noise (see Example 17.2) and therefore, has a zero mean and ACS $r_X[k] = \sigma^2 \delta[k]$. Then,

$$
P_X(f) = \lim_{M \to \infty} \frac{1}{2M+1} E\left[\sum_{n=-M}^{M} X[n] \exp(j2\pi f n) \sum_{m=-M}^{M} X[m] \exp(-j2\pi f m) \right]
$$

$$
= \lim_{M \to \infty} \frac{1}{2M+1} \sum_{n=-M}^{M} \sum_{m=-M}^{M} \underbrace{E[X[n]X[m]]}_{r_X[m-n]} \exp[-j2\pi f(m-n)] \quad (17.31)
$$

$$
= \lim_{M \to \infty} \frac{1}{2M+1} \sum_{n=-M}^{M} \sum_{m=-M}^{M} \sigma^2 \delta[m-n] \exp[-j2\pi f(m-n)]
$$

$$
= \lim_{M \to \infty} \frac{1}{2M+1} \sum_{n=-M}^{M} \sigma^2
$$

$$
= \lim_{M \to \infty} \sigma^2 = \sigma^2. \quad (17.32)
$$

Hence, for white noise the PSD is

$$
P_X(f) = \sigma^2 \qquad -1/2 \leq f \leq 1/2.
$$

As first mentioned in Chapter 16 white noise contains equal contributions of average power at all frequencies.

$\diamondsuit$

A more straightforward approach to obtaining the PSD is based on knowledge of the ACS. From (17.31) we see that

$$
P_X(f) = \lim_{M \to \infty} \frac{1}{2M+1} \sum_{n=-M}^{M} \sum_{m=-M}^{M} r_X[m-n] \exp[-j2\pi f(m-n)]. \quad (17.33)
$$

This can be simplified using the formula (see Problem 17.26)

$$
\sum_{n=-M}^{M} \sum_{m=-M}^{M} g[m-n] = \sum_{k=-2M}^{2M} (2M+1-|k|)g[k]
$$

which results from considering $g[m-n]$ as an element of the $(2M+1) \times (2M+1)$ matrix $\mathbf{G}$ with elements $[\mathbf{G}]_{mn} = g[m-n]$ for $m = -M, \ldots, M$ and $n = -M, \ldots, M$ and then summing all the elements. Using this relationship in (17.33) produces

$$
P_X(f) = \lim_{M \to \infty} \frac{1}{2M+1} \sum_{k=-2M}^{2M} (2M+1-|k|)r_X[k] \exp(-j2\pi f k)
$$

$$
= \lim_{M \to \infty} \sum_{k=-2M}^{2M} \left(1 - \frac{|k|}{2M+1}\right) r_X[k] \exp(-j2\pi f k).
$$

Assuming that $\sum_{k=-\infty}^{\infty} |r_X[k]| < \infty$, the limit can be shown to produce the final result (see Problem 17.27)

$$P_X(f) = \sum_{k=-\infty}^{\infty} r_X[k] \exp(-j2\pi fk) \qquad (17.34)$$

which says that *the power spectral density is the discrete-time Fourier transform of the ACS*. This relationship is known as the *Wiener-Khinchine* theorem. Some examples follow.

Example 17.9 – White noise

From Example 17.2 $r_X[k] = \sigma^2 \delta[k]$ and so

$$
\begin{aligned}
P_X(f) &= \sum_{k=-\infty}^{\infty} r_X[k] \exp(-j2\pi fk) \\
&= \sum_{k=-\infty}^{\infty} \sigma^2 \delta[k] \exp(-j2\pi fk) \\
&= \sigma^2.
\end{aligned}
$$

This is shown in Figure 17.10. Note that the total average power in $X[n]$, which is $r_X[0] = \sigma^2$, is given by the area under the PSD curve.

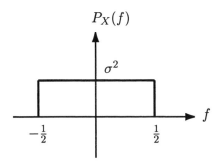

Figure 17.10: PSD of white noise.

◇

Example 17.10 – AR random process

From (17.21) we have that

$$r_X[k] = \frac{\sigma_U^2}{1 - a^2} a^{|k|} \qquad -\infty < k < \infty$$

and from (17.34)

$$
\begin{aligned}
P_X(f) &= \sum_{k=-\infty}^{\infty} r_X[k] \exp(-j2\pi f k) \\
&= \frac{\sigma_U^2}{1-a^2} \sum_{k=-\infty}^{\infty} a^{|k|} \exp(-j2\pi f k) \\
&= \frac{\sigma_U^2}{1-a^2} \left[\sum_{k=-\infty}^{-1} a^{-k} \exp(-j2\pi f k) + \sum_{k=0}^{\infty} a^k \exp(-j2\pi f k) \right] \\
&= \frac{\sigma_U^2}{1-a^2} \left[\sum_{k=1}^{\infty} [a\exp(j2\pi f)]^k + \sum_{k=0}^{\infty} [a\exp(-j2\pi f)]^k \right].
\end{aligned}
$$

Since $|a\exp(\pm j2\pi f)| = |a| < 1$, we can use the formula $\sum_{k=k_0}^{\infty} z^k = z^{k_0}/(1-z)$ for z a complex number with $|z| < 1$ to evaluate the sums. This produces

$$
\begin{aligned}
P_X(f) &= \frac{\sigma_U^2}{1-a^2} \left(\frac{a\exp(j2\pi f)}{1 - a\exp(j2\pi f)} + \frac{1}{1 - a\exp(-j2\pi f)} \right) \\
&= \frac{\sigma_U^2}{1-a^2} \frac{a\exp(j2\pi f)(1 - a\exp(-j2\pi f)) + (1 - a\exp(j2\pi f))}{(1 - a\exp(j2\pi f))(1 - a\exp(-j2\pi f))} \\
&= \frac{\sigma_U^2}{1-a^2} \frac{1-a^2}{|1 - a\exp(-j2\pi f)|^2} \\
&= \frac{\sigma_U^2}{|1 - a\exp(-j2\pi f)|^2}. \quad\quad (17.35)
\end{aligned}
$$

This can also be written in real form as

$$
P_X(f) = \frac{\sigma_U^2}{1 + a^2 - 2a\cos(2\pi f)} \quad\quad -1/2 \le f \le 1/2. \quad\quad (17.36)
$$

For $a = 0.25$ and $a = 0.98$ and $\sigma_U^2 = 1 - a^2$, the PSDs are plotted in Figure 17.11. Note that the total average power in each PSD is the same, being $r_X[0] = \sigma_U^2/(1-a^2) = 1$. As expected the more noise-like random process has a PSD (see Figure 17.11a) with more high frequency average power than the slowly varying random process (see Figure 17.11b) which has all its average power near $f = 0$ (or at DC).

$\Diamond$

From the previous example, we observe that the PSD exhibits the properties of being a real nonnegative function of frequency, consistent with our notion of power as a nonnegative physical quantity, of being symmetric about $f = 0$, and of being periodic with period one (see (17.36)). We next prove that these properties are true in general.

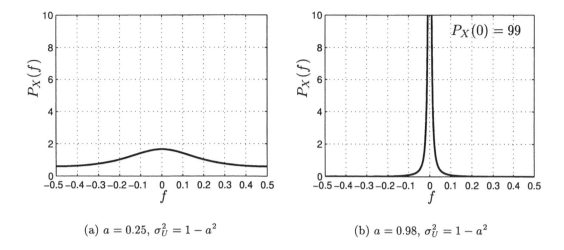

(a) $a = 0.25$, $\sigma_U^2 = 1 - a^2$ (b) $a = 0.98$, $\sigma_U^2 = 1 - a^2$

Figure 17.11: Power spectral densities for autoregressive random process with different parameters. The periodograms, which are estimated PSDs, were given in Figure 17.9.

Property 17.7 – PSD is a real function.

The PSD is also given by the real function

$$P_X(f) = \sum_{k=-\infty}^{\infty} r_X[k] \cos(2\pi f k). \tag{17.37}$$

Proof:

$$
\begin{aligned}
P_X(f) &= \sum_{k=-\infty}^{\infty} r_X[k] \exp(-j2\pi f k) \\
&= \sum_{k=-\infty}^{\infty} r_X[k](\cos(2\pi f k) - j \sin(2\pi f k)) \\
&= \sum_{k=-\infty}^{\infty} r_X[k] \cos(2\pi f k) - j \sum_{k=-\infty}^{\infty} r_X[k] \sin(2\pi f k).
\end{aligned}
$$

But

$$\sum_{k=-\infty}^{\infty} r_X[k] \sin(2\pi f k) = \sum_{k=-\infty}^{-1} r_X[k] \sin(2\pi f k) + \sum_{k=1}^{\infty} r_X[k] \sin(2\pi f k)$$

since the $k = 0$ term is zero, and letting $l = -k$ in the first sum we have

$$\sum_{k=-\infty}^{\infty} r_X[k]\sin(2\pi fk) = \sum_{l=1}^{\infty} r_X[-l]\sin(2\pi f(-l)) + \sum_{k=1}^{\infty} r_X[k]\sin(2\pi fk)$$

$$= \sum_{k=1}^{\infty} r_X[k](-\sin(2\pi fk) + \sin(2\pi fk)) = 0 \quad (r_X[-l] = r_X[l])$$

from which (17.37) follows.

$\square$

Property 17.8 – PSD is nonnegative.

$$P_X(f) \geq 0$$

Proof: Follows from (17.30) but can also be shown to follow from the positive semidefinite property of the ACS [Brockwell and Davis 1987]. (See also Problem 17.19.)

$\square$

Property 17.9 – PSD is symmetric about $f = 0$.

$$P_X(-f) = P_X(f)$$

Proof: Follows from (17.37).

$\square$

Property 17.10 – PSD is periodic with period one.

$$P_X(f + 1) = P_X(f)$$

Proof: From (17.37) we have

$$P_X(f+1) = \sum_{k=-\infty}^{\infty} r_X[k]\cos(2\pi(f+1)k)$$

$$= \sum_{k=-\infty}^{\infty} r_X[k]\cos(2\pi fk + 2\pi k)$$

$$= \sum_{k=-\infty}^{\infty} r_X[k]\cos(2\pi fk) \quad (\cos(2\pi k) = 1,\ \sin(2\pi k) = 0)$$

$$= P_X(f)$$

$\square$

Property 17.11 – ACS recovered from PSD using inverse Fourier transform

$$r_X[k] \;=\; \int_{-\frac{1}{2}}^{\frac{1}{2}} P_X(f)\exp(j2\pi fk)df \qquad -\infty < k < \infty \qquad (17.38)$$

$$=\; \int_{-\frac{1}{2}}^{\frac{1}{2}} P_X(f)\cos(2\pi fk)df \qquad -\infty < k < \infty \qquad (17.39)$$

<u>Proof</u>: (17.38) follows from properties of discrete-time Fourier transform [Jackson 1991]. (17.39) follows from Property 17.9 (see Appendix B.5 and also Problem 17.49).

□

Property 17.12 – PSD yields average power over band of frequencies.
To obtain the average power in the frequency band $f_1 \le f \le f_2$ we need only find the area under the PSD curve for this band. The average *physical* power is obtained as twice this area since the negative frequencies account for half of the average power (recall Property 17.9). Hence,

$$\text{Average physical power in } [f_1, f_2] = 2\int_{f_1}^{f_2} P_X(f)df. \qquad (17.40)$$

The proof of this property requires some concepts to be described in the next chapter, and thus, we defer the proof until Section 18.4. Note, however, that if $f_1 = 0$ and $f_2 = 1/2$, then the average power in this band is

$$\begin{aligned}
\text{Average physical power in } [0, 1/2] \;&=\; 2\int_{0}^{1/2} P_X(f)df \\
&=\; \int_{-\frac{1}{2}}^{\frac{1}{2}} P_X(f)df \quad \text{(due to symmetry of PSD)} \\
&=\; \int_{-\frac{1}{2}}^{\frac{1}{2}} P_X(f)\exp(j2\pi f(0))df \\
&=\; r_X[0] \quad \text{(from (17.38))}
\end{aligned}$$

which we have already seen yields the total average power since $r_X[0] = E[X^2[n]]$. Hence, we see that the *total average power* is obtained by integrating the PSD over all frequencies to yield

$$r_X[0] = \int_{-\frac{1}{2}}^{\frac{1}{2}} P_X(f)df. \qquad (17.41)$$

$\square$

 Definitions of PSD are not consistent.

In some texts, especially ones describing the use of the PSD for physical measurements, the definition of the PSD is slightly different. The alternative definition relies on the relationship of (17.40) to define the PSD as $G_X(f) = 2P_X(f)$. It is called the *one-sided PSD* and its advantage is that it yields directly the average power over a band when integrated over the band. As can be seen from (17.40)

$$\text{Average physical power in } [f_1, f_2] = \int_{f_1}^{f_2} G_X(f) df.$$

A final comment concerns the periodicity of the PSD. We have chosen the frequency interval $[-1/2, 1/2]$ over which to display the PSD. The rationale for this choice arises from the practical situation in which a *continuous-time* WSS random process (see Section 17.8) is sampled to produce a discrete-time WSS random process. Then, if the continuous-time random process $X(t)$ has a PSD that is bandlimited to W Hz and is sampled at F_s samples/sec, the discrete-time PSD $P_X(f)$ will have discrete-time frequency units of W/F_s. For Nyquist rate sampling of $F_s = 2W$, the maximum discrete-time frequency will be $f = W/F_s = 1/2$. Hence, our choice of the frequency interval $[-1/2, 1/2]$ corresponds to the continuous-time frequency interval of $[-W, W]$ Hz. The discrete-time frequency is also referred to as the *normalized frequency*, the normalizing factor being F_s.

17.7   Estimation of the ACS and PSD

Recall from our discussion of ergodicity that in the problem of mean estimation for a WSS random process, we were restricted to observing only a finite number of samples of *one realization* of the random process. If the random process is ergodic in the mean, then we saw that as the number of samples increases to infinity, the temporal average $\hat{\mu}_N$ will converge to the ensemble average μ. To apply this result to estimation of the ACS consider the problem of estimating the ACS for lag $k = k_0$ which is

$$r_X[k_0] = E[X[n]X[n + k_0]].$$

Then by defining the product random process $Y[n] = X[n]X[n + k_0]$ we see that

$$r_X[k_0] = E[Y[n]] \qquad -\infty < n < \infty$$

or the desired quantity to be estimated is just the mean of the random process $Y[n]$. The mean of $Y[n]$ does not depend on n. This suggests that we replace the observed values of $X[n]$ with those of $Y[n]$ by using $y[n] = x[n]x[n + k_0]$, and then use a temporal average to estimate the ensemble average. Hence, we have the temporal average estimate

$$
\begin{aligned}
\hat{r}_X[k_0] &= \frac{1}{N} \sum_{n=0}^{N-1} y[n] \\
&= \frac{1}{N} \sum_{n=0}^{N-1} x[n]x[n + k_0].
\end{aligned}
\tag{17.42}
$$

Also, since $r_X[-k] = r_X[k]$, we need only estimate the ACS for $k \geq 0$. There is one slight modification that we need to make to the estimate. Assuming that $\{x[0], x[1], \ldots, x[N-1]\}$ are observed, we must choose the upper limit on the summation in (17.42) to satisfy the constraint $n + k_0 \leq N - 1$. This is because $x[n + k_0]$ is unobserved for $n + k_0 > N - 1$. With this modification we have as our estimate of the ACS (and now replacing the specific lag of k_0 by the more general lag k)

$$
\hat{r}_X[k] = \frac{1}{N-k} \sum_{n=0}^{N-1-k} x[n]x[n + k] \qquad k = 0, 1, \ldots, N - 1.
\tag{17.43}
$$

We have also changed the $1/N$ averaging factor to $1/(N - k)$. This is because the number of terms in the sum is only $N - k$. For example, if $N = 4$ so that we observe $\{x[0], x[1], x[2], x[3]\}$, then (17.43) yields the estimates

$$
\begin{aligned}
\hat{r}_X[0] &= \frac{1}{4}(x^2[0] + x^2[1] + x^2[2] + x^2[3]) \\
\hat{r}_X[1] &= \frac{1}{3}(x[0]x[1] + x[1]x[2] + x[2]x[3]) \\
\hat{r}_X[2] &= \frac{1}{2}(x[0]x[2] + x[1]x[3]) \\
\hat{r}_X[3] &= x[0]x[3].
\end{aligned}
$$

As k increases, the distance between the samples increases and so there are less products available for averaging. In fact, for $k > N - 1$, we cannot estimate the value of the ACS at all. With the estimate given in (17.43) we see that $E[\hat{r}_X[k]] = r_X[k]$ for $k = 0, 1, \ldots, N - 1$. In order for the estimate to converge to the true value as $N \to \infty$, i.e, for the random process to be *ergodic in the autocorrelation* or

$$
\lim_{N \to \infty} \hat{r}_X[k] = \lim_{N \to \infty} \frac{1}{N-k} \sum_{n=0}^{N-1-k} x[n]x[n + k] = r_X[k] \qquad k = 0, 1, \ldots
$$

we require that $\text{var}(\hat{r}_X[k]) \to 0$ as $N \to \infty$. This will generally be true if $r_X[k] \to 0$ as $k \to \infty$ for a zero mean random process but see Problem 17.25 for a case where this is not required. To illustrate the estimation performance consider the AR random process described in Example 17.5. The true ACS and the estimated one using (17.43) and based on the realizations shown in Figure 17.5 are shown in Figure 17.12. The estimated ACS is shown as the dark lines while the true ACS as given by (17.21) is shown as light lines, which are slightly displaced to the right for easier viewing. Note that in Figure 17.12 the estimated values for k large exhibit

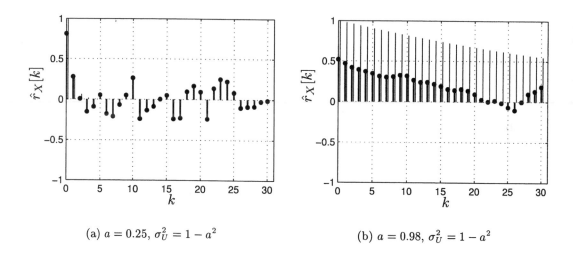

(a) $a = 0.25$, $\sigma_U^2 = 1 - a^2$ (b) $a = 0.98$, $\sigma_U^2 = 1 - a^2$

Figure 17.12: Estimated ACSs (dark lines) and the true ACSs given in Figure 17.6 (light lines) for the AR random process realizations shown in Figure 17.5.

a large error. This is due to the fewer number of products, i.e., $N - k = 31 - k$, that are available for averaging in (17.43). In the case of $k = 30$ the estimate is $\hat{r}_X[30] = x[0]x[30]$, which as you might expect is very poor since there is no averaging at all! Clearly, for accurate estimates of the ACS we require that $k_{\max} \ll N$. The MATLAB code used to estimate the ACS for Figure 17.12 is given below.

```
n=[0:30]';N=length(n);
a1=0.25;a2=0.98;
varu1=1-a1^2;varu2=1-a2^2;
r1true=(varu1/(1-a1^2))*a1.^n; % see (17.21)
r2true=(varu2/(1-a2^2))*a2.^n;
for k=0:N-1
    r1est(k+1,1)=(1/(N-k))*sum(x1(1:N-k).*x1(1+k:N));
    r2est(k+1,1)=(1/(N-k))*sum(x2(1:N-k).*x2(1+k:N));
end
```

To estimate the PSD requires somewhat more care than the ACS. We have already seen that the periodogram estimate of (17.29) is not suitable. There are many ways to estimate the PSD based on either (17.30) or (17.34). We illustrate one approach based on (17.30). Others may be found in [Jenkins and Watts 1968, Kay 1988]. Since we only have a segment of a single realization of the random process, we cannot implement the expectation operation required in (17.30). Note that the operation of $E[\cdot]$ represents an average down the ensemble or equivalently an average over multiple realizations. To obtain some averaging, however, we can break up the data $\{x[0], x[1], \ldots, x[N-1]\}$ into I nonoverlapping blocks, with each block having a total of L samples. We assume for simplicity that there is an integer number of blocks so that $N = IL$. The implicit assumption in doing so is that each block exhibits the statistical characteristics of a single realization and so we can mimic the averaging down the ensemble by averaging temporally across successive blocks of data. Once again, the assumption of ergodicity is being employed. Thus, we first break up the data set into the I nonoverlapping data blocks

$$y_i[n] = x[n+iL] \qquad n = 0, 1, \ldots, L-1; i = 0, 1, \ldots, I-1$$

where each data block has a length of L samples. Then, for each data block we compute a periodogram as

$$\hat{P}_X^{(i)}(f) = \frac{1}{L} \left| \sum_{n=0}^{L-1} y_i[n] \exp(-j2\pi f n) \right|^2 \tag{17.44}$$

and then average all the periodograms together to yield the final PSD estimate as

$$\hat{P}_{\mathrm{av}}(f) = \frac{1}{I} \sum_{i=0}^{I-1} \hat{P}_X^{(i)}(f). \tag{17.45}$$

This estimate is called the *averaged periodogram*. It can be shown that under some conditions, $\lim_{N \to \infty} \hat{P}_{\mathrm{av}}(f) = P_X(f)$. Once again we are calling upon an ergodicity type of property in that we are averaging the periodograms obtained in time instead of the theoretical ensemble averaging. Of course, for convergence to hold as $N \to \infty$, we must have $L \to \infty$ and $I \to \infty$ as well.

As an example, we examine the averaged periodogram estimates for the two AR processes whose PSDs are shown in Figure 17.11. The number of data samples was $N = 310$, which was broken up into $I = 10$ nonoverlapping blocks of data with $L = 31$ samples in each one. By comparing the spectral estimates in Figure 17.13 with those of Figure 17.9, it is seen that the averaging has yielded a better estimate. Of course, the price paid is that the data set needs to be $I = 10$ times as long! The MATLAB code used to implement the averaged periodogram estimate is given next. A fast Fourier transform (FFT) is used to compute the Fourier transform of the $y_i[n]$ sequences at the frequencies $f = -0.5 + k\Delta_f$, where $k = 0, 1, \ldots, 1023$ and $\Delta_f = 1/1024$ (see [Kay 1988] for a more detailed description).

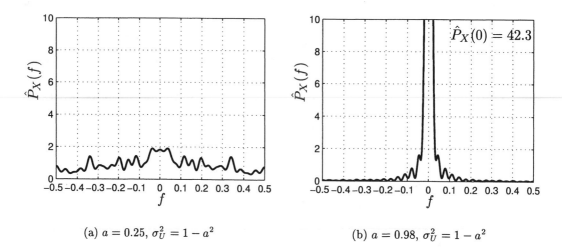

(a) $a = 0.25$, $\sigma_U^2 = 1 - a^2$ (b) $a = 0.98$, $\sigma_U^2 = 1 - a^2$

Figure 17.13: Power spectral density estimates using the averaged periodogram method for autoregressive processes with different parameters. The true PSDs are shown in Figure 17.11.

```
Nfft=1024; % set FFT size
Pav1=zeros(Nfft,1);Pav2=Pav1; % set up arrays with desired dimension
f=[0:Nfft-1]'/Nfft-0.5; % set frequencies for later plotting
                        % of PSD estimate
for i=0:I-1
   nstart=1+i*L;nend=L+i*L; % set up beginning and end points
                            % of ith block of data
   y1=x1(nstart:nend);
   y2=x2(nstart:nend);
% take FFT of block, since FFT outputs samples of Fourier
% transform over frequency range [0,1), must shift FFT outputs
% for [1/2,1) to [-1/2, 0), then take complex magnitude-squared,
% normalize by L and average
Pav1=Pav1+(1/(I*L))*abs(fftshift(fft(y1,Nfft))).^2;
Pav2=Pav2+(1/(I*L))*abs(fftshift(fft(y2,Nfft))).^2;
end
```

17.8 Continuous-Time WSS Random Processes

In this section we give the corresponding definitions and formulas for continuous-time WSS random processes. A more detailed description can be found in [Papoulis 1965]. Also, an important example is described to illustrate the use of these formulas.

A continuous-time random process $X(t)$ for $-\infty < t < \infty$ is defined to be WSS if the *mean function* $\mu_X(t)$ satisfies

$$\mu_X(t) = E[X(t)] = \mu \qquad -\infty < t < \infty \qquad (17.46)$$

which is to say it is constant in time and an *autocorrelation function* (ACF) can be defined as

$$r_X(\tau) = E[X(t)X(t+\tau)] \qquad -\infty < \tau < \infty \qquad (17.47)$$

which is not dependent on the value of t. Thus, $E[X(t_1)X(t_2)]$ depends only on $|t_2 - t_1|$. Note the use of the "parentheses" indicates that the argument of the ACF is continuous and serves to distinguish $r_X[k]$ from $r_X(\tau)$. The ACF has the following properties.

Property 17.13 – ACF is positive for the zero lag or $r_X(0) > 0$.
The total average power is $r_X(0) = E[X^2(t)]$.

$\square$

Property 17.14 – ACF is an even function or $r_X(-\tau) = r_X(\tau)$.

$\square$

Property 17.15 – Maximum value of ACF is at $\tau = 0$ or $|r_X(\tau)| \leq r_X(0)$.

$\square$

Property 17.16 – ACF measures the predictability of a random process.
The correlation coefficient for two samples of a zero mean WSS random process is

$$\rho_{X(t),X(t+\tau)} = \frac{r_X(\tau)}{r_X(0)}.$$

$\square$

Property 17.17 – ACF approaches μ^2 as $\tau \to \infty$.
This assumes that the samples become uncorrelated for large lags, which is usually the case.

$\square$

Property 17.18 – $r_X(\tau)$ is a positive semidefinite function.
See [Papoulis 1965] for the definition of a positive semidefinite function. This property assumes that the some samples of $X(t)$ may be perfectly predictable. If it is not, then the ACF is positive definite.

$\square$

The PSD is defined as

$$P_X(F) = \lim_{T\to\infty} \frac{1}{T} E\left[\left|\int_{-T/2}^{T/2} X(t)\exp(-j2\pi Ft)dt\right|^2\right] \quad -\infty < F < \infty \quad (17.48)$$

where F is the frequency in Hz. We use a capital F to denote continuous-time or analog frequency. By the Wiener-Khinchine theorem this is equivalent to the continuous-time Fourier transform of the ACF

$$P_X(F) = \int_{-\infty}^{\infty} r_X(\tau)\exp(-j2\pi F\tau)d\tau \quad (17.49)$$

$$= \int_{-\infty}^{\infty} r_X(\tau)\cos(2\pi F\tau)d\tau. \quad (17.50)$$

(See also Problem 17.49.) The PSD has the usual interpretation as the average power distribution with frequency. In particular, it is the average power per Hz. The average physical power in a frequency band $[F_1, F_2]$ is given by

$$\text{Average physical power in } [F_1, F_2] = 2\int_{F_1}^{F_2} P_X(F)dF$$

where again the 2 factor reflects the additional contribution of the negative frequencies. The properties of the PSD are as follows:

Property 17.19 – PSD is a real function.

The PSD is given by the real function

$$P_X(F) = \int_{-\infty}^{\infty} r_X(\tau)\cos(2\pi F\tau)d\tau$$

$\square$

Property 17.20 – PSD is nonnegative.

$$P_X(F) \geq 0$$

$\square$

Property 17.21 – PSD is symmetric about $F = 0$.

$$P_X(-F) = P_X(F)$$

$\square$

Property 17.22 – ACF recovered from PSD using inverse Fourier transform

$$r_X(\tau) = \int_{-\infty}^{\infty} P_X(F) \exp(j2\pi F\tau) dF \qquad -\infty < \tau < \infty \qquad (17.51)$$

$$= \int_{-\infty}^{\infty} P_X(F) \cos(2\pi F\tau) dF \qquad -\infty < \tau < \infty. \qquad (17.52)$$

(See also Problem 17.49.)

$\square$

Unlike the PSD for a discrete-time WSS random process, the PSD for a continuous-time WSS random process is *not* periodic. We next illustrate these definitions and formulas with an example of practical importance.

Example 17.11 – Obtaining discrete-time WGN from continuous-time WGN

A common model for a continuous-time noise random process $X(t)$ in a physical system is a WSS random process with a zero mean. In addition, due to the origin of noise as microscopic fluctuations of a large number of electrons, or molecules, etc., a central limit theorem can be employed to assert that $X(t)$ is a Gaussian random variable for all t. The average power of the noise in a band of frequencies is observed to be the same for all bands up to some upper frequency limit, at which the average power begins to decrease. For instance, consider thermal noise in a conductor due to random fluctuations of the electrons about some mean velocity. The average power versus frequency is predicted by physics to be constant until a cutoff frequency of about $F_c = 1000$ GHz at room temperature [Bell Telephone Labs 1970]. Hence, we can assume that the PSD of the noise has a PSD shown in Figure 17.14 as the true PSD. To further simplify the mathematically modeling without sacrificing the realism of the model, we can observe that all physical systems will only pass frequency components that are much lower than F_c—typically the bandwidth of the system is W Hz as shown in Figure 17.14. Any frequencies above W Hz will be cut off by the system. Therefore, the noise output of the system will be the same whether we use the true PSD or the modeled one shown in Figure 17.10. The modeled PSD is given by

$$P_X(F) = \frac{N_0}{2} \qquad -\infty < F < \infty.$$

This is clearly a physically impossible PSD in that the total average power is $r_X(0) = \int_{-\infty}^{\infty} P_X(F) df = \infty$. However, its use simplifies much systems analysis (see Problem 17.50). The corresponding ACF is from (17.51) the inverse Fourier transform, which is

$$r_X(\tau) = \frac{N_0}{2} \delta(\tau) \qquad (17.53)$$

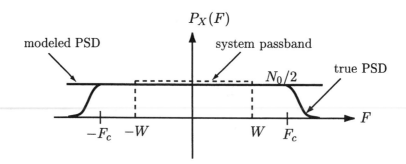

Figure 17.14: True and modeled PSDs for continuous-time white Gaussian noise.

and is seen to be an impulse at $\tau = 0$. Again the nonphysical nature of this model is manifest by the value $r_X(0) = \infty$. A continuous-time WSS Gaussian random process with zero mean and the ACF given by (17.53) is called continuous-time *white Gaussian noise* (WGN) (see also Example 20.6). It is a standard model in many disciplines.

Now as was previously mentioned, all physical systems are bandlimited to W Hz, which is typically chosen to ensure that a desired signal with a bandwidth of W Hz is not distorted. Modern signal processing hardware first bandlimits the continuous-time waveform to a maximum of W Hz using a lowpass filter and then samples the output of the filter at the Nyquist rate of $F_s = 2W$ samples/sec. The samples are then input into a digital computer. An important question to answer is: What are the statistical characteristics of the noise samples that are input to the computer? To answer this question we let Δ_t be the time interval between successive samples. Also, let $X(t)$ be the noise at the output of an ideal lowpass filter ($H(F) = 1$ for $|F| \leq W$ and $H(F) = 0$ for $|F| > W$) over the system passband shown in Figure 17.14. Then, the noise samples can be represented as

$$X(t)|_{t=n\Delta_t} = X[n] \qquad \text{for } -\infty < n < \infty.$$

Since $X(t)$ is bandlimited to W Hz and prior to filtering had the modeled PSD shown in Figure 17.14, its PSD is

$$P_X(F) = \begin{cases} \frac{N_0}{2} & |F| \leq W \\ 0 & |F| > W. \end{cases}$$

The noise samples $X[n]$ comprise a discrete-time random process. Its characteristics follow those of $X(t)$. Since $X(t)$ is Gaussian, then so is $X[n]$ (being just a sample). Also, since $X(t)$ is zero mean, so is $X[n]$ for all n. Finally, we inquire as to whether $X[n]$ is WSS, i.e., can we define an ACS? To answer this we first note that $X[n] = X(n\Delta_t)$ and recall that $X(t)$ is WSS. Then from the definition of the ACS

$$\begin{aligned} E[X[n]X[n+k]] &= E[X(n\Delta_t)X((n+k)\Delta_t)] \\ &= r_X(k\Delta_t) \quad \text{(definition of continuous-time ACF)} \end{aligned}$$

which does not depend on n, and so $X[n]$ is a zero mean discrete-time WSS random process with ACS

$$r_X[k] = r_X(k\Delta_t). \tag{17.54}$$

It is seen to be a sampled version of the continuous-time ACF. To explicitly evaluate the ACS we have from (17.51)

$$
\begin{aligned}
r_X(\tau) &= \int_{-\infty}^{\infty} P_X(F)\exp(j2\pi F\tau)dF \\
&= \int_{-W}^{W} \frac{N_0}{2}\exp(j2\pi F\tau)dF \\
&= \frac{N_0}{2}\int_{-W}^{W}\cos(2\pi F\tau)dF \quad \text{(sine component is odd function)} \\
&= \frac{N_0}{2}\frac{\sin(2\pi F\tau)}{2\pi\tau}\Big|_{-W}^{W} \\
&= N_0 W\frac{\sin(2\pi W\tau)}{2\pi W\tau} \tag{17.55}
\end{aligned}
$$

which is shown in Figure 17.15. Now since $r_X[k] = r_X(k\Delta_t) = r_X(k/(2W))$, we

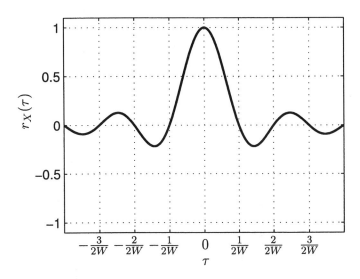

Figure 17.15: ACF for bandlimited continuous-time WGN with $N_0 W = 1$.

see from Figure 17.15 that for $k = \pm 1, \pm 2, \ldots$ the ACS is zero, being the result of sampling the continuous-time ACF at its zeros. The only nonzero value is for $k = 0$, which is $r_X[0] = r_X(0) = N_0 W$ from (17.55). Therefore, we finally observe that the ACS of the noise samples is

$$r_X[k] = N_0 W\delta[k]. \tag{17.56}$$

The discrete-time noise random process is indeed WSS and has the ACS of (17.56). The PSD corresponding to this ACS has already been found and is shown in Figure 17.10, where $\sigma^2 = N_0 W$. Therefore, $X[n]$ *is a discrete-time white Gaussian noise random process.* This example justifies the use of the WGN model for discrete-time systems analysis.

$\Diamond$

 Sampling faster gives only marginally better performance.

It is sometimes argued that by sampling the output of a system lowpass filter whose cutoff frequency is W Hz at a rate greater than $2W$, we can improve the performance of a signal processing system. For example, consider the estimation of the mean μ based on samples $Y[n] = \mu + X[n]$ for $n = 0, 1, \ldots, N - 1$ where $E[X[n]] = 0$, $\text{var}(X[n]) = \sigma^2$, and the $X[n]$ samples are uncorrelated. The obvious estimate is the sample mean or $(1/N) \sum_{n=0}^{N-1} Y[n]$, whose expectation is μ and whose variance is σ^2/N. Clearly, if we could increase N, then the variance could be reduced and a better estimate would result. This suggests sampling the continuous-time random process at a rate faster than $2W$ samples/sec. The fallacy, however, is that as the sampling rate increases, *the noise samples become correlated* as can be seen by considering a sampling rate of $4W$ for which the time interval between samples becomes $\tau = \Delta_t/2 = 1/(4W)$. Then, as observed from Figure 17.15, the correlation between successive samples is $r_X(1/(4W)) = 0.6$. In effect, by sampling faster we are not obtaining any new realizations of the noise samples but nearly repetitions of the same noise samples. As a result, the variance will *not* decrease as $1/N$ but at a slower rate (see also Problem 17.51).

17.9   Real-World Example – Random Vibration Testing

Anyone who has ever traveled in a jet knows that upon landing, the cabin can vibrate greatly. This is due to the air currents outside the cabin which interact with the metallic aircraft surface. These pressure variations give rise to vibrations which are referred to as *turbulent boundary layer noise*. A manufacturer that intends to attach an antenna or other device to an aircraft must be cognizant of this vibration and plan for it. It is customary then to subject the antenna to a random vibration test in the lab to make sure it is not adversely affected in flight [McConnell 1995]. To do so the antenna would be mounted on a shaker table and the table shaken in a manner to simulate the turbulent boundary layer (TBL) noise. The problem the manufacturer faces is how to provide the proper vibration signal to the table, which presumably will then be transmitted to the antenna. We now outline a possible solution to this problem.

The National Aeronautics and Space Administration (NASA) has determined PSD models for the TBL noise through physical modeling and experimentation. A reasonable model for the *one-sided* PSD of TBL noise upon reentry of a space vehicle, such as the space shuttle, into the earth's atmosphere is given by [NASA 2001]

$$G_X(F) = \begin{cases} G_X(500) & 0 \le F < 500 \text{ Hz} \\ \frac{9 \times 10^{14} r^2}{F + 11364} & 500 \le F \le 50000 \text{ Hz} \end{cases}$$

where r represents a reference value which is 20μPa. A μPa is a unit of pressure equal to 10^{-6} nt/m². This PSD is shown in Figure 17.16 referenced to the standard unit so that $r = 1$. Note that it has a lowpass type of characteristic. In order

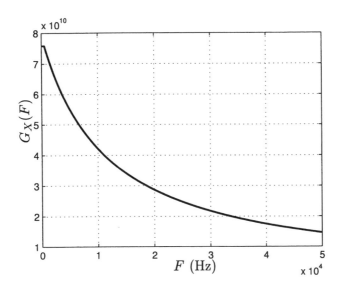

Figure 17.16: Continuous-time one-sided PSD for TBL noise.

to provide a signal to the shaker table that is random and has the PSD shown in Figure 17.16, we will assume that the signal is produced in a digital computer and then converted via a digital-to-analog convertor to a continuous-time signal. Hence, we need to produce a discrete-time WSS random process within the computer that has the proper PSD. Recalling our discussion in Section 17.8 we know that $r_X[k] = r_X(k\Delta_t)$ and since the highest frequency in the PSD is $W = 50,000$ Hz, we choose $\Delta_t = 1/(2W) = 1/100,000$. This produces the discrete-time PSD shown in Figure 17.17 and is given by $P_X(f) = (1/(2\Delta_t))G_X(f/\Delta_t)$. (We have divided by two to obtain the usual *two-sided* PSD. Also, the sampling operation introduces a factor of $1/\Delta_t$ [Jackson 1991].) To generate a realization of a discrete-time WSS random process with PSD given in Figure 17.17 we will use the AR model introduced in

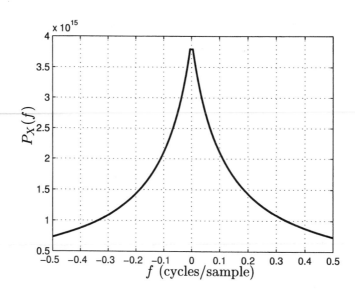

Figure 17.17: Discrete-time PSD for TBL noise.

Example 17.5. From the ACS we can determine values of a and σ_U^2 if we know $r_X[0]$ and $r_X[1]$ since

$$a = \frac{r_X[1]}{r_X[0]} \tag{17.57}$$

$$\sigma_U^2 = r_X[0](1 - a^2) = r_X[0]\left[1 - \left(\frac{r_X[1]}{r_X[0]}\right)^2\right]. \tag{17.58}$$

Knowing a and σ_U^2 will allow us to use the defining recursive difference equation, $X[n] = aX[n-1] + U[n]$, of an AR random process to generate the realization. To obtain the first two lags of the ACS we use (17.39)

$$r_X[0] = \int_{-\frac{1}{2}}^{\frac{1}{2}} P_X(f)df$$

$$r_X[1] = \int_{-\frac{1}{2}}^{\frac{1}{2}} P_X(f)\cos(2\pi f)df$$

where $P_X(f)$ is given in Figure 17.17. These can be evaluated numerically by replacing the integrals with approximating sums to yield $r_X[0] = 1.5169 \times 10^{15}$ and $r_X[1] = 4.8483 \times 10^{14}$. Then, using (17.57) and (17.58), we have the AR parameters $a = 0.3196$ and $\sigma_U^2 = 1.362 \times 10^{15}$. With these parameters the AR PSD (see (17.36)) and the true PSD (shown in Figure 17.17) are plotted in Figure 17.18. The agreement between them is fairly good except near $f = 0$. Hence, with these values of the parameters a random process realization could be synthesized within a digital computer and then converted to analog form to drive the shaker table.

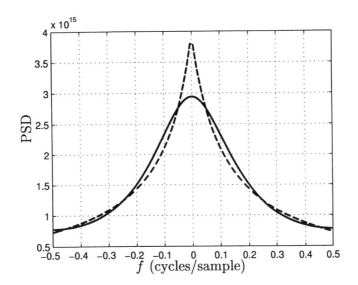

Figure 17.18: Discrete-time PSD and its AR PSD model for TBL noise. The true PSD is shown as the dashed line and the AR PSD model as the solid line.

References

Bell Telephone Laboratories, *Transmission Systems for Communications*, Western Electric Co, Winston-Salem, NC, 1970.

Box, E.P.G., G.M. Jenkins, *Time Series Analysis, Forecasting and Control*, Holden-Day, San Francisco, 1970.

Brockwell, P.J., R.A. Davis, *Time Series: Theory and Methods*, Springer-Verlag, New York, 1987.

Harvey, A.C., *Forecasting, Structural Time Series Models and the Kalman Filter*, Cambridge University Press, New York, 1989.

Jackson, L.B., *Signals, Systems, and Transforms*, Addison-Wesley, Reading, MA 1991.

Jenkins, G.M., D.G. Watts, *Spectral Analysis and Its Applications*, Holden-Day, San Francisco, 1968.

Kay, S., *Modern Spectral Estimation, Theory and Applications*, Prentice-Hall, Englewood Cliffs, NJ, 1988.

McConnell, K.G., *Vibration Testing*, John Wiley & Sons, New York, 1995.

NASA, *Dynamic Environmental Criteria, NASA Technical Handbook*, pp. 50–51, NASA-HDBK-7005, March 13, 2001.

Papoulis, A., *Probability, Random Variables, and Stochastic Processes*, McGraw-Hill, New York, 1965.

Robinson, E.A., *Multichannel Time Series Analysis with Digital Computer Programs*, Holden-Day, San Francisco, 1967.

Schuster, A., "On the Investigation of Hidden Periodicities with Applications to a Supposed 26 Day Period of Meterological Phenomena," *Terr. Magn.*, Vol. 3, pp. 13–41, March 1898.

Schwartz, M., L. Shaw, *Signal Processing, Discrete Spectral Analysis, Detection, and Estimation*, McGraw-Hill, New York, 1975.

Taylor, S., *Modelling Financial Time Series*, John Wiley & Sons, New York, 1986.

Tong, H., *Non-linear Time Series*, Oxford University Press, New York, 1990.

Problems

17.1 (☺) (w) A Bernoulli random process $X[n]$ for $-\infty < n < \infty$ consists of independent random variables with each random variable taking on the values $+1$ and -1 with probabilities p and $1-p$, respectively. Is this random process WSS? If it is WSS, find its mean sequence and autocorrelation sequence.

17.2 (w) Consider the random process defined as $X[n] = a_0 U[n] + a_1 U[n-1]$ for $-\infty < n < \infty$, where a_0 and a_1 are constants, and $U[n]$ is an IID random process with each $U[n]$ having a mean of zero and a variance of one. Is this random process WSS? If it is WSS, find its mean sequence and autocorrelation sequence.

17.3 (w) A sinusoidal random process is defined as $X[n] = A\cos(2\pi f_0 n)$ for $-\infty < n < \infty$, where $0 < f_0 < 0.5$ is a discrete-time frequency, and $A \sim \mathcal{N}(0,1)$. Is this random process WSS? If it is WSS, find its mean sequence and autocorrelation sequence.

17.4 (f) A WSS random process has $E[X[0]] = 1$ and a covariance sequence $c_X[n_1, n_2] = 2\delta[n_2 - n_1]$. Find the ACS and plot it.

17.5 (☺) (w) A random process $X[n]$ for $-\infty < n < \infty$ consists of independent random variables with

$$X[n] \sim \begin{cases} \mathcal{N}(0,1) & \text{for } n \text{ even} \\ \mathcal{U}(-\sqrt{3}, \sqrt{3}) & \text{for } n \text{ odd.} \end{cases}$$

Is this random process WSS? Is it stationary?

17.6 (w) The random processes $X[n]$ and $Y[n]$ are both WSS. Every sample of $X[n]$ is independent of every sample of $Y[n]$. Is $Z[n] = X[n] + Y[n]$ WSS? If it is WSS, find its mean sequence and autocorrelation sequence.

17.7 (w) The random processes $X[n]$ and $Y[n]$ are both WSS. Every sample of $X[n]$ is independent of every sample of $Y[n]$. Is $Z[n] = X[n]Y[n]$ WSS? If it is WSS, find its mean sequence and autocorrelation sequence.

17.8 (f) For the ACS $r_X[k] = (1/2)^k$ for $k \geq 0$ and $r_X[k] = (1/2)^{-k}$ for $k < 0$, verify that Properties 17.1–17.3 are satisfied.

17.9 (☺) (w) For the sequence $r_X[k] = ab^{|k|}$ for $-\infty < k < \infty$, determine the values of a and b that will result in a valid ACS.

17.10 (w) A periodic WSS random process with period P is defined to be a random process $X[n]$ whose ACS satisfies $r_X[k + P] = r_X[k]$ for all k. An example is the randomly phased sinusoid of Example 17.10 for which $P = 10$. Show that the correlation coefficient for two samples of a *zero mean* periodic random process that are separated by P samples is one. Comment on the predictability of $X[n + P]$ based on $X[n] = x[n]$.

17.11 (w) A WSS random process has an ACS $r_X[k]$ and mean μ. Find the correlation coefficient for two samples of the random process that are separated by k samples.

17.12 (☺) (w) Which of the sequences in Figure 17.19 cannot be valid ACSs? If the sequence cannot be an ACS, explain why not.

17.13 (w) For the randomly phased sinusoid described in Example 17.4 find the optimal linear prediction of $X[1]$ based on observing $X[0] = x[0]$, and also of $X[10]$ based on observing $X[0] = x[0]$. Can either of these samples be perfectly predicted? Explain why or why not.

17.14 (w) For the AR random process described in Example 17.10 find the optimal linear prediction of $X[n_0+k_0]$ based on observing $X[n_0] = x[n_0]$. How accurate is your prediction in terms of MSE as k_0 increases?

17.15 (t) In this problem we derive $r_X[0]$ for the AR random process described in Example 17.5. To do so assume that $X[n]$ can be written as

$$X[n] = \sum_{k=0}^{\infty} a^k U[n - k]. \tag{17.59}$$

This was shown to be true in Example 17.5. Then verify that $r_X[0]$ can be written as

$$r_X[0] = \sum_{k=0}^{\infty} \sum_{l=0}^{\infty} a^k a^l E[U[n - k]U[n - l]]$$

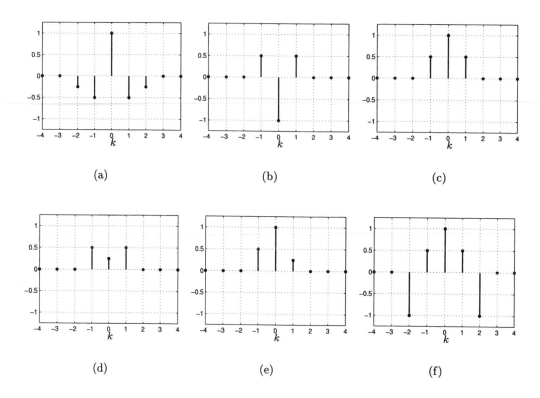

Figure 17.19: Possible ACSs for Problem 17.12.

and use the properties of the $U[n]$ random process to finish the derivation.

17.16 (t) Using a similar approach to the one used in Problem 17.15 derive the ACS for the AR random process described in Example 17.5. Hint: Start with the definition of the ACS and use (17.59).

17.17 (☺) (w) To generate a realization of an AR process on the computer we can use the recursive difference equation $X[n] = aX[n-1] + U[n]$ for $n \geq 0$. However, in doing so, we soon realize that the initial condition $X[-1]$ is required. Assume that we set $X[-1] = 0$ and use the recursion $X[0] = U[0], X[1] = aX[0] + U[1], \ldots$. Determine the mean and variance of $X[n]$ for $n \geq 0$, where as usual $U[n]$ consists of uncorrelated random variables with zero mean and variance σ_U^2. Does the mean depend on n? Does the variance depend on n? What happens as $n \to \infty$? Hint: First show that $X[n]$ can be written as $X[n] = \sum_{k=0}^{n} a^k U[n-k]$ for $n \geq 0$.

17.18 (w) This problem continues Problem 17.17. Instead of letting $X[-1] = 0$, set $X[-1]$ equal to a random variable with mean 0 and a variance of $\sigma_U^2/(1-a^2)$ and that is uncorrelated with $U[n]$ for $n \geq 0$. Find the mean and variance of

$X[0]$. Explain your results and why this makes sense.

17.19 ($\smile$) (w) An example of a sequence that is not positive semidefinite is $r[0] = 1$, $r[-1] = r[1] = -7/8$ and equals zero otherwise. Compute the determinant of the 1×1 principal minor, the 2×2 principal minor, and the 3×3 principal minor of the 3×3 autocorrelation matrix $\mathbf{R}_X$ using these values. Also, plot the discrete-time Fourier transform of $r[k]$. Why do you think the positive semidefinite property is important?

17.20 ($\smile$) (w) For the general MA random process of Example 17.6 show that the process is WSS.

17.21 (f) Use (17.28) to show that the MA random process defined in Example 17.6 is ergodic in the mean.

17.22 (t,f) Show that a WSS random process whose ACS satisfies $r_X[k] = \mu^2$ for $k > k_0 \geq 0$ must be ergodic in the mean.

17.23 (t) Prove (17.28) by using the relationship

$$\sum_{i=0}^{N-1} \sum_{j=0}^{N-1} g[i-j] = \sum_{k=-(N-1)}^{N-1} (N - |k|)g[k].$$

Try verifying this relationship for $N = 3$.

17.24 (f) For the random DC level defined in Example 17.7 prove that $\mathrm{var}(\hat{\mu}_N) = 1$.

17.25 (f) Explain why the randomly phased sinusoid defined in Example 17.4 is ergodic in the mean. Next show that it is ergodic in the ACS in that

$$\lim_{N \to \infty} \hat{r}_X[k] = \lim_{N \to \infty} \frac{1}{N-k} \sum_{n=0}^{N-1-k} X[n]X[n+k] = \frac{1}{2}\cos(2\pi(0.1)k) = r_X[k] \quad k \geq 0$$

by computing $\hat{r}_X[k]$ directly. Hint: Use the fact that $\lim_{N\to\infty}(1/(N-k)) \sum_{n=0}^{N-1-k} \cos(2\pi f n + \phi) = 0$ for any $0 < f < 1$ and any phase angle ϕ. This is because the temporal average of an infinite duration sinusoid is zero.

17.26 (t) Show that the formula

$$\sum_{m=-M}^{M} \sum_{n=-M}^{M} g[m-n] = \sum_{k=-2M}^{2M} (2M + 1 - |k|)g[k]$$

is true for $M = 1$.

17.27 (t) Argue that

$$\lim_{M\to\infty}\sum_{k=-2M}^{2M}\underbrace{\left(1-\frac{|k|}{2M+1}\right)}_{w[k]}r_X[k]\exp(-j2\pi fk)=\sum_{k=-\infty}^{\infty}r_X[k]\exp(-j2\pi fk)$$

by drawing pictures of $r_X[k]$, which decays to zero, and overlay it with $w[k]$ as M increases.

17.28 ($\smile$) (w) For the differenced random process defined in Example 17.1 determine the PSD. Explain your results.

17.29 (f) Determine the PSD for the randomly phased sinusoid described in Example 17.4. Is this result reasonable? Hint: The discrete-time Fourier transform of $\exp(j2\pi f_0 n)$ for $-1/2 < f_0 < 1/2$ is $\delta(f - f_0)$ over the frequency interval $-1/2 \le f \le 1/2$.

17.30 ($\smile$) (w) A random process is defined as $X[n] = AU[n]$, where $A \sim \mathcal{N}(0, \sigma_A^2)$ and $U[n]$ is white noise with variance σ_U^2. The random variable A is independent of all the samples of $U[n]$. Determine the PSD of $X[n]$.

17.31 (w) Find the PSD of the random process $X[n] = (1/2)^{|n|}U[n]$ for $-\infty < n < \infty$, where $U[n]$ is white noise with variance σ_U^2.

17.32 (w) Find the PSD of the random process $X[n] = a_0 U[n] + a_1 U[n-1]$, where a_0, a_1 are constants and $U[n]$ is white noise with variance $\sigma_U^2 = 1$.

17.33 (w) A Bernoulli random process consists of IID Bernoulli random variables taking on values $+1$ and -1 with equal probabilities. Determine the PSD and explain your results.

17.34 ($\smile$) (w) A random process is defined as $X[n] = U[n] + \mu$ for $-\infty < n < \infty$, where $U[n]$ is white noise with variance σ_U^2. Find the ACS and PSD and plot your results.

17.35 (w,c) Consider the AR random process defined in Example 17.5 and described further in Example 17.10 with $-1 < a < 0$ and for some $\sigma_U^2 > 0$. Plot the PSD for several values of a and explain your results.

17.36 (f,c) Plot the corresponding PSD for the ACS

$$r_X[k] = \begin{cases} 1 & k = 0 \\ 1/2 & k = \pm 1 \\ 1/4 & k = \pm 2 \\ 0 & \text{otherwise.} \end{cases}$$

17.37 (w) If a random process has the PSD $P_X(f) = 1 + \cos(2\pi f)$, are the samples of the random process uncorrelated?

17.38 ($\smile$) (f) If a random process has the PSD $P_X(f) = |1 + \exp(-j2\pi f) + (1/2)\exp(-j4\pi f)|^2$, determine the ACS.

17.39 (c) For the AR random processes whose ACSs are shown in Figure 17.6 generate a realization of $N = 2000$ samples for each process. Use the MATLAB code segment given in Section 17.4 to do this. Then, estimate the ACS for $k = 0, 1, \ldots, 30$ and plot the results. Compare your results to those shown in Figure 17.12 and explain.

17.40 ($\smile$) (w) A PSD is given as $P_X(f) = a + b\cos(2\pi f)$ for some constants a and b. What values of a and b will result in a valid PSD?

17.41 (f) A PSD is given as

$$P_X(f) = \begin{cases} 2 - 8f & 0 \le f \le 1/4 \\ 0 & 1/4 < f \le 1/2. \end{cases}$$

Plot the PSD and find the total average power in the random process.

17.42 ($\smile$) (c) Plot 50 realizations of the randomly phased sinusoid described in Example 17.4 with $N = 50$, and overlay the samples in a scatter diagram plot such as shown in Figure 16.15. Explain the results by referring to the PDF of Figure 16.12. . Next estimate the following quantities: $E[X[10]], E[X[12]]$, $E[X[10]X[12]], E[X[12]X[14]]$ by averaging down the ensemble, and compare your simulated results to the theoretical values.

17.43 (c) In this problem we support the results of Problem 17.18 by using a computer simulation. Specifically, generate $M = 10,000$ realizations of the AR random process $X[n] = 0.95X[n-1] + U[n]$ for $n = 0, 1, \ldots, 49$, where $U[n]$ is WGN with $\sigma_U^2 = 1$. Do so two ways: for the first set of realizations let $X[-1] = 0$ and for the second set of realizations let $X[-1] \sim \mathcal{N}(0, \sigma_U^2/(1-a^2))$, using a different random variable for each realization. Now estimate the variance for each sample time n, which is $r_X[0]$, by averaging $X^2[n]$ down the ensemble of realizations. Do you obtain the theoretical result of $r_X[0] = \sigma_U^2/(1-a^2)$?

17.44 ($\smile$) (c) Generate a realization of discrete-time white Gaussian noise with variance $\sigma_X^2 = 1$. For $N = 64$, $N = 128$, and $N = 256$, plot the periodogram. What is the true PSD? Does your estimated PSD get closer to the true PSD as N increases? If not, how could you improve your estimate?

17.45 (c) Generate a realization of an AR random process of length $N = 31,000$ with $a = 0.25$ and $\sigma_U^2 = 1-a^2$. Break up the data set into 1000 nonoverlapping blocks of data and compute the periodogram for each block. Finally, average the periodograms together for each point in frequency to determine the final averaged periodogram estimate. Compare your results to the theoretical PSD shown in Figure 17.11a.

17.46 (f) A continuous-time randomly phased sinusoid is defined by $X(t) = \cos(2\pi F_0 t + \Theta)$, where $\Theta \sim \mathcal{U}(0, 2\pi)$. Determine the mean function and ACF for this random process.

17.47 (☺) (f) For the PSD $P_X(F) = \exp(-|F|)$, determine the average power in the band $[10, 100]$ Hz.

17.48 (w) If a PSD is given as $P_X(F) = \exp(-|F/F_0|)$, what happens to the ACF as F_0 increases and also as $F_0 \to \infty$?

17.49 (t) Based on (17.49) derive (17.50), and also based on (17.51) derive (17.52).

17.50 (☺) (w) A continuous-time white noise random process $U(t)$ whose PSD is given as $P_U(F) = N_0/2$ is integrated to form the continuous-time MA random process

$$X(t) = \frac{1}{T} \int_{t-T}^{t} U(\xi) d\xi.$$

Determine the mean function and the variance of $X(t)$. Does $X(t)$ have infinite total average power?

17.51 (☺) (w,c) Consider a continuous-time random process $X(t) = \mu + U(t)$, where $U(t)$ is zero mean and has the ACF given in Figure 17.15. If $X(t)$ is sampled at twice the Nyquist rate, which is $F_s = 4W$, determine the ACS of $X[n]$. Next using (17.28) find the variance of the sample mean estimator $\hat{\mu}_N$ for $N = 20$. Is it half of the variance of the sample mean estimator if we had sampled at the Nyquist rate and used $N = 10$ samples in our estimate? Note that in either case the total length of the data interval in seconds is the same, which is $20/(4W) = 10/(2W)$.

17.52 (f) A PSD is given as

$$P_X(f) = \left| 1 + \frac{1}{2} \exp(-j2\pi f) \right|^2.$$

Model this PSD by using an AR PSD as was done in Section 17.9. Plot the true PSD and the AR model PSD.

Chapter 18

Linear Systems and Wide Sense Stationary Random Processes

18.1 Introduction

Most physical systems are conveniently modeled by a *linear system*. These include electrical circuits, mechanical machines, human biological functions, and chemical reactions, just to name a few. When the system is capable of responding to a continuous-time input, its effect can be described using a linear differential equation. For a system that responds to a discrete-time input a linear difference equation can be used to characterize the effect of the system. Furthermore, for systems whose characteristics do not change with time, the coefficients of the differential or difference equation are constants. Such a system is termed a *linear time invariant* (LTI) system for continuous-time inputs/outputs and a *linear shift invariant* (LSI) system for discrete-time inputs/outputs. In this chapter we explore the effect of these systems on wide sense stationary (WSS) random process inputs. The reader who is unfamiliar with the basic concepts of linear systems should first read Appendix D for a brief introduction. Many excellent books are available to supplement this material [Jackson 1991, Oppenheim, Willsky, and Nawab 1997, Poularikas and Seely 1985]. We will now consider only discrete-time systems and discrete-time WSS random processes. A summary of the analogous concepts for the continuous-time case is given in Section 18.6.

The importance of LSI systems is that they maintain the wide sense stationarity of the random process. That is to say, *if the input to an LSI system is a WSS random process, then the output is also a WSS random process.* The mean and ACS, or equivalently the PSD, however, are modified by the action of the system. We will be able to obtain simple formulas yielding these quantities at the system output. In effect, the linear system modifies the first two moments of the random process but in an easily determined and intuitively pleasing way. This allows us to assess the effect of a linear system on a WSS random process and therefore provides a means to produce a WSS random process at the output with some desired characteristics. Furthermore, the theory is easily extended to the case of multiple random processes and multiple linear systems as we will see in the next chapter.

18.2   Summary

For the linear shift invariant system shown in Figure 18.1 the output random process is given by (18.2). If the input random process is WSS, then the output random process is also WSS. The output random process has a mean given by (18.9), an ACS given by (18.10), and a PSD given by (18.11). If the input WSS random process is white noise, then the output random process has the ACS of (18.15). In Section 18.4 the PSD is interpreted, using the results of Theorem 18.3.1, as the average power in a narrow frequency band divided by the width of the frequency band. The application of discrete-time linear systems to estimation of samples of a random process is explored in Section 18.5. Generically known as Wiener filtering, there are four separate problems defined, of which the smoothing and prediction problems are solved. For smoothing of a random process signal in noise the estimate is given by (18.20) and the optimal filter has the frequency response of (18.25). A specific application is given in Example 18.4 to estimation of an AR signal that has been corrupted by white noise. The minimum MSE of the optimal Wiener smoother is given by (18.27). One-step linear prediction of a random process sample based on the current and all past samples as given by (18.21) leads to the optimal filter impulse response satisfying the infinite set of linear equations of (18.28). The general solution is summarized in Section 18.5.2 and then illustrated in Example 18.6. For linear prediction based on the current sample and a finite number of past samples the optimal impulse response is given by the solution of the Wiener-Hopf equations of (18.36). The corresponding minimum MSE is given by (18.37). In particular, if the random process is an AR random process of order p, the Wiener-Hopf equations are the same as the Yule-Walker equations of (18.38) and the minimum mean square error equation of (18.37) is the same as for the white noise variance of (18.39). In Section 18.6 the corresponding formulas for a continuous-time random process that is input to a linear time invariant system are summarized. The mean at the output is given by (18.40), the ACF is given by (18.41), and the PSD is given by (18.42). Example 18.7 illustrates the use of these formulas. In Section 18.7 the application of AR random process modeling to speech synthesis is described. In particular, it is shown how a segment of speech can first be modeled, and then how for an actual segment of speech, the parameters of the model can be extracted. The model with its estimated parameters can then be used for speech synthesis.

18.3   Random Process at Output of Linear System

We wish to consider the effect of an LSI system on a discrete-time WSS random process. We will from time to time refer to the linear system as a *filter*, a term that is synonomous. In Section 18.6 we summarize the results for a continuous-time WSS random process that is input to an LTI system. To proceed, let $U[n]$ be the WSS random process input and $X[n]$ be the random process output of the system. We generally represent an LSI system schematically with its input and output as shown in Figure 18.1. Previously, in Chapters 16 and 17 we have seen several examples

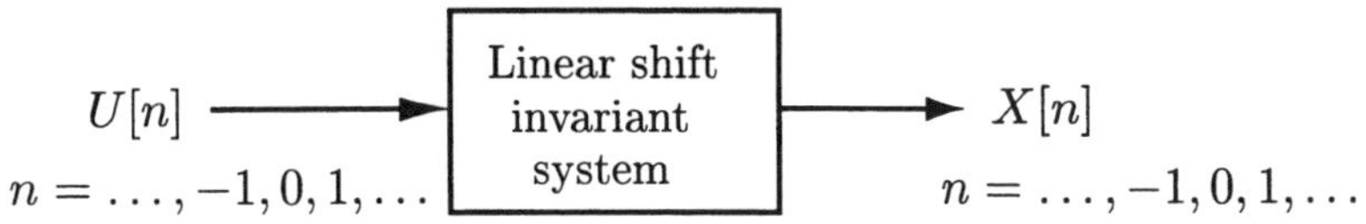

Figure 18.1: Linear shift invariant system with random process input and output.

of LSI systems with WSS random process inputs. One example is the MA random process (see Example 16.7) for which $X[n] = (U[n] + U[n-1])/2$, with $U[n]$ a white Gaussian noise process with variance σ_U^2. (Recall that discrete-time white noise is a zero mean WSS random process with ACS $r_U[k] = \sigma_U^2 \delta[k]$.) We may view the MA random process as the output $X[n]$ of an LSI filter excited at the input by the white Gaussian noise random process $U[n]$. (In this chapter we will be considering only the *first two moments* of $X[n]$. That $U[n]$ is a random process consisting of *Gaussian* random variables is of no consequence to these discussions. The same results are obtained for any white noise random process $U[n]$ irregardless of the marginal PDFs. In Chapter 20, however, we will consider the joint PDF of samples of $X[n]$, and in that case, the fact that $U[n]$ is white *Gaussian* noise will be very important.) The averaging operation can be thought of as a filtering by the LSI filter having an *impulse response*

$$h[k] = \begin{cases} \frac{1}{2} & k = 0 \\ \frac{1}{2} & k = 1 \\ 0 & \text{otherwise.} \end{cases} \tag{18.1}$$

(Recall that the impulse response $h[n]$ is the output of the LSI system when the input $u[n]$ is a unit impulse $\delta[n]$.) This is because the output of an LSI filter is obtained using the convolution sum formula

$$X[n] = \sum_{k=-\infty}^{\infty} h[k]U[n-k] \tag{18.2}$$

so that upon using (18.1) in (18.2) we have

$$\begin{aligned} X[n] &= h[0]U[n] + h[1]U[n-1] \\ &= \frac{1}{2}U[n] + \frac{1}{2}U[n-1] \\ &= \frac{1}{2}(U[n] + U[n-1]). \end{aligned}$$

In general, the LSI system will be specified by giving its impulse response $h[k]$ for $-\infty < k < \infty$ or equivalently by giving its *system function*, which is defined as the z-transform of the impulse response. The system function is thus given by

$$\mathcal{H}(z) = \sum_{k=-\infty}^{\infty} h[k]z^{-k}. \tag{18.3}$$

In addition, we will have need for the *frequency response* of the LSI system, which is defined as the discrete-time Fourier transform of the impulse response. It is therefore given by

$$H(f) = \sum_{k=-\infty}^{\infty} h[k]\exp(-j2\pi fk). \tag{18.4}$$

This function assesses the effect of the system on a complex sinusoidal input sequence $u[n] = \exp(j2\pi f_0 n)$ for $-\infty < n < \infty$. It can be shown that the response of the system to this input is $x[n] = H(f_0)\exp(j2\pi f_0 n) = H(f_0)u[n]$ (use (18.2) with the deterministic input $u[n] = \exp(j2\pi f_0 n)$). Hence, its name derives from the fact that the system action is to modify the amplitude of the complex sinusoid by $|H(f_0)|$ and the phase of the complex sinusoid by $\angle H(f_0)$, but otherwise retains the complex sinusoidal sequence. It should also be noted that the frequency response is easily obtained from the system function as $H(f) = \mathcal{H}(\exp(j2\pi f))$. For the MA random process we have upon using (18.1) in (18.3) that the system function is

$$\mathcal{H}(z) = \frac{1}{2} + \frac{1}{2}z^{-1}$$

and the frequency response is the system function when z is replaced by $\exp(j2\pi f)$, yielding

$$H(f) = \frac{1}{2} + \frac{1}{2}\exp(-j2\pi f).$$

It is said that the system function has been evaluated "on the unit circle in the z-plane".

We next give an example to determine the characteristics of the output random process of an LSI system with a WSS input random process. The previous example is generalized slightly to prepare for the theorem to follow.

Example 18.1 – Output random process characteristics

Let $U[n]$ be a WSS random process with mean μ_U and ACS $r_U[k]$. This random process is input to an LSI system with impulse response

$$h[k] = \begin{cases} h[0] & k = 0 \\ h[1] & k = 1 \\ 0 & \text{otherwise.} \end{cases}$$

This linear system is called a finite impulse response (FIR) filter since its impulse response has only a finite number of nonzero samples. We wish to determine if

a. the output random process is WSS and if so

b. its mean sequence and ACS.

The output of the linear system is from (18.2)

$$X[n] = h[0]U[n] + h[1]U[n-1].$$

The mean sequence is found as

$$\begin{aligned}
E[X[n]] &= h[0]E[U[n]] + h[1]E[U[n-1]] \\
&= h[0]\mu_U + h[1]\mu_U \\
&= (h[0] + h[1])\mu_U
\end{aligned}$$

so that the mean is constant with time and is given by

$$\mu_X = (h[0] + h[1])\mu_U.$$

It can also be written from (18.4) as

$$\mu_X = \left. \sum_{k=-\infty}^{\infty} h[k]\exp(-j2\pi fk) \right|_{f=0} \mu_U = H(0)\mu_U.$$

The mean at the output of the LSI system is seen to be modified by the frequency response evaluated at $f = 0$. Does this seem reasonable? Next, if $E[X[n]X[n+k]]$ is found not to depend on n, we will be able to conclude that $X[n]$ is WSS. Continuing we have

$$\begin{aligned}
E[X[n]X[n+k]] &= E[(h[0]U[n] + h[1]U[n-1])(h[0]U[n+k] + h[1]U[n+k-1])] \\
&= h^2[0]E[U[n]U[n+k]] + h[0]h[1]E[U[n]U[n+k-1]] \\
&\quad + h[1]h[0]E[U[n-1]U[n+k]] + h^2[1]E[U[n-1]U[n+k-1]] \\
&= (h^2[0] + h^2[1])r_U[k] + h[0]h[1]r_U[k-1] + h[1]h[0]r_U[k+1]
\end{aligned}$$

and is seen not to depend on n. Hence, $X[n]$ is WSS and its ACS is

$$r_X[k] = (h^2[0] + h^2[1])r_U[k] + h[0]h[1]r_U[k-1] + h[1]h[0]r_U[k+1]. \qquad (18.5)$$

$\Diamond$

Using the previous example for sake of illustration, we next show that the ACS of the output random process of an LSI system can be written as a multiple convolution of sequences. To do so consider (18.5) and let

$$\begin{aligned}
g[0] &= h^2[0] + h^2[1] \\
g[1] &= h[0]h[1] \\
g[-1] &= h[1]h[0]
\end{aligned}$$

and zero otherwise. Then

$$
\begin{aligned}
r_X[k] &= g[0]r_U[k] + g[1]r_U[k-1] + g[-1]r_U[k+1] \\
&= \sum_{j=-1}^{1} g[j]r_U[k-j] \\
&= g[k] \star r_U[k] \quad \text{(definition of convolution sum)} \quad (18.6)
\end{aligned}
$$

where $\star$ denotes convolution. Also, it is easily shown by direct computation that

$$
\begin{aligned}
g[k] &= \sum_{j=-1}^{0} h[-j]h[k-j] \\
&= h[-k] \star h[k] \quad\quad\quad\quad\quad\quad\quad\quad (18.7)
\end{aligned}
$$

and therefore from (18.6) and (18.7) we have the final result

$$
\begin{aligned}
r_X[k] &= (h[-k] \star h[k]) \star r_U[k] \\
&= h[-k] \star h[k] \star r_U[k]. \quad\quad\quad\quad (18.8)
\end{aligned}
$$

The parentheses can be omitted in (18.8) since the order in which the convolutions are carried out is immaterial (due to associative and commutative property of convolution).

To find the PSD of $X[n]$ we note from (18.4) that the Fourier transform of the impulse response is the frequency response and therefore

$$
\begin{aligned}
\mathcal{F}\{h[k]\} &= H(f) \\
\mathcal{F}\{h[-k]\} &= H^*(f)
\end{aligned}
$$

where $\mathcal{F}$ indicates the discrete-time Fourier transform. Fourier transforming (18.8) produces

$$
P_X(f) = H^*(f)H(f)P_U(f)
$$

or finally we have

$$
P_X(f) = |H(f)|^2 P_U(f).
$$

This is the fundamental relationship for the PSD at the output of an LSI system—*the output PSD is the input PSD multiplied by the magnitude-squared of the frequency response.* We summarize the foregoing results in a theorem.

Theorem 18.3.1 (Random Process Characteristics at LSI System Output)
If a WSS random process $U[n]$ with mean μ_U and ACS $r_U[k]$ is input to an LSI system which has an impulse response $h[k]$ and frequency response $H(f)$, then the output random process $X[n] = \sum_{k=-\infty}^{\infty} h[k]U[n-k]$ is also WSS and

$$
\mu_X = \sum_{k=-\infty}^{\infty} h[k]\mu_U = H(0)\mu_U \quad\quad\quad (18.9)
$$

$$
r_X[k] = h[-k] \star h[k] \star r_U[k] \quad\quad\quad (18.10)
$$

$$
P_X(f) = |H(f)|^2 P_U(f). \quad\quad\quad\quad (18.11)
$$

Proof: The mean sequence at the output is

$$
\begin{aligned}
\mu_X[n] &= E[X[n]] = E\left[\sum_{k=-\infty}^{\infty} h[k]U[n-k]\right] \\
&= \sum_{k=-\infty}^{\infty} h[k]E[U[n-k]] \\
&= \sum_{k=-\infty}^{\infty} h[k]\mu_U = H(0)\mu_U \qquad (U[n] \text{ is WSS})
\end{aligned}
$$

and is not dependent on n. To determine if an ACS can be defined, we consider $E[X[n]X[n+k]]$. This becomes

$$
\begin{aligned}
E[X[n]X[n+k]] &= E\left[\sum_{i=-\infty}^{\infty} h[i]U[n-i]\sum_{j=-\infty}^{\infty} h[j]U[n+k-j]\right] \\
&= \sum_{i=-\infty}^{\infty}\sum_{j=-\infty}^{\infty} h[i]h[j]\underbrace{E[U[n-i]U[n+k-j]]}_{r_U[k-j+i]}
\end{aligned}
$$

since $U[n]$ was assumed to be WSS. It is seen that there is no dependence on n and hence $X[n]$ is WSS. The ACS is

$$
\begin{aligned}
r_X[k] &= \sum_{i=-\infty}^{\infty}\sum_{j=-\infty}^{\infty} h[i]h[j]r_U[(k+i)-j] \\
&= \sum_{i=-\infty}^{\infty} h[i]\underbrace{\sum_{j=-\infty}^{\infty} h[j]r_U[(k+i)-j]}_{g[k+i]}
\end{aligned}
$$

where

$$
g[m] = h[m] \star r_U[m]. \tag{18.12}
$$

Now we have

$$
\begin{aligned}
r_X[k] &= \sum_{i=-\infty}^{\infty} h[i]g[k+i] \\
&= \sum_{l=-\infty}^{\infty} h[-l]g[k-l] \qquad (\text{let } l = -i) \\
&= h[-k] \star g[k].
\end{aligned}
$$

But from (18.12) $g[k] = h[k] \star r_U[k]$ and therefore

$$
\begin{aligned}
r_X[k] &= h[-k] \star (h[k] \star r_U[k]) \\
&= h[-k] \star h[k] \star r_U[k] \tag{18.13}
\end{aligned}
$$

due to the associate and commutative properties of convolution. The last result of (18.11) follows by taking the Fourier transform of (18.13) and noting that $\mathcal{F}\{h[-k]\} = H^*(f)$.

△

A special case of particular interest occurs when the input to the system is white noise. Then using $P_U(f) = \sigma_U^2$ in (18.11), the output PSD becomes

$$P_X(f) = |H(f)|^2 \sigma_U^2. \tag{18.14}$$

Using $r_U[k] = \sigma_U^2 \delta[k]$ in (18.10), the output ACS becomes

$$r_X[k] = h[-k] \star h[k] \star \sigma_U^2 \delta[k]$$

and noting that $h[k] \star \delta[k] = h[k]$

$$
\begin{aligned}
r_X[k] &= \sigma_U^2 h[-k] \star h[k] \\
&= \sigma_U^2 \sum_{i=-\infty}^{\infty} h[-i]h[k-i].
\end{aligned}
$$

Finally, letting $m = -i$ we have the result

$$r_X[k] = \sigma_U^2 \sum_{m=-\infty}^{\infty} h[m]h[m+k] \qquad -\infty < k < \infty. \tag{18.15}$$

This formula is useful for determining the output ACS, as is illustrated next.

Example 18.2 – AR random process

In Examples 17.5 and 17.10 we derived the ACS and PSD for an AR random process. We now rederive these quantities using the linear systems concepts just described. Recall that an AR random process is defined as $X[n] = aX[n-1] + U[n]$ and can be viewed as the output of an LSI filter with system function

$$\mathcal{H}(z) = \frac{1}{1 - az^{-1}}$$

with white Gaussian noise $U[n]$ at the input. This is shown in Figure 18.2 and follows from the definition of the system function $\mathcal{H}(z)$ as the z-transform of the output sequence divided by the z-transform of the input sequence. To see this let $u[n]$ be a deterministic input sequence with z-transform $\mathcal{U}(z)$ and $x[n]$ be the corresponding deterministic output sequence with z-transform $\mathcal{X}(z)$. Then we have by the definition of the system function

$$\mathcal{H}(z) = \frac{\mathcal{X}(z)}{\mathcal{U}(z)}$$

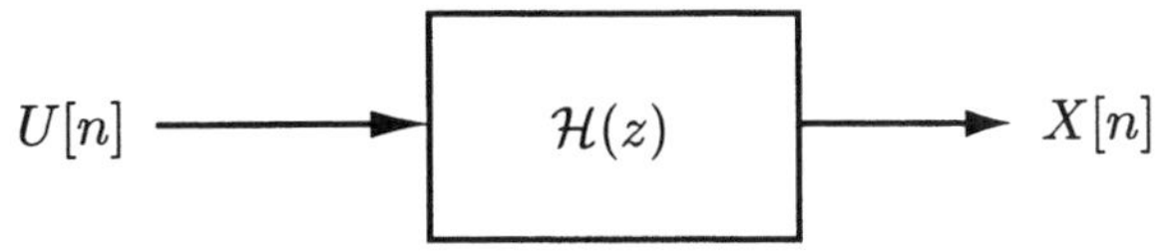

$$\mathcal{H}(z) = \frac{1}{1-az^{-1}}$$

Figure 18.2: Linear system model for AR random process. The input random process $U[n]$ is white Gaussian noise with variance σ_U^2.

and therefore for the given system function

$$\begin{aligned} \mathcal{X}(z) &= \mathcal{H}(z)\mathcal{U}(z) \\ &= \frac{1}{1-az^{-1}}\mathcal{U}(z). \end{aligned}$$

Thus,

$$\mathcal{X}(z) - az^{-1}\mathcal{X}(z) = \mathcal{U}(z)$$

and taking the inverse z-transform yields the recursive difference equation

$$x[n] - ax[n-1] = u[n] \tag{18.16}$$

which is equivalent to our AR random process definition when the input and output sequences are replaced by random processes.

The output PSD is now found by using (18.14) to yield

$$\begin{aligned} P_X(f) &= |\mathcal{H}(\exp(j2\pi f))|^2 \sigma_U^2 \\ &= \frac{\sigma_U^2}{|1 - a\exp(-j2\pi f)|^2} \end{aligned} \tag{18.17}$$

which agrees with our previous results. To determine the ACS we can either take the inverse Fourier transform of (18.17) or use (18.15). The latter approach is generally easier. To find the impulse response we can use (18.16) with the input set to $\delta[n]$ so that the output is by definition $h[n]$. Since the LSI system is assumed to be causal, we need to determine the solution of the difference equation $h[n] = ah[n-1] + \delta[n]$ for $n \geq 0$ with initial condition $h[-1] = 0$. The reason that the initial condition is set equal to zero is *our assumption that the LSI system is causal*. A causal system cannot produce an output which is nonzero, in this case $h[-1]$, before the input is applied, in this case at $n = 0$ since the input is $\delta[n]$. This produces $h[n] = a^n u_s[n]$, where we now use $u_s[n]$ to denote the unit step in order to avoid confusion with the random process realization $u[n]$ (see Appendix D.3). Thus, (18.15) becomes for

$k \geq 0$

$$r_X[k] = \sigma_U^2 \sum_{m=-\infty}^{\infty} a^m u_s[m] a^{m+k} u_s[m+k]$$

$$= \sigma_U^2 a^k \sum_{m=0}^{\infty} a^{2m} \quad (m \geq 0 \text{ and } m + k \geq 0 \text{ for nonzero term in sum})$$

$$= \sigma_U^2 \frac{a^k}{1 - a^2} \quad (\text{since } |a| < 1)$$

and therefore for all k

$$r_X[k] = \sigma_U^2 \frac{a^{|k|}}{1 - a^2}.$$

Again the ACS is in agreement with our previous results. Note that the linear system shown in Figure 18.2 is called an *infinite impulse response* (IIR) filter. This is because the impulse response $h[n] = a^n u_s[n]$ is infinite in length.

◇

 Fourier and z-transforms of WSS random process don't exist.

To determine the system function in the previous example we assumed the input to the linear system was a deterministic sequence $u[n]$. The corresponding output $x[n]$, therefore, was also a deterministic sequence. This is because formally the z-transform (and also the Fourier transform) cannot exist for a WSS random process. Existence requires the sequence to decay to zero as time becomes large. But of course if the random process is WSS, then we know that $E[X^2[n]]$ is constant as $n \to \pm\infty$ and so we cannot have $|X[n]| \to 0$ as $n \to \pm\infty$.

Example 18.3 – MA random process

In Example 17.3 we derived the ACS for an MA random process. We now show how to accomplish this more easily using (18.15). Recall the definition of the MA random process in Example 17.3 as $X[n] = (U[n] + U[n-1])/2$, with $U[n]$ being white Gaussian noise. This may be interpreted as the output of an LSI filter with white Gaussian noise at the input. In fact, it should now be obvious that the system function is $\mathcal{H}(z) = 1/2 + (1/2)z^{-1}$ and therefore the impulse response is $h[m] = 1/2$

for $m = 0, 1$ and zero otherwise. Using (18.15) we have

$$r_X[k] = \sigma_U^2 \sum_{m=-\infty}^{\infty} h[m]h[m+k]$$

$$= \sigma_U^2 \sum_{m=0}^{1} h[m]h[m+k]$$

and so for $k \geq 0$

$$r_X[k] = \begin{cases} \sigma_U^2 \sum_{m=0}^{1} h^2[m] & k = 0 \\ \sigma_U^2 \sum_{m=0}^{1} h[m]h[m+1] & k = 1 \\ 0 & k \geq 2. \end{cases}$$

Finally, we have

$$r_X[k] = \begin{cases} \sigma_U^2[(\frac{1}{2})^2 + (\frac{1}{2})^2] = \sigma_U^2/2 & k = 0 \\ \sigma_U^2(\frac{1}{2})(\frac{1}{2}) = \sigma_U^2/4 & k = 1 \\ 0 & k \geq 2 \end{cases}$$

which is the same as previously obtained.

18.4 Interpretation of the PSD

We are now in a position to prove that *the PSD, when integrated over a band of frequencies yields the average power within that band.* In doing so, the PSD may then be interpreted as the average power per unit frequency. We next consider a method to measure the average power of a WSS random process within a very narrow band of frequencies. To do so we filter the random process with an ideal narrowband filter whose frequency response is

$$H(f) = \begin{cases} 1 & -f_0 - \frac{\Delta f}{2} \leq f \leq -f_0 + \frac{\Delta f}{2}, \ f_0 - \frac{\Delta f}{2} \leq f \leq f_0 + \frac{\Delta f}{2} \\ 0 & \text{otherwise} \end{cases}$$

and which is shown in Figure 18.3a. The width of the passband of the filter Δf is assumed to be very small. If a WSS random process $X[n]$ is input to this filter, then the output WSS random process $Y[n]$ will be composed of frequency components within the Δf frequency band, the remaining ones having been "filtered out". The total average power in the output random process $Y[n]$ (which is WSS by Theorem 18.3.1) is $r_Y[0]$ and represents the sum of the average powers in $X[n]$ within the bands $[-f_0 - \Delta f/2, -f_0 + \Delta f/2]$ and $[f_0 - \Delta f/2, f_0 + \Delta f/2]$. It can be found from

$$r_Y[0] = \int_{-\frac{1}{2}}^{\frac{1}{2}} P_Y(f)df \qquad \text{(from (17.38)).}$$

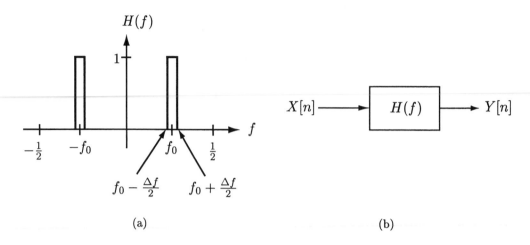

(a)   (b)

Figure 18.3: Narrowband filtering of random process to measure power within a band of frequencies.

Now using (18.11) and the definition of the narrowband filter frequency response we have

$$r_Y[0] = \int_{-\frac{1}{2}}^{\frac{1}{2}} P_Y(f)df$$

$$= \int_{-\frac{1}{2}}^{\frac{1}{2}} |H(f)|^2 P_X(f)df \qquad \text{(from (18.11))}$$

$$= \int_{-f_0-\Delta f/2}^{-f_0+\Delta f/2} 1 \cdot P_X(f)df + \int_{f_0-\Delta f/2}^{f_0+\Delta f/2} 1 \cdot P_X(f)df$$

$$= 2\int_{f_0-\Delta f/2}^{f_0+\Delta f/2} 1 \cdot P_X(f)df \qquad \text{(since } P_X(-f) = P_X(f)\text{).}$$

If we let $\Delta f \to 0$, so that $P_X(f) \to P_X(f_0)$ within the integration interval, this becomes approximately

$$r_Y[0] = 2P_X(f_0)\Delta f$$

or

$$P_X(f_0) = \frac{1}{2}\frac{r_Y[0]}{\Delta f}.$$

Since $r_Y[0]$ is the total average power due to the frequency components within the bands shown in Figure 18.3a, which is twice the total average power in the positive frequency band, we have that

$$P_X(f_0) = \frac{\text{Total average power in band } [f_0 - \Delta f/2, f_0 + \Delta f/2]}{\Delta f}. \qquad (18.18)$$

This says that *the PSD $P_X(f_0)$ is the average power of $X[n]$ in a small band of frequencies about $f = f_0$ divided by the width of the band.* It justifies the name of power spectral *density*. Furthermore, to obtain the average power within a frequency band from knowledge of the PSD, we can reverse (18.18) to obtain

$$\text{Total average power in band } [f_0 - \Delta f/2, f_0 + \Delta f/2] = P_X(f_0)\Delta f$$

which is the *area* under the PSD curve. More generally, we have for an arbitrary frequency band

$$\text{Total average power in band } [f_1, f_2] = \int_{f_1}^{f_2} P_X(f)df$$

which was previously asserted.

18.5 Wiener Filtering

Armed with the knowledge of the mean and ACS or equivalently the mean and PSD of a WSS random process, there are several important problems that can be solved. Because the required knowledge consists of only the first two moments of the random process (which in practice can be estimated), the solutions to these problems have found widespread application. The generic approach that results is termed *Wiener filtering*, although there are actually four slightly different problems and corresponding solutions. These problems are illustrated in Figure 18.4 and are referred to as *filtering, smoothing, prediction*, and *interpolation* [Wiener 1949]. In the filtering problem (see Figure 18.4a) it is assumed that a signal $S[n]$ has been corrupted by additive noise $W[n]$ so that the observed random process is $X[n] = S[n] + W[n]$. It is desired to estimate $S[n]$ by filtering $X[n]$ with an LSI filter having an impulse response $h[k]$. The filter will hopefully reduce the noise but pass the signal. The filter estimates a particular sample of the signal, say $S[n_0]$, by processing the current data sample $X[n_0]$ and the past data samples $\{X[n_0 - 1], X[n_0 - 2], \ldots\}$. Hence, the filter is assumed to be causal with an impulse response $h[k] = 0$ for $k < 0$. This produces the estimator

$$\hat{S}[n_0] = \sum_{k=0}^{\infty} h[k]X[n_0 - k] \tag{18.19}$$

which depends on the current sample, containing the signal sample of interest, and past observed data samples. Presumably, the past signal samples are correlated with the present signal sample and hence the use of past samples of $X[n]$ should enhance the estimation performance. This type of processing is called *filtering* and can be implemented in *real time*.

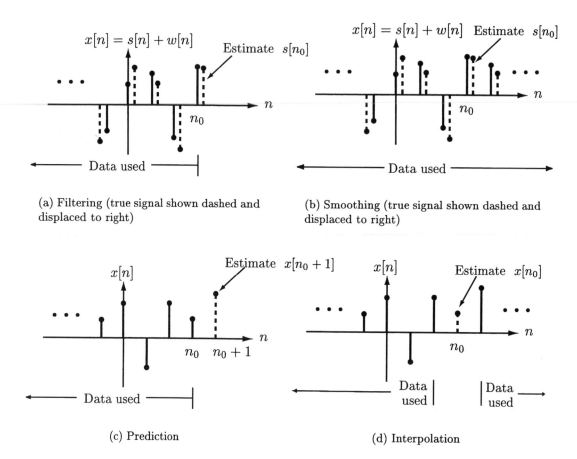

(a) Filtering (true signal shown dashed and displaced to right)

(b) Smoothing (true signal shown dashed and displaced to right)

(c) Prediction

(d) Interpolation

Figure 18.4: Definition of Wiener "filtering" problems.

 What are we really estimating here?

In Section 7.9 we attempted to estimate the outcome of a random variable, which was unobserved, based on the outcome of another random variable, which was observed. The correlation between the two random variables allowed us to do this. Here we have essentially the same problem, except that the outcome of interest to us is of the random variable $S[n_0]$. The random variables that are observed are $\{X[n_0], X[n_0 - 1], \ldots\}$ or we have access to the *realization* (another name for outcome) $\{x[n_0], x[n_0 - 1], \ldots\}$. Thus, we are attempting to estimate the *realization* of $S[n_0]$ based on the realization $\{x[n_0], x[n_0 - 1], \ldots\}$. This should be kept in mind since our notation of $\hat{S}[n_0] = \sum_{k=0}^{\infty} h[k]X[n_0 - k]$ seems to indicate that we are attempting to estimate a random variable $S[n_0]$ based on other random variables $\{X[n_0], X[n_0 - 1], \ldots\}$. What we are actually trying to accomplish is a *procedure* of estimating a realization of a random variable based on realizations of other random variables that will work *for all realizations*. Hence, we employ the capital letter notation for random variables to indicate our interest in all realizations and to allow us to employ expectation operations on the random variables.

The second problem is called smoothing (see Figure 18.4b). It differs from filtering in that the filter is not constrained to be causal. Therefore, the estimator becomes

$$\hat{S}[n_0] = \sum_{k=-\infty}^{\infty} h[k]X[n_0 - k] \tag{18.20}$$

where $\hat{S}[n_0]$ now depends on present, past, and *future* samples of $X[n]$. Clearly, this is not realizable in real time but can be approximated if we allow a delay before determining the estimate. The delay is necessary to accumulate the samples $\{X[n_0 + 1], X[n_0 + 2], \ldots\}$ before computing $\hat{S}[n_0]$. Within a digital computer we would store these "future" samples.

For problems three and four we observe samples of the WSS random process $X[n]$ and wish to estimate an unobserved sample. For *prediction*, which is also called *extrapolation* and *forecasting*, we observe the current and past samples $\{X[n_0], X[n_0 - 1], \ldots\}$ and wish to estimate a future sample, $X[n_0 + L]$, for some positive integer L. The prediction is called an *L-step prediction*. We will only consider one-step prediction or $L = 1$ (see Figure 18.4c). The reader should see [Yaglom 1962] for the more general case and also Problem 18.26 for an example. The predictor then becomes

$$\hat{X}[n_0 + 1] = \sum_{k=0}^{\infty} h[k]X[n_0 - k] \tag{18.21}$$

which of course uses a causal filter. For *interpolation* (see Figure 18.4d) we observe samples $\{\ldots, X[n_0 - 1], X[n_0 + 1], \ldots\}$ and wish to estimate $X[n_0]$. The interpolator then becomes

$$\hat{X}[n_0] = \sum_{\substack{k=-\infty \\ k\neq 0}}^{\infty} h[k]X[n_0 - k] \tag{18.22}$$

which is a noncausal filter. For practical implementation of (18.19)–(18.22) we must truncate the impulse responses to some finite number of samples.

To determine the optimal filter impulse responses we adopt the mean square error (MSE) criterion. Estimators that consist of LSI filters whose impulses are chosen to minimize a MSE are generically referred to as *Wiener filters* [Wiener 1949]. Of the four problems mentioned, we will solve the smoothing and prediction problems. The solution for the filtering problem can be found in [Orfanidis 1985] while that for the interpolation problem is described in [Yaglom 1962] (see also Problem 18.27).

18.5.1   Wiener Smoothing

We observe $X[n] = S[n] + W[n]$ for $-\infty < n < \infty$ and wish to estimate $S[n_0]$ using (18.20). It is assumed that $S[n]$ and $W[n]$ are both zero mean WSS random processes with *known ACSs (PSDs)*. Also, since there is usually no reason to assume otherwise, we assume that the signal and noise random processes are uncorrelated. This means that any sample of $S[n]$ is uncorrelated with any sample of $W[n]$ or $E[S[n_1]W[n_2]] = 0$ for all n_1 and n_2. The MSE for this problem is defined as

$$\text{mse} = E[\epsilon^2[n_0]] = E[(S[n_0] - \hat{S}[n_0])^2]$$

where $\epsilon[n_0] = S[n_0] - \hat{S}[n_0]$ is the error. To minimize the MSE we utilize the orthogonality principle described in Section 14.7 which states that the error should be orthogonal, i.e., uncorrelated, with the data. Since the data consists of $X[n]$ for all n, the orthogonality principle produces the requirement

$$E[\epsilon[n_0]X[n_0 - l]] = 0 \qquad -\infty < l < \infty.$$

Thus, we have that

$$E[(S[n_0] - \hat{S}[n_0])X[n_0 - l]] = 0$$

$$E\left[\left(S[n_0] - \sum_{k=-\infty}^{\infty} h[k]X[n_0 - k]\right)X[n_0 - l]\right] = 0 \qquad \text{(from (18.20))}$$

which results in

$$E[S[n_0]X[n_0 - l]] = \sum_{k=-\infty}^{\infty} h[k]E[X[n_0 - k]X[n_0 - l]]. \qquad (18.23)$$

But

$$
\begin{aligned}
E[S[n_0]X[n_0 - l]] &= E[S[n_0](S[n_0 - l] + W[n_0 - l])] \\
&= E[S[n_0]S[n_0 - l]] \qquad (S[n] \text{ and } W[n] \text{ are} \\
&\qquad\qquad\qquad\qquad\qquad \text{uncorrelated and zero mean}) \\
&= r_S[l]
\end{aligned}
$$

and

$$
\begin{aligned}
E[X[n_0 - k]X[n_0 - l]] &= E[(S[n_0 - k] + W[n_0 - k])(S[n_0 - l] + W[n_0 - l])] \\
&= E[S[n_0 - k]S[n_0 - l]] + E[W[n_0 - k]W[n_0 - l]] \\
&= r_S[l - k] + r_W[l - k].
\end{aligned}
$$

The infinite set of simultaneous linear equations becomes from (18.23)

$$r_S[l] = \sum_{k=-\infty}^{\infty} h[k](r_S[l - k] + r_W[l - k]) \qquad -\infty < l < \infty. \qquad (18.24)$$

Note that the equations *do not depend on* n_0 and therefore the solution for the optimal impulse response is the same for any n_0. This is due to the WSS assumption coupled with the LSI assumption for the estimator, which together imply that a shift in the sample to be estimated results in the same filtering operation but shifted. To solve this set of equations we can use transform techniques since the right-hand side of (18.24) is seen to be a discrete-time convolution. It follows then that

$$r_S[l] = h[l] \star (r_S[l] + r_W[l])$$

and taking Fourier transforms of both sides yields

$$P_S(f) = H(f)(P_S(f) + P_W(f))$$

or finally *the frequency response of the optimal Wiener smoothing filter* is

$$H_{\text{opt}}(f) = \frac{P_S(f)}{P_S(f) + P_W(f)}. \tag{18.25}$$

The optimal impulse response can be found by taking the inverse Fourier transform of (18.25). We next give an example.

Example 18.4 – Wiener smoother for AR signal in white noise

Consider a signal that is an AR random process corrupted by additive white noise with variance σ_W^2. Then, the PSDs are

$$
\begin{aligned}
P_S(f) &= \frac{\sigma_U^2}{|1 - a\exp(-j2\pi f)|^2} \\
P_W(f) &= \sigma_W^2.
\end{aligned}
$$

The PSDs and corresponding Wiener smoother frequency responses are shown in Figure 18.5. In both cases the white noise variance is the same, $\sigma_W^2 = 1$, and the AR input noise variance is the same, $\sigma_U^2 = 0.5$, but the AR filter parameter a has been chosen to yield a wide PSD and a very narrow PSD. As an example, consider the case of $a = 0.9$, which results in a lowpass signal random process as shown in Figure 18.5b. Then, the results of a computer simulation are shown in Figure 18.6. In Figure 18.6a the signal realization $s[n]$ is shown as the dashed curve and the noise corrupted signal realization $x[n]$ is shown as the solid curve. The points have been connected by straight lines for easier viewing. Applying the Wiener smoother results in the estimated signal shown in Figure 18.6b as the solid curve. Once again the true signal realization is shown as dashed. Note that the estimated signal shown in Figure 18.6b exhibits less noise fluctuations but having been smoothed, also exhibits a reduced ability to follow the signal when the signal changes rapidly (see the estimated signal from $n = 25$ to $n = 35$). This is a standard tradeoff in that noise smoothing is obtained at the price of poorer signal following dynamics.

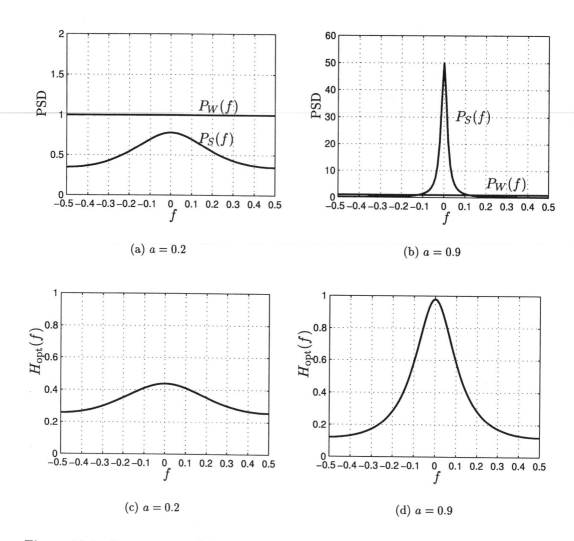

Figure 18.5: Power spectral densities of the signal and noise and corresponding frequency responses of Wiener smoother.

In order to implement the Wiener smoother for the previous example the data was filtered in the frequency domain and converted back into the time domain. This was done using the inverse discrete-time Fourier transform

$$\hat{s}[n] = \int_{-\frac{1}{2}}^{\frac{1}{2}} \frac{P_S(f)}{P_S(f) + \sigma_W^2} X_N(f) \exp(j2\pi f n) df \qquad n = 0, 1, \ldots, N - 1$$

where $X_N(f)$ is the Fourier transform of the available data $\{x[0], x[1], \ldots, x[N-1]\}$,

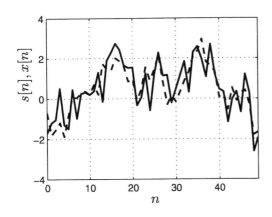

(a) True (dashed) and noisy (solid) signal

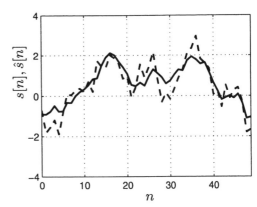

(b) True (dashed) and estimated (solid) signal

Figure 18.6: Example of Wiener smoother for additive noise corrupted AR signal. The true PSDs are shown in Figure 18.5b. In a) the true signal is shown as the dashed curve and the noisy signal as the solid curve and in b) the true signal is shown as the dashed curve and the Wiener smoothed signal estimate (using the Wiener smoother shown in Figure 18.5d) as the solid curve.

which is

$$X_N(f) = \sum_{n=0}^{N-1} x[n] \exp(-j2\pi fn)$$

($N = 50$ for the previous example). The actual implementation used an inverse FFT to approximate the integral as is shown in the MATLAB code given next. Note that in using the FFT and inverse FFT to calculate the Fourier transform and inverse Fourier transform, respectively, the frequency interval has been changed to $[0, 1]$. Because the Fourier transform is periodic with period one, however, this will not affect the result.

```
clear all
randn('state',0)
a=0.9;varu=0.5;vars=varu/(1-a^2);varw=1;N=50; % set up parameters
for n=0:N-1 % generate signal realization
    nn=n+1;
    if n==0 % use Gaussian random processes
        s(nn,1)=sqrt(vars)*randn(1,1); % initialize first sample
                                        % to avoid transient
    else
        s(nn,1)=a*s(nn-1)+sqrt(varu)*randn(1,1);
```

```
      end
end
x=s+sqrt(varw)*randn(N,1); % add white Gaussian noise
Nfft=1024; % set up FFT length
% compute PSD of signal, frequency interval is [0,1]
Ps=varu./(abs(1-a*exp(-j*2*pi*[0:Nfft-1]'/Nfft)).^2);
Hf=Ps./(Ps+varw); % form Wiener smoother
sestf=Hf.*fft(x,Nfft);   % signal estimate in frequency domain,
                         % frequency interval is [0,1]
sest=real(ifft(sestf,Nfft)); % inverse Fourier transform
```

One can also determine the minimum MSE to assess how well the smoother performs. This is

$$
\begin{aligned}
\text{mse}_{\min} &= E[(S[n_0] - \hat{S}[n_0])^2] \\
&= E[(S[n_0] - \hat{S}[n_0])S[n_0]] - E[(S[n_0] - \hat{S}[n_0])\hat{S}[n_0]].
\end{aligned}
$$

But the second term is zero since by the orthogonality principle

$$
\begin{aligned}
E[(S[n_0] - \hat{S}[n_0])\hat{S}[n_0]] &= E\left[\epsilon[n_0]\sum_{k=-\infty}^{\infty} h_{\text{opt}}[k]X[n_0 - k]\right] \\
&= \sum_{k=-\infty}^{\infty} h_{\text{opt}}[k]\underbrace{E[\epsilon[n_0]X[n_0 - k]]}_{=0} = 0.
\end{aligned}
$$

Thus, we have

$$
\begin{aligned}
\text{mse}_{\min} &= E[(S[n_0] - \hat{S}[n_0])S[n_0]] \\
&= r_S[0] - E\left[\sum_{k=-\infty}^{\infty} h_{\text{opt}}[k]X[n_0 - k]S[n_0]\right] \\
&= r_S[0] - \sum_{k=-\infty}^{\infty} h_{\text{opt}}[k]\underbrace{E[(S[n_0 - k] + W[n_0 - k])S[n_0]]}_{=E[S[n_0-k]S[n_0]]=r_S[k]}
\end{aligned}
$$

since $S[n_1]$ and $W[n_2]$ are uncorrelated for all n_1 and n_2 and also are zero mean. The minimum MSE becomes

$$
\text{mse}_{\min} = r_S[0] - \sum_{k=-\infty}^{\infty} h_{\text{opt}}[k]r_S[k]. \tag{18.26}
$$

This can also be written in the frequency domain by using Parseval's theorem to yield

$$\text{mse}_{\min} = \int_{-\frac{1}{2}}^{\frac{1}{2}} P_S(f)df - \int_{-\frac{1}{2}}^{\frac{1}{2}} H_{\text{opt}}(f)P_S(f)df \qquad ((17.38) \text{ and Parseval})$$

$$= \int_{-\frac{1}{2}}^{\frac{1}{2}} (1 - H_{\text{opt}}(f))P_S(f)df$$

$$= \int_{-\frac{1}{2}}^{\frac{1}{2}} \left(1 - \frac{P_S(f)}{P_S(f) + P_W(f)} \right) P_S(f)df$$

$$= \int_{-\frac{1}{2}}^{\frac{1}{2}} \frac{P_W(f)}{P_S(f) + P_W(f)} P_S(f)df$$

and finally letting $\rho(f) = P_S(f)/P_W(f)$ be the signal-to-noise ratio in the frequency domain we have

$$\text{mse}_{\min} = \int_{-\frac{1}{2}}^{\frac{1}{2}} \frac{P_S(f)}{1 + \rho(f)} df. \qquad (18.27)$$

It is seen that the frequency bands for which the contribution to the minimum MSE is largest, are the bands for which the signal-to-noise ratio is smallest or for which $\rho(f) \ll 1$.

18.5.2 Prediction

We consider only the case of $L = 1$ or *one-step prediction*. The more general case can be found in [Yaglom 1962] (see also Problem 18.26). As before, the criterion of MSE is used to design the predictor so that from (18.21)

$$\text{mse} = E[(X[n_0 + 1] - \hat{X}[n_0 + 1])^2]$$

$$= E\left[\left(X[n_0 + 1] - \sum_{k=0}^{\infty} h[k]X[n_0 - k] \right)^2 \right]$$

is to be minimized over $h[k]$ for $k \geq 0$. Invoking the orthogonality principle leads to the infinite set of simultaneous linear equations

$$E\left[\left(X[n_0 + 1] - \sum_{k=0}^{\infty} h[k]X[n_0 - k] \right) X[n_0 - l] \right] = 0 \qquad l = 0, 1, \ldots .$$

These equations become

$$E[X[n_0 + 1]X[n_0 - l]] = \sum_{k=0}^{\infty} h[k]E[X[n_0 - k]X[n_0 - l]]$$

or finally

$$r_X[l+1] = \sum_{k=0}^{\infty} h[k] r_X[l-k] \qquad l = 0, 1, \ldots . \tag{18.28}$$

Note that once again the optimal impulse response does not depend upon n_0 so that we obtain the same predictor for any sample. Although it appears that we should be able to solve these simultaneous linear equations using the previous Fourier transform approach, this is not so. Because the equations are only valid for $l \geq 0$ and not for $l < 0$, a z-transform cannot be used. Consider forming the z-transform of the left-hand-side as $\sum_{l=0}^{\infty} r_X[l+1] z^{-l}$ and note that it is *not* equal to $z\mathcal{P}(z)$. (See also Problem 18.15 to see what would happen if we blindly went ahead with this approach.)

The minimum MSE is evaluated by using a similar argument as for the Wiener smoother

$$
\begin{aligned}
\text{mse}_{\min} &= E\left[\left(X[n_0+1] - \sum_{k=0}^{\infty} h_{\text{opt}}[k] X[n_0-k]\right) X[n_0+1]\right] \\
&= r_X[0] - \sum_{k=0}^{\infty} h_{\text{opt}}[k] r_X[k+1]
\end{aligned}
\tag{18.29}
$$

where $h_{\text{opt}}[k]$ is the impulse response solution from (18.28). A simple example for which the equations of (18.28) can be solved is given next.

Example 18.5 – Prediction of AR random process

Consider an AR random process for which the ACS is given by $r_X[k] = (\sigma_U^2/(1 - a^2)) a^{|k|} = r_X[0] a^{|k|}$. Then from (18.28)

$$r_X[0] a^{|l+1|} = \sum_{k=0}^{\infty} h[k] r_X[0] a^{|l-k|} \qquad l = 0, 1, \ldots$$

and if we let $h[k] = 0$ for $k \geq 1$, we have

$$a^{|l+1|} = h[0] a^{|l|} \qquad l = 0, 1, \ldots .$$

Since $l \geq 0$, the solution is easily seen to be

$$h_{\text{opt}}[0] = \frac{a^{l+1}}{a^l} = a$$

or finally

$$\hat{X}[n_0+1] = aX[n_0].$$

Also, since this is true for any n_0, we can replace the specific sample by a more general sample by replacing n_0 by $n-1$. This results in

$$\hat{X}[n] = aX[n-1]. \tag{18.30}$$

Recalling that the AR random process is defined as $X[n] = aX[n-1] + U[n]$, it is now seen that the optimal one-step linear predictor is obtained from the definition by ignoring the term $U[n]$. This is because $U[n]$ cannot be predicted from the past samples $\{X[n-1], X[n-2], \ldots\}$, which are uncorrelated with $U[n]$ (see also Example 17.5). Furthermore, the prediction error is $\epsilon[n] = X[n] - \hat{X}[n] = X[n] - aX[n-1] = U[n]$. Finally, note that the prediction only depends on the most recent sample and not on the past samples of $X[n]$. In effect, to predict $X[n_0 + 1]$ all the past information of the random process is embodied in the sample $X[n_0]$. To illustrate the prediction solution consider the AR random process whose parameters and realizations were shown in Figure 17.5. The realizations, along with the one-step predictions, shown as the "*"s, are given in Figure 18.7. Note the good predictions

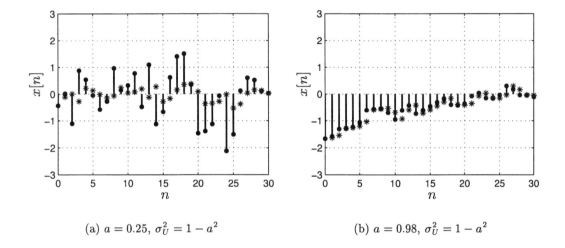

(a) $a = 0.25$, $\sigma_U^2 = 1 - a^2$ (b) $a = 0.98$, $\sigma_U^2 = 1 - a^2$

Figure 18.7: Typical realizations of autoregressive random process with different parameters and their one-step linear predictions indicated by the "*"s as $\hat{X}[n+1] = ax[n]$.

for the AR random process with $a = 0.98$ but the relatively poor ones for the AR random process with $a = 0.25$. Can you justify these results by comparing the minimum MSEs? (See Problem 18.17.)

$\diamond$

The general solution of (18.28) is fairly complicated. The details are given in Appendix 18A. We now summarize the solution and then present an example.

1. Assume that the z-transform of the ACS, which is

$$\mathcal{P}_X(z) = \sum_{k=-\infty}^{\infty} r_X[k] z^{-k}$$

can be written as

$$\mathcal{P}_X(z) = \frac{\sigma_U^2}{\mathcal{A}(z)\mathcal{A}(z^{-1})} \tag{18.31}$$

where

$$\mathcal{A}(z) = 1 - \sum_{k=1}^{\infty} a[k]z^{-k}.$$

It is required that $\mathcal{A}(z)$ have all its zeros inside the unit circle of the z-plane, i.e., the filter with z-transform $1/\mathcal{A}(z)$ is a stable and causal filter [Jackson 1991].

2. The solution of (18.28) for the impulse response is

$$h_{\text{opt}}[k] = a[k+1] \qquad k = 0, 1, \ldots$$

and the minimum MSE is

$$\text{mse}_{\min} = E[(X[n_0 + 1] - \hat{X}[n_0 + 1])^2] = \sigma_U^2.$$

3. The optimal linear predictor becomes from (18.21)

$$\hat{X}[n_0 + 1] = \sum_{k=0}^{\infty} a[k+1]X[n_0 - k] \tag{18.32}$$

and has the minimum MSE, $\text{mse}_{\min} = \sigma_U^2$.

Clearly, the most difficult part of the solution is putting $\mathcal{P}_X(z)$ into the required form of (18.31). In terms of the PSD the requirement is

$$
\begin{aligned}
P_X(f) = \mathcal{P}_X(\exp(j2\pi f)) &= \frac{\sigma_U^2}{\mathcal{A}(\exp(j2\pi f))\mathcal{A}(\exp(-j2\pi f))} \\
&= \frac{\sigma_U^2}{\mathcal{A}(\exp(j2\pi f))\mathcal{A}^*(\exp(j2\pi f))} \\
&= \frac{\sigma_U^2}{|\mathcal{A}(\exp(j2\pi f))|^2} \\
&= \frac{\sigma_U^2}{\left|1 - \sum_{k=1}^{\infty} a[k]\exp(-j2\pi f k)\right|^2}.
\end{aligned}
$$

But the form of the PSD is seen to be a generalization of the PSD for the AR random process. In fact, if we truncate the sum so that the required PSD becomes

$$P_X(f) = \frac{\sigma_U^2}{\left|1 - \sum_{k=1}^{p} a[k]\exp(-j2\pi f k)\right|^2}$$

then we have the PSD of what is referred to as an AR random process of *order p*, which is also denoted by the symbolism AR(p). In this case, the random process is defined as

$$X[n] = \sum_{k=1}^{p} a[k]X[n-k] + U[n] \qquad (18.33)$$

where as usual $U[n]$ is white Gaussian noise with variance σ_U^2. Of course, for $p = 1$ we have our previous definition of the AR random process, which is an AR(1) random process with $a[1] = a$. Assuming an AR(p) random process so that $a[l] = 0$ for $l > p$, the solution for the optimal one-step linear predictor is from (18.32)

$$\hat{X}[n_0 + 1] = \sum_{l=0}^{p-1} a[l+1]X[n_0 - l]$$

and letting $k = l + 1$ produces

$$\hat{X}[n_0 + 1] = \sum_{k=1}^{p} a[k]X[n_0 + 1 - k] \qquad (18.34)$$

and the minimum MSE is σ_U^2. Another example follows.

Example 18.6 – One-step linear prediction of MA random process
Consider the zero mean WSS random process given by $X[n] = U[n] - bU[n-1]$, where $|b| < 1$ and $U[n]$ is white Gaussian noise with variance σ_U^2 (also called an MA random process). This random process is a special case of that used in Example 18.1 for which $h[0] = 1$ and $h[1] = -b$ and $U[n]$ is white Gaussian noise. To find the optimal linear predictor we need to put the z-transform of the ACS into the required form. First we determine the PSD. Since the system function is easily shown to be $\mathcal{H}(z) = 1 - bz^{-1}$, the frequency response follows as $H(f) = 1 - b\exp(-j2\pi f)$. From (18.14) the PSD becomes

$$P_X(f) = H(f)H^*(f)\sigma_U^2 = (1 - b\exp(-j2\pi f))(1 - b\exp(j2\pi f))\sigma_U^2$$

and hence replacing $\exp(j2\pi f)$ by z, we have

$$\mathcal{P}_X(z) = (1 - bz^{-1})(1 - bz)\sigma_U^2. \qquad (18.35)$$

By equating (18.35) to the required form for $\mathcal{P}_X(z)$ given in (18.31) we have

$$\mathcal{A}(z) = \frac{1}{1 - bz^{-1}}.$$

To convert this to $1 - \sum_{k=1}^{\infty} a[k]z^{-k}$, we take the inverse z-transform, assuming a stable and causal sequence, to yield

$$\mathcal{Z}^{-1}\{\mathcal{A}(z)\} = \begin{cases} b^k & k \geq 0 \\ 0 & k < 0 \end{cases}$$

and so $a[k] = -b^k$ for $k \geq 1$. (Note why $|b| < 1$ is required or else $a[n]$ would not be stable.) The optimal predictor is from (18.32)

$$
\begin{aligned}
\hat{X}[n_0 + 1] &= \sum_{k=0}^{\infty} a[k+1] X[n_0 - k] \\
&= \sum_{k=0}^{\infty} (-b^{k+1}) X[n_0 - k] \\
&= -bX[n_0] - b^2 X[n_0 - 1] - b^3 X[n_0 - 2] - \cdots
\end{aligned}
$$

and the minimum MSE is

$$
\text{mse}_{\min} = \sigma_U^2.
$$

$\diamondsuit$

As a special case of practical interest, we next consider a *finite length* one-step linear predictor. By finite length we mean that the prediction can only depend on the present sample and past $M - 1$ samples. In a derivation similar to the infinite length predictor it is easy to show (see the discussion in Section 14.8 and also Problem 18.20) that if the predictor is given by

$$
\hat{X}[n_0 + 1] = \sum_{k=0}^{M-1} h[k] X[n_0 - k]
$$

which is just (18.21) with $h[k] = 0$ for $k \geq M$, then the optimal impulse response satisfies the M simultaneous linear equations

$$
r_X[l+1] = \sum_{k=0}^{M-1} h[k] r_X[l-k] \qquad l = 0, 1, \ldots, M - 1.
$$

(If $M \to \infty$, these equations are identical to (18.28)). The equations can be written in vector/matrix form as

$$
\underbrace{\begin{bmatrix} r_X[0] & r_X[1] & \cdots & r_X[M-1] \\ r_X[1] & r_X[0] & \cdots & r_X[M-2] \\ \vdots & \vdots & \ddots & \vdots \\ r_X[M-1] & r_X[M-2] & \cdots & r_X[0] \end{bmatrix}}_{\mathbf{R}_X} \begin{bmatrix} h[0] \\ h[1] \\ \vdots \\ h[M-1] \end{bmatrix} = \begin{bmatrix} r_X[1] \\ r_X[2] \\ \vdots \\ r_X[M] \end{bmatrix}.
$$

$$(18.36)$$

The corresponding minimum MSE is given by

$$
\text{mse}_{\min} = r_X[0] - \sum_{k=0}^{M-1} h_{\text{opt}}[k] r_X[k+1]. \tag{18.37}
$$

These equations are called the *Wiener-Hopf equations*. In general, they must be solved numerically but there are many efficient algorithms to do so [Kay 1988]. The algorithms take advantage of the structure of the matrix which is seen to be an autocorrelation matrix $\mathbf{R}_X$ as first described in Section 17.4. As such, it is symmetric, positive definite, and has the *Toeplitz* property. The Toeplitz property asserts that the elements along each northwest-southeast diagonal are identical. Another important connection between the linear prediction equations and an AR(p) random process is made by letting $M = p$ in (18.36). Then, since for an AR(p) process, we have that $h[n] = a[n+1]$ for $n = 0, 1, \ldots, p-1$ (recall from (18.34) that $\hat{X}[n_0 + 1] = \sum_{k=1}^{p} a[k]X[n_0 + 1 - k]$) the Wiener-Hopf equations become

$$\begin{bmatrix} r_X[0] & r_X[1] & \cdots & r_X[p-1] \\ r_X[1] & r_X[0] & \cdots & r_X[p-2] \\ \vdots & \vdots & \ddots & \vdots \\ r_X[p-1] & r_X[p-2] & \cdots & r_X[0] \end{bmatrix} \begin{bmatrix} a[1] \\ a[2] \\ \vdots \\ a[p] \end{bmatrix} = \begin{bmatrix} r_X[1] \\ r_X[2] \\ \vdots \\ r_X[p] \end{bmatrix}. \tag{18.38}$$

It is important to note that for *an AR(p) random process*, the optimal one-step linear predictor based on the infinite number of samples $\{X[n_0], X[n_0 - 1], \ldots\}$ is the same as that based on only the finite number of samples $\{X[n_0], X[n_0 - 1], \ldots, X[n_0 - (p-1)]\}$ [Kay 1988]. The equations of (18.38) are now referred to as the *Yule-Walker equations*. In this form they relate the ACS samples $\{r_X[0], r_X[1], \ldots r_X[p]\}$ to the AR filter parameters $\{a[1], a[2], \ldots, a[p]\}$. If the ACS samples are known, then the AR filter parameters can be obtained by solving the equations. Furthermore, once the filter parameters have been found from (18.38), the variance of the white noise random process $U[n]$ is found from

$$\sigma_U^2 = \text{mse}_{\min} = r_X[0] - \sum_{k=1}^{p} a[k]r_X[k] \tag{18.39}$$

which follows by letting $h_{\text{opt}}[k] = a[k+1]$ with $M = p$ in (18.37). In the real-world example of Section 18.7 we will see how these equations can provide a method to synthesize speech.

18.6 Continuous-Time Definitions and Formulas

For a continuous-time WSS random process as defined in Section 17.8 the linear system of interest is a linear time invariant (LTI) system. It is characterized by its impulse response $h(\tau)$. If a random process $U(t)$ is input to an LTI system with impulse response $h(\tau)$, the output random process $X(t)$ is

$$X(t) = \int_{-\infty}^{\infty} h(\tau)U(t - \tau)d\tau.$$

The integral is referred to as a *convolution integral* and in shorthand notation the output is given by $X(t) = h(t) \star U(t)$. If $U(t)$ is WSS with constant mean μ_U and ACF $r_U(\tau)$, then the output random process $X(t)$ is also WSS. It has a mean function

$$\mu_X = \left(\int_{-\infty}^{\infty} h(\tau)d\tau \right) \mu_U = H(0)\mu_U \tag{18.40}$$

where

$$H(F) = \int_{-\infty}^{\infty} h(\tau)\exp(-j2\pi F\tau)d\tau$$

is the frequency response of the LTI system. The ACF of the output random process $X(t)$ is

$$r_X(\tau) = h(-\tau) \star h(\tau) \star r_U(\tau) \tag{18.41}$$

and therefore the PSD becomes

$$P_X(F) = |H(F)|^2 P_U(F). \tag{18.42}$$

An example follows.

Example 18.7 – Inteference rejection filter

A signal, which is modeled as a WSS random process $S(t)$, is corrupted by an additive interference $I(t)$, which can be modeled as a randomly phased sinusoid with a frequency of $F_0 = 60$ Hz. The corrupted signal is $X(t) = S(t) + I(t)$. It is desired to filter out the interference but if possible, to avoid altering the PSD of the signal due to the filtering. Since the sinusoidal interference has a period of $T = 1/F_0 = 1/60$ seconds, it is proposed to filter $X(t)$ with the differencing filter

$$Y(t) = X(t) - X(t - T). \tag{18.43}$$

The motivation for choosing this type of filter is that a periodic signal with period T will have the same value at any two time instants separated by T seconds. Hence, the difference should be zero for all t. We wish to determine the PSD at the filter output. We will assume that the interference is uncorrelated with the signal. This assumption means that the ACF of $X(t)$ is the sum of the ACFs of $S(t)$ and $I(t)$ and consequently the PSDs sum as well (see Problem 18.33). The differencing filter is an LTI system and so its output can be written as

$$Y(t) = \int_{-\infty}^{\infty} h(\tau)X(t - \tau)d\tau \tag{18.44}$$

for the appropriate choice of the impulse response. The impulse response is obtained by equating (18.44) to (18.43) from which it follows that

$$h(\tau) = \delta(\tau) - \delta(\tau - T) \tag{18.45}$$

as can easily be verified. By taking the Fourier transform, the frequency response becomes

$$
\begin{aligned}
H(F) &= \int_{-\infty}^{\infty} (\delta(\tau) - \delta(\tau - T)) \exp(-j2\pi F\tau)d\tau \\
&= 1 - \exp(-j2\pi FT).
\end{aligned}
\tag{18.46}
$$

To determine the PSD at the filter output we use (18.42) and note that for the randomly phased sinusoid with amplitude A and frequency F_0, the ACF is (see Problem 17.46)

$$
r_I(\tau) = \frac{A^2}{2} \cos(2\pi F_0 \tau)
$$

and therefore its PSD, which is the Fourier transform, is given by

$$
P_I(F) = \frac{A^2}{4} \delta(F + F_0) + \frac{A^2}{4} \delta(F - F_0).
$$

The PSD at the filter input is $P_X(F) = P_S(F) + P_I(F)$ (the PSDs add due to the uncorrelated assumption) and therefore the PSD at the filter output is

$$
\begin{aligned}
P_Y(F) &= |H(F)|^2 P_X(F) = |H(F)|^2 (P_S(F) + P_I(F)) \\
&= |1 - \exp(-j2\pi FT)|^2 (P_S(F) + P_I(F)).
\end{aligned}
$$

The magnitude-squared of the frequency response of (18.46) can also be written in real form as

$$
|H(F)|^2 = 2 - 2\cos(2\pi FT)
$$

and is shown in Figure 18.8. Note that it exhibits zeros at multiples of $F = 1/T =$

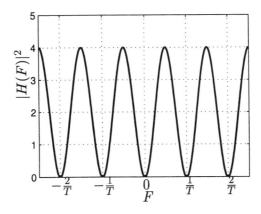

Figure 18.8: Magnitude-squared frequency response of interference canceling filter with $F_0 = 1/T$.

F_0. Hence, $|H(F_0)|^2 = 0$ and so the interfering sinusoid is filtered out. The PSD at the filter output then becomes

$$
\begin{aligned}
P_Y(F) &= |H(F)|^2 P_S(F) \\
&= 2(1 - \cos(2\pi FT)) P_S(F).
\end{aligned}
$$

Unfortunately, the signal PSD has also been modified. What do you think would happen if the signal were periodic with period $1/(2F_0)$?

18.7   Real-World Example – Speech Synthesis

It is commonplace to hear computer generated speech when asking for directory assistance in obtaining telephone numbers, in using text to speech conversion programs in computers, and in playing with a multitude of children's toys. One of the earliest applications of computer speech synthesis was the Texas Instruments Speak and Spell[1]. The approach to producing intelligible, if not exactly human sounding, speech, is to mimic the human speech production process. A speech production model is shown in Figure 18.9 [Rabiner and Schafer 1978]. It is well known that speech sounds can be delineated into two classes—*voiced speech* such as a vowel sound and *unvoiced speech* such as a consonant sound. A voiced sound such as "ahhh" (the o in "lot" for example) is produced by the vibration of the vocal cords, while an unvoiced sound such as "sss" (the s in "runs" for example) is produced by passing air over a constriction in the mouth. In either case, the sound is the output of the vocal tract with the difference being the excitation sound and the subsequent filtering of that sound. For voiced sounds the excitation is modeled as a train of impulses to produce a periodic sound while for an unvoiced sound it is modeled as white noise to produce a noise-like sound (see Figure 18.9). The excita-

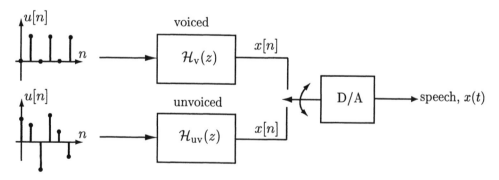

Figure 18.9: Speech production model.

tion is modified by the vocal tract, which can be modeled by an LSI filter. Knowing

---

[1]Registered trademark of Texas Instruments

the excitation waveform and the vocal tract system function allows us to synthesize speech. For the unvoiced sound we pass discrete white Gaussian noise through an LSI filter with system function $\mathcal{H}_{\mathrm{uv}}(z)$. We next concentrate on the synthesis of unvoiced sounds with the synthesis of voiced sounds being similar.

It has been found that a good model for the vocal tract is the LSI filter with system function

$$\mathcal{H}_{\mathrm{uv}}(z) = \frac{1}{1 - \sum_{k=1}^{p} a[k] z^{-k}}$$

which is an *all-pole filter*. Typically, the order of the filter p, which is the number of poles, is chosen to be $p = 12$. The output of the filter $X[n]$ for a white Gaussian noise random process input $U[n]$ with variance σ_U^2 is given as the WSS random process

$$X[n] = \sum_{k=1}^{p} a[k] X[n - k] + U[n]$$

which is recognized as the defining difference equation for an AR(p) random process. Hence, unvoiced speech sounds can be synthesized using this difference equation for an appropriate choice of the parameters $\{a[1], a[2], \ldots, a[p], \sigma_U^2\}$. The parameters will be different for each unvoiced sound to be synthesized. To determine the parameters for a given sound, a segment of the target speech sound is used to estimate the ACS. Estimation of the ACS was previously described in Section 17.7. Then, the parameters $a[k]$ for $k = 1, 2, \ldots, p$ can be obtained by solving the Yule-Walker equations (same as Wiener-Hopf equations). The theoretical ACS samples required are replaced by estimated ones to yield the set of simultaneous linear equations from (18.38) as

$$\begin{bmatrix} \hat{r}_X[0] & \hat{r}_X[1] & \cdots & \hat{r}_X[p-1] \\ \hat{r}_X[1] & \hat{r}_X[0] & \cdots & \hat{r}_X[p-2] \\ \vdots & \vdots & \ddots & \vdots \\ \hat{r}_X[p-1] & \hat{r}_X[p-2] & \cdots & \hat{r}_X[0] \end{bmatrix} \begin{bmatrix} a[1] \\ a[2] \\ \vdots \\ a[p] \end{bmatrix} = \begin{bmatrix} \hat{r}_X[1] \\ \hat{r}_X[2] \\ \vdots \\ \hat{r}_X[p] \end{bmatrix} \quad (18.47)$$

which are solved to yield the $\hat{a}[k]$'s. Then, the white noise variance estimate is found from (18.39) as

$$\hat{\sigma}_U^2 = \hat{r}_X[0] - \sum_{k=1}^{p} \hat{a}[k] \hat{r}_X[k] \quad (18.48)$$

where $\hat{a}[k]$ is given by the solution of the Yule-Walker equations of (18.47). Hence, we estimate the ACS for lags $k = 0, 1, \ldots, p$ based on an actual speech sound and then solve the equations of (18.47) to obtain $\{\hat{a}[1], \hat{a}[2], \ldots, \hat{a}[p]\}$ and finally, determine $\hat{\sigma}_U^2$ using (18.48). The only modification that is commonly made is to the ACS estimate, which is chosen to be

$$\hat{r}_X[k] = \frac{1}{N} \sum_{n=0}^{N-1-k} x[n] x[n+k] \qquad k = 0, 1, \ldots, p \quad (18.49)$$

and which differs from the one given in Section 17.7 in that the normalizing factor is N instead of $N - k$. For $N \gg p$ this will have minimal effect on the parameter estimates but has the benefit of ensuring a stable filter estimate, i.e., the poles of $\hat{\mathcal{H}}_{uv}(z)$ will lie inside the unit circle [Kay 1988]. This method of estimating the AR parameters is called the *autocorrelation method of linear prediction*. The entire procedure of modeling speech by an AR(p) model is referred to as *linear predictive coding* (LPC). The name originated with the connection of (18.47) as a set of linear prediction equations, although the ultimate goal here is not linear prediction but speech modeling [Makhoul 1975].

To demonstrate the modeling of an unvoiced sound consider the spoken word "seven" shown in Figure 18.10. A portion of the "sss" utterance is shown in Figure

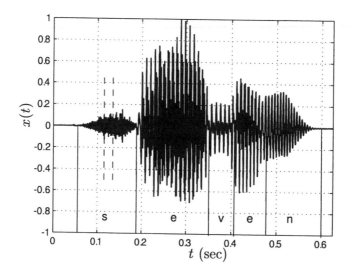

Figure 18.10: Waveform for the utterance "seven" [Allu 2005].

18.11 and as expected is noise-like. It is composed of the samples indicated between the dashed vertical lines in Figure 18.10. Typically, in analyzing speech sounds to estimate its AR parameters, we sample at 8 KHz and use a block of data 20 msec (about 160 samples) in length. The samples of $x(t)$ in Figure 18.10 from $t = 115$ msec to $t = 135$ msec are shown in Figure 18.11. With a model order of $p = 12$ we use (18.49) to estimate the ACS lags and then solve the Yule-Walker equations of (18.47) and also use (18.48) to yield the estimated parameters $\{\hat{a}[1], \hat{a}[2], \ldots, \hat{a}[p], \hat{\sigma}_U^2\}$. If the model is reasonably accurate, then the synthesized sound should be perceived as being similar to the original sound. It has been found through experimentation that if the PSDs are similar, then this will be the case. Hence, the estimated PSD

$$\hat{P}_X(f) = \frac{\hat{\sigma}_U^2}{\left|1 - \sum_{k=1}^{p} \hat{a}[k] \exp(-j2\pi f k)\right|^2} \tag{18.50}$$

should be a good match to the normalized and squared-magnitude of the Fourier

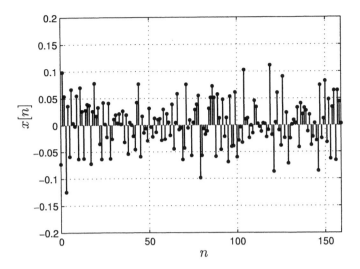

Figure 18.11: A 20 msec segment of the waveform for "sss". See Figure 18.10 for segment extracted as indicated by the vertical dashed lines.

transform of the speech sound. The latter is of course the periodogram. We need only consider the match in power since it is well known that the ear is relatively insensitive to the phase of the speech waveform [Rabiner and Schafer 1978].

As an example, for the portion of the "sss" sound shown in Figure 18.11 a periodogram as well as the AR PSD model of (18.50), is compared in Figure 18.12. Both PSDs are plotted in dB quantities, which is obtained by taking $10 \log_{10}$ of the PSD. Note that the resonances, i.e., the portions of the PSD that are large and which are most important for intelligibility, are well matched by the model. This verifies the validity of the AR model. Finally, to synthesize the "sss" sound we compute

$$x[n] = \sum_{k=1}^{p} \hat{a}[k]x[n-k] + u[n]$$

where $u[n]$ is a *pseudorandom* Gaussian noise sequence [Knuth 1981] with variance $\hat{\sigma}_U^2$, for a total of about 20 msec. Then, the samples are converted to an analog sound using a digital-to-analog (D/A) convertor (see Figure 18.9). The TI Speak and Spell used $p = 10$ and stored the AR parameters in memory for each sound. The MATLAB code used to generate Figure 18.12 is given below.

```
N=length(xseg); % xseg is the data shown in Figure 18.11
Nfft=1024; % set up FFT length for Fourier transforms
freq=[0:Nfft-1]'/Nfft-0.5; % PSD frequency points to be plotted
P_per=(1/N)*abs(fftshift(fft(xseg,Nfft))).^2; % compute periodogram
p=12; % dimension of autocorrelation matrix
for k=1:p+1 % estimate ACS for k=0,1,...,p (MATLAB indexes
```

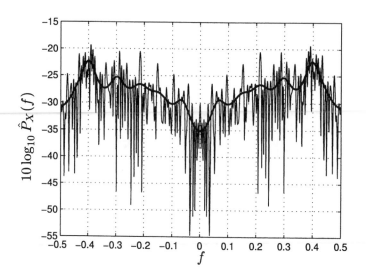

Figure 18.12: Periodogram, shown as the light line, and AR PSD model, shown as the darker line for speech segment of Figure 18.11.

```
            % must start at 1)
    rX(k,1)=(1/N)*sum(xseg(1:N-k+1).*xseg(k:N));
end
r=rX(2:p+1); % fill in right-hand-side vector
for i=1:p % fill in autocorrelation matrix
    for j=1:p
        R(i,j)=rX(abs(i-j)+1);
    end
end
a=inv(R)*r; % solve linear equations to find AR filter parameters
varu=rX(1)-a'*r; % find excitation noise variance
den=abs(fftshift(fft([1;-a],Nfft))).^2; % compute denominator of AR PSD
P_AR=varu./den; % compute AR PSD
```

See also Problem 18.34 for an application of AR modeling to spectral estimation [Kay 1988].

References

Allu, G., personal communication of speech data, 2005.

Jackson, L.B., *Signals, Systems, and Transforms*, Addison-Wesley, Reading, MA, 1991.

Kay, S., *Modern Spectral Estimation: Theory and Application*, Prentice-Hall, Englewood Cliffs, NJ, 1988.

Knuth, D.E., *The Art of Computer Programming, Vol. 2*, Addison-Wesley, Reading, MA, 1981.

Makhoul, J., "Linear Prediction: A Tutorial Review," *IEEE Proceedings*, Vol. 63, pp. 561–580, 1975.

Oppenheim, A.V., A.S. Willsky, S.H. Nawab, *Signal and Systems*, Prentice-Hall, Upper Saddle River, NJ, 1997.

Orfanidis, S.J., *Optimum Signal Processing, An Introduction*, Macmillan, New York, 1985.

Poularikas, A.D., S. Seely, *Signals and Systems*, PWS, Boston, 1985.

Rabiner, L.R., R.W. Schafer, *Digital Processing of Speech Signals*, Prentice-Hall, Englewood Cliffs, NJ, 1978.

Wiener, N., *Extrapolation, Interpolation, and Smoothing of Stationary Time Series*, MIT Press, Cambridge, MA, 1949.

Yaglom, A.M., *An Introduction to the Theory of Stationary Random Functions*, Dover, New York, 1962.

Problems

18.1 (☺) **(f)** An LSI system with system function $\mathcal{H}(z) = 1 - z^{-1} - z^{-2}$ is used to filter a discrete-time white noise random process with variance $\sigma_U^2 = 1$. Determine the ACS and PSD of the output random process.

18.2 (f) A discrete-time WSS random process with mean $\mu_U = 2$ is input to an LSI system with impulse response $h[n] = (1/2)^n$ for $n \geq 0$ and $h[n] = 0$ for $n < 0$. Find the mean sequence at the system output.

18.3 (w) A discrete-time white noise random process $U[n]$ is input to a system to produce the output random process $X[n] = a^{|n|}U[n]$ for $|a| < 1$. Determine the output PSD.

18.4 (☺) **(w)** A randomly phased sinusoid $X[n] = \cos(2\pi(0.25)n + \Theta)$ with $\Theta \sim \mathcal{U}(0, 2\pi)$ is input to an LSI system with system function $\mathcal{H}(z) = 1 - b_1 z^{-1} - b_2 z^{-2}$. Determine the filter coefficients b_1, b_2 so that the sinusoid will have zero power at the filter output.

18.5 (f,c) A discrete-time WSS random process $X[n]$ is defined by the difference equation $X[n] = aX[n-1] + U[n] - bU[n-1]$, where $U[n]$ is a discrete-time white noise random process with variance $\sigma_U^2 = 1$. Plot the PSD of $X[n]$ if $a = 0.9, b = 0.2$ and also if $a = 0.2, b = 0.9$ and explain your results.

18.6 (f) A discrete-time WSS random process $X[n]$ is defined by the difference equation $X[n] = 0.5X[n-1] + U[n] - 0.5U[n-1]$, where $U[n]$ is a discrete-time white noise random process with variance $\sigma_U^2 = 1$. Find the ACS and PSD of $X[n]$ and explain your results.

18.7 (⌣) (f) A differencer is given by $X[n] = U[n] - U[n-1]$. If the input random process $U[n]$ has the PSD $P_U(f) = 1 - \cos(2\pi f)$, determine the ACS and PSD at the output of the differencer.

18.8 (t) Verify that the discrete-time Fourier transform of $r_X[k]$ given in (18.15) is $\sigma_U^2 |H(f)|^2$.

18.9 (w) A discrete-time white noise random process is input to an LSI system which has $h[0] = 1$ with all the other impulse response samples nonzero. Can the output power of the filter ever be less than the input power?

18.10 (w) A random process with PSD

$$P_X(f) = \frac{1}{\left|1 - \frac{1}{2}\exp(-j2\pi f)\right|^2}$$

is to be filtered with an LSI system to produce a white noise random process $U[n]$ with variance $\sigma_U^2 = 4$ at the output. What should the difference equation of the LSI system be?

18.11 (w,c) An AR random process of order 2 is given by the recursive difference equation $X[n] = 2r\cos(2\pi f_0)X[n-1] - r^2 X[n-2] + U[n]$, where $U[n]$ is white Gaussian noise with variance $\sigma_U^2 = 1$. For $r = 0.7, f_0 = 0.1$ and also for $r = 0.95, f_0 = 0.1$ plot the PSD of $X[n]$. Can you explain your results? Hint: Determine the pole locations of $\mathcal{H}(z)$.

18.12 (w) A signal, which is bandlimited to B cycles/sample with $B < 1/2$, is modeled as a WSS random process with zero mean and PSD $P_S(f)$. If white noise is added to the signal with $\sigma_W^2 = 1$, find the frequency response of the optimal Wiener smoother. Explain your results.

18.13 (⌣) (f,c) A zero mean signal with PSD $P_S(f) = 2 - 2\cos(2\pi f)$ is embedded in white noise with variance $\sigma_W^2 = 1$. Plot the frequency response of the optimal Wiener smoother. Also, compute the minimum MSE. Hint: For the MSE use a "sum" approximation to the integral (see Problem 1.14).

18.14 (c) In this problem we simulate the Wiener smoother. First generate $N = 50$ samples of a signal $S[n]$, which is an AR random process (assumes that $U[n]$ is white Gaussian noise) with $a = 0.25$ and $\sigma_U^2 = 0.5$. Remember to set the initial condition $S[-1] \sim \mathcal{N}(0, \sigma_U^2/(1 - a^2))$. Next add white Gaussian noise $W[n]$ with $\sigma_W^2 = 1$ to the AR random process realization. Finally, use the MATLAB code in the chapter to smooth the noise-corrupted signal. Plot the true signal and the smoothed signal. How well does the smoother perform?

18.15 (w) To see that the linear prediction equations of (18.28) cannot be solved directly using z-transforms, take the z-transform of both sides of the equation. Next solve for $\mathcal{H}(z) = \mathcal{Z}\{h[k]\}$. Explain why the solution for the predictor cannot be correct.

18.16 (t) In this problem we rederive the optimal one-step linear predictor for the AR random process of Example 18.5. Assume that $X[n_0 + 1]$ is to be predicted based on observing the realization of $\{X[n_0], X[n_0 - 1], \ldots\}$. The random process $X[n]$ is assumed to be an AR random process described in Example 18.5. Prove that $\hat{X}[n_0 + 1] = aX[n_0]$ satisfies the orthogonality principle, making use of the result that $E[U[n_0 + 1]X[n_0 - k]] = 0$ for $k = 0, 1, \ldots$. The latter result says that "future" samples of $U[n]$ must be uncorrelated with the present and past samples of $X[n]$. Explain why this is true. Hint: Recall that for an AR random process $X[n]$ can be rewritten as $X[n] = \sum_{l=0}^{\infty} a^l U[n - l]$.

18.17 (w) For the AR random process described in Example 18.5 show that the minimum MSE for the optimal predictor $\hat{X}[n_0 + 1] = aX[n_0]$ is given by $\text{mse}_{\min} = r_X[0](1 - a^2)$. Use this to explain why the results shown in Figure 18.7 are reasonable.

18.18 (☺) (w) Express the minimum MSE given in the previous problem in terms of $r_X[0]$ and the correlation coefficient between $X[n_0]$ and $X[n_0 + 1]$. What happens to the minimum MSE if the correlation coefficient magnitude approaches one and also if it is zero?

18.19 (c) Consider an AR(2) random process given by $X[n] = -r^2 X[n - 2] + U[n]$, where $U[n]$ is white Gaussian noise with variance σ_U^2 and $0 < r < 1$. This random process follows from (18.33) with $p = 2$ and $a[1] = 0$, $a[2] = -r^2$. The ACS for this random process can be shown to be $r_X[k] = (\sigma_U^2/(1 - r^4))r^{|k|} \cos(k\pi/2)$ [Kay 1988]. Find the optimal one-step linear predictor based on the present and past samples of $X[n]$. Next perform a computer simulation to see how the predictor performs. Consider the two cases $r = 0.5, \sigma_U^2 = 1 - r^4$ and $r = 0.95, \sigma_U^2 = 1 - r^4$ so that the average power in each case is the same $(r_X[0] = 1)$. Generate 150 samples of each process and discard the first 100 samples to make sure the generated samples are WSS. Then, plot the realization and its predicted values for each case. Which value of r results in a more predictable process?

18.20 (t) Derive the Wiener-Hopf equations given by (18.36) and the resulting minimum MSE given by (18.37) for the finite length predictor.

18.21 (f) For $M = 1$ solve the Wiener-Hopf equations given by (18.36) to find $h[0]$. Relate this to $\text{cov}(X, Y)/\text{var}(X)$ used in the prediction of Y given $X = x$.

18.22 (☺) (f) The MA random process described in Example 18.6 and given by $X[n] = U[n] - bU[n-1]$ has as its ACS for $\sigma_U^2 = 1$

$$r_X[k] = \begin{cases} 1 + b^2 & k = 0 \\ -b & k = 1 \\ 0 & k \geq 2. \end{cases}$$

For $M = 2$ solve the Wiener-Hopf equations to find this finite length predictor and then determine the minimum MSE. Compare this minimum MSE to that of the infinite length predictor given in Example 18.6.

18.23 (f) It is desired to predict white noise. Solve the Wiener-Hopf equations for $r_X[k] = \sigma_X^2 \delta[k]$ and explain your results.

18.24 (☺) (f,c) For the MA random process $X[n] = U[n] - \frac{1}{2}U[n-1]$ where $U[n]$ is white Gaussian noise with $\sigma_U^2 = 1$ find the optimal finite length predictor $\hat{X}[n_0 + 1] = h[0]X[n_0] + h[1]X[n_0 - 1]$ and the corresponding minimum MSE. Next simulate the random process and compare the estimated minimum MSE with the theoretical one. Hint: Use your results from Problem 18.22.

18.25 (f) Consider the prediction of a randomly phased sinusoid whose ACS is $r_X[k] = \cos(2\pi f_0 k)$. For $M = 2$ solve the Wiener-Hopf equations to determine the optimal linear predictor and also the minimum MSE. Hint: You should be able to show that the minimum MSE is zero. Use the trigonometric identity $\cos(2\theta) = 2\cos^2(\theta) - 1$.

18.26 (t) In this problem we consider the L-step infinite length predictor of an AR random process. Let the predictor be given as

$$\hat{X}[n_0 + L] = \sum_{k=0}^{\infty} h[k]X[n_0 - k]$$

and show that the linear equations to be solved to determine the optimal $h[k]$'s are

$$r_X[l + L] = \sum_{k=0}^{\infty} h[k]r_X[l - k] \qquad l = 0, 1, \ldots .$$

Next show that the minimum MSE is

$$\text{mse}_{\min} = r_X[0] - \sum_{k=0}^{\infty} h_{\text{opt}}[k]r_X[k + L].$$

Finally, for an AR random process with ACS $r_X[k] = (\sigma_U^2/(1-a^2))a^{|k|}$ show that

$$\hat{X}[n_0 + L] = a^L X[n_0]$$
$$\text{mse}_{\min} = r_X[0](1 - a^{2L})$$

for a predictor based on $\{X[n_0], X[n_0 - 1], \ldots\}$. To do so assume that $h[k] = 0$ for $k \geq 1$ and show that the equations can be satisfied by choosing $h[0]$. Explain what happens to the quality of the prediction as L increases and why.

18.27 ($\smile$) (t) In this problem we consider the interpolation of a random process using a sample on either side of the sample to be interpolated. We wish to estimate or interpolate $X[n_0]$ using $\hat{X}[n_0] = h[-1]X[n_0+1]+h[1]X[n_0-1]$ for some impulse response values $h[-1], h[1]$. Find the optimal impulse response values by minimizing the MSE of the interpolated sample if $X[n]$ is the AR random process given by $X[n] = aX[n-1] + U[n]$. Does your interpolator average the samples on either side of $X[n_0]$? What happens as $a \to 1$ and as $a \to 0$?

18.28 (f) An LTI system has the impulse response $h(\tau) = \exp(-\tau)$ for $\tau \geq 0$ and is zero for $\tau < 0$. If continuous-time white noise with ACF $r_U(\tau) = (N_0/2)\delta(\tau)$ is input to the system, what is the PSD of the output random process? Sketch the PSD.

18.29 ($\smile$) (f) An LTI system has the impulse response $h(\tau) = 1$ for $0 \leq \tau \leq T$ and is zero otherwise. If continuous-time white noise with ACF $r_U(\tau) = (N_0/2)\delta(\tau)$ is input to the system, what is the PSD of the output random process? Sketch the PSD.

18.30 (f) A filter with frequency response $H(F) = \exp(-j2\pi F\tau_0)$ is used to filter a WSS random process with PSD $P_X(F)$. What is the PSD at the filter output and why?

18.31 (t) Prove that if a continuous-time white noise random process with ACF $r_U(\tau) = (N_0/2)\delta(\tau)$ is input to an LTI system with impulse response $h(\tau)$, then the ACF of the output random process is

$$r_X(\tau) = \frac{N_0}{2} \int_{-\infty}^{\infty} h(t)h(t + \tau)dt.$$

18.32 ($\smile$) (w) An RC electrical circuit with frequency response

$$H(F) = \frac{1/RC}{1/RC + j2\pi F}$$

is used to filter a white noise random process with ACF $r_U(\tau) = (N_0/2)\delta(\tau)$. Find the total average power at the filter output. Is it infinite? Hint: See previous problem.

18.33 (t) Two continuous-time WSS zero mean random processes $X(t)$ and $Y(t)$ are uncorrelated, which means that $E[X(t_1)Y(t_2)] = 0$ for all t_1 and t_2. Is the sum random process $Z(t) = X(t) + Y(t)$ also WSS, and if so, what is its ACF and PSD?

18.34 (c) In this problem we compare the periodogram spectral estimator to one based on an AR(2) model. This assumes, however, that the AR model is an accurate one for the random process. First generate $N = 50$ samples of a realization of the AR(2) random process described in Problem 18.19 with $r = 0.5$ and $\sigma_U^2 = 1 - r^4$. Next plot the periodogram of the realization (see Section 17.6). Using the estimate of the ACS given in (18.49) solve the Yule-Walker equations of (18.47) for $p = 2$ and then find $\hat{\sigma}_U^2$ from (18.48). Finally, plot the estimated PSD given by (18.50) and compare it to the periodogram as well as the true PSD. You may also wish to print out $\hat{a}[1]$ and $\hat{a}[2]$ and compare them to the theoretical values of $a[1] = 0$ and $a[2] = -r^2 = -0.25$. Hint: You can use the MATLAB code given in Section 18.7.

Appendix 18A

Solution for Infinite Length Predictor

The equations to be solved for the one-step predictor are from (18.28)

$$r_X[l+1] = \sum_{k=0}^{\infty} h[k]r_X[l-k] \qquad l = 0, 1, \dots \qquad (18A.1)$$

and the minimum MSE can be written from (18.29) as

$$\text{mse}_{\min} = r_X[0] - \sum_{k=0}^{\infty} h_{\text{opt}}[k]r_X[-1-k]. \qquad (18A.2)$$

Now let $n = l + 1$ in (18A.1) so that

$$
\begin{aligned}
r_X[n] &= \sum_{k=0}^{\infty} h[k]r_X[n-1-k] \qquad n = 1, 2, \dots \\
&= \sum_{j=1}^{\infty} h[j-1]r_X[n-j] \qquad (\text{let } j = k+1) \qquad (18A.3)
\end{aligned}
$$

and also let $j = k + 1$ in (18A.2) to yield

$$r_X[0] = \sum_{j=1}^{\infty} h[j-1]r_X[-j] + \text{mse}_{\min} \qquad (18A.4)$$

where we drop the "opt" on $h_{\text{opt}}[k]$ since $h[k]$ and $\text{mse}_{\min}$ are unknowns that we wish to solve for. Then combining (18A.3) and (18A.4) we have

$$r_X[n] = \sum_{j=1}^{\infty} h[j-1]r_X[n-j] + \text{mse}_{\min}\delta_{n0} \qquad n = 0, 1, \dots$$

where $\delta_{n0} = 1$ for $n = 0$ and $\delta_{n0} = 0$ for $n \geq 1$. Next divide both sides by $\mathrm{mse}_{\min}$ to yield

$$\frac{r_X[n]}{\mathrm{mse}_{\min}} = \sum_{j=1}^{\infty} \frac{h[j-1]}{\mathrm{mse}_{\min}} r_X[n-j] + \delta_{n0} \qquad n = 0, 1, \ldots .$$

Let

$$g[j] = \begin{cases} 1/\mathrm{mse}_{\min} & j = 0 \\ -h[j-1]/\mathrm{mse}_{\min} & j = 1, 2, \ldots \end{cases} \qquad (18A.5)$$

so that the equations become

$$r_X[n]g[0] = -\sum_{j=1}^{\infty} g[j] r_X[n-j] + \delta_{n0}$$

or

$$\sum_{j=0}^{\infty} g[j] r_X[n-j] = \delta_{n0} \quad n = 0, 1, \ldots . \qquad (18A.6)$$

Now if (18A.6) can be solved for $g[j]$, then $h[j], \mathrm{mse}_{\min}$ can then be found from (18A.5). Note that (18A.6) is a discrete-time convolution that holds for $n \geq 0$. We therefore need to find a *causal sequence* $g[n]$ (since the sum in (18A.6) is only over $j \geq 0$), which when convolved with $r_X[n]$ yields 1 for $n = 0$ and 0 for $n > 0$. Note that the values of $g[n] \star r_X[n]$ for $n < 0$ are unspecified by the equations. Hence, $g[n] \star r_X[n]$ must be an *anticausal sequence* to be a solution of (18A.6). This can easily be solved if

$$\mathcal{P}_X(z) = \sum_{k=-\infty}^{\infty} r_X[n] z^{-n}$$

can be written as

$$\mathcal{P}_X(z) = \frac{\sigma_U^2}{\mathcal{A}(z)\mathcal{A}(z^{-1})} \qquad (18A.7)$$

where

$$\mathcal{A}(z) = 1 - \sum_{k=1}^{\infty} a[k] z^{-k}$$

has all its zeros within the unit circle of the z plane. Now $1/\mathcal{A}(z)$ is the z-transform of a causal sequence. This is because if all the zeros of $\mathcal{A}(z)$ are within the unit circle, then all the poles of $1/\mathcal{A}(z)$ are within the unit circle. Thus, the z-transform $1/\mathcal{A}(z)$ must converge on and outside of the unit circle. Also, then $1/\mathcal{A}(z^{-1})$ is the z-transform of an anticausal sequence. Assuming this is possible (18A.6) becomes

$$\mathcal{Z}^{-1}\{\mathcal{G}(z)\mathcal{P}_X(z)\} = \mathcal{Z}^{-1}\left\{\mathcal{G}(z)\frac{\sigma_U^2}{\mathcal{A}(z)\mathcal{A}(z^{-1})}\right\}$$

$$= \begin{cases} 1 & n = 0 \\ 0 & n > 0 \end{cases}$$

where $\mathcal{G}(z)$ is the z-transform of $g[n]$ and $\mathcal{Z}^{-1}$ denotes the inverse z-transform. Now if we choose

$$\mathcal{G}(z) = \frac{\mathcal{A}(z)}{\sigma_U^2} \qquad (18A.8)$$

then

$$\mathcal{Z}^{-1}\left\{\mathcal{G}(z)\frac{\sigma_U^2}{\mathcal{A}(z)\mathcal{A}(z^{-1})}\right\} = \mathcal{Z}^{-1}\left\{\frac{1}{\mathcal{A}(z^{-1})}\right\}$$

$$= \begin{cases} 1 & n = 0 \\ 0 & n > 0 \end{cases}$$

since $1/\mathcal{A}(z^{-1})$ is the z-transform of an anticausal sequence, and the equations are satisfied. The inverse z-transform for $n = 0$ has been obtained by using the initial value theorem [Jackson 1991] which says that for an anticausal sequence $x[n]$

$$\mathcal{Z}^{-1}\left\{\sum_{n=-\infty}^{0} x[n]z^{-n}\right\}\Bigg|_{n=0} = \lim_{z \to 0} \sum_{n=-\infty}^{0} x[n]z^{-n} = x[0].$$

Therefore, we have that

$$\mathcal{Z}^{-1}\left\{\frac{1}{\mathcal{A}(z^{-1})}\right\}\Bigg|_{n=0} = \lim_{z \to 0} \frac{1}{\mathcal{A}(z^{-1})} = 1.$$

The solution for $g[n]$ is from (18A.8)

$$g[n] = \mathcal{Z}^{-1}\left\{\frac{\mathcal{A}(z)}{\sigma_U^2}\right\} = \begin{cases} 1/\sigma_U^2 & n = 0 \\ -a[n]/\sigma_U^2 & n \geq 1 \end{cases}$$

and using (18A.5)

$$\frac{1}{\text{mse}_{\min}} = g[0] = \frac{1}{\sigma_U^2}$$

$$-\frac{h[j-1]}{\text{mse}_{\min}} = g[j] = -\frac{a[j]}{\sigma_U^2} \qquad j \geq 1.$$

Finally, we have the result that

$$h[n] = a[n+1] \qquad n = 0, 1, \ldots$$

$$\text{mse}_{\min} = \sigma_U^2.$$

Chapter 19

Multiple Wide Sense Stationary Random Processes

19.1 Introduction

In Chapters 7 and 12 we defined multiple random variables X and Y as a mapping from the sample space S of the experiment to a *point* (x, y) in the x-y plane. We now extend that definition to be a mapping from S to a point in the x-y plane that evolves with time, and denote that point as $(x[n], y[n])$ for $-\infty < n < \infty$. The mapping, denoted either by $(X[n], Y[n])$ or equivalently by $[X[n]\, Y[n]]^T$, is called a *jointly distributed random process*. An example is the mapping from a point at some geographical location, where the possible choices for the location constitute S, to the daily temperature and pressure at that point or $(T[n], P[n])$. Instead of treating the random processes, which describe temperature and pressure, separately, it makes more sense to analyze them jointly. This is especially true if the random processes are correlated. For example, a drop in barometric pressure usually indicates the onset of a storm, which in turn will cause a drop in the temperature. Another example of great interest is the effect of a change in the Federal Reserve discount rate, which is the percentage interest charged to banks by the federal government, on the rate of job creation. It is generally assumed that by lowering the discount rate, companies can borrow money more cheaply and thus invest in new products and services, thus increasing the demand for labor. The jointly distributed random processes describing this situation are $I[n]$, the daily discount interest rate, and $J[n]$, the daily number of employed Americans. Many other examples are possible, encompassing a wide range of disciplines.

In this chapter we extend the concept of a wide sense stationary (WSS) random process to *two jointly distributed WSS random processes*. The extension to any number of WSS random processes can be found in [Bendat and Piersol 1971, Jenkins and Watts 1968, Kay 1988, Koopmans 1974, Robinson 1967]. Multiple random process theory is known by the synonymous terms *multivariate random processes, multichannel random processes*, and *vector random processes*. Also, the characterization of the random processes at the input and output of an LSI system is explored. We will find that the extensions are much the same as in going from a single random *variable* to two random *variables*, especially since the definitions are based on samples of the random process, which themselves are random variables. As in previous chapters our focus will be on discrete-time random processes but the analogous concepts and formulas for continuous-time random processes will be summarized later.

19.2   Summary

Two random processes are jointly WSS if they are individually WSS (satisfy (19.1)–(19.4)) and also the cross-correlation given by (19.5) does not depend on n. The sequence given by (19.5) is called the cross-correlation sequence. The cross-correlation sequence has the properties given in Property 19.1–19.4, which differ from those of the ACS. Jointly WSS random processes are defined to be uncorrelated if (19.12) holds. The cross-power spectral density is defined by (19.13) and is evaluated using (19.14). It has the properties given by Property 19.5–19.9, which differ from those of the PSD. The correlation between two jointly WSS random processes can be measured in the frequency domain using the coherence function defined in (19.20). The ACS and PSD for the sum of two jointly distributed WSS random processes is given in Section 19.5. If the random processes are uncorrelated, then the ACS and PSD of the sum random process are given by (19.25) and (19.26), respectively. For the filtering operation shown in Figure 19.2a the cross-correlation sequence is given by (19.27) and the cross-power spectral density by (19.28). For the filtering operation shown in Figure 19.2b the cross-correlation sequence is given by (19.29) and the cross-power spectral density by (19.30). The corresponding definitions and formulas for continuous-time random processes are given in Section 19.6. Estimation of the cross-correlation sequence is discussed in Section 19.7 with the estimate given by (19.46). Finally, an application of cross-correlation to brain physiology research is described in Section 19.8.

19.3   Jointly Distributed WSS Random Processes

We will denote the two discrete-time random processes by $X[n]$ and $Y[n]$ for $-\infty < n < \infty$. Of particular interest will be the extension of the concept of wide sense stationarity from one to two random processes. To do so we first assume that each random process is *individually WSS*, which is to say that

$$\mu_X[n] \;=\; E[X[n]] = \mu_X \qquad\qquad (19.1)$$
$$r_X[k] \;=\; E[X[n]X[n+k]] \qquad\qquad (19.2)$$
$$\mu_Y[n] \;=\; E[Y[n]] = \mu_Y \qquad\qquad (19.3)$$
$$r_Y[k] \;=\; E[Y[n]Y[n+k]] \qquad\qquad (19.4)$$

or the first two moments do not depend on n. For the concept of wide sense stationarity to be useful in the context of two random processes, we require a further definition. To motivate it, consider the situation in which we add the two random processes together and wish to determine the overall average power. Then, if $Z[n] = X[n] + Y[n]$, we need to find $E[Z^2[n]]$. Proceeding we have

$$
\begin{aligned}
E[Z^2[n]] &= E[(X[n] + Y[n])^2] \\
&= E[X^2[n]] + E[X[n]Y[n]] + E[Y[n]X[n]] + E[Y^2[n]] \\
&= r_X[0] + 2E[X[n]Y[n]] + r_Y[0].
\end{aligned}
$$

To complete the calculation we require knowledge of the joint moment $E[X[n]Y[n]]$. If it does not depend on n, then $E[Z^2[n]]$ will likewise not depend on n. More generally, if we were to compute $E[Z[n]Z[n+k]]$, then we would require knowledge of $E[X[n]Y[n+k]]$ and so we will *assume* that the latter does not depend on n. Therefore, with this assumption we can now define

$$
r_{X,Y}[k] = E[X[n]Y[n+k]] \qquad k = \ldots, -1, 0, 1, \ldots . \tag{19.5}
$$

This new sequence is called the *cross-correlation sequence* (CCS). Returning to our average power computation we can now write that

$$
E[Z^2[n]] = r_X[0] + 2r_{X,Y}[0] + r_Y[0]
$$

and the average power is seen not to depend on n. Note also from the definition of the CCS, that the ACS is just $r_{X,X}[k]$.

If $X[n]$ and $Y[n]$ are WSS random processes and a CCS can be defined ($E[X[n]Y[n+k]]$ not dependent on n), then the random processes are said to be *jointly wide sense stationary*. In summary, for the two random processes to be jointly WSS we require the conditions (19.1)–(19.5) to hold. An example follows.

Example 19.1 – CCS for WSS random processes delayed with respect to each other

Let $X[n]$ be a WSS random process and let $Y[n]$ be a delayed version of $X[n]$ so that $Y[n] = X[n - n_0]$. Then, to determine if the random processes are jointly WSS we have

$$
\begin{aligned}
E[X[n]] &= \mu_X \\
E[Y[n]] &= E[X[n - n_0]] = \mu_X \\
E[X[n]X[n+k]] &= r_X[k] \\
E[Y[n]Y[n+k]] &= E[X[n - n_0]X[n + k - n_0]] = r_X[k] \\
E[X[n]Y[n+k]] &= E[X[n]X[n + k - n_0]] = r_X[k - n_0] \tag{19.6}
\end{aligned}
$$

all of which follow from our definition of $Y[n]$ and the assumption that $X[n]$ is WSS. Note that $E[X[n]Y[n+k]]$ does not depend on n and so a CCS can be defined. It is given by (19.6) as

$$
r_{X,Y}[k] = r_X[k - n_0]. \tag{19.7}
$$

Since all the first and second moments do not depend on n, the random processes are jointly WSS.

◇

We will henceforth assume that $X[n]$ and $Y[n]$ are jointly WSS unless stated otherwise. From the previous example it is observed that the CCS has very different properties than the ACS. Unlike the ACS, *the CCS does not necessarily have its maximum value at $k = 0$.* In the previous example, the maximum of the CCS occurs at $k = n_0$ (see (19.7)). Also, in general *we do not have $r_{X,Y}[-k] = r_{X,Y}[k]$ or the CCS is not symmetric about $k = 0$.* In the previous example, we have from (19.7)

$$
\begin{aligned}
r_{X,Y}[-k] &= r_X[-k - n_0] \\
&= r_X[k + n_0] \neq r_X[k - n_0] = r_{X,Y}[k].
\end{aligned}
$$

Furthermore, even though the CCS is symmetric about $k = n_0$ in the previous example, it need not be symmetric at all.

 CCS asymmetry requires vigilance.

Since the CCS is not symmetric, in contrast to the ACS, one must be careful. The cross-second moment $E[X[m]Y[n]]$, where $X[n]$ and $Y[n]$ are jointly WSS, is expressed in terms of the CCS as $r_{X,Y}[n - m]$, *not* $r_{X,Y}[m - n]$. To determine the argument k of the CCS for $r_{X,Y}[k]$, always take the index of the Y random variable and subtract the index of the X random variable. For example, $E[X[3]Y[1]] = r_{X,Y}[1 - 3] = r_{X,Y}[-2]$. This is especially important in light of the fact that the definition of the CCS is not standard. Some authors use $r_{X,Y}[k] = E[X[n]Y[n - k]]$, which will produce a CCS that is "flipped around" in k, relative to our definition.

We give one more example and then summarize the properties of the CCS.

Example 19.2 – Another calculation of the CCS

Assume that $X[n] = U[n]$ and $Y[n] = U[n] + 2U[n - 1]$, where $U[n]$ is white noise with variance $\sigma_U^2 = 1$. Thus, $X[n]$ is a white noise random process and $Y[n]$ is a general MA random process, i.e., no Gaussian assumption is made. Then, it is easily shown that $\mu_X[n] = \mu_Y[n] = 0$, $r_X[k] = \delta[k]$, and

$$
r_Y[k] = \begin{cases} 5 & k = 0 \\ 2 & k = \pm 1 \\ 0 & \text{otherwise} \end{cases}
$$

so that $X[n]$ and $Y[n]$ are individually WSS. Now computing the cross-second moment, we have

$$
\begin{aligned}
E[X[n]Y[n+k]] &= E[U[n](U[n+k]+2U[n+k-1])] \\
&= r_U[k] + 2r_U[k-1] \\
&= \delta[k] + 2\delta[k-1]
\end{aligned}
$$

and it is seen to be independent of n. Hence, the CCS is

$$
r_{X,Y}[k] = \delta[k] + 2\delta[k-1]
$$

and the random processes are jointly WSS. The ACSs and the CCS are shown in Figure 19.1. We observe that $r_{X,Y}[-k] \neq r_{X,Y}[k]$ and that the maximum does not occur at $k = 0$. We can assert, however, that the maximum must be less than or equal to $\sqrt{r_X[0]r_Y[0]}$ since by the Cauchy-Schwarz inequality (see Appendix 7A)

$$
\begin{aligned}
|r_{X,Y}[k]| &= |E[X[n]Y[n+k]]| \\
&\leq \sqrt{E[X^2[n]]E[Y^2[n+k]]} \\
&= \sqrt{r_X[0]r_Y[0]}.
\end{aligned}
$$

For this example we see that

$$
|r_{X,Y}[k]| \leq \sqrt{1 \cdot 5} = \sqrt{5}.
$$

$\Diamond$

We now summarize the properties (or more appropriately the *non*properties) of the CCS.

Property 19.1 – CCS is not necessarily symmetric.

$$
r_{X,Y}[-k] \neq r_{X,Y}[k] \tag{19.8}
$$

$\square$

Property 19.2 – The maximum of the CCS can occur for any value of k.

$\square$

Property 19.3 – The maximum value of the CCS is bounded.

$$
|r_{X,Y}[k]| \leq \sqrt{r_X[0]r_Y[0]} \tag{19.9}
$$

$\square$

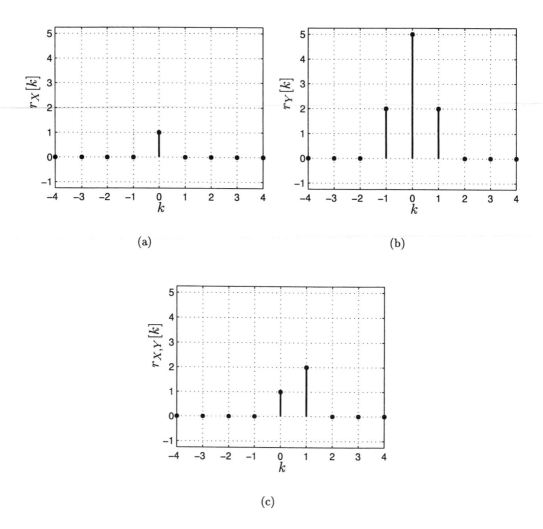

(a)   (b)

(c)

Figure 19.1: Autocorrelation and cross-correlation sequences for Example 19.2.

A fourth property that is useful arises by considering $E[Y[n]X[n + k]]$, which is the cross-second moment with $X[n]$ and $Y[n]$ interchanged. Assuming jointly WSS random processes, this moment becomes

$$
\begin{aligned}
E[Y[n]X[n + k]] &= E[X[n + k]Y[n]] \\
&= E[X[m]Y[m - k]] \qquad (\text{let } m = n + k) \\
&= r_{X,Y}[-k] \qquad\qquad (\text{from definition of CCS}).
\end{aligned}
$$

Therefore, $E[Y[n]X[n + k]]$ does not depend on n and so we can define another cross-correlation sequence as

$$
r_{Y,X}[k] = E[Y[n]X[n + k]] \tag{19.10}
$$

and it is seen to be equal to $r_{X,Y}[-k]$. Thus, as our last property we have

Property 19.4 – Interchanging $X[n]$ and $Y[n]$ flips the CCS about $k = 0$.

$$r_{Y,X}[k] = r_{X,Y}[-k] \tag{19.11}$$

□

Next, we define the concept of *uncorrelated jointly WSS random processes*. Two *zero mean* jointly WSS random processes are said to be uncorrelated if

$$r_{X,Y}[k] = 0 \qquad \text{for } -\infty < k < \infty \tag{19.12}$$

including $k = 0$. (For nonzero mean random processes the definition of uncorrelated random processes is that $r_{X,Y}[k] = \mu_X\mu_Y$ for $-\infty < k < \infty$.) Of course, if the random processes are independent so that $E[X[n]Y[n+k]] = 0$ does not depend on n, then they must be jointly WSS as well. It also follows from Property 19.4 that if the random processes are uncorrelated, then $r_{Y,X}[k] = 0$ for all k. An example follows.

Example 19.3 – Uncorrelated sinusoidal random processes

Let $X[n] = \cos(2\pi f_0 n + \Theta_1)$ and $Y[n] = \cos(2\pi f_0 n + \Theta_2)$, where $\Theta_1 \sim \mathcal{U}(0, 2\pi)$, $\Theta_2 \sim \mathcal{U}(0, 2\pi)$, and Θ_1 and Θ_2 are independent random variables. Then, we have seen previously that $X[n]$ and $Y[n]$ are individually WSS (see Example 17.4) and

$$
\begin{aligned}
r_{X,Y}[k] &= E[X[n]Y[n+k]] \\
&= E_{\Theta_1,\Theta_2}[\cos(2\pi f_0 n + \Theta_1)\cos(2\pi f_0(n+k) + \Theta_2)] \\
&= E_{\Theta_1}[\cos(2\pi f_0 n + \Theta_1)]E_{\Theta_2}[\cos(2\pi f_0(n+k) + \Theta_2)] \text{ (independent random} \\
&\qquad\qquad\qquad\qquad\qquad\qquad\qquad\qquad\qquad\qquad \text{variables and (12.30))} \\
&= 0
\end{aligned}
$$

since each random sinusoid has a zero mean (see Example 16.11). Thus, the random processes are uncorrelated and jointly WSS. Can you interpret this result physically?

◊

19.4 The Cross-Power Spectral Density

The PSD of a WSS random process was seen earlier to describe the distribution of average power with frequency. Also, the average power of the random process in a band of frequencies is obtained by integrating the PSD over that frequency band. In a similar vein to the definition of the PSD, we can define the *cross-power spectral density* (CPSD) of two jointly WSS random processes as

$$P_{X,Y}(f) = \lim_{M \to \infty} \frac{1}{2M+1} E\left[\left(\sum_{n=-M}^{M} X[n]\exp(-j2\pi fn)\right)^* \left(\sum_{n=-M}^{M} Y[n]\exp(-j2\pi fn)\right)\right]$$

$$\tag{19.13}$$

which results in the usual PSD if $Y[n] = X[n]$. Using a similar derivation as the one that resulted in the Wiener-Khinchine theorem, it can be shown that (see Problem 19.8)

$$P_{X,Y}(f) = \sum_{k=-\infty}^{\infty} r_{X,Y}[k] \exp(-j2\pi f k). \qquad (19.14)$$

It is less clear than for the PSD what the physical significance of the CPSD is. From (19.13) it appears that the CPSD will be large when the Fourier transforms of $X[n]$ and $Y[n]$ at a given frequency are large and *are in phase*. Conversely, when the Fourier transforms are either small or out of phase, the CPSD will be small. This is confirmed by the results of Example 19.3 in which the sinusoidal processes have all their power at $f = \pm f_0$ since $P_X(f) = P_Y(f) = \frac{1}{2}\delta(f + f_0) + \frac{1}{2}\delta(f - f_0)$. However, because they have phases that are independent of each other and can take on values in $(0, 2\pi)$ uniformly, $r_{X,Y}[k] = 0$ and therefore, $P_{X,Y}(f) = 0$. On the other hand, if the phase random variables were statistically dependent, say $\Theta_1 = \Theta_2$, then the CPSD would be large (see Problem 19.9). Another example follows.

Example 19.4 – CCS for WSS random processes delayed with respect to each other (continued)

We continue Example 19.1 in which $Y[n] = X[n - n_0]$ and $X[n]$ is WSS. We saw that the CCS is given by $r_{X,Y}[k] = r_X[k - n_0]$. Using (19.14) the CPSD is

$$P_{X,Y}(f) = \sum_{k=-\infty}^{\infty} r_X[k - n_0] \exp(-j2\pi f k)$$

and letting $l = k - n_0$ produces

$$
\begin{aligned}
P_{X,Y}(f) &= \sum_{l=-\infty}^{\infty} r_X[l] \exp[-j2\pi f(l + n_0)] \\
&= \sum_{l=-\infty}^{\infty} r_X[l] \exp(-j2\pi f l) \exp(-j2\pi f n_0) \\
&= P_X(f) \exp(-j2\pi f n_0).
\end{aligned}
$$

It is seen that the CPSD is a complex function and that $P_{X,Y}(-f) \neq P_{X,Y}(f)$. It does appear, however, that $P_{X,Y}(-f) = P_{X,Y}^*(f)$ so that it has the symmetry properties

$$
\begin{aligned}
|P_{X,Y}(-f)| &= |P_{X,Y}(f)| \\
\angle P_{X,Y}(-f) &= -\angle P_{X,Y}(f)
\end{aligned}
\qquad (19.15)
$$

or the magnitude of the CPSD is an even function and the phase of the CPSD is an odd function. This result is indeed true as we will prove in Property 19.6.

$\Diamond$

One way to think about the CPSD is as a correlation between the *normalized* Fourier transforms of $X[n]$ and $Y[n]$ *at a given frequency*. From (19.13) we see that if

$$X_{2M+1}(f) = \frac{1}{\sqrt{2M+1}} \sum_{n=-M}^{M} X[n] \exp(-j2\pi fn)$$

$$Y_{2M+1}(f) = \frac{1}{\sqrt{2M+1}} \sum_{n=-M}^{M} Y[n] \exp(-j2\pi fn) \qquad (19.16)$$

then

$$P_{X,Y}(f) = \lim_{M\to\infty} E[X_{2M+1}^*(f)Y_{2M+1}(f)]. \qquad (19.17)$$

This is a correlation between the two *complex random variables* $X_{2M+1}(f)$ and $Y_{2M+1}(f)$. In fact, a normalized version of the CPSD is a *complex correlation coefficient*. Indeed, from the Cauchy-Schwarz inequality for complex random variables (see Appendix 7A for real random variables), we have that (recall that if $X = U + jV$, then $E[X]$ is defined as $E[X] = E[U] + jE[V]$)

$$|E[X_{2M+1}^*(f)Y_{2M+1}(f)]| \le \sqrt{E[|X_{2M+1}(f)|^2]E[|Y_{2M+1}(f)|^2]} \qquad (19.18)$$

and therefore as $M \to \infty$, this becomes from (19.17) and (17.30)

$$|P_{X,Y}(f)| \le \sqrt{P_X(f)P_Y(f)}. \qquad (19.19)$$

Thus, if we normalize the CPSD to form the complex function of frequency

$$\gamma_{X,Y}(f) = \frac{P_{X,Y}(f)}{\sqrt{P_X(f)P_Y(f)}} \qquad (19.20)$$

then we have that $|\gamma_{X,Y}(f)| \le 1$. The complex function of frequency $\gamma_{X,Y}(f)$ is called the *coherence function* and it is a *complex correlation coefficient*. It measures the correlation between the Fourier transforms of two jointly WSS random processes at a given frequency. As an example, consider the random processes of Example 19.4. Then

$$\begin{aligned}
\gamma_{X,Y}(f) &= \frac{P_{X,Y}(f)}{\sqrt{P_X(f)P_Y(f)}} \\
&= \frac{P_X(f)\exp(-j2\pi fn_0)}{\sqrt{P_X(f)P_X(f)}} \\
&= \exp(-j2\pi fn_0) \qquad \text{(since } P_X(f) \ge 0\text{)}.
\end{aligned}$$

The magnitude of the coherence is unity for all frequencies, meaning that the Fourier transform of $Y[n]$ at a given frequency can be perfectly predicted from the Fourier transform of $X[n]$ at the same frequency since $Y[n] = X[n - n_0]$. It follows that

$Y_{2M+1}(f) = \exp(-j2\pi f n_0)X_{2M+1}(f)$ and therefore $Y_{2M+1}(f) = \gamma_{X,Y}(f)X_{2M+1}(f)$ for all f. Furthermore, since the coherence magnitude is unity for *all* frequencies, the prediction of the frequency component of $Y[n]$ is perfect for all frequencies as well. This says finally that $Y[n]$ can be perfectly predicted from $X[n]$. To do so just let $\hat{Y}[n] = X[n + n_0]$. In general, we will see later that if $Y[n]$ is the output of an LSI system whose input is $X[n]$, then the coherence magnitude is always unity. Can you interpret $Y[n] = X[n - n_0]$ as the action of an LSI system? Finally, in contrast to perfect prediction, consider the CPSD if $X[n]$ and $Y[n]$ are zero mean and uncorrelated. Then since $r_{X,Y}[k] = 0$, we have that $P_{X,Y}(f) = 0$ for f, and of course the coherence will be zero as well. We now summarize the properties of the CPSD.

Property 19.5 – CPSD is Fourier transform of the CCS.

$$P_{X,Y}(f) = \sum_{k=-\infty}^{\infty} r_{X,Y}[k]\exp(-j2\pi fk)$$

Proof: See Problem 19.8.

$\square$

Property 19.6 – CPSD is a hermitian function.

A complex function $g(f)$ is hermitian if its real part is an even function and its imaginary part is an odd function about $f = 0$. This is equivalent to saying that $g(-f) = g^*(f)$. Thus,

$$P_{X,Y}(-f) = P_{X,Y}^*(f) \tag{19.21}$$

(see also (19.15) which is valid for a hermitian function).
Proof:

$$
\begin{aligned}
P_{X,Y}(f) &= \sum_{k=-\infty}^{\infty} r_{X,Y}[k]\exp(-j2\pi fk) \\
P_{X,Y}(-f) &= \sum_{k=-\infty}^{\infty} r_{X,Y}[k]\exp(j2\pi fk) \\
&= \sum_{k=-\infty}^{\infty} r_{X,Y}[k]\cos(2\pi fk) + j\sum_{k=-\infty}^{\infty} r_{X,Y}[k]\sin(2\pi fk) \\
&= \left(\sum_{k=-\infty}^{\infty} r_{X,Y}[k]\cos(2\pi fk) - j\sum_{k=-\infty}^{\infty} r_{X,Y}[k]\sin(2\pi fk)\right)^* \\
&= \left(\sum_{k=-\infty}^{\infty} r_{X,Y}[k]\exp(-j2\pi fk)\right)^* \\
&= P_{X,Y}^*(f)
\end{aligned}
$$

$\square$

Property 19.7 – CPSD is bounded.

$$|P_{X,Y}(f)| \leq \sqrt{P_X(f)P_Y(f)} \qquad (19.22)$$

Proof: See argument leading to (19.20).

$\square$

Property 19.8 – CPSD is zero for zero mean uncorrelated random processes.

If $X[n]$ and $Y[n]$ are jointly WSS random processes that are zero mean and uncorrelated, then $P_{X,Y}(f) = 0$ for all f.

Proof: Since the random processes are zero mean and uncorrelated, $r_{X,Y}[k] = 0$ by definition. Hence, the CPSD is zero as well, being the Fourier transform of the CCS.

$\square$

Property 19.9 – CPSD of $(Y[n], X[n])$ is the complex conjugate of the CPSD of $(X[n], Y[n])$.

$$P_{Y,X}(f) = P_{X,Y}^*(f) \qquad (19.23)$$

Proof:

$$
\begin{aligned}
P_{Y,X}(f) &= \sum_{k=-\infty}^{\infty} r_{Y,X}[k] \exp(-j2\pi f k) \\
&= \sum_{k=-\infty}^{\infty} r_{X,Y}[-k] \exp(-j2\pi f k) \qquad \text{(using (19.11))} \\
&= \sum_{l=-\infty}^{\infty} r_{X,Y}[l] \exp(j2\pi f l) \qquad \text{(let } l = -k) \\
&= P_{X,Y}(-f) \\
&= P_{X,Y}^*(f) \qquad \text{(using (19.21))}
\end{aligned}
$$

$\square$

We conclude this section with one more example.

Example 19.5 – MA Random Process

Let $Y[n] = X[n] - bX[n-1]$, where $X[n]$ is white Gaussian noise with variance σ_X^2. We wish to determine the CPSD between the input $X[n]$ and output $Y[n]$ random processes (assuming they are jointly WSS, which will be borne out shortly). (Are $X[n]$ and $Y[n]$ individually WSS?) To do so we first find the CCS and then take the

Fourier transform of it. Proceeding we have

$$
\begin{aligned}
r_{X,Y}[k] &= E[X[n]Y[n+k]] \\
&= E[X[n](X[n+k] - bX[n+k-1])] \\
&= E[X[n]X[n+k]] - bE[X[n]X[n+k-1]] \\
&= r_X[k] - br_X[k-1]
\end{aligned}
$$

which does not depend on n and hence $X[n]$ and $Y[n]$ are jointly WSS with the CCS

$$
r_{X,Y}[k] = \sigma_X^2 \delta[k] - b\sigma_X^2 \delta[k-1].
$$

The CPSD is found as the Fourier transform to yield

$$
\begin{aligned}
P_{X,Y}(f) &= \sigma_X^2 - b\sigma_X^2 \exp(-j2\pi f) \\
&= \sigma_X^2(1 - b\exp(-j2\pi f)).
\end{aligned}
$$

$\diamondsuit$

Note that in the previous example we can view $Y[n]$ as the output of an LSI filter with frequency response $H(f) = 1 - b\exp(-j2\pi f)$. Therefore, we have the result

$$
P_{X,Y}(f) = H(f)\sigma_X^2. \tag{19.24}
$$

More generally, we will prove in the next section that if $X[n]$ is the input to an LSI system with $Y[n]$ as its corresponding output, then $X[n]$ and $Y[n]$ are jointly WSS and $P_{X,Y}(f) = H(f)P_X(f)$. As an application note, if the input to the LSI system is white noise with $\sigma_X^2 = 1$, then $P_{X,Y}(f) = H(f)$. To measure the frequency response of an unknown LSI system one can input white noise with a variance equal to one and then estimate the CCS from the input and observed output (see Section 19.7). Upon Fourier transforming that estimate one obtains an estimate of the frequency response. Lastly, since $P_{X,Y}(f) = H(f)$ for $P_X(f) = 1$, it is clear that the properties of the CPSD should mirror those of a frequency response, i.e., complex in general, hermitian, etc.

19.5 Transformations of Multiple Random Processes

We now consider the effect of some transformations on jointly WSS random processes. As a simple first example, we add the two random processes together. Hence, assume $X[n]$ and $Y[n]$ are jointly WSS random processes, and $Z[n] = X[n] + Y[n]$. We next compute the first two moments. Clearly, we will have $\mu_Z[n] = \mu_X[n] + \mu_Y[n] = \mu_X + \mu_Y$ and

$$
\begin{aligned}
r_Z[k] &= E[Z[n]Z[n+k]] \\
&= E[(X[n] + Y[n])(X[n+k] + Y[n+k])] \\
&= r_X[k] + r_{X,Y}[k] + r_{Y,X}[k] + r_Y[k] \qquad \text{(assumed jointly WSS)}
\end{aligned}
$$

and hence $Z[n]$ is a WSS random process. Its PSD is found by taking the Fourier transform of the ACS to yield

$$P_Z(f) = P_X(f) + P_{X,Y}(f) + P_{Y,X}(f) + P_Y(f).$$

If in particular $X[n]$ and $Y[n]$ are zero mean and uncorrelated, so that $r_{X,Y}[k] = 0$ and hence $r_{Y,X}[k] = r_{X,Y}[-k] = 0$ as well, we have

$$r_Z[k] = r_X[k] + r_Y[k] \tag{19.25}$$
$$P_X(f) = P_X(f) + P_Y(f). \tag{19.26}$$

Another frequently encountered transformation is that due to filtering of a WSS random process by one or two LSI filters. These transformations are shown in Figure 19.2. For the transformation shown in Figure 19.2a we already know from Chapter

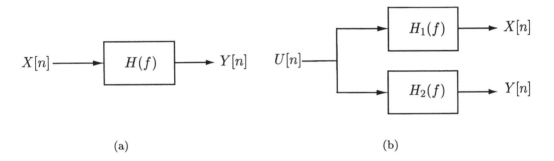

$$
\begin{array}{cc}
\text{(a)} & \text{(b)}
\end{array}
$$

Figure 19.2: Common filtering operations.

18 that if $X[n]$ is WSS, then $Y[n]$ is also WSS and its mean and ACS are easily found. The question arises, however, as to whether $X[n]$ and $Y[n]$ are *jointly* WSS. To answer this we compute $E[X[n]Y[n+k]]$ to see if it depends on n. Proceeding, we have for the filtering operation shown in Figure 19.2a with $h[k]$ denoting the impulse response

$$
\begin{aligned}
E[X[n]Y[n+k]] &= E\left[X[n]\sum_{l=-\infty}^{\infty} h[l]X[n+k-l]\right] \\
&= \sum_{l=-\infty}^{\infty} h[l]E[X[n]X[n+k-l]] \\
&= \sum_{l=-\infty}^{\infty} h[l]r_X[k-l]
\end{aligned}
$$

and we see that it does not depend on n. Hence, *if $X[n]$ is the input to an LSI system and $Y[n]$ is its corresponding output, then $X[n]$ and $Y[n]$ are jointly WSS.*

Also, we have for the CCS

$$r_{X,Y}[k] = \sum_{l=-\infty}^{\infty} h[l] r_X[k-l] \qquad (19.27)$$

which can be seen to be a discrete convolution or $r_{X,Y}[k] = h[k] \star r_X[k]$. As a result, by Fourier transforming the CCS we obtain the CPSD as

$$P_{X,Y}(f) = H(f) P_X(f) \qquad (19.28)$$

which agrees with our earlier result of (19.24). As previously asserted, we can also now prove that if $X[n]$ is a WSS random process that is input to an LSI system and $Y[n]$ is the output random process, then the coherence magnitude is one. This says that $Y[n]$ is perfectly predictable from $X[n]$, which upon reflection just says that to predict $Y[n]$ we need only pass $X[n]$ through the same filter! To verify the assertion about the coherence magnitude

$$\begin{aligned}
\gamma_{X,Y}(f) &= \frac{P_{X,Y}(f)}{\sqrt{P_X(f) P_Y(f)}} \\
&= \frac{H(f) P_X(f)}{\sqrt{P_X(f) |H(f)|^2 P_X(f)}} \qquad \text{(using (19.28) and (18.11))} \\
&= \frac{H(f)}{|H(f)|} \\
&= \exp(j\phi(f))
\end{aligned}$$

where $\phi(f)$ is the phase response of the LSI system or $\phi(f) = \angle H(f)$ Thus, $|\gamma_{X,Y}(f)| = 1$ (assuming $H(f) \neq 0$) and $Y[n]$ is perfectly predictable from $X[n]$ as

$$Y[n] = \sum_{k=-\infty}^{\infty} h[k] X[n-k] \qquad \text{for all } n$$

where $h[k]$ is the impulse response of $H(f)$. Also, $X[n]$ can be perfectly predicted from $Y[n]$ as one might expect from the analogous result of the symmetry of the correlation coefficient, which is $\rho_{X,Y} = \rho_{Y,X}$ (see Problem 19.21).

Next consider the transformation depicted in Figure 19.2b. The input random process $U[n]$ is WSS so that $X[n]$ and $Y[n]$ are individually WSS according to Theorem 18.3.1. To determine if they are jointly WSS we again compute $E[X[n]Y[n+k]]$ to see if it depends on n. Therefore, with $h_1[k]$, $h_2[k]$ denoting the impulse responses, and $H_1(f)$, $H_2(f)$ denoting the corresponding frequency responses

$$
\begin{aligned}
E[X[n]Y[n+k]] &= E\left[\sum_{i=-\infty}^{\infty} h_1[i]U[n-i] \sum_{j=-\infty}^{\infty} h_2[j]U[n+k-j]\right] \\
&= \sum_{i=-\infty}^{\infty}\sum_{j=-\infty}^{\infty} h_1[i]h_2[j]E[U[n-i]U[n+k-j]] \\
&= \sum_{i=-\infty}^{\infty}\sum_{j=-\infty}^{\infty} h_1[i]h_2[j]r_U[k+i-j]
\end{aligned}
$$

and does not depend on n. Hence, $X[n]$ and $Y[n]$ are jointly WSS and the CCS is

$$
r_{X,Y}[k] = \sum_{i=-\infty}^{\infty} h_1[i] \underbrace{\sum_{j=-\infty}^{\infty} h_2[j]r_U[k+i-j]}_{g[k+i]}
$$

where $g[n] = h_2[n] \star r_U[n]$. Continuing we have

$$
\begin{aligned}
r_{X,Y}[k] &= \sum_{i=-\infty}^{\infty} h_1[i]g[k+i] \\
&= \sum_{l=-\infty}^{\infty} h_1[-l]g[k-l] \qquad (\text{let } l = -i) \\
&= h_1[-k] \star g[k]
\end{aligned}
$$

so that

$$
r_{X,Y}[k] = h_1[-k] \star h_2[k] \star r_U[k] \tag{19.29}
$$

(this should be reminiscent of another relationship that results if $h_1[k] = h_2[k] = h[k]$). Upon Fourier transforming both sides we have the CPSD

$$
P_{X,Y}(f) = H_1^*(f)H_2(f)P_U(f). \tag{19.30}
$$

An interesting observation from (19.30) is that if the two filters have nonoverlapping passbands, as shown in Figure 19.3, then

$$
H_1^*(f)H_2(f) = 0 \qquad -\tfrac{1}{2} \le f \le \tfrac{1}{2}
$$

and $P_{X,Y}(f) = 0$. Taking the inverse Fourier transform of the CPSD produces the CCS, which is $r_{X,Y}[k] = 0$ for all k. Hence, for nonoverlapping passband filters as shown in Figure 19.3 the $X[n]$ and $Y[n]$ random processes are *uncorrelated*. (Note that because of the nonoverlapping passbands we must have $\mu_X = 0$ or $\mu_Y = 0$.) Since this holds for any filters satisfying the nonoverlapping constraint, it also holds

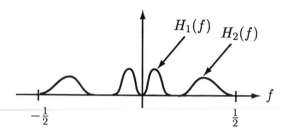

Figure 19.3: Nonoverlapping passband filters.

in particular for any narrowband filters with nonoverlapping passbands. What this says is that the *the Fourier transform of a WSS random process U[n] is uncorrelated at two different frequencies.* (Actually, it is the *truncated* Fourier transform or $U_{2M+1}(f) = (1/\sqrt{2M+1}) \sum_{n=-M}^{M} U[n] \exp(-j2\pi fn)$, which is required for existence, that is uncorrelated at different frequencies as $M \to \infty$.) This is because the Fourier transform can be thought of as resulting from filtering the random process with a narrowband filter and then determining the amplitude and phase of the resulting sinusoidal output. The *spectral representation* of a WSS random process is based upon this interpretation (see [Brockwell and Davis 1987] and also Problem 19.22).

$\triangle\!\!\!!!$ **Two random processes can be individually WSS but not jointly WSS.**

All the examples thus far of individually WSS random processes have also resulted in jointly WSS random processes. To dispel the notion that this is true in general consider the following example. Let $X[n] = A$ and $Y[n] = (-1)^n A$, where A is a random variable with $E[A] = 0$ and $\text{var}(A) = 1$. Then, $\mu_X[n] = \mu_Y[n] = 0$ and it is easily shown that $r_X[k] = 1$ for all k and $r_Y[k] = (-1)^k$ for all k. Therefore, $X[n]$ and $Y[n]$ are individually WSS random processes but they are not jointly WSS since

$$E[X[n]Y[n+k]] = E[A^2(-1)^{n+k}] = (-1)^n(-1)^k$$

which depends on n. For example, since $X[0] = Y[2] = A$ and $X[1] = -Y[3] = A$, we have that

$$E[X[0]Y[2]] = E[A^2] = 1$$
$$E[X[1]Y[3]] = E[A(-A)] = -1$$

so that the cross-correlation between two samples spaced two units apart depends on n.

19.6 Continuous-Time Definitions and Formulas

Two continuous-time random processes $X(t)$ and $Y(t)$ for $-\infty < t < \infty$ are *jointly WSS* if $X(t)$ is WSS, $Y(t)$ is WSS, and we can define the cross-correlation function (CCF) as

$$r_{X,Y}(\tau) = E[X(t)Y(t+\tau)] \qquad -\infty < \tau < \infty \qquad (19.31)$$

which does not depend on t. Some properties (actually *non*properties) of the CCF are CCF is not necessarily symmetric about $\tau = 0$.

$$r_{X,Y}(\tau) \neq r_{X,Y}(-\tau) \qquad (19.32)$$

□

Property 19.10 – The maximum of the CCF can occur for any value of τ.

□

Property 19.11 – The maximum value of the CCF is bounded.

$$|r_{X,Y}(\tau)| \leq \sqrt{r_X(0)r_Y(0)} \qquad (19.33)$$

□

Property 19.12 – Interchanging $X(t)$ and $Y(t)$ flips the CCF about $\tau = 0$.

$$r_{Y,X}(\tau) = r_{X,Y}(-\tau) \qquad (19.34)$$

□

Two zero mean jointly WSS continuous random processes are said to be uncorrelated if $r_{X,Y}(\tau) = 0$ for $-\infty < \tau < \infty$.

The CPSD for two jointly WSS random processes is defined as

$$P_{X,Y}(F) = \lim_{T \to \infty} \frac{1}{T} E\left[\left(\int_{-T/2}^{T/2} X(t)\exp(-j2\pi Ft)dt\right)^* \left(\int_{-T/2}^{T/2} Y(t)\exp(-j2\pi Ft)dt\right)\right] \qquad (19.35)$$

and is evaluated as

$$P_{X,Y}(F) = \int_{-\infty}^{\infty} r_{X,Y}(\tau)\exp(-j2\pi F\tau)d\tau. \qquad (19.36)$$

Some properties of the CPSD follow. The proofs are similar to those for the discrete-time case.

Property 19.13 – CPSD is a complex and hermitian function.

The hermitian property is

$$P_{X,Y}(-F) = P_{X,Y}^*(F) \tag{19.37}$$

$\square$

Property 19.14 – CPSD is bounded.

$$|P_{X,Y}(F)| \leq \sqrt{P_X(F)P_Y(F)} \tag{19.38}$$

$\square$

Property 19.15 – CPSD of $(Y(t), X(t))$ is the complex conjugate of the CPSD of $(X(t), Y(t))$.

$$P_{Y,X}(F) = P_{X,Y}^*(F) \tag{19.39}$$

$\square$

The formulas for the linear system configuration corresponding to that shown in Figure 19.2a are (continuous-time system is assumed to be LTI with impulse response $h(\tau)$ and frequency response $H(f)$)

$$
\begin{aligned}
r_{X,Y}(\tau) &= h(\tau) \star r_X(\tau) & (19.40) \\
P_{X,Y}(F) &= H(F)P_X(F) & (19.41)
\end{aligned}
$$

and for the configuration of Figure 19.2b (continuous-time systems are assumed to be LTI with impulse responses $h_1(\tau)$, $h_2(\tau)$, and corresponding frequency responses $H_1(f)$, $H_2(f)$)

$$
\begin{aligned}
r_{X,Y}(\tau) &= h_1(-\tau) \star h_2(\tau) \star r_X(\tau) & (19.42) \\
P_{X,Y}(F) &= H_1^*(F)H_2(F)P_U(F). & (19.43)
\end{aligned}
$$

An example of great practical importance is given next to illustrate the concepts and formulas.

Example 19.6 – Measurement of Channel Delay

It is frequently of interest to be able to measure the propagation time of a signal through a channel. This allows one to determine distance if the speed of propagation is known. This idea forms the basis for the global positioning system (GPS) [Hofmann-Wellenhof, Lichtenegger, Collins 1992]. See also Problem 19.28 for another application. To do so we transmit a WSS random process $X(t)$, that is bandlimited to W Hz (meaning that $P_X(F) = 0$ for $|F| > W$) through a channel and observe the output of the channel $Y(t)$. We furthermore assume that the channel is modeled as an LTI system with frequency response

$$H(F) = \frac{\exp(-j2\pi F t_0)}{1 + j2\pi F}. \tag{19.44}$$

Note that the numerator term represents a delay of t_0 seconds, sometimes called the propagation or bulk delay, and the term $H_{\text{LP}}(F) = 1/(1 + j2\pi F)$ represents a low-pass filter response since $H_{\text{LP}}(0) = 1$ and $H_{\text{LP}}(F) \to 0$ as $F \to \infty$. A question arises as to how to choose the transmit random process $X(t)$ so that we can accurately measure the delay t_0 through the channel. In the ideal case in which $Y(t)$ is just a delayed replica of $X(t)$ or $Y(t) = X(t - t_0)$, we know that the CCF is

$$\begin{aligned} r_{X,Y}(\tau) &= E[X(t)Y(t + \tau)] \\ &= E[X(t)X(t + \tau - t_0)] \\ &= r_X(\tau - t_0). \end{aligned}$$

Since the ACF has a maximum at lag zero, there will be maximum of $r_{X,Y}(\tau)$ at $\tau = t_0$, suggesting that the location of this maximum can be used to measure the delay. But when the channel has the frequency response given by (19.44) the maximum of the CCF may no longer be located at $\tau = t_0$. To see why, first compute the CCF as

$$\begin{aligned} r_{X,Y}(\tau) &= \int_{-\infty}^{\infty} P_{X,Y}(F) \exp(j2\pi F\tau) dF &&\text{(inverse Fourier transform)} \\ &= \int_{-\infty}^{\infty} H(F) P_X(F) \exp(j2\pi F\tau) dF &&\text{(from (19.41))} \\ &= \int_{-\infty}^{\infty} \frac{\exp(-j2\pi F t_0)}{1 + j2\pi F} P_X(F) \exp(j2\pi F\tau) dF &&\text{(from (19.44))} \\ &= \int_{-\infty}^{\infty} \frac{1}{1 + j2\pi F} P_X(F) \exp(j2\pi F(\tau - t_0)) dF \end{aligned}$$

and since $X(t)$ is assumed to be bandlimited to W Hz, we have

$$r_{X,Y}(\tau) = \int_{-W}^{W} \frac{1}{1 + j2\pi F} P_X(F) \exp(j2\pi F(\tau - t_0)) dF. \tag{19.45}$$

If, as an example, we choose $X(t)$ to be bandlimited white noise (see Example 17.11) or $P_X(F) = N_0/2$ for $|F| \leq W$ and $P_X(F) = 0$ for $|F| > W$, then

$$r_{X,Y}(\tau) = \frac{N_0}{2} \int_{-W}^{W} \frac{1}{1 + j2\pi F} \exp(j2\pi F(\tau - t_0)) dF.$$

To evaluate this we first note that

$$\mathcal{F}^{-1}\left\{\frac{1}{1+j2\pi F}\right\} = \int_{-\infty}^{\infty} \frac{1}{1+j2\pi F} \exp(j2\pi Ft)dF = \exp(-t)u(t)$$

so that if we define the *frequency window* function

$$G(F) = \left\{ \begin{array}{ll} 1 & |F| \leq W \\ 0 & |F| > W \end{array} \right.$$

then

$$
\begin{aligned}
r_{X,Y}(\tau) &= \frac{N_0}{2} \int_{-\infty}^{\infty} G(F)\frac{1}{1+j2\pi F} \exp(j2\pi F(\tau - t_0))dF \\
&= \frac{N_0}{2} g(t) \star \exp(-t)u(t)|_{t=\tau-t_0} \qquad \text{(convolution in time yields} \\
&\qquad\qquad\qquad\qquad\qquad\qquad\qquad\qquad \text{multiplication in frequency).}
\end{aligned}
$$

where $g(t)$ is the inverse Fourier transform of $G(F)$. We have chosen to express the integral in the time domain since its physical significance becomes clearer. In particular, note that the convolution in time results in a wider pulse. But

$$g(t) = 2W\frac{\sin(2\pi Wt)}{2\pi Wt} \qquad \text{(see Example 17.11)}$$

so that using a convolution integral, we have

$$
\begin{aligned}
r_{X,Y}(\tau) &= N_0 W \int_{-\infty}^{\infty} \exp(-\xi)u(\xi)\frac{\sin[2\pi W((\tau - t_0) - \xi)]}{2\pi W((\tau - t_0) - \xi)}d\xi \\
&= N_0 W \int_0^{\infty} \exp(-\xi)\frac{\sin[2\pi W((\tau - t_0) - \xi)]}{2\pi W((\tau - t_0) - \xi)}d\xi.
\end{aligned}
$$

This is shown in Figure 19.4 for the case when $W = 1$ and $t_0 = 2$ as the light line and has been normalized to have a maximum value of 1. The integral has been evaluated numerically. Note that the maximum does not occur at $t_0 = 2$ because the phase response of the channel has added a time delay. To remedy this problem we can insert an *equalizing filter* at the channel output whose frequency response is

$$H_{\text{eq}}(F) = \left\{ \begin{array}{ll} 1 + j2\pi F & |F| \leq W \\ 0 & |F| > W. \end{array} \right.$$

Then, we have for the CPSD between the input $X(t)$ and output random process of the equalizer $Y(t)$

$$P_{X,Y}(F) = H_{\text{eq}}(F)H(F)P_X(F) = \left\{ \begin{array}{ll} \frac{N_0}{2} \exp(-j2\pi Ft_0) & |F| \leq W \\ 0 & |F| > W. \end{array} \right.$$

The CCF is found as before by using the inverse Fourier transform

$$r_{X,Y}(\tau) = \int_{-\infty}^{\infty} P_{X,Y}(F) \exp(j2\pi F\tau)dF$$

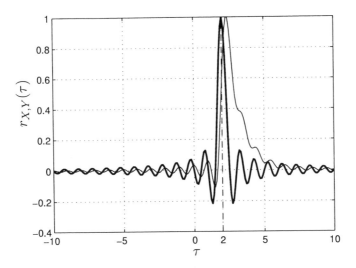

Figure 19.4: Cross-correlation functions for $W = 1$ and $t_0 = 2$. Both curves are normalized to yield one at their peak. The light line is for no equalization while the dark line incorporates equalization. The dashed line indicates $\tau = 2$, the true delay.

$$
\begin{aligned}
&= \int_{-W}^{W} \frac{N_0}{2} \exp(j2\pi F(\tau - t_0))dF \\
&= N_0 W \frac{\sin(2\pi W(\tau - t_0))}{2\pi W(\tau - t_0)}
\end{aligned}
$$

which is shown in Figure 19.4 as the dark line. As before it has been normalized to yield one at its peak. Note that the maximum now occurs at the correct location and also the width of the maximum peak is narrower. This allows a better location of the maximum in the presence of noise.

19.7 Cross-correlation Sequence Estimation

The estimation of the CCS is similar to that for the ACS (see Section 17.7). The main difference between the two stems from the fact that the ACS is guaranteed to have a maximum at $k = 0$ while the maximum for the CCS can be located anywhere. Furthermore, two samples of a WSS random process tend to become less correlated as the spacing between them increases. This implies that it is only necessary to estimate the ACS $r_X[k]$ for $k = 0, 1, \ldots, M$ if we assume that $r_X[k] \approx 0$ for $k > M$. For the CCS, however, we must estimate $r_{X,Y}[k]$ for $k = -M_1, \ldots, 0, \ldots, M_2$ (recall that $r_{X,Y}[-k] \neq r_{X,Y}[k]$) for which $r_{X,Y}[k] \approx 0$ if $k < -M_1$ or $k > M_2$. In practice, it is not clear how M_1 and M_2 should be chosen. Frequently, a preliminary estimate of $r_{X,Y}[k]$ is made, followed by a search for the maximum location. Then, the data records used to estimate the CCS are *shifted* relative to each other to place the maximum at $k = 0$. This is called *time alignment* [Jenkins and Watts

1968]. We assume that this has already been done. Then, we estimate the CCS for $|k| \leq M$ assuming that we have observed the realizations for $X[n]$ and $Y[n]$, both for $n = 0, 1, \ldots, N - 1$. The estimate becomes

$$\hat{r}_{X,Y}[k] = \begin{cases} \frac{1}{N-k} \sum_{n=0}^{N-1-k} x[n]y[n+k] & k = 0, 1, \ldots, M \\ \frac{1}{N-|k|} \sum_{n=|k|}^{N-1} x[n]y[n+k] & k = -M, -(M-1), \ldots, -1. \end{cases} \tag{19.46}$$

Note that the summation limits have been chosen to make sure that all the available products $x[n]y[n+k]$ are used. Similar to the estimation of the ACS, there will be a different number of products for each k. For example, if $N = 4$ so that $\{x[0], x[1], x[2], x[3]\}$ and $\{y[0], y[1], y[2], y[3]\}$ are observed, and we wish to compute the CCS estimate for $|k| \leq M = 2$, we will have

$$\hat{r}_{X,Y}[-2] = \frac{1}{2} \sum_{n=2}^{3} x[n]y[n-2] = \frac{1}{2}(x[2]y[0] + x[3]y[1])$$

$$\hat{r}_{X,Y}[-1] = \frac{1}{3} \sum_{n=1}^{3} x[n]y[n-1] = \frac{1}{3}(x[1]y[0] + x[2]y[1] + x[3]y[2])$$

$$\hat{r}_{X,Y}[0] = \frac{1}{4} \sum_{n=0}^{3} x[n]y[n] = \frac{1}{4}(x[0]y[0] + x[1]y[1] + x[2]y[2] + x[3]y[3])$$

$$\hat{r}_{X,Y}[1] = \frac{1}{3} \sum_{n=0}^{2} x[n]y[n+1] = \frac{1}{3}(x[0]y[1] + x[1]y[2] + x[2]y[3])$$

$$\hat{r}_{X,Y}[2] = \frac{1}{2} \sum_{n=0}^{1} x[n]y[n+2] = \frac{1}{2}(x[0]y[2] + x[1]y[3]).$$

As an example, consider the jointly WSS random processes described in Example 19.2, where $X[n] = U[n]$, $Y[n] = U[n] + 2U[n-1]$ and $U[n]$ is white noise with variance $\sigma_U^2 = 1$. We further assume that $U[n]$ has a Gaussian PDF for each n for the purpose of computer simulation (although we could use any PDF or PMF). Recall that the theoretical CCS is $r_{X,Y}[k] = \delta[k] + 2\delta[k-1]$. The estimated CCS using $N = 1000$ data samples is shown in Figure 19.5. The MATLAB code used to estimate the CCS is given below.

```
% assume realizations are x[n], y[n] for n=1,2,...,N
for k=0:M % compute zero and positive lags, see (19.46)
  % compute values for k=0,1,...,M
  rxypos(k+1,1)=(1/(N-k))*sum(x(1:N-k).*y(1+k:N));
end
for k=1:M % compute negative lags, see (19.46)
  % compute values for k=-M,-(M-1),...,-1
  rxyneg(k+1,1)=(1/(N-k))*sum(x(k+1:N).*y(1:N-k));
end
rxy=[flipud(rxyneg(2:M+1,1));rxypos]; % arrange values from k=-M to k=M
```

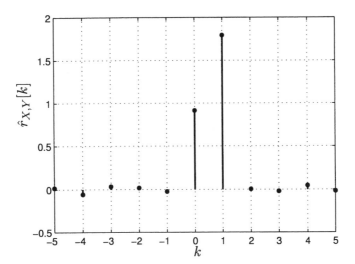

Figure 19.5: Estimated CCS using realizations for $X[n]$ and $Y[n]$ with $N = 1000$ samples. The theoretical CCS is $r_{X,Y}[k] = \delta[k] + 2\delta[k-1]$.

Finally, we note that estimation of the CPSD is more difficult and so we refer the interested reader to [Jenkins and Watts 1968, Kay 1988].

19.8 Real-World Example – Brain Physiology Research

Understanding the operation of the human brain is one of the most important goals of physiological research. Currently, there is an enormous effort to decipher its inner workings. At a very fundamental level is the study of its cells or *neurons*, which when working in unison form the basis for our behavior. Their electrical activity and the transmission of that activity to neighboring neurons yields clues as to the brain's operation. When an individual neuron "fires" it produces a *spike* or electrical pulse that propagates to nearby neurons. The connections between the neurons that allow this propagation to occur are called *synapses* and it is this connectivity that is the focus of much research. A typical spike train that might be recorded is shown in Figure 19.6a for a neuron at rest and in Figure 19.6b for a neuron that has been excited by some stimulus. Clearly, the firing rate increases in response to a stimulus. The model used to produce this figure is an IID Bernoulli random process with $p = p_q = 0.1$ for Figure 19.6a and $p = p_s = 0.6$ for Figure 19.6b. The subscripts "q" and "s" are meant to indicate the state of the neuron, either *quiescent* or *stimulated*. Now consider the question of whether two neurons are connected via a synapse. If they are, and a stimulus is applied to the first neuron, then the electrical pulse will propagate to the second neuron and appear some time later. Then, we would expect the second neuron electrical activity to change from that in Figure 19.6a to that in Figure 19.6b. It would be fairly simple

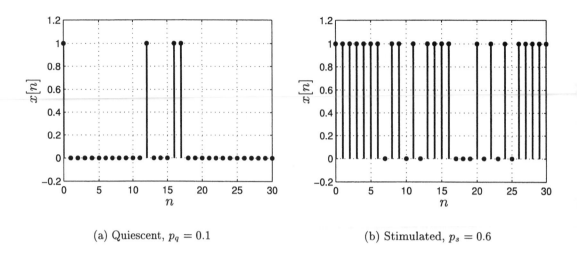

(a) Quiescent, $p_q = 0.1$ (b) Stimulated, $p_s = 0.6$

Figure 19.6: Typical spike trains for neurons.

then to estimate the p for each possible connected neuron and choose the neuron or neurons (there may be multiple connections with the stimulated neuron) for which p is large. Unfortunately, it is not easy to stimulate a single neuron so that when a stimulus is applied, many neurons may be activated. Thus, we need a method to associate one stimulated neuron with its connected ones. Ideally, if we record the electrical activity at two neurons under consideration, denoted by $X_1[n]$ and $X_2[n]$, then for connected neurons $X_2[n] = X_1[n - n_0]$. Since we have assumed that the spike train for the first neuron $X_1[n]$ is an IID random process, it is therefore WSS and we know from Example 19.1, the two random processes are jointly WSS. Therefore, we have as before

$$
\begin{aligned}
r_{X_1,X_2}[k] &= E[X_1[n]X_2[n+k]] \\
&= E[X_1[n]X_1[n - n_0 + k]] \\
&= r_{X_1}[k - n_0]
\end{aligned}
$$

and therefore the CCS will exhibit a maximum at $k = n_0$. Otherwise, if the neurons are not connected, we would expect a much smaller value of the maximum or no discernible maximum at all. For example, for unconnected but simultaneously stimulated neurons it is reasonable to assume that $X_1[n]$ and $X_2[n]$ are uncorrelated and hence $r_{X_1,X_2}[k] = E[X_1[n]]E[X_2[n+k]] = p_s^2$, which presumably will be less than $r_{X_1}[k - n_0]$ at its peak. Note that for connected neurons

$$
r_{X_1,X_2}[k] = \text{cov}(X_1[n], X_2[n+k]) + E[X_1[n]]E[X_2[n+k]] > E[X_1[n]]E[X_2[n+k]]
$$

if the covariance is positive.

Specifically, we assume that a neuron output is modeled as an IID Bernoulli random process that takes on the values 1 and 0 with probabilities p_s and $1 - p_s$, respectively. For two neurons that are connected we have that $r_{X_1,X_2}[k] = r_{X_1}[k - n_0]$. But

$$
\begin{aligned}
r_{X_1}[k] &= E[X_1[n]X_1[n+k]] \\
&= \begin{cases} E[X_1^2[n]] & k = 0 \\ E[X_1[n]]E[X_1[n+k]] & k \neq 0 \end{cases} \\
&= \begin{cases} p_s & k = 0 \\ p_s^2 & k \neq 0 \end{cases} \\
&= p_s(1 - p_s)\delta[k] + p_s^2.
\end{aligned}
$$

Hence, for two connected neurons the CCS is

$$
r_{X_1,X_2}[k] = p_s(1 - p_s)\delta[k - n_0] + p_s^2.
$$

For two neurons that are not connected, so that their outputs are uncorrelated (even if both are stimulated), the CCS is

$$
\begin{aligned}
r_{X_1,X_2}[k] &= E[X_1[n]]X_2[n+k]] \\
&= E[X_1[n]]E[X_2[n+k]] \\
&= p_s^2 \quad \text{for all } k.
\end{aligned}
$$

As a result, the maximum is p_s^2 for unconnected neurons but $p_s(1-p_s)+p_s^2 = p_s > p_s^2$ for connected neurons. The two different CCSs are shown in Figure 19.7 for $p_s = 0.6$ and $n_0 = 2$. As an example, for $p_s = 0.6$ we show realizations of three neuron

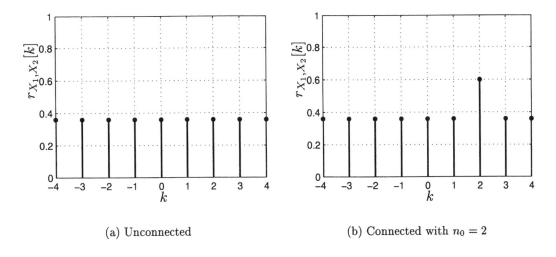

(a) Unconnected (b) Connected with $n_0 = 2$

Figure 19.7: CCS for unconnected and connected stimulated neurons with $p_s = 0.6$.

outputs in Figure 19.8, where only neuron 1 and neuron 3 are connected. There is

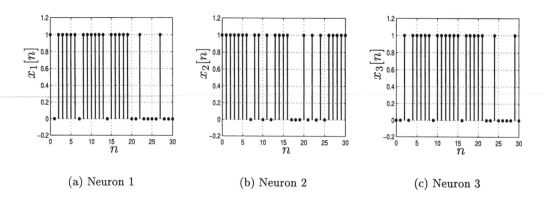

(a) Neuron 1   (b) Neuron 2   (c) Neuron 3

Figure 19.8: Spike trains for three neurons with neuron 1 connected to neuron 3 with a delay of two samples. The spike train of neuron 2 is uncorrelated with those for neurons 1 and 3.

a two sample delay between neurons 1 and 3. Neuron 2 is not connected to either of the other neurons and hence its spike train is uncorrelated with the others. The theoretical CCS between neurons 1 and 2 is given in Figure 19.7a while that between neurons 1 and 3 is given in Figure 19.7b. The estimated CCS for the spike trains shown in Figure 19.8 and based on the estimate of (19.46) is shown in Figure 19.9.

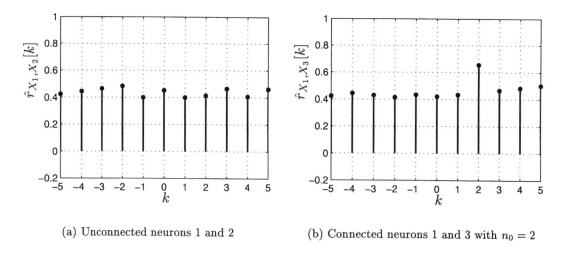

(a) Unconnected neurons 1 and 2   (b) Connected neurons 1 and 3 with $n_0 = 2$

Figure 19.9: Estimated CCS for unconnected and connected stimulated neurons with $p_s = 0.6$.

It is seen that as expected there is a maximum at $k = n_0 = 2$. The interested reader should consult [Univ. Pennsylvannia 2005] for further details.

References

Bendat, J.S., A.G. Piersol, *Random Data: Analysis and Measurement Procedures*, Wiley-Interscience, New York, 1971.

Brockwell, P.J., R.A. Davis, *Time Series: Theory and Methods*, Springer-Verlag, New York, 1987.

Hofmann-Wellenhof, H. Lichtenegger, J. Collins, *Global Positioning System: Theory and Practice*, Springer-Verlag, New York, 1992.

Jenkins, G.M., D.G. Watts, *Spectral Analysis and Its Applications*, Holden-Day, San Francisco, 1968.

Kay, S., *Modern Spectral Estimation: Theory and Application*, Prentice Hall, Englewood Cliffs, NJ, 1988.

Koopmans, L.H., *The Spectral Analysis of Time Series*, Academic Press, New York, 1974.

Robinson, E.A., *Multichannel Time Series Analysis with Digital Computer Programs*, Holden-Day, San Francisco, 1967.

University of Pennsylvannia, Multiple Unit Laboratory,
http://mulab.physiol.upenn.edu/analysis.html

Problems

19.1 ($\smile$) (w) Two discrete-time random processes are defined as $X[n] = U[n]$ and $Y[n] = (-1)^n U[n]$ for $-\infty < n < \infty$, where $U[n]$ is white noise with variance σ_U^2. Are the random processes $X[n]$ and $Y[n]$ jointly WSS?

19.2 (w) Two discrete-time random processes are defined as $X[n] = a_1 U_1[n] + a_2 U_2[n]$ and $Y[n] = b_1 U_1[n] + b_2 U_2[n]$ for $-\infty < n < \infty$, where $U_1[n]$ and $U_2[n]$ are jointly WSS and a_1, a_2, b_1, b_2 are constants. Are the random processes $X[n]$ and $Y[n]$ jointly WSS?

19.3 (f) If the CCS is given as $r_{X,Y}[k] = (1/2)^{|k-1|}$ for $-\infty < k < \infty$, plot it and describe which properties are the same or different from an ACS.

19.4 (f) If $Y[n] = X[n] + W[n]$, where $X[n]$ and $W[n]$ are jointly WSS, find $r_{X,Y}[k]$ and $P_{X,Y}(f)$.

19.5 ($\smile$) (w) A discrete-time random process is defined as $Y[n] = X[n]W[n]$, where $X[n]$ is WSS and $W[n]$ is an IID Bernoulli random process that takes on values ± 1 with equal probability. The random processes $X[n]$ and $W[n]$ are independent of each other, which means that $X[n_1]$ is independent of $W[n_2]$ for all n_1 and n_2. Find $r_{X,Y}[k]$ and explain your results.

19.6 (☺) (w) In this problem we show that for the AR random process $X[n] = aX[n-1] + U[n]$, which was described in Example 17.5, the cross-correlation sequence $E[X[n]U[n+k]] = 0$ for $k > 0$. Do so by evaluating $E[X[n](X[n+k] - aX[n+k-1])]$. Determine and plot the CCS $r_{X,U}[k]$ for $-\infty < k < \infty$ if $a = 0.5$ and $\sigma_U^2 = 1$. Hint: Refer back to Example 17.5 for the ACS of an AR random process.

19.7 (f) If $X[n]$ and $Y[n]$ are jointly WSS with ACSs

$$r_X[k] = 4 \left(\frac{1}{2} \right)^{|k|}$$
$$r_Y[k] = 3\delta[k] + 2\delta[k+1] + 2\delta[k-1]$$

determine the maximum possible value of $r_{X,Y}[k]$.

19.8 (t) Derive (19.14). To do so use the relationship $\sum_{m=-M}^{M} \sum_{n=-M}^{M} g[m-n] = \sum_{k=-2M}^{2M} (2M+1-|k|)g[k]$.

19.9 (f) For the two sinusoidal random processes $X[n] = \cos(2\pi f_0 n + \Theta_1)$ and $Y[n] = \cos(2\pi f_0 n + \Theta_2)$, where $\Theta_1 = \Theta_2 \sim \mathcal{U}(0, 2\pi)$ find the CPSD and explain your results versus the case when Θ_1 and Θ_2 are independent random variables.

19.10 (☺) (f,c) If $r_{X,Y}[k] = \delta[k] + 2\delta[k-1]$, plot the magnitude and phase of the CPSD. You will need a computer to do this.

19.11 (f) For the random processes $X[n] = U[n]$ and $Y[n] = U[n] - bU[n-1]$, where $U[n]$ is discrete white noise with variance $\sigma_U^2 = 1$, find the CPSD and explain what happens as $b \to 0$.

19.12 (☺) (w) If a random process is defined as $Z[n] = X[n] - Y[n]$, where $X[n]$ and $Y[n]$ are jointly WSS, determine the ACS and PSD of $Z[n]$.

19.13 (w) For the random processes $X[n]$ and $Y[n]$ defined in Problem 19.11 find the coherence function. Explain what happens as $b \to 0$.

19.14 (f) Determine the CPSD for two jointly WSS random processes if $r_{X,Y}[k] = \delta[k] - \delta[k-1]$. Also, explain why the coherence function at $f = 0$ is zero. Hint: The random processes $X[n]$ and $Y[n]$ are those given in Problem 19.11 if $b = 1$.

19.15 (☺) (f) If $Y[n] = -X[n]$ for $-\infty < n < \infty$, determine the coherence function and relate it to the predictability of $Y[n_0]$ based on observing $X[n]$ for $-\infty < n < \infty$.

19.16 (t) A *cross-spectral matrix* is defined as

$$\begin{bmatrix} P_X(f) & P_{X,Y}(f) \\ P_{Y,X}(f) & P_Y(f) \end{bmatrix}.$$

Prove that the cross-spectral matrix is positive semidefinite for all f. Hint: Show that the principal minors of the matrix are all nonnegative (see Appendix C for the definition of principal minors). To do so use the properties of the coherence function.

19.17 (w) The random processes $X[n]$ and $Y[n]$ are zero mean jointly WSS and are uncorrelated with each other. If $r_X[k] = 2\delta[k]$ and $r_Y[k] = (1/2)^{|k|}$ for $-\infty < k < \infty$, find the PSD of $X[n] + Y[n]$.

19.18 (⌣) (t) In this problem we derive an extension of the Wiener smoother (see Section 18.5.1). We consider the problem of estimating $Y[n_0]$ based on observing $X[n]$ for $-\infty < n < \infty$. To do so we use the linear estimator

$$\hat{Y}[n_0] = \sum_{k=-\infty}^{\infty} h[k]X[n_0 - k].$$

To find the optimal impulse response we employ the orthogonality principle to yield the infinite set of simultaneous linear equations

$$E\left[(Y[n_0] - \sum_{k=-\infty}^{\infty} h[k]X[n_0 - k])X[n_0 - l]\right] = 0 \qquad -\infty < l < \infty.$$

Assuming that $X[n]$ and $Y[n]$ are jointly WSS random processes, determine the frequency response of the optimal Wiener estimator. Then, show how the Wiener smoother, where $Y[n]$ represents the signal $S[n]$ and $X[n]$ represents the signal $S[n]$ plus noise $W[n]$ (recall that $S[n]$ and $W[n]$ are zero mean and uncorrelated random processes), arises as a special case of this solution.

19.19 (f) For the random processes defined in Example 19.2 determine the CPSD. Next, find the optimal Wiener smoother for $Y[n_0]$ based on the realization of $X[n]$ for $-\infty < n < \infty$.

19.20 (t) Prove that if $X[n]$ is a WSS random process that is input to an LSI system and $Y[n]$ is the corresponding random process output, then the coherence function between the input and output has a magnitude of one.

19.21 (t) Consider a WSS random process $X[n]$ that is input to an LSI system with frequency response $H(f)$, where $H(f) \neq 0$ for $|f| \leq 1/2$, and let $Y[n]$ be the corresponding random process output. It is desired to predict $X[n_0]$ based on observing $Y[n]$ for $-\infty < n < \infty$. Draw a linear filtering diagram (similar to that shown in Figure 19.2) to explain why $X[n_0]$ is perfectly predictable by passing $Y[n]$ through a filter with frequency response $1/H(f)$.

19.22 (t) In this problem we argue that a Fourier transform is actually a narrow-band filtering operation. First consider the Fourier transform at $f = f_0$ for the truncated random process $X[n]$, $n = -M, \ldots, 0, \ldots, M$ which is $\hat{X}(f_0) = \sum_{k=-M}^{M} X[k] \exp(-j2\pi f_0 k)$. Next show that this may be written as

$$\hat{X}(f_0) = \left. \sum_{k=-\infty}^{\infty} X[k]h[n-k] \right|_{n=0}$$

where

$$h[k] = \begin{cases} \exp(j2\pi f_0 k) & k = -M, \ldots, 0, \ldots, M \\ 0 & |k| > M. \end{cases}$$

Notice that this is a convolution sum so that $h[k]$ can be considered as the impulse response, although a complex one, of an LSI filter. Finally, find and plot the frequency response of this filter. Hint: You will need

$$\sum_{k=-M}^{M} \exp(jk\theta) = \frac{\sin((2M+1)\theta/2)}{\sin(\theta/2)}.$$

19.23 (☺) (w) Consider the continuous-time averager

$$Y(t) = \frac{1}{T} \int_{t-T}^{t} X(\xi)d\xi$$

where the random process $X(t)$ is continuous-time white noise with PSD $P_X(F) = N_0/2$ for $-\infty < F < \infty$. Determine the CCF $r_{X,Y}(\tau)$ and show that it is zero for τ outside the interval $[0, T]$. Explain why it is zero outside this interval.

19.24 (f) If a continuous-time white noise process $X(t)$ with ACF $r_X(\tau) = (N_0/2)\delta(\tau)$ is input to an LTI system with impulse response $h(\tau) = \exp(-\tau)u(\tau)$, determine $r_{X,Y}(\tau)$.

19.25 (t) Can the CPSD ever have the same properties as the PSD in terms of being real and symmetric? If so, give an example. Hint: Consider the relationship given in (19.43).

19.26 (☺) (f,c) Consider the random processes $X[n] = U[n]$ and $Y[n] = U[n] - bU[n-1]$, where $U[n]$ is white Gaussian noise with variance $\sigma_U^2 = 1$. Find $r_{X,Y}[k]$ and then to verify your results perform a computer simulation. To do so first generate $N = 1000$ samples of $X[n]$ and $Y[n]$. Then, estimate the CCS for $b = -0.1$ and $b = -1$. Explain your results.

19.27 (f,c) An AR random process is given by $X[n] = aX[n-1]+U[n]$, where $U[n]$ is white Gaussian noise with variance σ_U^2. Find the CCS $r_{X,U}[k]$ and then to verify your results perform a computer simulation using $a = 0.5$ and $\sigma_U^2 = 1$. To do so first generate $N = 1000$ samples of $U[n]$ and $X[n]$. Then, estimate the CCS. Hint: Remember to set the initial condition $X[-1] \sim \mathcal{N}(0, \sigma_U^2/(1-a^2))$.

19.28 (w) In this problem we explore the use of the CCF to determine the direction of arrival of a sound source. Referring to Figure 19.10, a sound source emits a pulse that propagates to a set of two receivers. Because the distance from the source to the receivers is large, it is assumed that the wavefronts are planar as shown. If the source has the angle θ with respect to the x axis as shown, it first reaches receiver 2 and then reaches receiver 1 at a time $t_0 = d\cos(\theta)/c$ seconds later, where d is the distance between receivers and c is the propagation speed. Assume that the received signal at receiver 2 is a WSS random process $X_2(t) = U(t)$ with a PSD

$$P_U(F) = \begin{cases} N_0/2 & |F| \leq W \\ 0 & |F| > W \end{cases}$$

and therefore the received signal at receiver 1 is $X_1(t) = U(t - t_0)$. Determine the CCF $r_{X_1,X_2}(\tau)$ and describe how it could be used to find the arrival angle θ.

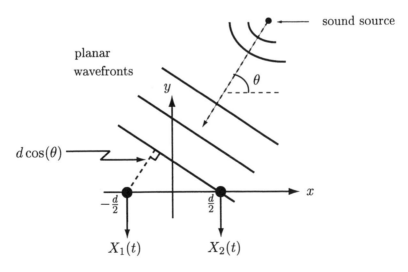

Figure 19.10: Geometry for sound source arrival angle measurement (figure for Problem 19.28).

Chapter 20

Gaussian Random Processes

20.1 Introduction

There are several types of random processes that have found wide application because of their realistic physical modeling yet relative mathematical simplicity. In this and the next two chapters we describe these important random processes. They are the Gaussian random process, the subject of this chapter; the Poisson random process, described in Chapter 21; and the Markov chain, described in Chapter 22. Concentrating now on the Gaussian random process, we will see that it has many important properties. These properties have been inherited from those of the N-dimensional Gaussian PDF, which was discussed in Section 14.3. Specifically, the important characteristics of a Gaussian random process are:

1. It is physically motivated by the central limit theorem (see Chapter 15).

2. It is a mathematically tractable model.

3. The joint PDF of any set of samples is a multivariate Gaussian PDF, which enjoys many useful properties (see Chapter 14).

4. Only the first two moments, the mean sequence and the covariance sequence, are required to completely describe it. As a result,

   a. In practice the joint PDF can be estimated by estimating only the first two moments.

   b. If the Gaussian random process is wide sense stationary, then it is also stationary.

5. The processing of a Gaussian random process by a linear filter does not alter its Gaussian nature, but only modifies the first two moments. The modified moments are easily found.

In effect, the Gaussian random process has so many useful properties that it is always the first model to be proposed in the solution of a problem. It finds application as a model for electronic noise [Bell Labs 1970], ambient ocean noise [Urick 1975], scattering phenomena such as reverberation of sound in the ocean or electromagnetic clutter in the atmosphere [Van Trees 1971], and financial time series [Taylor 1986], just to name a few. Any time a random process can be modeled as due to the sum of a large number of independent and similar type effects, a Gaussian random process results due to the central limit theorem. One example that we will explore in detail is the use of the scattering of a sound pulse from a school of fish to determine their numbers (see Section 20.9). In this case, the received waveform is the sum of a large number of scattered pulses that have been added together. The addition occurs because the leading edge of a pulse that is reflected from a fish farther away will coincide in time with the trailing edge of the pulse that is reflected from a fish that is nearer (see Figure 20.14). If the fish are about the same size and type, then the *average intensity* of the returned echos will be relatively constant. However, the echo amplitudes will be different due to the different reflection characteristics of each fish, i.e., its exact position, orientation, and motion will all determine how the incoming pulse is scattered. These characteristics cannot be predicted in advance and so the amplitudes are modeled as random variables. When overlapped in time, these random echos are well modeled by a Gaussian random process. As an example, consider a transmitted pulse $s(t) = \cos(2\pi F_0 t)$, where $F_0 = 10$ Hz, over the time interval $0 \leq t \leq 1$ second as shown in Figure 20.1. Assuming a single reflection

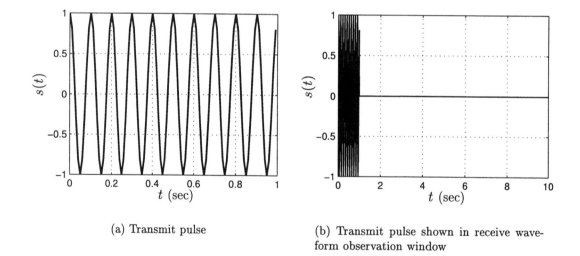

(a) Transmit pulse

(b) Transmit pulse shown in receive waveform observation window

Figure 20.1: Transmitted sinusoidal pulse.

for every 0.1 second interval with the starting time being a uniformly distributed random variable within the interval and an amplitude A that is a random variable with $A \sim \mathcal{U}(0,1)$ to account for the unknown reflection coefficient of each fish, a typical received waveform is shown in Figure 20.2. If we now estimate the marginal

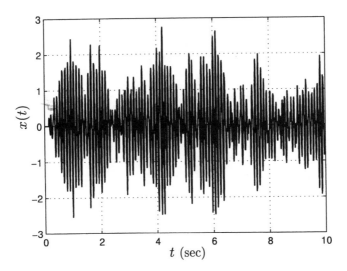

Figure 20.2: Received waveform consisting of many randomly overlapped and random amplitude echos.

PDF for $x(t)$ as shown in Figure 20.2 by assuming that each sample has the same marginal PDF, we have the estimated PDF shown in Figure 20.3 (see Section 10.9 on how to estimate the PDF). Also shown is the Gaussian PDF with its mean and variance estimated from uniformly spaced samples of $x(t)$. It is seen that the Gaussian PDF is very accurate as we would expect from the central limit theorem. The MATLAB code used to generate Figure 20.2 is given in Appendix 20A. In Section 20.3 we formally define the Gaussian random process.

20.2   Summary

Section 20.1 gives an example of why the Gaussian random process arises quite frequently in practice. The discrete-time Gaussian random process is defined in Section 20.3 as one whose samples comprise a Gaussian random vector as characterized by the PDF of (20.1). Also, some examples are given and are shown to exhibit two important properties, which are summarized in that section. Any linear transformation of a Gaussian random process produces another Gaussian random process. In particular for a discrete-time WSS Gaussian random process that is filtered by an LSI filter, the output random process is Gaussian with PDF given in Theorem 20.4.1. A nonlinear transformation does not maintain the Gaussian random process but its effect can be found in terms of the output moments using (20.12). An example of a squaring operation on a discrete-time WSS Gaussian random process produces an output random process that is still WSS with moments

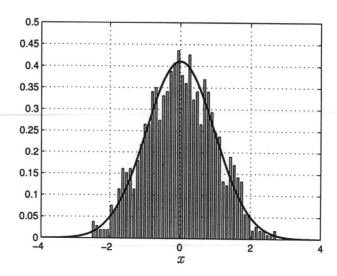

Figure 20.3: Marginal PDF of samples of received waveform shown in Figure 20.2 and Gaussian PDF fit.

given by (20.14). A continuous-time Gaussian random process is defined in Section 20.6 and examples are given. An important one is the Wiener random process examined in Example 20.7. Its covariance matrix is found using (20.16). Some special continuous-time Gaussian random processes are described in Section 20.7. The Rayleigh fading sinusoid is described in Section 20.7.1. It has the ACF given by (20.17) and corresponding PSD given by (20.18). A continuous-time bandpass Gaussian random process is described in Section 20.7.2. It has an ACF given by (20.21) and a corresponding PSD given by (20.22). The important example of bandpass "white" Gaussian noise is discussed in Example 20.8. The computer generation of a discrete-time WSS Gaussian random process realization is described in Section 20.8. Finally, an application of the theory to estimating fish populations using a sonar is the subject of Section 20.9.

20.3   Definition of the Gaussian Random Process

We will consider here the *discrete-time* Gaussian random process, an example of which was given in Figure 16.5b as the discrete-time/continuous-valued (DTCV) random process. The continuous-time/continuous-valued (CTCV) Gaussian random process, an example of which was given in Figure 16.5d, will be discussed in Section 20.6. Before defining the Gaussian random process we briefly review the N-dimensional multivariate Gaussian PDF as described in Section 14.3. An $N \times 1$ random vector $\mathbf{X} = [X_1 \, X_2 \ldots X_N]^T$ is defined to be a Gaussian random vector if its joint PDF is given by the multivariate Gaussian PDF

$$p_{\mathbf{X}}(\mathbf{x}) = \frac{1}{(2\pi)^{N/2}\det^{1/2}(\mathbf{C})} \exp\left[-\frac{1}{2}(\mathbf{x}-\boldsymbol{\mu})^T \mathbf{C}^{-1}(\mathbf{x}-\boldsymbol{\mu})\right] \qquad (20.1)$$

where $\boldsymbol{\mu} = [\mu_1\,\mu_2\ldots\mu_N]^T$ is the mean vector defined as

$$\boldsymbol{\mu} = E_{\mathbf{X}}[\mathbf{X}] = \begin{bmatrix} E_{X_1}[X_1] \\ E_{X_2}[X_2] \\ \vdots \\ E_{X_N}[X_N] \end{bmatrix} \qquad (20.2)$$

and $\mathbf{C}$ is the $N \times N$ covariance matrix defined as

$$\mathbf{C} = \begin{bmatrix} \text{var}(X_1) & \text{cov}(X_1,X_2) & \ldots & \text{cov}(X_1,X_N) \\ \text{cov}(X_2,X_1) & \text{var}(X_2) & \ldots & \text{cov}(X_2,X_N) \\ \vdots & \vdots & \ddots & \vdots \\ \text{cov}(X_N,X_1) & \text{cov}(X_N,X_2) & \ldots & \text{var}(X_N) \end{bmatrix}. \qquad (20.3)$$

In shorthand notation $\mathbf{X} \sim \mathcal{N}(\boldsymbol{\mu},\mathbf{C})$. The important properties of a Gaussian random vector are:

1. Only the first two moments $\boldsymbol{\mu}$ and $\mathbf{C}$ are required to specify the entire PDF.

2. If all the random variables are uncorrelated so that $[\mathbf{C}]_{ij} = 0$ for $i \neq j$, then they are also independent.

3. A linear transformation of $\mathbf{X}$ produces another Gaussian random vector. Specifically, if $\mathbf{Y} = \mathbf{GX}$, where $\mathbf{G}$ is an $M \times N$ matrix with $M \leq N$, then $\mathbf{Y} \sim \mathcal{N}(\mathbf{G}\boldsymbol{\mu}, \mathbf{GCG}^T)$.

Now we consider a discrete-time random process $X[n]$, where $n \geq 0$ for a semi-infinite random process and $-\infty < n < \infty$ for an infinite random process. *The random process is defined to be a Gaussian random process if all finite sets of samples have a multivariate Gaussian PDF as per (20.1).* Mathematically, if $\mathbf{X} = [X[n_1]\,X[n_2]\ldots X[n_K]]^T$ has a multivariate Gaussian PDF (given in (20.1) with N replaced by K) for all $\{n_1, n_2, \ldots, n_K\}$ and all K, then $X[n]$ is said to be a Gaussian random process. Some examples follow.

Example 20.1 – White Gaussian noise

White Gaussian noise was first introduced in Example 16.6. We revisit that example in light of our formal definition of a Gaussian random process. First recall that discrete-time white noise is a WSS random process $X[n]$ for which $E[X[n]] = \mu = 0$ for $-\infty < n < \infty$ and $r_X[k] = \sigma^2\delta[k]$. This says that all the samples are zero mean, uncorrelated with each other, and have the same variance σ^2. If we now furthermore assume that the samples are also *independent* and each sample has a *Gaussian*

PDF, then $X[n]$ is a Gaussian random process. It is referred to as *white Gaussian noise* (WGN). To verify this we need to show that any set of samples has a multivariate Gaussian PDF. Let $\mathbf{X} = [X[n_1]\, X[n_2] \ldots X[n_K]]^T$ and note that the joint K-dimensional PDF is the product of the marginal PDFs due to the independence assumption. Also, each marginal PDF is $X[n_i] \sim \mathcal{N}(0, \sigma^2)$ by assumption. This produces the joint PDF

$$
\begin{aligned}
p_{\mathbf{X}}(\mathbf{x}) &= \prod_{i=1}^{K} p_{X[n_i]}(x[n_i]) \\
&= \prod_{i=1}^{K} \frac{1}{\sqrt{2\pi\sigma^2}} \exp\left(-\frac{1}{2\sigma^2} x^2[n_i]\right) \\
&= \frac{1}{(2\pi\sigma^2)^{K/2}} \exp\left(-\frac{1}{2\sigma^2} \mathbf{x}^T \mathbf{x}\right) \\
&= \frac{1}{(2\pi)^{K/2} \det^{1/2}(\sigma^2 \mathbf{I})} \exp\left(-\tfrac{1}{2}\mathbf{x}^T (\sigma^2 \mathbf{I})^{-1} \mathbf{x}\right)
\end{aligned}
$$

or $\mathbf{X} \sim \mathcal{N}(\mathbf{0}, \sigma^2 \mathbf{I})$, where $\mathbf{I}$ is the $K \times K$ identity matrix. Note also that since WGN is an IID random process, it is also stationary (see Example 16.3).

$\Diamond$

Example 20.2 – Moving average random process

Consider the MA random process $X[n] = (U[n] + U[n-1])/2$, where $U[n]$ is WGN with variance σ_U^2. Then, $X[n]$ is a Gaussian random process. This is because $U[n]$ is a Gaussian random process (from previous example) and $X[n]$ is just a linear transformation of $U[n]$. For instance, if $K = 2$, and $n_1 = 0$, $n_2 = 1$, then

$$
\underbrace{\begin{bmatrix} X[0] \\ X[1] \end{bmatrix}}_{\mathbf{X}} = \underbrace{\begin{bmatrix} \frac{1}{2} & \frac{1}{2} & 0 \\ 0 & \frac{1}{2} & \frac{1}{2} \end{bmatrix}}_{\mathbf{G}} \underbrace{\begin{bmatrix} U[-1] \\ U[0] \\ U[1] \end{bmatrix}}_{\mathbf{U}}
$$

and thus $\mathbf{X} \sim \mathcal{N}(\mathbf{0}, \mathbf{G}\mathbf{C}_U \mathbf{G}^T) = \mathcal{N}(\mathbf{0}, \sigma_U^2 \mathbf{G}\mathbf{G}^T)$. The same argument applies to any number of samples K and any samples times $n_1, n_2, \ldots, n_K$. Note here that the MA random process is also stationary. If we were to change the two samples to $n_1 = n_0$ and $n_2 = n_0 + 1$, then

$$
\begin{bmatrix} X[n_0] \\ X[n_0 + 1] \end{bmatrix} = \begin{bmatrix} \frac{1}{2} & \frac{1}{2} & 0 \\ 0 & \frac{1}{2} & \frac{1}{2} \end{bmatrix} \begin{bmatrix} U[n_0 - 1] \\ U[n_0] \\ U[n_0 + 1] \end{bmatrix}
$$

and the joint PDF will be the same since the **U** vector has the same PDF. Again this result remains the same for any number of samples and sample times. We will see shortly that a Gaussian random process that is WSS is also stationary. Here, the $U[n]$ random process is WSS and hence $X[n]$ is WSS, being the output of an LSI filter (see Theorem 18.3.1).

As a typical probability calculation let $\sigma_U^2 = 1$ and determine $P[X[1] - X[0] > 1]$. We would expect this to be less than $P[U[1] - U[0] > 1] = Q(1/\sqrt{2})$ (since $U[1] - U[0] \sim \mathcal{N}(0, 2)$) due to the smoothing effect of the filter ($\mathcal{H}(z) = \frac{1}{2} + \frac{1}{2}z^{-1}$). Thus, let $Y = X[1] - X[0]$ or

$$Y = \underbrace{[\ -1 \quad 1\ ]}_{\mathbf{A}} \underbrace{\begin{bmatrix} X[0] \\ X[1] \end{bmatrix}}_{\mathbf{X}}.$$

Then, since Y is a linear transformation of **X**, we have $Y \sim \mathcal{N}(0, \mathrm{var}(Y))$, where $\mathrm{var}(Y) = \mathbf{ACA}^T$. Thus,

$$
\begin{aligned}
\mathrm{var}(Y) &= [\ -1 \quad 1\ ] \, \mathbf{C} \begin{bmatrix} -1 \\ 1 \end{bmatrix} \\
&= [\ -1 \quad 1\ ] \, \mathbf{GG}^T \begin{bmatrix} -1 \\ 1 \end{bmatrix} \qquad (\mathbf{C} = \sigma_U^2 \mathbf{GG}^T = \mathbf{GG}^T) \\
&= [\ -1 \quad 1\ ] \begin{bmatrix} \frac{1}{2} & \frac{1}{2} & 0 \\ 0 & \frac{1}{2} & \frac{1}{2} \end{bmatrix} \begin{bmatrix} \frac{1}{2} & 0 \\ \frac{1}{2} & \frac{1}{2} \\ 0 & \frac{1}{2} \end{bmatrix} \begin{bmatrix} -1 \\ 1 \end{bmatrix} \\
&= \frac{1}{2}
\end{aligned}
$$

so that $Y \sim \mathcal{N}(0, 1/2)$. Therefore,

$$P[X[1] - X[0] > 1] = Q\left(\frac{1}{\sqrt{1/2}}\right) = Q(\sqrt{2}) = 0.0786 < Q\left(\frac{1}{\sqrt{2}}\right) = 0.2398$$

and is consistent with our notion of smoothing.

Example 20.3 – Discrete-time Wiener random process or Brownian motion

This random process is basically a *random walk* with Gaussian "steps" or more specifically the sum process (see also Example 16.4)

$$X[n] = \sum_{i=0}^{n} U[i] \qquad n \geq 0$$

where $U[n]$ is WGN with variance σ_U^2. Note that the increments $X[n_2] - X[n_1]$ are independent and stationary (why?). As in the previous example, any set of samples of $X[n]$ is a linear transformation of the $U[i]$'s and hence has a multivariate Gaussian PDF. For example,

$$
\left[\begin{array}{c} X[0] \\ X[1] \end{array} \right] = \underbrace{\left[\begin{array}{cc} 1 & 0 \\ 1 & 1 \end{array} \right]}_{\mathbf{G}} \left[\begin{array}{c} U[0] \\ U[1] \end{array} \right]
$$

and therefore the Wiener random process is a Gaussian random process. It is clearly nonstationary, since, for example, the variance increases with n (recall from Example 16.4 that $\mathrm{var}(X[n]) = (n+1)\sigma_U^2$).

$\Diamond$

In Example 20.1 we saw that if the samples are uncorrelated, and the random process is Gaussian and hence the multivariate Gaussian PDF applies, then the samples are also *independent*. In Examples 20.1 and 20.2, the random processes were WSS but due to the fact that they are also Gaussian random processes, they are also *stationary*. We summarize and then prove these two properties next.

Property 20.1 – A Gaussian random process with uncorrelated samples has independent samples.

Proof:

Since the random process is Gaussian, the PDF of (20.1) applies for any set of samples. But for uncorrelated samples, the covariance matrix is diagonal and hence the joint PDF factors into the product of its marginal PDFs. Hence, the samples are independent.

□

Property 20.2 – A WSS Gaussian random process is also stationary.

Proof:

Since the random process is Gaussian, the PDF of (20.1) applies for any set of samples. But if $X[n]$ is also WSS, then for any n_0

$$
E[X[n_i + n_0]] = \mu_X[n_i + n_0] = \mu \qquad i = 1, 2, \ldots, K
$$

and

$$
\begin{aligned}
[\mathbf{C}]_{ij} &= \mathrm{cov}(X[n_i + n_0], X[n_j + n_0]) \\
&= E[X[n_i + n_0]X[n_j + n_0]] - E[X[n_i + n_0]]E[X[n_j + n_0]] \\
&= r_X[n_j - n_i] - \mu^2 \qquad \text{(due to WSS)}
\end{aligned}
$$

for $i = 1, 2, \ldots, K$ and $j = 1, 2, \ldots, K$. Since the mean vector and the covariance matrix do not depend on n_0, the joint PDF also does not depend on n_0. Hence, the WSS Gaussian random process is also stationary.

□

20.4 Linear Transformations

Any linear transformation of a Gaussian random process produces another Gaussian random process. In Example 20.2 the white noise random process $U[n]$ was Gaussian, and the MA random process $X[n]$, which was the result of a linear transformation, is another Gaussian random process. The MA random process in that example can be viewed as the output of the LSI filter with system function $\mathcal{H}(z) = 1/2 + (1/2)z^{-1}$ whose input is $U[n]$. This result, that if the input to an LSI filter is a Gaussian random process, then the output is also a Gaussian random process, is true in general. The random processes described by the linear difference equations

$$
\begin{aligned}
X[n] &= aX[n-1] + U[n] & &\text{AR random process (see Example 17.5)} \\
X[n] &= U[n] - bU[n-1] & &\text{MA random process (see Example 18.6)} \\
X[n] &= aX[n-1] + U[n] - bU[n-1] & &\text{ARMA random process} \\
& & &\text{(This is the definition.)}
\end{aligned}
$$

can also be viewed as the outputs of LSI filters with respective system functions

$$
\begin{aligned}
\mathcal{H}(z) &= \frac{1}{1 - az^{-1}} \\
\mathcal{H}(z) &= 1 - bz^{-1} \\
\mathcal{H}(z) &= \frac{1 - bz^{-1}}{1 - az^{-1}}.
\end{aligned}
$$

As a result, since the input $U[n]$ is a Gaussian random process, they are all Gaussian random processes. Furthermore, since it is only necessary to know the first two moments to specify the joint PDF of a set of samples of a Gaussian random process, the PDF for the output random process of an LSI filter is easily found. In particular, assume we are interested in the filtering of a WSS Gaussian random process by an LSI filter with frequency response $H(f)$. *Then, if the input to the filter is the WSS Gaussian random process $X[n]$, which has a mean of μ_X and an ACS of $r_X[k]$, then we know from Theorem 18.3.1 that the output random process $Y[n]$ is also WSS and its mean and ACS are*

$$
\begin{aligned}
\mu_Y &= \mu_X H(0) & &(20.4) \\
P_Y(f) &= |H(f)|^2 P_X(f) & &(20.5)
\end{aligned}
$$

and furthermore $Y[n]$ is a Gaussian random process (and is stationary according to Property 20.2). (See also Problem 20.7.) The joint PDF for any set of samples of $Y[n]$ is found from (20.1) by using (20.4) and (20.5). An example follows.

Example 20.4 – A differencer

A WSS Gaussian random process $X[n]$ with mean μ_X and ACS $r_X[k]$ is input to a differencer. The output random process is defined to be $Y[n] = X[n] - X[n-1]$.

What is the PDF of two successive output samples? To solve this we first note that the output random process is Gaussian and also WSS since a differencer is just an LSI filter whose system function is $\mathcal{H}(z) = 1 - z^{-1}$. We need only find the first two moments of $Y[n]$. The mean is

$$E[Y[n]] = E[X[n]] - E[X[n-1]] = \mu_X - \mu_X = 0$$

and the ACS can be found as the inverse Fourier transform of $P_Y(f)$. But from (20.5) with $H(f) = \mathcal{H}(\exp(j2\pi f)) = 1 - \exp(-j2\pi f)$, we have

$$
\begin{aligned}
P_Y(f) &= H(f)H^*(f)P_X(f) \\
&= [1 - \exp(-j2\pi f)][1 - \exp(j2\pi f)]P_X(f) \\
&= 2P_X(f) - \exp(j2\pi f)P_X(f) - \exp(-j2\pi f)P_X(f).
\end{aligned}
$$

Taking the inverse Fourier transform produces

$$r_Y[k] = 2r_X[k] - r_X[k+1] - r_X[k-1]. \tag{20.6}$$

For two successive samples, say $Y[0]$ and $Y[1]$, we require the covariance matrix of $\mathbf{Y} = [Y[0]\, Y[1]]^T$. Since $Y[n]$ has a zero mean, this is just

$$\mathbf{C}_Y = \begin{bmatrix} r_Y[0] & r_Y[1] \\ r_Y[1] & r_Y[0] \end{bmatrix}$$

and thus using (20.6), it becomes

$$\mathbf{C}_Y = \begin{bmatrix} 2(r_X[0] - r_X[1]) & 2r_X[1] - r_X[2] - r_X[0] \\ 2r_X[1] - r_X[2] - r_X[0] & 2(r_X[0] - r_X[1]) \end{bmatrix}.$$

The joint PDF is then

$$p_{Y[0],Y[1]}(y[0], y[1]) = \frac{1}{2\pi \det^{1/2}(\mathbf{C}_Y)} \exp(-\tfrac{1}{2}\mathbf{y}^T\mathbf{C}_Y^{-1}\mathbf{y})$$

where $\mathbf{y} = [y[0]\, y[1]]^T$. See also Problem 20.5.

$\Diamond$

We now summarize the foregoing results in a theorem.

Theorem 20.4.1 (Linear filtering of a WSS Gaussian random process)
Suppose that $X[n]$ is a WSS Gaussian random process with mean μ_X and ACS $r_X[k]$ that is input to an LSI filter with frequency response $H(f)$. Then, the PDF of N successive output samples $\mathbf{Y} = [Y[0]\, Y[1]\ldots Y[N-1]]^T$ is given by

$$p_{\mathbf{Y}}(\mathbf{y}) = \frac{1}{(2\pi)^{N/2} \det^{1/2}(\mathbf{C}_Y)} \exp\left[-\tfrac{1}{2}(\mathbf{y} - \boldsymbol{\mu}_Y)^T\mathbf{C}_Y^{-1}(\mathbf{y} - \boldsymbol{\mu}_Y)\right] \tag{20.7}$$

where

$$\boldsymbol{\mu}_Y = \begin{bmatrix} \mu_X H(0) \\ \vdots \\ \mu_X H(0) \end{bmatrix} \tag{20.8}$$

$$
\begin{aligned}
{[\mathbf{C}_Y]}_{mn} &= r_Y[m-n] - (\mu_X H(0))^2 \tag{20.9} \\
&= \int_{-\frac{1}{2}}^{\frac{1}{2}} |H(f)|^2 P_X(f) \exp(j2\pi f(m-n))df - (\mu_X H(0))^2
\end{aligned}
$$

$$\tag{20.10}$$

for $m = 1, 2, \ldots, N; n = 1, 2, \ldots, N$. The same PDF is obtained for any shifted set of successive samples since $Y[n]$ is stationary.

Note that in the preceding theorem the covariance matrix is a symmetric *Toeplitz* matrix (all elements along each northwest-southeast diagonal are the same) due to the assumption of successive samples (see also Section 17.4).

Another transformation that occurs quite frequently is the sum of two independent Gaussian random processes. If $X[n]$ is a Gaussian random process and $Y[n]$ is another Gaussian random process, *and $X[n]$ and $Y[n]$ are independent*, then $Z[n] = X[n] + Y[n]$ is a Gaussian random process (see Problem 20.9). By independence of two random processes we mean that all sets of samples of $X[n]$ or $\{X[n_1], X[n_2], \ldots, X[n_K]\}$ and of $Y[n]$ or $\{Y[m_1], Y[m_2], \ldots, Y[m_L]\}$ are independent of each other. This must hold for all $n_1, \ldots, n_K, m_1, \ldots, m_L$ and for all K and L. If this is the case then the PDF of the entire set of samples can be written as the product of the PDFs of each set of samples.

20.5 Nonlinear Transformations

The Gaussian random process is one of the few random processes for which the moments at the output of a nonlinear transformation can easily be found. In particular, a polynomial transformation lends itself to output moment evaluation. This is because the *higher-order* joint moments of a multivariate Gaussian PDF can be expressed in terms of *first- and second-order* moments. In fact, this is not surprising in that the multivariate Gaussian PDF is characterized by its first- and second-order moments. As a result, in computing the joint moments, any integral of the form $\int_{-\infty}^{\infty} \cdots \int_{-\infty}^{\infty} x_1^{l_1} \ldots x_N^{l_N} p_{X_1, \ldots, X_N}(x_1, \ldots, x_N) dx_1 \ldots dx_N$ must be a function of the mean vector and covariance matrix. Hence, the joint moments must be a function of the first- and second-order moments. As a particular case of interest, consider the fourth-order moment $E[X_1 X_2 X_3 X_4]$ for $\mathbf{X} = [X_1 \, X_2 \, X_3 \, X_4]^T$ a zero mean Gaussian random vector. Then, it can be shown that (see Problem 20.12)

$$E[X_1X_2X_3X_4] = E[X_1X_2]E[X_3X_4] + E[X_1X_3]E[X_2X_4] + E[X_1X_4]E[X_2X_3]$$
$$(20.11)$$

and this holds even if some of the random variables are the same (try $X_1 = X_2 = X_3 = X_4$ and compare it to $E[X^4]$ for $X \sim \mathcal{N}(0,1)$). It is seen that the fourth-order moment is expressible as the sum of products of the second-order moments, which are found from the covariance matrix. Now if $X[n]$ is a Gaussian random process with zero mean, then we have for any four samples (which by the definition of a Gaussian random process has a fourth-order Gaussian PDF)

$$
\begin{aligned}
E[X[n_1]X[n_2]X[n_3]X[n_4]] &= E[X[n_1]X[n_2]]E[X[n_3]X[n_4]] \\
&\quad + E[X[n_1]X[n_3]]E[X[n_2]X[n_4]] \\
&\quad + E[X[n_1]X[n_4]]E[X[n_2]X[n_3]] \quad (20.12)
\end{aligned}
$$

and if furthermore, $X[n]$ is WSS, then this reduces to

$$
\begin{aligned}
E[X[n_1]X[n_2]X[n_3]X[n_4]] &= r_X[n_2 - n_1]r_X[n_4 - n_3] + r_X[n_3 - n_1]r_X[n_4 - n_2] \\
&\quad + r_X[n_4 - n_1]r_X[n_3 - n_2]. \quad (20.13)
\end{aligned}
$$

This formula allows us to easily calculate the effect of a polynomial transformation on the moments of a WSS Gaussian random process. An example follows.

Example 20.5 – Effect of squaring WSS Gaussian random process

Assuming that $X[n]$ is a zero mean WSS Gaussian random process, we wish to determine the effect of squaring it to form $Y[n] = X^2[n]$. Clearly, $Y[n]$ will no longer be a Gaussian random process since it can only take on nonnegative values (see also Example 10.8). We can, however, show that $Y[n]$ is still WSS. To do so we calculate the mean as

$$E[Y[n]] = E[X^2[n]] = r_X[0]$$

which does not depend on n, and the covariance sequence as

$$
\begin{aligned}
E[Y[n]Y[n+k]] &= E[X^2[n]X^2[n+k]] \\
&= r_X^2[0] + 2r_X^2[k] \qquad \text{(using } n_1 = n_2 = n \\
&\qquad\qquad\qquad\qquad \text{and } n_3 = n_4 = n+k \text{ in (20.13))}
\end{aligned}
$$

which also does not depend on n. Thus, at the output of the squarer the random process is WSS with

$$
\begin{aligned}
\mu_Y &= r_X[0] \\
r_Y[k] &= r_X^2[0] + 2r_X^2[k]. \quad (20.14)
\end{aligned}
$$

Note that if the PSD at the input to the squarer is $P_X(f)$, then the output PSD is obtained by taking the Fourier transform of (20.14) to yield

$$P_Y(f) = r_X^2[0]\delta(f) + 2P_X(f) \star P_X(f) \tag{20.15}$$

where

$$P_X(f) \star P_X(f) = \int_{-\frac{1}{2}}^{\frac{1}{2}} P_X(\nu)P_X(f-\nu)d\nu$$

is a convolution integral. As a specific example, consider the MA random process $X[n] = (U[n]+U[n-1])/2$, where $U[n]$ is WGN with variance $\sigma_U^2 = 1$. Then, typical realizations of $X[n]$ and $Y[n]$ are shown in Figure 20.4. The MA random process

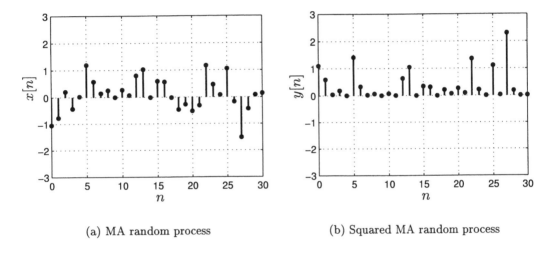

(a) MA random process   (b) Squared MA random process

Figure 20.4: Typical realization of a Gaussian MA random process and its squared realization.

has a zero mean and ACS $r_X[k] = (1/2)\delta[k] + (1/4)\delta[k+1] + (1/4)\delta[k-1]$ (see Example 17.3). Because of the squaring, the output mean is $E[Y[n]] = r_X[0] = 1/2$. The PSD of $X[n]$ can easily be shown to be $P_X(f) = (1+\cos(2\pi f))/2$ and the PSD of $Y[n]$ follows most easily by taking the Fourier transform of $r_Y[k]$. From (20.14) we have

$$\begin{aligned}
r_Y[k] &= r_X^2[0] + 2r_X^2[k] \\
&= \frac{1}{4} + 2\left(\frac{1}{2}\delta[k] + \frac{1}{4}\delta[k+1] + \frac{1}{4}\delta[k-1]\right)^2 \\
&= \frac{1}{4} + 2\left(\frac{1}{4}\delta[k] + \frac{1}{16}\delta[k+1] + \frac{1}{16}\delta[k-1]\right)
\end{aligned}$$

since all the cross-terms must be zero and $\delta^2[k-k_0] = \delta[k-k_0]$. Thus, we have

$$r_Y[k] = \frac{1}{4} + \frac{1}{2}\delta[k] + \frac{1}{8}\delta[k+1] + \frac{1}{8}\delta[k-1]$$

and taking the Fourier transform produces the PSD as

$$P_Y(f) = \frac{1}{4}\delta(f) + \frac{1}{2} + \frac{1}{4}\cos(2\pi f).$$

The PSDs are shown in Figure 20.5. Note that the squaring has produced an impulse

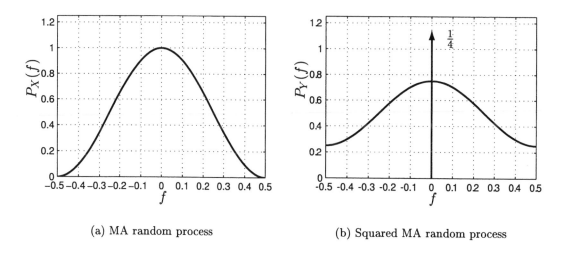

(a) MA random process   (b) Squared MA random process

Figure 20.5: PSDs of Gaussian MA random process and the squared random process.

at $f = 0$ of strength 1/4 that is due to the nonzero mean of the $Y[n]$ random process. Also, the squaring has "widened" the PSD, the usual consequence of a convolution in frequency.

$\Diamond$

20.6   Continuous-Time Definitions and Formulas

A continuous-time random process is defined to be a Gaussian random process if the random vector $\mathbf{X} = [X(t_1)\, X(t_2) \ldots X(t_K)]^T$ has a multivariate Gaussian PDF for all $\{t_1, t_2, \ldots, t_K\}$ and all K. The properties of a continuous-time Gaussian random process are identical to those for the discrete-time random process as summarized in Properties 20.1 and 20.2. Therefore, we will proceed directly to some examples of interest.

Example 20.6 – Continuous-time WGN

The continuous-time version of discrete-time WGN as defined in Example 20.1 is a *continuous-time* Gaussian random process $X(t)$ that has a zero mean and an ACF $r_X(\tau) = (N_0/2)\delta(\tau)$. The factor of $N_0/2$ is customarily used, since it is the level of the corresponding PSD (see Example 17.11). The random process is called *continuous-time white Gaussian noise (WGN)*. This was previously described in

Example 17.11. Note that in addition to the samples being uncorrelated (since $r_X(\tau) = 0$ for $\tau \neq 0$), they are also *independent* because of the Gaussian assumption. Unfortunately, for continuous-time WGN, it is not possible to explicitly write down the multivariate Gaussian PDF since $r_X(0) \to \infty$. Instead, as explained in Example 17.11 we use continuous-time WGN only as a model, reserving any probability calculations for the random process at the output of some filter, whose input is WGN. This is illustrated next.

$\diamondsuit$

Example 20.7 – Continuous-time Wiener random process or Brownian motion

Let $U(t)$ be WGN and define the semi-infinite random process

$$X(t) = \int_0^t U(\xi)d\xi \qquad t \geq 0.$$

This random process is called the *Wiener random process* and is often used as a model for Brownian motion. It is the continuous-time equivalent of the discrete-time random process of Example 20.3. A typical realization of the Wiener random process is shown in Figure 20.6 (see Problem 20.18 on how this was done). Note that

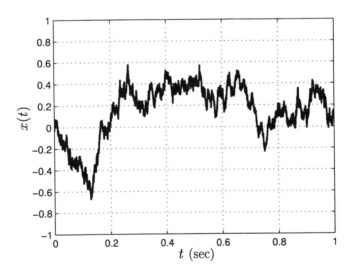

Figure 20.6: Typical realization of the Wiener random process.

because of its construction as the "sum" of independent and identically distributed random variables (the $U(t)$'s), the increments are also independent and stationary. To prove that $X(t)$ is a Gaussian random process is somewhat tricky in that it is an uncountable "sum" of independent random variables $U(\xi)$ for $0 \leq \xi \leq t$. We will take it on faith that any integral, which is a linear transformation, of a continuous-time Gaussian random process produces another continuous-time Gaussian random process (see also Problem 20.16 for a heuristic proof). As such, we need only determine the mean and covariance functions. These are found as

$$E[X(t)] = E\left[\int_0^t U(\xi)d\xi\right]$$

$$= \int_0^t E[U(\xi)]d\xi = 0$$

$$E[X(t_1)X(t_2)] = E\left[\int_0^{t_1} U(\xi_1)d\xi_1 \int_0^{t_2} U(\xi_2)d\xi_2\right]$$

$$= \int_0^{t_1}\int_0^{t_2} \underbrace{E[U(\xi_1)U(\xi_2)]}_{r_U(\xi_2-\xi_1)=(N_0/2)\delta(\xi_2-\xi_1)} d\xi_1 d\xi_2$$

$$= \frac{N_0}{2}\int_0^{t_1}\left(\int_0^{t_2}\delta(\xi_2-\xi_1)d\xi_2\right)d\xi_1.$$

To evaluate the double integral we first examine the inner integral and assume that $t_2 > t_1$. Then, the function $\delta(\xi_2 - \xi_1)$ with ξ_1 *fixed* is integrated over the interval $0 \le \xi_2 \le t_2$ as shown in Figure 20.7. It is clear from the figure that if we fix ξ_1

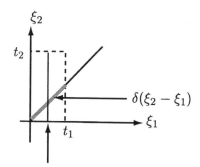

integrate along here first

Figure 20.7: Evaluation of double integral of Dirac delta function for the case of $t_2 > t_1$.

and integrate along ξ_2, then we will include the impulse in the inner integral for all ξ_1. (This would not be the case if $t_2 < t_1$ as one can easily verify by redrawing the rectangle for this condition.) As a result, if $t_2 > t_1$, then

$$\int_0^{t_2}\delta(\xi_2-\xi_1)d\xi_2 = 1 \qquad \text{for all } 0 \le \xi_1 \le t_1$$

and therefore

$$E[X(t_1)X(t_2)] = \frac{N_0}{2}\int_0^{t_1}d\xi_1 = \frac{N_0}{2}t_1$$

and similarly if $t_2 < t_1$, we will have $E[X(t_1)X(t_2)] = (N_0/2)t_2$. Combining the two results produces

$$E[X(t_1)X(t_2)] = \frac{N_0}{2}\min(t_1, t_2) \qquad (20.16)$$

which should be compared to the discrete-time result obtained in Problem 16.26. Hence, the joint PDF of the samples of a Wiener random process is a multivariate Gaussian PDF with mean vector equal to zero and covariance matrix having as its (i,j)th element

$$[\mathbf{C}]_{ij} = \frac{N_0}{2}\min(t_i, t_j).$$

Note that from (20.16) with $t_1 = t_2 = t$, the PDF of $X(t)$ is $\mathcal{N}(0, (N_0/2)t)$. Clearly, *the Wiener random process is a nonstationary correlated random process whose mean is zero, variance increases with time, and marginal PDF is Gaussian.*

$\diamondsuit$

In the next section we explore some other important continuous-time Gaussian random processes often used as models in practice.

20.7 Special Continuous-Time Gaussian Random Processes

20.7.1 Rayleigh Fading Sinusoid

In Example 16.11 we studied a discrete-time randomly phased sinusoid. Here we consider the continuous-time equivalent for that random process, which is given by $X(t) = A\cos(2\pi F_0 t + \Theta)$, where $A > 0$ is the amplitude, F_0 is the frequency in Hz, and Θ is the random phase with PDF $\mathcal{U}(0, 2\pi)$. We now further assume that the *amplitude* is also a random variable. This is frequently a good model for a sinusoidal signal that is subject to *multipath fading*. It occurs when a sinusoidal signal propagates through a medium, e.g., an electromagnetic pulse in the atmosphere or a sound pulse in the ocean, and reaches its destination by several different paths. The constructive and destructive interference of several overlapping sinusoids causes the received waveform to exhibit amplitude fluctuations or fading. An example of this was given in Figure 20.2. However, over any short period of time, say $5 \leq t \leq 5.5$ seconds, the waveform will have approximately a constant amplitude and a constant phase as shown in Figure 20.8. Because the amplitude and phase are not known in advance, we model them as realizations of random variables. That the waveform does not maintain the constant amplitude level and phase outside of the small interval will be of no consequence to us if we are only privy to observing the waveform over a small time interval. Hence, a reasonable model for the random process (over the small time interval) is to assume a random amplitude and random phase so that $X(t) = A\cos(2\pi F_0 t + \Theta)$, where A and Θ are random variables. A more convenient

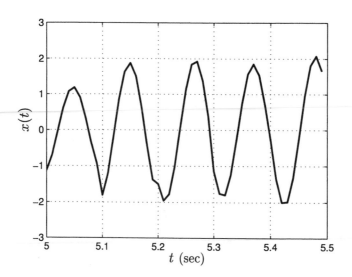

Figure 20.8: Segment of waveform shown in Figure 20.2 for $5 \leq t \leq 5.5$ seconds.

form is obtained by expanding the sinusoid as

$$
\begin{aligned}
X(t) &= A\cos(2\pi F_0 t + \Theta) \\
&= A\cos(\Theta)\cos(2\pi F_0 t) - A\sin(\Theta)\sin(2\pi F_0 t) \\
&= U\cos(2\pi F_0 t) - V\sin(2\pi F_0 t)
\end{aligned}
$$

where we have let $A\cos(\Theta) = U$, $A\sin(\Theta) = V$. Clearly, since A and Θ are random variables, so are U and V. Since the physical waveform is due to the sum of many sinusoids, we once again use a central limit theorem argument to assume that U and V are Gaussian. Furthermore, if we assume that they are independent and have the same PDF of $\mathcal{N}(0, \sigma^2)$, we will obtain PDFs for the amplitude and phase which are found to be valid in practice. With the Gaussian assumptions for U and V, the random amplitude becomes a Rayleigh distributed random variable, the random phase becomes a uniformly distributed random variable, and the amplitude and phase random variables are independent of each other. To see this note that since $U = A\cos(\Theta)$, $V = A\sin(\Theta)$, we have $A = \sqrt{U^2 + V^2}$ and $\Theta = \arctan(V/U)$. It was shown in Example 12.12 that if $X \sim \mathcal{N}(0, \sigma^2)$, $Y \sim \mathcal{N}(0, \sigma^2)$, and X and Y are independent, then $R = \sqrt{X^2 + Y^2}$ is a Rayleigh random variable, $\Theta = \arctan(Y/X)$ is a uniformly distributed random variable, and R and Θ are independent. Hence, we have that for the random amplitude/random phase sinusoid $X(t) = A\cos(2\pi F_0 t + \Theta)$, the amplitude has the PDF

$$
p_A(a) = \begin{cases} \frac{a}{\sigma^2}\exp\left(-\frac{1}{2}\frac{a^2}{\sigma^2}\right) & a \geq 0 \\ 0 & a < 0 \end{cases}
$$

and the phase has the PDF $\Theta \sim \mathcal{U}(0, 2\pi)$, and A and Θ are independent. This model is usually referred to as the *Rayleigh fading sinusoidal* model. It is also a

Gaussian random process since all sets of K samples can be written as

$$\begin{bmatrix} X(t_1) \\ X(t_2) \\ \vdots \\ X(t_K) \end{bmatrix} = \begin{bmatrix} \cos(2\pi F_0 t_1) & -\sin(2\pi F_0 t_1) \\ \cos(2\pi F_0 t_2) & -\sin(2\pi F_0 t_2) \\ \vdots & \vdots \\ \cos(2\pi F_0 t_K) & -\sin(2\pi F_0 t_K) \end{bmatrix} \begin{bmatrix} U \\ V \end{bmatrix}$$

which is a linear transformation of the Gaussian random vector $[U\ V]^T$, and so has a multivariate Gaussian PDF. (For $K > 2$ the covariance matrix will be singular, so that to be more rigorous we would need to modify our definition of the Gaussian random process. This would involve the characteristic function which exists even for a singular covariance matrix.) Furthermore, $X(t)$ is WSS, as we now show. Its mean is zero since $E[U] = E[V] = 0$ and its ACF is

$$r_X(\tau)$$

$$\begin{aligned}
&= E[X(t)X(t+\tau)] \\
&= E[[U\cos(2\pi F_0 t) - V\sin(2\pi F_0 t)][U\cos(2\pi F_0(t+\tau)) - V\sin(2\pi F_0(t+\tau))]] \\
&= E[U^2]\cos(2\pi F_0 t)\cos(2\pi F_0(t+\tau)) + E[V^2]\sin(2\pi F_0 t)\sin(2\pi F_0(t+\tau)) \\
&= \sigma^2[\cos(2\pi F_0 t)\cos(2\pi F_0(t+\tau)) + \sin(2\pi F_0 t)\sin(2\pi F_0(t+\tau))] \\
&= \sigma^2\cos(2\pi F_0\tau) \qquad\qquad\qquad\qquad\qquad\qquad\qquad\qquad (20.17)
\end{aligned}$$

where we have used $E[UV] = E[U]E[V] = 0$ due to independence. Its PSD is obtained by taking the Fourier transform to yield

$$P_X(F) = \frac{\sigma^2}{2}\delta(F + F_0) + \frac{\sigma^2}{2}\delta(F - F_0) \qquad\qquad (20.18)$$

and it is seen that all its power is concentrated at $F = F_0$ as expected.

20.7.2 Bandpass Random Process

The Rayleigh fading sinusoid model assumed that our observation time was short. Within that time window, the sinusoid exhibits approximately constant amplitude and phase. If we observe a longer time segment of the random process whose typical realization is shown in Figure 20.2, then the constant in time (but random amplitude/random phase) sinusoid is not a good model. A more realistic but more complicated model is to let both the amplitude and phase be random processes so that they vary in time. As such, the random process will be made up of many frequencies, although they will be concentrated about $F = F_0$. Such a random process is usually called a *narrowband random process*. Our model, however, will actually be valid for a bandpass random process whose PSD is shown in Figure 20.9. Hence, we will assume that the bandpass random process can be represented as

$$\begin{aligned}
X(t) &= A(t)\cos(2\pi F_0 t + \Theta(t)) \\
&= A(t)\cos(\Theta(t))\cos(2\pi F_0 t) - A(t)\sin(\Theta(t))\sin(2\pi F_0 t)
\end{aligned}$$

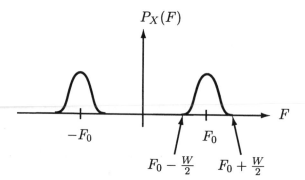

Figure 20.9: Typical PSD for bandpass random process. The PSD is assumed to be symmetric about $F = F_0$ and also that $F_0 > W/2$.

where $A(t)$ and $\Theta(t)$ are now random processes. As before we let

$$\begin{aligned} U(t) &= A(t)\cos(\Theta(t)) \\ V(t) &= A(t)\sin(\Theta(t)) \end{aligned}$$

so that we have as our model for a bandpass random process

$$X(t) = U(t)\cos(2\pi F_0 t) - V(t)\sin(2\pi F_0 t). \tag{20.19}$$

The $X(t)$ random process is seen to be a modulated version of $U(t)$ and $V(t)$ (modulation meaning that $U(t)$ and $V(t)$ are multiplied by $\cos(2\pi F_0 t)$ and $\sin(2\pi F_0 t)$, respectively). This modulation shifts the PSD of $U(t)$ and $V(t)$ to be centered about $F = F_0$. Therefore, $U(t)$ and $V(t)$ must be slowly varying or *lowpass* random processes. As a suitable description of $U(t)$ and $V(t)$ we assume that *they are each zero mean lowpass Gaussian random processes, independent of each other, and jointly WSS (see Chapter 19) with the same ACF,* $r_U(\tau) = r_V(\tau)$. Then, as before $X(t)$ is a zero mean Gaussian random process, which as we now show is also WSS. Clearly, since both $U(t)$ and $V(t)$ are zero mean, from (20.19) so is $X(t)$, and the ACF is

$$\begin{aligned} r_X(\tau) &= E[X(t)X(t+\tau)] & (20.20) \\ &= E[[U(t)\cos(2\pi F_0 t) - V(t)\sin(2\pi F_0 t)] \\ &\quad \cdot [U(t+\tau)\cos(2\pi F_0(t+\tau)) - V(t+\tau)\sin(2\pi F_0(t+\tau))] \\ &= r_U(\tau)\cos(2\pi F_0 t)\cos(2\pi F_0(t+\tau)) + r_V(\tau)\sin(2\pi F_0 t)\sin(2\pi F_0(t+\tau)) \\ &= r_U(\tau)\cos(2\pi F_0\tau) & (20.21) \end{aligned}$$

since $E[U(t_1)V(t_2)] = 0$ for all t_1 and t_2 due to the independence assumption, and $r_U(\tau) = r_V(\tau)$ by assumption. Note that this extends the previous case in which $U(t) = U$ and $V(t) = V$ and $r_U(\tau) = \sigma^2$ (see (20.17)). The PSD is found by taking the Fourier transform of the ACF to yield

$$P_X(F) = \frac{1}{2}P_U(F + F_0) + \frac{1}{2}P_U(F - F_0). \tag{20.22}$$

If $U(t)$ and $V(t)$ have the lowpass PSD shown in Figure 20.10, then in accordance with (20.22) $P_X(F)$ is given by the dashed curve. As desired, we now have a repre-

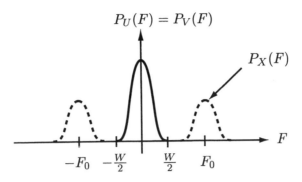

Figure 20.10: PSD for lowpass random processes $U(t)$ and $V(t)$. The PSD for the bandpass random process $X(t)$ is shown as the dashed curve.

sentation for a bandpass random process. It is obtained by *modulating* two lowpass random processes $U(t)$ and $V(t)$ up to a center frequency of F_0 Hz. Hence, (20.19) is called the *bandpass random process representation* and since the random process may either represent a signal or noise, it is also referred to as the bandpass signal representation or the bandpass noise representation. Note that because $P_U(F)$ is symmetric about $F = 0$, $P_X(F)$ must be symmetric about $F = F_0$. To represent bandpass PSDs that are not symmetric requires the assumption that $U(t)$ and $V(t)$ are correlated [Van Trees 1971].

In summary, to model a WSS Gaussian random process $X(t)$ that has a zero mean and a bandpass PSD given by

$$P_X(F) = \frac{1}{2}P_U(F + F_0) + \frac{1}{2}P_U(F - F_0)$$

where $P_U(F) = 0$ for $|F| > W/2$ as shown in Figure 20.10 by the dashed curve, we use

$$X(t) = U(t)\cos(2\pi F_0 t) - V(t)\sin(2\pi F_0 t).$$

The assumptions are that $U(t), V(t)$ are each Gaussian random processes with zero mean, independent of each other and each is WSS with PSD $P_U(F)$. The random processes $U(t), V(t)$ are lowpass random processes and are sometimes referred to as the *in phase* and *quadrature* components of $X(t)$. This is because the "carrier" sinusoid $\cos(2\pi F_0 t)$ is in phase with the sinusoidal carrier in $U(t)\cos(2\pi F_0 t)$ and 90° out of phase with the sinusoidal carrier in $V(t)\sin(2\pi F_0 t)$. See Problem 20.24 on how to extract the lowpass random processes from $X(t)$. In addition, the amplitude of $X(t)$, which is $\sqrt{U^2(t) + V^2(t)}$ is called the *envelope* of $X(t)$. This is because if $X(t)$ is written as $X(t) = \sqrt{U^2(t) + V^2(t)}\cos(2\pi F_0 t + \arctan(V(t)/U(t)))$ (see Problem 12.42) the envelope consists of the maximums of the waveform. An example of a *deterministic* bandpass signal, given for sake of illustration, is $s(t) = 3t\cos(2\pi 20t) - 4t\sin(2\pi 20t)$ for $0 \le t \le 1$, and is shown in Figure 20.11. Note that the envelope is $\sqrt{(3t)^2 + (4t)^2} = 5|t|$. For a bandpass *random process* the envelope will also be a

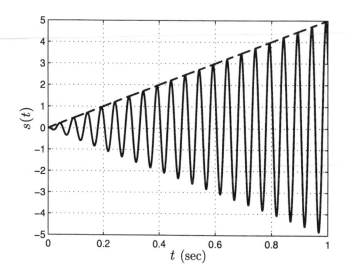

Figure 20.11: Plot of the deterministic bandpass signal $s(t) = 3t\cos(2\pi 20t) - 4t\sin(2\pi 20t)$ for $0 \le t \le 1$. The envelope is shown as the dashed line.

random process. Since $U(t)$ and $V(t)$ both have the same ACF, the characteristics of the envelope depend directly on $r_U(\tau)$. An illustration is given in the next example.

Example 20.8 – Bandpass random process envelope

Consider the bandpass Gaussian random process whose PSD is shown in Figure 20.12. This is often used as a model for bandpass "white" Gaussian noise. It results from having filtered WGN with a bandpass filter. Note that from (20.22) the PSD

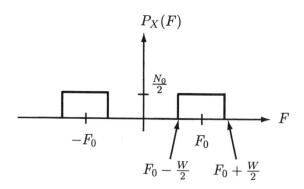

Figure 20.12: PSD for bandpass "white" Gaussian noise.

of $U(t)$ and $V(t)$ must be

$$P_U(F) = P_V(F) = \begin{cases} N_0 & |F| \leq \frac{W}{2} \\ 0 & |F| > \frac{W}{2} \end{cases}$$

and therefore by taking the inverse Fourier transform, the ACF becomes (see also (17.55) for a similar calculation)

$$r_U(\tau) = r_V(\tau) = N_0 W \frac{\sin(\pi W \tau)}{\pi W \tau}. \tag{20.23}$$

The correlation between two samples *of the envelope* will be approximately zero when $\tau > 1/W$ since then $r_U(\tau) = r_V(\tau) \approx 0$. Examples of some bandpass realizations are shown for $F_0 = 20$ Hz, $W = 1$ Hz in Figure 20.13a and for $F_0 = 20$ Hz, $W = 4$ Hz in Figure 20.13b. The time for which two samples must be separated before they become uncorrelated is called the *correlation time* τ_c. It is defined by $r_X(\tau) \approx 0$ for $\tau > \tau_c$. Here it is $\tau_c \approx 1/W$, and is shown in Figure 20.13.

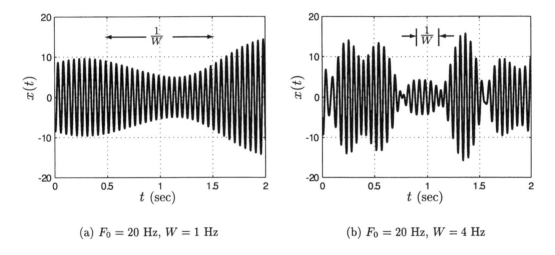

(a) $F_0 = 20$ Hz, $W = 1$ Hz $\qquad\qquad$ (b) $F_0 = 20$ Hz, $W = 4$ Hz

Figure 20.13: Typical realizations of bandpass "white" Gaussian noise. The PSD is given in Figure 20.12.

A typical probability calculation might be to determine the probability that the envelope at $t = t_0$ exceeds some threshold γ. Thus, we wish to find $P[A(t_0) > \gamma]$, where $A(t_0) = \sqrt{U^2(t_0) + V^2(t_0)}$. Since the $U(t)$ and $V(t)$ are independent Gaussian random processes with $U(t) \sim \mathcal{N}(0, \sigma^2)$ and $V(t) \sim \mathcal{N}(0, \sigma^2)$, it follows that $A(t_0)$ is a Rayleigh random variable. Hence, we have that

$$\begin{aligned} P[A(t_0) > \gamma] &= \int_\gamma^\infty \frac{a}{\sigma^2} \exp\left(-\tfrac{1}{2}\tfrac{a^2}{\sigma^2}\right) da \\ &= \exp\left(-\tfrac{1}{2}\tfrac{\gamma^2}{\sigma^2}\right). \end{aligned}$$

To complete the calculation we need to determine σ^2. But $\sigma^2 = E[U^2(t_0)] = r_U[0] = N_0 W$ from (20.23). Therefore, we have that

$$P[A(t_0) > \gamma] = \exp\left(-\tfrac{1}{2}\frac{\gamma^2}{N_0 W}\right).$$

$\diamondsuit$

20.8   Computer Simulation

We now discuss the generation of a realization of a discrete-time Gaussian random process. The generation of a continuous-time random process realization can be accomplished by approximating it by a discrete-time realization with a sufficiently small time interval between samples. We have done this to produce Figure 20.6 (see also Problem 20.18). In particular, we wish to generate a realization of a *WSS* Gaussian random process with mean zero and ACS $r_X[k]$ or equivalently a PSD $P_X(f)$. For nonzero mean random processes we need only add the mean to the realization. The method is based on Theorem 20.4.1, where we use a WGN random process $U[n]$ as the input to an LSI filter with frequency response $H(f)$. Then, we know that the output random process will be WSS and Gaussian with a PSD $P_X(f) = |H(f)|^2 P_U(f)$. Now assuming that $P_U(f) = \sigma_U^2 = 1$, so that $P_X(f) = |H(f)|^2$, we see that a filter whose frequency response magnitude is $|H(f)| = \sqrt{P_X(f)}$ and whose phase response is arbitrary (but must be an odd function) will be required. Finding the filter frequency response from the PSD is known as *spectral factorization* [Priestley 1981]. As special cases of this problem, if we wish to generate either the AR, MA, or ARMA Gaussian random processes described in Section 20.4, then the filters are already known and have been implemented as difference equations. For example, the MA random process is generated by filtering $U[n]$ with the LSI filter whose frequency response is $H(f) = 1 - b\exp(-j2\pi f)$. This is equivalent to the implementation using the difference equation $X[n] = U[n] - bU[n-1]$. For higher-order (more coefficients) AR, MA, and ARMA random processes, the reader should consult [Kay 1988] for how the appropriate coefficients can be obtained from the PSD. Also, note that the problem of designing a filter whose frequency response magnitude *approximates* a given one is called *digital filter design*. Many techniques are available to do this [Jackson 1996]. We next give a simple example of how to generate a realization of a WSS Gaussian random process with a given PSD.

Example 20.9 – Filter determination to produce Gaussian random process with given PSD

Assume we wish to generate a realization of a WSS Gaussian random process with zero mean and PSD $P_X(f) = (1 + \cos(4\pi f))/2$. Then, for $P_U(f) = 1$ the magnitude of the frequency response should be

$$|H(f)| = \sqrt{\tfrac{1}{2}(1 + \cos(4\pi f))}.$$

We will choose the phase response or $\angle H(f) = \theta(f)$ to be any convenient function. Thus, we wish to determine the impulse response $h[k]$ of the filter whose frequency response is

$$H(f) = \sqrt{\tfrac{1}{2}(1 + \cos(4\pi f))}\exp(j\theta(f))$$

since then we can generate the random process using a convolution sum as

$$X[n] = \sum_{k=-\infty}^{\infty} h[k]U[n-k]. \tag{20.24}$$

The impulse response is found as the inverse Fourier transform of the frequency response

$$
\begin{aligned}
h[n] &= \int_{-\frac{1}{2}}^{\frac{1}{2}} H(f)\exp(j2\pi fn)df \\
&= \int_{-\frac{1}{2}}^{\frac{1}{2}} \sqrt{\tfrac{1}{2}(1 + \cos(4\pi f))}\exp(j\theta(f))\exp(j2\pi fn)df \qquad -\infty < n < \infty.
\end{aligned}
$$

This can be evaluated by noting that $\cos(2\alpha) = \cos^2(\alpha) - \sin^2(\alpha)$ and therefore

$$
\begin{aligned}
\sqrt{\tfrac{1}{2}(1 + \cos(4\pi f))} &= \sqrt{\tfrac{1}{2}(1 + \cos^2(2\pi f) - \sin^2(2\pi f))} \\
&= \sqrt{\cos^2(2\pi f)} \\
&= |\cos(2\pi f)|.
\end{aligned}
$$

Thus,

$$h[n] = \int_{-\frac{1}{2}}^{\frac{1}{2}} |\cos(2\pi f)|\exp(j\theta(f))\exp(j2\pi fn)df$$

and we choose $\exp(j\theta(f)) = 1$ if $\cos(2\pi f) > 0$ and $\exp(j\theta(f)) = -1$ if $\cos(2\pi f) < 0$. This produces

$$h[n] = \int_{-\frac{1}{2}}^{\frac{1}{2}} \cos(2\pi f)\exp(j2\pi fn)df$$

which is easily shown to evaluate to

$$h[n] = \begin{cases} \tfrac{1}{2} & n = \pm 1 \\ 0 & \text{otherwise.} \end{cases}$$

Hence, from (20.24) we have that

$$X[n] = \frac{1}{2}U[n+1] + \frac{1}{2}U[n-1].$$

Note that the filter is noncausal. We could also use $X[n] = \frac{1}{2}U[n] + \frac{1}{2}U[n-2]$ if a causal filter is desired and still obtain the same PSD (see Problem 20.28).

$$\diamondsuit$$

Finally, it should be pointed out that an alternative means of generating successive samples of a zero mean Gaussian WSS random process is by applying a matrix transformation to a vector of independent $\mathcal{N}(0,1)$ samples. If a realization of $\mathbf{X} = [X[0]\, X[1] \ldots X[N-1]]^T$, where $\mathbf{X} \sim \mathcal{N}(\mathbf{0}, \mathbf{R}_X)$ and $\mathbf{R}_X$ is the $N \times N$ Toeplitz autocorrelation matrix given in (17.23) is desired, then the method described in Section 14.9 can be used. We need only replace $\mathbf{C}$ by $\mathbf{R}_X$. For a nonzero mean WSS Gaussian random process, we add the mean μ to each sample after this procedure is employed. The only drawback is that the realization is assumed to consist of a fixed number of samples N, and so for each value of N the procedure must be repeated. Filtering, as previously described, allows any number of samples to be easily generated.

20.9   Real-World Example – Estimating Fish Populations

Of concern to biologists, and to us all, is the fish population. Traditionally, the population has been estimated using a count produced by a net catch. However, this is expensive, time consuming, and relatively inaccurate. A better approach is therefore needed. In the introduction we briefly indicated how an echo sonar would produce a Gaussian random process as the reflected waveform from a school of fish. We now examine this in more detail and explain how estimation of the fish population might be done. The discussion is oversimplified so that the interested reader may consult [Ehrenberg and Lytle 1972, Stanton 1983, Stanton and Chu 1998] for more detail. Referring to Figure 20.14 a sound pulse, which is assumed to be sinusoidal, is transmitted from a ship. As it encounters a school of fish, it will be reflected from each fish and the entire waveform, which is the sum of all the reflections, will be received at the ship. The received waveform will be examined for the time interval from $t = 2R_{\min}/c$ to $t = 2R_{\max}/c$, where $R_{\min}$ and $R_{\max}$ are the minimum and maximum ranges of interest, respectively, and c is the speed of sound in the water. This corresponds to the time interval over which the reflections from the desired ranges will be present. Based on the received waveform we wish to estimate the number of fish in the vertical direction in the desired *range window* from $R_{\min}$ to $R_{\max}$. Note that only the fish within the nearly dashed vertical lines, which indicate the width of the transmitted sound energy, will produce reflections. For different angular regions other pulses must be transmitted. As discussed in the introduction, if there are a large number of fish producing reflections, then by the central limit theorem, the received waveform can be modeled as a Gaussian random process. As shown in Figure 20.14 the sinusoidal pulse first encounters the fish nearest in range, producing a reflection, while the fish farthest in range produces the last reflection. As a result, the many reflected pulses will overlap in time, with two of the reflected pulses shown in the figure. Hence, each reflected pulse can be

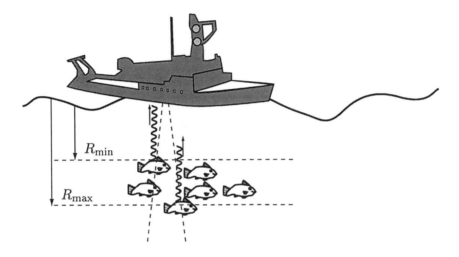

Figure 20.14: Fish counting by echo sonar.

represented as

$$X_i(t) = A_i \cos(2\pi F_0(t - \tau_i) + \Theta_i) \qquad (20.25)$$

where F_0 is the transmit frequency in Hz and $\tau_i = 2R_i/c$ is the time delay of the pulse reflected from the ith fish. As explained in the introduction, since A_i, Θ_i will depend upon the fish's position, orientation, and motion, which are not known a priori, we assume that they are realizations of random variables. Futhermore, since the ranges of the individual fish are unknown, we also do not know τ_i. Hence, we replace (20.25) by

$$X_i(t) = A_i \cos(2\pi F_0 t + \Theta_i')$$

where $\Theta_i' = \Theta_i - 2\pi F_0 \tau_i$ (which is reduced by multiples of 2π until it lies within the interval $(0, 2\pi)$), and model Θ_i' as a new random variable. Hence, for N reflections we have as our model

$$
\begin{aligned}
X(t) &= \sum_{i=1}^{N} X_i(t) \\
&= \sum_{i=1}^{N} A_i \cos(2\pi F_0 t + \Theta_i')
\end{aligned}
$$

and letting $U_i = A_i \cos(\Theta_i')$ and $V_i = A_i \sin(\Theta_i')$, we have

$$
\begin{aligned}
X(t) &= \sum_{i=1}^{N} (U_i \cos(2\pi F_0 t) - V_i \sin(2\pi F_0 t)) \\
&= \left(\sum_{i=1}^{N} U_i \right) \cos(2\pi F_0 t) - \left(\sum_{i=1}^{N} V_i \right) \sin(2\pi F_0 t) \\
&= U \cos(2\pi F_0 t) - V \sin(2\pi F_0 t)
\end{aligned}
$$

where $U = \sum_{i=1}^{N} U_i$ and $V = \sum_{i=1}^{N} V_i$. We assume that all the fish are about the same size and hence the echo amplitudes are about the same. Then, since U and V are the sums of random variables that we assume are independent (reflection from one fish does not affect reflection from any of the others) and identically distributed (fish are same size), we use a central limit theorem argument to postulate a Gaussian PDF for U and V. We furthermore assume that U and V are independent so that if $E[U_i] = E[V_i] = 0$ and $\text{var}(U_i) = \text{var}(V_i) = \sigma^2$, then we have that $U \sim \mathcal{N}(0, N\sigma^2)$, $V \sim \mathcal{N}(0, N\sigma^2)$, and U and V are independent. This is the Rayleigh fading sinusoid model discussed in Section 20.7. As a result, the envelope of the received waveform $X(t)$, which is given by $A = \sqrt{U^2 + V^2}$ has a Rayleigh PDF. Specifically, it is

$$
p_A(a) = \begin{cases} \frac{a}{N\sigma^2} \exp\left(-\frac{1}{2}\frac{a^2}{N\sigma^2}\right) & a \geq 0 \\ 0 & a < 0. \end{cases}
$$

Hence, if we have previously measured the reflection characteristics of a single fish, then we will know σ^2. To estimate N we recall that the mean of the Rayleigh random variable is

$$
E[A] = \sqrt{\frac{\pi}{2} N\sigma^2}
$$

so that upon solving for N, we have

$$
N = \frac{2}{\pi\sigma^2} E^2[A].
$$

To estimate the mean we can transmit a series of M pulses and measure the envelope for each received waveform $X_m(t)$ for $m = 1, 2 \ldots, M$. Calling the envelope measurement for the mth pulse $\hat{A}_m$, we can form the estimator for the number of fish as

$$
\hat{N} = \frac{2}{\pi\sigma^2} \left(\frac{1}{M} \sum_{m=1}^{M} \hat{A}_m \right)^2. \tag{20.26}
$$

See Problem 20.20 on how to obtain $\hat{A}_m = \sqrt{U_m^2 + V_m^2}$ from $X_m(t)$. It is shown there that $U_m = [2X_m(t)\cos(2\pi F_0 t)]_{\text{LPF}}$ and $V_m = [-2X_m(t)\sin(2\pi F_0 t)]_{\text{LPF}}$, where the designation "LPF" indicates that the time waveform has been lowpass filtered.

References

Bell Telephone Laboratories, *Transmission Systems for Communications*, Western Electric Co, Winston-Salem, NC, 1970.

Ehrenberg, J.E., D.W. Lytle, "Acoustic Techniques for Estimating Fish Abundance," *IEEE Trans. Geoscience Electronics*, pp. 138–145, 1972.

Jackson, L.B., *Digital Filters and Signal Processing: with MATLAB Exercises, 3rd Ed.*, Kluwer Academic Press, New York, 1996.

Kay, S., *Modern Spectral Estimation: Theory and Application*, Prentice-Hall, Englewood Cliffs, NJ, 1988.

Priestley, M.B., *Spectral Analysis and Time Series*, Academic Press, New York, 1981.

Stanton, T.K., "Multiple Scattering with Application to Fish-Echo Processing," *Journal Acoustical Soc. of America*, pp. 1164–1169, April 1983.

Stanton, T.K., D. Chu, "Sound Scattering by Several Zooplankton Groups. II. Scattering Models," *Journal Acoustical Soc. of America*, pp. 236–253, Jan. 1998.

Taylor, S., *Modelling Financial Time Series*, John Wiley & Sons, New York, 1986.

Urick, R.J., *Principles of Underwater Sound*, McGraw-Hill, New York, 1975.

Van Trees, H.L., *Detection, Estimation, and Modulation Theory, Part III*, John Wiley & Sons, New York, 1971.

Problems

20.1. (w) Determine the probability that 5 successive samples $\{X[0], X[1], X[2], X[3], X[4]\}$ of discrete-time WGN with $\sigma_U^2 = 1$ will all exceed zero. Then, repeat the problem if the samples are $\{X[10], X[11], X[12], X[13], X[14]\}$.

20.2 (☺) (w) If $X[n]$ is the random process described in Example 20.2, find $P[X[0] > 0, X[3] > 0]$ if $\sigma_U^2 = 1$.

20.3 (w) If $X[n]$ is a discrete-time Wiener random process with $\mathrm{var}(X[n]) = 2(n + 1)$, determine $P[-3 \leq X[5] \leq 3]$.

20.4 (w) A discrete-time Wiener random process $X[n]$ is input to a differencer to generate the output random process $Y[n] = X[n] - X[n-1]$. Describe the characteristics of the output random process.

20.5 (⌣) (w) If discrete-time WGN $X[n]$ with $\sigma_X^2 = 1$ is input to a differencer to generate the output random process $Y[n] = X[n] - X[n-1]$, find the PDF of the samples $Y[0], Y[1]$. Are the samples independent?

20.6 (w) If in Example 20.4 the input random process to the differencer is an AR random process with parameters a and $\sigma_U^2 = 1$, determine the PDF of $Y[0], Y[1]$. What happens as $a \to 1$? Explain your results.

20.7 (t) In this problem we argue that if $X[n]$ is a Gaussian random process that is input to an LSI filter so that the output random process is $Y[n] = \sum_{i=-\infty}^{\infty} h[i]X[n-i]$, then $Y[n]$ is also a Gaussian random process. To do so consider a finite impulse response filter so that $Y[n] = \sum_{i=0}^{I-1} h[i]X[n-i]$ with $I = 4$ (the infinite impulse response filter argument is a bit more complicated but is similar in nature) and choose to test the set of output samples $n_1 = 0, n_2 = 1, n_3 = 2$ so that $K = 3$ (again the more general case proceeds similarly). Now prove that the output samples have a 3-dimensional Gaussian PDF. Hint: Show that the samples of $Y[n]$ are obtained as a linear transformation of $X[n]$.

20.8 (w) A discrete-time WGN random process is input to an LSI filter with system function $\mathcal{H}(z) = z - z^{-1}$. Determine the PDF of the output samples $Y[n]$ for $n = 0, 1, \ldots, N-1$. Are any of these samples independent of each other?

20.9 (t) In this problem we prove that if $X[n]$ and $Y[n]$ are both Gaussian random processes that are independent of each other, then $Z[n] = X[n] + Y[n]$ is also a Gaussian random process. To do so we prove that the characteristic function of $\mathbf{Z} = [Z[n_1]\, Z[n_2] \ldots Z[n_K]]^T$ is that of a Gaussian random vector. First note that since $\mathbf{X} = [X[n_1]\, X[n_2] \ldots X[n_K]]^T$ and $\mathbf{Y} = [Y[n_1]\, Y[n_2] \ldots Y[n_K]]^T$ are both Gaussian random vectors (by definition of a Gaussian random process), then each one has the characteristic function

$$\phi(\boldsymbol{\omega}) = \exp\left(j\boldsymbol{\omega}^T\boldsymbol{\mu} - \tfrac{1}{2}\boldsymbol{\omega}^T\mathbf{C}\boldsymbol{\omega}\right)$$

where $\boldsymbol{\omega} = [\omega_1\, \omega_2 \ldots \omega_K]^T$. Next use the property that the characteristic function of a sum of independent random vectors is the product of the characteristic functions to show that $\mathbf{Z}$ has a K-dimensional Gaussian PDF.

20.10 (⌣) (w) Let $X[n]$ and $Y[n]$ be WSS Gaussian random processes with zero mean and independent of each other. It is known that $Z[n] = X[n]Y[n]$ is not a Gaussian random process. However, can we say that $Z[n]$ is a WSS random process, and if so, what is its mean and PSD?

20.11 (w) An AR random process is described by $X[n] = \tfrac{1}{2}X[n-1] + U[n]$, where $U[n]$ is WGN with $\sigma_U^2 = 1$. This random process is input to an LSI filter with system function $\mathcal{H}(z) = 1 - \tfrac{1}{2}z^{-1}$ to generate the output random process $Y[n]$. Find $P[Y^2[0] + Y^2[1] > 1]$. Hint: Consider $X[n]$ as the output of an LSI filter.

20.12 (t) We prove (20.11) in this problem by using the method of characteristic functions. Recall that for a multivariate zero mean Gaussian PDF the characteristic function is

$$\phi_{\mathbf{X}}(\boldsymbol{\omega}) = \exp\left(-\tfrac{1}{2}\boldsymbol{\omega}^T \mathbf{C}\boldsymbol{\omega}\right)$$

and the fourth-order moment can be found using (see Section 14.6)

$$E[X_1 X_2 X_3 X_4] = \left.\frac{\partial^4 \phi_{\mathbf{X}}(\boldsymbol{\omega})}{\partial \omega_1 \partial \omega_2 \partial \omega_3 \partial \omega_4}\right|_{\boldsymbol{\omega}=0}.$$

Although straightforward, the algebra is tedious (see also Example 14.5 for the second-order moment calculations). To avoid frustration (with $P[\text{frustration}] = 1$) note that

$$\boldsymbol{\omega}^T \mathbf{C}\boldsymbol{\omega} = \sum_{i=1}^{4}\sum_{j=1}^{4} \omega_i \omega_j E[X_i X_j]$$

and let $L_i = \sum_{j=1}^{4} \omega_j E[X_i X_j]$. Next show that

$$\begin{aligned}\frac{\partial \phi_{\mathbf{X}}(\boldsymbol{\omega})}{\partial \omega_k} &= -\phi_{\mathbf{X}}(\boldsymbol{\omega}) L_k \\ \frac{\partial L_i}{\partial \omega_k} &= E[X_i X_k]\end{aligned}$$

and finally note that $L_i|_{\boldsymbol{\omega}=0} = 0$ to avoid some algebra in the last differentiation.

20.13 (w) It is desired to estimate $r_X[0]$ for $X[n]$ being WGN. If we use the estimator, $\hat{r}_X[0] = (1/N)\sum_{n=0}^{N-1} X^2[n]$, determine the mean and variance of $\hat{r}_X[0]$. Hint: Use (20.13).

20.14 (⌣) (f) If $X[n] = U[n] + U[n-1]$, where $U[n]$ is a WGN random process with $\sigma_U^2 = 1$, find $E[X[0]X[1]X[2]X[3]]$.

20.15 (f) Find the PSD of $X^2[n]$ if $X[n]$ is WGN with $\sigma_X^2 = 2$.

20.16 (t) To argue that the continuous-time Wiener random process is a Gaussian random process, we replace $X(t) = \int_0^t U(\xi)d\xi$, where $U(\xi)$ is continuous-time WGN, by the approximation

$$\bar{X}(t) = \sum_{n=0}^{[t/\Delta t]} Z(n\Delta t)\Delta t$$

where $[x]$ indicates the largest integer less than or equal to x and $Z(t)$ is a zero mean WSS Gaussian random process. The PSD of $Z(t)$ is given by

$$P_Z(F) = \begin{cases} \frac{N_0}{2} & |F| \leq W \\ 0 & |F| > W \end{cases}$$

where $W = 1/(2\Delta t)$. Explain why $\bar{X}(t)$ is a Gaussian random process. Next let $\Delta t \to 0$ and explain why $\bar{X}(t)$ becomes a Wiener random process.

20.17 ($\smile$) **(w)** To extract A from a realization of the random process $X(t) = A + U(t)$, where $U(t)$ is WGN with PSD $P_U(F) = 1$ for all F, it is proposed to use

$$\hat{A} = \frac{1}{T} \int_0^T X(\xi)d\xi.$$

How large should T be chosen to ensure that $P[|\hat{A} - A| \leq 0.01] = 0.99$?

20.18 (w) To generate a realization of a continuous-time Wiener random process on a computer we must replace the continuous-time random process by a sampled approximation. To do so note that we can first describe the Wiener random process by breaking up the integral into integrals over smaller time intervals. This yields

$$
\begin{aligned}
X(t) &= \int_0^t U(\xi)d\xi \\
&= \sum_{i=1}^n \underbrace{\int_{t_{i-1}}^{t_i} U(\xi)d\xi}_{X_i}
\end{aligned}
$$

where $t_i = i\Delta t$ with Δt very small, and $t_n = n\Delta t = t$. It is assumed that $t/\Delta t$ is an integer. Thus, the samples of $X(t)$ are conveniently found as

$$X(t_n) = X(n\Delta t) = \sum_{i=1}^n X_i$$

and the approximation is completed by connecting the samples $X(t_n)$ by straight lines. Find the PDF of the X_i's to determine how they should be generated. Hint: The X_i's are increments of $X(t)$.

20.19 ($\smile$) **(f)** For a continuous-time Wiener random process with $\text{var}(X(t)) = t$, determine $P[|X(t)| > 1]$. Explain what happens as $t \to \infty$ and why.

20.20 (w) Show that if $X(t)$ is a Rayleigh fading sinusoid, the "demodulation" and lowpass filtering shown in Figure 20.15 will yield U and V, respectively. What should the bandwidth of the lowpass filter be?

20.21 (c) Generate 10 realizations of a Rayleigh fading sinusoid for $0 \leq t \leq 1$. Use $F_0 = 10$ Hz and $\sigma^2 = 1$ to do so. Overlay your realizations. Hint: Replace $X(t) = U\cos(2\pi F_0 t) - V\sin(2\pi F_0 t)$ by $X[n] = X(n\Delta t) = U\cos(2\pi F_0 n\Delta t) - V\sin(2\pi F_0 n\Delta t)$ for $n = 0, 1, \ldots, N\Delta t$, where $\Delta t = 1/N$ and N is large.

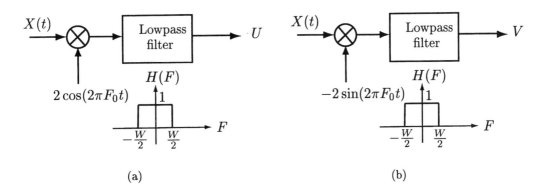

(a)   (b)

Figure 20.15: Extraction of Rayleigh fading sinusoid lowpass components for Problem 20.20.

20.22 (☺) (w) Consider $X_1(t)$ and $X_2(t)$, which are both Rayleigh fading sinusoids with frequency $F_0 = 1/2$ and which are independent of each other. Each random process has the total average power $\sigma^2 = 1$. If $Y(t) = X_1(t) + X_2(t)$, find the joint PDF of $Y(0)$ and $Y(1/4)$.

20.23 (f) A Rayleigh fading sinusoid has the PSD $P_X(F) = \delta(F + 10) + \delta(F - 10)$. Find the PSDs of $U(t)$ and $V(t)$ and plot them.

20.24 (w) Show that if $X(t)$ is a bandpass random process, the "demodulation" and lowpass filtering given in Figure 20.16 will yield $U(t)$ and $V(t)$, respectively.

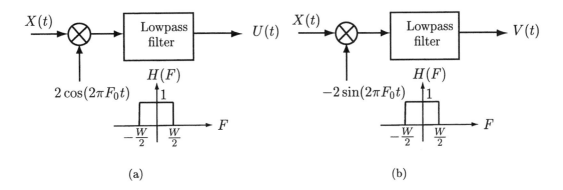

(a)   (b)

Figure 20.16: Extraction of bandpass random process lowpass components for Problem 20.24.

20.25 (☺) (f) If a bandpass random process has the PSD shown in Figure 20.17, find the PSD of $U(t)$ and $V(t)$.

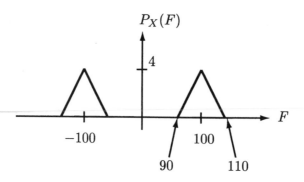

Figure 20.17: PSD for bandpass random process for Problem 20.25.

20.26 (c) The random process whose realization is shown in Figure 20.2 appears
to be similar in nature to the bandpass random processes shown in Figure
20.13b. We have already seen that the marginal PDF appears to be Gaussian
(see Figure 20.3). To see if it is reasonable to model it as a bandpass random
process we estimate the PSD. First run the code given in Appendix 20A to
produce the realization shown in Figure 20.2. Then, run the code given below
to estimate the PSD using an averaged periodogram (see also Section 17.7
for a description of this). Does the estimated PSD indicate that the random
process is a bandpass random process? If so, explain how you can give a
complete probabilistic model for this random process.

```
Fs=100;  % set sampling rate for later plotting
L=50;I=20; % L = length of block, I = number of blocks
n=[0:I*L-1]'; % set up time indices
Nfft=1024; % set FFT length for Fourier transform
Pav=zeros(Nfft,1);
f=[0:Nfft-1]'/Nfft-0.5; % set discrete-time frequencies
for i=0:I-1
    nstart=1+i*L;nend=L+i*L; % set start and end time indices
                             % of block
    y=x(nstart:nend); % extract block of data
Pav=Pav+(1/(I*L))*abs(fftshift(fft(y,Nfft))).^2;
                             % compute periodogram
                             % and add to average
                             % of periodograms
end
F=f*Fs; % convert to continuous-time (analog) frequency in Hz
Pest=Pav/Fs; % convert discrete-time PSD to continuous-time PSD
plot(F,Pest)
```

20.27 (f) For the Gaussian random process with mean zero and PSD

$$P_X(F) = \begin{cases} 4 & 90 \leq |F| \leq 110 \\ 0 & \text{otherwise} \end{cases}$$

find the probability that its envelope will be less than or equal to 10 at $t = 10$ seconds. Repeat the calculation if $t = 20$ seconds.

20.28 (w) Prove that $X_1[n] = \frac{1}{2}U[n+1]+\frac{1}{2}U[n-1]$ and $X_2[n] = \frac{1}{2}U[n]+\frac{1}{2}U[n-2]$, where $U[n]$ is WGN with $\sigma_U^2 = 1$, both have the same PSD given by $P_X(F) = \frac{1}{2}(1 + \cos(4\pi f))$.

20.29 (w) It is desired to generate a realization of a WSS Gaussian random process by filtering WGN with an LSI filter. If the desired PSD is $P_X(f) = |1 - \frac{1}{2}\exp(-j2\pi f)|^2$, explain how to do this.

20.30 (☼) (w) It is desired to generate a realization of a WSS Gaussian random process by filtering WGN with an LSI filter. If the desired PSD is $P_X(f) = 2 - 2\cos(2\pi f)$, explain how to do this.

20.31 (☼) (c) Using the results of Problem 20.30, generate a realization of $X[n]$. To verify that your data generation appears correct, estimate the ACS for $k = 0, 1, \ldots, 9$ and compare it to the theoretical ACS.

Appendix 20A

MATLAB Listing for Figure 20.2

```
clear all
rand('state',0)
t=[0:0.01:0.99]'; %  set up transmit pulse time interval
F0=10;
s=cos(2*pi*F0*t); %  transmit pulse
ss=[s;zeros(1000-length(s),1)];  % put transmit pulse in receive window
tt=[0:0.01:9.99]';  % set up receive window time interval
x=zeros(1000,1);
for i=1:100  % add up all echos, one for each 0.1 sec interval
tau=round(10*i+10*(rand(1,1)-0.5));  % time delay for each 0.1 sec interval
                                     % is uniformly distributed - round
                                     % time delay to integer
  x=x+rand(1,1)*shift(ss,tau);
end
```

shift.m subprogram

```
% shift.m
%
function y=shift(x,Ns)
%
%  This function subprogram shifts the given sequence by Ns points.
%  Zeros are shifted in either from the left or right.
%
%  Input parameters:
%    x  - array of dimension Lx1
%    Ns - integer number of shifts where Ns>0 means a shift to the
```

```
%                right and Ns<0 means a shift to the left and if Ns=0, then
%                the sequence is not shifted
%
%     Output parameters:
%       y   - array of dimension Lx1 containing the
%                    shifted sequence
L=length(x);
if abs(Ns)>L
   y=zeros(L,1);
   else
if Ns>0
y(1:Ns,1)=0;
y(Ns+1:L,1)=x(1:L-Ns);
elseif Ns<0
   y(L-abs(Ns)+1:L,1)=0;
   y(1:L-abs(Ns),1)=x(abs(Ns)+1:L);
else
   y=x;
end
end
```

Chapter 21

Poisson Random Processes

21.1 Introduction

A random process that is useful for modeling events occurring in time is the *Poisson random process*. A typical realization is shown in Figure 21.1 in which the events, indicated by the "x"s, occur randomly in time. The random process, whose real-

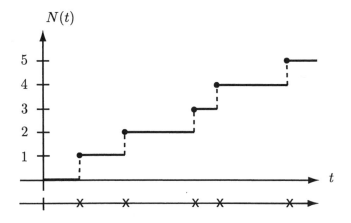

Figure 21.1: Poisson process events and the Poisson counting random process $N(t)$.

ization is a set of times, is called the *Poisson random process*. The random process that counts the number of events in the time interval $[0, t]$, and which is denoted by $N(t)$, is called the *Poisson counting random process*. It is clear from Figure 21.1 that the two random processes are equivalent descriptions of the same random phenomenon. Note that $N(t)$ is a continuous-time/discrete-valued (CTDV) random process. Also, because $N(t)$ counts the number of events from the initial time $t = 0$ up to and *including* the time t, the value of $N(t)$ at a jump is $N(t^+)$. Thus, $N(t)$ is *right-continuous* (the same property as for the CDF of a discrete random variable). The motivation for the widespread use of the Poisson random process is its ability to model a wide range of physical and man-made random phenomena. Some of these are the distribution in time of radioactive counts, the arrivals of customers at a cashier, requests for service in computer networks, and calls made to a central location, to name just a few. In Chapter 5 we gave an example of the application of the Poisson PMF to the servicing of customers at a supermarket checkout. Here we examine the characteristics of a Poisson random process in more detail, paying particular attention not only to the probability of a given number of events in a time interval but also to the probability for the arrival times of those events. In order to avoid confusing the probabilistic notion of an event with the common usage, we will refer to the events shown in Figure 21.1 as *arrivals*.

The Poisson random process is a natural extension of a sequence of independent and identically distributed Bernoulli trials (see Example 16.1). The Poisson counting random process $N(t)$ then becomes the extension of the binomial counting random process discussed in Example 16.5. To make this identification, consider a Bernoulli random process, which is defined as a sequence of IID Bernoulli trials, with $U[n] = 1$ with probability p and $U[n] = 0$ with probability $1 - p$. Now envision a Bernoulli trial for each small time slot of width Δt in the interval $[0, t]$ as shown in Figure 21.2. Thus, we will observe either a 1 with probability p or a 0 with probability

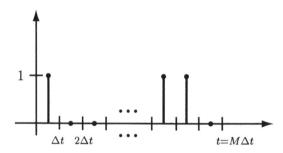

Figure 21.2: IID Bernoulli random process with one trial per time slot.

$1 - p$ for each of the $M = t/\Delta t$ time slots. Recall that on the average we will observe Mp ones. Now if $\Delta t \to 0$ and $M \to \infty$ with $t = M\Delta t$ held constant, we will obtain the Poisson random process as the limiting form of the Bernoulli random process. Also, recall that the number of ones in M IID Bernoulli trials is a binomial random variable. Hence, it seems reasonable that the number of arrivals in a Poisson random process should be a Poisson random variable in accordance with our results in Section 5.6. We next argue that this is indeed the case. For the binomial counting random process, thought of as one trial per time slot, we have that the number of ones in the interval $[0, t]$ has the PMF

$$P[N(t) = k] = \binom{M}{k} p^k (1 - p)^{M-k} \qquad k = 0, 1, \ldots, M.$$

But as $M \to \infty$ and $p \to 0$ with $E[N(t)] = Mp$ being fixed, the binomial PMF becomes the Poisson PMF or $N(t) \sim \text{Pois}(\lambda')$, where $\lambda' = E[N(t)] = Mp$. (Note that as the number of time slots M increases, we need to let $p \to 0$ in order to maintain an average number of arrivals in $[0, t]$.) Thus, replacing λ' by $E[N(t)]$, we write the Poisson PMF as

$$P[N(t) = k] = \exp(-E[N(t)])\frac{E^k[N(t)]}{k!} \qquad k = 0, 1, \ldots . \tag{21.1}$$

To determine $E[N(t)]$ for use in (21.1), where t may be arbitrary, we examine Mp in the limit. Thus,

$$
\begin{aligned}
E[N(t)] &= \lim_{\substack{M \to \infty \\ p \to 0}} Mp \\
&= \lim_{\substack{\Delta t \to 0 \\ p \to 0}} \frac{t}{\Delta t}p = t \lim_{\substack{\Delta t \to 0 \\ p \to 0}} \frac{p}{\Delta t} \\
&= \lambda t
\end{aligned}
$$

where we define λ as the limit of $p/\Delta t$. Since $\lambda = E[N(t)]/t$, we can interpret λ as the *average number of arrivals per second* or the *rate of the Poisson random process*. This is a parameter that is easily specified in practice. Using this definition we have that

$$P[N(t) = k] = \exp(-\lambda t)\frac{(\lambda t)^k}{k!} \qquad k = 0, 1, \ldots . \tag{21.2}$$

As mentioned previously, $N(t)$ is the Poisson counting random process and the probability of k arrivals from $t = 0$ up to and including t is given by (21.2). It is a semi-infinite random process with $N(0) = 0$ by definition.

It is possible to derive all the properties of a Poisson counting random process by employing the previous device of viewing it as the limiting form of a binomial counting random process as $\Delta t \to 0$. However, it is cumbersome to do so and therefore, we present an alternative derivation that is consistent with the same basic assumptions. One advantage of viewing the Poisson random process as a limiting form is that many of its properties become more obvious by consideration of a sequence of IID Bernoulli trials. These properties are inherited from the binomial, such as, for example, the increments $N(t_2) - N(t_1)$ must be independent. (Can you explain why this must be true for the binomial counting random process?)

21.2 Summary

The Poisson counting random process is introduced in Section 21.1. The probability of k arrivals in the time interval $[0, t]$ is given by (21.2). This probability is also derived in Section 21.3 based on a set of axioms that the Poisson random process should adhere to. Some examples of typical problems for which this probability is useful are also described in that section. The times between arrivals or interarrival times is shown in Section 21.4 to be independent and exponentially distributed as given by (21.6). The arrival times of a Poisson random process are described by an Erlang PDF given in (21.8). An extension of the Poisson random process that is useful is the compound Poisson random process described in Section 21.6. Moments of the random process can be found from the characteristic function of (21.12). In particular, the mean is given by (21.13). A Poisson random process is easily simulated on a computer using the MATLAB code listed in Section 21.7. Finally, an application of the compound Poisson random process to automobile traffic signal planning is the subject of Section 21.8.

21.3   Derivation of Poisson Counting Random Process

We next derive the Poisson counting random process by appealing to a set of axioms that are consistent with our previous assumptions. Clearly, since the random process starts at $t = 0$, we assume that $N(0) = 0$. Next, since the binomial counting random process has increments that are independent and stationary (Bernoulli trials are IID), we assume the same for the Poisson counting random process. Thus, for two increments we assume that the random variables $I_1 = N(t_2) - N(t_1)$ and $I_2 = N(t_4) - N(t_3)$ are independent if $t_4 > t_3 > t_2 > t_1$ and also have the same PDF if additionally $t_4 - t_3 = t_2 - t_1$. Likewise, we assume this is true for *all* possible sets of increments. Note that $t_4 > t_3 > t_2 > t_1$ corresponds to *nonoverlapping* time intervals. The increments will still be independent if $t_2 = t_3$ or the time intervals have a single point in common since the probability of $N(t)$ changing at a point is zero as we will see shortly. As for the Bernoulli random process, there can be at most one arrival in each time slot. Similarly, for the Poisson counting random process we allow at most one arrival for each time slot so that

$$P[N(t + \Delta t) - N(t) = k] = \begin{cases} 1 - p & k = 0 \\ p & k = 1 \end{cases}$$

and recall that

$$\lim_{\substack{\Delta t \to 0 \\ p \to 0}} \frac{p}{\Delta t} = \lambda$$

so that for Δt small, $p = \lambda \Delta t$ and

$$P[N(t + \Delta t) - N(t) = k] = \begin{cases} 1 - \lambda \Delta t & k = 0 \\ \lambda \Delta t & k = 1 \\ 0 & k \geq 2. \end{cases}$$

Therefore, our axioms become

Axiom 1 $N(0) = 0$.

Axiom 2 $N(t)$ has independent and stationary increments.

Axiom 3 $P[N(t + \Delta t) - N(t) = k] = \begin{cases} 1 - \lambda\Delta t & k = 0 \\ \lambda\Delta t & k = 1 \end{cases}$

for all t.

With these axioms we wish to prove that (21.2) follows. The derivation is indicative of an approach commonly used for analyzing *continuous-time Markov random processes* [Cox and Miller 1965] and so is of interest in its own right.

21.3.1   Derivation

To begin, consider the determination of $P[N(t) = 0]$ for an arbitrary $t > 0$. Then referring to Figure 21.3a we see that for no arrivals in $[0, t]$, there must be no arrivals in $[0, t - \Delta t]$ and also no arrivals in $(t - \Delta t, t]$. Therefore,

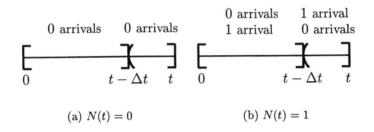

Figure 21.3: Possible number of arrivals in indicated time intervals.

$$\begin{aligned} P[N(t) = 0] &= P[N(t - \Delta t) = 0, N(t) - N(t - \Delta t) = 0] \\ &= P[N(t - \Delta t) = 0]P[N(t) - N(t - \Delta t) = 0] \quad \text{(Axiom 2 - independence)} \\ &= P[N(t - \Delta t) = 0]P[N(t + \Delta t) - N(t) = 0] \quad \text{(Axiom 2 - stationarity)} \\ &= P[N(t - \Delta t) = 0](1 - \lambda\Delta t) \quad \text{(Axiom 3).} \end{aligned}$$

If we let $P_0(t) = P[N(t) = 0]$, then

$$P_0(t) = P_0(t - \Delta t)(1 - \lambda\Delta t)$$

or

$$\frac{P_0(t) - P_0(t - \Delta t)}{\Delta t} = -\lambda P_0(t - \Delta t).$$

Now letting $\Delta t \to 0$, we arrive at the linear differential equation

$$\frac{dP_0(t)}{dt} = -\lambda P_0(t)$$

for which the solution is $P_0(t) = c\exp(-\lambda t)$, where c is an arbitrary constant. To evaluate the constant we invoke the initial condition that $P_0(0) = P[N(0) = 0] = 1$ by Axiom 1 to yield $c = 1$. Thus, we have finally that

$$P[N(t) = 0] = P_0(t) = \exp(-\lambda t).$$

Next we use the same argument to find a differential equation for $P_1(t) = P[N(t) = 1]$ by referring to Figure 21.3b. We can either have no arrivals in $[0, t - \Delta t]$ and one arrival in $(t - \Delta t, t]$ *or* one arrival in $[0, t - \Delta t]$ and no arrivals in $(t - \Delta t, t]$. These are the only possibilities since there can be at most one arrival in a time interval of length Δt. The two events are mutually exclusive so that

$$
\begin{aligned}
P[N(t) = 1] &= P[N(t - \Delta t) = 0, N(t) - N(t - \Delta t) = 1] \\
&\quad + P[N(t - \Delta t) = 1, N(t) - N(t - \Delta t) = 0] \\
&= P[N(t - \Delta t) = 0]P[N(t) - N(t - \Delta t) = 1] \\
&\quad + P[N(t - \Delta t) = 1]P[N(t) - N(t - \Delta t) = 0] \quad \text{(independence)} \\
&= P[N(t - \Delta t) = 0]P[N(t + \Delta t) - N(t) = 1] \\
&\quad + P[N(t - \Delta t) = 1]P[N(t + \Delta t) - N(t) = 0]. \quad \text{(stationarity)}
\end{aligned}
$$

Using the definition of $P_1(t)$ and Axiom 3,

$$
P_1(t) = P_0(t - \Delta t)\lambda \Delta t + P_1(t - \Delta t)(1 - \lambda \Delta t)
$$

or

$$
\frac{P_1(t) - P_1(t - \Delta t)}{\Delta t} = -\lambda P_1(t - \Delta t) + \lambda P_0(t - \Delta t)
$$

and as $\Delta t \to 0$, we have the differential equation

$$
\frac{dP_1(t)}{dt} + \lambda P_1(t) = \lambda P_0(t).
$$

In like fashion we can show (see Problem 21.1) that if $P_k(t) = P[N(t) = k]$, then

$$
\frac{dP_k(t)}{dt} + \lambda P_k(t) = \lambda P_{k-1}(t) \quad k = 1, 2, \ldots \tag{21.3}
$$

where we know that $P_0(t) = \exp(-\lambda t)$. This is a set of simultaneous linear differential equations that fortunately can be solved recursively. Since $P_0(t)$ is known, we can solve for $P_1(t)$. Once $P_1(t)$ has been found, then $P_2(t)$ can be solved for, etc. It is shown in Problem 21.2 that by using Laplace transforms, we can easily solve these equations. The result is

$$
P_k(t) = \exp(-\lambda t)\frac{(\lambda t)^k}{k!} \quad k = 0, 1, \ldots
$$

so that finally we have the desired result

$$
P[N(t) = k] = \exp(-\lambda t)\frac{(\lambda t)^k}{k!} \quad k = 0, 1, \ldots \tag{21.4}
$$

which is the usual Poisson PMF. The only difference from that described in Section 5.5.4 is that here λ represents an *arrival rate*. Since if $X \sim \text{Pois}(\lambda')$, then $E[X] = \lambda'$, we have $\lambda' = \lambda t$. Hence, $\lambda = \lambda'/t = E[N(t)]/t$, which is seen to be the *average number of arrivals per second*.

21.3.2 Some Examples

Before proceeding with some examples it should be pointed out that the Poisson counting random process is *not* stationary or even WSS. This is evident from the PMF of $N(t)$ since $E[N(t_2)] = \lambda t_2 \neq \lambda t_1 = E[N(t_1)]$ for $t_2 \neq t_1$. As its properties are inherited from the binomial counting random process, it exhibits the properties of a sum random process (see Section 16.4). Also, in determining probabilities of events, the fact that the *increments* are independent and stationary will greatly simplify our calculations.

Example 21.1 − Customer arrivals

Customers arrive at a checkout lane at the rate of 0.1 customers per second according to a Poisson random process. Determine the probability that 5 customers will arrive during the first minute the lane is open and also 5 customers will arrive the second minute it is open. During the time interval $[0, 60]$ the probability of 5 arrivals is from (21.4)

$$P[N(60) = 5] = \exp[-0.1(60)]\frac{[(0.1)(60)]^5}{5!} = 0.1606.$$

This will also be the probability of 5 customers arriving during the second minute interval or *for any one minute interval* $[t, t + 60]$ since

$$
\begin{aligned}
P[N(t+60) - N(t) = 5] &= P[N(60) - N(0) = 5] \quad &\text{(increment stationarity)}\\
&= P[N(60) = 5] \quad &(N(0) = 0)
\end{aligned}
$$

which is not dependent on t. Hence, the probability of 5 customers arriving in the first minute and 5 more arriving in the second minute is

$$P[N(60) - N(0) = 5, N(120) - N(60) = 5]$$

$$
\begin{aligned}
&= P[N(60) - N(0) = 5]P[N(120) - N(60) = 5] \quad &\text{(increment independence)}\\
&= P[N(60) - N(0) = 5]P[N(60) - N(0) = 5] \quad &\text{(increment stationarity)}\\
&= P^2[N(60) = 5] = 0.0258 \quad &(N(0) = 0)
\end{aligned}
$$

$\Diamond$

Example 21.2 − Traffic bursts

Consider the arrival of cars at an intersection. It is known that for any 5 minute interval 50 cars arrive on the average. For any 5 minute interval what is the probability of 20 cars in the first minute and 30 cars in the next 4 minutes? Since the probabilities of the increments do not change with the time origin due to stationarity, we can assume that the 5 minute interval in question starts at $t = 0$ and ends at $t = 300$ seconds. Thus, we wish to determine the probability of a traffic burst P_B, which is

$$P_B = P[N(60) = 20, N(300) - N(60) = 30].$$

Since the increments are independent, we have

$$P_B = P[N(60) = 20]P[N(300) - N(60) = 30]$$

and because they are also stationary

$$
\begin{aligned}
P_B &= P[N(60) = 20]P[N(240) - N(0) = 30] \\
&= P[N(60) = 20]P[N(240) = 30] \\
&= \exp(-60\lambda)\frac{(60\lambda)^{20}}{20!} \exp(-240\lambda)\frac{(240\lambda)^{30}}{30!}.
\end{aligned}
$$

Finally, since the arrival rate is given by $\lambda = 50/300 = 1/6$, the probability of a traffic burst is

$$P_B = \exp(-10)\frac{(10)^{20}}{20!} \exp(-40)\frac{(40)^{30}}{30!} = 3.4458 \times 10^{-5}.$$

$\Diamond$

In many applications it is important to assess not only the probability of a number of arrivals within a given time interval but also the distribution of these arrival times. Are they evenly spaced or can they bunch up as in the last example? In the next section we answer these questions.

21.4   Interarrival Times

Consider a typical realization of a Poisson random process as shown in Figure 21.4. The times $t_1, t_2, t_3, \ldots$ are called the *arrival times* while the time intervals

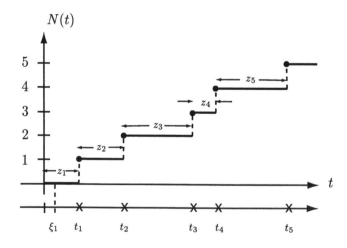

Figure 21.4: Definition of arrival times t_i's and interarrival times z_i's.

$z_1, z_2, z_3, \ldots$ are called the *interarrival times*. The interarrival times shown in Figure 21.4 are realizations of the random variables $Z_1, Z_2, Z_3, \ldots$. We wish to be able to compute probabilities for a finite set, say $Z_1, Z_2, \ldots, Z_K$. Since $N(t)$ is a continuous-time random process, the time between arrivals is also continuous and so a joint PDF is sought. To begin we first determine $p_{Z_1}(z_1)$. Note that $Z_1 = T_1$, where T_1 is the random variable denoting the first arrival. By the definition of the first arrival if $Z_1 > \xi_1$, then $N(\xi_1) = 0$ as shown in Figure 21.4. Conversely, if $N(\xi_1) = 0$, then the first arrival has not occurred as of time ξ_1 and so $Z_1 > \xi_1$. This argument shows that the events $\{Z_1 > \xi_1\}$ and $\{N(\xi_1) = 0\}$ are equivalent and therefore

$$
\begin{aligned}
P[Z_1 > \xi_1] &= P[N(\xi_1) = 0] \\
&= \exp(-\lambda\xi_1) \qquad \xi_1 \geq 0
\end{aligned}
\tag{21.5}
$$

where we have used (21.4). As a result, the PDF is for $z_1 \geq 0$

$$
\begin{aligned}
p_{Z_1}(z_1) &= \frac{d}{dz_1} F_{Z_1}(z_1) \\
&= \frac{d}{dz_1}(1 - P[Z_1 > z_1]) \\
&= \frac{d}{dz_1}[1 - \exp(-\lambda z_1)] \\
&= \lambda \exp(-\lambda z_1)
\end{aligned}
$$

and finally the PDF of the first arrival is

$$
p_{Z_1}(z_1) = \begin{cases} \lambda \exp(-\lambda z_1) & z_1 \geq 0 \\ 0 & z_1 < 0 \end{cases}
\tag{21.6}
$$

or $Z_1 \sim \exp(\lambda)$. An example follows.

Example 21.3 – Waiting for an arrival

Assume that at $t = 0$ we start to wait for an arrival. Then we know from (21.6) that the time we will have to wait is a random variable with $Z_1 \sim \exp(\lambda)$. On the average we will have to wait $E[Z_1] = 1/\lambda$ seconds. This is reasonable in that λ is average arrivals per second and therefore $1/\lambda$ is seconds per arrival. However, say we have already waited ξ_1 seconds—what is the probability that we will have to wait more than an additional ξ_2 seconds? In probabilistic terms we wish to compute the conditional probability $P[Z_1 > \xi_1 + \xi_2 | Z_1 > \xi_1]$. This is found as follows.

$$
\begin{aligned}
P[Z_1 > \xi_1 + \xi_2 | Z_1 > \xi_1] &= \frac{P[Z_1 > \xi_1 + \xi_2, Z_1 > \xi_1]}{P[Z_1 > \xi_1]} \\
&= \frac{P[Z_1 > \xi_1 + \xi_2]}{P[Z_1 > \xi_1]}
\end{aligned}
$$

since the arrival time will be greater than both $\xi_1 + \xi_2$ and ξ_1 only if it is greater than the former. Now using (21.5) we have that

$$P[Z_1 > \xi_1 + \xi_2 | Z_1 > \xi_1] = \frac{\exp[-\lambda(\xi_1 + \xi_2)]}{\exp(-\lambda\xi_1)}$$
$$= \exp(-\lambda\xi_2)$$
$$= P[Z_1 > \xi_2]. \qquad (21.7)$$

Hence, the conditional probability that we will have to wait more than an additional ξ_2 seconds given that we have already waited ξ_1 seconds is just the probability that we will have to wait more than ξ_2 seconds. The fact that we have already waited does not in any way affect the probability of the first arrival. Once we have waited and observed that no arrival has occured up to time ξ_1, then the random process in essence starts over as if it were at time $t = 0$. This property of the Poisson random process is referred to as the *memoryless property*. It is somewhat disconcerting to know that the chances your bus will arrive in the next 5 minutes, given that it is already 5 minutes late, is not any better than your chances it will be late by 5 minutes. However, this conclusion is consistent with the Poisson random process model. It is also evident by examining the similar result of waiting for a fair coin to comes up heads given that it has already exhibited 10 tails in a row. In Problem 21.12 an alternative derivation of the memoryless property is given which makes use of the geometric random variable.

$\diamondsuit$

We next give the joint PDF for two or more interarrival times. It is shown in Appendix 21A that the interarrival times $Z_1, Z_2, \ldots, Z_K$ are IID random variables with each one having $Z_1 \sim \exp(\lambda)$. This result may also be reconciled in light of the Poisson random process being the limiting form of a Bernoulli random process. Consider a Bernoulli random process $\{X[0] = 0, X[1], X[2], \ldots\}$, where $X[0] = 0$ by definition, and assume *interarrival* times of k_1 and k_2, where $k_1 \geq 1$, $k_2 \geq 1$. For example, if $X[1] = 0, X[2] = 1, X[3] = 0, X[4] = 0$, and $X[5] = 1$, then we would have $k_1 = 2$ and $k_2 = 3$. In general,

$$P[\text{first interarrival time} = k_1, \text{second interarrival time} = k_2]$$
$$= P[X[n] = 0 \text{ for } 1 \leq n \leq k_1 - 1, X[k_1] = 1, X[n] = 0$$
$$\text{for } k_1 + 1 \leq n \leq k_1 + k_2 - 1, X[k_1 + k_2] = 1]$$
$$= [(1-p)^{k_1-1}p][(1-p)^{k_2-1}p].$$

Hence, the joint PMF factors so that the interarrival times are independent and furthermore they are identically distributed (let $k_1 = k_2$). An example follows.

Example 21.4 – Expected time for calls

A customer call service center opens at 9 A.M. The calls received follow a Poisson random process at the average rate of 600 calls per hour. The 20th call comes in at 9:01 A.M. At what time can we expect the next call to come in? Let Z_{21} be the elapsed time from 9:01 A.M. until the next call comes in. Since the interarrival times are independent, they do not depend upon the past history of arrivals. Hence, $Z_{21} = T_{21} - T_{20} \sim \exp(\lambda)$. Since the mean of an exponential random variable Z is just $1/\lambda$ and from the information given $\lambda = 600/3600 = 1/6$ calls per second, we have that $E[Z_{21}] = 1/(1/6) = 6$ seconds. Hence, we can expect the next call to come in at 9:01:06 A.M.

$\diamondsuit$

21.5 Arrival Times

The kth arrival time T_k is defined as the time from $t = 0$ until the kth arrival occurs. The arrival times are illustrated in Figure 21.4, where T_k is also referred to as the *waiting time* until the kth arrival. In this section we will determine the PDF of T_k. It is seen from Figure 21.4 that $t_k = \sum_{i=1}^{k} z_i$ so that the random variable of interest is

$$T_k = \sum_{i=1}^{k} Z_i.$$

But we saw in the last section that the Z_i's are IID with $Z_1 \sim \exp(\lambda)$. Hence, the PDF of T_k is obtained by determining the PDF for a sum of IID random variables. This is a problem that has been studied in Section 14.6, and is solved most readily by the use of the characteristic function. Recall that if $X_1, X_2, \ldots, X_k$ are IID random variables, then the characteristic function for $Y = \sum_{i=1}^{k} X_i$ is $\phi_Y(\omega) = \phi_X^k(\omega)$. Thus, the PDF for Y, assuming that Y is a continuous random variable, is found from the continuous-time inverse Fourier transform (defined to correspond to the Fourier transform used in the characteristic function definition, and uses a $-j$ and radian frequency ω) as

$$p_Y(y) = \int_{-\infty}^{\infty} \phi_X^k(\omega) \exp(-j\omega y) \frac{d\omega}{2\pi}.$$

From Table 11.1 we have that $\phi_{Z_1}(\omega) = \lambda/(\lambda - j\omega)$ and therefore

$$\phi_{T_k}(\omega) = \left(\frac{\lambda}{\lambda - j\omega} \right)^k = \left(\frac{1}{1 - j\omega/\lambda} \right)^k.$$

Again referring to Table 11.1, we see that this is the characteristic function of a Gamma random variable with $\alpha = k$ so that $T_k \sim \Gamma(k, \lambda)$. Specifically, this is the Erlang random variable described in Section 10.5.6 . Hence, we have that

$$p_{T_k}(t) = \frac{\lambda^k}{(k-1)!} t^{k-1} \exp(-\lambda t). \tag{21.8}$$

(See also Problem 21.15 for the derivation for $k = 2$ using a convolution integral and Problem 21.16 for an alternative derivation for the general case.) Note that for a

$\Gamma(\alpha, \lambda)$ random variable the mean is α/λ so that with $\alpha = k$, we have the expected time for the kth arrival as

$$E[T_k] = \frac{k}{\lambda} \tag{21.9}$$

or equivalently

$$E[T_k] = kE[T_1]. \tag{21.10}$$

On the average the time to the kth arrival is just k times the time to the first arrival, a somewhat pleasing result. An example follows.

Example 21.5 – Computer servers

A computer server is designed to provide downloaded software when requested. It can honor a total of 80 requests in each hour before it becomes overloaded. If the requests are made in accordance with a Poisson random process at an average rate of 60 requests per hour, what is the probability that it will be overloaded in the first hour? We need to determine the probability that the 81st request will occur at a time $t \leq 3600$ seconds. Thus, from (21.8) with $k = 81$

$$
\begin{aligned}
P[\text{overloaded in first hour}] &= P[T_{81} \leq 3600] \\
&= \int_0^{3600} \frac{\lambda^{81}}{80!} t^{80} \exp(-\lambda t) dt.
\end{aligned}
$$

Here the arrival rate of the requests is $\lambda = 60/3600 = 1/60$ per second and therefore

$$P[\text{overloaded in first hour}] = \frac{1}{60} \int_0^{3600} \frac{1}{80!} \left(\frac{t}{60}\right)^{80} \exp(-t/60) dt$$

Using the result

$$\int \frac{(at)^n}{n!} \exp(-at) dt = -\frac{\exp(-at)}{a} \sum_{i=0}^{n} \frac{(at)^i}{i!}$$

it follows that

$$
\begin{aligned}
P[\text{overloaded in first hour}] &= \frac{1}{60} \left[-\frac{\exp(-t/60)}{1/60} \sum_{i=0}^{80} \frac{(t/60)^i}{i!} \Big|_0^{3600} \right] \\
&= -\left[\exp(-60) \sum_{i=0}^{80} \frac{(60)^i}{i!} - 1 \right] \\
&= 1 - \exp(-60) \sum_{i=0}^{80} \frac{(60)^i}{i!} = 0.0056.
\end{aligned}
$$

$\diamond$

21.6 Compound Poisson Random Process

A Poisson counting random process increments its value by one for each new arrival. In some applications we may not know the increment in advance. An example would be to determine the average amount of all transactions within a bank for a given day. In this case the amount obtained is the sum of all deposits and withdrawals. To model these transactions we could assume that customers arrive at the bank according to a Poisson random process. If, for example, each customer deposited one dollar, then at the end of the day, say at time t_0, the total amount of the transactions $X(t_0)$ could be written as

$$X(t_0) = \sum_{i=1}^{N(t_0)} 1 = N(t_0).$$

This is the standard Poisson counting random process. If, however, there are withdrawals, then this would no longer hold. Furthermore, if the deposits and withdrawals are unknown to us before they are made, then we would need to model each one by a random variable, say U_i. The random variable would take on positive values for deposits and negative values for withdrawals and probabilities could be assigned to the possible values of U_i. The total dollar amount of the transactions at the end of the day would be

$$\sum_{i=1}^{N(t_0)} U_i.$$

With this motivation we will consider the more general case in which the U_i's are either discrete or continuous random variables, and denote the total at time t by the random process $X(t)$. This random process is therefore given by

$$X(t) = \sum_{i=1}^{N(t)} U_i \qquad t \geq 0. \tag{21.11}$$

It is a continuous-time random process but can be either continuous-valued or discrete-valued depending upon whether the U_i's are continuous or discrete random variables. We furthermore assume that the U_i's are IID random variables. Hence, $X(t)$ is similar to the usual sum of IID random variables except that the *number of terms in the sum is random* and the number of terms is distributed according to a Poisson random process. This random process is called a *compound Poisson random process*.

   In summary, we let $X(t) = \sum_{i=1}^{N(t)} U_i$ for $t \geq 0$, where the U_i's are IID random variables and $N(t)$ is a Poisson counting random process with arrival rate λ. Also, we define $X(0) = 0$, and furthermore assume that the U_i's and $N(t)$ are independent of each other for all t.

We next determine the marginal PMF or PDF of $X(t)$. To do so we will use characteristic functions in conjunction with conditioning arguments. The key to success here is to turn the sum with a *random* number of terms into one with a *fixed* number by conditioning. Then, the usual characteristic function approach described in Section 14.6 will be applicable. Hence, consider for a fixed $t = t_0$ the random variable $X(t_0)$ and write its characteristic function as

$$
\begin{aligned}
\phi_{X(t_0)}(\omega) &= E[\exp(j\omega X(t_0))] && \text{(definition)} \\[2mm]
&= E\left[\exp\left(j\omega \sum_{i=1}^{N(t_0)} U_i\right)\right] \\[2mm]
&= E_{N(t_0)}\left[E_{U_1,\dots,U_k|N(t_0)}\left[\exp\left(j\omega \sum_{i=1}^{k} U_i\right)\Bigg| N(t_0) = k\right]\right] \\
&&& \text{(see Problem 21.18)} \\[2mm]
&= E_{N(t_0)}\left[E_{U_1,\dots,U_k}\left[\exp\left(j\omega \sum_{i=1}^{k} U_i\right)\right]\right] && (U_i\text{'s independent of } N(t_0)) \\[2mm]
&= E_{N(t_0)}\left[E_{U_1,\dots,U_k}\left[\prod_{i=1}^{k}\exp(j\omega U_i)\right]\right] \\[2mm]
&= E_{N(t_0)}\left[\prod_{i=1}^{k} E_{U_i}\left[\exp(j\omega U_i)\right]\right] && (U_i\text{'s are independent}) \\[2mm]
&= E_{N(t_0)}\left[\prod_{i=1}^{k}\phi_{U_i}(\omega)\right] && \text{(definition of char. function)} \\[2mm]
&= E_{N(t_0)}\left[\phi_{U_1}^{k}(\omega)\right] && (U_i\text{'s identically dist.}) \\[2mm]
&= \sum_{k=0}^{\infty}\phi_{U_1}^{k}(\omega)p_{N(t_0)}[k] \\[2mm]
&= \sum_{k=0}^{\infty}\phi_{U_1}^{k}(\omega)\exp(-\lambda t_0)\frac{(\lambda t_0)^{k}}{k!} \\[2mm]
&= \exp(-\lambda t_0)\sum_{k=0}^{\infty}\frac{(\lambda t_0\phi_{U_1}(\omega))^{k}}{k!} \\[2mm]
&= \exp(-\lambda t_0)\exp(\lambda t_0\phi_{U_1}(\omega))
\end{aligned}
$$

so that finally we have the characteristic function

$$
\phi_{X(t_0)}(\omega) = \exp[\lambda t_0(\phi_{U_1}(\omega) - 1)]. \tag{21.12}
$$

To determine the PMF or PDF of $X(t_0)$ we would need to take the inverse Fourier transform of the characteristic function. As a check, if we let $U_i = 1$ for all i so that from (21.11) $X(t_0) = N(t_0)$, then since

$$\phi_{U_1}(\omega) = E[\exp(j\omega U_1)] = \exp(j\omega)$$

we have the usual characteristic function of a Poisson random variable (see Table 6.1)

$$\phi_{X(t_0)}(\omega) = \exp[\lambda t_0(\exp(j\omega) - 1)].$$

(The derivation of (21.12) can be shown to hold for this choice of the U_i's, which are degenerate random variables.) An example follows.

Example 21.6 – Poisson random process with dropped arrivals

Consider a Poisson random process in which some of the arrivals are dropped. This means for example that a Geiger counter may not record radioactive particles if their intensity is too low. Assume that the probability of dropping an arrival is $1 - p$, and that this event is independent of the Poisson arrival process. Then, we wish to determine the PMF of the number of arrivals within the time interval $[0, t_0]$. Thus, the number of arrivals can be represented as

$$X(t_0) = \sum_{i=1}^{N(t_0)} U_i$$

where $U_i = 1$ if the ith arrival is counted and $U_i = 0$ if it is dropped. Assuming that the U_i's are IID, we have a compound Poisson random process. The characteristic function of $X(t_0)$ is found using (21.12) where we note that

$$\begin{aligned} \phi_{U_1}(\omega) &= E[\exp(j\omega U_1)] \\ &= p\exp(j\omega) + (1 - p) \end{aligned}$$

so that from (21.12)

$$\begin{aligned} \phi_{X(t_0)}(\omega) &= \exp[\lambda t_0(p\exp(j\omega) + (1 - p) - 1)] \\ &= \exp[p\lambda t_0(\exp(j\omega) - 1)]. \end{aligned}$$

But this is just the characteristic of a Poisson counting random process with arrival rate of $p\lambda$. Hence, by dropping arrivals the arrival rate is reduced but $X(t)$ is still a Poisson counting process, a very reasonable result.

$$\diamondsuit$$

Since the characteristic function of a compound Poisson random process is available, we can use it to easily find the moments of $X(t_0)$. In particular, we now determine the mean, leaving the variance as a problem (see Problem 21.22). Using (21.12) we have

$$E[X(t_0)] = \frac{1}{j} \left. \frac{d\phi_{X(t_0)}(\omega)}{d\omega} \right|_{\omega=0} \qquad \text{(using (6.13))}$$

$$= \frac{1}{j} \lambda t_0 \left. \frac{d\phi_{U_1}(\omega)}{d\omega} \exp[\lambda t_0(\phi_{U_1}(\omega) - 1)] \right|_{\omega=0}$$

$$= \lambda t_0 \frac{1}{j} \left. \frac{d\phi_{U_1}(\omega)}{d\omega} \right|_{\omega=0}$$

since $\phi_{U_1}(0) = 1$. But

$$E[U_1] = \frac{1}{j} \left. \frac{d\phi_{U_1}(\omega)}{d\omega} \right|_{\omega=0}$$

so that the average value is

$$E[X(t_0)] = \lambda t_0 E[U_1] = E[N(t_0)]E[U_1]. \qquad (21.13)$$

It is seen that the average value of $X(t_0)$ is just the average value of U_1 times the expected number of arrivals. This result also holds even if the U_i's only have the same mean, without the IID assumption (see Problem 21.25 and the real-world problem). An example follows.

Example 21.7 – Expected number of points scored in basketball game

A basketball player, dubbed the "Poisson pistol Pete" of college basketball, shoots the ball at an average rate of 1 shot per minute according to a Poisson random process. He shoots a 2 point shot with a probability of 0.6 and a 3 point shot with a probability of 0.4. If his 2 point field goal percentage is 50% and his 3 point field goal percentage is 30%, what is his expected total number of points scored in a 40 minute game? (We assume that the referees "let them play" so that no fouls are called and hence no free throw points.) The average number of points is $E[N(t_0)]E[U_1]$, where $t_0 = 2400$ seconds and U_1 is a random variable that denotes his points made for the first shot (the distribution for each shot is identical). We first determine the PMF for U_1, where we have implicitly assumed that the U_i's are IID random variables. From the problem description we have that

$$U_1 = \begin{cases} 2 & \text{if 2 point shot attempted and made} \\ 3 & \text{if 3 point shot attempted and made} \\ 0 & \text{otherwise.} \end{cases}$$

Hence,

$$\begin{aligned} p_{U_1}[2] &= P[\text{2 point shot attempted and made}] \\ &= P[\text{2 point shot made} \mid \text{2 point shot attempted}]P[\text{2 point shot attempted}] \\ &= 0.5(0.6) = 0.3 \end{aligned}$$

and similarly $p_{U_1}[3] = 0.3(0.4) = 0.12$ and therefore, $p_{U_1}[0] = 0.58$. The expected value becomes $E[U_1] = 2(0.3) + 3(0.12) = 0.96$ and therefore the expected number of points scored is

$$\begin{aligned} E[N(t_0)]E[U_1] &= \lambda t_0 E[U_1] \\ &= \frac{1}{60}(2400)(0.96) \\ &= 38.4 \text{ points per game.} \end{aligned}$$

21.7 Computer Simulation

To generate a realization of a Poisson random process on a computer is relatively simple. It relies on the property that the interarrival times are IID $\exp(\lambda)$ random variables. We observe from Figure 21.4 that the ith interarrival time is $Z_i = T_i - T_{i-1}$, where T_i is the ith arrival time. Hence,

$$T_i = T_{i-1} + Z_i \qquad i = 1, 2, \dots$$

where we define $T_0 = 0$. Each Z_i has the PDF $\exp(\lambda)$ and the Z_i's are IID. Hence, to generate a realization of each Z_i we use the inverse probability integral transformation technique (see Section 10.9) to yield

$$Z_i = \frac{1}{\lambda} \ln \frac{1}{1 - U_i}$$

where $U_i \sim \mathcal{U}(0, 1)$ and the U_i's are IID. A typical realization using the following MATLAB code is shown in Figure 21.5a for $\lambda = 2$. The arrivals are indicated now by $+$'s for easier viewing. If we were to increase the arrival rate to $\lambda = 5$, then a typical realization is shown in Figure 21.5b.

```
clear all
rand('state',0)
lambda=2; % set arrival rate
T=5; % set time interval in seconds
for i=1:1000
    z(i,1)=(1/lambda)*log(1/(1-rand(1,1))); % generate interarrival times
    if i==1 % generate arrival time
        t(i,1)=z(i);
    else
        t(i,1)=t(i-1)+z(i,1);
    end
    if t(i)>T % test to see if desired time interval has elapsed
```

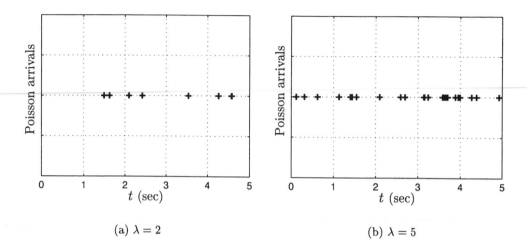

(a) $\lambda = 2$ (b) $\lambda = 5$

Figure 21.5: Realizations of Poisson random process.

```
        break
    end
end
M=length(t)-1; % number of arrivals in interval [0,T]
arrivals=t(1:M); % arrival times in interval [0,T]
```

21.8   Real-World Example – Automobile Traffic Signal Planning

An important responsibility of traffic engineers is to decide which intersections require traffic lights. Although general guidelines are available [Federal Highway Ad. 1988], new situations constantly arise that warrant a reassessment of the situation—principally an unusually high accident rate [Imada 2001]. In this example, we suppose that a particular intersection, which has two stop signs, is prone to accidents. The situation is depicted in Figure 21.6, where it is seen that the two intersecting streets are one-way streets with a stop sign at the corner of each one. A traffic engineer believes that the high accident rate is due to motorists who ignore the stop signs and proceed at full speed through the intersection. If this is indeed the case, then the installation of a traffic light is warranted. To determine if the accident rate is consistent with his belief that motorists are "running" the stop signs, he wishes to determine the average number of accidents that would occur if this is true. As shown in Figure 21.6, if 2 vehicles arrive at the intersection within a given time interval, an accident will occur. It is assumed the two cars are identical and move with the same speed. The traffic engineer then models the arrivals as two indepen-

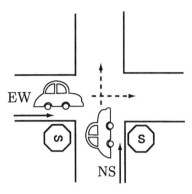

Figure 21.6: Intersection with two automobiles approaching at constant speed.

dent Poisson random processes, one for each direction of travel. A typical set of car arrivals based on this assumption is shown in Figure 21.7. Specifically, an accident

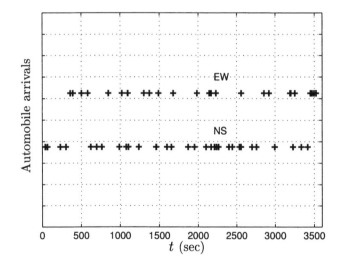

Figure 21.7: Automobile arrivals.

will occur if any two arrivals satisfy $|T^{\text{EW}} - T^{\text{NS}}| \leq \tau$, where T^{EW} and T^{NS} refer to the arrival time at the center of the intersection from the east-west direction and the north-south direction, respectively, and τ is some minimum time for which the cars can pass each other without colliding. The actual value of τ can be estimated using $\tau = d/c$, where d is the length of a car and c is its speed. As an example, if we assume that $d = 22$ ft and $c = 44$ ft/sec (about 30 mph), then $\tau = 0.5$ sec. An accident will occur if two arrivals are within one-half second of each other. In Figure 21.7 this does not occur, but there is a near miss as can be seen in Figure 21.8, which is an expanded version. The east-west car arrives at $t = 2167.5$ seconds while the north-south car arrives at $t = 2168.4$ seconds.

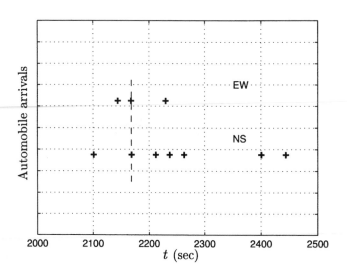

Figure 21.8: Automobile arrivals—expanded version of Figure 21.7. There is a near miss at $t = 2168$ seconds, shown by the dashed vertical line.

We now describe how to determine the average number of accidents per day. This can be obtained by defining a set of indicator random variables (see Example 11.4) as

$$
I_i = \begin{cases} 1 & \text{if there is at least one NS arrival with } |T_i^{\mathrm{EW}} - T^{\mathrm{NS}}| \leq \tau \\ 0 & \text{otherwise.} \end{cases}
$$

Here T^{NS} can be *any* NS arrival time and T_i^{EW} is the ith arrival time for the EW traffic. (More explicitly, the event for which the indicator random variable is 1 occurs when $\min_{j=1,2,\ldots} |T_i^{\mathrm{EW}} - T_j^{\mathrm{NS}}| \leq \tau$, where T_j^{NS} is the jth arrival for the NS traffic.) Now the number of accidents in the time interval $[0, t]$ is

$$
X(t) = \sum_{i=1}^{N(t)} I_i \tag{21.14}
$$

where $N(t)$ is the Poisson counting random process for the EW traffic. To find the expected value of $X(t)$ we note that the equation (21.13), although originally derived under the assumption that the U_i's are IID, is also valid under the weaker assumption that the means of the U_i's are the same as shown in Problem 21.25. Since the I_i's will be seen shortly to have the same mean, the expected value of (21.14) is from (21.13) with $U_1 = I_1$

$$
E[X(t)] = \lambda t E[I_1]. \tag{21.15}
$$

Now to evaluate $E[I_i]$, we note that

$$
E[I_i] = P[|T_i^{\mathrm{EW}} - T^{\mathrm{NS}}| \leq \tau]
$$

and the probability can be found using a conditioning approach (see (13.12)). This produces

$$P[|T_i^{\text{EW}} - T^{\text{NS}}| \leq \tau] = \int_0^\infty P[|T_i^{\text{EW}} - T^{\text{NS}}| \leq \tau | T_i^{\text{EW}} = t] p_{T_i}(t) dt.$$

Proceeding we have that

$$
\begin{aligned}
P[|T_i^{\text{EW}} - T^{\text{NS}}| \leq \tau] &= \int_0^\infty P[|t - T^{\text{NS}}| \leq \tau | T_i^{\text{EW}} = t] p_{T_i}(t) dt \\
&= \int_0^\infty P[|t - T^{\text{NS}}| \leq \tau] p_{T_i}(t) dt \quad (T_i^{\text{EW}}, T^{NS} \text{ are independent}) \\
&= \int_0^\infty P[t - \tau \leq T^{\text{NS}} \leq t + \tau] p_{T_i}(t) dt. \qquad (21.16)
\end{aligned}
$$

Note that $t - \tau \leq T^{\text{NS}} \leq t + \tau$ is the event that the NS traffic will have at least one arrival (and hence an accident) in the interval $[t - \tau, t + \tau]$. Its probability is just

$$
\begin{aligned}
P[t - \tau \leq T^{\text{NS}} \leq t + \tau] &= P[\text{one or more arrivals in } [t - \tau, t + \tau]] \\
&= 1 - P[\text{no arrival in } [t - \tau, t + \tau]] \\
&= 1 - P[\text{no arrivals in } [0, 2\tau]] \quad \text{(increment stationarity)} \\
&= 1 - P[N(2\tau) = 0] \\
&= 1 - \exp(-2\lambda\tau) \quad \text{(from (21.2))}
\end{aligned}
$$

and is not dependent on t. Thus,

$$
\begin{aligned}
E[I_i] &= P[|T_i^{\text{EW}} - T^{\text{NS}}| \leq \tau] \\
&= \int_0^\infty (1 - \exp(-2\lambda\tau)) p_{T_i}(t) dt \quad \text{(from (21.16))} \\
&= 1 - \exp(-2\lambda\tau)
\end{aligned}
$$

for all i, and therefore all the I_i's have the same mean. From (21.15)

$$E[X(t)] = \lambda t (1 - \exp(-2\lambda\tau)).$$

For the same example as before with $\tau = 0.5$, the average number of accidents per second is

$$\frac{E[X(t)]}{t} = \lambda(1 - \exp(-\lambda)).$$

For a more meaningful measure we convert this to the average number of accidents per hour, which is $(E[X(t)]/t)3600$. This is plotted versus λ', where λ' is in arrivals per hour, in Figure 21.9. Specifically, it is given by

$$
\begin{aligned}
3600\frac{E[X(t)]}{t} &= 3600\lambda(1 - \exp(-\lambda)) \\
&= \lambda'\left[1 - \exp\left(-\frac{\lambda'}{3600}\right)\right]
\end{aligned}
$$

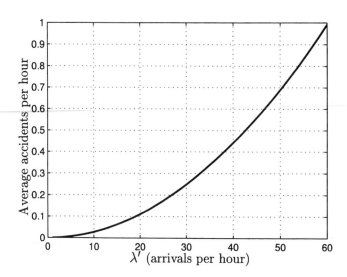

Figure 21.9: Average number of accidents per hour versus arrival rate (in per hour units).

where $\lambda' =$ arrivals per hour $= 3600\lambda$. As seen in Figure 21.9 for about 1 arrival every 3 minutes or 20 arrivals per hour, we will have an average of 0.1 accidents per hour or about an average of one accident every two days. This assumes a busy intersection for about 5 hours per day. Thus, if the traffic engineer notices an accident nearly every other day, he will request that a traffic light be put in.

References

Cox, D.R., H.D. Miller, *The Theory of Stochastic Processes*, John Wiley & Sons, New York, 1965.

Federal Highway Administration, *Manual on Uniform Traffic Control Devices*, U.S. Govt. Printing Office, 1988.

Imada, T., "Traffic Control of Closely Located Intersections: the U.S. 1 Busway Experience," *ITE Journal on the Web*, pp. 81–84, May 2001.

Problems

21.1 (t) Prove that the differential equation describing $P_k(t) = P[N(t) = k]$ for a Poisson counting random process is given by (21.3). To do so use Figure 21.3 with *either* k arrivals in $[0, t - \Delta t]$ and no arrivals in $(t - \Delta t, t]$ *or* $k - 1$ arrivals in $[0, t - \Delta t]$ and one arrival in $(t - \Delta t, t]$. Since there can be at most one arrival in a time interval of length Δt, these are the only possibilities.

21.2 (t) Solve the differential equation of (21.3) by taking the (one-sided) Laplace transform of both sides, noting that $P_k(0^+) = 0$. Explain why the latter condition is consistent with the assumptions of a Poisson random process. You should be able to show that the Laplace transform of $P_k(t)$ is

$$\mathcal{P}_k(s) = \frac{\lambda^k}{(s+\lambda)^{k+1}}$$

by finding $\mathcal{P}_1(s)$ from $\mathcal{P}_0(s)$, and then $\mathcal{P}_2(s)$ from $\mathcal{P}_1(s)$, etc. The desired inverse Laplace transform is found by referring to a table of Laplace transforms.

21.3 (⌣) (f) Find the probability of 6 arrivals of a Poisson random process in the time interval $[7, 12]$ if $\lambda = 1$. Next determine the average number of arrivals for the same time interval.

21.4 (w) For a Poisson random process with an arrival rate of 2 arrivals per second, find the probability of exactly 2 arrivals in 5 successive time intervals of length 1 second each.

21.5 (f) What is the probability of a single arrival for a Poisson random process with arrival rate λ in the time interval $[t, t + \Delta t]$ if $\Delta t \to 0$?

21.6 (w) Telephone calls come into a service center at an average rate of one per 5 seconds. What is the probability that there will be more than 12 calls in the first one minute?

21.7 (⌣) (f,c) For a Poisson random process with an arrival rate of λ use a computer simulation to estimate the arrival rate if $\lambda = 2$ and also if $\lambda = 5$. To do so relate λ to the average number of arrivals in $[0, t]$. Hint: Use the MATLAB code in Section 21.7.

21.8 (w) Two independent Poisson random processes both have an arrival rate of λ. What is the expected time of the first arrival observed from either of the two random processes? Explain your results. Hint: Let this time be denoted by T and note that $T = \min(T_1^{(1)}, T_1^{(2)})$, where $T_1^{(i)}$ is the first arrival time of the ith random process. Then, note that $P[T > t] = P[T_1^{(1)} > t, T_1^{(2)} > t]$.

21.9 (t) In this problem we prove that the sum of two independent Poisson counting random processes is another Poisson counting random process whose arrival rate is the sum of the arrival rates of the two random processes. Let the Poisson counting random processes be $N_1(t)$ and $N_2(t)$ and consider the increments $N(t_2) - N(t_1)$ and $N(t_4) - N(t_3)$ for nonoverlapping time intervals. Argue that the corresponding increments for the sum random process are independent and stationary, knowing that this is true for each individual random process. Then, use characteristic functions to prove that if $N_1(t) \sim \text{Pois}(\lambda_1 t)$ and

$N_2(t) \sim \text{Pois}(\lambda_2 t)$ and $N_1(t)$ and $N_2(t)$ are independent, then $N_1(t) + N_2(t) \sim \text{Pois}((\lambda_1 + \lambda_2)t)$.

21.10 (☺) (w) If $N(t)$ is a Poisson counting random process, determine $E[N(t_2) - N(t_1)]$ and $\text{var}(N(t_2) - N(t_1))$.

21.11 (w) Commuters arrive at a subway station that has 3 turnstyles with the arrivals at each turnstyle characterized by an independent Poisson random process with arrival rate of λ commuters per second. Determine the probability of a total of k arrivals in the time interval $[0, t]$. Hint: See Problem 21.9.

21.12 (t) In this problem we present an alternate proof that the Poisson random process has no memory as described by (21.7). It is based on the observation that a Poisson random process is the limiting form of a Bernoulli random process as explained in Section 21.1. Consider first the geometric PMF of the first success or arrival which is $P[X = k] = (1 - p)^{k-1}p$ for $k = 1, 2, \ldots$. Then show that

$$P[X > k_1 + k_2 | X > k_1] = (1 - p)^{k_2}.$$

Next let $p = \lambda \Delta t$ and $k_1 = \xi_1/\Delta t$ and $k_2 = \xi_2/\Delta t$ and prove that as $\Delta t \to 0$

$$P[X > k_1 + k_2 | X > k_1] = P[X\Delta t > k_1\Delta t + k_2\Delta t | X\Delta t > k_1\Delta t] \to \exp(-\lambda \xi_2).$$

Hint: As $x \to 0$, $(1 - ax)^{1/x} \to \exp(-a)$.

21.13 (☺) (w) Taxi cabs arrive at the rate of 1 per minute at a taxi stand. If a person has already waited 10 minutes for a cab, what is the probability that he will have to wait less than 1 additional minute?

21.14 (w) A computer memory has the capacity to store 10^6 words. If requests for word storage follow a Poisson random process with a request rate of 1 per millisecond, how long on average will it be before the memory capacity is exceeded?

21.15 (t) If $X_1 \sim \exp(\lambda)$, $X_2 \sim \exp(\lambda)$, and X_1 and X_2 are independent random variables, derive the PDF of the sum by using a convolution integral.

21.16 (t) We give an alternate derivation of the PDF for the kth arrival time of a Poisson random process. This PDF can be expressed as

$$\lim_{\Delta t \to 0} \frac{P[t - \Delta t \le T_k \le t]}{\Delta t}.$$

Use the fact that the event $\{t - \Delta t \le T_k \le t\}$ can only occur as $\Delta t \to 0$ if there are $k - 1$ arrivals in $[0, t - \Delta t]$ and 1 arrival in $(t - \Delta t, t]$.

21.17 ($\smile$) **(w)** People arrive at a football game at a rate of 100 per minute. If the 1000th person is to receive a seat at the 50th yard line (which is highly desirable), how long should you wait before entering the stadium?

21.18 (t) Prove that if X, Y, Z are jointly distributed continuous random variables, then $E_{X,Y,Z}[g(X,Y,Z)] = E_Z[E_{X,Y|Z}[g(X,Y,Z)|z]]$ by expressing the expectations using integrals. You may wish to refer back to Section 13.6.

21.19 (t) The Poisson random process exhibits the *Markov property*. This says that the conditional probability of $N(t)$ based on past samples of the random process only depends upon the most recent sample. Mathematically, if $t_3 > t_2 > t_1$, then

$$P[N(t_3) = k_3 | N(t_2) = k_2, N(t_1) = k_1] = P[N(t_3) = k_3 | N(t_2) = k_2].$$

Prove that this is true by making use of the property that the increments are independent. Specifically, consider the equivalent probability

$$P[N(t_3) - N(t_2) = k_3 - k_2 | N(t_2) = k_2, N(t_1) - N(0) = k_1]$$

and also explain why this probability is equivalent.

21.20 ($\smile$) **(c)** Use a computer simulation to generate multiple realizations of a Poisson random process with $\lambda = 1$. Then, use the simulation to estimate $P[T_2 \leq 1]$. Compare your result to the true value. Hint: Use the MATLAB code in Section 21.7.

21.21 (w) An airport has two security screening lines. An employee directs the incoming travelers to one of the two lines at random. If the incoming travelers arrive at the airport with a rate of λ travelers per second, what is the arrival rate at each of the two security screening lines? What assumptions are implicit in arriving at your answer?

21.22 (t) Prove that the variance of a compound Poisson random process is $\text{var}(X(t_0)) = \lambda t_0 E[U_1^2]$. If you guessed that the result would be $\lambda t_0 \text{var}(U_1)$, then evaluate your guess for a Poisson random process (let $U_i = 1$).

21.23 ($\smile$) **(f)** A compound Poisson random process $X(t)$ is composed of random variables U_i that can take on the values ± 1 with $P[U_i = 1] = p$. What is the expected value of $X(t)$?

21.24 (c) Perform a computer simulation to lend credibility to the expected number of points scored in the basketball game described in Example 21.7.

21.25 (t) Derive (21.13) for the case where the U_i's have the same mean and are independent of $N(t_0)$. Start your derivation with the expression

$$E[X(t_0)] = E_{N(t_0)} \left[E_{U_1,\ldots,U_k|N(t_0)} \left[\sum_{i=1}^{k} U_i \;\middle|\; N(t_0) = k \right] \right]$$

and then follow the same approach as given in Section 21.6. You do not need the characteristic function to do this.

Appendix 21A

Joint PDF for Interarrival Times

We prove in this appendix that the first two interarrival times Z_1, Z_2 are IID with $Z_i \sim \exp(\lambda)$. The general case of any number of interarrival times can similarly be proven to be IID with an $\exp(\lambda)$ PDF. We now refer to Figure 21.4 and prove that the joint CDF factors and each marginal CDF is that corresponding to the $\exp(\lambda)$ PDF. The joint CDF is given as

$$P[Z_1 \leq \xi_1, Z_2 \leq \xi_2] = \int_0^{\xi_1} P[Z_2 \leq \xi_2 | Z_1 = z_1] p_{Z_1}(z_1) dz_1 \qquad (21A.1)$$

which follows from (13.12) where $A = \{z_2 : z_2 \leq \xi_2\}$. But if $Z_1 = z_1$, then $Z_2 \leq \xi_2$ if and only if $N(z_1 + \xi_2) - N(z_1) \geq 1$ since an arrival must have occurred in $[z_1, z_1 + \xi_2]$. Hence,

$$P[Z_2 \leq \xi_2 | Z_1 = z_1] = P[N(z_1 + \xi_2) - N(z_1) \geq 1 | Z_1 = z_1]$$

and because the event $Z_1 = z_1$ is equivalent to the increment $N(z_1) - N(0) = 1$, and the increments are independent and stationary, we have

$$
\begin{aligned}
P[Z_2 \leq \xi_2 | Z_1 = z_1] &= P[N(z_1 + \xi_2) - N(z_1) \geq 1 | Z_1 = z_1] \\
&= P[N(z_1 + \xi_2) - N(z_1) \geq 1] \qquad \text{(independence)} \\
&= P[N(\xi_2) \geq 1] \qquad \text{(stationarity)}.
\end{aligned}
$$

Using this in (21A.1) produces

$$
\begin{aligned}
P[Z_1 \le \xi_1, Z_2 \le \xi_2] &= \int_0^{\xi_1} P[N(\xi_2) \ge 1] p_{Z_1}(z_1) dz_1 \\
&= \int_0^{\xi_1} [1 - P[N(\xi_2) < 1]] p_{Z_1}(z_1) dz_1 \\
&= \int_0^{\xi_1} (1 - P[N(\xi_2) = 0]) p_{Z_1}(z_1) dz_1 \\
&= \int_0^{\xi_1} (1 - \exp(-\lambda \xi_2)) p_{Z_1}(z_1) dz_1 \\
&= [1 - \exp(-\lambda \xi_2)] \int_0^{\xi_1} p_{Z_1}(z_1) dz_1 \\
&= [1 - \exp(-\lambda \xi_2)] P[Z_1 \le \xi_1] \\
&= [1 - \exp(-\lambda \xi_2)] P[N(\xi_1) \ge 1] \\
&= [1 - \exp(-\lambda \xi_2)][1 - P[N(\xi_1) < 1]] \\
&= [1 - \exp(-\lambda \xi_2)][1 - \exp(-\lambda \xi_1)] \\
&= P[Z_1 \le z_1] P[Z_2 \le z_2].
\end{aligned}
$$

It is seen that the joint CDF factors into the product of the marginal CDFs, where each marginal is the CDF of an $\exp(\lambda)$ random variable. Thus, the first two inter-arrival times are IID with PDF $\exp(\lambda)$.

Chapter 22

Markov Chains

22.1 Introduction

We have seen in Chapter 16 that an important random process is the IID random process. When applicable to a specific problem, it lends itself to a very simple analysis. A Bernoulli random process, which consists of independent Bernoulli trials, is the archetypical example of this. In practice, it is found, however, that there is usually some dependence between samples of a random process. In Chapters 17 and 18 we modeled this dependence using wide sense stationary random process theory, but restricted the modeling to only the first two moments. In an effort to introduce a more general dependence into the modeling of a random process, we now reconsider the Bernoulli random process but assume dependent samples. We briefly introduced this extension in Example 4.10 as a sequence of *dependent* Bernoulli trials. The dependence of the PMF that we will be interested in is dependence *on the previous trial only*. This type of dependence leads to what is generically referred to as a *Markov random process*. A special case of this for a *discrete-time/discrete-valued* (DTDV) random process is called a *Markov chain*. Specifically, it has the property that the probability of the random process $X[n]$ at time $n = n_0$ only depends upon the *outcome* or *realization* of the random process at the previous time $n = n_0 - 1$. It can then be viewed as the next logical step in extending an IID random process to a random process with statistical dependence. Recall from Chapter 8 that for discrete random variables statistical dependence is quantified using conditional probabilities. The reader should review Example 4.10 and also Chapter 8 in preparation for our discussion of Markov chains.

Although we will restrict our description to a DTDV Markov random process, i.e., the Markov chain, there are many generalizations that are important in practice. The interested reader can consult the excellent books by [Bharucha-Reid 1988], [Cox and Miller 1965], [Gallagher 1996] and [Parzen 1962] for these other random processes. Before proceeding with our discussion we present an example to illustrate typical concepts associated with a Markov chain.

In the game of golf it is very desirable to be a good putter. The best golfers in the world are able to hit a golf ball lying on the green into the hole using only a few strokes, called *putting the ball*. At times they can even "one-putt" the ball, in which they require only a single stroke to hit the ball into the hole. Of course, their chances of doing so rely heavily on how far the ball is from the hole when they first reach the green. If the ball is say 3 feet from the hole, then they will almost always one-putt. If, however, it is near the edge of the green, possibly 20 feet from the hole, then their chances are small. For our hypothetical golfer we will assume that her chance of a one-putt is 50% at the start of a round of golf, i.e., at hole one. If she one-putts on hole one, then her chances on hole two will remain at 50%. If not, she becomes somewhat discouraged which reduces her chances at hole two to only 25%. Hence, at each hole her chances of a one-putt are 50% if she has one-putted the previous hole and 25% if she has not. To model this situation we let $X[n] = 1$ for a one-putt at hole n and $X[n] = 0$ otherwise. We label hole one by $n = 0$. A round of golf, which consists of 18 holes, produces a sequence of 18 1's and 0's with a 1 indicating a one-putt. For the probabilities assumed a typical set of outcomes is shown in Figure 22.1. Note that she has played three rounds of

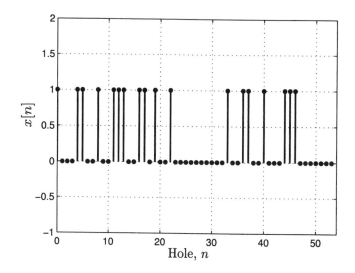

Figure 22.1: Outcomes of three rounds of golf. A 1 indicates a one-putt on hole n.

golf or 54 holes, of which 18 were one-putts. It appears that her probability of a one-putt is closer to 1/3 than either 1/2 or 1/4. Also, it is of interest to determine the average number of holes played between one-putts. The actual number varies as seen in Figure 22.1 and is $\{4, 1, 3, 3, 1, 1, 3, 1, 2, 3, 11, 3, 1, 3, 4, 1, 1\}$ for an average of $46/17 = 2.70$. It would seem that the expected number of holes played between one-putts, about 3, is the reciprocal of the probability of a one-putt, about 1/3. This suggests a geometric-type PMF, which we will confirm in Section 22.6.

Probabilistically, we are observing a sequence of *dependent Bernoulli trials*. The dependence arises (in contrast to the usual Bernoulli random process which had independent trials) due to the probability of a one-putt at hole n being dependent upon the outcome at hole $n - 1$. We can model this dependence using conditional probabilities to say that

$$P[\text{one-putt at hole } n | \text{no one-putt at hole } n - 1] \;=\; \frac{1}{4}$$

$$P[\text{one-putt at hole } n | \text{one-putt at hole } n - 1] \;=\; \frac{1}{2}$$

or

$$P[X[n] = 1 | X[n - 1] = 0] \;=\; \frac{1}{4}$$

$$P[X[n] = 1 | X[n - 1] = 1] \;=\; \frac{1}{2}.$$

Completing the conditional probability description, we have

$$P[X[n] = 0 | X[n - 1] = 0] \;=\; \frac{3}{4}$$

$$P[X[n] = 1 | X[n - 1] = 0] \;=\; \frac{1}{4}$$

$$P[X[n] = 0 | X[n - 1] = 1] \;=\; \frac{1}{2}$$

$$P[X[n] = 1 | X[n - 1] = 1] \;=\; \frac{1}{2}.$$

Note that we have assumed that the *conditional probabilities do not change with "time"* (actually hole number). Lastly, we require the initial probability of a one-putt for the first hole. We assign this to be $P[X[0] = 1] = 1/2$. In summary, we have two sets of conditional probabilities and one set of initial probabilities which can be arranged conveniently using a matrix and vector to be

$$\mathbf{P} \;=\; \begin{bmatrix} P[X[n] = 0 | X[n - 1] = 0] & P[X[n] = 1 | X[n - 1] = 0] \\ P[X[n] = 0 | X[n - 1] = 1] & P[X[n] = 1 | X[n - 1] = 1] \end{bmatrix} \quad (22.1)$$

$$=\; \begin{bmatrix} \frac{3}{4} & \frac{1}{4} \\ \frac{1}{2} & \frac{1}{2} \end{bmatrix}$$

and

$$\mathbf{p}[0] \;=\; \begin{bmatrix} P[X[0] = 0] \\ P[X[0] = 1] \end{bmatrix} \quad (22.2)$$

$$=\; \begin{bmatrix} \frac{1}{2} \\ \frac{1}{2} \end{bmatrix}.$$

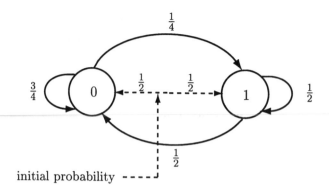

Figure 22.2: Markov state probability diagram for putting example.

The probabilities can also be summarized using the diagram shown in Figure 22.2, where for example the *conditional probability* of a one-putt on hole n given that the golfer has not one-putted on hole $n-1$ is 1/4. We may view this diagram as one in which we are in "state" 0, which corresponds to the previous outcome of no one-putt and will move to "state" 1, which corresponds to a one-putt, with a conditional probability of 1/4. If we do move to a new state, it means the outcome is a 1 and otherwise, the outcome is a 0. In interpreting the diagram one should visualize that a 0 or 1 is emitted as we *enter* the 0 or 1 state, respectively. Then, the current state becomes the last value emitted. Also, our initial *unconditional probabilities* of 1/2 and 1/2 of entering state 0 or state 1 are shown as dashed lines. The diagram is called the *Markov state probability diagram*. The use of the term "state" is derived from physics in that the future evolution (in terms of probabilities) of the process is only dependent upon the current state and not upon how the process arrived in that state. The probabilistic structure summarized in Figure 22.2 is called a *Markov chain*. As mentioned previously, it is a DTDV random process. Although we have used a dependent Bernoulli random process as an example, it easily generalizes to any finite number of states. It is common in the discussion of Markov chains to term the matrix of conditional probabilities **P** in (22.1) as the *state transition probability matrix* or more succinctly the *transition probability matrix*. The initial probability vector **p**[0] in (22.2) is called the *initial state probability vector* or more succinctly the *initial probability vector*. Note that in using the state probability diagram to summarize the Markov chain we will henceforth omit the initial probability assignment in the diagram but it should be kept in mind that it is necessary in order to complete the description.

   As an example of a typical probability computation, consider the probability of $X[0] = 0, X[1] = 1, X[2] = 1$ versus $X[0] = 1, X[1] = 1, X[2] = 1$. Then, using the chain rule (see (4.10)) we have

$$P[X[0] = 0, X[1] = 1, X[2] = 1] \quad = \quad P[X[2] = 1 | X[1] = 1, X[0] = 0]$$
$$\cdot P[X[1] = 1 | X[0] = 0] P[X[0] = 0].$$

But due to the assumption that the probability of $X[n]$ only depends upon the outcome at time $n-1$, which is called the *Markov property*, we have

$$P[X[2] = 1|X[1] = 1, X[0] = 0] = P[X[2] = 1|X[1] = 1]$$

and therefore

$$\begin{aligned} P[X[0] = 0, X[1] = 1, X[2] = 1] &= P[X[2] = 1|X[1] = 1]P[X[1] = 1|X[0] = 0] \\ &\quad \cdot P[X[0] = 0]. \end{aligned}$$

But from Figure 22.2 this is

$$P[X[0] = 0, X[1] = 1, X[2] = 1] = \left(\frac{1}{2}\right)\left(\frac{1}{4}\right)\left(\frac{1}{2}\right) = \frac{1}{16}.$$

Similary,

$$\begin{aligned} P[X[0] = 1, X[1] = 1, X[2] = 1] &= P[X[2] = 1|X[1] = 1]P[X[1] = 1|X[0] = 1] \\ &\quad \cdot P[X[0] = 1] = \left(\frac{1}{2}\right)\left(\frac{1}{2}\right)\left(\frac{1}{2}\right) = \frac{1}{8}. \end{aligned}$$

We see that joint probabilities are easily determined from the initial probabilities and the transition probabilities. If we are only interested in the marginal PMF at a given time say $P[X[n] = k]$ for $k = 0, 1$, as, for example, $P[X[2] = 1]$, we need only sum over the other variables of the joint PMF. This produces

$$\begin{aligned} P[X[2] = 1] &= \sum_{i=0}^{1}\sum_{j=0}^{1} P[X[0] = i, X[1] = j, X[2] = 1] \\ &= \sum_{i=0}^{1}\sum_{j=0}^{1} P[X[2] = 1|X[0] = i, X[1] = j]P[X[1] = j|X[0] = i] \\ &\qquad\qquad\qquad\qquad\qquad\qquad\qquad\qquad \cdot P[X[0] = i] \\ &= \sum_{i=0}^{1}\sum_{j=0}^{1} P[X[2] = 1|X[1] = j]P[X[1] = j|X[0] = i]P[X[0] = i] \\ &\qquad\qquad\qquad\qquad\qquad\qquad\qquad\qquad\qquad \text{(Markov property)} \\ &= \sum_{j=0}^{1} P[X[2] = 1|X[1] = j]\underbrace{\sum_{i=0}^{1} P[X[1] = j|X[0] = i]P[X[0] = i]}_{P[X[1]=j]}. \end{aligned}$$

Note that $P[X[1] = j]$ can be found and then used to find $P[X[2] = 1]$. Of course, this is getting somewhat messy algebraically but as shown in the next section the use of vectors and matrices will simplify the computation.

Finally, some questions of interest to the golfer are:

1. After playing many holes, will the probability of a one-putt settle down to some constant value? Mathematically, will $P[X[n] = k]$ converge to some constant PMF as $n \to \infty$?

2. Given that the golfer has just one-putted, how many holes on the average will she have to wait until the next one-putt? Or given that she has not one-putted, how many holes on the average will she have to wait until she one-putts? In the first case, mathematically we wish to determine if given $X[n_0] = 1$ and $X[n_0 + 1] = 0, \ldots, X[n_0 + N - 1] = 0, X[n_0 + N] = 1$, what is $E[N]$?

We will answer both these questions shortly, but before doing so some definitions are necessary.

22.2 Summary

A motivating example of a Markov chain is given in Section 22.1. A Markov chain is defined by the property of (22.3). The state transition probabilities, which describe the probabilities of movements between states, is given by (22.4). When arranged in a matrix it is equivalent to (22.5) for a two-state Markov chain and is called the transition probability matrix. The probabilities of the states are defined in (22.6) and succinctly summarized by the vector of (22.7) for a two-state Markov chain. Table 22.1 summarizes the notational conventions. The state probability vector can be found for any time by using (22.9). To evaluate a power of the transition probability matrix (22.12) can be used if the eigenvalues of the matrix are distinct. For a two-state Markov chain the state probabilities are explicitly found in Section 22.4 with the general transition probability matrix given by (22.14). For an ergodic Markov chain the state probabilities approach a constant value as time increases and this value is found by solving (22.17). Also, the value of the n-step transition probability matrix approaches the steady-state value given by (22.19). In Section 22.6 the occupation time of a state for an ergodic Markov chain is shown to be given by the steady-state probabilities and also, the mean recurrence time is the inverse of the occupation time. An explicit solution for the steady-state or stationary probabilities can be found using (22.22). The MATLAB code for a computer simulation of a 3-state Markov chain is given in Section 22.8 while a concluding real-world example is given in Section 22.9.

22.3 Definitions

We restrict ourselves to a discrete-time random process $X[n]$ with K possible values or states. In the introduction $K = 2$ and the values were $0, 1$. This is a DTDV random process that starts at $n = 0$ (semi-infinite). We define $X[n]$ as a Markov chain if *given* the entire past set of outcomes, the PMF of $X[n]$ depends on only the outcome of the previous sample $X[n-1]$ so that

$$P[X[n] = j | X[n-1] = i, X[n-2] = k, \ldots, X[0] = l] = P[X[n] = j | X[n-1] = i].$$

Using the concept of a PMF this is equivalent to

$$p_{X[n]|X[n-1],\ldots,X[0]} = p_{X[n]|X[n-1]}. \tag{22.3}$$

This implies that the joint PMF only depends on the product of the *first-order* conditional PMFs and the initial probabilities, for example

$$p_{X[0],X[1],X[2]} = p_{X[2]|X[1],X[0]} p_{X[1]|X[0]} p_{X[0]}$$

$$= \underbrace{p_{X[2]|X[1]}}_{\substack{\text{conditional} \\ \text{probability}}} \underbrace{p_{X[1]|X[0]}}_{\substack{\text{conditional} \\ \text{probability}}} \underbrace{p_{X[0]}}_{\substack{\text{initial} \\ \text{probability}}} .$$

As mentioned previously, this is an extension of the idea of independence in that it asserts a type of *conditional independence*. Most importantly, the *joint PMF* is obtained as the product of *first-order conditional PMFs*. An example follows.

Example 22.1 – A coin with memory

Assume that a coin is tossed three times with the outcome of a head represented by a 1 and a tail by a 0. If the coin has memory and is modeled by the state probability diagram of Figure 22.2, determine the probability of the sequence HTH. Note that the conditional probabilities are equivalent to those in Example 4.10. Writing the joint probability in the more natural order of increasing time, we have

$$P[X[0] = 1, X[1] = 0, X[2] = 1] = P[X[0] = 1]P[X[1] = 0 | X[0] = 1]$$

$$\cdot P[X[2] = 1 | X[1] = 0]$$

$$= \left(\frac{1}{2}\right)\left(\frac{1}{2}\right)\left(\frac{1}{4}\right) = \frac{1}{16}.$$

Hence, the sequence HTH is less probable than for a fair coin without memory for which 3 independent tosses would yield a probability of 1/8. Can you explain why this is less probable?

$$\diamond$$

We will now use the terminology of the introduction to refer to the conditional probabilities $P[X[n] = j | X[n-1] = i]$ as the *state transition probabilities*. Note that they are assumed not to depend on n and therefore the Markov chain is said to be *homogeneous*. To simplify the notation further and to prepare for subsequent probability calculations we denote the state transition probabilities as

$$P_{ij} = P[X[n] = j | X[n-1] = i] \qquad i = 0, 1, \ldots, K-1; j = 0, 1, \ldots, K-1. \tag{22.4}$$

This is the conditional probability of observing an outcome j given that the previous outcome was i. It is also said that P_{ij} is the probability of the chain moving from state i to state j, but keep in mind that it is a conditional probability. In the case of a two-state Markov chain or $K = 2$, we have $i = 0, 1; j = 0, 1$ and the state transition probabilities are most conveniently arranged in a matrix $\mathbf{P}$. From (22.1) we have

$$\mathbf{P} = \begin{bmatrix} P_{00} & P_{01} \\ P_{10} & P_{11} \end{bmatrix} \tag{22.5}$$

which as previously mentioned is the transition probability matrix. Note that the sum of the elements along each row must be one since they represent all the values of a conditional PMF. In accordance with the assumption of homogeneity $\mathbf{P}$ is a constant matrix. Finally, we define the state probabilities at time n as

$$p_i[n] = P[X[n] = i] \qquad i = 0, 1, \ldots, K - 1. \tag{22.6}$$

This is the probability of observing an outcome i at time n or equivalently the PMF of $X[n]$. This notation is somewhat at odds with our previous notation, which would be $p_{X[n]}[i]$, but is a standard one. The PMF depends on n and it is this PMF that we will be most concerned. In particular, how the PMF changes with n will be of interest. Hence, a Markov chain is in general *a nonstationary random process*. For ease of notation and later computation we also define the *state probability vector* for $K = 2$ as

$$\mathbf{p}[n] = \begin{bmatrix} p_0[n] \\ p_1[n] \end{bmatrix}. \tag{22.7}$$

A summary of these definitions and notation is given in Table 22.1. An example is given next to illustrate the utility of definitions (22.4) and (22.6) and their vector/matrix representations of (22.5) and (22.7).

Example 22.2 – Two-state Markov chain

Consider the computation of $P[X[n] = j]$ for a two-state Markov chain ($K = 2$). Then,

$$\begin{aligned} P[X[n] = j] &= \sum_{i=0}^{1} P[X[n-1] = i, X[n] = j] \\ &= \sum_{i=0}^{1} P[X[n] = j | X[n-1] = i] P[X[n-1] = i] \end{aligned}$$

which can now be written as

$$p_j[n] = \sum_{i=0}^{1} P_{ij} p_i[n-1] \qquad j = 0, 1.$$

In vector/matrix notation we have

$$\underbrace{\begin{bmatrix} p_0[n] & p_1[n] \end{bmatrix}}_{\mathbf{p}^T[n]} = \underbrace{\begin{bmatrix} p_0[n-1] & p_1[n-1] \end{bmatrix}}_{\mathbf{p}^T[n-1]} \underbrace{\begin{bmatrix} P_{00} & P_{01} \\ P_{10} & P_{11} \end{bmatrix}}_{\mathbf{P}}$$

Terminology	Description	Notation
Random process	DTDV	$X[n] \quad n = 0, 1, \ldots$
State	Sample space	$k = 0, 1, \ldots, K - 1$
State probability vector	PMF of $X[n]$	$\mathbf{p}[n] = [p_0[n] \ldots p_{K-1}[n]]^T$ $p_k[n] = P[X[n] = k]$
State transition probability matrix	Conditional prob.	$\mathbf{P} = \begin{bmatrix} P_{00} & P_{01} & \cdots & P_{0,K-1} \\ P_{10} & P_{11} & \cdots & P_{1,K-1} \\ \vdots & \vdots & \ddots & \vdots \\ P_{K-1,0} & P_{K-1,1} & \cdots & P_{K-1,K-1} \end{bmatrix}$ $P_{ij} = P[X[n] = j \mid X[n-1] = i]$
Initial state probability vector	PMF of $X[0]$	$\mathbf{p}[0]$

Table 22.1: Markov chain definitions and notation.

or

$$\mathbf{p}^T[n] = \mathbf{p}^T[n-1]\mathbf{P}. \tag{22.8}$$

The evolution of the state probability vector in time is easily found by post-multiplying the previous state probability vector (in row form) by the transition probability matrix.

$\diamond$

Note that we have defined $\mathbf{p}[n]$ as a column vector in accordance with our usual convention. Other textbooks may use row vectors. A numerical example follows.

Example 22.3 – Golfer one-putting

From Figure 22.2 we have the transition probability matrix and initial state probability vector as

$$\mathbf{P} = \begin{bmatrix} \frac{3}{4} & \frac{1}{4} \\ \frac{1}{2} & \frac{1}{2} \end{bmatrix}$$

$$\mathbf{p}^T[0] = \begin{bmatrix} \frac{1}{2} & \frac{1}{2} \end{bmatrix}.$$

To find $\mathbf{p}[1]$ we use (22.8) to yield

$$
\begin{aligned}
\mathbf{p}^T[1] &= \mathbf{p}^T[0]\mathbf{P} \\
&= \begin{bmatrix} \frac{1}{2} & \frac{1}{2} \end{bmatrix} \begin{bmatrix} \frac{3}{4} & \frac{1}{4} \\ \frac{1}{2} & \frac{1}{2} \end{bmatrix} \\
&= \begin{bmatrix} \frac{5}{8} & \frac{3}{8} \end{bmatrix}.
\end{aligned}
$$

As expected the elements of $\mathbf{p}[1]$ sum to one. Also, note that $p_1[1] = 3/8 < 1/2$, which means that initially the probability of a one-putt is $1/2$ but after the first hole, it is reduced to $3/8$. Can you explain why? We can continue in this manner to compute the state probability vector for $n = 2$ as

$$
\begin{aligned}
\mathbf{p}^T[2] &= \mathbf{p}^T[1]\mathbf{P} \\
&= \begin{bmatrix} \frac{5}{8} & \frac{3}{8} \end{bmatrix} \begin{bmatrix} \frac{3}{4} & \frac{1}{4} \\ \frac{1}{2} & \frac{1}{2} \end{bmatrix} \\
&= \begin{bmatrix} \frac{21}{32} & \frac{11}{32} \end{bmatrix}
\end{aligned}
$$

and so forth for all n.

$\diamondsuit$

22.4   Computation of State Probabilities

We are now in a position to determine $\mathbf{p}[n]$ for all n. The key of course is the recursion of (22.8). In a slightly more general form where we wish to go from $\mathbf{p}[n_1]$ to $\mathbf{p}[n_2]$, the resulting equations are known as the *Chapman-Kolmogorov* equations. For example, if $n_2 = n_1 + 2$, then

$$
\begin{aligned}
\mathbf{p}^T[n_2] &= \mathbf{p}^T[n_2 - 1]\mathbf{P} \\
&= (\mathbf{p}^T[n_2 - 2]\mathbf{P})\mathbf{P} \\
&= \mathbf{p}^T[n_1]\mathbf{P}^2.
\end{aligned}
$$

The matrix $\mathbf{P}^2$ is known as the *two-step transition probability matrix*. It allows the state probabilities for two steps into the future to be found if we know the state probabilities at the current time. In general, then we see that

$$
\mathbf{p}^T[n_1 + n] = \mathbf{p}^T[n_1]\mathbf{P}^n
$$

as is easily verified, where $\mathbf{P}^n$ is the *n-step transition probability matrix*. In particular, if $n_1 = 0$, then

$$
\mathbf{p}^T[n] = \mathbf{p}^T[0]\mathbf{P}^n \qquad n = 1, 2, \ldots \tag{22.9}
$$

which can be used to find the state probabilities for all time. These probabilities can exhibit markedly different behaviors depending upon the entries in $\mathbf{P}$. To illustrate

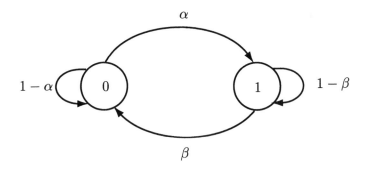

Figure 22.3: General two-state probability diagram.

this consider the two-state Markov chain with the state probability diagram shown in Figure 22.3. This corresponds to the transition probability matrix

$$\mathbf{P} = \begin{bmatrix} 1-\alpha & \alpha \\ \beta & 1-\beta \end{bmatrix} \tag{22.10}$$

where $0 \le \alpha \le 1$ and $0 \le \beta \le 1$. As always the rows sum to one. We give an example and then generalize the results.

Example 22.4 – State probability vector computation for all n

Let $\alpha = \beta = 1/2$ and $\mathbf{p}^T[0] = [1\,0]$ so that we are intially in state 0 and the transition to either of the states is equally probable. Then from (22.9) we have

$$\begin{aligned} \mathbf{p}^T[n] &= \mathbf{p}^T[0]\mathbf{P}^n \\ &= \begin{bmatrix} 1 & 0 \end{bmatrix} \begin{bmatrix} \frac{1}{2} & \frac{1}{2} \\ \frac{1}{2} & \frac{1}{2} \end{bmatrix}^n \\ \mathbf{p}^T[1] &= \begin{bmatrix} 1 & 0 \end{bmatrix} \begin{bmatrix} \frac{1}{2} & \frac{1}{2} \\ \frac{1}{2} & \frac{1}{2} \end{bmatrix} = \begin{bmatrix} \frac{1}{2} & \frac{1}{2} \end{bmatrix} \\ \mathbf{p}^T[2] &= \begin{bmatrix} 1 & 0 \end{bmatrix} \begin{bmatrix} \frac{1}{2} & \frac{1}{2} \\ \frac{1}{2} & \frac{1}{2} \end{bmatrix}^2 = \begin{bmatrix} \frac{1}{2} & \frac{1}{2} \end{bmatrix} \end{aligned}$$

Clearly, $\mathbf{p}^T[n] = \begin{bmatrix} \frac{1}{2} & \frac{1}{2} \end{bmatrix}$ for all $n \ge 1$. The Markov chain is said to be in *steady-state* for $n \ge 1$. In addition, for $n \ge 1$, the PMF $\mathbf{p}^T[n] = \begin{bmatrix} \frac{1}{2} & \frac{1}{2} \end{bmatrix}$ is called the *steady-state PMF*.

$\Diamond$

More generally, the state probabilities of a Markov chain may or may not approach a steady-state value. It depends upon the form of $\mathbf{P}$. To study the behavior more thoroughly we require a means of determining $\mathbf{P}^n$. To do so we next review the diagonalization of a matrix using an eigenanalysis (see also Appendix D).

Computing Powers of P

Assuming that the eigenvalues of $\mathbf{P}$ are distinct, it is possible to find eigenvectors $\mathbf{v}_i$ that are linearly independent. Arranging them as the columns of a matrix and assuming that $K = 2$, we have the modal matrix $\mathbf{V} = [\mathbf{v}_1 \ \mathbf{v}_2]$ which is a nonsingular matrix since the eigenvectors are linearly independent. Then we can write that

$$\mathbf{V}^{-1}\mathbf{P}\mathbf{V} = \mathbf{\Lambda} \tag{22.11}$$

where $\mathbf{\Lambda} = \text{diag}(\lambda_1, \lambda_2)$ and λ_i is the eigenvalue corresponding to the ith eigenvector of $\mathbf{P}$. Now from (22.11) we have that $\mathbf{P} = \mathbf{V}\mathbf{\Lambda}\mathbf{V}^{-1}$ and therefore, the powers of $\mathbf{P}$ can be found as follows.

$$\begin{aligned} \mathbf{P}^2 &= (\mathbf{V}\mathbf{\Lambda}\mathbf{V}^{-1})(\mathbf{V}\mathbf{\Lambda}\mathbf{V}^{-1}) = \mathbf{V}\mathbf{\Lambda}^2\mathbf{V}^{-1} \\ \mathbf{P}^3 &= \mathbf{P}^2\mathbf{P} = (\mathbf{V}\mathbf{\Lambda}^2\mathbf{V}^{-1})\mathbf{V}\mathbf{\Lambda}\mathbf{V}^{-1} = \mathbf{V}\mathbf{\Lambda}^3\mathbf{V}^{-1} \end{aligned}$$

and in general we have that

$$\mathbf{P}^n = \mathbf{V}\mathbf{\Lambda}^n\mathbf{V}^{-1}. \tag{22.12}$$

But since $\mathbf{\Lambda}$ is a diagonal matrix its powers are easily found as

$$\mathbf{\Lambda}^n = \left[\begin{array}{cc} \lambda_1^n & 0 \\ 0 & \lambda_2^n \end{array} \right]$$

and finally we have that

$$\mathbf{P}^n = \mathbf{V} \left[\begin{array}{cc} \lambda_1^n & 0 \\ 0 & \lambda_2^n \end{array} \right] \mathbf{V}^{-1}. \tag{22.13}$$

It should be observed that the eigenvectors need not be normalized to unity length for (22.13) to hold. As an example, if

$$\mathbf{P} = \left[\begin{array}{cc} \frac{1}{2} & \frac{1}{2} \\ 0 & 1 \end{array} \right]$$

then the eigenvalues are found from the characteristic equation as the solutions of $\det(\mathbf{P} - \lambda\mathbf{I}) = 0$. This yields the equation $(1/2 - \lambda)(1 - \lambda) = 0$ which produces $\lambda_1 = 1/2$ and $\lambda_2 = 1$. The eigenvectors are found from

$$(\mathbf{P} - \lambda_1\mathbf{I})\mathbf{v}_1 = \left[\begin{array}{cc} 0 & \frac{1}{2} \\ 0 & \frac{1}{2} \end{array} \right] \mathbf{v}_1 = \mathbf{0} \Rightarrow \mathbf{v}_1 = \left[\begin{array}{c} 1 \\ 0 \end{array} \right]$$

$$(\mathbf{P} - \lambda_2\mathbf{I})\mathbf{v}_2 = \left[\begin{array}{cc} -\frac{1}{2} & \frac{1}{2} \\ 0 & 0 \end{array} \right] \mathbf{v}_2 = \mathbf{0} \Rightarrow \mathbf{v}_2 = \left[\begin{array}{c} 1 \\ 1 \end{array} \right]$$

and hence the modal matrix and its inverse are

$$\mathbf{V} = [\ \mathbf{v}_1 \quad \mathbf{v}_2\ ] = \begin{bmatrix} 1 & 1 \\ 0 & 1 \end{bmatrix}$$

$$\mathbf{V}^{-1} = \begin{bmatrix} 1 & -1 \\ 0 & 1 \end{bmatrix}.$$

Finally, for $n \geq 1$ we can easily find the powers of $\mathbf{P}$ from (22.13) as

$$\mathbf{P}^n = \begin{bmatrix} 1 & 1 \\ 0 & 1 \end{bmatrix} \begin{bmatrix} \left(\frac{1}{2}\right)^n & 0 \\ 0 & 1 \end{bmatrix} \begin{bmatrix} 1 & -1 \\ 0 & 1 \end{bmatrix}$$

$$= \begin{bmatrix} \left(\frac{1}{2}\right)^n & 1 - \left(\frac{1}{2}\right)^n \\ 0 & 1 \end{bmatrix}.$$

This can easily be verified by direct multiplication of $\mathbf{P}$.

$\triangle$

Now returning to the problem of determining the state probability vector for the general two-state Markov chain, we need to first find the eigenvalues of (22.10). The characteristic equation is

$$\det(\mathbf{P} - \lambda\mathbf{I}) = \det \begin{bmatrix} 1 - \alpha - \lambda & \alpha \\ \beta & 1 - \beta - \lambda \end{bmatrix} = 0$$

which produces $(1 - \alpha - \lambda)(1 - \beta - \lambda) - \alpha\beta = 0$ or

$$\lambda^2 + (\alpha + \beta - 2)\lambda + (1 - \alpha - \beta) = 0.$$

Letting $r = \alpha + \beta$, which is nonnegative, we have that $\lambda^2 + (r - 2)\lambda + (1 - r) = 0$ for which the solution is

$$\lambda = \frac{-(r - 2) \pm \sqrt{(r - 2)^2 - 4(1 - r)}}{2}$$

$$= \frac{-(r - 2) \pm r}{2}$$

$$= 1 \text{ and } 1 - r.$$

Thus, the eigenvalues are $\lambda_1 = 1$ and $\lambda_2 = 1 - \alpha - \beta$. Next we determine the corresponding eigenvectors as

$$(\mathbf{P} - \lambda_1\mathbf{I})\mathbf{v}_1 = \begin{bmatrix} -\alpha & \alpha \\ \beta & -\beta \end{bmatrix} \mathbf{v}_1 = \mathbf{0} \Rightarrow \mathbf{v}_1 = \begin{bmatrix} 1 \\ 1 \end{bmatrix}$$

$$(\mathbf{P} - \lambda_2\mathbf{I})\mathbf{v}_2 = \begin{bmatrix} \beta & \alpha \\ \beta & \alpha \end{bmatrix} \mathbf{v}_2 = \mathbf{0} \Rightarrow \mathbf{v}_2 = \begin{bmatrix} 1 \\ -\frac{\beta}{\alpha} \end{bmatrix}$$

and therefore the modal matrix and its inverse are

$$\mathbf{V} = \begin{bmatrix} 1 & 1 \\ 1 & -\frac{\beta}{\alpha} \end{bmatrix}$$

$$\mathbf{V}^{-1} = -\frac{1}{1+\beta/\alpha} \begin{bmatrix} -\frac{\beta}{\alpha} & -1 \\ -1 & 1 \end{bmatrix}.$$

With the matrix

$$\mathbf{\Lambda}^n = \begin{bmatrix} 1 & 0 \\ 0 & (1-\alpha-\beta)^n \end{bmatrix}$$

we have

$$\mathbf{P}^n = -\frac{1}{1+\beta/\alpha} \begin{bmatrix} 1 & 1 \\ 1 & -\frac{\beta}{\alpha} \end{bmatrix} \begin{bmatrix} 1 & 0 \\ 0 & (1-\alpha-\beta)^n \end{bmatrix} \begin{bmatrix} -\frac{\beta}{\alpha} & -1 \\ -1 & 1 \end{bmatrix}$$

and after some algebra

$$\mathbf{P}^n = \begin{bmatrix} \frac{\beta}{\alpha+\beta} & \frac{\alpha}{\alpha+\beta} \\ \frac{\beta}{\alpha+\beta} & \frac{\alpha}{\alpha+\beta} \end{bmatrix} + (1-\alpha-\beta)^n \begin{bmatrix} \frac{\alpha}{\alpha+\beta} & -\frac{\alpha}{\alpha+\beta} \\ -\frac{\beta}{\alpha+\beta} & \frac{\beta}{\alpha+\beta} \end{bmatrix}. \tag{22.14}$$

We now examine three cases of interest. They are distinguished by the value that $\lambda_2 = 1-\alpha-\beta$ takes on. Clearly, as seen from (22.14) this is the factor that influences the behavior of $\mathbf{P}^n$ with n. Since α and β are both conditional probabilities we must have that $0 \le \alpha+\beta \le 2$ and hence $-1 \le \lambda_2 = 1-\alpha-\beta \le 1$. The cases are delineated by whether this eigenvalue is strictly less than one in magnitude or not.

Case 1. $-1 < 1-\alpha-\beta < 1$

Here $|1-\alpha-\beta| < 1$ and therefore from (22.14) as $n \to \infty$

$$\mathbf{P}^n \to \begin{bmatrix} \frac{\beta}{\alpha+\beta} & \frac{\alpha}{\alpha+\beta} \\ \frac{\beta}{\alpha+\beta} & \frac{\alpha}{\alpha+\beta} \end{bmatrix}. \tag{22.15}$$

As a result,

$$\mathbf{p}^T[n] = \mathbf{p}^T[0]\mathbf{P}^n \to \begin{bmatrix} p_0[0] & p_1[0] \end{bmatrix} \begin{bmatrix} \frac{\beta}{\alpha+\beta} & \frac{\alpha}{\alpha+\beta} \\ \frac{\beta}{\alpha+\beta} & \frac{\alpha}{\alpha+\beta} \end{bmatrix}$$

$$= \begin{bmatrix} \frac{\beta}{\alpha+\beta} & \frac{\alpha}{\alpha+\beta} \end{bmatrix}$$

for any $\mathbf{p}[0]$. Hence, the Markov chain approaches a steady-state irregardless of the initial state probabilities. It is said to be an *ergodic Markov chain*, the reason for which we will discuss later. Also, the state probability vector approaches the *steady-state probability vector* $\mathbf{p}^T[\infty]$, which is denoted by

$$\boldsymbol{\pi}^T = \begin{bmatrix} \pi_0 & \pi_1 \end{bmatrix} = \begin{bmatrix} \frac{\beta}{\alpha+\beta} & \frac{\alpha}{\alpha+\beta} \end{bmatrix}. \tag{22.16}$$

Finally, note that each row of $\mathbf{P}^n$ becomes the same as $n \to \infty$.

Case 2. $1 - \alpha - \beta = 1$ or $\alpha = \beta = 0$

If we draw the state probability diagram in this case, it should become clear what will happen. This is shown in Figure 22.4a, where the zero transition probability branches are omitted from the diagram. It is seen that there is no chance of leaving the initial state so that we should have $\mathbf{p}[n] = \mathbf{p}[0]$ for all n. To verify this, for $\alpha = \beta = 0$, the eigenvalues are both 1 and therefore $\mathbf{\Lambda} = \mathbf{I}$. Hence, $\mathbf{P} = \mathbf{I}$ and $\mathbf{P}^n = \mathbf{I}$. Here the Markov chain also attains steady-state and $\boldsymbol{\pi} = \mathbf{p}[0]$ but *the steady-state PMF depends upon the initial probability vector*, unlike in Case 1. Note that the only possible realizations are $0000\ldots$ and $1111\ldots$.

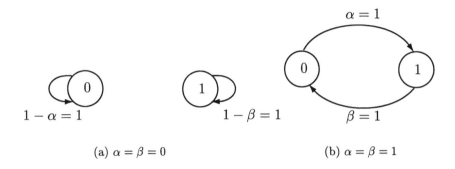

(a) $\alpha = \beta = 0$ (b) $\alpha = \beta = 1$

Figure 22.4: State probability diagrams for anomalous behaviors of two-state Markov chain.

Case 3. $1 - \alpha - \beta = -1$ or $\alpha = \beta = 1$

It is also easy to see what will happen in this case by referring to the state probability diagram in Figure 22.4b. The outcomes must alternate and thus the only realizations are $0101\ldots$ and $1010\ldots$, with the realization generated depending upon the initial state. Unlike the previous two cases, here there are no steady-state probabilities as we now show. From (22.14) we have

$$
\begin{aligned}
\mathbf{P}^n &= \begin{bmatrix} \frac{1}{2} & \frac{1}{2} \\ \frac{1}{2} & \frac{1}{2} \end{bmatrix} + (-1)^n \begin{bmatrix} \frac{1}{2} & -\frac{1}{2} \\ -\frac{1}{2} & \frac{1}{2} \end{bmatrix} \\
&= \begin{bmatrix} 1 & 0 \\ 0 & 1 \end{bmatrix} \quad \text{for } n \text{ even} \\
&= \begin{bmatrix} 0 & 1 \\ 1 & 0 \end{bmatrix} \quad \text{for } n \text{ odd.}
\end{aligned}
$$

Hence, the state probability vector is

$$\mathbf{p}^T[n] = \mathbf{p}^T[0]\mathbf{P}^n = \begin{bmatrix} p_0[0] & p_1[0] \end{bmatrix} \mathbf{P}^n$$

$$= \begin{cases} \begin{bmatrix} p_0[0] & p_1[0] \end{bmatrix} & \text{for } n \text{ even} \\ \begin{bmatrix} p_1[0] & p_0[0] \end{bmatrix} & \text{for } n \text{ odd.} \end{cases}$$

As an example, if $\mathbf{p}^T[0] = \begin{bmatrix} 1/4 & 3/4 \end{bmatrix}$, then

$$\mathbf{p}^T[n] = \begin{cases} \begin{bmatrix} \frac{1}{4} & \frac{3}{4} \end{bmatrix} & \text{for } n \text{ even} \\ \begin{bmatrix} \frac{3}{4} & \frac{1}{4} \end{bmatrix} & \text{for } n \text{ odd} \end{cases}$$

as shown in Figure 22.5. It is seen that the state probabilities cycle between two PMFs and hence there is no steady-state.

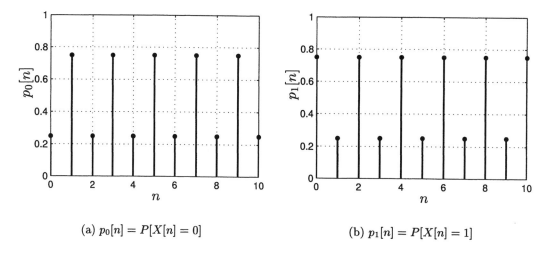

(a) $p_0[n] = P[X[n] = 0]$ (b) $p_1[n] = P[X[n] = 1]$

Figure 22.5: Cycling of state probability vector for Case 3.

The last two cases are of little practical importance for a two-state Markov chain since we usually have $0 < \alpha < 1$ and $0 < \beta < 1$. However, for a K-state Markov chain it frequently occurs that some of the transition probabilities are zero (corresponding to missing branches of the state probability diagram and an inability of the Markov chain to transition between certain states). Then, the dependence upon the initial state and cycling or periodic PMFs become quite important. The interested reader should consult [Gallagher 1996] and [Cox and Miller 1965] for further details. We next return to our golfing friend.

Example 22.5 – One-putting

Recall that our golfer had a transition probability matrix given by

$$\mathbf{P} = \begin{bmatrix} \frac{3}{4} & \frac{1}{4} \\ \frac{1}{2} & \frac{1}{2} \end{bmatrix}.$$

It is seen from (22.10) that $\alpha = 1/4$ and $\beta = 1/2$ and so this corresponds to Case 1 in which the same steady-state probability is reached regardless of the initial probability vector. Hence, as $n \to \infty$, $\mathbf{P}^n$ will converge to a constant matrix and therefore so will $\mathbf{p}[n]$. After many rounds of golf the probability of a one-putt or of going to state 1 is found from the second element of the stationary probability vector $\boldsymbol{\pi}$. This is from (22.16)

$$\boldsymbol{\pi}^T = \begin{bmatrix} \pi_0 & \pi_1 \end{bmatrix} = \begin{bmatrix} \frac{\beta}{\alpha+\beta} & \frac{\alpha}{\alpha+\beta} \end{bmatrix}$$
$$= \begin{bmatrix} \frac{1/2}{3/4} & \frac{1/4}{3/4} \end{bmatrix}$$
$$= \begin{bmatrix} \frac{2}{3} & \frac{1}{3} \end{bmatrix}$$

so that her probability of a one-putt is now only $1/3$ as we surmised by examination of Figure 22.1. At the first hole it was $p_1[0] = 1/2$. To determine how many holes she must play until this steady-state probability is attained we let this be $n = n_{ss}$ and determine from (22.14) when $(1 - \alpha - \beta)^{n_{ss}} = (1/4)^{n_{ss}} \approx 0$. This is about $n_{ss} = 10$ for which $(1/4)^{10} = 10^{-6}$. The actual state probability vector is shown in Figure 22.6 using an initial state probability of $\mathbf{p}^T[0] = [1/2 \; 1/2]$. The steady-state values of $\boldsymbol{\pi} = [2/3 \; 1/3]^T$ are also shown as dashed lines.

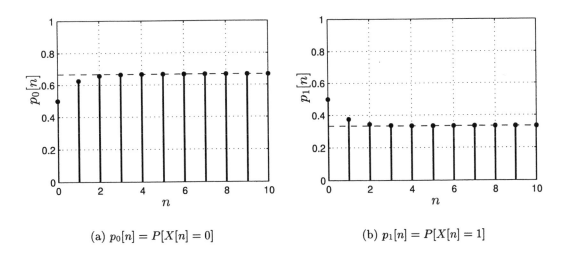

(a) $p_0[n] = P[X[n] = 0]$ (b) $p_1[n] = P[X[n] = 1]$

Figure 22.6: Convergence of state probability vector for Case 1 with $\alpha = 1/4$ and $\beta = 1/2$.

22.5   Ergodic Markov Chains

We saw in the previous section that as $n \to \infty$, then for some $\mathbf{P}$ the state probability vector approaches a steady-state value irregardless of the initial state probabilities. This was Case 1 for which each element of $\mathbf{P}$ was nonzero or

$$\mathbf{P} = \begin{bmatrix} 1 - \alpha & \alpha \\ \beta & 1 - \beta \end{bmatrix} > 0$$

where the "> 0" is meant to indicate that every element of $\mathbf{P}$ is greater than zero. Equivalently, all the branches of the state probability diagram were present. A Markov chain of this type is said to be ergodic in that a temporal average is equal to an ensemble average as we will later show. The key requirement for this to be true for any K-state Markov chain is that the $K \times K$ transition probability matrix satisfies $\mathbf{P} > 0$. The matrix $\mathbf{P}$ then has some special properties. We already have pointed out that the rows must sum to one; a matrix of this type is called a *stochastic matrix*, and for ergodicity, we must have $\mathbf{P} > 0$; a matrix satisfying this requirement is called an *irreducible stochastic matrix*. The associated Markov chain is known as an *ergodic or irreducible Markov chain*. A theorem termed the Perron-Frobenius theorem [Gallagher 1996] states that if $\mathbf{P} > 0$, then the transition probability matrix will always have one eigenvalue equal to 1 and the remaining eigenvalues will have magnitudes strictly less than 1. Such was the case for the two-state probability transition matrix of Case 1 for which $\lambda_1 = 1$ and $|\lambda_2| = |1 - \alpha - \beta| < 1$. This condition on $\mathbf{P}$ assures convergence of $\mathbf{P}^n$ to a constant matrix. Convergence may also occur if some of the elements of $\mathbf{P}$ are zero but it is not guaranteed. A slightly more general condition for convergence is that $\mathbf{P}^n > 0$ for some n (not necessarily $n = 1$). An example is

$$\mathbf{P} = \begin{bmatrix} \frac{1}{2} & \frac{1}{2} & 0 \\ 0 & \frac{1}{2} & \frac{1}{2} \\ \frac{1}{2} & 0 & \frac{1}{2} \end{bmatrix}$$

(see Problem 22.13).

We now assume that $\mathbf{P} > 0$ and determine the steady-state probabilities for a general K-state Markov chain. Since

$$\mathbf{p}^T[n] = \mathbf{p}^T[n-1]\mathbf{P}$$

and in steady-state we have that $\mathbf{p}^T[n-1] = \mathbf{p}^T[n] = \mathbf{p}^T[\infty]$, it follows that

$$\mathbf{p}^T[\infty] = \mathbf{p}^T[\infty]\mathbf{P}.$$

Letting the steady-state probability vector be $\boldsymbol{\pi} = \mathbf{p}[\infty]$, we have

$$\boldsymbol{\pi}^T = \boldsymbol{\pi}^T \mathbf{P} \tag{22.17}$$

and we need only solve for $\boldsymbol{\pi}$. An example follows.

Example 22.6 – Two-state Markov chain

We solve for the steady-state probability vector for Case 1. From (22.17) we have

$$\begin{bmatrix} \pi_0 & \pi_1 \end{bmatrix} = \begin{bmatrix} \pi_0 & \pi_1 \end{bmatrix} \begin{bmatrix} 1-\alpha & \alpha \\ \beta & 1-\beta \end{bmatrix}$$

so that

$$\begin{aligned} \pi_0 &= (1-\alpha)\pi_0 + \beta\pi_1 \\ \pi_1 &= \alpha\pi_0 + (1-\beta)\pi_1 \end{aligned}$$

or

$$\begin{aligned} 0 &= -\alpha\pi_0 + \beta\pi_1 \\ 0 &= \alpha\pi_0 - \beta\pi_1. \end{aligned}$$

The yields $\pi_1 = (\alpha/\beta)\pi_0$ since the two linear equations are identical. Of course, we also require that $\pi_0 + \pi_1 = 1$ and so this forms the second linear equation. The solution then is

$$\begin{aligned} \pi_0 &= \frac{\beta}{\alpha+\beta} \\ \pi_1 &= \frac{\alpha}{\alpha+\beta} \end{aligned} \tag{22.18}$$

and agrees with our previous results of (22.16).

$\Diamond$

It can further be shown that if a steady-state probability vector exists (which will be the case if $\mathbf{P} > 0$), then the solution for $\boldsymbol{\pi}$ is unique [Gallagher 1996]. Finally, note that if we intialize the Markov chain with $\mathbf{p}[0] = \boldsymbol{\pi}$, then since $\mathbf{p}^T[1] = \mathbf{p}^T[0]\mathbf{P} = \boldsymbol{\pi}^T\mathbf{P} = \boldsymbol{\pi}^T$, the state probability vector will be $\boldsymbol{\pi}$ for $n \geq 0$. The Markov chain is then stationary since the state probability vector is the same for all n and $\boldsymbol{\pi}$ is therefore referred to as the *stationary probability vector*. We will henceforth use this terminology for $\boldsymbol{\pi}$.

Another observation of importance is that if $\mathbf{P} > 0$, then $\mathbf{P}^n$ converges, and it converges to $\mathbf{P}^\infty$, whose rows are identical. This was borne out in (22.15) and is true in general (see Problem 22.17). (Note that this is not true for Case 2 in which although $\mathbf{P}^n$ converges, it converges to $\mathbf{I}$, whose rows are not the same.) As a result of this property, the steady-state value of the state probability vector does not depend upon the initial probabilities since

$$\begin{aligned} \mathbf{p}^T[n] &= \mathbf{p}^T[0]\mathbf{P}^n \\ &= \begin{bmatrix} p_0[0] & p_1[0] \end{bmatrix} \begin{bmatrix} \frac{\beta}{\alpha+\beta} & \frac{\alpha}{\alpha+\beta} \\ \frac{\beta}{\alpha+\beta} & \frac{\alpha}{\alpha+\beta} \end{bmatrix} + \underbrace{\mathbf{p}^T[0](1-\alpha-\beta)^n \begin{bmatrix} \frac{\alpha}{\alpha+\beta} & -\frac{\alpha}{\alpha+\beta} \\ -\frac{\beta}{\alpha+\beta} & \frac{\beta}{\alpha+\beta} \end{bmatrix}}_{\to \mathbf{0}^T \text{ as } n\to\infty} \\ &\to \begin{bmatrix} \frac{\beta}{\alpha+\beta} & \frac{\alpha}{\alpha+\beta} \end{bmatrix} = \boldsymbol{\pi}^T \end{aligned}$$

independent of $\mathbf{p}^T[0]$. Also, as previously mentioned, if $\mathbf{P} > 0$, then as $n \to \infty$

$$\mathbf{P}^n \to \begin{bmatrix} \frac{\beta}{\alpha+\beta} & \frac{\alpha}{\alpha+\beta} \\ \frac{\beta}{\alpha+\beta} & \frac{\alpha}{\alpha+\beta} \end{bmatrix}$$

whose rows are identical. As a result, we have that

$$\lim_{n\to\infty} [\mathbf{P}^n]_{ij} = \pi_j \qquad j = 0, 1, \ldots, K-1. \tag{22.19}$$

Hence, the stationary probabilities may be obtained either by solving the set of linear equations as was done for Example 22.6 or by examining a row of $\mathbf{P}^n$ as $n \to \infty$. In Section 22.7 we give the general solution for the stationary probabilities. We next give another example.

Example 22.7 – Machine failures

A machine is in operation at the beginning of day $n = 0$. It may break during operation that day in which case repairs will begin at the beginning of the next day ($n = 1$). In this case, the machine will not be in operation at the beginning of day $n = 1$. There is a probability of $1/2$ that the technician will be able to repair the machine that day. If it is repaired, then the machine will be in operation for day $n = 2$ and if not, the technician will again attempt to fix it the next day ($n = 2$). The probability that the machine will operate without a failure during the day is $7/8$. After many days of operation or failure what is the probability that the machine will be working at the *beginning* of a day? Here there are two states, either $X[n] = 0$ if the machine is not in operation at the beginning of day n, or $X[n] = 1$ if the machine is in operation at the beginning of day n. The transition probabilities are given as

$$p_{01} = P[\text{machine operational on day } n|\text{machine nonoperational on day } n-1] = \frac{1}{2}$$

$$p_{11} = P[\text{machine operational on day } n|\text{machine operational on day } n-1] = \frac{7}{8}$$

and so the state transition probability matrix is

$$\mathbf{P} = \begin{bmatrix} \frac{1}{2} & \frac{1}{2} \\ \frac{1}{8} & \frac{7}{8} \end{bmatrix}$$

noting that $p_{00} = 1 - p_{01} = 1/2$ and $p_{10} = 1 - p_{11} = 1/8$. This Markov chain is shown in Figure 22.7. Since $\mathbf{P} > 0$, a steady-state is reached and the stationary probabilities are from (22.18)

$$\pi_0 = \frac{\beta}{\alpha + \beta} = \frac{\frac{1}{8}}{\frac{1}{2} + \frac{1}{8}} = \frac{1}{5}$$

$$\pi_1 = 1 - \pi_0 = \frac{4}{5}.$$

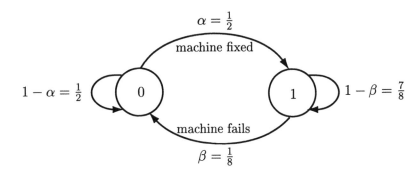

Figure 22.7: State probability diagram for Example 22.7.

The machine will be in operation at the beginning of a day with a probability of 0.8.

$\diamond$

Note that in the last example the states of 0 and 1 are arbitrary labels. They could just as well have been "nonoperational" and "operational". In problems such as these the state description is chosen to represent meaningful attributes of interest. One last comment concerns our apparent preoccupation with the steady-state behavior of a Markov chain. Although not always true, we are many times only interested in this because the choice of a starting time, i.e., at $n = 0$, is not easy to specify. In the previous example, it is conceivable that the machine in question has been in operation for a long time and it is only recently that a plant manager has become interested in its failure rate. Therefore, its initial starting time was probably some time in the past and we are now observing the states for some large n. We continue our discussion of steady-state characteristics in the next section.

22.6 Further Steady-State Characteristics

22.6.1 State Occupation Time

It is frequently of interest to be able to determine the percentage of time that a Markov chain is in a particular state, also called the *state occupation time*. Such was the case in Example 22.7, although a careful examination reveals that what we actually computed was the *probability* of being operational at the beginning of each day. In essence we are now asking for the *relative frequency* (or percentage of time) of the machine being operational. This is much the same as asking for the relative frequency of heads in a long sequence of independent fair coin tosses. We have proven by the law of large numbers (see Chapter 15) that this relative frequency must approach a probability of 1/2 as the number of coin tosses approaches infinity. For Markov chains the trials are not independent and so the law of large numbers does not apply directly. However, as we now show, if steady-state is attained, then the fraction of time the Markov chain spends in a particular state approaches the steady-state probability. This allows us to say that the fraction of time that the Markov chain spends in state j is just π_j.

Again consider a two-state Markov chain with states 0 and 1 and assume that $\mathbf{P} > 0$. We wish to determine the fraction of time spent in state 1. For some large n this is given by

$$\frac{1}{N} \sum_{j=n}^{n+N-1} X[j]$$

which is recognized as the sample mean of the N state outcomes for $\{X[n], X[n+1], \ldots, X[n+N-1]\}$. We first determine the expected value as

$$E\left[\frac{1}{N} \sum_{j=n}^{n+N-1} X[j]\right] = E_{X[0]}\left[E\left[\frac{1}{N} \sum_{j=n}^{n+N-1} X[j] \,\middle|\, X[0] = i\right]\right]$$

$$= E_{X[0]}\left[\frac{1}{N} \sum_{j=n}^{n+N-1} E[X[j]|X[0] = i]\right]. \qquad (22.20)$$

But

$$E[X[j]|X[0] = i] = P[X[j] = 1|X[0] = i]$$
$$= [\mathbf{P}^j]_{i1} \to \pi_1$$

as $j \geq n \to \infty$ which follows from (22.19). The expected value does not depend upon the initial state i. Therefore, we have from (22.20) that

$$E\left[\frac{1}{N} \sum_{j=n}^{n+N-1} X[j]\right] \to E_{X[0]}\left[\frac{1}{N} \sum_{j=n}^{n+N-1} \pi_1\right] = \pi_1.$$

Thus, as $n \to \infty$, the *expected* fraction of time in state 1 is π_1. Furthermore, although it is more difficult to show, the variance of the sample mean converges to zero as $N \to \infty$ so that the fraction of time (and not just the expected value) spent in state 1 will converge to π_1 or

$$\frac{1}{N} \sum_{j=n}^{n+N-1} X[j] \to \pi_1. \qquad (22.21)$$

This is the same result as for the repeated independent tossing of a fair coin. The result stated in (22.21) is that the *temporal mean* is equal to the ensemble mean which says that for large n, i.e., in steady-state, $\frac{1}{N} \sum_{j=n}^{n+N-1} X[j] \to \pi_1$ as $N \to \infty$. This is the property of *ergodicity* as previously described in Chapter 17. Thus, a Markov chain that achieves a steady-state irregardless of the initial state probabilities is called an *ergodic Markov chain*.

Returning to our golfing friend, we had previously questioned the fraction of the time she will achieve one-putts. We know that her stationary probability is $\pi_1 = 1/3$. Thus, after playing many rounds of golf, she will be one-putting about $1/3$ of the time.

22.6.2 Mean Recurrence Time

Another property of the ergodic Markov chain that is of interest is the average number of steps before a state is revisited. For example, the golfer may wish to know the average number of holes she will have to play before another one-putt occurs, *given that she has just one-putted*. This is equivalent to determining the average number of steps the Markov chain will undergo before it returns to state 1. The time between visits to the same state is called the *recurrence time* and the average of this is called the *mean recurrence time*. We next determine this average.

Let T_R denote the recurrence time and note that it is an integer random variable that can take on values in the sample space $\{1, 2, \ldots\}$. For the two-state Markov chain shown in Figure 22.3 we first assume that we are in state 1 at time $n = n_0$. Then, the value of the recurrence time will be 1, or 2, or 3, etc. if $X[n_0 + 1] = 1$, or $X[n_0 + 1] = 0, X[n_0 + 2] = 1$, or $X[n_0 + 1] = 0, X[n_0 + 2] = 0, X[n_0 + 3] = 1$, etc., respectively. The probabilities of these events are $1 - \beta$, $\beta\alpha$, and $\beta(1 - \alpha)\alpha$, respectively as can be seen by referring to Figure 22.3. In general, the PMF is given as

$$P[T_R = k|\text{initially in state 1}] = \begin{cases} 1 - \beta & k = 1 \\ \beta\alpha(1 - \alpha)^{k-2} & k \geq 2 \end{cases}$$

which is a geometric-type PMF (see Chapter 5). To find the mean recurrence time we need only determine the expected value of T_R. This is

$$\begin{aligned} E[T_R|\text{initially in state 1}] &= (1 - \beta) + \sum_{k=2}^{\infty} k\left[\beta\alpha(1 - \alpha)^{k-2}\right] \\ &= (1 - \beta) + \alpha\beta\sum_{l=1}^{\infty}(l + 1)(1 - \alpha)^{l-1} \quad (\text{let } l = k - 1) \\ &= (1 - \beta) + \left[\alpha\beta\sum_{l=1}^{\infty}(1 - \alpha)^{l-1} + \beta\sum_{l=1}^{\infty}l\underbrace{\alpha(1 - \alpha)^{l-1}}_{\text{geom}(\alpha)\text{ PMF}}\right] \\ &= (1 - \beta) + \alpha\beta\frac{1}{1 - (1 - \alpha)} + \beta\frac{1}{\alpha} \quad (\text{from Section 6.4.3}) \\ &= \frac{\alpha + \beta}{\alpha} \end{aligned}$$

so that we have finally

$$E[T_R|\text{initially in state 1}] = \frac{1}{\pi_1}.$$

It is seen that mean recurrence time is the reciprocal of the stationary state probability. This is much the same result as for a geometric PMF and is interpreted as the number of failures (not returning to state 1) before a success (returning to state 1). For our golfer, since she has a stationary probability of one-putting of 1/3, she must wait on the average $1/(1/3)=3$ holes between one-putts. This agrees with our simulation results shown in Figure 22.1.

22.7 K-State Markov Chains

Markov chains with more than two states are quite common and useful in practice but their analysis can be difficult. Most of the previous properties of a Markov chain apply to any finite number K of states. Computation of the n-step transition probability matrix is of course more difficult and requires computer evaluation. Most importantly, however, is that steady-state is still attained if $\mathbf{P} > 0$. The solution for the stationary probabilities is given next. It is derived in Appendix 22A.

The stationary probability vector for a K-state Markov chain is $\boldsymbol{\pi}^T = [\pi_0 \, \pi_1 \dots \pi_{K-1}]$. Its solution is given as

$$\boldsymbol{\pi} = (\mathbf{I} - \mathbf{P}^T + \mathbf{11}^T)^{-1}\mathbf{1} \tag{22.22}$$

where $\mathbf{I}$ is the $K \times K$ identity matrix and $\mathbf{1} = [1 \, 1 \dots 1]^T$, which is a $K \times 1$ vector of ones. We next give an example of a 3-state Markov chain.

Example 22.8 – Weather modeling

 Assume that the weather for each day can be classified as being either rainy (state 0), cloudy (state 1), or sunny (state 2). We wish to determine in the long run (steady-state) the percentage of sunny days. From the discussion in Section 22.6.1 this is the state occupation time, and is equal to the stationary probability π_2. To do so we assume the conditional probabilities

$$\text{currently raining (state 0)} \quad : \quad P_{00} = \frac{4}{8}, P_{01} = \frac{3}{8}, P_{02} = \frac{1}{8}$$

$$\text{currently cloudy (state 1)} \quad : \quad P_{10} = \frac{3}{8}, P_{11} = \frac{2}{8}, P_{12} = \frac{3}{8}$$

$$\text{currently sunny (state 2)} \quad : \quad P_{20} = \frac{1}{8}, P_{21} = \frac{3}{8}, P_{22} = \frac{4}{8}.$$

This says that if it is currently raining, then it is most probable that the next day will also have rain (4/8). The next most probable weather condition will be cloudy for the next day (3/8), and the least probable weather condition is sunny for the next day (1/8). See if you can rationalize the other entries in **P**. The complete state transition probability matrix is

$$\mathbf{P} = \begin{bmatrix} \frac{4}{8} & \frac{3}{8} & \frac{1}{8} \\ \frac{3}{8} & \frac{2}{8} & \frac{3}{8} \\ \frac{1}{8} & \frac{3}{8} & \frac{4}{8} \end{bmatrix}$$

and the state probability diagram is shown in Figure 22.8. We can use this to

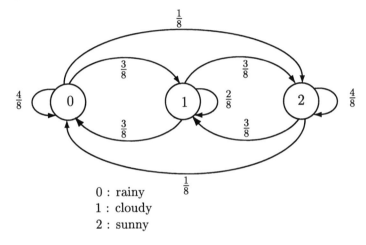

0 : rainy
1 : cloudy
2 : sunny

Figure 22.8: Three-state probability diagram for weather example.

determine the probability of the weather conditions on any day if we know the weather on day $n = 0$. For example, to find the probability of the weather on Saturday knowing that it is raining on Monday, we use

$$\mathbf{p}^T[n] = \mathbf{p}^T[0]\mathbf{P}^n$$

with $n = 5$ and $\mathbf{p}^T[0] = [1\,0\,0]$. Using a computer to evalute this we have that

$$\mathbf{p}[5] = \begin{bmatrix} 0.3370 \\ 0.3333 \\ 0.3296 \end{bmatrix}$$

and it appears that the possible weather conditions are nearly equiprobable. To find the stationary probabilities for the weather conditions we must solve $\boldsymbol{\pi}^T = \boldsymbol{\pi}^T\mathbf{P}$. Using the solution of (22.22), we find that

$$\boldsymbol{\pi} = \begin{bmatrix} \frac{1}{3} \\ \frac{1}{3} \\ \frac{1}{3} \end{bmatrix}.$$

As $n \to \infty$, it is equiprobable that the weather will be rainy, cloudy, or sunny. Furthermore, because of ergodicity the fraction of days that it will be rainy, or be cloudy, or be sunny will all be 1/3.

$$\diamond$$

The previous result that the stationary probabilities are equal is true in general for the type of transition probability matrix given. Note that $\mathbf{P}$ not only has all its rows summing to one but also its column entries sum to one for all the columns. This is called a *doubly stochastic matrix* and always results in equal stationary probabilities (see Problem 22.27).

22.8   Computer Simulation

The computer simulation of a Markov chain is very simple. Consider the weather example of the previous section. We first need to generate a realization of a random variable taking on the values $0, 1, 2$ with the PMF $p_0[0], p_1[0], p_2[0]$. This can be done using the approach of Section 5.9. Once the realization has been obtained, say $x[0] = i$, we continue the same procedure *but* must choose the next PMF, which is actually a *conditional PMF*. If $x[0] = i = 1$ for example, then we use the PMF $p[0|1] = P_{10}, p[1|1] = P_{11}, p[2|1] = P_{12}$, which are just the entries in the second row of $\mathbf{P}$. We continue this procedure for all $n \geq 1$. Some MATLAB code to generate a realization for the weather example is given below.

```
clear all
rand('state',0)
N=1000;   % set number of samples desired
p0=[1/3 1/3 1/3]'; % set initial probability vector
P=[4/8 3/8 1/8;3/8 2/8 3/8;1/8 3/8 4/8]; % set transition prob. matrix
xi=[0 1 2]'; % set values of PMF
X0=PMFdata(1,xi,p0); % generate X[0] (see Appendix 6B for PMFdata.m
                     % function subprogram)
i=X0+1; % choose appropriate row for PMF
X(1,1)=PMFdata(1,xi,P(i,:)); % generate X[1]
i=X(1,1)+1; % choose appropriate row for PMF
for n=2:N % generate X[n]
   i=X(n-1,1)+1; % choose appropriate row for PMF
   X(n,1)=PMFdata(1,xi,P(i,:));
end
```

The reader may wish to modify and run this program to gain some insight into the effect of the conditional probabilities on the predicted weather patterns.

22.9 Real-World Example – Strange Markov Chain Dynamics

It is probably fitting that as the last real-world example, we choose one that questions what the real-world actually is. Is it a place of determinism, however complex, or one that is subject to the whims of chance events? *Random*, as defined by Webster's dictionary, means "lacking a definite plan, purpose, or *pattern*". Is this a valid definition? We do not plan to answer this question, but only to present some "food for thought". The seemingly random Markov chain provides an interesting example.

Consider a square arrangement of 101×101 points and define a set of states as the locations of the integer points within this square. The points are therefore denoted by the integer coordinates (i, j), where $i = 0, 1, \ldots, 100; j = 0, 1, \ldots, 100$. The number of states is $K = 101^2$. Next define a Markov chain for this set of states such that the nth outcome is a realization of the random point $\mathbf{X}[n] = [I[n] \, J[n]]^T$, where $I[n]$ and $J[n]$ are random variables taking on integer values in the interval $[0, 100]$. The initial point is chosen to be $\mathbf{X}[0] = [10 \, 80]^T$ and succeeding points evolve according to the random process:

1. Choose at random one of the *reference points* $(0, 0), (100, 0), (50, 100)$.

2. Find the midpoint between the initial point and the chosen reference point and round it to the nearest integer coordinates (so that it becomes a state output).

3. Replace the initial point with the one found in step 2.

4. Go to step 1 and repeat the process, always using the previous point and one of the reference points chosen at random.

This procedure is equivalent to the formula

$$\mathbf{X}[n] = \left[\frac{1}{2}(\mathbf{X}[n-1] + \mathbf{R}[n]) \right]_{\text{round}} \qquad n \geq 1 \qquad (22.23)$$

where $\mathbf{R}[n] = [r_1[n] r_2[n]]^T$ is the reference point chosen at random and $[\cdot]_{\text{round}}$ denotes rounding of both elements of the vector to the nearest integer. Note that this is a Markov chain. The points generated must all lie within the square at integer coordinates due to the averaging and rounding that is ongoing. Also, the current output only depends upon the previous output $\mathbf{X}[n-1]$, i.e., justifying the claim of a Markov chain. The process is "random" due to our choice of $\mathbf{R}[n]$ from the sample space $\{(0, 0), (100, 0), (50, 100)\}$ with equal probabilities.

The behavior of this Markov chain is shown in Figure 22.9, where the successive output points have been plotted with the first few shown with their values of n. It appears that the chain attains a steady-state and its steady-state PMF is zero over many triangular regions. It is interesting to note that the pattern consists of 3 triangles—one with vertices $(0, 0), (50, 0), (25, 50)$, and the others with vertices

$(50, 0), (100, 0), (75, 50)$, and $(25, 50), (75, 50), (50, 100)$. Within each of these triangles resides an exact replica of the whole pattern and within each replica resides another replica, etc.! Such a figure is called a *fractal* with this particular one termed a *Sierpinski triangle*. The MATLAB code used to produce this figure is given below.

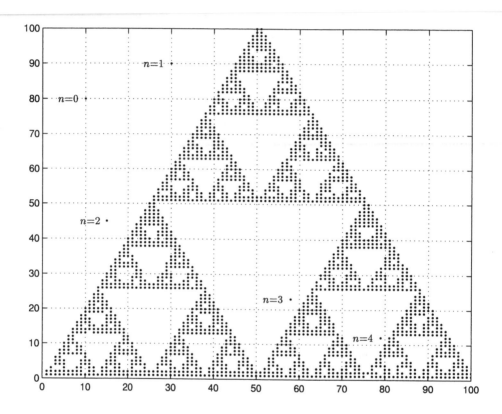

Figure 22.9: Steady-state Markov chain.

```
% sierpinski.m
%
clear all
rand('state',0)
r(:,1)=[0 0]'; % set up reference points
r(:,2)=[100 0]';
r(:,3)=[50 100]';
x0=[10 80]'; % set initial state
plot(x0(1),x0(2),'.') % plot state outcome as point
axis([0 100 0 100])
hold on
xn_1=x0;
```

```
for n=1:10000 % generate states
   j=floor(3*rand(1,1)+1); % choose at random one of three
                           % reference points
   xn=round(0.5*(r(:,j)+xn_1)); % generate new state
   plot(xn(1),xn(2),'.') % plot state outcome as point
   xn_1=xn; % make current state the previous one for
            % next transition
end
grid
hold off
```

The question arises as to whether the Markov chain is deterministic or random. We choose not to answer this question (because we don't know the answer!). Instead we refer the interested reader to the excellent book [Peitgen, Jurgens, and Saupe 1992] and also the popular layman's account [Gleick 1987] for further details. As a more practical application, it is observed that seemingly complex figures can be generated using a simple algorithm. This leads to the idea of data compression in which the only information needed to store a complex figure is the details of the algorithm. A field of sunflowers is such an example for which the reader should consult [Barnsley 1988] on how this is done.

References

Barnsley, M., *Fractals Everywhere*, Academic Press, New York, 1988.

Bharucha-Reid, A.T., *Elements of the Theory of Markov Processes and Their Applications*, Dover, Mineola, NY, 1988.

Cox, D.R., H.D. Miller, *The Theory of Stochastic Processes*, Methuen and Co, LTD, London, 1965.

Gallagher, R.G., *Discrete Stochastic Processes*, Kluwer Academic Press, Boston, 1996.

Gleick, J., *Chaos, Making a New Science*, Penguin Books, New York, 1987.

Parzen, E., *Stochastic Processes*, Holden-Day, San Francisco, 1962.

Peitgen, H., H. Jurgens, D. Saupe, *Chaos and Fractals*, Springer-Verlag, New York, 1992.

Problems

22.1 (w) A Markov chain has the states "A" and "B" or equivalently 0 and 1. If the conditional probabilities are $P[A|B] = 0.1$ and $P[B|A] = 0.4$, draw the state probability diagram. Also, find the transition probability matrix.

22.2 (∵) (f) For the state probability diagram shown in Figure 22.2 find the probability of obtaining the outcomes $X[n] = 0, 1, 0, 1, 1$ for $n = 0, 1, 2, 3, 4$, respectively.

22.3 (f) For the state probability diagram shown in Figure 22.3 find the probabilities of the outcomes $X[n] = 0, 1, 0, 1, 1, 1$ for $n = 0, 1, 2, 3, 4, 5$, respectively and also for $X[n] = 1, 1, 0, 1, 1, 1$ for $n = 0, 1, 2, 3, 4, 5$, respectively. Compare the two and explain the difference.

22.4 (w) In some communication systems it is important to determine the percentage of time a person is talking. From measurements it is found that if a person is talking in a given time interval, then he will be talking in the next time interval with a probability of 0.75. If he is not talking in a time interval, then he will be talking in the next time interval with a probability of 0.5. Draw the state probability diagram using the states "talking" and "not talking".

22.5 (∵) (t) In this problem we give an example of a random process that does *not* have the Markov property. The random process is defined as an *exclusive OR* logical function. This is $Y[n] = X[n] \oplus X[n-1]$ for $n \geq 0$, where $X[n]$ for $n \geq 0$ takes on values 0 and 1 with probabilities $1 - p$ and p, respectively. The $X[n]$'s are IID. Also, for $n = 0$ we define $Y[0] = X[0]$. The definition of this operation is that $Y[n] = 0$ only if $X[n]$ and $X[n-1]$ are the same (both equal to 0 or both equal to 1), and otherwise $Y[n] = 1$. Determine $P[Y[2] = 1 | Y[1] = 1, Y[0] = 0]$ and $P[Y[2] = 1 | Y[1] = 1]$ to show that they are not equal in general.

22.6 (f) For the transition probability matrix given below draw the corresponding state probability diagram.

$$\mathbf{P} = \begin{bmatrix} \frac{1}{2} & \frac{1}{4} & \frac{1}{4} \\ \frac{1}{3} & \frac{1}{3} & \frac{1}{3} \\ \frac{2}{3} & \frac{1}{6} & \frac{1}{6} \end{bmatrix}$$

22.7 (w) A fair die is tossed many times in succession. The tosses are independent of each other. Let $X[n]$ denote the maximum of the first $n + 1$ tosses. Determine the transition probability matrix. Hint: The maximum value cannot decrease as n increases.

22.8 (w) A particle moves along the circle shown in Figure 22.10 from one point to the other in a clockwise (CW) or counterclockwise (CCW) direction. At each step it can move either CW 1 unit or CCW 1 unit. The probabilities are $P[\text{CCW}] = p$ and $P[\text{CW}] = 1 - p$ and do not depend upon the current

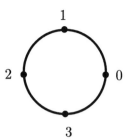

Figure 22.10: Movement of particle along a circle for Problem 22.8.

location of the particle. For the states $0, 1, 2, 3$ find the transition probability matrix.

22.9 (☺) (w,c) A digital communication system transmits a 0 or a 1. After 10 miles of cable a repeater decodes the bit and declares it either a 0 or a 1. The probability of a decoding error is 0.1 as shown schematically in Figure 22.11. It is then retransmitted to the next repeater located 10 miles away. If the repeaters are all located 10 miles apart and the communication system is 50 miles in length, find the probability of an error if a 0 is initially transmitted. Hint: You will need a computer to work this problem.

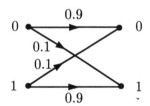

Figure 22.11: One section of a communication link.

22.10 (w,c) If $\alpha = \beta = 1/4$ for the state probability diagram shown in Figure 22.3, determine n so that the Markov chain is in steady-state. Hint: You will need a computer to work this problem.

22.11 (☺) (w) There are two urns filled with red and black balls. Urn 1 has 60% red balls and 40% black balls while urn 2 has 20% red balls and 80% black balls. A ball is drawn from urn 1, its color noted, and then replaced. If it is red, the next ball is also drawn from urn 1, its color noted and then replaced. If the ball is black, then the next ball is drawn from urn 2, its color noted and then replaced. This procedure is continued indefinitely. Each time a ball is drawn the next ball is drawn from urn 1 if the ball is red and from urn 2 if it is black. After many trials of this experiment what is the probability of drawing a red ball? Hint: Define the states 1 and 2 as urns 1 and 2 chosen. Also, note that $P[\text{red drawn}] = P[\text{red drawn}|\text{urn 1 chosen}]P[\text{urn 1 chosen}] + P[\text{red drawn}|\text{urn 2 chosen}]P[\text{urn 2 chosen}]$.

22.12 (⌣) (w) A contestant answers questions posed to him from a game show host. If his answer is correct, the game show host gives him a harder question for which his probability of answering correctly is 0.01. If however, his answer is incorrect, the contestant is given an easy question for which his probability of answering correctly is 0.99. After answering many questions, what is the probability of answering a question correctly?

22.13 (f) For the transition probability matrix

$$\mathbf{P} = \begin{bmatrix} \frac{1}{2} & \frac{1}{2} & 0 \\ 0 & \frac{1}{2} & \frac{1}{2} \\ \frac{1}{2} & 0 & \frac{1}{2} \end{bmatrix}$$

will $\mathbf{P}^n$ converge as $n \to \infty$? You should be able to answer this question without the use of a computer. Hint: Determine $\mathbf{P}^2$.

22.14 (⌣) (w,c) For the transition probability matrix

$$\mathbf{P} = \begin{bmatrix} \frac{1}{2} & \frac{1}{2} & 0 & 0 \\ \frac{1}{4} & \frac{3}{4} & 0 & 0 \\ \frac{1}{4} & \frac{1}{4} & \frac{1}{4} & \frac{1}{4} \\ \frac{1}{4} & \frac{1}{4} & \frac{1}{4} & \frac{1}{4} \end{bmatrix}$$

does the Markov chain attain steady-state? If it does, what are the steady-state probabilities? Hint: You will need a computer to evaluate the answer.

22.15 (w,c) There are three lightbulbs that are always on in a room. At the beginning of each day the custodian checks to see if at least one lightbulb is working. If all three lightbulbs have failed, then he will replace them all. During the day each lightbulb will fail with a probability of 1/2 and the failure is independent of the other lightbulbs failing. Letting the state be the number of working lightbulbs draw the state probability diagram and determine the transition probability matrix. Show that eventually all three bulbs must fail and the custodian will then have to replace them. Hint: You will need a computer to work this problem.

22.16 (f) Find the stationary probabilities for the transition probability matrix

$$\mathbf{P} = \begin{bmatrix} \frac{1}{3} & \frac{2}{3} \\ \frac{1}{4} & \frac{3}{4} \end{bmatrix}.$$

22.17 (t) In this problem we discuss the proof of the property that if $\mathbf{P} > 0$, the rows of $\mathbf{P}^n$ will all converge to the same values and that these values are the stationary probabilities. We consider the case of $K = 3$ for simplicity and assume distinct eigenvalues. Then, it is known from the Perron-Frobenius theorem that we will have the eigenvalues $\lambda_1 = 1$, $|\lambda_2| < 1$, and $|\lambda_3| < 1$. From (22.12) we have that $\mathbf{P}^n = \mathbf{V}\mathbf{\Lambda}^n\mathbf{V}^{-1}$ which for $K = 3$ is

$$
\mathbf{P}^n = \begin{bmatrix} \mathbf{v}_1 & \mathbf{v}_2 & \mathbf{v}_3 \end{bmatrix} \begin{bmatrix} 1 & 0 & 0 \\ 0 & \lambda_2^n & 0 \\ 0 & 0 & \lambda_3^n \end{bmatrix} \underbrace{\begin{bmatrix} \mathbf{w}_1^T \\ \mathbf{w}_2^T \\ \mathbf{w}_3^T \end{bmatrix}}_{\mathbf{W}}
$$

where $\mathbf{W} = \mathbf{V}^{-1}$ and $\mathbf{w}_i^T$ is the ith row of $\mathbf{W}$. Next argue that as $n \to \infty$, $\mathbf{P}^n \to \mathbf{v}_1\mathbf{w}_1^T$. Use the relation $\mathbf{P}^\infty \mathbf{1} = \mathbf{1}$ (why?) to show that $\mathbf{v}_1 = c\mathbf{1}$, where c is a constant. Next use $\boldsymbol{\pi}^T\mathbf{P}^\infty = \boldsymbol{\pi}^T$ (why?) to show that $\mathbf{w}_1 = d\boldsymbol{\pi}$, where d is a constant. Finally, use the fact that $\mathbf{w}_1^T\mathbf{v}_1 = 1$ since $\mathbf{W}\mathbf{V} = \mathbf{I}$ to show that $cd = 1$ and therefore, $\mathbf{P}^\infty = \mathbf{1}\boldsymbol{\pi}^T$. The latter is the desired result which can be verified by direct multiplication of $\mathbf{1}$ by $\boldsymbol{\pi}^T$.

22.18 (f,c) For the transition probability matrix

$$
\mathbf{P} = \begin{bmatrix} 0.1 & 0.4 & 0.5 \\ 0.2 & 0.5 & 0.3 \\ 0.3 & 0.3 & 0.4 \end{bmatrix}
$$

find $\mathbf{P}^{100}$ using a computer evaluation. Does the form of $\mathbf{P}^{100}$ agree with the theory?

22.19 ($\smile$) (f,c) Using the explicit solution for the stationary probability vector given by (22.22), determine its value for the transition probability matrix given in Problem 22.18. Hint: You will need a computer to evaluate the solution.

22.20 (w) The result of multiplying two identical matrices together produces the same matrix as shown below.

$$
\begin{bmatrix} 0.2 & 0.1 & 0.7 \\ 0.2 & 0.1 & 0.7 \\ 0.2 & 0.1 & 0.7 \end{bmatrix} \begin{bmatrix} 0.2 & 0.1 & 0.7 \\ 0.2 & 0.1 & 0.7 \\ 0.2 & 0.1 & 0.7 \end{bmatrix} = \begin{bmatrix} 0.2 & 0.1 & 0.7 \\ 0.2 & 0.1 & 0.7 \\ 0.2 & 0.1 & 0.7 \end{bmatrix}.
$$

Explain what this means for Markov chains.

22.21 (f) For the transition probability matrix

$$
\mathbf{P} = \begin{bmatrix} 0.99 & 0.01 \\ 0.01 & 0.99 \end{bmatrix}
$$

solve for the stationary probabilities. Compare your probabilities to those obtained if a fair headed coin is tossed repeatedly and the tosses are independent. Do you expect the realization for this Markov chain to be similar to that of the fair coin tossing?

22.22 (c) Simulate on the computer the Markov chain described in Problem 22.21. Use $\mathbf{p}^T[0] = [1/2 \ 1/2]$ for the initial probability vector. Generate a realization for $n = 0, 1, \ldots, 99$ and plot the results. What do you notice about the realization? Next generate a realization for $n = 0, 1, \ldots, 9999$ and estimate the stationary probability of observing 1 by taking the sample mean of the realization. Do you obtain the theoretical result found in Problem 22.21 (recall that this type of Markov chain is ergodic and so a temporal average is equal to an ensemble average).

22.23 (w) A person is late for work on his first day with a probability of 0.1. On succeeding days he is late for work with a probability of 0.2 if he was late the previous day and with a probability of 0.4 if he was on time the previous day. In the long run what percentage of time is he late to work?

22.24 (⌣) (f,c) Assume for the weather example of Example 22.8 that the transition probability matrix is

$$\mathbf{P} = \begin{bmatrix} \frac{6}{8} & \frac{1}{8} & \frac{1}{8} \\ \frac{5}{8} & \frac{2}{8} & \frac{1}{8} \\ \frac{4}{8} & \frac{3}{8} & \frac{1}{8} \end{bmatrix}.$$

What is the steady-state probability of rain? Compare your answer to that obtained in Example 22.8 and explain the difference. Hint: You will need a computer to find the solution.

22.25 (w,c) Three machines operate together on a manufacturing floor, and each day there is a possibility that any of the machines may fail. The probability of their failure depends upon how many other machines are still in operation. The number of machines in operation at the beginning of each day is represented by the state values of $0, 1, 2, 3$ and the corresponding state transition probability matrix is

$$\mathbf{P} = \begin{bmatrix} 1 & 0 & 0 & 0 \\ 0.5 & 0.5 & 0 & 0 \\ 0.1 & 0.3 & 0.6 & 0 \\ 0.4 & 0.3 & 0.2 & 0.1 \end{bmatrix}.$$

First explain why $\mathbf{P}$ has zero entries. Next determine how many days will pass before the probability of all 3 machines failing is greater than 0.8. Assume that intially all 3 machines are working. Hint: You will need a computer to find the solution.

22.26 (☺) **(w,c)** A pond holds 4 fish. Each day a fisherman goes fishing and his probability of catching $k = 0, 1, 2, 3, 4$ fish that day follows a binomial PDF with $p = 1/2$. How many days should he plan on fishing so that the probability of his catching all 4 fish exceeds 0.9? Note that initially, i.e., at $n = 0$, all 4 fish are present. Hint: You will need a computer to find the solution.

22.27 (t) In this problem we prove that a doubly stochastic transition probability matrix with $\mathbf{P} > 0$ produces equal stationary probabilities. First recall that since the columns of $\mathbf{P}$ sum to one, we have that $\mathbf{P}^T \mathbf{1} = \mathbf{1}$ and therefore argue that $\mathbf{P}^{\infty^T} \mathbf{1} = \mathbf{1}$. Next use the results of Problem 22.17 that $\mathbf{P}^\infty = \mathbf{1}\boldsymbol{\pi}^T$ to show that $\boldsymbol{\pi} = 1/K$.

22.28 (☺) **(c)** Use a computer simulation to generate a realization of the golf example for a large number of holes (very much greater than 18). Estimate the percentage of one-putts from your realization and compare it to the theoretical results.

22.29 (c) Repeat Problem 22.28 but now estimate the average time between one-putts. Compare your results to the theoretical value.

22.30 (c) Run the program sierpinski.m given in Section 22.9 but use instead the initial position $\mathbf{X}[0] = [50 \, 30]^T$. Do you obtain similar results to those shown in Figure 22.9? What is the difference, if any?

Appendix 22A

Solving for the Stationary PMF

We derive the formula of (22.22). The set of equations to be solved (after transposition) is $\mathbf{P}^T\boldsymbol{\pi} = \boldsymbol{\pi}$ or equivalently

$$(\mathbf{I} - \mathbf{P}^T)\boldsymbol{\pi} = \mathbf{0}. \tag{22A.1}$$

Since we have assumed a unique solution, it is clear that the matrix $\mathbf{I} - \mathbf{P}^T$ cannot be invertible or else we would have $\boldsymbol{\pi} = \mathbf{0}$. This is to say that the linear equations are not all independent. To make them independent we must add the constraint equation $\sum_{i=0}^{K-1} \pi_i = 1$ or in vector form this is $\mathbf{1}^T\boldsymbol{\pi} = 1$. Equivalently, the constraint equation is $\mathbf{11}^T\boldsymbol{\pi} = \mathbf{1}$. Adding this to (22A.1) produces

$$(\mathbf{I} - \mathbf{P}^T)\boldsymbol{\pi} + \mathbf{11}^T\boldsymbol{\pi} = \mathbf{1}$$

or

$$(\mathbf{I} - \mathbf{P}^T + \mathbf{11}^T)\boldsymbol{\pi} = \mathbf{1}.$$

It can be shown that the matrix $\mathbf{I} - \mathbf{P}^T + \mathbf{11}^T$ is now invertible and so the solution is

$$\boldsymbol{\pi} = (\mathbf{I} - \mathbf{P}^T + \mathbf{11}^T)^{-1}\mathbf{1}.$$

Appendix A

Glossary of Symbols and Abbrevations

Symbols

Boldface characters denote vectors or matrices. All others are scalars. All vectors are column vectors. Random variables are denoted by capital letters such as U, V, W, X, Y, Z and random vectors by $\mathbf{U}, \mathbf{V}, \mathbf{W}, \mathbf{X}, \mathbf{Y}, \mathbf{Z}$ and their values by corresponding lowercase letters.

$\angle$	angle of
$*$	complex conjugate
$\star$	convolution operator, either convolution sum or integral
$\hat{\ }$	denotes estimator
$\sim$	denotes *is distributed according to*
$[x]$	denotes the largest integer $\leq x$
x^+	denotes a number slightly larger than x
x^-	denotes a number slightly smaller than x
$A \times B$	cartesian product of sets A and B
$[\mathbf{A}]_{ij}$	(i,j)th element of $\mathbf{A}$
$\mathcal{A}(z)$	z-transform of $a[n]$ sequence
$[\mathbf{b}]_i$	ith element of $\mathbf{b}$
$\mathrm{Ber}(p)$	Bernoulli random variable
$\mathrm{bin}(M,p)$	binomial random variable
χ_N^2	chi-squared distribution with N degrees of freedom
$\binom{N}{k}$	number of combinations of N things taken k at a time
c	complement of set
$\mathrm{cov}(X,Y)$	covariance of X and Y
$\mathbf{C}$	covariance matrix

$\mathbf{C}_X$	covariance matrix of $\mathbf{X}$
$\mathbf{C}_{X,Y}$	covariance matrix of X and Y
$c_X[n_1, n_2]$	covariance sequence of discrete-time random process $X[n]$
$c_X(t_1, t_2)$	covariance function of continuous-time random process $X(t)$
$\delta(t)$	Dirac delta function or impulse function
$\delta[n]$	discrete-time unit impulse sequence
δ_{ij}	Kronecker delta
Δf	small interval in frequency f
Δt	small interval in t
Δx	small interval in x
Δ_t	time interval between samples
$\det(\mathbf{A})$	determinant of matrix $\mathbf{A}$
$\mathrm{diag}(a_{11}, \ldots, a_{NN})$	diagonal matrix with elements a_{ii} on main diagonal
$\mathbf{e}_i$	natural unit vector in ith direction
η	signal-to-noise ratio
$E[\cdot]$	expected value
$E[X^n]$	nth moment
$E[(X - E[X])^n]$	nth central moment
$E_X[\cdot]$	expected value with respect to PMF or PDF of X
$E_{X,Y}[\cdot]$	expected value with respect to joint PMF or joint PDF of (X, Y)
$E_{X_1, X_2, \ldots, X_N}[\cdot]$	expected value with respect to N-dimensional joint PMF or PDF
$E_{\mathbf{X}}[\cdot]$	shortened notation for $E_{X_1, X_2, \ldots, X_N}[\cdot]$
$E_{Y\|X}[Y\|X]$	conditional expected value considered as random variable
$E_{Y\|X}[Y\|x_i]$	expected value of PMF $p_{Y\|X}[y_j\|x_i]$
$E_{Y\|X}[Y\|x]$	expected value of PDF $p_{Y\|X}(y\|x)$
$E[\mathbf{X}]$	expected value of random vector $\mathbf{X}$
$\in$	element of set
$\exp(\lambda)$	exponential random variable
f	discrete-time frequency
F	continuous-time frequency
$F_X(x)$	cumulative distribution function of X
$F_X^{-1}(x)$	inverse cumulative distribution function of X
$F_{X,Y}(x, y)$	cumulative distribution function of X and Y
$F_{X_1, \ldots, X_N}(x_1, \ldots, x_N)$	cumulative distribution function of $X_1, \ldots, X_N$
$F_{Y\|X}(y\|x)$	cumulative distribution function of Y conditioned on $X = x$
$\mathcal{F}$	Fourier transform
$\mathcal{F}^{-1}$	inverse Fourier transform
$g(\cdot)$	general notation for function of real variable
$g^{-1}(\cdot)$	general notation for inverse function of $g(\cdot)$

$\Gamma(x)$	Gamma function
$\Gamma(\alpha, \lambda)$	Gamma random variable
$\gamma_{X,Y}(f)$	coherence function for discrete-time random processes $X[n]$ and $Y[n]$
$\text{geom}(p)$	geometric random variable
$h[n]$	impulse response of LSI system
$h(t)$	impulse response of LTI system
$H(f)$	frequency response of LSI system
$H(F)$	frequency response of LTI system
$\mathcal{H}(z)$	system function of LSI system
$I_A(x)$	indicator function for the set A
$\mathbf{I}$	identity matrix
$\cap$	intersection of sets
j	$\sqrt{-1}$
$\frac{\partial(w,z)}{\partial(x,y)}$	Jacobian matrix of transformation of $w = g(x,y), z = h(x,y)$
$\frac{\partial(x_1,...,x_N)}{\partial(y_1,...,y_N)}$	Jacobian matrix of transformation from $\mathbf{y}$ to $\mathbf{x}$
$\mathbf{\Lambda}$	diagonal matrix with eigenvalues on main diagonal
mse	mean square error
μ	mean
$\mu_X[n]$	mean sequence of discrete-time random process $X[n]$
$\mu_X(t)$	mean function of continuous-time random process $X(t)$
$\boldsymbol{\mu}$	mean vector
$\binom{M}{k_1,k_2,...,k_N}$	multinomial coefficient
n	discrete-time index
$N!$	N factorial
$(N)_r$	equal to $N(N-1)\cdots(N-r+1)$
N_A	number of elements in set A
$\mathcal{N}(\mu, \sigma^2)$	normal or Gaussian random variable with mean μ and variance σ^2
$\mathcal{N}(\boldsymbol{\mu}, \mathbf{C})$	multivariate normal or Gaussian random vector with mean $\boldsymbol{\mu}$ and covariance $\mathbf{C}$
$\|\mathbf{x}\|$	Euclidean norm or length of vector $\mathbf{x}$
$\emptyset$	null or empty set
opt	optimal value
$\mathbf{1}$	vector of all ones
$\text{Pois}(\lambda)$	Poisson random variable
$p_X[x_i]$	PMF of X
$p_X[k]$	PMF of integer-valued random variable X (or $p_X[i]$, $p_X[j]$)
$p_{X,Y}[x_i, y_j]$	joint PMF of X and Y
$p_{X_1,...,X_N}[x_1, \ldots, x_N]$	joint PMF of $X_1, \ldots, X_N$
$p_{\mathbf{x}}[\mathbf{x}]$	shortened notation for $p_{X_1,...,X_N}[x_1, \ldots, x_N]$
$p_{X_1,...,X_N}[k_1, \ldots, k_N]$	joint PMF of integer-valued random variables $X_1, \ldots, X_N$

$p_{Y\|X}[y_j\|x_i]$	conditional PMF of Y given $X = x_i$
$p_{X_N\|X_1,\ldots,X_{N-1}}[x_N\|$ $x_1,\ldots,x_{N-1}]$	conditional PMF of X_N given $X_1,\ldots,X_{N-1}$
$p_{X,Y}[i,j]$	joint PMF of integer-valued random variables X and Y
$p_{Y\|X}[j\|i]$	conditional PMF of integer-valued random variable Y given $X = i$
$p_X(x)$	PDF of X
$p_{X,Y}(x,y)$	joint PDF of X and Y
$p_{X_1,\ldots,X_N}(x_1,\ldots,x_N)$	joint PDF of $X_1,\ldots,X_N$
$p_{\mathbf{x}}(\mathbf{x})$	shortened notation for $p_{X_1,\ldots,X_N}(x_1,\ldots,x_N)$
$p_{Y\|X}(y\|x)$	conditional PDF of Y given $X = x$
$P[E]$	probability of the event E
P_e	probability of error
$P_X(f)$	power spectral density of discrete-time random process $X[n]$
$\mathcal{P}_X(z)$	z-transform of autocorrelation sequence $r_X[k]$
$P_X(F)$	power spectral density of continuous-time random process $X(t)$
$P_{X,Y}(f)$	cross-power spectral density of discrete-time random processes $X[n]$ and $Y[n]$
$P_{X,Y}(F)$	cross-power spectral density of continuous-time random processes $X(t)$ and $Y(t)$
$\phi_X(\omega)$	characteristic function of X
$\phi_{X,Y}(\omega_X,\omega_Y)$	joint characteristic function of X and Y
$\phi_{X_1,\ldots,X_N}(\omega_1,\ldots,\omega_N)$	joint characteristic function of $X_1,\ldots,X_N$
$\Phi(x)$	cumulative distribution function of $\mathcal{N}(0,1)$ random variable
$Q(x)$	probability that a $\mathcal{N}(0,1)$ random variable exceeds x
$Q^{-1}(u)$	value of $\mathcal{N}(0,1)$ random variable that is exceeded with probability of u
$\rho_{X,Y}$	correlation coefficient of X and Y
$r_X[k]$	autocorrelation sequence of discrete-time random process $X[n]$
$r_X(\tau)$	autocorrelation function of continuous-time random process $X(t)$
$r_{X,Y}[k]$	cross-correlation sequence of discrete-time random processes $X[n]$ and $Y[n]$
$r_{X,Y}(\tau)$	cross-correlation function of continuous-time random processes $X(t)$ and $Y(t)$
R or R^1	denotes real line
R^N	denotes N-dimensional Euclidean space
$\mathbf{R}_X$	autocorrelation matrix
$\mathcal{S}$	sample space

$\mathcal{S}_X$	sample space of random variable X
$\mathcal{S}_{X,Y}$	sample space of random variables X and Y
$\mathcal{S}_{X_1,X_2,\ldots,X_N}$	sample space of random variables $X_1, X_2, \ldots, X_N$
s_i	element of discrete sample space
s	element of continuous sample space
σ^2	variance
σ_X^2	variance of random variable X
$\sigma_X^2[n]$	variance sequence of discrete-time random process $X[n]$
$\sigma_X^2(t)$	variance function of continuous-time random process $X(t)$
$s[n]$	discrete-time signal
$\mathbf{s}$	vector of signal samples
$s(t)$	continuous-time signal
t	continuous time
T	transpose of matrix
$\mathcal{U}(a,b)$	uniform random variable over the interval (a,b)
$\cup$	union of sets
$u[n]$	discrete unit step function
$u(x)$	unit step function
$\mathcal{U}(z)$	z-transform of $u[n]$ sequence
$\mathbf{V}$	modal matrix
$\mathrm{var}(X)$	variance of X
$\mathrm{var}(Y|x_i)$	variance of conditional PMF or of $p_{Y|X}[y_j|x_i]$
x_i	value of discrete random variable
x	value of continuous random variable
X_s	standardized version of random variable X
x_s	value for X_s
$X[n]$	discrete-time random process
$x[n]$	realization of discrete-time random process
$X(t)$	continuous-time random process
$x(t)$	realization of continuous-time random process
$\mathcal{X}(z)$	z-transform of $x[n]$ sequence
$\bar{X}$	sample mean random variable
$\bar{x}$	value of $\bar{X}$
$\mathbf{X}$	random vector $(X_1, X_2, \ldots, X_N)$
$\mathbf{x}$	value $(x_1, x_2, \ldots, x_N)$ of random vector $\mathbf{X}$
$Y|(X = x_i)$	random variable Y conditioned on $X = x_i$
$\mathcal{Z}$	z-transform
$\mathcal{Z}^{-1}$	inverse z-transform
$\mathbf{0}$	vector or matrix of all zeros

Abbreviations

ACF	autocorrelation function
ACS	autocorrelation sequence
AR	autoregressive
AR(p)	autoregressive process of order p
ARMA	autoregressive moving average
CCF	cross-correlation function
CCS	cross-correlation sequence
CDF	cumulative distribution function
CPSD	cross-power spectral density
CTCV	continuous-time/continuous-valued
CTDV	continuous-time/discrete-valued
D/A	digital-to-analog
dB	decibel
DC	constant level (direct current)
DFT	discrete Fourier transform
DTCV	discrete-time/continuous-valued
DTDV	discrete-time/discrete-valued
FFT	fast Fourier transform
FIR	finite impulse response
GHz	giga-hertz
Hz	hertz
IID	independent and identically distributed
IIR	infinite impulse response
KHz	kilo-hertz
LSI	linear shift invariant
LTI	linear time invariant
MA	moving average
MHz	mega-hertz
MSE	mean square error
PDF	probability density function
PMF	probability mass function
PSD	power spectral density
SNR	signal-to-noise ratio
WGN	white Gaussian noise
WSS	wide sense stationary

Appendix B

Assorted Math Facts and Formulas

An extensive summary of math facts and formulas can be found in [Gradshteyn and Ryzhik 1994].

B.1 Proof by Induction

To prove that a statement is true, for example,

$$\sum_{i=1}^{N} i = \frac{N}{2}(N+1) \tag{B.1}$$

by mathematical induction we proceed as follows:

1. Prove the statement is true for $N = 1$.

2. *Assume* the statement is true $N = n$ and prove that it therefore must be true for $N = n + 1$.

Obviously, (B.1) is true for $N = 1$ since $\sum_{i=1}^{1} i = 1$ and $(N/2)(N+1) = (1/2)(2) = 1$. Now assume it is true for $N = n$. Then for $N = n + 1$ we have

$$
\begin{aligned}
\sum_{i=1}^{n+1} i &= \sum_{i=1}^{n} i + (n+1) \\
&= \frac{n}{2}(n+1) + (n+1) \qquad \text{(since it is true for } N = n\text{)} \\
&= \frac{n+1}{2}(n+2) \\
&= \frac{(n+1)}{2}[(n+1) + 1]
\end{aligned}
$$

which proves that it is also true for $N = n + 1$. By induction, since it is true for $N = n = 1$ from step 1, it must also be true for $N = (n + 1) = 2$ from step 2. And since it is true for $N = n = 2$, it must also be true for $N = n + 1 = 3$, etc.

B.2 Trigonometry

Some useful trigonometric identities are:

Fundamental

$$\sin^2 \alpha + \cos^2 \alpha = 1 \tag{B.2}$$

Sum of angles

$$\sin(\alpha + \beta) = \sin \alpha \cos \beta + \cos \alpha \sin \beta \tag{B.3}$$
$$\cos(\alpha + \beta) = \cos \alpha \cos \beta - \sin \alpha \sin \beta \tag{B.4}$$

Double angle

$$\sin(2\alpha) = 2 \sin \alpha \cos \alpha \tag{B.5}$$
$$\cos(2\alpha) = \cos^2 \alpha - \sin^2 \alpha = 2 \cos^2 \alpha - 1 \tag{B.6}$$

Squared sine and cosine

$$\sin^2 \alpha = \frac{1}{2} - \frac{1}{2} \cos(2\alpha) \tag{B.7}$$
$$\cos^2 \alpha = \frac{1}{2} + \frac{1}{2} \cos(2\alpha) \tag{B.8}$$

Euler identities For $j = \sqrt{-1}$

$$\exp(j\alpha) = \cos \alpha + j \sin \alpha \tag{B.9}$$

$$\cos \alpha = \frac{\exp(j\alpha) + \exp(-j\alpha)}{2} \tag{B.10}$$
$$\sin \alpha = \frac{\exp(j\alpha) - \exp(-j\alpha)}{2j} \tag{B.11}$$

B.3 Limits

Alternative definition of exponential function

$$\lim_{M \to \infty} \left(1 + \frac{x}{M}\right)^M = \exp(x) \tag{B.12}$$

Taylor series expansion about the point $x = x_0$

$$g(x) = \sum_{i=0}^{\infty} \frac{g^{(i)}(x_0)}{i!} (x - x_0)^i \qquad \text{(B.13)}$$

where $g^{(i)}(x_0)$ is the ith derivative of $g(x)$ evaluated at $x = x_0$ and $g^{(0)}(x_0) = g(x_0)$. As an example, consider $g(x) = \exp(x)$, which when expanded about $x = x_0 = 0$ yields

$$\exp(x) = \sum_{i=0}^{\infty} \frac{x^i}{i!}$$

B.4   Sums

Integers

$$\sum_{i=0}^{N-1} i = \frac{N(N-1)}{2} \qquad \text{(B.14)}$$

$$\sum_{i=0}^{N-1} i^2 = \frac{N(N-1)(2N-1)}{6} \qquad \text{(B.15)}$$

Real geometric series

$$\sum_{i=k}^{N-1} x^i = \frac{x^k(1 - x^{N-k})}{1 - x} \qquad (x \text{ is real}) \qquad \text{(B.16)}$$

If $|x| < 1$, then

$$\sum_{i=k}^{\infty} x^i = \frac{x^k}{1 - x} \qquad \text{(B.17)}$$

Complex geometric series

$$\sum_{i=k}^{N-1} z^i = \frac{z^k(1 - z^{N-k})}{1 - z} \qquad (z \text{ is complex}) \qquad \text{(B.18)}$$

A special case is when $z = \exp(j\theta)$. Then

$$\sum_{i=0}^{N-1} \exp(j\theta) = \frac{1 - \exp(jN\theta)}{1 - \exp(j\theta)}$$

$$= \exp\left[j\left(\frac{N-1}{2}\right)\theta\right] \frac{\sin\left(\frac{N\theta}{2}\right)}{\sin\left(\frac{\theta}{2}\right)} \qquad \text{(B.19)}$$

If $|z| = |x + jy| = \sqrt{x^2 + y^2} < 1$, then as $N \to \infty$ (B.18) becomes

$$\sum_{i=k}^{\infty} z^i = \frac{z^k}{1-z} \tag{B.20}$$

Double sums

$$\sum_{i=1}^{M} \sum_{j=1}^{M} (x_i y_j) = \left(\sum_{i=1}^{M} x_i \right) \left(\sum_{j=1}^{M} y_j \right) \tag{B.21}$$

B.5   Calculus

Convergence of sum to integral

If $g(x)$ is a continuous function over $[a, b]$, then

$$\lim_{\Delta x \to 0} \sum_{i=0}^{M} g(x_i) \Delta x = \int_a^b g(x) dx \tag{B.22}$$

where $x_i = a + i\Delta x$ and $x_M = b$. Also, this shows how to approximate an integral by a sum.

Approximation of integral over small interval

$$\int_{x_0 - \Delta x/2}^{x_0 + \Delta x/2} g(x) dx \approx g(x_0) \Delta x \tag{B.23}$$

Differentiation of composite function

$$\left. \frac{dg(h(x))}{dx} \right|_{x=x_0} = \left. \frac{dg(u)}{du} \right|_{u=h(x_0)} \left. \frac{dh(x)}{dx} \right|_{x=x_0} \qquad \text{(chain rule)} \tag{B.24}$$

Change of integration variable

If $u = h(x)$, then

$$\int_a^b g(u) du = \int_{h^{-1}(a)}^{h^{-1}(b)} g(h(x)) h'(x) dx \tag{B.25}$$

where $h'(x)$ is the derivative of $h(x)$ and $h^{-1}(\cdot)$ denotes the inverse function. This assumes that there is one solution to the equation $u = h(x)$ over the interval $a \le u \le b$.

Fundamental theorem of calculus

$$\frac{d}{dx} \int_{-\infty}^{x} g(t) dt = g(x) \tag{B.26}$$

Leibnitz's rule

$$\frac{d}{dy}\int_{h_1(y)}^{h_2(y)} g(x,y)dx = \int_{h_1(y)}^{h_2(y)} \frac{\partial}{\partial y}g(x,y)dx + g(h_2(y),y)\frac{dh_2(y)}{dy} - g(h_1(y),y)\frac{dh_1(y)}{dy}$$

(B.27)

Integration of even and odd functions An even function is defined as having the property $g(-x) = g(x)$, while an odd function has the property $g(-x) = -g(x)$. As a result,

$$\int_{-M}^{M} g(x)dx = 2\int_{0}^{M} g(x)dx \qquad \text{for } g(x) \text{ an even function}$$

$$\int_{-M}^{M} g(x)dx = 0 \qquad \text{for } g(x) \text{ an odd function}$$

Integration by parts
   If U and V are both functions of x, then

$$\int U dV = UV - \int V dU$$

(B.28)

Dirac delta "function" or impulse
   Denoted by $\delta(x)$ it is not really a function but a symbol that has the definition

$$\delta(x) = \begin{cases} 0 & x \neq 0 \\ \infty & x = 0 \end{cases}$$

and

$$\int_{a}^{b} \delta(x)dx = \begin{cases} 1 & 0 \in [a^-, b^+] \\ 0 & \text{otherwise} \end{cases}$$

Some properties are for $u(x)$ the unit step function

$$\frac{du(x)}{dx} = \delta(x)$$

$$\int_{-\infty}^{x} \delta(t)dt = u(x)$$

Double integrals

$$\int_{c}^{d}\int_{a}^{b} g(x)h(y)dxdy = \left(\int_{a}^{b} g(x)dx\right)\left(\int_{c}^{d} h(y)dy\right)$$

(B.29)

References

Gradshteyn, I.S., I.M. Ryzhik, *Tables of Integrals, Series, and Products*, Fifth Ed., Academic Press, New York, 1994.

Appendix C

Linear and Matrix Algebra

Important results from linear and matrix algebra theory are reviewed in this appendix. It is assumed that the reader has had some exposure to matrices. For a more comprehensive treatment the books [Noble and Daniel 1977] and [Graybill 1969] are recommended.

C.1 Definitions

Consider an $M \times N$ matrix $\mathbf{A}$ with elements a_{ij}, $i = 1, 2, \ldots, M$; $j = 1, 2, \ldots, N$. A shorthand notation for describing $\mathbf{A}$ is

$$[\mathbf{A}]_{ij} = a_{ij}.$$

Likewise a shorthand notation for describing an $N \times 1$ vector $\mathbf{b}$ is

$$[\mathbf{b}]_i = b_i.$$

An $M \times N$ matrix $\mathbf{A}$ may multiply an $N \times 1$ vector $\mathbf{b}$ to yield a new $M \times 1$ vector $\mathbf{c}$ whose ith element is

$$c_i = \sum_{j=1}^{N} a_{ij} b_j \qquad i = 1, 2, \ldots, M.$$

Similarly, an $M \times N$ matrix $\mathbf{A}$ can multiply an $N \times L$ matrix $\mathbf{B}$ to yield an $M \times L$ matrix $\mathbf{C} = \mathbf{AB}$ whose (i, j) element is

$$c_{ij} = \sum_{k=1}^{N} a_{ik} b_{kj} \qquad i = 1, 2, \ldots, M; j = 1, 2, \ldots, L.$$

Vectors and matrices that can be multiplied together are said to be *conformable*.

The *transpose* of $\mathbf{A}$, which is denoted by $\mathbf{A}^T$, is defined as the $N \times M$ matrix with elements a_{ji} or

$$[\mathbf{A}^T]_{ij} = a_{ji}.$$

A *square* matrix is one for which $M = N$. A square matrix is *symmetric* if $\mathbf{A}^T = \mathbf{A}$ or $a_{ji} = a_{ij}$.

The *inverse* of a square $N \times N$ matrix is the square $N \times N$ matrix $\mathbf{A}^{-1}$ for which

$$\mathbf{A}^{-1}\mathbf{A} = \mathbf{A}\mathbf{A}^{-1} = \mathbf{I}$$

where $\mathbf{I}$ is the $N \times N$ identity matrix. If the inverse does not exist, then $\mathbf{A}$ is *singular*. Assuming the existence of the inverse of a matrix, the unique solution to a set of N simultaneous linear equations given in matrix form by $\mathbf{A}\mathbf{x} = \mathbf{b}$, where $\mathbf{A}$ is $N \times N$, $\mathbf{x}$ is $N \times 1$, and $\mathbf{b}$ is $N \times 1$, is $\mathbf{x} = \mathbf{A}^{-1}\mathbf{b}$.

The *determinant* of a square $N \times N$ matrix is denoted by $\det(\mathbf{A})$. It is computed as

$$\det(\mathbf{A}) = \sum_{j=1}^{N} a_{ij}C_{ij}$$

where

$$C_{ij} = (-1)^{i+j}D_{ij}.$$

D_{ij} is the determinant of the submatrix of $\mathbf{A}$ obtained by deleting the ith row and jth column and is termed the *minor* of a_{ij}. C_{ij} is the *cofactor* of a_{ij}. Note that any choice of i for $i = 1, 2, \ldots, N$ will yield the same value for $\det(\mathbf{A})$. A square $N \times N$ matrix is nonsingular if and only if $\det(\mathbf{A}) \neq 0$.

A *quadratic form* Q, which is a *scalar*, is defined as

$$Q = \sum_{i=1}^{N} \sum_{j=1}^{N} a_{ij}x_i x_j.$$

In defining the quadratic form it is assumed that $a_{ji} = a_{ij}$. This entails no loss in generality since any quadratic function may be expressed in this manner. Q may also be expressed as

$$Q = \mathbf{x}^T \mathbf{A} \mathbf{x}$$

where $\mathbf{x} = [x_1 \, x_2 \ldots x_N]^T$ and $\mathbf{A}$ is a square $N \times N$ matrix with $a_{ji} = a_{ij}$ or $\mathbf{A}$ is a symmetric matrix.

A square $N \times N$ matrix $\mathbf{A}$ is *positive semidefinite* if $\mathbf{A}$ is symmetric and

$$Q = \mathbf{x}^T \mathbf{A} \mathbf{x} \geq 0$$

for all $\mathbf{x}$. If the quadratic form is strictly positive for $\mathbf{x} \neq \mathbf{0}$, then $\mathbf{A}$ is *positive definite*. When referring to a matrix as positive definite or positive semidefinite, it is always assumed that the matrix is symmetric.

A *partitioned* $M \times N$ matrix $\mathbf{A}$ is one that is expressed in terms of its submatrices. An example is the 2×2 partitioning

$$\mathbf{A} = \left[\begin{array}{cc} \mathbf{A}_{11} & \mathbf{A}_{12} \\ \mathbf{A}_{21} & \mathbf{A}_{22} \end{array} \right].$$

Each "element" $\mathbf{A}_{ij}$ is a submatrix of $\mathbf{A}$. The dimensions of the partitions are given as

$$\left[\begin{array}{cc} K \times L & K \times (N-L) \\ (M-K) \times L & (M-K) \times (N-L) \end{array} \right].$$

C.2 Special Matrices

A *diagonal* matrix is a square $N \times N$ matrix with $a_{ij} = 0$ for $i \neq j$ or all elements not on the *principal diagonal* (the diagonal containing the elements a_{ii}) are zero. The elements a_{ij} for which $i \neq j$ are termed the *off-diagonal* elements. A diagonal matrix appears as

$$\mathbf{A} = \left[\begin{array}{cccc} a_{11} & 0 & \cdots & 0 \\ 0 & a_{22} & \cdots & 0 \\ \vdots & \vdots & \ddots & \vdots \\ 0 & 0 & \cdots & a_{NN} \end{array} \right].$$

A diagonal matrix will sometimes be denoted by $\mathrm{diag}(a_{11}, a_{22}, \ldots, a_{NN})$. The inverse of a diagonal matrix is found by simply inverting each element on the principal diagonal, assuming that $a_{ii} \neq 0$ for $i = 1, 2, \ldots, N$ (which is necessary for invertibility).

A square $N \times N$ matrix is *orthogonal* if

$$\mathbf{A}^{-1} = \mathbf{A}^T.$$

For a matrix to be orthogonal the columns (and rows) must be orthonormal or if

$$\mathbf{A} = \left[\begin{array}{cccc} \mathbf{a}_1 & \mathbf{a}_2 & \cdots & \mathbf{a}_N \end{array} \right]$$

where $\mathbf{a}_i$ denotes the ith column, the conditions

$$\mathbf{a}_i^T \mathbf{a}_j = \left\{ \begin{array}{ll} 0 & \text{for } i \neq j \\ 1 & \text{for } i = j \end{array} \right.$$

must be satisfied. Other "matrices" that can be constructed from vector operations on the $N \times 1$ vectors $\mathbf{x}$ and $\mathbf{y}$ are the *inner product*, which is defined as the scalar

$$\mathbf{x}^T \mathbf{y} = \sum_{i=1}^{N} x_i y_i$$

and the *outer product*, which is defined as the $N \times N$ matrix

$$\mathbf{x}\mathbf{y}^T = \begin{bmatrix} x_1 y_1 & x_1 y_2 & \cdots & x_1 y_N \\ x_2 y_1 & x_2 y_2 & \cdots & x_2 y_N \\ \vdots & \vdots & \ddots & \vdots \\ x_N y_1 & x_N y_2 & \cdots & x_N y_N \end{bmatrix}.$$

C.3   Matrix Manipulation and Formulas

Some useful formulas for the algebraic manipulation of matrices are summarized in this section. For $N \times N$ matrices $\mathbf{A}$ and $\mathbf{B}$ the following relationships are useful.

$$\begin{aligned} (\mathbf{A}^T)^{-1} &= (\mathbf{A}^{-1})^T \\ (\mathbf{AB})^{-1} &= \mathbf{B}^{-1}\mathbf{A}^{-1} \\ \det(\mathbf{A}^T) &= \det(\mathbf{A}) \\ \det(c\mathbf{A}) &= c^N \det(\mathbf{A}) \quad (c \text{ a scalar}) \\ \det(\mathbf{AB}) &= \det(\mathbf{A})\det(\mathbf{B}) \\ \det(\mathbf{A}^{-1}) &= \frac{1}{\det(\mathbf{A})}. \end{aligned}$$

Also, for any conformable matrices (or vectors) we have

$$(\mathbf{AB})^T = \mathbf{B}^T \mathbf{A}^T.$$

It is frequently necessary to determine the inverse of a matrix analytically. To do so one can make use of the following formula. The inverse of a square $N \times N$ matrix is

$$\mathbf{A}^{-1} = \frac{\mathbf{C}^T}{\det(\mathbf{A})}$$

where $\mathbf{C}$ is the square $N \times N$ matrix of cofactors of $\mathbf{A}$. The cofactor matrix is defined by

$$[\mathbf{C}]_{ij} = (-1)^{i+j} D_{ij}$$

where D_{ij} is the minor of a_{ij} obtained by deleting the ith row and jth column of $\mathbf{A}$.

Partitioned matrices may be manipulated according to the usual rules of matrix algebra by considering each submatrix as an element. For multiplication of partitioned matrices the submatrices that are multiplied together must be conformable. As an illustration, for 2×2 partitioned matrices

$$\begin{aligned} \mathbf{AB} &= \begin{bmatrix} \mathbf{A}_{11} & \mathbf{A}_{12} \\ \mathbf{A}_{21} & \mathbf{A}_{22} \end{bmatrix} \begin{bmatrix} \mathbf{B}_{11} & \mathbf{B}_{12} \\ \mathbf{B}_{21} & \mathbf{B}_{22} \end{bmatrix} \\ &= \begin{bmatrix} \mathbf{A}_{11}\mathbf{B}_{11} + \mathbf{A}_{12}\mathbf{B}_{21} & \mathbf{A}_{11}\mathbf{B}_{12} + \mathbf{A}_{12}\mathbf{B}_{22} \\ \mathbf{A}_{21}\mathbf{B}_{11} + \mathbf{A}_{22}\mathbf{B}_{21} & \mathbf{A}_{21}\mathbf{B}_{12} + \mathbf{A}_{22}\mathbf{B}_{22} \end{bmatrix}. \end{aligned}$$

Other useful relationships for partitioned matrices for an $M \times N$ matrix $\mathbf{A}$ and $N \times 1$ vectors $\mathbf{x}_i$ are

$$\begin{bmatrix} \mathbf{Ax}_1 & \mathbf{Ax}_2 & \ldots & \mathbf{Ax}_N \end{bmatrix} = \mathbf{A} \begin{bmatrix} \mathbf{x}_1 & \mathbf{x}_2 & \ldots & \mathbf{x}_N \end{bmatrix} \tag{C.1}$$

which is a $M \times N$ matrix and

$$\begin{bmatrix} a_{11}\mathbf{x}_1 & a_{22}\mathbf{x}_2 & \ldots & a_{NN}\mathbf{x}_N \end{bmatrix} = \begin{bmatrix} \mathbf{x}_1 & \mathbf{x}_2 & \ldots & \mathbf{x}_N \end{bmatrix} \begin{bmatrix} a_{11} & 0 & \ldots & 0 \\ 0 & a_{22} & \ldots & 0 \\ \vdots & \vdots & \ddots & \vdots \\ 0 & 0 & \ldots & a_{NN} \end{bmatrix} \tag{C.2}$$

which is an $N \times N$ matrix.

C.4 Some Properties of Positive Definite (Semidefinite) Matrices

Some useful properties of positive definite (semidefinite) matrices are:

1. A square $N \times N$ matrix $\mathbf{A}$ is positive definite if and only if the principal minors are all positive. (The ith principal minor is the determinant of the submatrix formed by deleting all rows and columns with an index greater than i.) If the principal minors are only nonnegative, then $\mathbf{A}$ is positive semidefinite.

2. If $\mathbf{A}$ is positive definite (positive semidefinite), then

   a. $\mathbf{A}$ is invertible (singular).

   b. the diagonal elements are positive (nonnegative).

   c. the determinant of $\mathbf{A}$, which is a principal minor, is positive (nonnegative).

C.5 Eigendecomposition of Matrices

An *eigenvector* of a square $N \times N$ matrix $\mathbf{A}$ is an $N \times 1$ vector $\mathbf{v}$ satisfying

$$\mathbf{Av} = \lambda \mathbf{v} \tag{C.3}$$

for some scalar λ, which may be complex. λ is the *eigenvalue* of $\mathbf{A}$ corresponding to the eigenvector $\mathbf{v}$. To determine the eigenvalues we must solve for the N λ's in $\det(\mathbf{A} - \lambda\mathbf{I}) = 0$, which is an Nth order polynomial in λ. Once the eigenvalues are found, the corresponding eigenvectors are determined from the equation $(\mathbf{A} - \lambda\mathbf{I})\mathbf{v} = \mathbf{0}$. It is assumed that the eigenvector is normalized to have unit length or $\mathbf{v}^T\mathbf{v} = 1$.

If $\mathbf{A}$ is symmetric, then one can always find N linearly independent eigenvectors, although they will not in general be unique. An example is the identity matrix for which any vector is an eigenvector with eigenvalue 1. If $\mathbf{A}$ is symmetric, then the eigenvectors corresponding to distinct eigenvalues are orthonormal or $\mathbf{v}_i^T \mathbf{v}_j = 0$ for $i \neq j$ and $\mathbf{v}_i^T \mathbf{v}_j = 1$ for $i = j$, and the eigenvalues are real. If, furthermore, the matrix is positive definite (positive semidefinite), then the eigenvalues are positive (nonnegative).

The defining relation of (C.3) can also be written as (using (C.1) and (C.2))

$$\left[\begin{array}{cccc} \mathbf{A}\mathbf{v}_1 & \mathbf{A}\mathbf{v}_2 & \ldots & \mathbf{A}\mathbf{v}_N \end{array}\right] = \left[\begin{array}{cccc} \lambda_1\mathbf{v}_1 & \lambda_2\mathbf{v}_2 & \ldots & \lambda_N\mathbf{v}_n \end{array}\right]$$

or

$$\mathbf{A}\mathbf{V} = \mathbf{V}\mathbf{\Lambda} \qquad\qquad (C.4)$$

where

$$\begin{aligned} \mathbf{V} &= \left[\begin{array}{cccc} \mathbf{v}_1 & \mathbf{v}_2 & \ldots & \mathbf{v}_n \end{array}\right] \\ \mathbf{\Lambda} &= \operatorname{diag}(\lambda_1, \lambda_2, \ldots, \lambda_n). \end{aligned}$$

If $\mathbf{A}$ is symmetric so that the eigenvectors corresponding to distinct eigenvalues are orthonormal and the remaining eigenvectors are chosen to yield an orthonormal eigenvector set, then $\mathbf{V}$ is an orthogonal matrix. As such, its inverse is $\mathbf{V}^T$, so that (C.4) becomes

$$\mathbf{A} = \mathbf{V}\mathbf{\Lambda}\mathbf{V}^T$$

Also, the inverse is easily determined as

$$\begin{aligned} \mathbf{A}^{-1} &= \mathbf{V}^{T^{-1}}\mathbf{\Lambda}^{-1}\mathbf{V}^{-1} \\ &= \mathbf{V}\mathbf{\Lambda}^{-1}\mathbf{V}^T. \end{aligned}$$

References

Graybill, F.A., *Introduction to Matrices with Applications in Statistics*, Wadsworth, Belmont, CA, 1969.

Noble, B., Daniel, J.W., *Applied Linear Algebra*, Prentice-Hall, Englewood Cliffs, NJ, 1977.

Appendix D

Summary of Signals, Linear Transforms, and Linear Systems

In this appendix we summarize the important concepts and formulas for discrete-time signal and system analysis. This material is used in Chapters 18–20. Some examples are given so that the reader unfamiliar with this material should try to verify the example results. For a more comprehensive treatment the books [Jackson 1991], [Oppenheim, Willsky, and Nawab 1997], [Poularikis and Seeley 1985] are recommended.

D.1 Discrete-Time Signals

A discrete-time signal is a *sequence* $x[n]$ for $n = \ldots, -1, 0, 1, \ldots$. It is defined only for the integers. Some important signals are:

a. *Unit impulse* – $x[n] = 1$ for $n = 0$ and $x[n] = 0$ for $n \neq 0$. It is also denoted by $\delta[n]$.

b. *Unit step* – $x[n] = 1$ for $n \geq 0$ and $x[n] = 0$ for $n < 0$. It is also denoted by $u[n]$.

c. *Real sinusoid* – $x[n] = A\cos(2\pi f_0 n + \theta)$ for $-\infty < n < \infty$, where A is the amplitude (must be nonnegative), f_0 is the frequency in cycles per sample and must be in the interval $0 < f_0 < 1/2$, and θ is the phase in radians.

d. *Complex sinusoid* – $x[n] = A\exp(j2\pi f_0 n + \theta)$ for $-\infty < n < \infty$, where A is the amplitude (must be nonnegative), f_0 is the frequency in cycles per sample and must be in the interval $-1/2 < f_0 < 1/2$, and θ is the phase in radians.

e. *Exponential* – $x[n] = a^n u[n]$

Note that any sequence can be written as a linear combination of unit impulses that are weighted by $x[k]$ and shifted in time as $\delta[n-k]$ to form

$$x[n] = \sum_{k=-\infty}^{\infty} x[k]\delta[n-k]. \tag{D.1}$$

For example, $a^n u[n] = \delta[n] + a\delta[n-1] + a^2\delta[n-2] + \cdots$.

Some special signals are defined next.

a. A signal is *causal* if $x[n] = 0$ for $n < 0$, for example, $x[n] = u[n]$.

b. A signal is *anticausal* if $x[n] = 0$ for $n > 0$, for example, $x[n] = u[-n]$.

c. A signal is *even* if $x[-n] = x[n]$ or it is symmetric about $n = 0$, for example, $x[n] = \cos(2\pi f_0 n)$.

d. A signal is *odd* if $x[-n] = -x[n]$ or it is antisymmetric about $n = 0$, for example, $x[n] = \sin(2\pi f_0 n)$.

e. A signal is *stable* if $\sum_{n=-\infty}^{\infty} |x[n]| < \infty$ (also called *absolutely summable*), for example, $x[n] = (1/2)^n u[n]$.

D.2   Linear Transforms

D.2.1   Discrete-Time Fourier Transforms

The *discrete-time Fourier transform* $X(f)$ of a discrete-time signal $x[n]$ is defined as

$$X(f) = \sum_{n=-\infty}^{\infty} x[n]\exp(-j2\pi f n) \qquad -1/2 \le f \le 1/2. \tag{D.2}$$

An example is $x[n] = (1/2)^n u[n]$ for which $X(f) = 1/(1 - (1/2)\exp(-j2\pi f))$. It converts a discrete-time signal into a complex function of f, where f is called the *frequency* and is measured in cycles per sample. The operation of taking the Fourier transform of a signal is denoted by $\mathcal{F}\{x[n]\}$ and the signal and its Fourier transform are referred to as a *Fourier transform pair*. The latter relationship is usually denoted by $x[n] \Leftrightarrow X(f)$. The discrete-time Fourier transform is *periodic in frequency with period one* and for this reason we need only consider the frequency interval $[-1/2, 1/2]$. Since the Fourier transform is a complex function of frequency, it can be represented by the two real functions

$$|X(f)| = \sqrt{\left(\sum_{n=-\infty}^{\infty} x[n]\cos(2\pi f n)\right)^2 + \left(\sum_{n=-\infty}^{\infty} x[n]\sin(2\pi f n)\right)^2}$$

$$\phi(f) = \arctan \frac{-\sum_{n=-\infty}^{\infty} x[n]\sin(2\pi f n)}{\sum_{n=-\infty}^{\infty} x[n]\cos(2\pi f n)}$$

Signal name	$x[n]$	$X(f)\ (-\frac{1}{2} \le f \le \frac{1}{2})$
Unit impulse	$\delta[n] = \begin{cases} 1 & n = 0 \\ 0 & n \ne 0 \end{cases}$	1
Real sinusoid	$\cos(2\pi f_0 n)$	$\frac{1}{2}\delta(f + f_0) + \frac{1}{2}\delta(f - f_0)$
Complex sinusoid	$\exp(j2\pi f_0 n)$	$\delta(f - f_0)$
Exponential	$a^n u[n]$	$\frac{1}{1 - a\exp(-j2\pi f)}$ $\qquad |a| < 1$
Double-sided exponential	$a^{|n|}$	$\frac{1 - a^2}{1 + a^2 - 2a\cos(2\pi f)}$ $\qquad |a| < 1$

Table D.1: Discrete-time Fourier transform pairs.

which are called the *magnitude* and *phase*, respectively. For example, if $x[n] = (1/2)^n u[n]$, then

$$|X(f)| = \frac{1}{\sqrt{5/4 - \cos(2\pi f)}}$$

$$\phi(f) = -\arctan \frac{\frac{1}{2}\sin(2\pi f)}{1 - \frac{1}{2}\cos(2\pi f)}.$$

Note that the magnitude is an even function or $|X(-f)| = |X(f)|$ and the phase is an odd function or $\phi(-f) = -\phi(f)$. Some Fourier transform pairs are given in Table D.1. Some important properties of the discrete-time Fourier transform are:

a. Linearity – $\mathcal{F}\{ax[n] + by[n]\} = aX(f) + bY(f)$

b. Time shift – $\mathcal{F}\{x[n - n_0]\} = \exp(-j2\pi f n_0)X(f)$

c. Modulation – $\mathcal{F}\{\cos(2\pi f_0 n)x[n]\} = \frac{1}{2}X(f + f_0) + \frac{1}{2}X(f - f_0)$

d. Time reversal – $\mathcal{F}\{x[-n]\} = X^*(f)$

e. Symmetry – if $x[n]$ is even, then $X(f)$ is even and real, and if $x[n]$ is odd, then $X(f)$ is odd and purely imaginary.

f. Energy – the energy defined as $\sum_{n=-\infty}^{\infty} x^2[n]$ can be found from the Fourier transform using *Parseval's theorem*

$$\sum_{n=-\infty}^{\infty} x^2[n] = \int_{-\frac{1}{2}}^{\frac{1}{2}} |X(f)|^2 df.$$

g. Inner product – as an extension of Parseval's theorem we have

$$\sum_{n=-\infty}^{\infty} x[n]y[n] = \int_{-\frac{1}{2}}^{\frac{1}{2}} X^*(f)Y(f)df.$$

Two signals $x[n]$ and $y[n]$ are said to be *convolved* together to yield a new signal $z[n]$ if

$$z[n] = \sum_{k=-\infty}^{\infty} x[k]y[n-k] \qquad -\infty < n < \infty.$$

As an example, if $x[n] = u[n]$ and $y[n] = u[n]$, then $z[n] = (n+1)u[n]$. The operation of convolving two signals together is called *convolution* and is implemented using a *convolution sum*. It is denoted by $x[n] \star y[n]$. The operation is commutative in that $x[n] \star y[n] = y[n] \star x[n]$ so that an equivalent form is

$$z[n] = \sum_{k=-\infty}^{\infty} y[k]x[n-k] \qquad -\infty < n < \infty.$$

As an example, if $y[n] = \delta[n - n_0]$, then it is easily shown that $x[n] \star \delta[n - n_0] = \delta[n - n_0] \star x[n] = x[n - n_0]$. The most important property of convolution is that two signals that are convolved together produce a signal whose Fourier transform is the product of the signals' Fourier transforms or

$$\mathcal{F}\{x[n] \star y[n]\} = X(f)Y(f).$$

Two signals $x[n]$ and $y[n]$ are said to be *correlated together* to yield a new signal $z[n]$ if

$$z[n] = \sum_{k=-\infty}^{\infty} x[k]y[k+n] \qquad -\infty < n < \infty.$$

The Fourier transform of $z[n]$ is $X^*(f)Y(f)$. The sequence $z[n]$ is also called the *deterministic cross-correlation*. If $x[n] = y[n]$, then $z[n]$ is called the *deterministic autocorrelation* and its Fourier transform is $|X(f)|^2$.

The discrete-time signal may be recovered from its Fourier transform by using the *discrete-time inverse Fourier transform*

$$x[n] = \int_{-\frac{1}{2}}^{\frac{1}{2}} X(f) \exp(j2\pi fn)df \qquad -\infty < n < \infty. \tag{D.3}$$

As an example, if $X(f) = \frac{1}{2}\delta(f + f_0) + \frac{1}{2}\delta(f - f_0)$, then the integral yields $x[n] = \cos(2\pi f_0 n)$. It also has the interpretation that a discrete-time signal $x[n]$ may be thought of as a sum of complex sinusoids $X(f)\exp(j2\pi fn)\Delta f$ for $-1/2 \leq f \leq 1/2$ with amplitude $|X(f)|\Delta f$ and phase $\angle X(f)$. There is a separate sinusoid for each frequency f, and the total number of sinusoids is uncountable.

D.2.2 Numerical Evaluation of Discrete-Time Fourier Transforms

The discrete-time Fourier transform of a signal $x[n]$, which is nonzero only for $n = 0, 1, \ldots, N-1$, is given by

$$X(f) = \sum_{n=0}^{N-1} x[n] \exp(-j2\pi fn) \qquad -1/2 \leq f \leq 1/2. \qquad (D.4)$$

Such a signal is said to be *time-limited*. Since the Fourier transform is periodic with period one, we can equivalently evaluate it over the interval $0 \leq f \leq 1$. Then, if we desire the Fourier transform for $-1/2 \leq f' < 0$, we use the previously evaluated $X(f)$ with $f = f' + 1$. To numerically evaluate the Fourier transform we therefore can use the frequency interval $[0, 1]$ and compute samples of $X(f)$ for $f = 0, 1/N, 2/N, \ldots, (N-1)/N$. This yields the *discrete Fourier transform* (DFT) which is defined as

$$X[k] = X(f)|_{f=k/N} = \sum_{n=0}^{N-1} x[n] \exp\left(-j2\pi(k/N)n\right) \qquad k = 0, 1, \ldots, N-1.$$

Since there are only N time samples, we may wish to compute more *frequency samples* since $X(f)$ is a *continuous* function of frequency. To do so we can *zero pad* the time samples with zeros to yield a new signal $x'[n]$ of length $M > N$ with samples $\{x[0], x[1], \ldots, x[N-1], 0, 0, \ldots, 0\}$. This new signal $x'[n]$ will consist of N time samples and $M - N$ zeros so that the DFT will compute more finely spaced frequency samples as

$$
\begin{aligned}
X[k] &= X(f)|_{f=k/M} = \sum_{n=0}^{M-1} x'[n] \exp\left(-j2\pi(k/M)n\right) & k = 0, 1, \ldots, M-1 \\
&= \sum_{n=0}^{N-1} x[n] \exp\left(-j2\pi(k/M)n\right) & k = 0, 1, \ldots, M-1.
\end{aligned}
$$

The actual DFT is computed using the fast Fourier transform (FFT), which is an algorithm used to reduce the computation.

The inverse Fourier transform of an infinite length causal sequence can be *approximated* using an *inverse DFT* as

$$
\begin{aligned}
x[n] &= \int_{-\frac{1}{2}}^{\frac{1}{2}} X(f) \exp(j2\pi fn) df = \int_0^1 X(f) \exp(j2\pi fn) df \\
&\approx \frac{1}{M} \sum_{k=0}^{M-1} X[k] \exp\left(j2\pi(k/M)n\right) & n = 0, 1, \ldots, M-1. \qquad (D.5)
\end{aligned}
$$

One should choose M large. The actual inverse DFT is computed using the inverse FFT.

D.2.3 z-Transforms

The z-transform of a discrete-time signal $x[n]$ is defined as

$$X(z) = \sum_{n=-\infty}^{\infty} x[n]z^{-n} \qquad (D.6)$$

where z is a complex variable that takes on values for which $|X(z)| < \infty$. As an example, if $x[n] = (1/2)^n u[n]$, then

$$X(z) = \frac{1}{1 - \frac{1}{2}z^{-1}} \qquad |z| > \frac{1}{2}. \qquad (D.7)$$

The operation of taking the z-transform is indicated by $\mathcal{Z}\{x[n]\}$. Some important properties of the z-transform are:

a. Linearity – $\mathcal{Z}\{ax[n] + by[n]\} = a X(z) + b \mathcal{Y}(z)$

b. Time shift – $\mathcal{Z}\{x[n - n_0]\} = z^{-n_0} X(z)$

c. Convolution – $\mathcal{Z}\{x[n] \star y[n]\} = X(z)\mathcal{Y}(z)$.

Assuming that the z-transform converges on the unit circle, the discrete-time Fourier transform is given by

$$X(f) = X(z)|_{z=\exp(j2\pi f)} \qquad (D.8)$$

as is seen by comparing (D.6) to (D.2). As an example, if $x[n] = (1/2)^n u[n]$, then from (D.7)

$$X(f) = \frac{1}{1 - \frac{1}{2}\exp(-j2\pi f)}$$

since $X(z)$ converges for $|z| = |\exp(j2\pi f)| = 1 > 1/2$.

D.3   Discrete-Time Linear Systems

A discrete-time system takes an input signal $x[n]$ and produces an output signal $y[n]$. The transformation is symbolically represented as $y[n] = \mathcal{L}\{x[n]\}$. The system is *linear* if $\mathcal{L}\{ax[n] + by[n]\} = a\mathcal{L}\{x[n]\} + b\mathcal{L}\{y[n]\}$. A system is defined to be *shift invariant* if $\mathcal{L}\{x[n - n_0]\} = y[n - n_0]$. If the system is *linear and shift invariant* (LSI), then the output is easily found if we know the output to a unit impulse. To see this we compute the output of the system as

$$
\begin{aligned}
y[n] &= \mathcal{L}\{x[n]\} \\
&= \mathcal{L}\left\{ \sum_{k=-\infty}^{\infty} x[k]\delta[n-k] \right\} && \text{(using (D.1))} \\
&= \sum_{k=-\infty}^{\infty} x[k]\mathcal{L}\{\delta[n-k]\} && \text{(linearity)} \\
&= \sum_{k=-\infty}^{\infty} x[k]\ \mathcal{L}\{\delta[n]\}|_{n\to n-k} && \text{(shift invariance)} \\
&= \sum_{k=-\infty}^{\infty} x[k]h[n-k]
\end{aligned}
$$

where $h[n] = \mathcal{L}\{\delta[n]\}$ is called the *impulse response* of the system. Note that $y[n] = x[n] \star h[n] = h[n] \star x[n]$ and so the output of the LSI system is also given by the convolution sum

$$
y[n] = \sum_{k=-\infty}^{\infty} h[k]x[n-k]. \tag{D.9}
$$

A *causal* system is defined as one for which $h[k] = 0$ for $k < 0$ since then the output depends only on the present input $x[n]$ and the past inputs $x[n-k]$ for $k \geq 1$. The system is said to be *stable* if

$$
\sum_{k=-\infty}^{\infty} |h[k]| < \infty.
$$

If this condition is satisfied, then a bounded input signal or $|x[n]| < \infty$ for $-\infty < n < \infty$ will always produce a bounded output signal or $|y[n]| < \infty$ for $-\infty < n < \infty$. As an example, the LSI system with impulse response $h[k] = (1/2)^k u[k]$ is stable but not the one with impulse response $h[k] = u[k]$. The latter system will produce the unbounded output $y[n] = (n+1)u[n]$ for the bounded input $x[n] = u[n]$ since $u[n] \star u[n] = (n+1)u[n]$.

Since for an LSI system $y[n] = h[n] \star x[n]$, it follows from the properties of z-transforms that $\mathcal{Y}(z) = \mathcal{H}(z)\mathcal{X}(z)$, where $\mathcal{H}(z)$ is the z-transform of the impulse response. As a result, we have that

$$
\mathcal{H}(z) = \frac{\mathcal{Y}(z)}{\mathcal{X}(z)} = \frac{\text{Output } z\text{-transform}}{\text{Input } z\text{-transform}} \tag{D.10}
$$

and $\mathcal{H}(z)$ is called the *system function*. Note that since it is the z-transform of the impulse response $h[n]$ we have

$$
\mathcal{H}(z) = \sum_{k=-\infty}^{\infty} h[n]z^{-n}. \tag{D.11}
$$

If the input to an LSI system is a complex sinusoid, $x[n] = \exp(j2\pi f_0 n)$, then the output is from (D.9)

$$
\begin{aligned}
y[n] &= \sum_{k=-\infty}^{\infty} h[k]\exp[j2\pi f_0(n-k)] \\
&= \underbrace{\sum_{k=-\infty}^{\infty} h[k]\exp(-j2\pi f_0 k)}_{H(f_0)} \exp(j2\pi f_0 n). \qquad (D.12)
\end{aligned}
$$

It is seen that the output is also a complex sinusoid with the same frequency but multiplied by the Fourier transform of the impulse response evaluated at the sinusoidal frequency. Hence, $H(f)$ is called the *frequency response*. Also, from (D.12) the frequency response is obtained from the system function (see (D.11)) by letting $z = \exp(j2\pi f)$. Finally, note that the frequency response is the discrete-time Fourier transform of the impulse response. As an example, if $h[n] = (1/2)^n u[n]$, then

$$
\mathcal{H}(z) = \frac{1}{1 - \frac{1}{2}z^{-1}}
$$

and

$$
H(f) = \mathcal{H}(\exp(j2\pi f)) = \frac{1}{1 - \frac{1}{2}\exp(-j2\pi f)}.
$$

The *magnitude response* of the LSI system is defined as $|H(f)|$ and the *phase response* as $\angle H(f)$.

As we have seen, LSI systems can be characterized by the equivalent descriptions: impulse response, system function, or frequency response. This means that *given one of these descriptions the output can be determined for any input.* LSI systems can also be characterized by linear difference equations with constant coefficients. Some examples are

$$
\begin{aligned}
y_1[n] &= x[n] - bx[n-1] \\
y_2[n] &= ay_2[n-1] + x[n] \\
y_3[n] &= ay_3[n-1] + x[n] - bx[n-1]
\end{aligned}
$$

and more generally

$$
y[n] = \sum_{k=1}^{p} a[k]y[n-k] + x[n] - \sum_{k=1}^{q} b[k]x[n-k]. \qquad (D.13)
$$

The system function is found by taking the z-transform of both sides of the difference equations and using (D.10) to yield

$$
\begin{aligned}
\mathcal{Y}_1(z) &= \mathcal{X}(z) - bz^{-1}\mathcal{X}(z) \Rightarrow \mathcal{H}_1(z) = 1 - bz^{-1} \\
\mathcal{Y}_2(z) &= az^{-1}\mathcal{Y}_2(z) + \mathcal{X}(z) \Rightarrow \mathcal{H}_2(z) = \frac{1}{1 - az^{-1}} \\
\mathcal{Y}_3(z) &= az^{-1}\mathcal{Y}_3(z) + \mathcal{X}(z) - bz^{-1}\mathcal{X}(z) \Rightarrow \mathcal{H}_3(z) = \frac{1 - bz^{-1}}{1 - az^{-1}}
\end{aligned}
$$

and the frequency response is obtained using $H(f) = \mathcal{H}(\exp(j2\pi f))$. More generally, for the LSI system whose difference equation description is given by (D.13) we have

$$
\mathcal{H}(z) = \frac{1 - \sum_{k=1}^{q} b[k]z^{-k}}{1 - \sum_{k=1}^{p} a[k]z^{-k}}. \tag{D.14}
$$

The impulse response is obtained by taking the inverse z-transform of the system function to yield for the previous examples

$$
h_1[n] = \begin{cases} 1 & n = 0 \\ -b & n = 1 \\ 0 & \text{otherwise} \end{cases}
$$

$$
\begin{aligned}
h_2[n] &= a^n u[n] & \text{(assuming system is causal)} \\
h_3[n] &= a^n u[n] - ba^{n-1}u[n-1] & \text{(assuming system is causal).}
\end{aligned}
$$

The impulse response could also be obtained by letting $x[n] = \delta[n]$ in the difference equations and setting $y[-1] = 0$, due to causality, and *recursing* the difference equation. For example, if the difference equation is $y[n] = (1/2)y[n-1] + x[n]$, then by definition the impulse response satisfies the equation $h[n] = (1/2)h[n-1] + \delta[n]$. By recursing this we obtain

$$
\begin{aligned}
h[0] &= \tfrac{1}{2}h[-1] + \delta[0] = 1 & \text{(since } h[-1] = 0 \text{ due to causality)} \\
h[1] &= \tfrac{1}{2}h[0] + \delta[1] = \tfrac{1}{2} & \text{(since } \delta[n] = 0 \text{ for } n \geq 1) \\
h[2] &= \tfrac{1}{2}h[1] = \tfrac{1}{4} \\
& \quad \text{etc.}
\end{aligned}
$$

and so in general we have the impulse response $h[n] = (1/2)^n u[n]$. The system with impulse response $h_1[n]$ is called a *finite impulse response* (FIR) system while those of $h_2[n]$ and $h_3[n]$ are called *infinite impulse response* (IIR) systems. The terminology refers to the number of nonzero samples of the impulse response.

For the system function $H_3(z) = (1 - bz^{-1})/(1 - az^{-1})$, the value of z for which the numerator is zero is called a *zero* and the value of z for which the denominator is zero is called a *pole*. In this case the system function has one zero at $z = b$ and one pole at $z = a$. For the system to be stable, assuming it is causal, *all the poles of the system function must be within the unit circle of the z-plane.* Hence, for stability we require $|a| < 1$. The zeros may lie anywhere in the z-plane. For a *second-order* system function (let $p = 2$ and $q = 0$ in (D.14)) given as

$$H(z) = \frac{1}{1 - a[1]z^{-1} - a[2]z^{-2}}$$

the poles, assuming they are complex, are located at $z = r\exp(\pm j\theta)$. Hence, for stability we require $r < 1$ and we note that since the poles are the z values for which the denominator polynomial is zero, we have

$$1 - a[1]z^{-1} - a[2]z^{-2} = z^{-2}(z - r\exp(j\theta))(z - r\exp(-j\theta)).$$

Therefore, the coefficients are related to the complex poles as

$$\begin{aligned} a[1] &= 2r\cos(\theta) \\ a[2] &= -r^2 \end{aligned}$$

which puts restrictions on the possible values of $a[1]$ and $a[2]$. As an example, the coefficients $a[1] = 0$, $a[2] = -1/4$ produce a stable filter but not $a[1] = 0$, $a[2] = -2$.

An LSI system whose frequency response is

$$H(f) = \begin{cases} 1 & |f| \leq B \\ 0 & |f| > B \end{cases}$$

is said to be an *ideal lowpass filter*. It passes complex sinusoids undistorted if their frequency is $|f| \leq B$ but nullifies ones with a higher frquency. The band of positive frequencies from $f = 0$ to $f = B$ is called the *passband* and the band of positive frequencies for which $f > B$ is called the *stopband*.

D.4   Continuous-Time Signals

A continuous-time signal is a *function of time* $x(t)$ for $-\infty < t < \infty$. Some important signals are:

a. *Unit impulse* – It is denoted by $\delta(t)$. An impulse $\delta(t)$, also called the *Dirac delta function*, is defined as the limit of a very narrow pulse as the pulsewidth goes to zero and the pulse amplitude goes to infinity, such that the overall area remains at one. Therefore, if we define a very narrow pulse as

$$x_T(t) = \begin{cases} \frac{1}{T} & |t| \leq T/2 \\ 0 & |t| > T/2 \end{cases}$$

then the unit impulse is defined as

$$\delta(t) = \lim_{T \to 0} x_T(t).$$

The impulse has the important *sifting property* that if $x(t)$ is continuous at $t = t_0$, then

$$\int_{-\infty}^{\infty} x(t)\delta(t - t_0)dt = x(t_0).$$

b. *Unit step* – $x(t) = 1$ for $t \geq 0$ and $x(t) = 0$ for $t < 0$. It is also denoted by $u(t)$.

c. *Real sinusoid* – $x(t) = A\cos(2\pi F_0 t + \theta)$ for $-\infty < t < \infty$, where A is the amplitude (must be nonnegative), F_0 is the frequency in Hz (cycles per second), and θ is the phase in radians.

d. *Complex sinusoid* – $x(t) = A\exp(j2\pi F_0 t + \theta)$ for $-\infty < t < \infty$, with the amplitude, frequency, and phase taking on same values as for real sinusoid.

e. *Exponential* – $x(t) = \exp(at)u(t)$

f. *Pulse* – $x(t) = 1$ for $|t| \leq T/2$ and $x(t) = 0$ for $|t| > T/2$.

Some special signals are defined next.

a. A signal is *causal* if $x(t) = 0$ for $t < 0$, for example, $x(t) = u(t)$.

b. A signal is *anticausal* if $x(t) = 0$ for $t > 0$, for example, $x(t) = u(-t)$.

c. A signal is *even* if $x(-t) = x(t)$ or it is symmetric about $t = 0$, for example, $x(t) = \cos(2\pi F_0 t)$.

d. A signal is *odd* if $x(-t) = -x(t)$ or it is antisymmetric about $t = 0$, for example, $x(t) = \sin(2\pi F_0 t)$.

e. A signal is *stable* if $\int_{-\infty}^{\infty} |x(t)|dt < \infty$ (also called *absolutely integrable*), for example, $x(t) = \exp(-t)u(t)$.

D.5 Linear Transforms

D.5.1 Continuous-Time Fourier Transforms

The *continuous-time Fourier transform* $X(F)$ of a continuous-time signal $x(t)$ is defined as

$$X(F) = \int_{-\infty}^{\infty} x(t)\exp(-j2\pi Ft)dt \qquad -\infty < F < \infty. \tag{D.15}$$

An example is $x(t) = \exp(-t)u(t)$ for which $X(F) = 1/(1 + j2\pi F)$. It converts a continuous-time signal into a complex function of F, where F is called the *frequency* and is measured in Hz (cycles per second). The operation of taking the Fourier transform of a signal is denoted by $\mathcal{F}\{x(t)\}$ and the signal and its Fourier transform are referred to as a *Fourier transform pair*. The latter relationship is usually

Signal name	$x(t)$	$X(F)$
Unit impulse	$\delta(t) = \begin{cases} \infty & t = 0 \\ 0 & t \neq 0 \end{cases}$	1
Real sinusoid	$\cos(2\pi F_0 t)$	$\frac{1}{2}\delta(F + F_0) + \frac{1}{2}\delta(F - F_0)$
Complex sinusoid	$\exp(j2\pi F_0 t)$	$\delta(F - F_0)$
Exponential	$\exp(-at)u(t)$	$\frac{1}{a+j2\pi F}$ $a > 0$
Pulse	$= \begin{cases} 1 & |t| \leq T/2 \\ 0 & |t| > T/2 \end{cases}$	$T\frac{\sin(\pi FT)}{\pi FT}$

Table D.2: Continuous-time Fourier transform pairs.

denoted by $x(t) \Leftrightarrow X(F)$. Note that the magnitude of $X(F)$ is an even function or $|X(-F)| = |X(F)|$ and the phase is an odd function or $\phi(-F) = -\phi(F)$. Some Fourier transform pairs are given in Table D.2.

Some important properties of the continuous-time Fourier transform are:

a. Linearity – $\mathcal{F}\{ax(t) + by(t)\} = aX(F) + bY(F)$

b. Time shift – $\mathcal{F}\{x(t - t_0)\} = \exp(-j2\pi F t_0)X(F)$

c. Modulation – $\mathcal{F}\{\cos(2\pi F_0 t)x(t)\} = \frac{1}{2}X(F + F_0) + \frac{1}{2}X(F - F_0)$

d. Time reversal – $\mathcal{F}\{x(-t)\} = X^*(F)$

e. Symmetry – if $x(t)$ is even, then $X(F)$ is even and real, and if $x(t)$ is odd, then $X(F)$ is odd and purely imaginary.

f. Energy – the energy defined as $\int_{-\infty}^{\infty} x^2(t)dt$ can be found from the Fourier transform using *Parseval's theorem*

$$\int_{-\infty}^{\infty} x^2(t)dt = \int_{-\infty}^{\infty} |X(F)|^2 dF.$$

g. Inner product – as an extension of Parseval's theorem we have

$$\int_{-\infty}^{\infty} x(t)y(t)dt = \int_{-\infty}^{\infty} X^*(F)Y(F)dF.$$

Two signals $x(t)$ and $y(t)$ are said to be *convolved* together to yield a new signal $z(t)$ if

$$z(t) = \int_{-\infty}^{\infty} x(\tau)y(t - \tau)d\tau \qquad -\infty < t < \infty.$$

As an example, if $x(t) = u(t)$ and $y(t) = u(t)$, then $z(t) = tu(t)$. The operation of convolving two signals together is called *convolution* and is implemented using a *convolution integral*. It is denoted by $x(t) \star y(t)$. The operation is commutative in that $x(t) \star y(t) = y(t) \star x(t)$ so that an equivalent form is

$$z(t) = \int_{-\infty}^{\infty} y(\tau) x(t - \tau) d\tau \qquad -\infty < t < \infty.$$

As an example, if $y(t) = \delta(t - t_0)$, then it is easily shown that $x(t) \star \delta(t - t_0) = \delta(t - t_0) \star x(t) = x(t - t_0)$. The most important property of convolution is that two signals that are convolved together produce a signal whose Fourier transform is the product of the signals' Fourier transforms or

$$\mathcal{F}\{x(t) \star y(t)\} = X(F)Y(F).$$

The continuous-time signal may be recovered from its Fourier transform by using the *continuous-time inverse Fourier transform*

$$x(t) = \int_{-\infty}^{\infty} X(F) \exp(j2\pi Ft) dF \qquad -\infty < t < \infty. \tag{D.16}$$

As an example, if $X(F) = \frac{1}{2}\delta(F + F_0) + \frac{1}{2}\delta(F - F_0)$, then the integral yields $x(t) = \cos(2\pi F_0 t)$. It also has the interpretation that a continuous-time signal $x(t)$ may be thought of as a sum of complex sinusoids $X(F) \exp(j2\pi Ft)\Delta F$ for $-\infty < F < \infty$ with amplitude $|X(F)|\Delta F$ and phase $\angle X(F)$. There is a separate sinusoid for each frequency F, and the total number of sinusoids is uncountable.

D.6 Continuous-Time Linear Systems

A continuous-time system takes an input signal $x(t)$ and produces an output signal $y(t)$. The transformation is symbolically represented as $y(t) = \mathcal{L}\{x(t)\}$. The system is *linear* if $\mathcal{L}\{ax(t) + by(t)\} = a\mathcal{L}\{x(t)\} + b\mathcal{L}\{y(t)\}$. A system is defined to be *time invariant* if $\mathcal{L}\{x(t-t_0)\} = y(t-t_0)$. If the system is *linear and time invariant* (LTI), then the output is easily found if we know the output to a unit impulse. It is given by the convolution integral

$$y(t) = \int_{-\infty}^{\infty} h(\tau) x(t - \tau) d\tau \tag{D.17}$$

where $h(t) = \mathcal{L}\{\delta(t)\}$ is called the *impulse response* of the system. A *causal* system is defined as one for which $h(\tau) = 0$ for $\tau < 0$ since then the output depends only on the present input $x(t)$ and the past inputs $x(t - \tau)$ for $\tau > 0$. The system is said to be *stable* if

$$\int_{-\infty}^{\infty} |h(\tau)| d\tau < \infty.$$

If this condition is satisfied, then a bounded input signal or $|x(t)| < \infty$ for $-\infty < t < \infty$ will always produce a bounded output signal or $|y(t)| < \infty$ for $-\infty < t < \infty$. As an example, the LTI system with impulse response $h(\tau) = \exp(-\tau)u(\tau)$ is stable but not the one with impulse response $h(\tau) = u(\tau)$. The latter system will produce the unbounded output $y(t) = tu(t)$ for the bounded input $x(t) = u(t)$ since $u(t) \star u(t) = tu(t)$.

If the input to an LTI system is a complex sinusoid, $x(t) = \exp(j2\pi F_0 t)$, then the output is from (D.17)

$$
\begin{aligned}
y(t) &= \int_{-\infty}^{\infty} h(\tau) \exp[j2\pi F_0(t - \tau)]d\tau \\
&= \underbrace{\int_{-\infty}^{\infty} h(\tau) \exp(-j2\pi F_0 \tau)d\tau}_{H(F_0)} \exp(j2\pi F_0 t).
\end{aligned}
\tag{D.18}
$$

It is seen that the output is also a complex sinusoid with the same frequency but multiplied by the Fourier transform of the impulse response evaluated at the sinusoidal frequency. Hence, $H(F)$ is called the *frequency response*. Finally, note that the frequency response is the continuous-time Fourier transform of the impulse response. As an example, if $h(t) = \exp(-at)u(t)$, then for $a > 0$

$$
H(F) = \frac{1}{a + j2\pi F}.
$$

The *magnitude response* of the LSI system is defined as $|H(F)|$ and the *phase response* as $\angle H(F)$.

An LTI system whose frequency response is

$$
H(F) = \begin{cases} 1 & |F| \le W \\ 0 & |F| > W \end{cases}
$$

is said to be an *ideal lowpass filter*. It passes complex sinusoids undistorted if their frequency is $|F| \le W$ Hz but nullifies ones with a higher frquency. The band of positive frequencies from $F = 0$ to $F = W$ is called the *passband* and the band of positive frequencies for which $F > W$ is called the *stopband*.

References

Jackson, L.B., *Signals, Systems, and Transforms*, Addison-Wesley, Reading, MA, 1991.

Oppenheim, A.V., A.S. Willsky, S.H. Nawab, *Signal and Systems*, Prentice-Hall, Upper Saddle River, NJ, 1997.

Poularikas, A.D., S. Seely, *Signals and Systems*, PWS, Boston, 1985.

Appendix E

Answers to Selected Problems

Note: For problems based on computer simulations the number of realizations used in the computer simulation will affect the numerical results. In the results listed below the number of realizations is denoted by N_{real}. Also, each result assumes that rand('state',0) and/or randn('state',0) have been used to initialize the random number generator (see Appendix 2A for further details).

Chapter 1

1. experiment: toss a coin; outcomes: {head, tail}; probabilities: $1/2, 1/2$

5. a. continuous; b. discrete; c. discrete; d. continuous; e. discrete

7. yes, yes

10. $P[k = 9] = 0.0537$, probably not

13. $1/2$

14. 0.9973 for $\Delta = 0.001$

Chapter 2

1. $\hat{P}[Y = 0] = 0.7490$, $\hat{P}[Y = 1] = 0.2510$ ($N_{real} = 1000$)

3. via simulation: $\hat{P}[-1 \leq X \leq 1] = 0.6863$; via numerical integration with $\Delta = 0.01$, $P[-1 \leq X \leq 1] = 0.6851$ ($N_{real} = 10,000$)

6. values near zero

8. estimated mean = 0.5021; true mean = $1/2$ ($N_{real} = 1000$)

11. estimated mean = 1.0042; true mean = 1 ($N_{real} = 1000$)

13. 1.2381 ($N_{real} = 1000$)

14. no; via simulation: mean of $\sqrt{U} = 0.6589$; via simulation: $\sqrt{\text{mean of } U} = 0.7125$ ($N_{\text{real}} = 1000$)

Chapter 3

1. a. $A^c = \{x : x \leq 1\}$, $B^c = \{x : x > 2\}$
   b. $A \cup B = \{x : -\infty < x < \infty\} = S$, $A \cap B = \{x : 1 < x < 2\}$
   c. $A - B = \{x : x > 2\}$, $B - A = \{x : x \leq 1\}$

7. $A = \{1, 2, 3\}$, $B = \{4, 5\}$, $C = \{1, 2, 3\}$, $D = \{4, 5, 6\}$

10. $A \cup B \cup C = (A^c \cap B^c \cap C^c)^c$, $A \cap B \cap C = (A^c \cup B^c \cup C^c)^c$

12. a. 10^7, discrete b. 1, discrete c. ∞ (uncountable), continuous d. ∞ (uncountable), continuous e. 2, discrete f. ∞ (countable), discrete

14. a. $S = \{t : 30 \leq t \leq 100\}$ b. outcomes are all t in interval $[30, 100]$ c. set of outcomes having no elements, i.e., {negative temperatures} d. $A = \{t : 40 \leq t \leq 60\}$, $B = \{t : 40 \leq t \leq 50 \text{ or } 60 \leq t \leq 70\}$, $C = \{100\}$ (simple event) e. $A = \{t : 40 \leq t \leq 60\}$, $B = \{t : 60 \leq t \leq 70\}$

18. a. 1/2 b. 1/2 c. 6/36 d. 24/36

19. $P_{\text{even}} = 1/2$, $\hat{P}_{\text{even}} = 0.5080$ ($N_{\text{real}} = 1000$)

21. a. even, 2/3 b. odd, 1/3 c. even or odd, 1 d. even and odd, 0

23. 1/56

25. 10/36

27. no

33. 90/216

35. 676,000

38. 0.00183

40. total number $= 16$, two-toppings $= 6$

44. a. 4 of a kind
$$\frac{13 \cdot 48}{\binom{52}{5}}$$

   b. flush
$$\frac{4 \cdot \binom{13}{5}}{\binom{52}{5}}$$

49. $P[k \geq 95] = 0.4407$, $\hat{P}[k \geq 95] = 0.4430$ ($N_{\text{real}} = 1000$)

Chapter 4

2. 1/4

5. 1/4

7. a. 0.53 b. 0.34

11. 0.5

14. yes

19. 0.03

21. a. no b. no

22. 4

26. 0.0439

28. 5/16

33. $P[k] = (k-1)(1-p)^{k-2}p^2$, $k = 2, 3, \ldots$,

38. 2 red, 2 black, 2 white

40. 3/64

43. 165/512

Chapter 5

4. $\mathcal{S}_X = \{0, 1, 4, 9\}$

$$p_X[x_i] = \begin{cases} \frac{1}{7} & x_i = 0 \\ \frac{2}{7} & x_i = 1 \\ \frac{2}{7} & x_i = 4 \\ \frac{2}{7} & x_i = 9 \end{cases}$$

6. $0 < p < 1$, $\alpha = (1-p)/p^2$

8. 0.99^{19}

13. Average value $= 5.0310$, true value shown in Chapter 6 to be $\lambda = 5$ ($N_{\text{real}} = 1000$)

14. $p_X[5] = 0.0029$, $\hat{p}_X[5] = 0.0031$ (from Poisson approximation)

18. $P[X = 3] = 0.0613$, $\hat{P}[X = 3] = 0.0607$ $(N_{\text{real}} = 10,000)$

20. $p_Y[k] = \exp(-\lambda)\lambda^{k/2}/k!$ for $k = 0, 2, 4, \ldots$

26. $p_X[k] = 1/5$ for $k = 1, 2, 3, 4, 5$

28. 0.4375

31. 8.68×10^{-7}

Chapter 6

2. 9/2

4. 2/3

8. geometric PMF

12. $(2/p^2) - 1/p$

13. yes, if $X = $ constant

14. predictor $= E[X] = 21/8$, $\text{mse}_{\min} = 47/64 = 0.7343$

15. estimated $\text{mse}_{\min} = 0.7371$ $(N_{\text{real}} = 10,000)$

20. $\lambda^2 + \lambda$

26. $\sum_{k=0}^{n}(-1)^{n-k}\binom{n}{k}E^{n-k}[X]E[X^k]$

27. $\phi_Y(\omega) = \exp(j\omega b)\phi_X(a\omega)$

28. $(1 + 2\cos(\omega) + 2\cos(2\omega))/5$

32. true mean $= 1/2$, true variance $= 3/4$; estimated mean $= 0.5000$, estimated variance $= 0.7500$ $(N_{\text{real}} = 1000)$

Chapter 7

3. $S = \{(\text{p,n}), (\text{p,d}), (\text{n,p}), (\text{n,d}), (\text{d,p}), (\text{d,n})\}$
$S_{X,Y} = \{(1, 5), (1, 10), (5, 1), (5, 10), (10, 1), (10, 5)\}$

8.
$$p_{X,Y}[i,j] = \begin{cases} 1/4 & (i,j) = (0,0) \\ 1/4 & (i,j) = (1,-1) \\ 1/4 & (i,j) = (1,1) \\ 1/4 & (i,j) = (2,0) \end{cases}$$

10. 1/5

13. $p_X[i] = (1-p)^{i-1}p$ for $i = 1, 2, \ldots$ and same for $p_Y[j]$

16. $p_{X,Y}[0,0] = 1/4$, $p_{X,Y}[0,1] = 0$, $p_{X,Y}[1,0] = 1/8$, $p_{X,Y}[1,1] = 5/8$

19. no

23. yes, $X \sim \text{bin}(10, 1/2)$, $Y \sim \text{bin}(11, 1/2)$

27. $p_Z[0] = 1/4$, $p_Z[1] = 1/2$, $p_Z[2] = 1/4$, variance always increases when uncorrelated random variables are added

33. $1/8$

37. 0

38. $3/22$

40. minimum MSE prediction $= E_Y[Y] = 5/8$ and minimum MSE $= \text{var}(Y) = 15/64$ for no knowledge
minimum MSE prediction $= \hat{Y} = -(1/15)x + 2/3$ and minimum MSE $= \text{var}(Y)(1 - \rho_{X,Y}^2) = 7/30$ based on observing outcome of X

41. $\hat{W} = 0.1063h + 164.6$

43. $\rho_{W,Z} = \sqrt{\eta/(\eta+1)}$, where $\eta = E_X[X^2]/E_N[N^2]$

46. see solution for Problem 7.27

48. $p_{X,Y}[0,0] = 0.1190$, $p_{X,Y}[0,1] = 0.1310$, $p_{X,Y}[1,0] = 0.2410$, $p_{X,Y}[1,1] = 0.5090$
($N_{\text{real}} = 1000$)

49. $\rho_{X,Y} = \sqrt{5}/15 = 0.1490$, $\hat{\rho}_{X,Y} = 0.1497$ ($N_{\text{real}} = 100,000$)

Chapter 8

2. $p_{Y|X}[j|0] = 1$ for $j = 0$
$p_{Y|X}[j|1] = 1/6$ for $j = 1, 2, 3, 4, 5, 6$
$P[Y = 1] = 1/12$

5. no, no, no

6. $p_{Y|X}[j|0] = 1/3$ for $j = 0$ and $= 2/3$ for $j = 1$
$p_{Y|X}[j|1] = 2/3$ for $j = 0$ and $= 1/3$ for $j = 1$
$p_{X|Y}[i|0] = 1/3$ for $i = 0$ and $= 2/3$ for $i = 1$
$p_{X|Y}[i|1] = 2/3$ for $i = 0$ and $= 1/3$ for $i = 1$

8. $p_{Y|X}[j|i] = 1/5$ for $j = 0, 1, 2, 3, 4; i = 1, 2$
$p_{X|Y}[i|j] = 1/2$ for $i = 1, 2; j = 0, 1, 2, 3, 4$

11. 0.4535

13. a. $p_{Y|X}[y_j|0] = 0, 1, 0$ for $y_j = -1/\sqrt{2}, 0, 1/\sqrt{2}$, respectively
$p_{Y|X}[y_j|1/\sqrt{2}] = 1/2, 0, 1/2$ for $y_j = -1/\sqrt{2}, 0, 1/\sqrt{2}$, respectively
$p_{Y|X}[y_j|\sqrt{2}] = 0, 1, 0$ for $y_j = -1/\sqrt{2}, 0, 1/\sqrt{2}$, respectively
not independent (conditional PMF depends on x_i)
b. $p_{Y|X}[y_j|0] = 1/2, 1/2$ for $y_j = 0, 1$, respectively
$p_{Y|X}[y_j|1] = 1/2, 1/2$ for $y_j = 0, 1$, respectively
independent

17. $p_Z[k] = p_X[k] \sum_{j=k}^{\infty} p_Y[j] + p_Y[k] \sum_{j=k+1}^{\infty} p_X[j]$

21. $E_{Y|X}[Y|0] = 0$, $E_{Y|X}[Y|1] = 1/2$, $E_{Y|X}[Y|2] = 1$

22. $\text{var}(Y|0) = 0$, $\text{var}(Y|1) = 1/4$, $\text{var}(Y|2) = 2/3$

28. optimal predictor: $\hat{Y} = 0$ for $x = -1$, $\hat{Y} = 1/2$ for $x = 0$, and $\hat{Y} = 0$ for $x = 1$
optimal linear predictor: $\hat{Y} = 1/4$ for $x = -1, 0, 1$

30. $\widehat{E_{Y|X}}[Y|0] = 0.5204$, $\widehat{E_{Y|X}}[Y|1] = 0.6677$ ($N_{\text{real}} = 10,000$)

Chapter 9

1. 0.0567

4. yes

6. (X_1, X_2) independent of X_3

10. $E[\bar{X}] = E_X[X]$, $\text{var}(\bar{X}) = \text{var}(X)/N$

13. $\mathbf{C}_X = \begin{bmatrix} 1 & 2 \\ 2 & 4 \end{bmatrix}$, $\det(\mathbf{C}_X) = 0$, no

17. a. no, b. no , c. yes, d. no

20. $\mathbf{C}_X = \begin{bmatrix} 2 & 3 \\ 3 & 6 \end{bmatrix}$

26. $\mathbf{A} = \begin{bmatrix} 0.9056 & 0.4242 \\ -0.4242 & 0.9056 \end{bmatrix}$ for MATLAB 5.2
$\mathbf{A} = \begin{bmatrix} -0.9056 & 0.4242 \\ 0.4242 & 0.9056 \end{bmatrix}$ for MATLAB 6.5, R13
$\text{var}(Y_1) = 7.1898$, $\text{var}(Y_2) = 22.8102$

35. $\mathbf{B} = \begin{bmatrix} \sqrt{3/2} & \sqrt{5/2} \\ -\sqrt{3/2} & \sqrt{5/2} \end{bmatrix}$

36. $\hat{\mathbf{C}}_X = \begin{bmatrix} 4.0693 & 0.9996 \\ 0.9996 & 3.9300 \end{bmatrix}$ $(N_{\text{real}} = 1000)$

Chapter 10

2. 1/80

4. a. no b. yes c. no

6. $\alpha_1 \geq 0$, $\alpha_2 \geq 0$, and $\alpha_1 + \alpha_2 = 1$

12. 0.0252

14. Gaussian: 0.0013 Laplacian: 0.0072

17. first person probability = 0.393, first two persons probability = 0.090

19. $F_X(x) = 1/2 + (1/\pi)\arctan(x)$

22. $F_X(x) = \Phi\left(\frac{x-\mu}{\sigma}\right)$

28. 2.28%

30. eastern U.S.

33. yes

36. $c \approx 14$

40.
$$p_Y(y) = \begin{cases} \frac{\lambda}{4(y-1)^{3/4}}\exp[-\lambda(y-1)^{1/4}] & y \geq 1 \\ 0 & y < 1 \end{cases}$$

43. $p_Y(y) = p_X(y) + p_X(-y)$

46.
$$p_Y(y) = \begin{cases} \frac{1}{2\sqrt{y}} & 0 < y < 1 \\ 0 & \text{otherwise} \end{cases}$$

51.
$$\begin{aligned}
P[-2 \leq X \leq 2] &= 1 - \tfrac{1}{2}\exp(-2) \\
P[-1 \leq X \leq 1] &= 1 - \tfrac{1}{2}\exp(-1) \\
P[-1 < X \leq 1] &= \tfrac{3}{4} - \tfrac{1}{2}\exp(-1) \\
P[-1 < X < 1] &= \tfrac{1}{2} - \tfrac{1}{2}\exp(-1) \\
P[-1 \leq X < 1] &= \tfrac{3}{4} - \tfrac{1}{2}\exp(-1)
\end{aligned}$$

54. $g(U) = \sqrt{2\ln(1/(1-U))}$

Chapter 11

1. $7/6$

10. ± 9.12

11. 0.1353

14. N

19. 0.0078

21. $\sqrt{E[U]} = \sqrt{1/2}$, $E[\sqrt{U}] = 2/3$

22. $E[s(t_0)] = 0$, $E[s^2(t_0)] = 1/2$

26. $\sigma^2/2$

27. $\sigma^2/2$

30. $T_{\min} = 5.04$, $T_{\max} = 8.96$

35. $E[X^3] = 3\mu\sigma^2 + \mu^3$, $E[(X-\mu)^3] = 0$

38. $E[X^n] = 0$ for n odd, $E[X^n] = n!$ for n even

42. $\delta(x-\mu)$

44. $\sqrt{2\mathrm{var}(X)}$

46. $E[X] = 1.2533$, $\widehat{E[X]} = 1.2538$; $\mathrm{var}(X) = 0.4292$, $\widehat{\mathrm{var}(X)} = 0.4269$ ($N_{\mathrm{real}} = 1000$)

Chapter 12

1. $7/16$

3. no, probability is $1/4$

5. $\pi = 4P[X^2 + Y^2 \le 1]$, $\hat{\pi} = 3.1140$ ($N_{\mathrm{real}} = 10,000$)

7. $1/4$

10. $P = 0.19$, $\hat{P} = 0.1872$ ($N_{\mathrm{real}} = 10,000$)

11. 0

15. $p_X(x) = 2x$ for $0 < x < 1$ and zero otherwise, $p_Y(y) = 2(1-y)$ for $0 < y < 1$ and zero otherwise

18.

$$F_{X,Y}(x,y) = \begin{cases} 0 & x < 0 \text{ or } y < 0 \\ \frac{1}{8}xy & 0 \le x < 2, 0 \le y < 4 \\ \frac{1}{4}y & x \ge 2, 0 \le y < 4 \\ \frac{1}{2}x & 0 \le x < 2, y \ge 4 \\ 1 & x \ge 2, y \ge 4 \end{cases}$$

23. $(1 - \exp(-2))^2$

25. no

26. $Q(2)$

30. $P[\text{bullseye}] = 1 - \exp(-2) = 0.8646$, $\hat{P}[\text{bullseye}] = 0.8730$ ($N_{\text{real}} = 1000$)

36. $W \sim \mathcal{N}(\mu_W, \sigma_W^2)$, $Z \sim \mathcal{N}(\mu_Z, \sigma_Z^2)$

38. $\begin{bmatrix} W \\ Z \end{bmatrix} \sim \mathcal{N}(\boldsymbol{\mu}, \mathbf{C})$, where

$$\boldsymbol{\mu} = \begin{bmatrix} 3 \\ 8 \end{bmatrix}$$

$$\mathbf{C} = \begin{bmatrix} 2 & 5 \\ 5 & 14 \end{bmatrix}$$

43. $\sqrt{5\pi}$

45. uncorrelated but not necessarily independent

47.

$$\mathbf{G} = \begin{bmatrix} \frac{1}{\sqrt{2}} & -\frac{1}{\sqrt{2}} \\ \frac{1}{\sqrt{2}} & \frac{1}{\sqrt{2}} \end{bmatrix}$$

52. $Q(1)$

Chapter 13

2. yes, $c = 1/x$

4. $p_{Y|X}(y|x) = \exp(-y)/(1 - \exp(-x))$ for $0 \le y \le x$, $x \ge 0$

8. $p_{X,Y}(x,y) = 1/x$ for $0 < y < x$, $0 < x < 1$; $p_Y(y) = -\ln y$ for $0 < y < 1$

10. $p_{Y|X}(y|x) = 1/x$ for $0 < y < x$, $0 < x < 1$; $p_{X|Y}(x|y) = 1/(1-y)$ for $y < x < 1$, $0 < y < 1$

14. Use $P = \int_{-\frac{1}{2}}^{\frac{1}{2}} P[|X_2| - |X_1| < 0|X_1 = x_1]p_{X_1}(x_1)dx_1$ and note independence of X_1 and X_2 so that $P = \int_{-\frac{1}{2}}^{\frac{1}{2}} P[|X_2| \le x_1]dx_1$

16. $Q(-1)$, assume R and E are independent

21. $1/2$

24. Use $E_{(X+Y)|X}[X+Y|x] = E_{Y|X}[Y|x] + x$ to yield $E_{(X+Y)|X}[X+Y|X = 50] = 77.45$ and $E_{(X+Y)|X}[X+Y|X = 75] = 84.57$

Chapter 14

1. $E_Y[Y] = 6$, $\text{var}(Y) = 11/2$

6. $1/16$

9. $Y \sim \mathcal{N}(0, \sigma_1^2 + \sigma_2^2 + \sigma_3^2)$

12. no since $\text{var}(\bar{X}) \to \sigma^2/2$ as $N \to \infty$

19. $E_Y[Y] = 0$, $\text{var}(Y) = 1$

21. $\hat{X}_3 = 7/5$

24. $\text{mse}_{\min} = 8/15 = 0.5333$

25. $\widehat{\text{mse}}_{\min} = 0.5407$ ($N_{\text{real}} = 5000$)

Chapter 15

4. $\lim_{N\to\infty} \sum_{i=1}^{N} \alpha_i^2 = 0$, $\alpha_i = 1/N^{3/4}$

7. no since the variance does not converge to zero

13. $Y \sim \mathcal{N}(2000, 1000/3)$

19. $N = 5529$

20. $1 - Q(-77.78) \approx 0$

22. Gaussian, "converges" for all $N \ge 1$

23. no since approximate 95% confidence interval is $[0.723, 0.777]$

26. drug group has approximate 95% confidence interval of $[0.69, 0.91]$ and placebo group has $[0.47, 0.73]$. Can't say if drug is effective since true value of p could be 0.7 for either group.

Chapter 16

1. a. temperature at noon b. expense in dollars and cents c. time left in hours and minutes

4. $p^{50}(1-p)^{50}$, 0

7. $E[X[n]] = (n+1)/2$, $\text{var}(X[n]) = (3/4)(n+1)$

9. $\exp(-3)$

13. independent but not stationary

16. $\mu_X[n] = 0$, $c_X[n_1, n_2] = \delta[n_2 - n_1]$, exactly the same as for WGN with $\sigma_U^2 = 1$

18. $P[X[n] > 3] = 0.000011$, $P[U[n] > 3] = 0.0013$

22. $E[X(t)] = 0$, $c_X(t_1, t_2) = \cos(2\pi t_1)\cos(2\pi t_2)$

24. $E[Y[n]] = 0$, $\text{cov}(Y[0], Y[1]) = -1$, not IID since samples are not independent

26. $E[X[n]] = 0$, $c_X[n_1, n_2] = \sigma_U^2 \min(n_1, n_2)$

27. $c_X[1,1] = 1/2$, $c_X[1,2] = 1/4$, $c_X[1,3] = 0$, $\hat{c}_X[1,1] = 0.5057$, $\hat{c}_X[1,2] = 0.2595$, $\hat{c}_X[1,3] = -0.0016$ ($N_{\text{real}} = 10,000$)

31. $\mu_X[n] = 0$, $c_X[n_1, n_2] = \sigma_A^2 \sigma_U^2 \delta[n_2 - n_1]$, white noise

34. $N = 209$

Chapter 17

1. yes, $\mu_X[n] = \mu = 2p - 1$, $r_X[k] = 1$ for $k = 0$ and $r_X[k] = \mu^2$ for $k \neq 0$

5. WSS but not stationary since $p_{X[0]} \neq p_{X[1]}$

9. $a > 0$, $|b| \leq 1$

12. b,d,e

17. $E[X[n]] = 0$, $\text{var}(X[n]) = \sigma_U^2(1 - a^{2(n+1)})/(1 - a^2)$; as $n \to \infty$, $\text{var}(X[n]) \to \sigma_u^2/(1 - a^2)$

19. Principal minors are 1, 15/64, and $-17/32$ for 1×1, 2×2 and 3×3, respectively. Fourier transform is $1 - (7/4)\cos(2\pi f)$ which can be negative.

20. $\mu_X[n] = \mu$, $r_X[k] = (1/2)r_U[k] + (1/4)r_U[k-1] + (1/4)r_U[k+1]$

28. $P_X(f) = 2\sigma_U^2(1 - \cos(2\pi f))$

30. $P_X(f) = \sigma_A^2 \sigma_U^2$

34. $r_X[k] = \sigma_U^2\delta[k] + \mu^2$, $P_X(f) = \sigma_U^2 + \mu^2\delta(f)$

38. $r_X[k] = 9/4, 3/2, 1/2$ for $k = 0, 1, 2$, respectively, and zero otherwise

40. $a \geq |b|$

42. $E[X[n]] = 0$, $\widehat{E}[X[10]] = -0.0105$, $\widehat{E}[X[12]] = 0.0177$; $r_X[2] = 0.1545$,
$\widehat{E}[X[10]X[12]] = 0.1501$, $\widehat{E}[X[12]X[14]] = 0.1533$ ($N_{\text{real}} = 1000$)

44. $P_X(f) = 1$, increasing N does not improve estimate – must average over ensemble of periodograms

47. $2(\exp(-10) - \exp(-100))$

50. $\mu_X(t) = 0$, $\text{var}(X(t)) = N_0/(2T)$, no

51. $\text{var}(\hat{\mu}_N) = (1/N)\sum_{k=-(N-1)}^{N-1}(1 - |k|/N)N_0W\sin(\pi k/2)/(\pi k/2)$ for $N = 20$.
It is 0.9841 times that of the variance of the sample mean for Nyquist rate sampled data.

Chapter 18

1. $r_X[k] = 3$ for $k = 0$, $r_X[k] = -1$ for $k = \pm 2$, and equals zero otherwise; $P_X(f) = 3 - 2\cos(4\pi f)$

4. $b_1 = 0$, $b_2 = -1$

7. $r_X[k] = 3$ for $k = 0$, $r_X[k] = -2$ for $k = \pm 1$, $r_X[k] = 1/2$ for $k = \pm 2$, and equals zero otherwise; $P_X(f) = 3 - 4\cos(2\pi f) + \cos(4\pi f)$

13. $H_{\text{opt}} = (2 - 2\cos(2\pi f))/(3 - 2\cos(2\pi f))$; $\text{mse}_{\text{min}} = 0.5552$

18. $\text{mse}_{\text{min}} = r_X[0](1 - \rho_{X[n_0],X[n_0+1]}^2)$

22. $\hat{X}[n_0 + 1] = -[b(1 + b^2)/(1 + b^2 + b^4)]X[n_0] - [b^2/(1 + b^2 + b^4)]X[n_0 - 1]$;
$\text{mse}_{\text{min}} = 1 + b^6/(1 + b^2 + b^4)$

24. $\text{mse}_{\text{min}} = 1 + [b^6/(1 + b^2 + b^4)] = 85/84 = 1.0119$, $\widehat{\text{mse}}_{\text{min}} = 1.0117$ ($N_{\text{real}} = 10,000$)

27. $\hat{X}[n_0] = [a/(1 + a^2)](X[n_0 + 1] + X[n_0 - 1])$

29. $P_X(F) = (N_0T^2/2)[(\sin(\pi FT))/(\pi FT)]^2$

32. $r_X(0) = N_0/(4RC)$, no

Chapter 19

1. $E[X[n]Y[n+k]] = (-1)^{n+k}\sigma_U^2\delta[k]$, no

5. $r_{X,Y}[k] = 0$

6. $r_{X,U}[k] = 0$ for $k > 0$ and $r_{X,Y}[k] = (1/2)^{(-k)}$ for $k < 0$

10. $|P_{X,Y}(f)| = \sqrt{5 + 4\cos(2\pi f)}$, $\angle P_{X,Y}(f) = \arctan\frac{-2\sin(2\pi f)}{1+2\cos(2\pi f)}$

12. $r_Z[k] = r_X[k] - r_{X,Y}[k] - r_{Y,X}[k] + r_Y[k]$, $P_Z(f) = P_X(f) - P_{X,Y}(f) - P_{Y,X}(f) + P_Y(f)$

15. $\gamma_{X,Y}(f) = -1$, perfectly predictable using $\hat{Y}[n_0] = -X[n_0]$

18. $H_{\text{opt}}(f) = P_{X,Y}(f)/P_X(f)$

23. $r_{X,Y}(\tau) = N_0/(2T)$ for $0 \leq \tau \leq T$ and zero otherwise

26. $r_{X,Y}[k] = \delta[k] - b\delta[k-1]$, for $b = -1$

k	$r_{X,Y}[k]$	$\hat{r}_{X,Y}[k]$	k	$r_{X,Y}[k]$	$\hat{r}_{X,Y}[k]$
-5	0	-0.0077	0	1	0.9034
-4	0	-0.0242	1	1	0.9031
-3	0	0.0259	2	0	-0.0064
-2	0	0.0004	3	0	-0.0007
-1	0	-0.0062	4	0	0.0267
			5	0	-0.0238

Chapter 20

2. $1/4$

5. $\mathbf{Y} = [Y[0]\, Y[1]]^T \sim \mathcal{N}(\mathbf{0}, \mathbf{C}_Y)$, where

$$\mathbf{C}_Y = \begin{bmatrix} 2 & -1 \\ -1 & 2 \end{bmatrix}$$

not independent

10. WSS with $\mu_Z = \mu_X\mu_Y$, $P_Z(f) = P_X(f) \star P_Y(f)$

14. 1

17. $T = 66,347$

19. $2Q(1/\sqrt{T})$

22. $\mathbf{Y} = [Y(0)\, Y(1/4)]^T \sim \mathcal{N}(\mathbf{0}, \mathbf{C}_Y)$, where

$$\mathbf{C}_Y = \begin{bmatrix} 2 & \sqrt{2} \\ \sqrt{2} & 2 \end{bmatrix}$$

25. $P_U(F) = P_V(F) = 8(1 - |F|/10)$ for $|F| \le 10$ and zero otherwise

30. $X[n] = U[n] - U[n-1]$, where $U[n]$ is WGN with $\sigma_U^2 = 1$

31. $r_X[0] = 2$, $\hat{r}_X[0] = 1.9591$; $r_X[1] = -1$, $\hat{r}_X[1] = -0.9614$; $r_X[2] = 0$, $\hat{r}_X[2] = -0.0195$; $r_X[3] = 0$, $\hat{r}_X[3] = -0.0154$

Chapter 21

3. probability $= 0.1462$, average $= 5$

7. $\lambda = 2, \hat{\lambda} = 1.9629$; $\lambda = 5, \hat{\lambda} = 4.9072$ (based on 10,000 arrivals with $\hat{\lambda} = \hat{E}[N(t)]/t$

10. $E[N(t_2) - N(t_1)] = \text{var}(N(t_2) - N(t_1)) = \lambda(t_2 - t_1)$,

13. 0.6321

17. 10 minutes

20. $P[T_2 \le 1] = 1 - 2\exp(-1) = 0.2642$, $\hat{P}[T_2 \le 1] = 0.2622$ ($N_{\text{real}} = 10,000$)

23. $\lambda t_0(2p - 1)$

Chapter 22

2. $1/128$

5. $P[Y[2] = 1|Y[1] = 1, Y[0] = 0] = 1 - p$, $P[Y[2] = 1|Y[1] = 1] = 1/2$ for all p

9. $P_e = 0.3362$

11. $P[\text{red drawn}] = 1/3$

12. $1/2$

14. yes, $\boldsymbol{\pi}^T = [\frac{1}{3}\ \frac{2}{3}\ 0\ 0]$

19. $\boldsymbol{\pi}^T = [0.2165\ 0.4021\ 0.3814]$

24. $P[\text{rain}] = 0.6964$

26. $n = 6$

28. $\pi_1 = 1/3$, $\hat{\pi}_1 = 0.3240$ (based on playing 1000 holes)

Index